1990

Third Edition

BASIC COLLEGE MATHEMATICS

A TEXT/WORKBOOK

Charles D. Miller

Stanley A. Salzman

American River College

Margaret L. Lial

American River College

HarperCollins*Publishers*

TO THE STUDENT

If you need further help with mathematics, you may want to obtain a copy of the *Student's Solutions Manual* that goes with this book. It contains solutions to the odd-numbered exercises that are not already solved at the back of the textbook, plus solutions to all chapter review exercises, cumulative review exercises, and chapter tests. Your college bookstore either has this book or can order it for you.

Sponsoring Editor:	Bill Poole
Developmental Editor:	Linda Bieze/Linda Youngman
Project Editor:	Janet Tilden/Ginny Guerrant
Art Direction:	Julie Anderson
Text and Cover Design:	Lucy Lesiak Design/Cynthia Crampton
Cover Photo:	Front Cover left, M. Angelo/West Light; right, H. D. Thoreau/West Light; Front illustration inset reprinted with permission from: Peitgen, H.-O./Richter, P.H.: *The Beauty of Fractals.* 1986 © 1986 Springer-Verlag Berlin-Heidelberg.
Photo Research:	Judy Ladendorf
Director of Production:	Jeanie A. Berke
Production Assistant:	Linda Murray
Compositor:	The Clarinda Company
Printer and Binder:	Courier Corporation
Cover Printer:	Lehigh Press Lithographers

Basic College Mathematics: A Text/Workbook, Third Edition

ISBN 0-673-46279-X

The pattern in the small insert on the cover of this text is an example of the types of structures that can be computer generated in accordance with the principles of fractal geometry. Fractal patterns also exist in nature; examples include the shapes of mountain ranges and the winding of coastlines.

90 91 92 93 9 8 7 6 5 4 3 2 1

PREFACE

Basic College Mathematics: A Text/Workbook, Third Edition, covers the topics needed for success in a developmental mathematics program. The book is comprehensive, providing the necessary background and review in whole numbers, fractions, decimals, ratio and proportion, and measurement, as well as an introduction to algebra and geometry, and a preview of statistics and consumer mathematics.

The book's clear explanations, precise learning objectives, approximately 675 detailed examples, carefully graded exercise sets, and open, accessible design make this an ideal text for either a traditional lecture class or individualized instruction.

NEW CONTENT FEATURES

Changes in the third edition include the following.

- New examples and exercises have been added, and the applications have been extensively rewritten and updated. New applications in science, medicine, business, and the technical fields are included throughout.
- Chapter 2 includes an increased emphasis on applications of multiplication and division of fractions.
- More thorough explanations and illustrations of finding the least common denominator are given in Chapter 3.
- Completely updated examples and applications on percent in Chapter 6 reflect current student interests and career goals.
- Chapter 8 on geometry has been substantially revised and the coverage expanded. It now includes thorough coverage of basic geometric terms and a discussion of angles and their relationships. These additions will fulfill mathematics competency requirements in many states.
- A reworked treatment of basic algebra topics in Chapter 9 ensures a mathematically sound yet unintimidating introduction to this subject.
- New figures, graphs, and charts have been added throughout the text to promote student understanding.
- A new appendix on inductive and deductive reasoning has been included. This material also satisfies mathematics competency requirements in many states.

NEW FEATURES

All the successful features of the workbook format in the previous edition are carried over in the new edition: Learning objectives for each section, careful exposition and fully developed examples, sample problems in the margins for immediate feedback, carefully graded exercises with work space, and a clear, functional design guide students through the text. Screened boxes that set off important definitions, formulas, rules, and procedures further aid in learning and reviewing the course material.

In addition, the following new features have been developed to enhance the pedagogy and usefulness of the text.

Example Titles

Each example now has a title to help students see the purpose of the example. The titles also facilitate working the exercises and studying for examinations.

Helpful Remarks

Reminders are used to emphasize key principles and important facts throughout the text. They are highlighted graphically and identified with the heading "Note."

Challenge Exercises

These new problems are grouped near the end of exercise sets and present additional challenges for the motivated student.

Review Exercises

These short sets of review exercises at the end of each exercise set beginning with Chapter 2 provide students with a quick review of skills that will be needed in the next section.

Chapter Summaries

This new study aid at the end of each chapter provides a glossary of key terms with section references and a "Quick Review" featuring section-referenced capsule summaries of key ideas accompanied by worked-out examples.

Chapter Review Exercises

A thorough set of chapter review exercises follows each chapter summary. Most of these exercises are organized and referenced by section, but a concluding section, titled "Mixed Review Exercises," includes randomly mixed exercises from the entire chapter. This helps students practice identifying problems by type. As in the previous edition, a sample test concludes each chapter.

Success in Mathematics

This foreword to the students provides additional support by offering suggestions for successful study of the course material.

Section and Chapter References in Answer Section

As another study aid, the answer section at the end of the book includes section references in the answers to the chapter tests, and chapter references in the answers to the final examination. Students can easily locate and review material that gave them difficulty when they took the sample chapter test or final examination.

Glossary

A comprehensive glossary is placed at the end of the book. Each term in the glossary is defined and then referenced to the appropriate section in the text, where students may find a more detailed explanation of the term.

SUPPLEMENTS

Our extensive supplemental package includes an annotated instructor's edition, testing materials, solutions, software, videotapes, and audiotapes.

Annotated Instructor's Edition

This edition provides instructors with immediate access to the answers to every exercise in the text; each answer is printed in color next to the corresponding text exercise.

Instructor's Testing Manual

The Instructor's Testing Manual includes suggestions for using the textbook in a mathematics laboratory; short-answer and multiple-choice versions of a pretest in basic mathematics; six forms of chapter tests for each chapter, including four open-response and two multiple-choice forms; two forms of a final examination; and an extensive set of additional exercises, providing 10 to 20 exercises for each textbook objective, which can be used as an additional source of questions for tests, quizzes, or student review of difficult topics.

Student's Solutions Manual

This book contains solutions to half of the odd-numbered section exercises (those not included at the back of the textbook) as well as solutions to all chapter review exercises, chapter tests, and cumulative review exercises.

Instructor's Solutions Manual

Available at no charge to instructors, this book includes solutions to all the margin problems in the textbook and solutions to the even-numbered section exercises. The two solutions manuals plus the solutions given at the back of the textbook provide detailed, worked-out solutions to each exercise and margin problem in the book.

HarperCollins Test Generator for Mathematics

Available in Apple, IBM, and Macintosh versions, the test generator enables instructors to select questions by objective, section, or chapter, or to use a ready-made test for each chapter. Instructors may generate tests in multiple-choice or open-response formats, scramble the order of questions while printing, and produce multiple versions of each test (up to 9 with Apple, up to 25 with IBM and Macintosh). The system features printed graphics and accurate mathematics symbols. It also features a preview option that allows instructors to view questions before printing, to regenerate variables, and to replace or skip questions if desired. The IBM version includes an editor that allows instructors to add their own problems to existing data disks.

Interactive Tutorial Software

This innovative package is also available in Apple, IBM, and Macintosh versions. It offers interactive modular units, specifically linked to the text, for reinforcement of selected topics. The tutorial is self-paced and provides unlimited opportunities to review lessons and to practice problem solving. When students give a wrong answer, they can request to see the problem worked out. The program is menu-driven for ease of use, and on-screen help can be obtained at any time with a single keystroke. Students' scores are automatically recorded and can be printed for a permanent record.

Audiotapes

A set of audiotapes, one per chapter, is available for the text. The tapes guide students through each topic, allowing individualized study and additional practice for troublesome areas. They are especially helpful for visually impaired students.

Videotapes

A new videotape series, *Algebra Connection: The Basic Mathematics Course,* has been developed to accompany *Basic College Mathematics,* Third Edition. Produced by an Emmy Award–winning team in consultation with a task force of academicians from both two-year and four-year colleges, the tapes cover all objectives, topics, and problem-solving techniques within the text. In addition, each lesson is preceded by motivational ''launchers'' that connect classroom activity to real-world applications.

ACKNOWLEDGMENTS

We wish to thank the many users of the second edition for their insightful suggestions on improvements for this book.

We also wish to thank our reviewers for their contributions: Carol Atnip, University of Louisville; Linda Barton, Ball State University; Solveig R. Bender, William Rainey Harper College; Ernest Berman, City Colleges of Chicago Harry S Truman College; Laurence Chernoff, Miami-Dade Community College; Cheryl S. Cummings, Cincinnati Technical College; Geraldine M. Curley, Bunker Hill Community College; Barbara Davis, Hillsborough Community College; Jeanne Fitzgerald, Phoenix College; Irwin Goldfine, City Colleges of Chicago Harry S Truman College; James R. Griffiths, College of the Mainland; Marilyn S. Hamilton, Indiana Vocational Technical College–Central Indiana; Robert Kaiden, Lorain County Community College; Betty W. Ledford, Cleveland State Community College; Gwen Magee, Jones County Junior College; Gael T. Mericle, Mankato State University; Anthony Monteith, College of Marin; Jack Preston, Canada College; Mark Rajai, Cleveland State Community College; Bobby Righi, Seattle Central Community College; Frances Rosamond, National University; Carolyn Ruegger, Pearl River Community College; Jane E. Sieberth, Franklin University; Faye Tamakawa, Honolulu Community College; George Thomas, College of the Mainland; Mary A. Wilber, Pasco-Hernando Community College; Lloyd F. Woodling, Williamsport Area Community College.

We also would like to thank our colleagues at American River College: William Edgar, Kathy Monaghan, and Paul Van Erden. We would be remiss if we did not also thank our typists, Judy Martinez, Sheri Minkner, and Sandy Yost, for their fine work.

Paul J. Eldersveld, College of DuPage, deserves heartfelt thanks for undertaking the enormous job of coordinating all of the print ancillaries for us.

We also want to thank Tommy Thompson, Seminole Community College, for his suggestions about the essay to the students on studying mathematics at the beginning of the book.

Our special thanks go to our editors, Bill Poole, Linda Bieze, Linda Youngman, and Janet Tilden. As always, they have done a wonderful job under difficult circumstances.

Stanley A. Salzman
Margaret L. Lial

CONTENTS

TO THE STUDENT: SUCCESS IN MATHEMATICS

The main reason students have difficulty with mathematics is that they don't know how to study it. Studying mathematics *is* different from studying subjects like English or history. The key to success is regular practice.

This should not be surprising. After all, can you learn to play the piano or to ski well without a lot of regular practice? The same thing is true for learning mathematics. Working problems nearly every day is the key to becoming successful. Here are suggestions to help you succeed in studying mathematics.

1. Pay attention in class to what your instructor says and does, and make careful notes. Note the problems the instructor works on the board and copy the complete solutions. Keep these notes separate from your homework to avoid confusion when you read them later.
2. Don't hesitate to ask questions in class. It is not a sign of weakness, but of strength. There are always other students with the same question who are too shy to ask.
3. Before you start on your homework assignment, rework the problems the instructor worked in class. This will reinforce what you have learned. Many students say, "I understand it perfectly when you do it, but I get stuck when I try to work the problem myself."
4. *Read your text carefully*. Many students read only enough to get by, usually only the examples. Reading the complete section will help you to be successful with the homework problems. As a bonus you will be able to do the problems more quickly if you have read the text first. As you read the text, work the sample problems given in the margin and check the answers to see if you have worked them correctly. This will test your understanding of what you have read.
5. Do your homework assignment only *after* reading the text and reviewing your notes from class. Check your work with the answers in the back of the book. If you get a problem wrong and are unable to see why, mark that problem and ask your instructor about it.
6. Work as neatly as you can. Write your symbols clearly, and make sure the problems are clearly separated from each other. Working neatly will help you to think clearly and also make it easier to review the homework before a test.
7. After you have completed a homework assignment, look over the text again. Try to decide what the main ideas are in the lesson. Often they are clearly highlighted or boxed in the text.
8. Keep any quizzes and tests that are returned to you for studying for future tests and the final exam. These quizzes and tests indicate what your instructor considers most important. Be sure to correct any problems on these tests that you missed, so you will have the corrected work to study.
9. Don't worry if you do not understand a new topic right away. As you read more about it and work through the problems, you will gain understanding. Each time you look back at a topic you will understand it a little better. No one understands each topic completely right from the start.
10. What a great feeling you will experience when, after a lot of time and hard work, you can say, "I really do understand this now."

name date hour

DIAGNOSTIC PRETEST

Ask your instructor whether or not you should work this pretest. It is designed to tell whether material in the course may already be familiar to you. The actual course begins on page 1.

[1.1]

1. Identify the place value of the 5 and the 8 in the number 725,283.

1. ______________________

____

2. Write 874 in words.

2. ______________________

____

[1.2]

3. Add.

$$\begin{array}{r} 324 \\ 7855 \\ 23 \\ 7 \\ +\quad 86 \\ \hline \end{array}$$

3. ______________________

[1.3]

4. Subtract.

$$\begin{array}{r} 7000 \\ -\ 4999 \\ \hline \end{array}$$

4. ______________________

[1.4]

5. Multiply.

$$\begin{array}{r} 724 \\ \times\quad 5 \\ \hline \end{array}$$

5. ______________________

[1.6]

6. Divide. $58\overline{)2730}$

6. ______________________

[1.7]

7. Round 37,892 to the nearest ten thousand.

7. ______________________

8. ______________________

9. ______________________

10. ______________________

11. ______________________

12. ______________________

13. ______________________

14. ______________________

[1.8]

8. Simplify $12 \cdot \sqrt{16} - 8 \cdot 4$.

[1.9]

9. At one college, the number of students enrolled this year is 1842 fewer than the number enrolled last year. If the enrollment last year was 12,622, find the enrollment this year.

[2.2]

10. Write $5\frac{9}{10}$ as an improper fraction.

[2.5]

11. Multiply. $\frac{4}{7} \cdot \frac{2}{5}$

[2.6]

12. Of the 42 students in a biology class, $\frac{2}{3}$ go on a field trip. How many go on the trip?

[2.8]

13. One tent requires $7\frac{1}{4}$ yards of nylon cloth. How many tents can be made from $65\frac{1}{4}$ yards of the cloth?

14. Write $\frac{5}{6}$ with a denominator of 42.

name date hour

DIAGNOSTIC PRETEST

[3.3]

15. Add. $\frac{3}{8} + \frac{7}{12}$

16. Subtract. $\frac{3}{4} - \frac{5}{9}$

[3.4]

17. Add $9\frac{5}{8}$ and $13\frac{7}{8}$.

18. Subtract. $137\frac{1}{3} - 72\frac{3}{4}$

[4.2]

19. Round .69413 to the nearest hundredth.

[4.3]

20. Add. $6.42 + 9.3 + 2.576$

[4.4]

21. Subtract. $59.7 - 38.914$

[4.5]

22. Multiply. $.042 \cdot .03$

[4.6]

23. Divide. $.3\overline{)27.69}$

24. Work this problem.
$1.82 + (6.7 - 5.2) \cdot 5.8$

15. ______

16. ______

17. ______

18. ______

19. ______

20. ______

21. ______

22. ______

23. ______

24. ______

[5.1]

25. ______________________

25. Write the ratio of 5 yards to 8 feet.

[5.2]

26. ______________________

26. Write 337.5 miles on 13.5 gallons of gas as a unit rate.

[5.3]

27. ______________________

27. Is the proportion $\frac{5}{9} = \frac{18}{27}$ true?

[5.4]

28. ______________________

28. Find the missing number. $\frac{8}{y} = \frac{12}{8}$

[6.1]

29. ______________________

29. Write 76% as a decimal.

[6.2]

30. ______________________

30. Write $\frac{3}{5}$ as a percent.

[6.5] *Find each of the following.*

31. ______________________

32. ______________________

31. 8 is 4% of what number?

32. 13 is what percent of 52?

[6.7]

33. ______________________

33. A furniture store has a sofa with an original price of $470, on sale at 15% off. Find the sale price of the sofa.

[6.8]

34. ______________________

34. Find the interest on $2200 at 11% for three and a half years.

name date hour

[7.2]

35. Simplify 5 yards 4 feet 17 inches.

35. ______________

[7.3]

36. Convert 37 km to meters.

36. ______________

[8.3]

37. Find the perimeter and area of this figure.

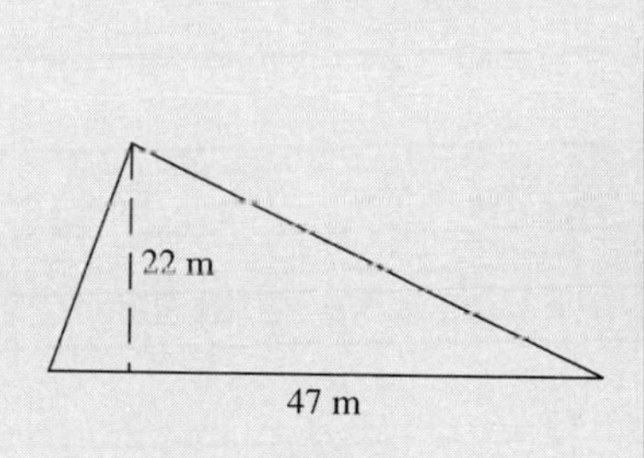

37. ______________

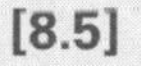

[8.5]

38. Find the area of this triangle.

22 m

47 m

38. ______________

[8.6]

39. Find the area of a circle with radius 8.4 cm. (Use 3.14 as an approximation for π.) Round to the nearest tenth.

39. ______________

[8.7]

40. Find the volume of this cone. (Use 3.14 as an approximation for π.) Round to the nearest tenth.

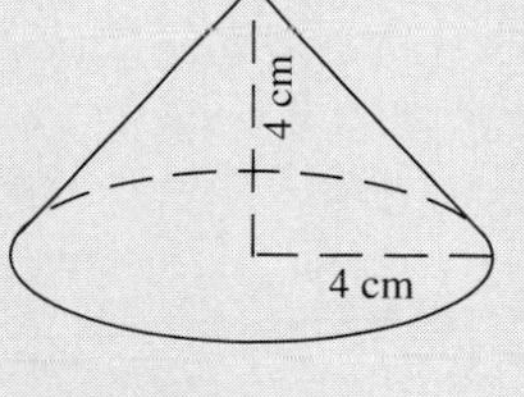

40. ______________

[8.9]

41. Find the missing lengths in the pair of similar triangles.

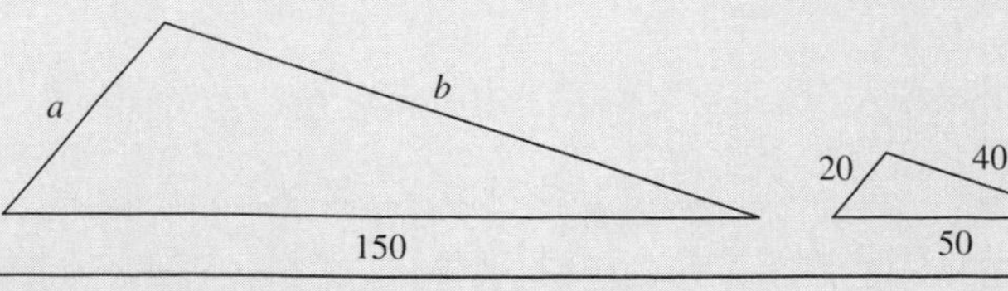

41. ______________

42. ______

43. ______

44. ______

45. ______

46. ______

47. ______

48. ______

49. ______

50. ______

[9.2]

42. Add. $-43 + (-15) + 25$

43. Subtract. $-4 - (-9)$

[9.3]

44. Multiply. $-10 \cdot (-5)$

45. Divide. $\frac{-80}{-4}$

[9.5]

46. Find the value of $7m - 8n + p$ if $m = -2$, $n = 4$, and $p = 3$.

[9.7]

47. Solve this equation. $2k - 5 = 5k - 11$

[10.4]

48. Find the weighted mean for the following. Round to the nearest tenth.

Value	*Frequency*
8	3
9	2
11	6
14	2
16	4
19	1

49. Find the median for the following list of numbers.

24, 32, 36, 102, 116, 184, 212

50. Find the mode or modes for the following list of numbers.

8, 18, 16, 12, 18, 4, 2, 6

WHOLE NUMBERS

1.1 READING AND WRITING WHOLE NUMBERS

OBJECTIVES

1. Identify whole numbers.
2. Give the place value of a digit.
3. Write a number in words or digits.

1 The **decimal system** of writing numbers uses the ten **digits**

0, 1, 2, 3, 4, 5, 6, 7, 8, 9

to write any number. For example, these digits can be used to write the **whole numbers:**

0, 1, 2, 3, 4, 5, 6, 7, 8, 9, 10, 11, 12, 13

and so on.

2 Each digit in a whole number has a **place value.** The following **place value chart** shows the names of the different places used most often.

hundred trillions	ten trillions	trillions	hundred billions	ten billions	billions	hundred millions	ten millions	millions	hundred thousands	ten thousands	thousands	hundreds	tens	ones

EXAMPLE 1 Identifying Whole Numbers

(a) In the whole number 42, the 2 is in the ones place and has a value of 2 **ones.**

(b) In 29, the 2 is in the tens place and has a value of 2 **tens.**

(c) In 281, the 2 is in the hundreds place and has a value of 2 **hundreds.** ■

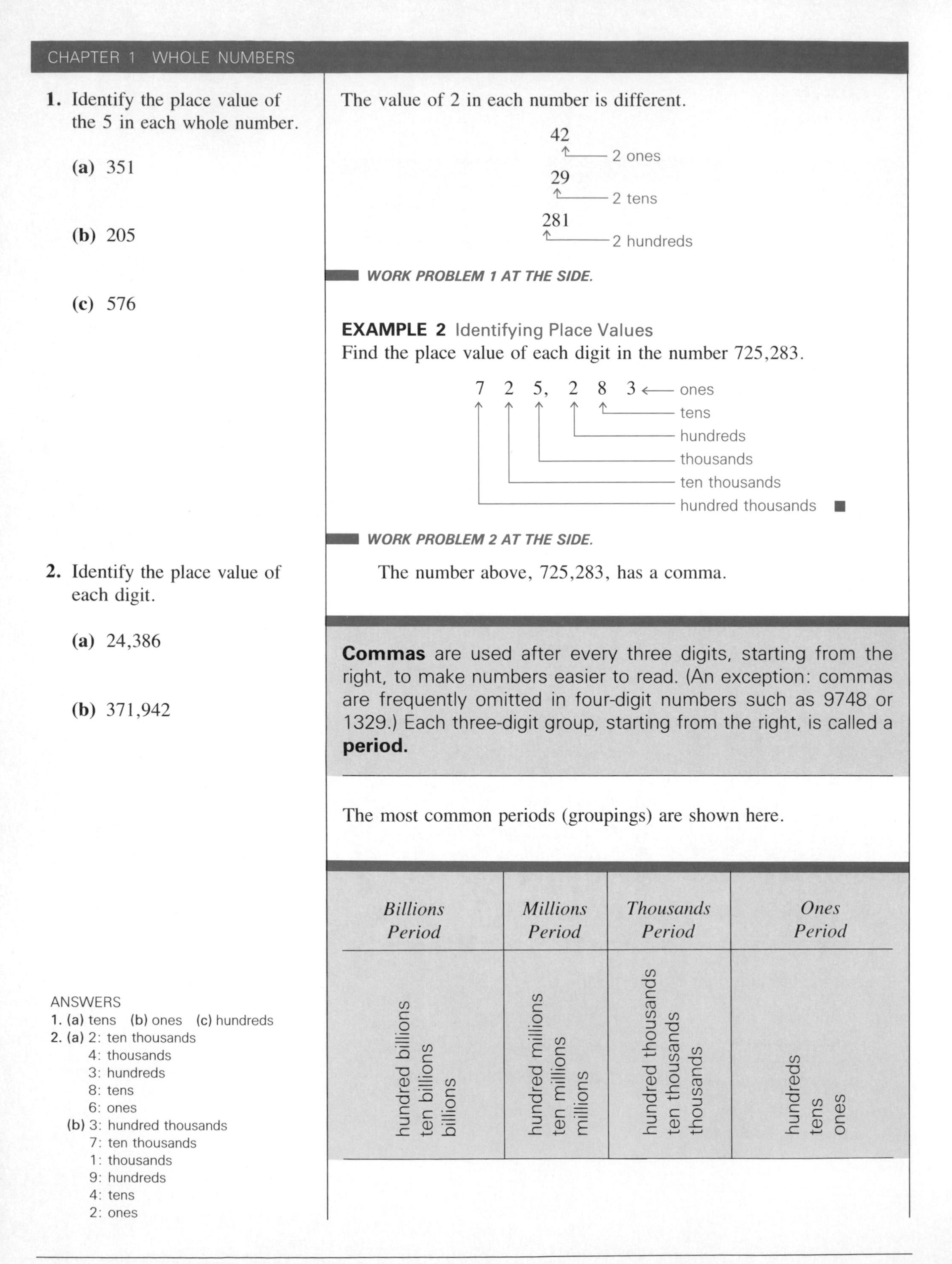

The value of 2 in each number is different.

42 ← 2 ones

29 ← 2 tens

281 ← 2 hundreds

WORK PROBLEM 1 AT THE SIDE.

EXAMPLE 2 Identifying Place Values

Find the place value of each digit in the number 725,283.

7 2 5, 2 8 3

- 3 ← ones
- 8 ← tens
- 2 ← hundreds
- 5 ← thousands
- 2 ← ten thousands
- 7 ← hundred thousands ■

WORK PROBLEM 2 AT THE SIDE.

The number above, 725,283, has a comma.

Commas are used after every three digits, starting from the right, to make numbers easier to read. (An exception: commas are frequently omitted in four-digit numbers such as 9748 or 1329.) Each three-digit group, starting from the right, is called a **period.**

The most common periods (groupings) are shown here.

Billions Period	*Millions Period*	*Thousands Period*	*Ones Period*
hundred billions ten billions billions	hundred millions ten millions millions	hundred thousands ten thousands thousands	hundreds tens ones

1. Identify the place value of the 5 in each whole number.

(a) 351

(b) 205

(c) 576

2. Identify the place value of each digit.

(a) 24,386

(b) 371,942

ANSWERS

1. (a) tens (b) ones (c) hundreds
2. (a) 2: ten thousands
   4: thousands
   3: hundreds
   8: tens
   6: ones
   (b) 3: hundred thousands
   7: ten thousands
   1: thousands
   9: hundreds
   4: tens
   2: ones

EXAMPLE 3 Knowing the Period Names
Write the digits in each period of 8,321,456,795.

billions period	8 ←	8,321,456,795
millions period	321 ←	
thousands period	456 ←	
ones period	795 ←	

■

WORK PROBLEM 3 AT THE SIDE.

Use the following rule to name a number with more than three digits.

> Start at the left when naming a number with more than three digits. Name the digits in each period, followed by the name of the period. The word "and" is *not* used in writing and reading whole numbers.

3 The following examples show how to write names for whole numbers.

EXAMPLE 4 Writing Numbers in Words
Write each number in words.

(a) 57

This number means 5 tens and 7 ones, or 50 ones and 7 ones. Write the number as

fifty-seven.

(b) 94

ninety-four

(c) 874

eight hundred seventy-four ■

WORK PROBLEM 4 AT THE SIDE.

EXAMPLE 5 Writing Numbers in Words by Using Period Names
Write each number in words.

(a) 725,283

seven hundred twenty-five	thousand,	two hundred eighty-three
number in period	name of period	number in period (not necessary to write "ones")

(b) 7835

seven thousand, eight hundred thirty-five

3. In the number 6,728,576,321 identify the digits in each of the following periods.

(a) billions period

(b) millions period

(c) thousands period

(d) ones period

4. Write each number in words.

(a) 27

(b) 68

(c) 293

ANSWERS
3. (a) 6 (b) 728 (c) 576 (d) 321
4. (a) twenty-seven (b) sixty-eight
(c) two hundred ninety-three

(c) 111,356,075

one hundred eleven million, three hundred fifty-six thousand, seventy-five

(d) 6,000,005,000

six million, five thousand ■

NOTE Do not use the word "and" when writing whole numbers.

WORK PROBLEM 5 AT THE SIDE.

EXAMPLE 6 Writing Numbers in Digits
Write each of the following numbers by using digits.

(a) seven thousand, eighty-five

7085

(b) two hundred fifty-six thousand, six hundred twelve

256,612

(c) nine million, seven hundred twenty-one thousand, five hundred fifty-nine

9,721,559 ■

WORK PROBLEM 6 AT THE SIDE.

5. Write each number in words.

(a) 7309

(b) 95,372

(c) 100,075,002

(d) 17,022,040,000

6. Write each of the following numbers by using digits.

(a) three thousand, eight hundred sixty-three

(b) nine hundred, seventy-one thousand, three hundred six

(c) twenty-two million, five hundred sixty thousand, eight hundred twenty-five

ANSWERS
5. (a) seven thousand, three hundred nine (b) ninety-five thousand, three hundred seventy-two (c) one hundred million, seventy-five thousand, two (d) seventeen billion, twenty-two million, forty thousand
6. (a) 3863 (b) 971,306 (c) 22,560,825

name date hour

1.1 EXERCISES

*Fill in the digit for the given **place value** in each of the following whole numbers.*

Example:		**Solution:**
564	hundreds	**5**
	ones	**4**

1. 1786
thousands
tens

2. 9225
thousands
ones

3. 18,015
ten thousands
hundreds

4. 75,229
ten thousands
ones

5. 4,702,201
millions
thousands

6. 1,700,225,016
billions
millions

*Fill in the number for the given **period** in each of the following whole numbers.*

Example:		**Solution:**
58,618	thousands	**58**
	ones	**618**

7. 71,105
thousands
ones

8. 28,785,203
millions
thousands
ones

9. 60,000,502,109
billions
millions
thousands
ones

10. 100,258,100,006
billions
millions
thousands
ones

Rewrite the following numbers in words.

Example: 1,630,254 **Solution: one million, six hundred thirty thousand, two hundred fifty-four**

11. 9613

12. 37,886

13. 725,659

14. 3,218,933

15. 75,756,665

16. 199,993,101

Rewrite each of the following numbers by using digits.

Example: three thousand, four hundred twenty **Solution: 3420**

17. three thousand, one hundred thirty-five

18. fifty-eight thousand, six hundred eight

19. seven hundred eighty-five thousand, two hundred twenty-three

20. one hundred million, two hundred

Rewrite the numbers from the following sentences by using digits.

Example: Her income tax refund was one thousand, five hundred twelve dollars.
Solution: In digits, the number is **1512.**

21. A bottle of a certain vaccine will give two thousand seventy injections.

22. Every year, nine hundred seventy-two thousand, six hundred fifty people visit a certain historical area.

23. A supermarket has thirteen thousand, one hundred twelve different items for sale.

24. The Chicago area has six million, two hundred seven thousand telephones.

CHALLENGE EXERCISES

25. Write eight hundred billion, six hundred twenty-one million, twenty thousand, two hundred fifteen by using digits.

26. Write 306,735,620,102 in words.

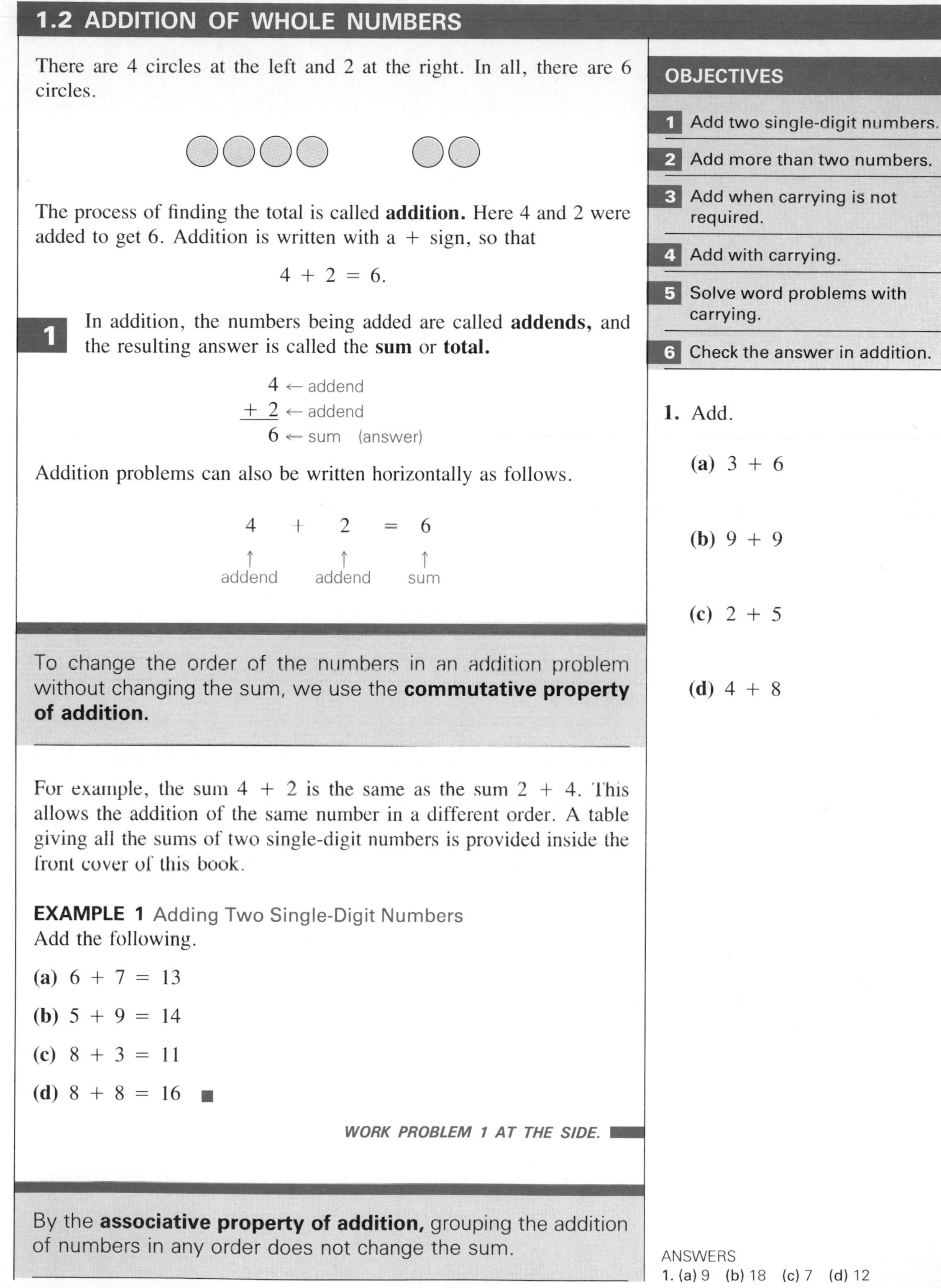

1.2 ADDITION OF WHOLE NUMBERS

OBJECTIVES

1. Add two single-digit numbers.
2. Add more than two numbers.
3. Add when carrying is not required.
4. Add with carrying.
5. Solve word problems with carrying.
6. Check the answer in addition.

There are 4 circles at the left and 2 at the right. In all, there are 6 circles.

The process of finding the total is called **addition.** Here 4 and 2 were added to get 6. Addition is written with a + sign, so that

$$4 + 2 = 6.$$

1 In addition, the numbers being added are called **addends,** and the resulting answer is called the **sum** or **total.**

$$\begin{array}{r l} 4 & \leftarrow \text{addend} \\ +\ 2 & \leftarrow \text{addend} \\ \hline 6 & \leftarrow \text{sum (answer)} \end{array}$$

Addition problems can also be written horizontally as follows.

$$\underset{\text{addend}}{\underset{\uparrow}{4}} \quad + \quad \underset{\text{addend}}{\underset{\uparrow}{2}} \quad = \quad \underset{\text{sum}}{\underset{\uparrow}{6}}$$

To change the order of the numbers in an addition problem without changing the sum, we use the **commutative property of addition.**

For example, the sum 4 + 2 is the same as the sum 2 + 4. This allows the addition of the same number in a different order. A table giving all the sums of two single-digit numbers is provided inside the front cover of this book.

EXAMPLE 1 Adding Two Single-Digit Numbers

Add the following.

(a) 6 + 7 = 13

(b) 5 + 9 = 14

(c) 8 + 3 = 11

(d) 8 + 8 = 16 ■

WORK PROBLEM 1 AT THE SIDE.

By the **associative property of addition,** grouping the addition of numbers in any order does not change the sum.

1. Add.

(a) 3 + 6

(b) 9 + 9

(c) 2 + 5

(d) 4 + 8

ANSWERS
1. (a) 9 (b) 18 (c) 7 (d) 12

2. Add the following columns of numbers.

(a)
$$\begin{array}{r} 7 \\ 5 \\ 2 \\ 3 \\ +\ 8 \\ \hline \end{array}$$

(b)
$$\begin{array}{r} 6 \\ 2 \\ 1 \\ 5 \\ +\ 7 \\ \hline \end{array}$$

(c)
$$\begin{array}{r} 9 \\ 2 \\ 1 \\ 3 \\ +\ 4 \\ \hline \end{array}$$

(d)
$$\begin{array}{r} 5 \\ 4 \\ 8 \\ 3 \\ +\ 6 \\ \hline \end{array}$$

ANSWERS
2. (a) 25 (b) 21 (c) 19 (d) 26

For example, the sum of $3 + 5 + 6$ may be found as follows.

$(3 + 5) + 6 = 8 + 6 = 14$ Parentheses tell what to do first.

Another way to add the same numbers is shown below.

$$3 + (5 + 6) = 3 + 11 = 14$$

Either method gives the answer 14.

2 To add several numbers, first write them in a column. Add the first number to the second. Add this sum to the third digit; continue until all the digits are used.

EXAMPLE 2 Adding More Than Two Numbers
Add 2, 5, 6, 1, and 4.

SOLUTION

$$\begin{array}{r} 2 \\ 5 \\ 6 \\ 1 \\ +\ 4 \\ \hline 18 \end{array}$$

■

NOTE By the commutative property of addition, numbers may be added from the bottom.

WORK PROBLEM 2 AT THE SIDE.

3 If numbers have two or more digits, first arrange the numbers in columns so that the ones digits are in the same column, tens are in the same column, hundreds are in the same column, and so on. Next, add.

EXAMPLE 3 Adding without Carrying
Add 511, 23, 154, and 10.

SOLUTION
First arrange the numbers in a column, with ones digits at the right.

$$\begin{array}{r} 511 \\ 23 \\ 154 \\ +\ 10 \\ \hline \end{array}$$

Start at the right and add the ones digits. Add the tens digits next, and finally, the hundreds digits.

```
   5 1 1
     2 3
   1 5 4
+    1 0
---------
   6 9 8
   ↑ ↑ ↑
   │ │ └──── sum of ones
   │ └────── sum of tens
   └──────── sum of hundreds
```

The sum of the four numbers is 698. ■

WORK PROBLEM 3 AT THE SIDE.

4 If the sum of the digits in a column is more than 9, use **carrying.**

EXAMPLE 4 Adding with Carrying

Add 27 and 49.

SOLUTION

Add ones.

```
   27
+  49
    ↑── Sum of ones is 16.
```

Since 16 is 1 ten plus 6 ones, place 6 in the ones column and carry 1 to the tens column.

```
   1 ←───────┐
   2 7       │  7 + 9 = 16
+  4 9       │
     6 ←─────┘
```

Add in the tens column.

```
   1
   27
+  49
   76
   ↑── sum of digits in tens column ■
```

WORK PROBLEM 4 AT THE SIDE.

3. Add.

(a) 12 + 57

(b) 153 + 346

(c) 42,305 + 11,563

4. Add by using carrying.

(a) 48 + 34

(b) 66 + 28

(c) 39 + 43

(d) 27 + 9

ANSWERS
3. (a) 69 (b) 499 (c) 53,868
4. (a) 82 (b) 94 (c) 82 (d) 36

5. Add by carrying as necessary.

(a)
```
   576
    92
    43
+  274
```

(b)
```
  4271
   372
  8976
+  162
```

(c)
```
    57
     4
   392
   804
    51
+   27
```

(d)
```
  3214
   762
    98
   302
+  154
```

ANSWERS
5. (a) 985 (b) 13,781 (c) 1335 (d) 4530

EXAMPLE 5 Adding with Carrying

Add 324 + 7855 + 23 + 7 + 86.

SOLUTION

Step 1 Add the digits in the ones column.

```
     2
   324
  7855
    23
     7
+   86
     5
```

Carry to the tens column.

Sum of the ones column is 25.

Write in the ones column.

In 25, the 5 represents 5 ones and is written in the ones column, while 2 represents 2 tens and is carried to the tens column.

Step 2 Now add the digits in the tens column, including the carried 2.

```
    12
   324
  7855
    23
     7
+   86
    95
```

Carry to the hundreds column.

Sum of the tens column is 19.

Write in the tens column.

Step 3

```
   112
   324
  7855
    23
     7
+   86
   295
```

Carry to the thousands column.

Sum of the hundreds column is 12.

Write in the hundreds column.

Step 4

```
  1 12
   324
  7855
   823
     7
+   86
  8295
```

Sum of the thousands column is 8.

Write in the thousands column.

Finally, 324 + 7855 + 23 + 7 + 86 = 8295. ■

WORK PROBLEM 5 AT THE SIDE.

NOTE For additional speed, try to carry mentally. Do not write the number carried, but just carry the number mentally to the top of the next column being added.

WORK PROBLEM 6 AT THE SIDE.

5 In Section 1.9 we will describe how to solve word problems in more detail. The next two examples have word problems that require adding.

EXAMPLE 6 Applying Addition Skills

On this map, the distance in miles from one town to another is written alongside the road. Find the shortest distance from Auburn to Folsom.

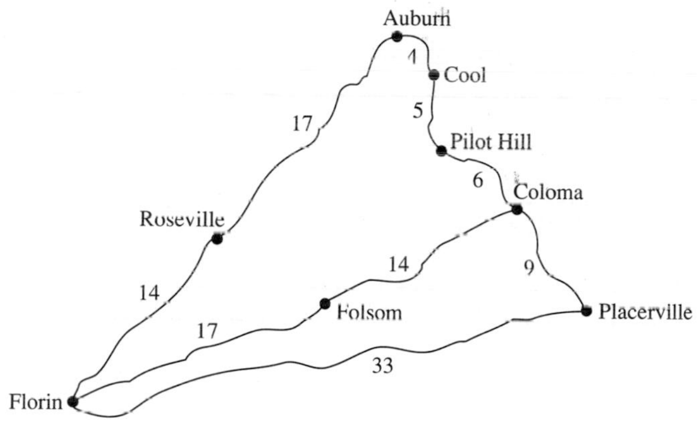

Auburn
4
Cool
5
17
Pilot Hill
6
Coloma
Roseville
14
9
14
Folsom
Placerville
17
33
Florin

SOLUTION

One way from Auburn to Folsom is through Coloma. Add the mileage numbers along this route.

4	Auburn to Cool
5	Cool to Pilot Hill
6	Pilot Hill to Coloma
+ 14	Coloma to Folsom
29	miles from Auburn to Folsom, going through Coloma

Another way is through Roseville and Florin. Add the mileage numbers along this route.

17	Auburn to Roseville
14	Roseville to Florin
+ 17	Florin to Folsom
48	miles from Auburn to Folsom, going through Roseville and Florin

The shortest way from Auburn to Folsom is through Coloma. ■

WORK PROBLEM 7 AT THE SIDE.

6. Add with mental carrying.

(a)
$$\begin{array}{r} 315 \\ 986 \\ 17 \\ 2 \\ 5 \\ +\ 8224 \\ \hline \end{array}$$

(b)
$$\begin{array}{r} 5280 \\ 407 \\ 805 \\ 31 \\ 20 \\ 4 \\ +\ \ \ 2 \\ \hline \end{array}$$

(c)
$$\begin{array}{r} 15{,}829 \\ 765 \\ 78 \\ 15 \\ 9 \\ 7 \\ +\ 13{,}179 \\ \hline \end{array}$$

7. Use the map and find the shortest distance from Florin to Coloma.

ANSWERS
6. (a) 9549 (b) 6549 (c) 29,882
7. 31 miles

8. The bridge between Cool and Auburn has been closed, so this route cannot be used. Use the map to find the next shortest distance from Cool to Auburn.

9. Check the following additions. If an answer is incorrect, give the correct answer.

(a)
```
   16
    3
    5
 + 27
   51
```

(b)
```
  715
  622
   38
+ 198
 1573
```

(c)
```
   79
  218
    7
+ 639
  953
```

(d)
```
  21,892
  11,746
+ 43,925
  79,563
```

ANSWERS

8.
Cool to Pilot Hill	5
Pilot Hill to Coloma	6
Coloma to Folsom	14
Folsom to Florin	17
Florin to Roseville	14
Roseville to Auburn	17
Cool to Auburn	73 miles

9. (a) correct (b) correct (c) incorrect, should be 943 (d) incorrect, should be 77,563

EXAMPLE 7 Finding a Total

Find the total distance from Placerville to Auburn to Florin and back to Placerville.

SOLUTION

Use the numbers from the map.

9	Placerville to Coloma
6	Coloma to Pilot Hill
5	Pilot Hill to Cool
4	Cool to Auburn
17	Auburn to Roseville
14	Roseville to Florin
+ 33	Florin to Placerville
88	miles from Placerville to Auburn to Florin to Placerville ■

WORK PROBLEM 8 AT THE SIDE.

6 Checking the answer is an important part of problem solving. A common method for checking addition is to re-add from bottom to top.

EXAMPLE 8 Checking Addition

Check the following addition.

```
(add down)      1428   ↑  Adding down and
    |            738   |    adding up should give
    |             63   |    the same answer.
    |            125   |
    |             17   |
    |          + 485   |  (Add up.)
    ↓           1428   └─ check
```

Here the answers agree, so the sum is probably correct. ■

EXAMPLE 9 Checking Addition

Check the following additions.

```
                 1033  ↑  correct, since both
(a)   785         785  |    answers are the same
       63          63  |  (Add up.)
    + 185       + 185  └─ check
     1033        1033

                 2454  ↑  error, since answers
(b)   635         635  |    are different
       73          73  |  (Add up.)
      831         831  |
    + 915       + 915  └─ check
     2444        2444
```

Re-add to find that the correct answer is 2454. ■

WORK PROBLEM 9 AT THE SIDE.

name date hour

1.2 EXERCISES

Add.

Examples:

$$\begin{array}{r} 57 \\ +\ 42 \\ \hline \end{array}$$

23 + 721 + 834

Solutions:

Line up numbers in a column, if necessary.

$$\begin{array}{r} 57 \\ +\ 42 \\ \hline \mathbf{99} \end{array}$$

(ones added; tens added)

$$\begin{array}{r} 23 \\ 721 \\ +\ 834 \\ \hline \mathbf{1578} \end{array}$$

1. $\begin{array}{r} 15 \\ +\ 24 \\ \hline \end{array}$ **2.** $\begin{array}{r} 10 \\ +\ 17 \\ \hline \end{array}$ **3.** $\begin{array}{r} 18 \\ +\ 61 \\ \hline \end{array}$ **4.** $\begin{array}{r} 87 \\ +\ 12 \\ \hline \end{array}$ **5.** $\begin{array}{r} 17 \\ +\ 52 \\ \hline \end{array}$

6. $\begin{array}{r} 651 \\ +\ 228 \\ \hline \end{array}$ **7.** $\begin{array}{r} 158 \\ +\ 340 \\ \hline \end{array}$ **8.** $\begin{array}{r} 135 \\ 253 \\ +\ 410 \\ \hline \end{array}$ **9.** $\begin{array}{r} 6310 \\ 252 \\ +\ 1223 \\ \hline \end{array}$ **10.** $\begin{array}{r} 121 \\ 5705 \\ +\ 3163 \\ \hline \end{array}$

11. 412 + 234 + 143

12. 131 + 242 + 311

13. 1251 + 4311 + 2114

14. 3241 + 1513 + 2014

15. 12,142 + 43,201 + 23,103

16. 41,124 + 12,302 + 23,500

17. 8642 + 7134

18. 9258 + 3421

19. 38,204 + 91,020

20. 53,251 + 96,305

Add the following numbers by carrying as necessary.

Example:

185 + 769

Solution:

1 1

185 + 769 = **954**

21. 89 + 38

22. 76 + 35

23. 65 + 77

24. 29 + 98

25. 47 + 74

26. 99 + 99

27. 78 + 83

28. 92 + 39

29. 73 + 29

30. 68 + 37

31. 906 + 875

32. 621 + 359

33. 201 + 769

34. 364 + 158

35. 278 + 135

36. 172 + 156

37. 928 + 843

38. 686 + 726

39. 157 + 968

40. 448 + 665

41. 7968 + 1285

42. 1768 + 8275

43. 2896 + 3728

44. 9382 + 7586

45. 1625 + 7986

46. 6829 + 6076 + 8218

47. 2056 + 78 + 5089 + 731

48. 6022 + 709 + 1621 + 37

49. 18 + 708 + 9286 + 636

50. 1708 + 321 + 61 + 8926

51. 218 + 7022 + 335 + 9283

52. 2505 + 173 + 6044 + 168

53. 1321 + 603 + 8 + 21 + 1604

54. 5631 + 7983 + 7 + 36 + 505

55. 2109 + 63 + 16 + 3 + 9887

56.	57.	58.	59.	60.
244	553	3187	413	576
67	97	810	85	7934
7076	2772	527	9919	60
13	437	76	602	781
618	63	2665	31	5968
+ 3005	+ 328	+ 317	+ 1218	+ 371

Check the following additions. If an answer is incorrect, give the correct answer.

Example:	**Solution:** 1535	correct
635	635	
378	378	
+ 522	+ 522	
1535	1535	

61.	62.	63.	64.	65.
729	371	179	17	713
372	820	214	296	28
+ 159	+ 914	+ 376	713	615
1260	2105	759	+ 94	+ 64
			1220	1420

66.	67.	68.	69.	70.
6 215	678	516	4 714	6715
744	7 952	8 760	27	283
36	56	24	77	617
+ 4 284	718	189	8 878	13
11,279	+ 2 173	+ 1 723	+ 636	+ 81
	11,377	11,212	14,332	7719

Using the map above, find the shortest distance between the following cities.

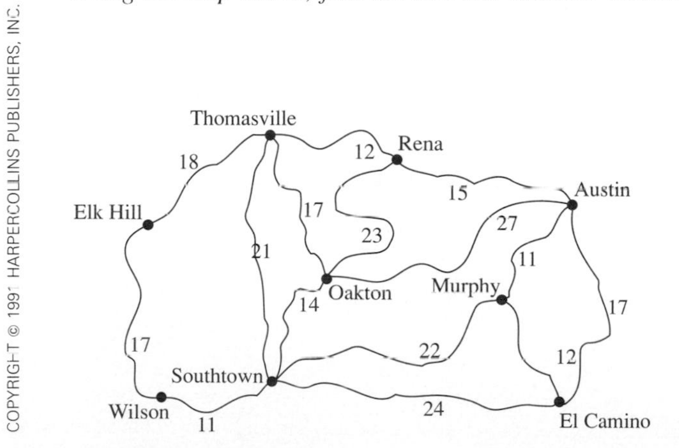

71. Wilson and Thomasville

72. Wilson and Rena

73. Southtown and Austin

74. El Camino and Thomasville

Solve the following word problems by using addition.

75. A box of apples costs \$19 and a box of peaches \$17. Find the total cost for a box of each.

76. Bill Poole has 55 nickels, 74 dimes, and 106 quarters. How many coins does he have altogether?

77. There were 275 women and 345 men at the city council meeting. How many people were at the meeting?

78. One floor of a hospital has 155 patients, while another floor has 173 patients. How many patients are on the two floors?

79. At a charity bazaar, a church has a total of 1792 books for sale, while a lodge has 3259 books for sale. How many books are for sale?

80. A plane is flying at an altitude of 5924 feet. It then increases its altitude by 7284 feet. Find its new altitude.

CHALLENGE EXERCISES

Find the perimeter or total distance around each of the following figures.

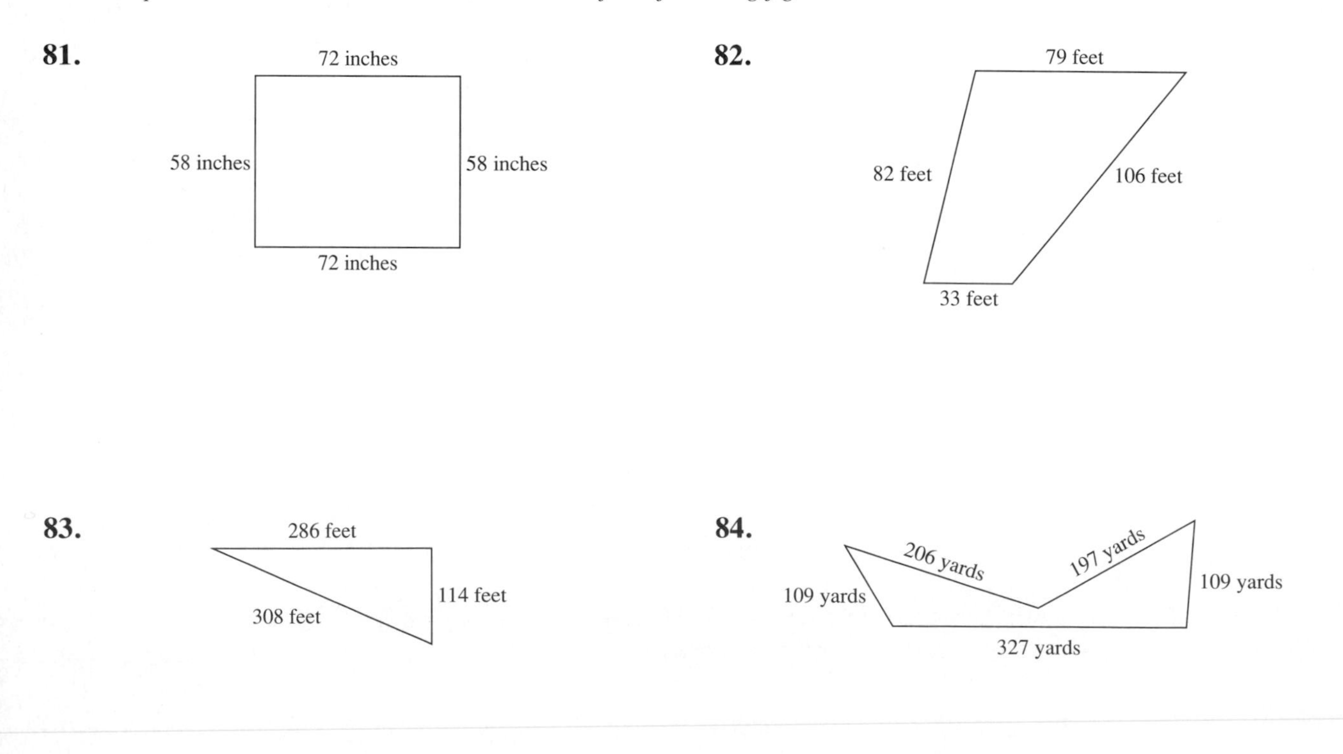

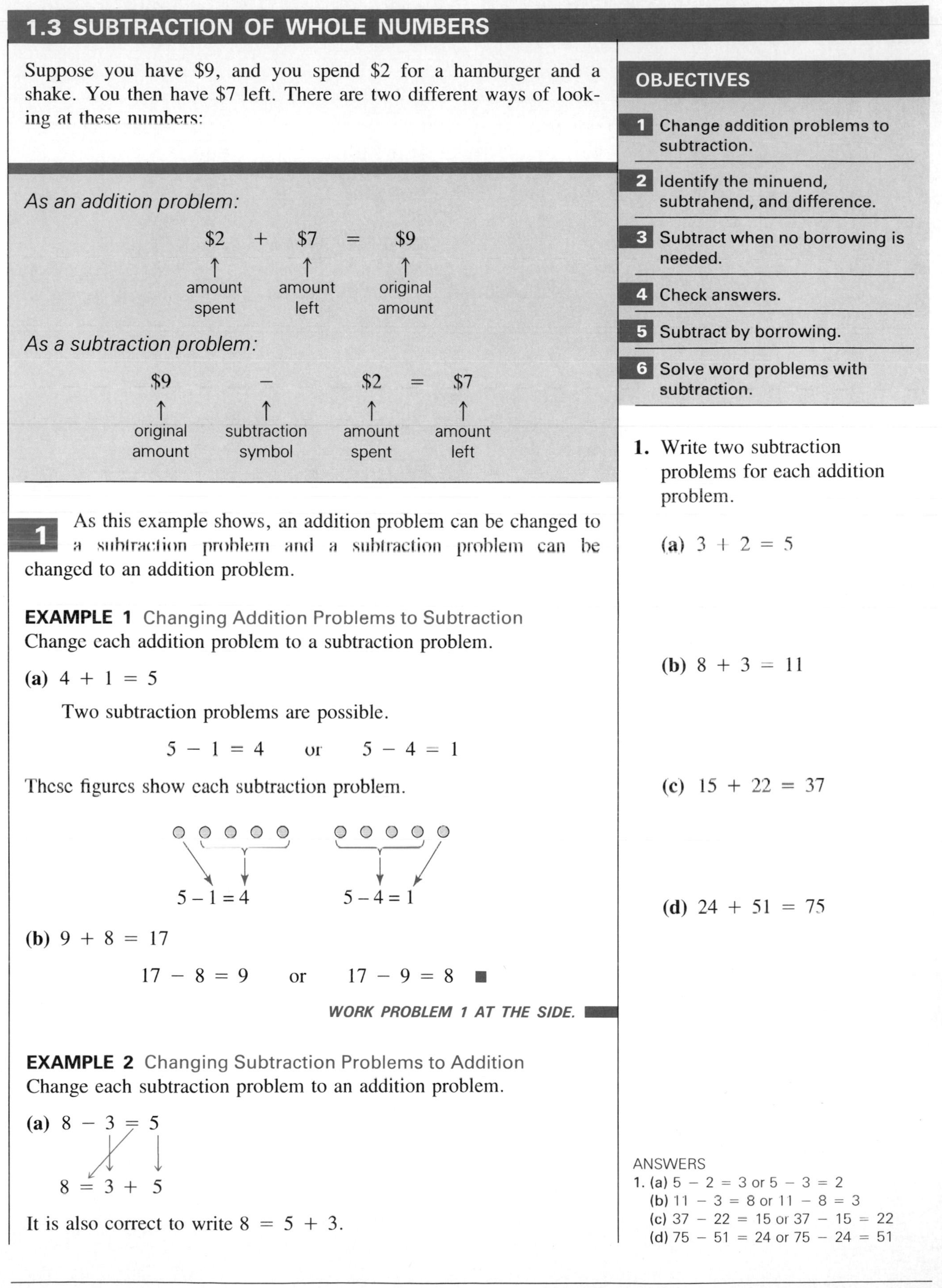

1.3 SUBTRACTION OF WHOLE NUMBERS

Suppose you have \$9, and you spend \$2 for a hamburger and a shake. You then have \$7 left. There are two different ways of looking at these numbers:

As an addition problem:

\$2	+	\$7	=	\$9
↑		↑		↑
amount spent		amount left		original amount

As a subtraction problem:

\$9	−	\$2	=	\$7
↑	↑	↑		↑
original amount	subtraction symbol	amount spent		amount left

OBJECTIVES

1. Change addition problems to subtraction.
2. Identify the minuend, subtrahend, and difference.
3. Subtract when no borrowing is needed.
4. Check answers.
5. Subtract by borrowing.
6. Solve word problems with subtraction.

1 As this example shows, an addition problem can be changed to a subtraction problem and a subtraction problem can be changed to an addition problem.

EXAMPLE 1 Changing Addition Problems to Subtraction
Change each addition problem to a subtraction problem.

(a) $4 + 1 = 5$

Two subtraction problems are possible.

$$5 - 1 = 4 \quad \text{or} \quad 5 - 4 = 1$$

These figures show each subtraction problem.

$5 - 1 = 4$ $\qquad$ $5 - 4 = 1$

(b) $9 + 8 = 17$

$$17 - 8 = 9 \quad \text{or} \quad 17 - 9 = 8 \ \blacksquare$$

WORK PROBLEM 1 AT THE SIDE.

EXAMPLE 2 Changing Subtraction Problems to Addition
Change each subtraction problem to an addition problem.

(a) $8 - 3 = 5$

$8 = 3 + 5$

It is also correct to write $8 = 5 + 3$.

1. Write two subtraction problems for each addition problem.

(a) $3 + 2 = 5$

(b) $8 + 3 = 11$

(c) $15 + 22 = 37$

(d) $24 + 51 = 75$

ANSWERS
1. (a) $5 - 2 = 3$ or $5 - 3 = 2$
(b) $11 - 3 = 8$ or $11 - 8 = 3$
(c) $37 - 22 = 15$ or $37 - 15 = 22$
(d) $75 - 51 = 24$ or $75 - 24 = 51$

(b) $17 - 15 = 2$
$17 = 15 + 2$

(c) $29 - 13 = 16$
$29 = 13 + 16$ ■

WORK PROBLEM 2 AT THE SIDE.

2 In subtraction, as in addition, the numbers in a problem have names. For example, in the box above, $9 - 2 = 7$, the number 9 is the **minuend,** 2 is the **subtrahend,** and 7 is the **difference** or answer.

$9 - 2 = 7$ ← difference
↑ minuend ↑ subtrahend

9 ← minuend
$-\ 2$ ← subtrahend
7 ← difference

3 Subtract two numbers by lining up the numbers, so the digits in the ones place are in the same column. Next, subtract by columns, starting at the right with the ones column.

EXAMPLE 3 Subtracting Two Numbers
Subtract.

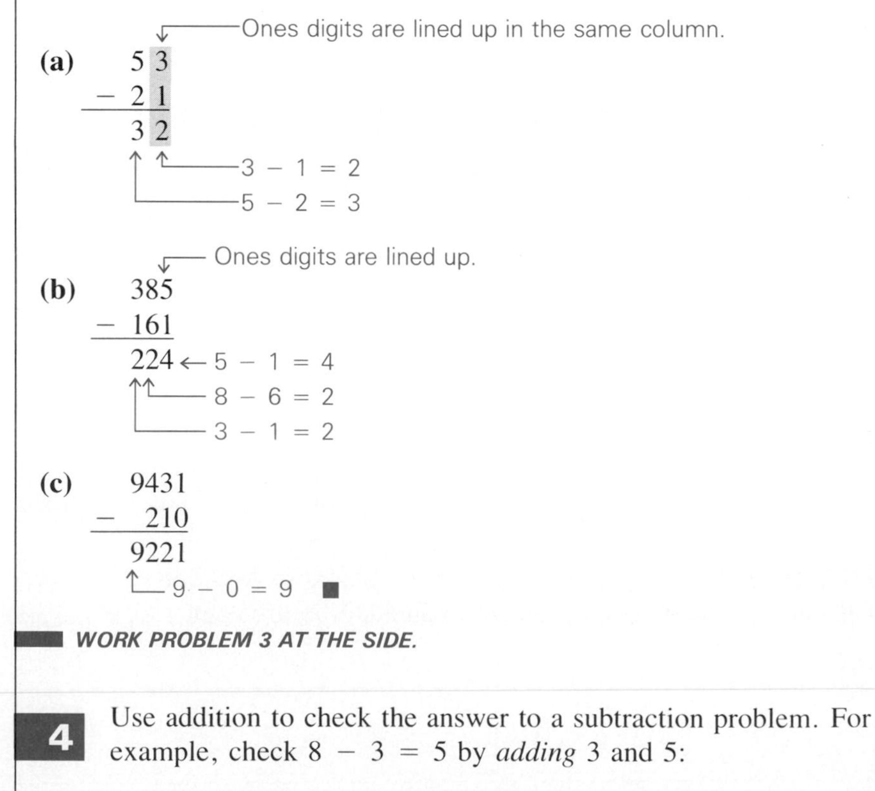

WORK PROBLEM 3 AT THE SIDE.

4 Use addition to check the answer to a subtraction problem. For example, check $8 - 3 = 5$ by *adding* 3 and 5:

$3 + 5 = 8$, so $8 - 3 = 5$ is correct.

2. Write an addition problem for each subtraction problem.

(a) $4 - 1 = 3$

(b) $7 - 2 = 5$

(c) $21 - 15 = 6$

(d) $58 - 42 = 16$

3. Subtract.

(a) $94 - 73$

(b) $57 - 25$

(c) $378 - 235$

(d) $4348 - 1237$

(e) $5464 - 324$

ANSWERS
2. (a) $4 = 1 + 3$ (b) $7 = 2 + 5$ (c) $21 = 15 + 6$ (d) $58 = 42 + 16$
3. (a) 21 (b) 32 (c) 143 (d) 3111 (e) 5140

EXAMPLE 4 Checking Subtraction
Check each answer.

(a)
$$\begin{array}{r} 89 \\ -\ 47 \\ \hline 42 \end{array}$$

Rewrite as an addition problem, as shown in Section 1.2.

subtraction problem
$$\begin{array}{r} 89 \\ -\ 47 \\ \hline 42 \\ 89 \end{array}$$
addition problem
$$\begin{array}{r} 47 \\ +\ 42 \\ \hline 89 \end{array}$$

Since 47 + 42 = 89, the answer is 42.

(b) 72 − 41 = 21

Rewrite as an addition problem.

$$72 = 41 + 21$$

Since 41 + 21 = 62, 21 does *not* check as the answer to 72 − 41. The answer is incorrect. Rework the original subtraction to get the correct answer, 31.

(c)
$$\begin{array}{r} 374 \\ 141 \\ \hline 233 \end{array}$$
374 ← match → 141 + 233 = 374

The answer checks. ■

WORK PROBLEM 4 AT THE SIDE.

5 If a digit in subtraction is larger than the one directly above, **borrowing** will be necessary.

EXAMPLE 5 Subtracting with Borrowing
Subtract 19 from 57.

SOLUTION
Write the problem.
$$\begin{array}{r} 57 \\ -\ 19 \\ \hline \end{array}$$

In the ones column, 9 is larger than 7, so borrow a 10 from the 5 (which represents 5 tens, or 50).

50 − 10 = 40 → 4   17 ← 10 + 7 = 17
$$\begin{array}{r} \overset{4}{\not 5}\ \overset{17}{\not 7} \\ -\ 1\ 9 \\ \hline \end{array}$$

Now subtract 9 from 17, and then 1 from 4.

$$\begin{array}{r} \overset{4}{\not 5}\ \overset{17}{\not 7} \\ -\ 1\ 9 \\ \hline 3\ 8 \end{array}$$
difference

Finally, 57 − 19 = 38. Check by adding 19 and 38. ■

WORK PROBLEM 5 AT THE SIDE.

4. Decide whether these answers are correct. If incorrect, what should they be?

(a)
$$\begin{array}{r} 52 \\ -\ 32 \\ \hline 20 \end{array}$$

(b)
$$\begin{array}{r} 46 \\ -\ 32 \\ \hline 24 \end{array}$$

(c)
$$\begin{array}{r} 587 \\ -\ 342 \\ \hline 345 \end{array}$$

(d)
$$\begin{array}{r} 9742 \\ -\ 8131 \\ \hline 1611 \end{array}$$

5. Subtract.

(a)
$$\begin{array}{r} 72 \\ -\ 38 \\ \hline \end{array}$$

(b)
$$\begin{array}{r} 54 \\ -\ 37 \\ \hline \end{array}$$

(c)
$$\begin{array}{r} 21 \\ -\ 7 \\ \hline \end{array}$$

(d)
$$\begin{array}{r} 863 \\ -\ 47 \\ \hline \end{array}$$

(e)
$$\begin{array}{r} 951 \\ -\ 39 \\ \hline \end{array}$$

ANSWERS
4. **(a)** correct **(b)** incorrect, should be 14 **(c)** incorrect, should be 245
5. **(a)** 34 **(b)** 17 **(c)** 14 **(d)** 816 **(e)** 912

6. Subtract.

(a) $\begin{array}{r} 129 \\ -\ 53 \\ \hline \end{array}$

(b) $\begin{array}{r} 457 \\ -\ 68 \\ \hline \end{array}$

(c) $\begin{array}{r} 653 \\ -\ 379 \\ \hline \end{array}$

(d) $\begin{array}{r} 1328 \\ -\ 879 \\ \hline \end{array}$

(e) $\begin{array}{r} 8739 \\ -\ 3892 \\ \hline \end{array}$

ANSWERS
6. (a) 76 (b) 389 (c) 274 (d) 449 (e) 4847

EXAMPLE 6 Subtracting with Borrowing
Subtract by borrowing as necessary.

(a) $\begin{array}{r} 7856 \\ -\ 137 \\ \hline \end{array}$

There is no need to borrow, since 4 is greater than 3.

10 + 6 = 16

$$\begin{array}{rrrrr} & & & 4 & 16 \\ & 7 & 8 & \not{5} & \not{6} \\ - & & 1 & 3 & 7 \\ \hline & 7 & 7 & 1 & 9 \end{array} \quad \text{difference}$$

(b) $\begin{array}{r} 635 \\ -\ 546 \\ \hline \end{array}$

100 + 20 = 120

600 − 100 = 500

10 + 5 = 15

$$\begin{array}{rrrr} & 5 & 12 & 15 \\ & \not{6} & \not{3} & \not{5} \\ - & 5 & 4 & 6 \\ \hline & & 8 & 9 \end{array} \quad \text{difference}$$

(c) $\begin{array}{r} 412 \\ -\ 225 \\ \hline \end{array}$

$$\begin{array}{rrrr} & 3 & 10 & 12 \\ & \not{4} & \not{1} & \not{2} \\ - & 2 & 2 & 5 \\ \hline & 1 & 8 & 7 \end{array}$$ ■

WORK PROBLEM 6 AT THE SIDE.

Sometimes a minuend has zeros in some of the positions. In such cases, borrowing may be a little more complicated than what we have shown so far.

EXAMPLE 7 Borrowing with Zeros
Subtract.

$$\begin{array}{r} 4607 \\ -\ 3168 \\ \hline \end{array}$$

SOLUTION
It is not possible to borrow from the tens position. Instead, first borrow from the hundreds position.

600 − 100 = 500 100 + 0 = 100

$$\begin{array}{rrrrr} & & 5 & 10 & \\ & 4 & \not{6} & \not{0} & 7 \\ - & 3 & 1 & 6 & 8 \\ \hline \end{array}$$

Now we may borrow from the tens position.

100 − 10 = 90

10 + 7 = 17

$$\begin{array}{rrrrr} & & & 9 & \\ & & 5 & \not{10} & 17 \\ & 4 & \not{6} & \not{0} & \not{7} \\ - & 3 & 1 & 6 & 8 \\ \hline & & & & 9 \end{array}$$

Complete the problem.

$$\begin{array}{r} 4\ \overset{5}{\cancel{6}}\ \overset{\overset{9}{\cancel{10}}}{\cancel{0}}\ \overset{17}{\cancel{7}} \\ -\ 3\ 1\ 6\ 8 \\ \hline 1\ 4\ 3\ 9 \end{array} \quad \text{difference}$$

As above, check by adding 1439 and 3168; you should get 4607. ■

WORK PROBLEM 7 AT THE SIDE.

EXAMPLE 8 Borrowing with Zeros

Subtract.

(a) $\begin{array}{r} 708 \\ -\ 149 \\ \hline \end{array}$

100 + 0 = 100 100 − 10 = 90

700 − 100 = 600 10 + 8 = 18

$$\begin{array}{r} \overset{6}{\cancel{7}}\ \overset{\overset{9}{\cancel{10}}}{\cancel{0}}\ \overset{18}{\cancel{8}} \\ -\ 1\ 4\ 9 \\ \hline 5\ 5\ 9 \end{array}$$

(b) $\begin{array}{r} 380 \\ -\ 276 \\ \hline \end{array}$

80 − 10 = 70 10 + 0 = 10

$$\begin{array}{r} 3\ \overset{7}{\cancel{8}}\ \overset{10}{\cancel{0}} \\ -\ 2\ 7\ 6 \\ \hline 1\ 0\ 4 \end{array}$$

(c) $\begin{array}{r} 7000 \\ -4999 \\ \hline \end{array}$

$$\begin{array}{r} \overset{6}{\cancel{7}}\ \overset{\overset{9}{\cancel{10}}}{\cancel{0}}\ \overset{\overset{9}{\cancel{10}}}{\cancel{0}}\ \overset{10}{\cancel{0}} \\ -\ 4\ 9\ 9\ 9 \\ \hline 2\ 0\ 0\ 1 \end{array}$$ ■

WORK PROBLEM 8 AT THE SIDE.

As explained above, an answer to a subtraction problem can be checked by adding.

EXAMPLE 9 Checking Subtraction

Check the following answers.

(a) $\begin{array}{r} 613 \\ -\ 275 \\ \hline 338 \end{array}$ check $\begin{array}{r} 338 \\ +\ 275 \\ \hline 613 \end{array}$

match correct

7. Subtract.

(a) $\begin{array}{r} 508 \\ -\ 352 \\ \hline \end{array}$

(b) $\begin{array}{r} 305 \\ -\ 247 \\ \hline \end{array}$

(c) $\begin{array}{r} 7045 \\ -\ 1324 \\ \hline \end{array}$

8. Subtract.

(a) $\begin{array}{r} 402 \\ -\ 378 \\ \hline \end{array}$

(b) $\begin{array}{r} 570 \\ -\ 163 \\ \hline \end{array}$

(c) $\begin{array}{r} 1570 \\ -\ 983 \\ \hline \end{array}$

(d) $\begin{array}{r} 9001 \\ -\ 7193 \\ \hline \end{array}$

(e) $\begin{array}{r} 6000 \\ -\ 2768 \\ \hline \end{array}$

ANSWERS
7. (a) 156 (b) 58 (c) 5721
8. (a) 24 (b) 407 (c) 587 (d) 1808 (e) 3232

(b)

```
   1915        check
 - 1635
    280          280
  match       + 1635
              → 1915     correct
```

(c)

```
  15,803       check
 -  7 325
    8 578         8 578
  no match      + 7 325
              → 15,903    error
```

Rework the original problem to get the correct answer, 8478. ■

WORK PROBLEM 9 AT THE SIDE.

As shown in the next example, subtraction can be used to solve a word problem.

EXAMPLE 10 Applying Subtraction Skills

Using the table below, decide how many more miles Scott traveled on Saturday than on Friday.

SCOTT'S VACATION TRAVEL

Day	*Number of Miles Traveled*
Sunday	572
Monday	385
Tuesday	78
Wednesday	22
Thursday	446
Friday	98
Saturday	275

SOLUTION

Scott traveled only 98 miles on Friday, but he traveled 275 miles on Saturday. Find how many more miles were traveled on Saturday than on Friday by subtracting 98 from 275.

```
   275    miles on Saturday
 -  98    miles on Friday
   177
```

Scott traveled 177 more miles on Saturday than on Friday. ■

WORK PROBLEM 10 AT THE SIDE.

9. Check the answers in the following problems. If the answer is incorrect, give the correct answer.

(a)

```
   218
 - 176
    42
```

(b)

```
   670
 - 439
   241
```

(c)

```
  11,263
 -  7 784
    3 479
```

10. Using the table from Example 10, how many more miles did Scott travel on

(a) Thursday than on Tuesday?

(b) Sunday than on Monday?

ANSWERS
9. (a) correct (b) incorrect, should be 231 (c) correct
10. (a) 368 (b) 187

name date hour

1.3 EXERCISES

Solve the following subtraction problems. Check each answer.

Example:

$$\begin{array}{r} 3722 \\ -\ 1610 \\ \hline \end{array}$$

Solution:

$$\begin{array}{r} 3722 \\ -\ 1610 \\ \hline \mathbf{2112} \end{array} \quad \text{check} \quad \begin{array}{r} 1610 \\ +\ 2112 \\ \hline \mathbf{3722} \end{array}$$

match

The answer, 2112, checks.

1. 37 − 12

2. 17 − 12

3. 65 − 24

4. 43 − 31

5. 77 − 60

6. 98 − 77

7. 79 − 17

8. 55 − 33

9. 73 − 61

10. 18 − 11

11. 445 − 323

12. 315 − 104

13. 602 − 301

14. 552 − 451

15. 777 − 112

16. 7317 − 4206

17. 2318 − 1207

18. 7911 − 710

19. 4420 − 310

20. 6639 − 5415

21. 1875 − 1362

22. 6259 − 4148

23. 7526 − 5313

24. 9988 − 677

25. 8625 − 311

26. 7428 − 3117

27. 24,392 − 11,232

28. 57,921 − 34,801

29. 46,253 − 5 143

30. 75,904 − 3 702

Check the following subtractions. If an answer is not correct, give the correct answer.

Example:

$$\begin{array}{r} 725 \\ -\ 413 \\ \hline 212 \end{array}$$

Add. 413 + 212 = 625, which is not 725.
The answer does not check. It should be **312.**

31. 58 − 24 = 34

32. 76 − 41 = 35

33. 89 − 27 = 63

34. 54 − 42 = 13

35. 274 − 153 = 131

36. $\begin{array}{r} 515 \\ -\ 304 \\ \hline 211 \end{array}$ **37.** $\begin{array}{r} 2984 \\ -\ 1321 \\ \hline 1663 \end{array}$ **38.** $\begin{array}{r} 5217 \\ -\ 4105 \\ \hline 1132 \end{array}$ **39.** $\begin{array}{r} 8643 \\ -\ 1421 \\ \hline 7212 \end{array}$ **40.** $\begin{array}{r} 9428 \\ -\ 3124 \\ \hline 6324 \end{array}$

Subtract by borrowing as necessary.

> **Example:** $\begin{array}{r} 92 \\ -\ 78 \\ \hline \end{array}$ **Solution:** $\begin{array}{rr} \overset{8}{\cancel{9}} & \overset{12}{\cancel{2}} \\ -\ 7 & 8 \\ \hline \mathbf{1} & \mathbf{4} \end{array}$

41. $\begin{array}{r} 32 \\ -\ 25 \\ \hline \end{array}$ **42.** $\begin{array}{r} 84 \\ -\ 75 \\ \hline \end{array}$ **43.** $\begin{array}{r} 61 \\ -\ 32 \\ \hline \end{array}$ **44.** $\begin{array}{r} 78 \\ -\ 49 \\ \hline \end{array}$ **45.** $\begin{array}{r} 45 \\ -\ 29 \\ \hline \end{array}$

46. $\begin{array}{r} 93 \\ -\ 37 \\ \hline \end{array}$ **47.** $\begin{array}{r} 613 \\ -\ 251 \\ \hline \end{array}$ **48.** $\begin{array}{r} 916 \\ -\ 618 \\ \hline \end{array}$ **49.** $\begin{array}{r} 326 \\ -\ 158 \\ \hline \end{array}$ **50.** $\begin{array}{r} 729 \\ -\ 635 \\ \hline \end{array}$

51. $\begin{array}{r} 581 \\ -\ 122 \\ \hline \end{array}$ **52.** $\begin{array}{r} 973 \\ -\ 788 \\ \hline \end{array}$ **53.** $\begin{array}{r} 8852 \\ -\ 573 \\ \hline \end{array}$ **54.** $\begin{array}{r} 6171 \\ -\ 1182 \\ \hline \end{array}$ **55.** $\begin{array}{r} 9988 \\ -\ 2399 \\ \hline \end{array}$

56. $\begin{array}{r} 3576 \\ -\ 1658 \\ \hline \end{array}$ **57.** $\begin{array}{r} 22{,}618 \\ -\ 6\ 719 \\ \hline \end{array}$ **58.** $\begin{array}{r} 75{,}261 \\ -\ 18{,}374 \\ \hline \end{array}$ **59.** $\begin{array}{r} 38{,}335 \\ -\ 29{,}476 \\ \hline \end{array}$ **60.** $\begin{array}{r} 61{,}278 \\ -\ 3\ 559 \\ \hline \end{array}$

61. $\begin{array}{r} 30 \\ -\ 23 \\ \hline \end{array}$ **62.** $\begin{array}{r} 90 \\ -\ 82 \\ \hline \end{array}$ **63.** $\begin{array}{r} 60 \\ -\ 37 \\ \hline \end{array}$ **64.** $\begin{array}{r} 70 \\ -\ 27 \\ \hline \end{array}$ **65.** $\begin{array}{r} 100 \\ -\ 38 \\ \hline \end{array}$

66. $\begin{array}{r} 108 \\ -\ 69 \\ \hline \end{array}$ **67.** $\begin{array}{r} 500 \\ -\ 189 \\ \hline \end{array}$ **68.** $\begin{array}{r} 603 \\ -\ 474 \\ \hline \end{array}$ **69.** $\begin{array}{r} 4041 \\ -\ 1208 \\ \hline \end{array}$ **70.** $\begin{array}{r} 5905 \\ -\ 3096 \\ \hline \end{array}$

71. $\begin{array}{r} 6036 \\ -\ 5822 \\ \hline \end{array}$ **72.** $\begin{array}{r} 2102 \\ -\ 1099 \\ \hline \end{array}$ **73.** $\begin{array}{r} 9305 \\ -\ 1530 \\ \hline \end{array}$ **74.** $\begin{array}{r} 7120 \\ -\ 6033 \\ \hline \end{array}$ **75.** $\begin{array}{r} 1580 \\ -\ 1077 \\ \hline \end{array}$

76. $\begin{array}{r} 3068 \\ -\ 2105 \\ \hline \end{array}$ **77.** $\begin{array}{r} 2006 \\ -\ 1850 \\ \hline \end{array}$ **78.** $\begin{array}{r} 8203 \\ -\ 5365 \\ \hline \end{array}$ **79.** $\begin{array}{r} 6020 \\ -\ 4078 \\ \hline \end{array}$ **80.** $\begin{array}{r} 7050 \\ -\ 6045 \\ \hline \end{array}$

81. $\begin{array}{r} 8503 \\ -\ 2816 \\ \hline \end{array}$ **82.** $\begin{array}{r} 16{,}004 \\ -\ 5\ 087 \\ \hline \end{array}$ **83.** $\begin{array}{r} 80{,}705 \\ -\ 61{,}667 \\ \hline \end{array}$ **84.** $\begin{array}{r} 36{,}000 \\ -\ 22{,}117 \\ \hline \end{array}$ **85.** $\begin{array}{r} 33{,}000 \\ -\ 17{,}222 \\ \hline \end{array}$

86. $\begin{array}{r} 71{,}080 \\ -\ 65{,}308 \\ \hline \end{array}$ **87.** $\begin{array}{r} 20{,}206 \\ -\ 18{,}077 \\ \hline \end{array}$ **88.** $\begin{array}{r} 90{,}056 \\ -\ 83{,}507 \\ \hline \end{array}$ **89.** $\begin{array}{r} 19{,}080 \\ -\ 13{,}496 \\ \hline \end{array}$ **90.** $\begin{array}{r} 80{,}056 \\ -\ 23{,}869 \\ \hline \end{array}$

Check the following subtractions. If an answer is incorrect, give the correct answer.

Example: $\begin{array}{r} 3084 \\ -\ 1278 \\ \hline 1806 \end{array}$ **Solution:** $\begin{array}{r} 1806 \\ +\ 1278 \\ \hline \mathbf{3084} \end{array}$ **match** correct

91. $\begin{array}{r} 7582 \\ -\ 1628 \\ \hline 5954 \end{array}$ **92.** $\begin{array}{r} 1671 \\ -\ 1325 \\ \hline 1346 \end{array}$ **93.** $\begin{array}{r} 1829 \\ -\ 1638 \\ \hline 191 \end{array}$ **94.** $\begin{array}{r} 3486 \\ -\ 1340 \\ \hline 2146 \end{array}$

95. $\begin{array}{r} 78{,}213 \\ -\ 17{,}346 \\ \hline 60{,}867 \end{array}$ **96.** $\begin{array}{r} 8235 \\ -\ 1439 \\ \hline 6896 \end{array}$ **97.** $\begin{array}{r} 2768 \\ -\ 2230 \\ \hline 538 \end{array}$ **98.** $\begin{array}{r} 34{,}821 \\ -\ 17{,}735 \\ \hline 17{,}735 \end{array}$

99. $\begin{array}{r} 17{,}005 \\ -\ 14{,}552 \\ \hline 2\ 553 \end{array}$ **100.** $\begin{array}{r} 18{,}039 \\ -\ 16{,}187 \\ \hline 1\ 852 \end{array}$ **101.** $\begin{array}{r} 47{,}900 \\ -\ 18{,}679 \\ \hline 29{,}231 \end{array}$ **102.** $\begin{array}{r} 73{,}800 \\ -\ 25{,}957 \\ \hline -\ 47{,}853 \end{array}$

Solve the following word problems.

103. A kennel has 62 dogs. It sells 40. How many dogs are left?

104. A bottle has 36 doses of a medicine. How many doses are left in the bottle after 12 doses have been given?

105. An airplane is carrying 254 passengers. When it lands in Atlanta 133 passengers get off the plane. How many passengers are left on the plane?

106. Paul Kramer has $553 in his checking account. He writes a check for $112. How much is left in his account?

107. Sandy Yost drove her truck 829 miles, while Lloyd Fleek drove his car 517 miles. How many more miles did Ms. Yost drive?

108. On Tuesday, 5822 people went to a soccer game, and on Friday, 7994 people went to a soccer game. How many more people went to the game on Friday?

109. In last fall's election, 3754 people voted. In this fall's election, 2511 people voted. How many more people voted in last fall's election?

110. One bid for painting a house was $1954. A second bid was $1742. How much would be saved by using the second bid?

CHALLENGE EXERCISES

Solve each of the following word problems. Add or subtract as necessary.

111. Rudy Tafoya had $1523 withheld from his paychecks last year for income tax, but he owes only $1379 in tax. What refund should he receive?

112. A retired couple used to receive a social security payment of $879 per month. Recently, benefits were increased by $118 per month. How much money does the couple now receive?

113. The Smiths now make a house payment of $439 per month. If they buy a larger house, the payment will increase by $263 per month. How much will they pay each month?

114. A car now goes 374 miles on a tank of gas. After a tune-up, the same car will go 401 miles on a tank of gas. How many additional miles will it go on a tank of gas after the tune-up?

115. On Monday, 11,594 people visited Arcade Amusement Park, and 12,352 people visited the park on Tuesday. How many more people visited the park on Tuesday?

116. At People's Bank, Sheri Minkner's account will earn $1468 per year in interest, and her account at Farmer's Bank will pay her $1543 in interest. Find the total interest earned at the two banks.

1.4 MULTIPLICATION OF WHOLE NUMBERS

Adding the number 3 a total of 4 times gives 12.

$$3 + 3 + 3 + 3 = 12$$

This result can also be shown with a figure.

3 + 3 + 3 + 3

3 items

4 times

OBJECTIVES

1. Know the parts of a multiplication problem.
2. Do chain multiplications.
3. Multiply by single-digit numbers.
4. Multiply quickly by numbers ending in zeros.
5. Multiply by numbers having more than one digit.
6. Solve word problems with multiplication.

1 **Multiplication** is a shortcut for this repeated addition. The numbers being multiplied are called **factors.** The answer is called the **product.** For example, the product of 3 and 4 can be written with the symbol ×, a raised dot, or parentheses, as follows.

3	factor (also called *multiplicand*)
× 4	factor (also called *multiplier*)
12	product

$3 \times 4 = 12$ $\quad$ $3 \cdot 4 = 12$ $\quad$ $(3)(4) = 12$

Computers use the symbol * for multiplication. A computer would show $3 \times 4 = 12$ as 3*4 = 12.

WORK PROBLEM 1 AT THE SIDE.

1. Identify the factors and the product in each multiplication problem.

(a) $5 \times 9 = 45$

(b) $6 \cdot 3 = 18$

(c) $(2)(8) = 16$

By the **commutative property of multiplication,** the answer or product remains the same when the order of the factors is changed.

For example,

$$3 \times 4 = 12 \quad \text{and} \quad 4 \times 3 = 12.$$

NOTE Recall that addition also has a commutative property.

All possible combinations and products of single-digit numbers are shown in a table provided inside the front cover.

ANSWERS

1. (a) factors: 5, 9; product: 45
   (b) factors: 6, 3; product: 18
   (c) factors: 2, 8; product: 16

2. Multiply.

(a) 5×7

(b) 0×9

(c) $8 \cdot 5$

(d) $4 \cdot 4$

(e) $(2)(7)$

3. Multiply.

(a) $3 \times 3 \times 2$

(b) $2 \times 1 \times 4$

(c) $7 \times 2 \times 0$

ANSWERS
2. (a) 35 (b) 0 (c) 40 (d) 16 (e) 14
3. (a) 18 (b) 8 (c) 0

EXAMPLE 1 Multiplying Two Numbers
Multiply. (Remember that a raised dot means to multiply.)

(a) $3 \times 4 = 12$

(b) $6 \cdot 0 = 0$ (The product of any number and 0 is 0; if you give no money to each of 6 relatives, you give no money.)

(c) $(4)(8) = 32$ ■

WORK PROBLEM 2 AT THE SIDE.

2 Some multiplications contain more than two factors.

By the **associative property of multiplication,** the grouping of numbers in any order gives the same product.

EXAMPLE 2 Multiplying Three Numbers
Multiply: $2 \times 3 \times 5$.

SOLUTION

$(2 \times 3) \times 5$ Parentheses tell what to do first.

$6 \times 5 = 30$

Also,

$2 \times (3 \times 5)$

$2 \times 15 = 30$

Either grouping results in the same answer. ■

A problem with more than two factors, such as the one in Example 2, is called a **chain multiplication.**

WORK PROBLEM 3 AT THE SIDE.

3 Carrying may be needed in multiplication problems having two or more digits in a factor.

EXAMPLE 3 Carrying with Multiplication
Multiply.

(a)
$$\begin{array}{r} 53 \\ \times\ 4 \\ \hline \end{array}$$

Start by multiplying in the ones column.

$$\begin{array}{r} {}^{1} \\ 53 \\ \times\ 4 \\ \hline 2 \end{array}$$

$4 \times 3 = \mathbf{12}$ Carry the 1 to the tens column. Place the 2 in the ones column.

Next, multiply 4 ones and 5 tens.

$$\begin{array}{r} {}^{1}\\ 53 \\ \times\ \ 4 \\ \hline 2 \end{array} \qquad 4 \times 5 = \mathbf{20}$$

Add the 1 that was carried to the tens column.

$$\begin{array}{r} {}^{1}\\ 53 \\ \times\ \ 4 \\ \hline 212 \end{array} \qquad 20 + 1 = \mathbf{21}$$

(b) $\begin{array}{r} 724 \\ \times\ \ 5 \\ \hline \end{array}$

Work as shown.

$$\begin{array}{r} {}^{1\,2}\\ 724 \\ \times\ \ \ 5 \\ \hline 3620 \end{array}$$

- 5 × 4 = **20** ones; write 0 ones and carry 2 tens.
- 5 × 2 = **10** tens; add the 2 tens to get 12; write 2 tens and carry 1 hundred.
- 5 × 7 = **35** hundreds; add the 1 hundred to get 36 ■

WORK PROBLEM 4 AT THE SIDE.

4 The product of two whole-number factors is also called a **multiple** of either factor. For example, since 4 · 2 = 8, the whole number 8 is a multiple of both 4 and 2. *Multiples of 10* are very useful when multiplying. A **multiple of 10** is a whole number that ends in zero, such as 10, 20, or 30; 100, 200, or 300; 1000, 2000, or 3000. There is a short way to multiply by these multiples of 10. For example,

$$8 \times 10 = 80, \quad 74 \times 100 = 7400,$$
$$\text{and}$$
$$953 \times 1000 = 953{,}000.$$

These examples suggest the following rule.

Multiply a whole number by 10, 100, or 1000, by attaching zeros to the right of the whole number.

EXAMPLE 4 Using Multiples of 10 to Multiply

Multiply.

(a) 59 × 10 = 590 ← Attach 0.

4. Multiply.

(a) $\begin{array}{r} 72 \\ \times\ \ 5 \\ \hline \end{array}$

(b) $\begin{array}{r} 54 \\ \times\ \ 0 \\ \hline \end{array}$

(c) $\begin{array}{r} 862 \\ \times\ \ \ 9 \\ \hline \end{array}$

(d) $\begin{array}{r} 3251 \\ \times\ \ \ \ 6 \\ \hline \end{array}$

(e) $\begin{array}{r} 5146 \\ \times\ \ \ \ 8 \\ \hline \end{array}$

ANSWERS

4. (a) 360 (b) 0 (c) 7758 (d) 19,506 (e) 41,168

5. Multiply.

(a) 83×10

(b) 102×100

(c) 647×1000

6. Multiply.

(a) 27×80

(b) 64×300

(c) $\begin{array}{r} 220 \\ \times\ 40 \\ \hline \end{array}$

(d) $\begin{array}{r} 6100 \\ \times\ 90 \\ \hline \end{array}$

(e) $\begin{array}{r} 800 \\ \times\ 200 \\ \hline \end{array}$

ANSWERS
5. (a) 830 (b) 10,200 (c) 647,000
6. (a) 2160 (b) 19,200 (c) 8800
(d) 549,000 (e) 160,000

(b) $74 \times 100 = 7400$ ← Attach 00.

(c) $803 \times 1000 = 803{,}000$ ← Attach 000. ■

WORK PROBLEM 5 AT THE SIDE.

Find the product of other multiples of ten by also attaching zeros.

EXAMPLE 5 Multiplying by Using Other Multiples
Multiply.

(a) 46×3000

Multiply 46 by 3 and attach 3 zeros.

$\begin{array}{r} 46 \\ \times\ 3 \\ \hline 138 \end{array}$ $\quad 46 \times 3\,000 = 138{,}000$ ← Attach 000.

(b) 150×70

Multiply 15 by 7, and then attach 2 zeros.

$\begin{array}{r} 15 \\ \times\ 7 \\ \hline 105 \end{array}$ $\quad 15\,0 \times 7\,0 = 10{,}5\,00$ ← Attach 00. ■

WORK PROBLEM 6 AT THE SIDE.

5 The next example shows multiplication when a factor has more than one digit.

EXAMPLE 6 Multiplying with More Than One Digit
Multiply 46 and 23.

SOLUTION
First multiply 46 by 3.

$$\begin{array}{r} ^{1} \\ 46 \\ \times\ 3 \\ \hline 138 \end{array} \leftarrow 46 \times 3 = 138$$

Now multiply 46 by 20.

$$\begin{array}{r} ^{1} \\ 46 \\ \times\ 20 \\ \hline 920 \end{array} \leftarrow 46 \times 20 = 920$$

Add the results.

$$\begin{array}{rl} 46 & \\ \times\ 23 & \\ \hline 138 & \leftarrow 46 \times 3 \\ +\ 920 & \leftarrow 46 \times 20 \\ \hline 1058 & \leftarrow \text{Add.} \end{array}$$

Both 138 and 920 are called **partial products.** To save time, the zero in 920 is often left off.

```
  46
× 23
 138
 92   ← 0 left off. Be very careful to
        place the 2 in the tens column. ■
```

WORK PROBLEM 7 AT THE SIDE.

EXAMPLE 7 Using Partial Products

Multiply.

(a)

```
    233
  × 132
    466
   6 99     (tens lined up)
  23 3      (hundreds lined up)
 30,756     product
```

(b)

```
  538
×  46
```

Multiply by 6.

```
   24
  538
×  46     Carrying is
 3228       needed here.
```

Multiply by 4, being careful to line up the tens.

```
    13
    24
   538
 ×  46
  3 228
 21 52
 24,748  ■
```

WORK PROBLEM 8 AT THE SIDE.

When zero appears in a factor that is multiplying, be sure to move the partial products to the left to account for the position held by the zero.

EXAMPLE 8 Multiplication with Zeros

Multiply.

(a)

```
    137
  × 306
    822
   0 00     (tens lined up)
  41 1      (hundreds lined up)
 41,922
```

7. Complete each multiplication.

(a)

```
  54
× 35
 270
162
```

(b)

```
  76
× 49
 684
304
```

8. Multiply.

(a)

```
  34
× 12
```

(b)

```
  213
× 323
```

(c)

```
  67
× 59
```

(d)

```
  234
×  73
```

(e)

```
  364
× 272
```

ANSWERS
7. (a) 1890 (b) 3724
8. (a) 408 (b) 68,799 (c) 3953
(d) 17,082 (e) 99,008

9. Multiply.

(a)
```
   28
× 60
```

(b)
```
  932
×  50
```

(c)
```
  481
× 206
```

(d)
```
  3526
× 6002
```

10. Find the total cost of the following items.

(a) 12 trucks at $6215 per truck

(b) 289 books at $14 per book

(c) 274 tickets at $35 per ticket

ANSWERS
9. (a) 1680 (b) 46,600 (c) 99,086 (d) 21,163,052
10. (a) $74,580 (b) $4046 (c) $9590

(b)
```
       1406
    × 2001
       1406
    281200  ←——(0 to line up tens)——→  0000
 2,813,406  ←——(0 to line up hundreds)——→  0000
                                      2812
                                   2813406
```

Zeros are inserted to move the partial products to the left.

Right-hand computation:
```
    1406
  × 2001
    1406
    0000
   0000
  2812
 2813406
```
■

WORK PROBLEM 9 AT THE SIDE.

6 The next example shows how multiplication can be used to solve a word problem.

EXAMPLE 9 Applying Multiplication Skills
Find the total cost of 54 pieces of lumber that cost $24 each.

APPROACH
To find the cost of all the lumber, multiply the number of pieces (54) by the cost per piece ($24).

SOLUTION
Multiply 54 by 24.

```
    54
  × 24
   216
  108
  1296
```

The total cost of the lumber is $1296. ■

WORK PROBLEM 10 AT THE SIDE.

name date hour

1.4 EXERCISES

Work each of the following chain multiplications.

Example: $3 \times 2 \times 9$

Solution: $3 \times 2 = 6 \longrightarrow 6 \times 9 = \mathbf{54}$

1. $2 \times 3 \times 4$ **2.** $3 \times 3 \times 7$ **3.** $7 \times 1 \times 8$ **4.** $2 \times 4 \times 5$

5. $8 \cdot 9 \cdot 0$ **6.** $7 \cdot 0 \cdot 9$ **7.** $3 \cdot 1 \cdot 9$ **8.** $1 \cdot 4 \cdot 8$

9. (2)(3)(6) **10.** (4)(1)(9) **11.** (3)(0)(7) **12.** (9)(0)(4)

Multiply.

Example: 24×2

Solution: **48** ← $2 \times 4 = 8$; $2 \times 2 = 4$

Example: 37×8 (carry 5)

Solution: **296** ← $8 \times 7 = 56$; write 6 and carry 5. $8 \times 3 = 24$, $24 + 5 = 29$

13. 55×8 **14.** 42×9 **15.** 28×9 **16.** 214×3

17. 512×4 **18.** 472×4 **19.** 624×3 **20.** 852×7

21. 2153×4 **22.** 1137×3 **23.** 2521×4 **24.** 2544×3

25. 7212×5 **26.** 9582×4 **27.** $21{,}835 \times 6$ **28.** $34{,}572 \times 5$

29. $81{,}259 \times 4$ **30.** $76{,}892 \times 6$

Multiply.

Example:

$$\begin{array}{r} 110 \\ \times\ 50 \\ \hline \end{array}$$

Solution:

First $\begin{array}{r} 11 \\ \times\ 5 \\ \hline 55 \end{array} \rightarrow \begin{array}{r} 110 \\ \times\ 50 \\ \hline \mathbf{5500} \end{array}$, then attach 00.

31. $\begin{array}{r} 20 \\ \times\ 4 \\ \hline \end{array}$

32. $\begin{array}{r} 30 \\ \times\ 5 \\ \hline \end{array}$

33. $\begin{array}{r} 50 \\ \times\ 7 \\ \hline \end{array}$

34. $\begin{array}{r} 90 \\ \times\ 8 \\ \hline \end{array}$

35. $\begin{array}{r} 740 \\ \times\ 3 \\ \hline \end{array}$

36. $\begin{array}{r} 200 \\ \times\ 9 \\ \hline \end{array}$

37. $\begin{array}{r} 500 \\ \times\ 4 \\ \hline \end{array}$

38. $\begin{array}{r} 72 \\ \times\ 20 \\ \hline \end{array}$

39. $\begin{array}{r} 125 \\ \times\ 30 \\ \hline \end{array}$

40. $\begin{array}{r} 246 \\ \times\ 50 \\ \hline \end{array}$

41. $\begin{array}{r} 1255 \\ \times\ 20 \\ \hline \end{array}$

42. $\begin{array}{r} 8522 \\ \times\ 50 \\ \hline \end{array}$

43. $\begin{array}{r} 800 \\ \times\ 400 \\ \hline \end{array}$

44. $\begin{array}{r} 300 \\ \times\ 600 \\ \hline \end{array}$

45. $\begin{array}{r} 43{,}000 \\ \times\ 2\ 000 \\ \hline \end{array}$

46. $\begin{array}{r} 11{,}000 \\ \times\ 9\ 000 \\ \hline \end{array}$

47. $970 \cdot 50$

48. $800 \cdot 300$

49. $500 \cdot 900$

50. $5600 \cdot 800$

51. $9700 \cdot 200$

52. $10{,}050 \cdot 300$

Multiply.

Example: $\begin{array}{r} 63 \\ \times\ 28 \\ \hline \end{array}$

Solution: $\begin{array}{r} 63 \\ \times\ 28 \\ \hline 504 \\ 126\ \ \\ \hline \mathbf{1764} \end{array}$

53. $\begin{array}{r} 29 \\ \times\ 27 \\ \hline \end{array}$

54. $\begin{array}{r} 19 \\ \times\ 36 \\ \hline \end{array}$

55. $\begin{array}{r} 72 \\ \times\ 33 \\ \hline \end{array}$

56. $\begin{array}{r} 79 \\ \times\ 49 \\ \hline \end{array}$

57. $\begin{array}{r} 83 \\ \times\ 45 \\ \hline \end{array}$

58. (43)(27)

59. (58)(41)

60. (82)(67)

61. (72)(85)

62. (38)(22)

name date hour

63. 758×24

64. 152×71

65. 631×35

66. 453×62

67. 331×44

68. 332×772

69. 735×112

70. 231×318

71. 638×555

72. 658×731

73. 1233×951

74. 6289×632

75. 4355×615

76. 2988×138

77. 1629×478

78. 426×103

79. 135×401

80. 305×101

81. 321×203

82. 780×105

83. 837×708

84. 781×630

85. 219×404

86. 6310×3078

87. 3533×5001

88. 2195×1038

89. 1592×2009

90. 7051×6060

91. 3789×2205

92. 6381×7009

Solve the following word problems by using multiplication.

93. An encyclopedia has 30 volumes. Each volume has 800 pages. What is the total number of pages in the encyclopedia?

94. A hospital has 20 bottles of thyroid medication, with each bottle containing 2500 tablets. How many of these tablets does the hospital have in all?

95. There are 9 assembly workers at a company. Each brings 12 cans of beverage to a company picnic. Find the total number of cans brought to the picnic.

96. A rental car company buys 7 cars at $9817 per car. Find the total cost of the cars.

97. A Ford Escort gets 27 miles per gallon in town. How many miles can it go in town on 8 gallons of gas?

98. A small clinic has 8 patients. Each patient pays $144 per day. How much does the clinic collect for these patients in one day?

Find the total cost of each of the following.

Example: 54 hammers at $8 per hammer
Solution: Multiply.

$$\begin{array}{r} 54 \\ \times \ 8 \\ \hline 432 \end{array}$$

The total cost is **$432.**

99. 14 pictures at $25 per picture

100. 23 bicycles at $72 per bicycle

101. 58 lockers at $53 per locker

102. 76 flats of flowers at $22 per flat

103. 108 boxes of fruit at $37 per box

104. 305 tons of dirt at $12 per ton

CHALLENGE EXERCISES

Multiply.

105. $21 \cdot 43 \cdot 56$

106. (600)(8)(75)(40)

Use addition, subtraction, or multiplication, as needed, to solve each of the following.

107. Sarah Stolz drove 36 miles one day and 104 miles the next day. How many miles did she drive altogether during the two days?

108. Find the cost of twelve shovels at $21 per shovel.

109. A large meal contains 1406 calories, while a small meal contains 348 calories. How many more calories are in the large meal than the small one?

110. One company pays its shop workers $9 for every assembly. How much would be paid to a person who had 104 assemblies?

111. A company has 22 clerical employees, 9 management employees, 16 technical employees, and 8 employees in the warehouse. How many employees does the company have altogether?

112. The distance from a certain city to the Atlantic Ocean is 1326 miles, while the distance to the Pacific Ocean is 825 miles. How much farther is it to the Atlantic Ocean than it is to the Pacific Ocean?

1.5 DIVISION OF WHOLE NUMBERS

Suppose \$12 must be divided into 3 equal parts. Each part would be \$4, as shown here.

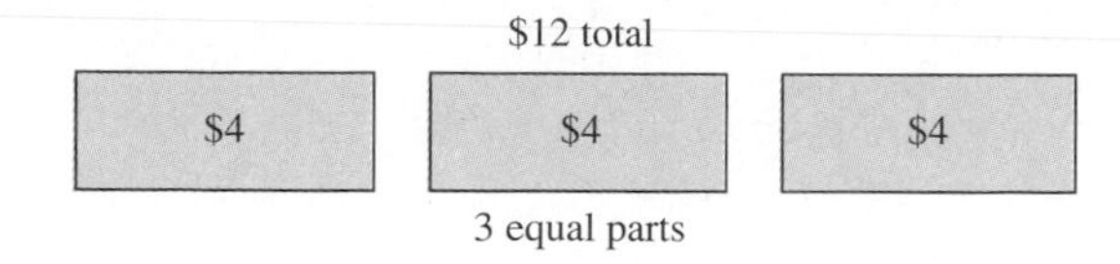

OBJECTIVES

1. Write division problems three ways.
2. Identify the parts of a division problem.
3. Divide zero by a number.
4. Divide a number by itself.
5. Use short division.
6. Check the answer to a division problem.
7. Use tests for divisibility.

1 Just as $3 \cdot 4$, 3×4, and $(3)(4)$ are different ways of indicating the multiplication of 3 and 4, there are several ways to write 12 divided by 3.

$$12 \div 3 = 4 \qquad \begin{array}{r}4\\3\overline{)12}\end{array} \qquad \frac{12}{3} = 4$$

The division symbols used above, $\div$, $\overline{)\quad}$, and —, are commonly used in mathematics. In more advanced courses such as algebra, a slash symbol, /, or a bar, —, are most often used.

EXAMPLE 1 Using Division Symbols
Write each division by using two other symbols.

(a) $12 \div 4 = 3$

This division can also be written as

$$\begin{array}{r}3\\4\overline{)12}\end{array} \quad \text{or} \quad \frac{12}{4} = 3.$$

(b) $\frac{15}{5} = 3$

$$15 \div 5 = 3 \quad \text{or} \quad \begin{array}{r}3\\5\overline{)15}\end{array}$$

(c) $\begin{array}{r}4\\5\overline{)20}\end{array}$

$$20 \div 5 = 4 \quad \text{or} \quad \frac{20}{5} = 4 \ ■$$

WORK PROBLEM 1 AT THE SIDE.

1. Write each division problem using two other symbols.

(a) $50 \div 5 = 10$

(b) $24 \div 6 = 4$

(c) $\begin{array}{r}4\\8\overline{)32}\end{array}$

(d) $\frac{42}{6} = 7$

2 In division, the number being divided is the **dividend,** the number divided by is the **divisor,** and the answer is the **quotient.**

$$\text{dividend} \div \text{divisor} = \text{quotient}$$

$$\begin{array}{r}\text{quotient}\\\text{divisor}\overline{)\text{dividend}}\end{array} \qquad \frac{\text{dividend}}{\text{divisor}} = \text{quotient}$$

ANSWERS

1. (a) $\begin{array}{r}10\\5\overline{)50}\end{array}$ and $\frac{50}{5} = 10$

(b) $\begin{array}{r}4\\6\overline{)24}\end{array}$ and $\frac{24}{6} = 4$

(c) $32 \div 8 = 4$ and $\frac{32}{8} = 4$

(d) $\begin{array}{r}7\\6\overline{)42}\end{array}$ and $42 \div 6 = 7$

2. Identify the dividend, divisor, and quotient.

(a) $12 \div 2 = 6$

(b) $30 \div 10 = 3$

(c) $\frac{28}{7} = 4$

(d) $2\overline{)36}$ with quotient 18

3. Divide.

(a) $0 \div 20$

(b) $\frac{0}{6}$

(c) $\frac{0}{28}$

(d) $57\overline{)0}$

ANSWERS

2. (a) dividend: 12; divisor: 2; quotient: 6 (b) dividend: 30; divisor: 10; quotient: 3 (c) dividend: 28; divisor: 7; quotient: 4 (d) dividend: 36; divisor: 2; quotient: 18

3. all 0

EXAMPLE 2 Identifying the Parts in a Division Problem

Identify the dividend, divisor, and quotient.

(a) $35 \div 7 = 5$

$35 \div 7 = 5$ ← quotient, where 35 is the dividend and 7 is the divisor.

(b) $\frac{100}{20} = 5$

$\frac{100}{20} = 5$ ← quotient, where 100 is the dividend and 20 is the divisor.

(c) $12\overline{)72}$ with quotient 6

6 ← quotient; 72 ← dividend; 12 ← divisor ■

WORK PROBLEM 2 AT THE SIDE.

3 No money, or \$0, can be divided equally among five people; each person gets \$0. As a general rule,

Zero divided by any nonzero number is **zero.**

EXAMPLE 3 Dividing Zero by a Number

Divide.

(a) $0 \div 12 = 0$

(b) $0 \div 1728 = 0$

(c) $\frac{0}{375} = 0$

(d) $129\overline{)0}$ with quotient 0 ■

WORK PROBLEM 3 AT THE SIDE.

Just as a subtraction such as $8 - 3 = 5$ can be written as the addition $8 = 3 + 5$, any division can be written as a multiplication. For example, $12 \div 3 = 4$ can be written as

$$12 = 3 \times 4.$$

EXAMPLE 4 Converting Division to Multiplication
Convert each division to a multiplication.

(a) $\frac{20}{4} = 5$ becomes $4 \cdot 5 = 20$

(b) $8\overline{)48}$ with quotient 6 becomes $8 \cdot 6 = 48$

(c) $72 \div 9 = 8$ becomes $9 \cdot 8 = 72$ ■

WORK PROBLEM 4 AT THE SIDE.

Division by zero is not possible. To see why, try to find

$$5 \div 0.$$

Suppose the answer to this problem is some number called x.

$$5 \div 0 = x$$

Change this division problem to a multiplication problem.

$$5 = 0 \cdot x$$

Since the product of any number and zero is zero, $0 \cdot x$ must be zero. But

$$5 = 0 \text{ is } \textit{false}.$$

Therefore, the number x cannot exist.

Division by **zero** is impossible.

EXAMPLE 5 Dividing by Zero Is Impossible
All the following are impossible.

(a) $\frac{6}{0}$

(b) $0\overline{)21}$

(c) $18 \div 0$ ■

Division involving 0 is summarized below.

$$\frac{0}{\text{nonzero number}} = \mathbf{0}$$

$$\frac{\text{number}}{0} \text{ is } \textbf{meaningless}$$

WORK PROBLEM 5 AT THE SIDE.

4. Write each division problem as a multiplication problem.

(a) $3\overline{)15}$ with quotient 5

(b) $\frac{28}{4} = 7$

(c) $42 \div 6 = 7$

5. Work the following problems whenever possible.

(a) $\frac{12}{0}$

(b) $\frac{0}{12}$

(c) $0\overline{)32}$

(d) $32\overline{)0}$

(3) $100 \div 0$

ANSWERS
4. (a) $3 \cdot 5 = 15$ (b) $4 \cdot 7 = 28$
(c) $6 \cdot 7 = 42$
5. (a) meaningless (b) 0
(c) meaningless (d) 0
(e) meaningless

6. Divide.

(a) $12 \div 12$

(b) $24\overline{)24}$

(c) $\frac{9}{9}$

7. Divide.

(a) $4\overline{)84}$

(b) $2\overline{)68}$

(c) $3\overline{)39}$

(d) $2\overline{)462}$

ANSWERS
6. all 1
7. (a) 21 (b) 34 (c) 13 (d) 231

4 The next rule tells the quotient when a number is divided by itself.

Any nonzero number divided by itself is **one.**

EXAMPLE 6 Any Nonzero Number Divided by Itself Equals One
Divide.

(a) $54 \div 54 = 1$

(b) $\overset{1}{32\overline{)32}}$

(c) $\frac{18}{18} = 1$ ■

WORK PROBLEM 6 AT THE SIDE.

5 **Short division** is a method of dividing a number by a one-digit divisor.

EXAMPLE 7 Using Short Division
Divide: $3\overline{)96}$.

SOLUTION
Divide 9 by 3.

$$\overset{3}{3\overline{)96}} \longleftarrow \frac{9}{3} = 3$$

Next, divide 6 by 3.

$$\overset{32}{3\overline{)96}} \leftarrow \frac{6}{3} = 2$$ ■

WORK PROBLEM 7 AT THE SIDE.

When two numbers do not divide exactly, a number called the **remainder** is left over.

EXAMPLE 8 Using Short Division with a Remainder
Divide 147 by 4.

SOLUTION
Write the problem.

$$4\overline{)147}$$

Since 1 cannot be divided by 4, divide 14 by 4.

$$\overset{3}{4\overline{)14^{2}7}} \qquad \frac{14}{4} = 3 \text{ with 2 left over}$$

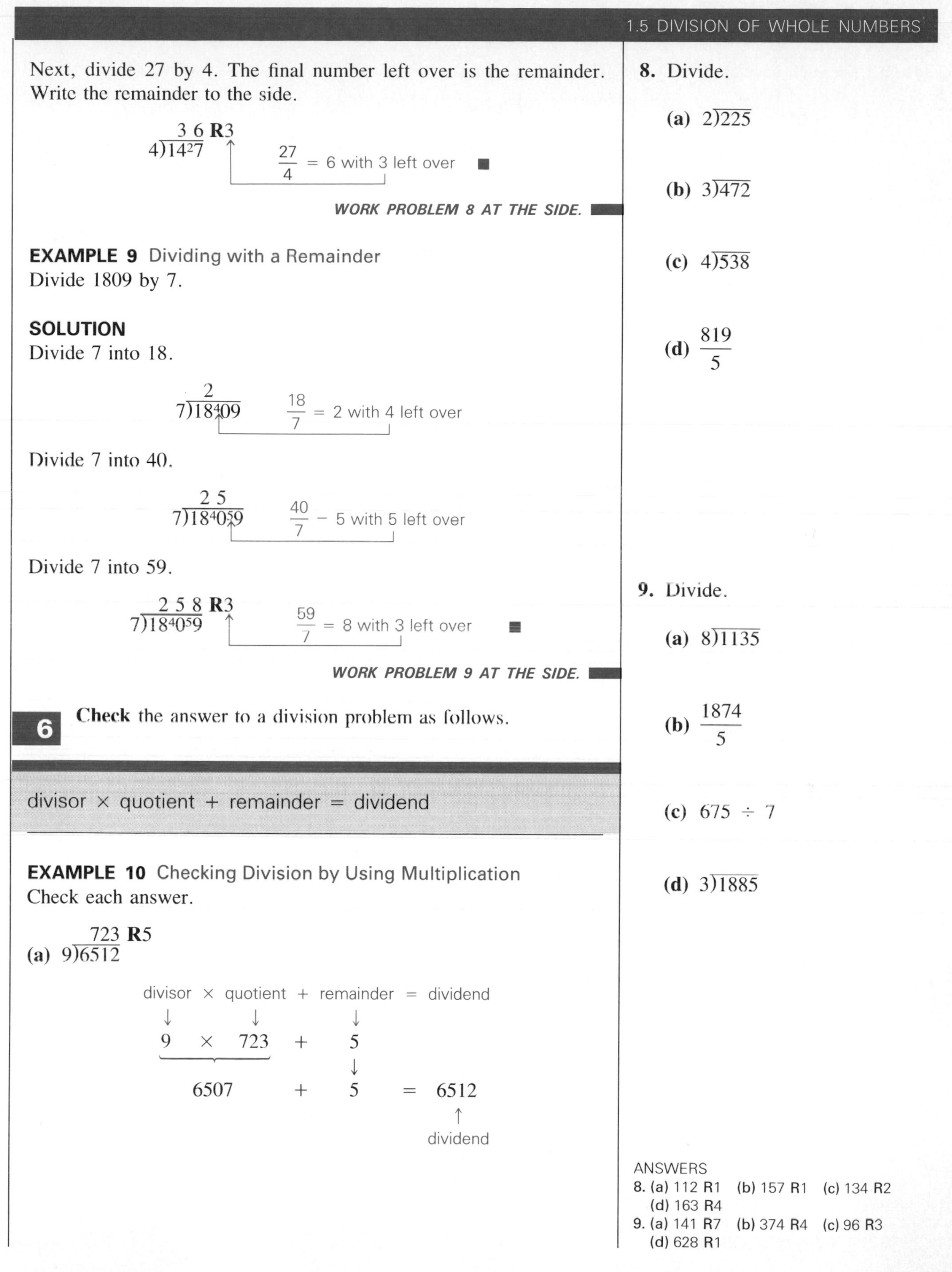

Next, divide 27 by 4. The final number left over is the remainder. Write the remainder to the side.

$$\begin{array}{r} 3\ 6\ \mathbf{R}3 \\ 4\overline{)14^27} \end{array} \qquad \frac{27}{4} = 6 \text{ with 3 left over}$$ ■

WORK PROBLEM 8 AT THE SIDE.

EXAMPLE 9 Dividing with a Remainder
Divide 1809 by 7.

SOLUTION
Divide 7 into 18.

$$\begin{array}{r} 2 \\ 7\overline{)18^409} \end{array} \qquad \frac{18}{7} = 2 \text{ with 4 left over}$$

Divide 7 into 40.

$$\begin{array}{r} 2\ 5 \\ 7\overline{)18^40^59} \end{array} \qquad \frac{40}{7} = 5 \text{ with 5 left over}$$

Divide 7 into 59.

$$\begin{array}{r} 2\ 5\ 8\ \mathbf{R}3 \\ 7\overline{)18^40^59} \end{array} \qquad \frac{59}{7} = 8 \text{ with 3 left over}$$ ■

WORK PROBLEM 9 AT THE SIDE.

6 **Check** the answer to a division problem as follows.

divisor × quotient + remainder = dividend

EXAMPLE 10 Checking Division by Using Multiplication
Check each answer.

(a) $\begin{array}{r} 723\ \mathbf{R}5 \\ 9\overline{)6512} \end{array}$

divisor × quotient + remainder = dividend

$$\underbrace{9 \times 723}_{6507} + 5 = 6512 \text{ (dividend)}$$

8. Divide.

(a) $2\overline{)225}$

(b) $3\overline{)472}$

(c) $4\overline{)538}$

(d) $\dfrac{819}{5}$

9. Divide.

(a) $8\overline{)1135}$

(b) $\dfrac{1874}{5}$

(c) $675 \div 7$

(d) $3\overline{)1885}$

ANSWERS
8. (a) 112 R1 (b) 157 R1 (c) 134 R2 (d) 163 R4
9. (a) 141 R7 (b) 374 R4 (c) 96 R3 (d) 628 R1

10. Check each answer.

(a) $4\overline{)234}$ quotient 58 **R**2

(b) $7\overline{)684}$ quotient 97 **R**3

(c) $8\overline{)2624}$ quotient 328

(d) $5\overline{)2383}$ quotient 476 **R**3

ANSWERS
10. (a) correct (b) incorrect, should be 97 R5 (c) correct (d) correct

The answer checks.

(b) $6\overline{)1437}$ quotient 239 **R**4

divisor × quotient + remainder = dividend

$6 \times 239 + 4$

$1434 + 4 = 1438$

↑ not the dividend

The answer does not check. Rework the original problem to get the correct answer, 239 **R**3. ■

WORK PROBLEM 10 AT THE SIDE.

7 It is often important to know whether a number is divisible by another number.

One whole number is **divisible** by another if the remainder is zero.

Decide whether one number is exactly divisible by another by using the following **tests for divisibility.**

A number is divisible by

2 if it ends in 0, 2, 4, 6, or 8.
3 if the sum of its digits is divisible by 3.
4 if the last two digits make a number that is divisible by 4.
5 if it ends in 0 or 5.
6 if it is divisible by both 2 and 3.
8 if the last three digits make a number that is divisible by 8.
9 if the sum of its digits is divisible by 9.
10 if it ends in 0.

The most commonly used tests are those for 2, 3, 5, and 10.

A number is divisible by **2** if the number ends in 0, 2, 4, 6, or 8.

EXAMPLE 11 Testing for Divisibility by 2
Are the following numbers divisible by 2?

(a) 986
↑ends in 6

Since the number ends in 6, which is in the list above, the number 986 is divisible by 2.

(b) 3255 is not divisible by 2.
↑ends in 5, and not in 0, 2, 4, 6, or 8 ■

WORK PROBLEM 11 AT THE SIDE.

11. Which are divisible by 2?

(a) 714

(b) 9003

(c) 5122

(d) 21,000

A number is divisible by **3** if the sum of its digits is divisible by 3.

EXAMPLE 12 Testing for Divisibility by 3
Are the following numbers divisible by 3?

(a) 4251

Add the digits. 4 2 5 1

$$4 + 2 + 5 + 1 = 12$$

Since 12 is divisible by 3, the number 4251 is divisible by 3.

(b) 29,806

Add the digits.

$$2 + 9 + 8 + 0 + 6 = 25$$

Since 25 is not divisible by 3, the number 29,806 is not divisible by 3. ■

WORK PROBLEM 12 AT THE SIDE.

12. Which are divisible by 3?

(a) 7813

(b) 3474

(c) 242,913

(d) 508,396

ANSWERS
11. all but (b)
12. (b) and (c)

13. Which are divisible by 5 or 10?

(a) 320

(b) 545

(c) 4980

(d) 32,705

A number is divisible by **5** if it ends in 0 or 5.
A number is divisible by **10** if it ends in 0.

EXAMPLE 13 Determining Divisibility by 5
Are the following numbers divisible by 5?

(a) 12,900 ends in 0 and is divisible by 5.

(b) 4325 ends in 5 and is divisible by 5.

(c) 392 ends in 2 and is not divisible by 5. ■

EXAMPLE 14 Determining Divisibility by 10
Are the following numbers divisible by 10?

(a) 80, 700, and 9140 all end in 0 and are divisible by 10.

(b) 29, 355, and 18,743 do not end in 0 and are not divisible by 10. ■

WORK PROBLEM 13 AT THE SIDE.

ANSWERS
13. all are divisible by 5; (a) and (c) are divisible by 10

name date hour

1.5 EXERCISES

Write each division problem by using two other symbols.

Example: $14 \div 2 = 7$

Solution: $\frac{14}{2} = 7$ $\quad 2\overline{)14}$ with quotient 7

Notice the three ways to write the same problem.

1. $21 \div 7 = 3$

2. $18 \div 9 = 2$

3. $\frac{45}{9} = 5$

4. $\frac{56}{8} = 7$

5. $2\overline{)16}$ with quotient 8

6. $8\overline{)48}$ with quotient 6

Divide.

Examples:

$21 \div 3$ $\quad \frac{0}{5}$ $\quad 18 \div 0$

Solutions:

$21 \div 3 = \mathbf{7}$ $\quad \frac{0}{5} = \mathbf{0}$ $\quad 18 \div 0$ is **meaningless**

7. $5 \div 5$

8. $35 \div 7$

9. $\frac{10}{2}$ **5**

10. $\frac{15}{0}$

11. $18 \div 0$

12. $2 \div 2$

13. $\frac{0}{4}$

14. $\frac{24}{8}$

15. $12\overline{)0}$

16. $\frac{0}{8}$

17. $0\overline{)15}$

18. $\frac{1}{0}$

19. $\frac{0}{4}$

20. $0\overline{)4}$

21. $\frac{8}{1}$

22. $\frac{0}{1}$

Divide by using short division. Check each answer.

Examples:

$8\overline{)376}$ $\qquad$ $6\overline{)1487}$

Solutions:

$\begin{array}{r}\mathbf{4\ 7}\\ 8\overline{)37^{5}6}\end{array}$ $\qquad$ $\begin{array}{r}\mathbf{2\ 4\ 7\ R5}\\ 6\overline{)14^{2}8^{4}7}\end{array}$

Check: $8 \times 47 = \mathbf{376}$ $\qquad$ $6 \times 247 + 5 = 1482 + 5 = \mathbf{1487}$

23. $5\overline{)130}$ **24.** $4\overline{)232}$ **25.** $7\overline{)322}$ **26.** $6\overline{)228}$

27. $3\overline{)1341}$ **28.** $7\overline{)952}$ **29.** $5\overline{)2005}$ **30.** $4\overline{)3208}$

31. $4\overline{)2509}$ **32.** $8\overline{)1335}$ **33.** $9\overline{)2024}$ **34.** $7\overline{)3739}$

35. $6\overline{)9137}$ **36.** $9\overline{)8371}$ **37.** $6\overline{)1854}$ **38.** $8\overline{)856}$

39. $9054 \div 9$ **40.** $14{,}049 \div 7$ **41.** $30{,}036 \div 6$ **42.** $32{,}008 \div 4$

43. $4867 \div 6$ **44.** $5993 \div 7$ **45.** $12{,}947 \div 5$ **46.** $33{,}285 \div 9$

47. $29{,}357 \div 3$ **48.** $29{,}419 \div 8$ **49.** $14{,}757 \div 4$ **50.** $46{,}560 \div 7$

51. $\frac{10{,}980}{4}$ **52.** $\frac{23{,}667}{7}$ **53.** $\frac{52{,}569}{9}$ **54.** $\frac{52{,}696}{8}$

55. $\frac{74{,}751}{6}$ **56.** $\frac{72{,}543}{5}$ **57.** $\frac{92{,}327}{9}$ **58.** $\frac{83{,}257}{4}$

59. $\frac{71{,}776}{7}$ **60.** $\frac{77{,}621}{3}$ **61.** $\frac{78{,}785}{7}$ **62.** $\frac{118{,}315}{8}$

name date hour

1.5 EXERCISES

Check each answer. If an answer is incorrect, give the correct answer.

Example: 9)1609 with answer **178 R7**

Solution: divisor × quotient + remainder = dividend

9 × 178 + 7

1602 + 7 = **1609**

The answer checks.

63. 69 R1
4)277

64. 74 R7
8)599

65. 1908 R2
3)5725

66. 432 R3
5)2158

67. 650 R2
7)4692

68. 663 R5
9)5974

69. 3 568 R2
6)21,409

70. 25,879
4)103,516

71. 20,763
8)166,104

72. 5 302
7)37,114

73. 11,523 R2
6)69,140

74. 27,532 R1
3)82,598

75. 9 628 R7
9)86,655

76. 7 258 R4
7)50,809

77. 27,822
8)222,576

78. 77,804
4)311,216

79. 52,136 R5
6)312,820

80. 74,361 R1
9)669,250

Solve each word problem.

81. An 8-ounce can of oil costs 96¢. Find the cost per ounce.

82. Eight people invested a total of $244,224 to buy a condominium. Each person invested the same amount of money. How much did each person invest?

83. If 7 identical cars cost \$57,764, find the cost of each car.

84. One gallon of beverage will serve 9 people. How many gallons are needed for 3483 people?

85. A teacher is paid \$29,493 for a 9-month school year. How much is this per month?

86. How many 5-pound bags of rice can be filled from 8750 pounds of rice?

Put a √ mark in the blank if the number at the left is divisible by the number at the top. Put an X in the blank if the number is not divisible by the number at the top.

> **Example:** 40
>
> The number 40 can be divided by 2, 5, and 10 but not by 3.
>
> | **Solution:** | **2** | **3** | **5** | **10** |
> |---|---|---|---|---|
> | 40 | √ | **X** | √ | √ |

	2	**3**	**5**	**10**
87. 30	_____	_____	_____	_____
89. 184	_____	_____	_____	_____
91. 355	_____	_____	_____	_____
93. 903	_____	_____	_____	_____
95. 2583	_____	_____	_____	_____
97. 21,763	_____	_____	_____	_____

	2	**3**	**5**	**10**
88. 36	_____	_____	_____	_____
90. 192	_____	_____	_____	_____
92. 897	_____	_____	_____	_____
94. 1000	_____	_____	_____	_____
96. 15,502	_____	_____	_____	_____
98. 32,472	_____	_____	_____	_____

CHALLENGE EXERCISES

99. An employee earns \$16,200 per year. Find the amount of earnings in 3 months.

100. A worker assembles 168 light diffusers in an 8-hour shift. Find the number assembled in 2 hours.

1.6 LONG DIVISION

OBJECTIVES

1. Do long division.
2. Divide numbers ending in zeros by numbers ending in zero.
3. Check answers.

Long division is used to divide by a number with more than one digit.

1 In long division, estimate the various numbers by using a **trial divisor,** which is used to get a **trial quotient.**

EXAMPLE 1 Using a Trial Divisor and a Trial Quotient
Divide: $42\overline{)3066}$.

SOLUTION
Since 42 is closer to 40 than to 50, use the first digit of the divisor as a trial divisor.

42 (4 ← trial divisor)

Try to divide the first digit of the dividend by 4. Since 3 cannot be divided by 4, use the first *two* digits, 30.

$\frac{30}{4}$ — 7 with remainder 2

$$\begin{array}{r} 7 \\ 42\overline{)3066} \end{array}$$

7 ← trial quotient

7 goes over the 6, since $\frac{306}{42}$ is about 7.

Multiply 7 and 42 to get 294; next, subtract 294 from 306.

$$\begin{array}{rl} 7 & \\ 42\overline{)3066} & \\ \underline{294} & \leftarrow 7 \times 42 \\ 12 & \leftarrow 306 - 294 \end{array}$$

Bring down the 6 at the right.

$$\begin{array}{rl} 7 & \\ 42\overline{)3066} & \\ \underline{294}\downarrow & \\ 126 & \leftarrow 6 \text{ brought down} \end{array}$$

Use the trial divisor, 4.

first digits of 126 → $\frac{12}{4} = 3$

$$\begin{array}{rl} 73 & \\ 42\overline{)3066} & \\ \underline{294} & \\ 126 & \\ \underline{126} & \\ 0 & \leftarrow 3 \times 42 = 126 \end{array}$$

Check the answer by multiplying 42 and 73. The product should be 3066. ■

WORK PROBLEM 1 AT THE SIDE.

1. Divide.

(a) $31\overline{)1643}$

(b) $52\overline{)4264}$

(c) $81\overline{)2835}$

(d) $\frac{3276}{42}$

ANSWERS
1. (a) 53 (b) 82 (c) 35 (d) 78

2. Divide.

(a) $39\overline{)1053}$

(b) $48\overline{)2691}$

(c) $78\overline{)7218}$

(d) $89\overline{)6649}$

ANSWERS
2. (a) 27 (b) 56 R3 (c) 92 R42
(d) 74 R63

EXAMPLE 2 Dividing to Find a Trial Quotient
Divide: $58\overline{)2730}$.

SOLUTION
Use 6 as a trial divisor, since 58 is closer to 60 than to 50.

$$\frac{27}{6} = 4 \text{ with 3 left over}$$

(27: first digits of dividend)

$$\begin{array}{r} 4 \\ 58\overline{)2730} \\ \underline{232} \\ 41 \end{array}$$

4 ← trial quotient
232 ← 4 × 58 = 232
41 ← 273 − 232 = 41 (smaller than 58, the divisor)

Bring down the 0.

$$\begin{array}{r} 4 \\ 58\overline{)2730} \\ \underline{232} \\ 410 \end{array}$$

410 ← 0 brought down

$$\frac{41}{6} = 6 \text{ with 5 left over}$$

(41: first digits of 410)

$$\begin{array}{r} 46 \\ 58\overline{)2730} \\ \underline{232} \\ 410 \\ \underline{348} \\ 62 \end{array}$$

46 ← trial quotient
348 ← 6 × 58 = 348
62 ← greater than 58

The remainder of 62 is greater than the divisor of 58, so 7 should have been used as a trial quotient instead of 6.

$$\begin{array}{r} 47 \textbf{ R4} \\ 58\overline{)2730} \\ \underline{232} \\ 410 \\ \underline{406} \\ 4 \end{array}$$

406 ← 7 × 58 = 406
4 ← 410 − 406 (the remainder, R4) ■

WORK PROBLEM 2 AT THE SIDE.

Sometimes it is necessary to insert a zero in the quotient.

EXAMPLE 3 Inserting Zeros in the Quotient
Divide: $42\overline{)8734}$.

SOLUTION
Start as above.

$$\begin{array}{r} 2 \\ 42\overline{)8734} \\ \underline{84} \\ 3 \end{array}$$

84 ← 2 × 42 = 84
3 ← 87 − 84 = 3

Bring down the 3.

```
      2
42)8734
   84
   33  ← 3 brought down
```

Since 33 cannot be divided by 42, place a 0 in the quotient as a placeholder.

```
     20  ← 0 in quotient
42)8734
   84
    33
```

Bring down the final digit, the 4.

```
     20
42)8734
   84
    334 ← 4 brought down
```

Complete the problem.

```
    207 R40
42)8734
   84
    334
    294
     40
```

The answer is 207 **R**40. ■

WORK PROBLEM 3 AT THE SIDE.

2 If the divisor and dividend both contain zeros at the far right, recall that these numbers are multiples of 10. There is a short way to divide these multiples of 10. For example,

$90 \div 10 = 9, \quad 6400 \div 100 = 64, \quad 857{,}000 \div 1000 = 875.$

These examples suggest the following rule

> Divide a whole number by 10, 100, or 1000 by dropping zeros from the whole number.

EXAMPLE 4 Using Multiples of 10 to Divide

Divide.

(a) $60 \div 10 = 6$ (1 zero in divisor; 0 dropped)

(b) $3500 \div 100 = 35$ (2 zeros in divisor; 00 dropped)

(c) $915{,}000 \div 1000 = 915$ (3 zeros in divisor; 000 dropped) ■

WORK PROBLEM 4 AT THE SIDE.

3. Divide.

(a) 63)12,663

(b) 24)3127

(c) 39)15,933

(d) 78)23,462

4. Divide.

(a) $50 \div 10$

(b) $3200 \div 100$

(c) $712{,}000 \div 1000$

ANSWERS
3. (a) 201 (b) 130 R7 (c) 408 R21 (d) 300 R62
4. (a) 5 (b) 32 (c) 712

5. Divide.

(a) $90\overline{)15{,}300}$

(b) $130\overline{)143{,}130}$

(c) $2600\overline{)195{,}000}$

6. Decide whether the following divisions are correct.

(a)
```
     37
23)851
   69
   161
   161
     0
```

(b)
```
          45
426)19,170
    17 04
     2 130
     2 130
         0
```

(c)
```
          57 R18
514)29,316
    25 70
     3 616
     3 598
        18
```

ANSWERS
5. (a) 170 (b) 1101 (c) 75
6. (a) correct (b) correct (c) correct

Find the quotient for other multiples of 10 by dropping zeros.

EXAMPLE 5 Dividing by Multiples of 10

Divide.

(a) $40\overline{)11{,}000}$ Drop 1 zero from the divisor and the dividend.

```
    275
4)1100
  8
  30
  28
   20
   20
    0
```

(b) $3500\overline{)31{,}500}$

```
     9
35)315     2 zeros dropped
   315
     0
```

■

NOTE Dropping zeros when dividing by multiples of 10 does not change the answer (quotient).

WORK PROBLEM 5 AT THE SIDE.

3 Answers in long division can be checked just as answers in short division were checked.

EXAMPLE 6 Checking Division

Check each answer.

(a) $48\overline{)5324}$ with quotient 114 **R**43

```
   114
×   48     Multiply the quotient and the divisor.
   912
  456
  5472
+   43  ← Add the remainder.
  5515  ← Result does not match dividend.
```

The answer does not check. Rework the original problem to get 110 **R**44.

(b)

```
          37             716
716)26,492             ×  37
    21 48               5 012
     5 012             21 48
     5 012             26,492     correct ■
         0
```

WORK PROBLEM 6 AT THE SIDE.

name date hour

1.6 EXERCISES

Divide by using long division. Check each answer.

Example: $41\overline{)2388}$

Solution:

$$\begin{array}{r} \textbf{58 R10} \\ 41\overline{)2388} \\ \underline{205} \\ 338 \\ \underline{328} \\ 10 \end{array}$$

Check:

$$\begin{array}{rl} 58 & \\ \times\ 41 & \\ \hline 58 & \\ \underline{232} & \\ 2378 & \\ +\quad 10 & \leftarrow \text{remainder} \\ \hline \textbf{2388} & \leftarrow \text{matches dividend} \end{array}$$

1. $21\overline{)735}$

2. $32\overline{)1696}$

3. $53\overline{)2501}$

4. $72\overline{)6308}$

5. $59\overline{)2190}$

6. $78\overline{)8190}$

7. $23\overline{)7065}$

8. $42\overline{)8699}$

9. $58\overline{)2204}$

10. $77\overline{)1650}$

11. $94\overline{)25{,}789}$

12. $83\overline{)39{,}692}$

13. $26\overline{)62{,}583}$

14. $28\overline{)84{,}249}$

15. $63\overline{)78{,}072}$

16. $86\overline{)10{,}327}$

17. $38\overline{)24{,}328}$

18. $52\overline{)68{,}025}$

19. $12\overline{)116{,}953}$

20. $21\overline{)149{,}826}$

21. $32\overline{)247{,}892}$

22. $343\overline{)315{,}764}$

23. $153\overline{)509{,}725}$

24. $308\overline{)18{,}788}$

25. $821\overline{)17{,}241}$

26. $523\overline{)638{,}075}$

27. $657\overline{)732{,}094}$

28. $360\overline{)25{,}920}$

29. $900\overline{)153{,}000}$

30. $230\overline{)253{,}230}$

Check each answer. If an answer is incorrect, give the correct answer.

31. $56\overline{)5943}$ 106 **R**17

32. $87\overline{)3254}$ 37 **R**37

33. $28\overline{)18,424}$ 658 **R**9

34. $191\overline{)88,604}$ 463 **R**171

35. $614\overline{)38,068}$ 62 **R**3

36. $557\overline{)97,286}$ 174 **R**368

Solve each word problem by using addition, subtraction, multiplication, or division as needed.

37. A car travels 1350 miles at 54 miles per hour. How many hours did it travel?

38. The total cost for 27 baseball uniforms is $2106. Find the cost of each uniform.

39. A person borrows $3200 and pays interest of $706. Find the total amount that must be repaid.

40. A small car goes 792 miles on 24 gallons of gas. How many miles per gallon does the car get?

41. Judy Martinez owes $3888 on a loan. Find her monthly payment, if the loan is paid off in 36 months.

42. Two sisters share a legal bill of $1806. One sister pays $954 toward the bill. How much must the other sister pay?

CHALLENGE EXERCISES

43. Bill Poole can assemble 42 circuits in 1 hour. How many circuits can he assemble in a 5-day work week of 8 hours per day.

44. The total receipts at a concert were $28,017. Each ticket cost $11. How many people attended the concert?

45. A car loan can be paid off in 3 years with monthly payments of $292 each. Find the total amount repaid.

46. A medical technician earns $27,996 per year. Find the earnings per month.

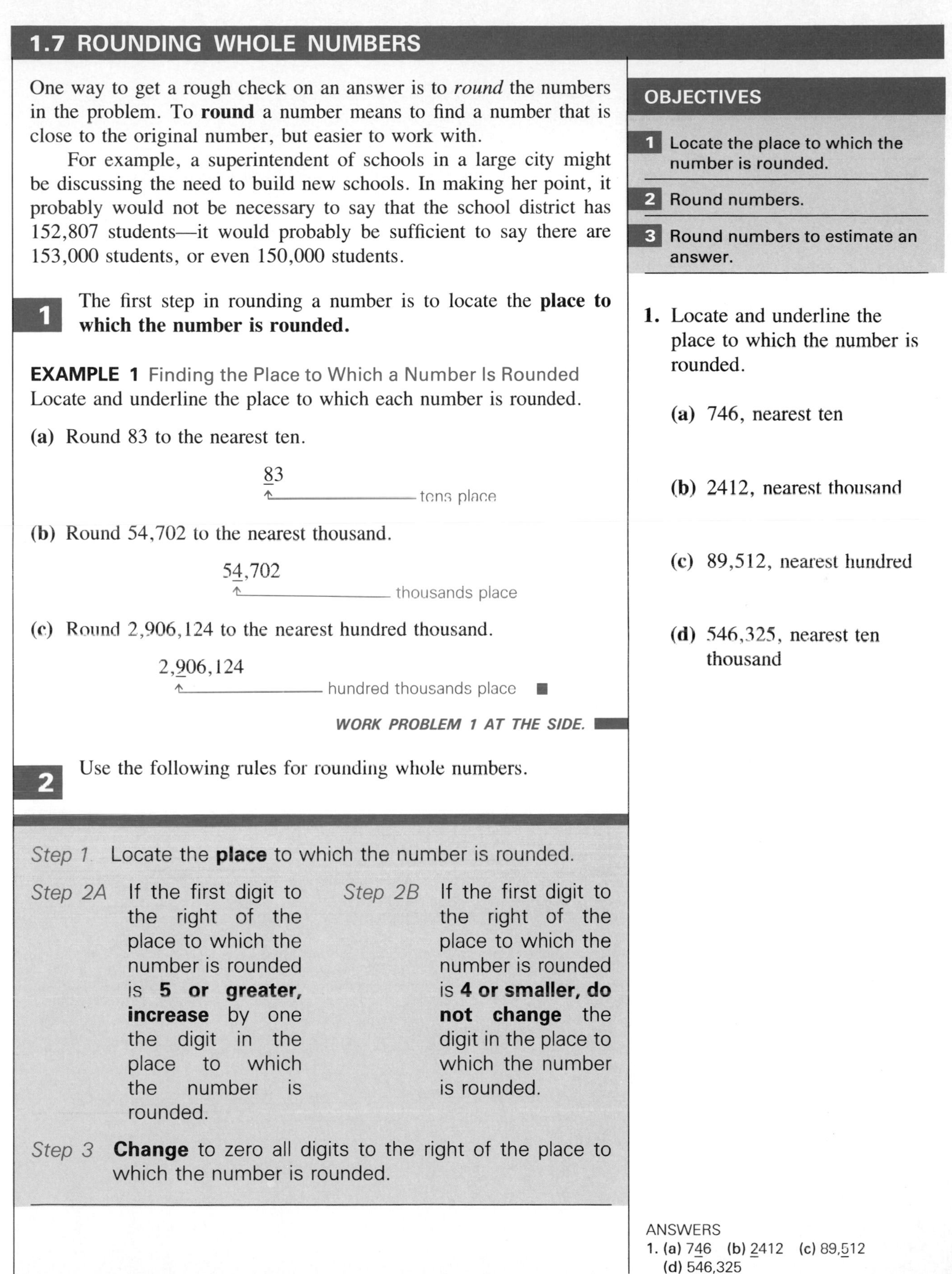

1.7 ROUNDING WHOLE NUMBERS

One way to get a rough check on an answer is to *round* the numbers in the problem. To **round** a number means to find a number that is close to the original number, but easier to work with.

For example, a superintendent of schools in a large city might be discussing the need to build new schools. In making her point, it probably would not be necessary to say that the school district has 152,807 students—it would probably be sufficient to say there are 153,000 students, or even 150,000 students.

OBJECTIVES

1. Locate the place to which the number is rounded.
2. Round numbers.
3. Round numbers to estimate an answer.

1 The first step in rounding a number is to locate the **place to which the number is rounded.**

EXAMPLE 1 Finding the Place to Which a Number Is Rounded

Locate and underline the place to which each number is rounded.

(a) Round 83 to the nearest ten.

83 ← tens place

(b) Round 54,702 to the nearest thousand.

54,702 ← thousands place

(c) Round 2,906,124 to the nearest hundred thousand.

2,906,124 ← hundred thousands place ■

WORK PROBLEM 1 AT THE SIDE.

1. Locate and underline the place to which the number is rounded.

(a) 746, nearest ten

(b) 2412, nearest thousand

(c) 89,512, nearest hundred

(d) 546,325, nearest ten thousand

2 Use the following rules for rounding whole numbers.

Step 1 Locate the **place** to which the number is rounded.

Step 2A If the first digit to the right of the place to which the number is rounded is **5 or greater, increase** by one the digit in the place to which the number is rounded.

Step 2B If the first digit to the right of the place to which the number is rounded is **4 or smaller, do not change** the digit in the place to which the number is rounded.

Step 3 **Change** to zero all digits to the right of the place to which the number is rounded.

ANSWERS
1. (a) 746 (b) 2412 (c) 89,512 (d) 546,325

2. Round to the nearest ten.

(a) 73

(b) 87

(c) 155

(d) 9839

3. Round to the nearest thousand.

(a) 92,405

(b) 86,032

(c) 5547

(d) 3594

ANSWERS
2. (a) 70 (b) 90 (c) 160 (d) 9840
3. (a) 92,000 (b) 86,000 (c) 6000 (d) 4000

EXAMPLE 2 Using Rounding Rules

Round 349 to the nearest hundred.

SOLUTION

Step 1 Locate the place to which the number is rounded.

349
↑ place to which number is rounded

Step 2 Since the first digit to the right of this place is 4, which is 4 or smaller, do not change the digit in the place to which the number is rounded.

3 remains 3

Step 3 Change to zeros all digits to the right of the 3.

349 rounded to the nearest hundred is 300. ■

WORK PROBLEM 2 AT THE SIDE.

EXAMPLE 3 Rounding Rules for 5 or Greater

Round 36,833 to the nearest thousand.

SOLUTION

Step 1 Find the place to which the number is rounded.

36,833
↑ place to which number is rounded

Step 2 Since the first digit to the right of this place is 8, which is 5 or greater, 1 must be added to the digit in the place to which the number is rounded.

Change 6 to 7. $(6 + 1 = 7)$

Step 3 All digits to the right of this 7 must be changed to zeros. Therefore,

36,833 rounded to the nearest thousand is 37,000. ■

WORK PROBLEM 3 AT THE SIDE.

EXAMPLE 4 Using Rules for Rounding

Round as shown.

(a) 2382 to the nearest ten

Step 1 2382
↑ place to which number is rounded

Step 2 The first digit to the right is a 2.

238 2 — 4 or smaller

Step 3 23 8 2 — Change to 0. Leave as 8.

2382 rounded to the nearest ten is 2380.

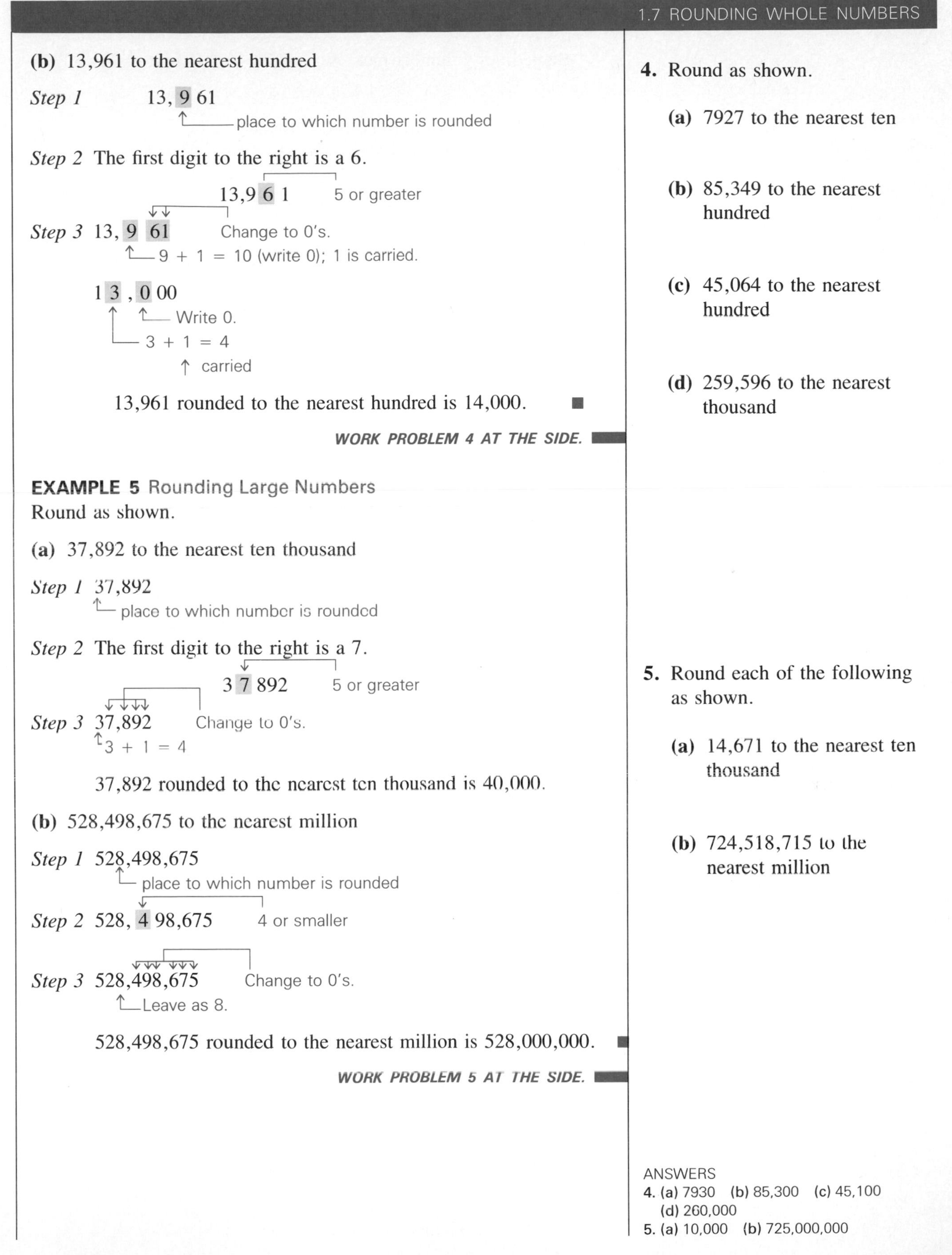

(b) 13,961 to the nearest hundred

Step 1 13,9 61
place to which number is rounded

Step 2 The first digit to the right is a 6.

13,961 5 or greater

Step 3 13,9 61 Change to 0's.
9 + 1 = 10 (write 0); 1 is carried.

13,0 00
Write 0.
3 + 1 = 4
carried

13,961 rounded to the nearest hundred is 14,000. ■

WORK PROBLEM 4 AT THE SIDE.

EXAMPLE 5 Rounding Large Numbers

Round as shown.

(a) 37,892 to the nearest ten thousand

Step 1 37,892
place to which number is rounded

Step 2 The first digit to the right is a 7.

3 7 892 5 or greater

Step 3 37,892 Change to 0's.
3 + 1 = 4

37,892 rounded to the nearest ten thousand is 40,000.

(b) 528,498,675 to the nearest million

Step 1 528,498,675
place to which number is rounded

Step 2 528, 4 98,675 4 or smaller

Step 3 528,498,675 Change to 0's.
Leave as 8.

528,498,675 rounded to the nearest million is 528,000,000. ■

WORK PROBLEM 5 AT THE SIDE.

4. Round as shown.

(a) 7927 to the nearest ten

(b) 85,349 to the nearest hundred

(c) 45,064 to the nearest hundred

(d) 259,596 to the nearest thousand

5. Round each of the following as shown.

(a) 14,671 to the nearest ten thousand

(b) 724,518,715 to the nearest million

ANSWERS
4. (a) 7930 (b) 85,300 (c) 45,100 (d) 260,000
5. (a) 10,000 (b) 725,000,000

Sometimes a number must be rounded to different places.

EXAMPLE 6 Rounding to Different Places
Round 648 **(a)** to the nearest ten and **(b)** to the nearest hundred.

SOLUTION

(a) to the nearest ten

6 4 8 — 5 or greater
tens position (4 + 1 = 5)

648 to the nearest ten is 650.

(b) to the nearest hundred

6 4 8 — 4 or smaller
hundreds position

648 to the nearest hundred is 600. ■

If 648 is rounded to the nearest ten at 650, and then 650 is rounded to the nearest hundred, the result is 700. If, however, 648 is rounded to the nearest hundred, the result is 600.

NOTE Always go back to the original number when rounding to a different place.

EXAMPLE 7 Applying Rounding Rules
Round each of the following to the nearest ten, nearest hundred, and nearest thousand.

(a) 4358 **(b)** 680,914

SOLUTION
(a) First round 4358 to the nearest ten.

435 8 — 5 or greater
tens position (5 + 1 = 6)

4358 rounded to the nearest ten is 4360.

Now go back to the original number to round to the nearest hundred.

43 5 8 — 5 or greater
hundreds position (3 + 1 = 4)

4358 rounded to the nearest hundred is 4400.

Again, go back to the original number to round, this time to the nearest thousand.

4 3 58   4 or smaller
↑ thousands position

4358 rounded to the nearest thousand is 4000.

(b) First, round to the nearest ten.

680,91 4   4 or smaller
↑ tens position

680,914 rounded to the nearest ten is 680,910.

Go back to the original number to round to the nearest hundred.

680,9 1 4   4 or smaller
↑ hundreds position

680,914 rounded to the nearest hundred is 680,900.

Go back to the original number to round to the nearest thousand.

680, 9 14   5 or greater
↑ thousands position (0 + 1 = 1)

680,914 rounded to the nearest thousand is 681,000. ■

WORK PROBLEM 6 AT THE SIDE.

3 Numbers are rounded to estimate an answer. An estimated answer is one that is close to the exact answer and may be used as a check when the exact answer is found.

EXAMPLE 8 Using Rounding to Estimate an Answer
Estimate the following answers by rounding to the nearest ten.

(a)

	Rounded	
76	80	rounded to the nearest ten
53	50	
38	40	
+ 91	+ 90	
	260	estimated answer

(b)

	Rounded	
27	30	rounded to the nearest ten
− 14	− 10	
	20	estimated answer

(c)

	Rounded	
16	20	rounded to the nearest ten
× 21	× 20	
	400	estimated answer ■

WORK PROBLEM 7 AT THE SIDE.

6. Round each of the following numbers to the nearest ten, nearest hundred, and nearest thousand.

(a) 7065

(b) 178,419

(c) 54,653

7. Estimate the following answers by rounding to the nearest ten.

(a) 24 + 68 + 82 + 56

(b) 33 − 17

(c) 43 × 68

ANSWERS
6. (a) 7070; 7100; 7000 (b) 178,420; 178,400; 178,000 (c) 54,650; 54,700; 55,000
7. (a) 230 (b) 10 (c) 2800

8. Estimate the following answers by rounding to the nearest hundred.

(a)
$$\begin{array}{r} 375 \\ 738 \\ 827 \\ +\ 673 \\ \hline \end{array}$$

(b)
$$\begin{array}{r} 792 \\ -\ 438 \\ \hline \end{array}$$

(c)
$$\begin{array}{r} 723 \\ \times\ 478 \\ \hline \end{array}$$

EXAMPLE 9 Using Rounding to Estimate an Answer

Estimate the following answers by rounding to the nearest hundred.

(a)
$$\begin{array}{r} 152 \\ 749 \\ 576 \\ +\ 819 \\ \hline \end{array} \qquad \left.\begin{array}{r} 200 \\ 700 \\ 600 \\ +\ 800 \\ \hline \end{array}\right\} \text{rounded to the nearest hundred}$$
$$2300 \quad \text{estimated answer}$$

(b)
$$\begin{array}{r} 780 \\ -\ 536 \\ \hline \end{array} \qquad \left.\begin{array}{r} 800 \\ -\ 500 \\ \hline \end{array}\right\} \text{rounded to the nearest hundred}$$
$$300 \quad \text{estimated answer}$$

(c)
$$\begin{array}{r} 664 \\ \times\ 843 \\ \hline \end{array} \qquad \left.\begin{array}{r} 700 \\ \times\ 800 \\ \hline \end{array}\right\} \text{rounded to the nearest hundred}$$
$$560{,}000 \quad \text{estimated answer}$$ ■

WORK PROBLEM 8 AT THE SIDE.

ANSWERS
8. (a) 2600 (b) 400 (c) 350,000

name date hour

1.7 EXERCISES

Round as shown.

Example: 6237 to the nearest ten **6240**

Solution: 6237 — 5 or greater, so add 1 to 3.
↑ tens place — Change 7 to 0.

1. 7862 to the nearest hundred

2. 674 to the nearest ten

3. 1276 to the nearest ten

4. 3928 to the nearest ten

5. 714 to the nearest ten

6. 16,215 to the nearest ten

7. 86,813 to the nearest ten

8. 17,211 to the nearest hundred

9. 42,495 to the nearest hundred

10. 8273 to the nearest hundred

11. 7998 to the nearest hundred

12. 12,303 to the nearest hundred

13. 15,758 to the nearest hundred

14. 28,065 to the nearest hundred

15. 31,052 to the nearest hundred

16. 4213 to the nearest thousand

17. 5847 to the nearest thousand

18. 49,706 to the nearest thousand

19. 53,182 to the nearest thousand

20. 13,124 to the nearest thousand

21. 496,192 to the nearest ten thousand

22. 515,035 to the nearest ten thousand

23. 8,906,422 to the nearest million

24. 13,713,409 to the nearest million

Round each of the following to the nearest ten, nearest hundred, and nearest thousand.

Example: 6375

Solution:

ten	***hundred***	***thousand***
637⑤ ← Change to 0. (5 or greater)	63⑦5 Change to 0's. (5 or greater)	6③75 Change to 0's. (4 or smaller)
↑ Add 1 to 7.	↑ Add 1 to 3.	↑ Leave original number.
6380	**6400**	**6000**

Remember to round from the original number.

	Ten	*Hundred*	*Thousand*
25. 7459	________	________	________
26. 8493	________	________	________
27. 4283	________	________	________
28. 6714	________	________	________
29. 5049	________	________	________
30. 7065	________	________	________
31. 3132	________	________	________
32. 7456	________	________	________

	Ten	*Hundred*	*Thousand*
33. 3645	______	______	______
34. 49,906	______	______	______
35. 28,171	______	______	______
36. 78,519	______	______	______
37. 23,502	______	______	______

Estimate the following answers by rounding to the nearest ten.

38. $\begin{array}{r} 26 \\ 72 \\ 84 \\ +\ 67 \\ \hline \end{array}$ ___ estimate

39. $\begin{array}{r} 39 \\ 14 \\ 56 \\ +\ 73 \\ \hline \end{array}$ ___ estimate

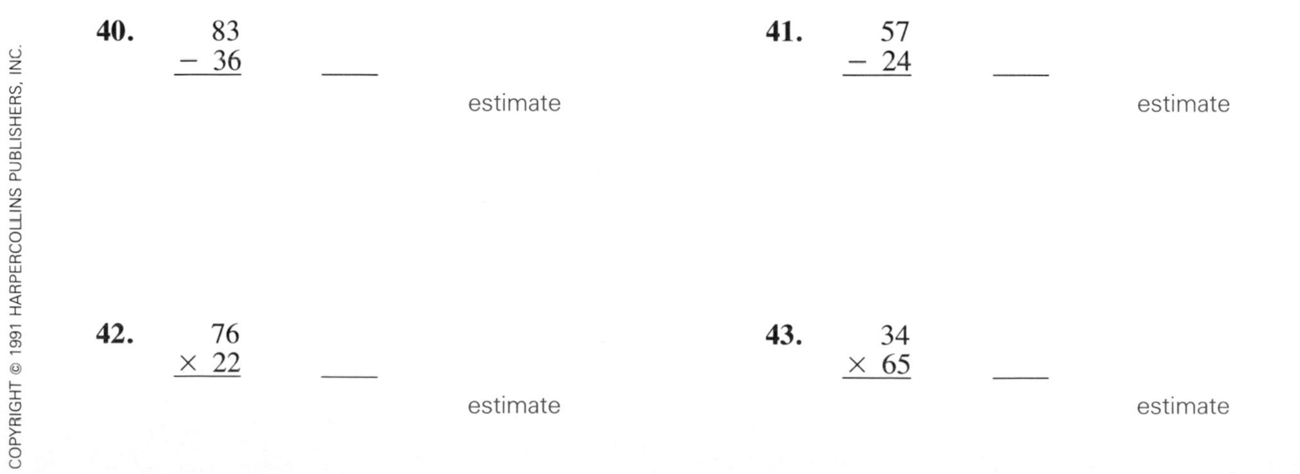

40. $\begin{array}{r} 83 \\ -\ 36 \\ \hline \end{array}$ ___ estimate

41. $\begin{array}{r} 57 \\ -\ 24 \\ \hline \end{array}$ ___ estimate

42. $\begin{array}{r} 76 \\ \times\ 22 \\ \hline \end{array}$ ___ estimate

43. $\begin{array}{r} 34 \\ \times\ 65 \\ \hline \end{array}$ ___ estimate

Estimate the following answers by rounding to the nearest hundred.

44.
$$\begin{array}{r} 786 \\ 823 \\ 342 \\ +\ 684 \\ \hline \end{array}$$
_____ estimate

45.
$$\begin{array}{r} 623 \\ 362 \\ 189 \\ +\ 736 \\ \hline \end{array}$$
_____ estimate

46.
$$\begin{array}{r} 936 \\ -\ 781 \\ \hline \end{array}$$
_____ estimate

47.
$$\begin{array}{r} 614 \\ -\ 276 \\ \hline \end{array}$$
_____ estimate

48.
$$\begin{array}{r} 279 \\ \times\ 518 \\ \hline \end{array}$$
_____ estimate

49.
$$\begin{array}{r} 649 \\ \times\ 594 \\ \hline \end{array}$$
_____ estimate

CHALLENGE EXERCISES

50. Round 70,987,652 to the nearest ten, nearest ten thousand, and nearest ten million.

51. Round 621,999,652 to the nearest thousand, nearest ten thousand, and nearest hundred thousand.

52. Round 357,654,218,862 to the nearest hundred thousand, nearest hundred million, and nearest hundred billion.

1.8 ROOTS AND ORDER OF OPERATIONS

1 The product $3 \cdot 3$ can be written as 3^2 (read as "3 squared"). The small raised number 2, called an exponent, says to use 2 factors of 3. The number 3 is called the **base.** Writing 3^2 as 9 is called **simplifying the expression.**

EXAMPLE 1 Simplifying an Expression
Identify the exponent and the base. Simplify each expression.

(a) 4^3

base → 4^3 ← exponent $\qquad 4^3 = 4 \times 4 \times 4 = 64$

(b) $2^5 = 2 \times 2 \times 2 \times 2 \times 2 = 32$

The base is 2 and the exponent is 5. ■

WORK PROBLEM 1 AT THE SIDE.

2 Since $3^2 = 9$, the number 3 is called the **square root** of 9. Square roots of numbers are written with the symbol $\sqrt{}$, so

$$\sqrt{9} = 3.$$

By definition,

$$\sqrt{\text{number}} \cdot \sqrt{\text{number}} = \text{number}.$$

Find the square root of 64 by asking, "What number can be multiplied by itself (that is, *squared*) to give 64?" The answer is 8, so $\sqrt{64} = 8$.

A **perfect square** is a number that is the square of a whole number. The first few perfect squares are listed here.

$0 = 0^2$	$16 = 4^2$	$64 = 8^2$	$144 = 12^2$
$1 = 1^2$	$25 = 5^2$	$81 = 9^2$	$169 = 13^2$
$4 = 2^2$	$36 = 6^2$	$100 = 10^2$	$196 = 14^2$
$9 = 3^2$	$49 = 7^2$	$121 = 11^2$	$225 = 15^2$

EXAMPLE 2 Using Perfect Squares
Find each square root.

(a) $\sqrt{16}$ Since $4^2 = 16$, $\sqrt{16} = 4$.

(b) $\sqrt{49} = 7$

(c) $\sqrt{0} = 0$

(d) $\sqrt{169} = 13$ ■

WORK PROBLEM 2 AT THE SIDE.

3 Frequently problems may have parentheses, exponents, and square roots, and may involve more than one operation. Work these problems with the following **order of operations.**

OBJECTIVES

1. Identify an exponent and a base.
2. Find the square root of a number.
3. Use the order of operations.

1. Identify the exponent and the base. Simplify each expression.

(a) 5^2

(b) 6^3

(c) 2^4

(d) 3^4

2. Find each square root.

(a) $\sqrt{4}$

(b) $\sqrt{36}$

(c) $\sqrt{81}$

(d) $\sqrt{196}$

(e) $\sqrt{1}$

ANSWERS
1. (a) 2; 5; 25 (b) 3; 6; 216 (c) 4; 2; 16 (d) 4; 3; 81
2. (a) 2 (b) 6 (c) 9 (d) 14 (e) 1

3. Work each problem.

(a) $3 + 9 + 4^2$

(b) $2^3 + 3^2$

(c) $5 \cdot 8 \div 20 - 1$

(d) $50 \div 2 \div 5$

(e) $8 + (14 \div 2) \cdot 6$

4. Work each problem.

(a) $8 - 5 + 2^2$

(b) $2^2 + 3^2 - (5 \cdot 2)$

(c) $5 \cdot \sqrt{100} - 7 \cdot 1$

(d) $16 \div 2 + (7 - 5)$

(e) $15 \cdot \sqrt{9} - 8 \cdot \sqrt{4}$

ANSWERS
3. (a) 28 (b) 17 (c) 1 (d) 5 (e) 50
4. (a) 7 (b) 3 (c) 43 (d) 10 (e) 29

ORDER OF OPERATIONS

1. Do all operations inside **parentheses.**
2. Simplify any expressions with **exponents** and find any **square roots.**
3. **Multiply** or **divide,** proceeding from left to right.
4. **Add** or **subtract,** proceeding from left to right.

EXAMPLE 3 Understanding Order of Operations
Work each problem.

(a) $8^2 + 5 + 2$

2 factors of 8

$$\begin{aligned} 8^2 + 5 + 2 &= 8 \cdot 8 + 5 + 2 && \text{Evaluate exponent first.} \\ &= 64 + 5 + 2 \\ &= 69 + 2 && \leftarrow \text{Add from the left.} \\ &= 71 \end{aligned}$$

(b)
$$\begin{aligned} 35 \div 5 \cdot 6 &= 7 \cdot 6 && \text{Divide first (start at left).} \\ &= 42 && \text{Multiply.} \end{aligned}$$

(c)
$$\begin{aligned} 9 + (20 - 4) \cdot 3 &= 9 + 16 \cdot 3 && \text{Work inside parentheses first.} \\ &= 9 + 48 && \text{Multiply.} \\ &= 57 && \text{Add.} \end{aligned}$$

(d)
$$\begin{aligned} 12 \cdot \sqrt{16} - 8 \cdot 4 &= 12 \cdot 4 - 8 \cdot 4 && \text{Find square root first.} \\ &= 48 - 32 && \text{Multiply from the left.} \\ &= 16 && \text{Subtract.} \end{aligned}$$
■

WORK PROBLEM 3 AT THE SIDE.

EXAMPLE 4 Using Order of Operations
Work each problem.

(a)
$$\begin{aligned} 15 - 4 + 2 &= 11 && \text{Subtract first (start at left).} \\ &= 13 && \text{Add.} \end{aligned}$$

(b)
$$\begin{aligned} 8 + (7 - 3) \div 2 &= 8 + 4 \div 2 && \text{Work inside parentheses first.} \\ &= 8 + 2 && \text{Divide.} \\ &= 10 && \text{Add.} \end{aligned}$$

(c)
$$\begin{aligned} 4^2 \cdot 2^2 + (7 + 3) \cdot 2 &= 4^2 \cdot 2^2 + 10 \cdot 2 && \text{Parentheses} \\ &= 16 \cdot 4 + 10 \cdot 2 && \text{Exponents} \\ &= 64 + 20 && \text{Multiply.} \\ &= 84 && \text{Add.} \end{aligned}$$

(d)
$$\begin{aligned} 4 \cdot \sqrt{25} - 7 \cdot 2 &= 4 \cdot 5 - 7 \cdot 2 && \text{Find square root first.} \\ &= 20 - 14 && \text{Multiply from the left.} \\ &= 6 && \text{Subtract.} \end{aligned}$$
■

NOTE Getting a correct answer depends on using the order of operations.

WORK PROBLEM 4 AT THE SIDE.

name date hour

1.8 EXERCISES

Use the table provided inside the back cover to find each square root.

Example: $\sqrt{169}$

Solution:
From the table, $13^2 = 169$, so $\sqrt{169} = \mathbf{13.}$

1. $\sqrt{4}$ **2.** $\sqrt{25}$ **3.** $\sqrt{16}$ **4.** $\sqrt{9}$

5. $\sqrt{144}$ **6.** $\sqrt{100}$ **7.** $\sqrt{121}$ **8.** $\sqrt{225}$

Identify the exponent and the base. Simplify each expression.

Example: 4^2

Solution:
4^2 ← exponent
↑ base

Simplify. $4^2 = 4 \times 4 = \mathbf{16}$

9. 2^3 **10.** 5^3

11. 6^2 **12.** 7^3

13. 12^2 **14.** 10^3

15. 15^2 **16.** 11^3

Complete each blank.

Example: $23^2 =$ so $\sqrt{\qquad} = 23$

Solution:
$23^2 = 23 \cdot 23 = \mathbf{529}$, so $\sqrt{\mathbf{529}} = 23$

17. $18^2 =$ so $\sqrt{\qquad} = 18$ **18.** $25^2 =$ so $\sqrt{\qquad} = 25$

19. $28^2 =$ so $\sqrt{\qquad} = 28$ **20.** $30^2 =$ so $\sqrt{\qquad} = 30$

21. $35^2 =$ so $\sqrt{\qquad} = 35$ **22.** $38^2 =$ so $\sqrt{\qquad} = 38$

23. $40^2 =$ so $\sqrt{} = 40$

24. $46^2 =$ so $\sqrt{} = 46$

25. $54^2 =$ so $\sqrt{} = 54$

26. $60^2 =$ so $\sqrt{} = 60$

Work each problem by using the order of operations.

Examples: $15 - 9 + 4 = 6 + 4$ $28 \div 7 + 3^2 = 28 \div 7 + 9$

Solutions: $= \mathbf{10}$ $= 4 + 9$ $= \mathbf{13}$

27. $9^2 + 5 - 2$

28. $4^3 + 12 - 9$

29. $6 \cdot 4 - \frac{9}{3}$

30. $2 \cdot 7 - 4$

31. $15 \cdot 2 \div 6$

32. $8 \cdot 9 \div 6$

33. $25 \div 5(8 - 4)$

34. $36 \div 18(7 - 3)$

35. $6 \cdot 2^2 + \frac{0}{6}$

36. $8 \cdot 3^2 - \frac{10}{2}$

37. $4 \cdot 1 + 8(9 - 2) + 3$

38. $3 \cdot 2 + 7(3 + 1) + 5$

39. $8 + 9 \div (5 - 2) + 6 \cdot 2$

40. $7 + 12 \div (3 \cdot 2) + \frac{0}{2}$

41. $2^3 \cdot 3^2 + (15 - 10)$

42. $4^2 \cdot 5^2 + (20 - 9) \cdot 3$

name date hour

43. $6 \cdot \sqrt{144} - 8 \cdot 6$

44. $9 \cdot \sqrt{100} - 3 \cdot 9$

45. $6 \cdot 2 + 9 \cdot 3 - 5$

46. $6 \cdot 3 + 4 \cdot 5 + 7$

47. $6 + 10 \div 2 + 3 \cdot 3$

48. $3^2 \cdot 4^2 + (15 - 6) \cdot 2$

49. $8 + 10 \div 5 + \frac{0}{3}$

50. $5^2 + 3^2 - (18 - 15)$

51. $3^2 + 6^2 + (30 - 21) \cdot 2$

52. $5 \cdot \sqrt{144} - 5 \cdot 7$

53. $7 \cdot \sqrt{81} - 5 \cdot 6$

54. $8 + 9 \div 3 + \sqrt{36}$

55. $6 \cdot 3 - 8 \cdot 2 + 4$

56. $5 \cdot 2 + 3(5 + 3) - 6$

57. $4 \cdot \sqrt{49} - 7(5 - 2)$

58. $8 \cdot \sqrt{25} - 4 \cdot 7$

59. $6 \cdot (5 - 1) + \sqrt{4}$

60. $7 \cdot 4 \div 2 + 8$

61. $6^2 + 2^2 - 6 \cdot 2$

62. $3^2 - 2^2 + 3 - 2$

63. $5^2 \cdot 2^2 + (8 - 4)$

64. $5^2 \cdot 3^2 + (30 - 20) \cdot 2$

65. $7 + 6 \div 3 + 5 \cdot 2$

66. $8 + 3 \div 3 + 6 \cdot 3$

67. $5 \cdot \sqrt{36} - 7(7 - 4)$

68. $9 \cdot \sqrt{64} - 5(4 + 2)$

69. $6^2 - 2^2 + 3 \cdot 4$

70. $4^2 + 3^2 - 5 \cdot 3$

71. $8 + 5 \div 5 + 7 \cdot 3$

72. $2 + 12 \div 6 + 5 + \frac{0}{5}$

73. $3 \cdot \sqrt{16} - 7 \cdot 1$

74. $7 \cdot \sqrt{81} - 3 \cdot 6$

75. $3 \cdot \sqrt{25} - 4 \cdot \sqrt{9}$

76. $8 \cdot \sqrt{100} - 6 \cdot \sqrt{36}$

77. $7 \div 1 \cdot 8 \cdot 2 \div (21 - 5)$

78. $12 \div 4 \cdot 5 \cdot 4 \div 2$

79. $15 \div 3 \cdot 2 \cdot 6 \div 3$

80. $9 \div 1 \cdot 4 \cdot 2 \div (11 - 5)$

81. $5 \cdot \sqrt{9} - 2 \cdot \sqrt{4}$

82. $12 \cdot \sqrt{81} - 7 \cdot \sqrt{25}$

83. $5 \div 1 \cdot 10 \cdot 4 \div (17 - 9)$

84. $15 \div 3 \cdot 8 \cdot 9 \div 4$

CHALLENGE EXERCISES

85. $8 \cdot 9 \div \sqrt{36} - 4 \div 2 + (14 - 8)$

86. $3 - 2 + 5 \cdot 4 \cdot \sqrt{144} \div \sqrt{36}$

87. $1 + 3 - 2 \cdot \sqrt{1} + 3 \cdot \sqrt{121} - 5 \cdot 3$

88. $6 - 4 + 2 \cdot 9 - 3 \cdot \sqrt{225} \div \sqrt{25}$

89. $6 \cdot \sqrt{25} \cdot \sqrt{100} \div 3 \cdot \sqrt{4} + 9$

90. $9 \cdot \sqrt{36} \cdot \sqrt{81} \div 2 + 6 - 3 - 5$

1.9 SOLVING WORD PROBLEMS

OBJECTIVES

1. Find indicator words in word problems.
2. Solve word problems.
3. Estimate an answer.

Problems involving applications of mathematics are usually presented in word problems. You must read the words carefully to decide how to solve the problem.

1 Look for **indicators** in the word problem—words that indicate the necessary operations—either addition, subtraction, multiplication, or division. Some of these word indicators appear below.

Addition	Subtraction
plus	less
more	subtract
more than	subtracted from
added to	difference
increased by	less than
sum	fewer
total	decreased by
sum of	loss of
increase of	minus
gain of	take away

Multiplication	Division
product	divided by
double	divided into
triple	quotient
times	goes into
of	divide
twice	divided equally
twice as much	per

Equals

is
the same as
equals
equal to

2 Solve word problems by using the following steps.

1. Pick the most reasonable answer for each problem.

(a) an hourly wage
$2; $6; $60

(b) the cost of a hamburger
$2; $8; $20

(c) a score on a test
6; 20; 74; 109

2. (a) On a recent geology field trip, 84 fossils were collected. If the fossils are divided equally among Gilbert, Sean, Noella, and Sue, how many fossils will each receive?

(b) A gasoline storage tank holds 295 gallons of gasoline. If the gas is divided equally among five people, how much does each person get?

ANSWERS
1. (a) $6 (b) $2 (c) 74
2. (a) 21 fossils
(b) 59 gallons

Step 1 Read the problem carefully and be certain you **understand** what the problem is asking. It may be necessary to read the problem several times.

Step 2 Before doing any calculations, work out a **plan.** Know which facts are given and which must be found. Use **indicators** to help decide on the plan.

Step 3 Estimate a **reasonable answer** by using rounding.

Step 4 **Solve** the problem by using the facts given and your plan. If the answer is reasonable, **check** your work. If the answer is not reasonable, begin again by rereading the problem.

3 These steps give a systematic approach for solving word problems. Each of the steps is important, but special emphasis should be placed on Step 3, estimating a *reasonable answer*. Many times an ''answer'' just does not fit the problem.

What is a reasonable answer? Read the problem and try to determine the approximate size of the answer. Should the answer be part of a dollar, a few dollars, hundreds, thousands, or even millions of dollars? For example, if a problem asks for the cost of a man's shirt, would an answer of $15 be reasonable? $1000? $65? $80?

Always make an estimate of a reasonable answer; then check the answer you get to see if it is close to your estimate.

WORK PROBLEM 1 AT THE SIDE.

EXAMPLE 1 Applying Division

At a recent garage sale, the total sales were $116. If the money was divided equally among Tom, Roietta, Maryann, and Jose, how much did each person get?

APPROACH

To find the amount received by each person, divide the total amount of sales by the number of people.

SOLUTION

Step 1 A reading of the problem shows that the four members in the group divided $116 equally.

Step 2 The word indicators, ***divided equally,*** show that the amount each received can be found by dividing $116 by 4.

Step 3 A reasonable answer would be a little more than $25 each, since $100 ÷ 4 = $25.

Step 4 Find the actual answer by dividing $116 by 4.

$$4\overline{)116}\quad 29$$

Each person should get $29.

The answer $29 is reasonable, since $29 is close to the estimated answer of $25.

Is the answer $29 correct? Check the work.

$$\begin{array}{r} \$29 \\ \times \quad 4 \\ \hline \$116 \end{array}$$ ■

WORK PROBLEM 2 AT THE SIDE.

EXAMPLE 2 Applying Addition

Larry earns $23 on Monday, $18 on Tuesday, $24 on Wednesday, $25 on Thursday, and $16 on Friday. Find his total earnings.

APPROACH

To find the total for the week, add the earnings for each day.

SOLUTION

Step 1 In this problem, the earnings for each day are given and the total earnings for the week must be found.

Step 2 Add the daily earnings to arrive at the weekly total.

Step 3 Since the earnings were about $20 per day for a week of 5 days, a reasonable estimate would be around $100 (5 × $20 = $100).

Step 4 Find the actual answer by adding the earnings for the 5 days.

$$\begin{array}{rl} \$106 & \text{check} \\ \$23 & \\ \$18 & \\ \$24 & \\ \$25 & \\ + \$16 & \\ \hline \$106 & \text{earnings for the week} \end{array}$$

This answer is reasonable. ■

WORK PROBLEM 3 AT THE SIDE.

EXAMPLE 3 Determining Whether Subtraction Is Necessary

The number of students enrolled in Peterson College this year is 1842 fewer than the number enrolled last year. Enrollment last year was 12,622. Find the enrollment this year.

APPROACH

To find the number of students enrolled this year, the enrollment decrease (fewer students) must be subtracted from last year.

SOLUTION

Step 1 In this problem, the enrollment has decreased from last year to this year. The enrollment last year and the decrease in enrollment are given. This year's enrollment must be found.

Step 2 The word indicator, ***fewer***, shows that subtraction must be used to find the number of students enrolled this year.

3. (a) During the semester, Cindy receives the following points on examinations and quizzes: 92, 81, 83, 98, 15, 14, 15, and 12. Find her total points for the semester.

(b) Marie Perino, a salesperson for a greeting card company, drives 73 miles on Monday, 117 miles on Tuesday, 235 miles on Wednesday, 94 miles on Thursday, and spends Friday in her office. Find her total mileage for the week.

ANSWERS

3. (a) 410 points
(b) 519 miles

Step 3 Since the enrollment was about 13,000 students, and the decrease in enrollment is about 2000 students, a reasonable estimate would be 11,000 students (13,000 − 2000 = 11,000).

Step 4 Find the actual answer by subtracting 1842 from 12,622.

$$\begin{array}{r} 12{,}622 \\ -\ \ 1\,842 \\ \hline 10{,}780 \end{array}$$

The enrollment this year is 10,780.

The answer 10,780 is reasonable, since it is close to the estimate of 11,000. Check.

10,780	enrollment this year
+ 1 842	decrease in enrollment
12,622	enrollment last year ■

WORK PROBLEM 4 AT THE SIDE.

EXAMPLE 4 Solving a Two-Step Problem

A landlord receives $680 from each of five tenants. After paying $1880 in expenses, how much rent money does the landlord have left?

APPROACH

To find the amount remaining, first find the total rent received. Next, subtract the expenses paid to find the amount remaining.

SOLUTION

Step 1 There are five tenants and each pays the same.

Step 2 The word *each* indicates that the five rents must be totaled. Since the rents are all the same, use multiplication to find the total rent received. Finally, subtract expenses.

Step 3 The amount of rent is about $700, making the total rent received about $3500 ($700 × 5). The expenses are about $1900. A reasonable estimate of the amount remaining is $1600 ($3500 − $1900).

Step 4 Find the exact amount by first multiplying $680 by 5 (the number of tenants).

$$\begin{array}{r} \$680 \\ \times\ \ \ \ 5 \\ \hline \$3400 \end{array}$$

Finally, subtract the $1880 in expenses from $3400.

$$\$3400 - \$1880 = \$1520$$

The amount remaining is $1520.

The answer $1520 is reasonable, since it is close to the estimated answer of $1600. Check the amount by adding the expenses and then dividing by 5.

$$\$1520 + \$1880 = \$3400$$

$$5\overline{)3400}\ \ \ \$680$$ ■

WORK PROBLEM 5 AT THE SIDE.

4. (a) One home occupies 1858 square feet, while another occupies 1640 square feet. Find the difference in the number of square feet of the two homes.

(b) A library had 25,622 books. After a loss of 1367 books, how many books were left?

5. (a) Mary receives four checks amounting to $160 each. If she spends $200, find the amount that remains.

(b) During a 4-hour period, 125 cars enter a parking lot each hour. In the same time period, 271 cars leave the lot. Find the number of cars remaining in the lot.

ANSWERS
4. (a) 218 square feet (b) 24,255 books
5. (a) $440 (b) 229

name date hour

1.9 EXERCISES

Solve each of the following word problems.

1. Sales at the Computer Center were $6975 today. If this is $1630 less than the sales yesterday, what were yesterday's sales?

2. Ted Slauson, coordinator of Toys for Tots, has collected 2628 toys. If his group can give the same number of toys to each of 657 children, how many toys will each child receive?

3. If 375 ski lift tickets are sold per day, how many tickets will be sold in a 7-day period?

4. If profits of $680,000 are divided evenly among a firm's 1000 employees, how much money will each employee receive?

5. Mark Steffans owes $3815 plus $268 in interest on his credit union loan. If he wishes to pay the loan in full, how much must he pay?

6. To qualify for a real estate loan at Uptown Bank, a borrower must have a monthly income of at least 4 times the monthly payment. For a monthly payment of $675, what must the borrower's minimum monthly income be?

7. In The Fine Leather Shop, 783 yards of leather are used from a supply of 938 yards. How many yards are not used?

8. Management reports that 73 window frames came off the assembly line each hour. At this rate, how many frames came off the assembly line in a 24-hour period?

9. Lois Stevens knows that her Escort gets 36 miles per gallon in town. How many miles can she travel in town on 37 gallons?

10. Erich Means completed a 2146-mile trip on his motorcycle and used 37 gallons of gasoline. How many miles did he travel on each gallon?

11. Beth Anderson spent \$286 on tuition, \$148 on books, and \$12 on supplies. If this money was withdrawn from her checking account, which had a balance of \$698, what is her new balance?

12. The Natural Chocolate Works melts 385 pounds of light chocolate, 100 pounds of dark chocolate, and 22 pounds of peanut butter. They next add 18 pounds of almonds and 1 pound of confectioner's wax. What is the total weight of the candy made from these ingredients?

13. A truck weighs 9250 pounds when empty. After being loaded with firewood, it weighs 21,375 pounds. What is the weight of the firewood?

14. How many 2-inch strips of leather can be cut from a piece of leather 1 foot wide? (*Hint:* 1 foot = 12 inches.)

15. If there are 43,560 square feet in an acre, how many square feet are there in 4 acres?

16. Find the total cost if a college bookstore buys 17 computers at \$506 each and 13 printers at \$482 each.

17. The number of gallons of water polluted each day in an industrial area is 209,670. How many gallons of water are polluted each year? (Use a 365-day year.)

18. Travel Rent-A-Car owns 385 compact cars, 483 full-size cars, 115 luxury cars, and 71 vans and trucks. How many vehicles does it have in all?

19. The total number of miles covered on a cross-country bicycle trip was 567. If the trip took 9 days and the same number of miles was traveled each day, how many miles were traveled per day?

20. Tom earned $14 on Monday, $23 on Tuesday, and $62 on Friday. Find the total amount he earned.

21. If beachfront property sells for $1018 per foot, what is the value of a 60-foot beachfront lot?

22. On a cross-country pigeon race, the total number of miles covered was 324. If the trip took 9 hours, find the number of miles flown per hour.

23. A biology class found 24 deer in one area, 232 in another, and 512 in a third. How many deer did the class find?

24. Barbara Salzman, Blue Bird leader, estimates that each of her Blue Birds will eat three cookies, and she and her assistant, Lana Meehan, will each eat four cookies. If she expects 14 Blue Birds and her assistant at the meeting, how many cookies will she need?

25. Edward Biondi has $3010 in his checking account. He writes checks for $280 for tires, $620 for equipment repairs, and $178 for fuel and oil. Find the balance remaining in his account.

26. Sherry Patterson can file 43 folders per hour. How many folders can she file in 3 hours?

27. A package of 5 shirts cost $18, and a package of 3 pairs of socks cost $4. Find the total cost of 25 shirts and 9 pairs of socks.

28. A car weighs 2425 pounds. If its 582-pound engine is removed and replaced with a 634-pound engine, find the weight of the car after the engine change.

29. Tino Santana has annual payroll deductions of $2184 for federal taxes and $372 for state taxes. If equal deductions are made each month, find his total monthly tax deductions.

30. In one week, Bob earned $5 per hour for working 37 hours. Russell earned $6 per hour for working 39 hours. Find their total combined income.

CHALLENGE EXERCISES

31. Jim Peppa's vending machine company had 325 machines on hand at the beginning of the month. At different times during the month, machines were distributed to new locations; 35 machines were taken at one time, then 23 machines, and then 76 machines. During the same month additional machines were returned; 15 machines were returned at one time, then 38 machines, and then 108 machines. How many machines were on hand at the end of the month?

32. Mike Fitzgerald owns 70 acres of land that he leases to an alfalfa farmer for $150 per acre per year. If property taxes are $28 per acre per year, find the total amount of yearly lease income he has left after taxes are paid.

33. A theater owner wants to provide enough seating for 1250 people. The main floor has 30 rows of 25 seats in each row. If the balcony has 25 rows, how many seats must be in each row to satisfy the owner's seating requirements?

34. Jennie makes 24 grapevine wreaths per week to sell to gift shops. She works 30 weeks a year and packages six wreaths per box. If she ships equal quantities to each of five shops, find the number of boxes each store will receive.

CHAPTER 1 SUMMARY

KEY TERMS

1.1	**whole numbers**	A number made up of digits to the left of the decimal point is called a whole number. Examples are 0, 1, 2, 3, and 4.
1.2	**sum**	The answer in an addition problem is called the sum.
	commutative property of addition	The commutative property of addition states that the order of numbers in an addition problem can be changed without changing the sum.
	carrying	The process of carrying is used in an addition problem when the sum of the digits in a column is greater than 9.
1.3	**difference**	The answer in a subtraction problem is called the difference.
	borrowing	The method of borrowing is used in subtraction if a digit is greater than the one directly above.
1.4	**factors**	Parts of a product are called factors. For example, since $3 \times 4 = 12$, both 3 and 4 are factors of 12.
	product	The answer in a multiplication problem is called the product.
	commutative property of multiplication	The commutative property of multiplication states that the product in a multiplication problem remains the same when the order of the factors is changed.
	multiple	The product of two whole-number factors is a multiple of those numbers.
1.5	**quotient**	The answer in a division problem is called the quotient.
	remainder	The remainder is the number left over when two numbers do not divide exactly.
1.6	**long division**	The process of long division is used to divide by a number with more than one digit.
1.7	**rounding**	To find a number that is close to the original number, but easier to work with, we use rounding.
	perfect square	A number that is the square of a whole number is a perfect square.
	indicators	Words that suggest the operations needed for solving a problem are called indicators.
1.8	**square root of a number**	The square root of a given number is the number that can be multiplied by itself to produce the given (larger) number.

QUICK REVIEW

Section Number and Topic	Approach	Example
1.1 Reading and Writing Whole Numbers	Do not use the word "and" with a whole number. Commas help divide the periods for ones, thousands, millions, and billions. A comma is not needed with a number having four digits or fewer.	795 is written *seven hundred ninety-five*. 9,768,002 is written *nine million, seven hundred sixty-eight thousand, two*.
1.2 Addition of Whole Numbers	Add from top to bottom, starting with units and working left. To check, add from bottom to top.	Problem (add down): 1 1 4 0 (carries); 687 + 26 + 9 + 418 = 1140; (add up) Check
1.2 Commutative Property of Addition	The order of numbers in an addition problem can be changed without changing the sum.	$4 + 2 = 6$ $2 + 4 = 6$ By the commutative property, the sum is the same.
1.3 Subtraction of Whole Numbers	Subtract the subtrahend from minuend to get the difference by borrowing when necessary. To check, add the difference to the subtrahend to get the minuend.	Problem: 4738 (minuend; borrowing marks 6 2 1) − 649 (subtrahend) = 4089 (difference) Check: 4089 + 649 = 4738
1.4 Multiplication of Whole Numbers	The numbers being multiplied are called *factors*. The multiplicand is being multiplied by the multiplier, giving the product. When the multiplier has more than one digit, partial products must be used and added to find the product.	78 multiplicand × 24 multiplier 312 partial product 156 partial product (one position left) 1872 product

Section Number and Topic	Approach	Example
1.4 Commutative Property of Multiplication	The answer or product in multiplication remains the same when the order of the factors is changed.	$3 \times 4 = 12$ $4 \times 3 = 12$ By the commutative property, the product is the same.
1.5 Division of Whole Numbers	$\div$ and $\overline{)\quad}$ mean divide. Also a —, as in $\frac{25}{5}$, means to divide the top number (dividend) by the bottom number (divisor).	22 quotient divisor $4\overline{)88}$ dividend 88 0 If the answer is rounded, a check will not match the dividend perfectly.
1.7 Rounding Whole Numbers	Rules for rounding: 1. Identify the position to be rounded. 2. If the digit to the right is 5 or greater, increase by 1; if 4 or smaller, do not change. 3. Change to zero all digits to the right of the place being rounded.	Round 726 to the nearest ten. ↑ tens position; 5 or greater, so add 1 to tens position. 726 rounds to 730. Round 1,498, 586 to the nearest million. (millions position); 4 or smaller, so do not change. 1,498,586 rounds to 1,000,000.
1.8 Order of Operations	Problems may have several operations. Work these problems with the following order of operations. 1. Do all operations inside parentheses. 2. Simplify any expressions with exponents and find any square roots. 3. Multiply or divide from left to right. 4. Add or subtract from left to right.	Solve, using the order of operations. $7 \cdot \sqrt{9} - 4 \cdot 5 = 7 \cdot 3 - 4 \cdot 5$ $= 7 \cdot 3 - 4 \cdot 5$ Find square root. $= 21 - 20$ Multiply from left to right. $= 1$ Subtract.

Section Number and Topic	Approach	Example
1.9 Word Problems	Follow these steps.	Manuel earns $118 on Sunday, $87 on Monday, and $63 on Tuesday. Find total earnings for the three days. *Total* means to add.
	1. Read the problem carefully.	1. The earnings for each day are given, and the total for the 3 days must be found.
	2. Work out a plan before starting.	2. Add the daily earnings to find the total.
	3. Estimate a reasonable answer.	3. Since the earnings were about $90 per day for 3 days, a reasonable estimate would be approximately $270 ($90 × 3 = $270).
	4. Solve the problem. If the answer is reasonable, check; if not, start over.	4. $\underline{\$268}$ Check $118 87 + 63 $268 total earnings

name date hour

CHAPTER 1 REVIEW EXERCISES

If you need help with any of these review exercises, look in the section indicated in brackets.

[1.1] *Fill in the digits for the given period in each of the following numbers.*

1. 7816
thousands
ones

2. 78,915
thousands
ones

3. 206,792
thousands
ones

4. 1,768,710,618
billions
millions
thousands
ones

Rewrite the following numbers in words.

5. 725

6. 17,615

7. 62,500,005

Rewrite each of the following numbers in digits.

8. eight thousand, one hundred twenty

9. six hundred million, fifteen thousand, seven hundred fifty-nine

[1.2] *Add.*

10. $\begin{array}{r} 74 \\ +\ 29 \\ \hline \end{array}$

11. $\begin{array}{r} 43 \\ +\ 77 \\ \hline \end{array}$

12. $\begin{array}{r} 778 \\ +\ 459 \\ \hline \end{array}$

13. $\begin{array}{r} 914 \\ 3708 \\ +\ \ \ 34 \\ \hline \end{array}$

14. $\begin{array}{r} 8215 \\ 9 \\ +\ 7433 \\ \hline \end{array}$

15. $\begin{array}{r} 1108 \\ 566 \\ 7201 \\ +\ \ 304 \\ \hline \end{array}$

16. $\begin{array}{r} 187 \\ 5543 \\ 246 \\ +\ 1003 \\ \hline \end{array}$

17. $\begin{array}{r} 5\ 732 \\ 11{,}069 \\ 37 \\ 1\ 595 \\ +\ 22{,}169 \\ \hline \end{array}$

[1.3] *Subtract.*

18. $\begin{array}{r} 26 \\ -\ 15 \\ \hline \end{array}$ **19.** $\begin{array}{r} 79 \\ -\ 57 \\ \hline \end{array}$ **20.** $\begin{array}{r} 238 \\ -\ 199 \\ \hline \end{array}$ **21.** $\begin{array}{r} 4380 \\ -\ 577 \\ \hline \end{array}$

22. $\begin{array}{r} 5210 \\ -\ 883 \\ \hline \end{array}$ **23.** $\begin{array}{r} 2215 \\ -\ 1198 \\ \hline \end{array}$ **24.** $\begin{array}{r} 2210 \\ -\ 1986 \\ \hline \end{array}$ **25.** $\begin{array}{r} 99{,}704 \\ -\ 73{,}838 \\ \hline \end{array}$

[1.4] *Multiply.*

26. $\begin{array}{r} 8 \\ \times\ 8 \\ \hline \end{array}$ **27.** $\begin{array}{r} 7 \\ \times\ 0 \\ \hline \end{array}$ **28.** $\begin{array}{r} 4 \\ \times\ 6 \\ \hline \end{array}$ **29.** 6×9

30. 6×6 **31.** 4×9 **32.** $7 \cdot 8$ **33.** $9 \cdot 9$

Work the following chain multiplications.

34. $2 \times 4 \times 6$ **35.** $9 \times 1 \times 5$ **36.** $3 \times 3 \times 8$ **37.** $2 \times 2 \times 2$

38. $8 \cdot 0 \cdot 6$ **39.** $9 \cdot 9 \cdot 1$ **40.** $5 \cdot 1 \cdot 7$ **41.** $7 \cdot 7 \cdot 0$

Multiply.

42. $\begin{array}{r} 65 \\ \times\ 2 \\ \hline \end{array}$ **43.** $\begin{array}{r} 92 \\ \times\ 7 \\ \hline \end{array}$ **44.** $\begin{array}{r} 24 \\ \times\ 3 \\ \hline \end{array}$ **45.** $\begin{array}{r} 89 \\ \times\ 1 \\ \hline \end{array}$

46. $\begin{array}{r} 39 \\ \times\ 6 \\ \hline \end{array}$ **47.** $\begin{array}{r} 781 \\ \times\ 7 \\ \hline \end{array}$ **48.** $\begin{array}{r} 349 \\ \times\ 4 \\ \hline \end{array}$ **49.** $\begin{array}{r} 9163 \\ \times\ 5 \\ \hline \end{array}$

50. $\begin{array}{r} 7259 \\ \times\ 2 \\ \hline \end{array}$ **51.** $\begin{array}{r} 4480 \\ \times\ 5 \\ \hline \end{array}$ **52.** $\begin{array}{r} 93{,}105 \\ \times\ 5 \\ \hline \end{array}$ **53.** $\begin{array}{r} 21{,}873 \\ \times\ 8 \\ \hline \end{array}$

name date hour

54. $\begin{array}{r} 22 \\ \times\ 15 \\ \hline \end{array}$

55. $\begin{array}{r} 52 \\ \times\ 36 \\ \hline \end{array}$

56. $\begin{array}{r} 98 \\ \times\ 12 \\ \hline \end{array}$

57. $\begin{array}{r} 708 \\ \times\ 65 \\ \hline \end{array}$

58. $\begin{array}{r} 655 \\ \times\ 21 \\ \hline \end{array}$

59. $\begin{array}{r} 392 \\ \times\ 77 \\ \hline \end{array}$

60. $\begin{array}{r} 5032 \\ \times\ 48 \\ \hline \end{array}$

61. $\begin{array}{r} 543 \\ \times\ 658 \\ \hline \end{array}$

Find the total cost of each of the following.

62. 18 chairs at $32 per chair

63. 24 soccer balls at $13 per ball

64. 278 tires at $48 per tire

65. 168 welders masks at $9 per mask

Multiply by using multiples of ten.

66. $\begin{array}{r} 50 \\ \times\ 7 \\ \hline \end{array}$

67. $\begin{array}{r} 380 \\ \times\ 80 \\ \hline \end{array}$

68. $\begin{array}{r} 752 \\ \times\ 400 \\ \hline \end{array}$

69. $\begin{array}{r} 16{,}000 \\ \times\ 8\ 000 \\ \hline \end{array}$

70. $\begin{array}{r} 43{,}000 \\ \times\ 2\ 100 \\ \hline \end{array}$

71. $\begin{array}{r} 30{,}200 \\ \times\ 20{,}000 \\ \hline \end{array}$

[1.5]

Divide whenever possible.

72. $6 \div 3$

73. $35 \div 7$

74. $42 \div 6$

75. $36 \div 4$

76. $\frac{72}{8}$

77. $\frac{27}{9}$

78. $\frac{54}{6}$

79. $\frac{0}{6}$

80. $\frac{125}{0}$

81. $\frac{0}{35}$

[1.5–1.6] *Divide.*

82. $8\overline{)648}$

83. $5\overline{)180}$

84. $9\overline{)56{,}259}$

85. $76\overline{)26{,}752}$

86. $576 \div 6$

87. $2704 \div 18$

88. $15{,}525 \div 125$

[1.7] *Round as shown.*

89. 118 to the nearest ten

90. 16,701 to the nearest hundred

91. 19,721 to the nearest thousand

92. 67,485 to the nearest ten thousand

Round each of the following to the nearest ten, nearest hundred, and nearest thousand.

	ten	***hundred***	***thousand***
93. 1496	______	______	______
94. 10,056	______	______	______
95. 98,201	______	______	______
96. 352,118	______	______	______

[1.8] *Find each square root by using the table provided inside the back cover of the book.*

97. $\sqrt{4}$

98. $\sqrt{144}$

99. $\sqrt{81}$

100. $\sqrt{196}$

name date hour

Identify the exponent and the base. Simplify each expression.

101. 3^2

102. 5^3

103. 3^5

104. 4^5

Work each problem by using the order of operations.

105. $9^2 - 9$

106. $2 \cdot 7 + 6$

107. $2 \cdot 3^2 \div 2$

108. $9 \div 1 \cdot 2 \cdot 2 \div (11 - 2)$

109. $\sqrt{9} + 2 \cdot 3$

110. $6 \cdot \sqrt{16} - 6 \cdot \sqrt{9}$

[1.9] *Solve each of the following word problems.*

111. Find the cost of 15 cash registers at $1356 per cash register.

112. A pulley turns 1400 revolutions per minute. How many revolutions will the pulley turn in 60 minutes?

113. Sheri types 120 words in 1 minute. Find the number of words she can type in 12 minutes.

114. A keg contains 8000 nails. How many nails are in 40 kegs?

115. It takes 2000 hours of work to build 1 home. How many hours of work are needed to build 12 homes?

116. A train travels 80 miles in 1 hour. Find the number of miles traveled in 5 hours.

117. There are 24 cans of soft drink in 1 case. How many cans are in 24 cases?

118. A newspaper girl has 56 customers who take the paper daily and 23 customers who take the paper on weekends only. A daily customer pays $6 per month and a weekend-only customer pays $3 per month. Find the total monthly collections.

119. Sam Lee had $1279 withheld from his paycheck last year for income tax. If he owed only $1080 in taxes, find the amount of refund he should receive.

120. A stamping machine produces 936 license plates per hour. How long will it take to produce 30,888 license plates?

121. A food canner uses 1 pound of pork for every 175 cans of pork and beans. How many pounds of pork are needed for 8750 cans of pork and beans?

122. Susan Hessney has $382 in her checking account. She writes a check for $135. How much does she have left in her account?

123. If an acre needs 250 pounds of fertilizer, how many acres can be fertilized with 5750 pounds of fertilizer?

124. A contract calls for 3415 feet of fencing. If 1786 feet have been installed, how many more feet are still needed?

MIXED REVIEW EXERCISES

Solve each of the following as indicated.

125. $\begin{array}{r} 64 \\ \times\ 5 \\ \hline \end{array}$

126. $81 \div 9$

127. $\begin{array}{r} 179 \\ -\ 64 \\ \hline \end{array}$

128. 8×5

129. $\begin{array}{r} 662 \\ +\ 379 \\ \hline \end{array}$

130. $\begin{array}{r} 72 \\ \times\ 29 \\ \hline \end{array}$

131. $\begin{array}{r} 38{,}140 \\ -\ 6\ 078 \\ \hline \end{array}$

132. $21 \div 7$

133. $\begin{array}{r} 6 \\ \times\ 5 \\ \hline \end{array}$

134. $\dfrac{42}{6}$

name date hour

135.
```
   7 218
       3
      18
   1 791
  82,623
+  1 982
```

136.
```
  623
×   9
```

137. $\frac{2}{0}$

138. $4\overline{)552}$

139. 49,509 ÷ 9

140.
```
  8430
×  128
```

141. $\frac{5}{1}$

142. $34\overline{)3672}$

143.
```
  38,571
×      3
```

144. Rewrite 286,753 in words.

145. Round 7245 to the nearest hundred.

146. Round 500,196 to the nearest million.

Find each square root.

147. $\sqrt{36}$

148. $\sqrt{100}$

Find the total cost of each of the following.

149. 56 skateboards at $38 per board

150. 72 ice machines at $435 per machine

151. 185 baseball bats at $12 per bat

152. 607 boxes of avocados at $26 per box

Solve each of the following word problems.

153. There are 52 cards in a deck. How many cards are there in 9 decks.

154. Each Brownie sells 20 boxes of cookies. How many boxes are sold by 500 Brownies?

155. If a family watches television 5 hours each day, how many hours of television will the family watch in 2 years of 365 days each?

156. A charity wants to raise \$84,235. If \$34,872 has already been raised, how much more must be raised to reach the goal?

157. If four people divided \$3156 worth of gold dust equally, how much would each receive.

158. John Miller travels 635 miles on a tank of Super Unleaded and 583 miles on a tank of Regular Unleaded. How many more miles does he travel on a tank of Super Unleaded?

159. Building a lighted tennis court costs \$26,950. If seven families divide the cost of a court equally, how much will each family pay?

160. An employee earns \$108 for a 9-hour work day. How much is this employee paid per hour?

name date hour

CHAPTER 1 TEST

Write the following numbers in words.

1. 3022

2. 52,008

3. Use digits to write "one hundred thirty-eight thousand, eight."

1. ______
2. ______
3. ______

Add the following.

4.
$$\begin{array}{r} 729 \\ 83 \\ 9821 \\ +\ 6073 \\ \hline \end{array}$$

5.
$$\begin{array}{r} 17{,}063 \\ 7 \\ 12 \\ 1\ 505 \\ 93{,}710 \\ +\ \ 333 \\ \hline \end{array}$$

4. ______
5. ______

Subtract.

6.
$$\begin{array}{r} 7005 \\ -\ 4889 \\ \hline \end{array}$$

7.
$$\begin{array}{r} 6202 \\ -\ 3660 \\ \hline \end{array}$$

6. ______
7. ______

Multiply.

8. $6 \times 5 \times 4$

9. 49×3000

10.
$$\begin{array}{r} 75 \\ \times\ 18 \\ \hline \end{array}$$

11.
$$\begin{array}{r} 7381 \\ \times\ 603 \\ \hline \end{array}$$

8. ______
9. ______
10. ______
11. ______

Divide whenever possible.

12. $16\overline{)123{,}952}$

13. $\dfrac{791}{0}$

14. $84\overline{)38{,}472}$

15. $450\overline{)76{,}500}$

12. ______
13. ______
14. ______
15. ______

Round as shown.

16. 7854 to the nearest ten

17. 76,671 to the nearest hundred

18. 45,698 to the nearest thousand

16. ______
17. ______
18. ______

Work each problem.

19. ______________

19. $6^2 + 8 + 7$

20. ______________

20. $7 \cdot \sqrt{64} - 14 \cdot 2$

Solve each of the following word problems.

21. ______________

21. The monthly rents collected from the four units in an apartment building are $485, $500, $515, and $425. After expenses of $785 are paid, find the amount that remains.

22. ______________

22. A factory produces 288 automobiles per day. How many days would be needed to manufacture 35,424 autos?

23. ______________

23. Kenée Shadbourne paid $690 for tuition, $185 on books, and $68 on supplies. If this money was withdrawn from her checking account, which had a balance of $1108, find her new balance.

24. ______________

24. Cooperative Almond Growers has 873 production employees, 74 office and clerical workers, and 22 management personnel. Find the total number of employees.

25. ______________

25. An appliance manufacturer assembles 118 self-cleaning ovens per hour for 4 hours and 139 standard ovens per hour for the next 4 hours. Find the total number of ovens assembled in the 8-hour period.

2

MULTIPLYING AND DIVIDING FRACTIONS

Chapter 1 discussed whole numbers. Many times, however, parts of whole numbers are considered. One way to write parts of a whole is with **fractions.** (Another way is with decimals, which is discussed in a later chapter.)

2.1 BASICS OF FRACTIONS

1 The number $\frac{1}{8}$ is a fraction that represents 1 of 8 equal parts. Read $\frac{1}{8}$ as "one eighth."

EXAMPLE 1 Identifying Fractions
Use a fraction to represent the shaded portions.

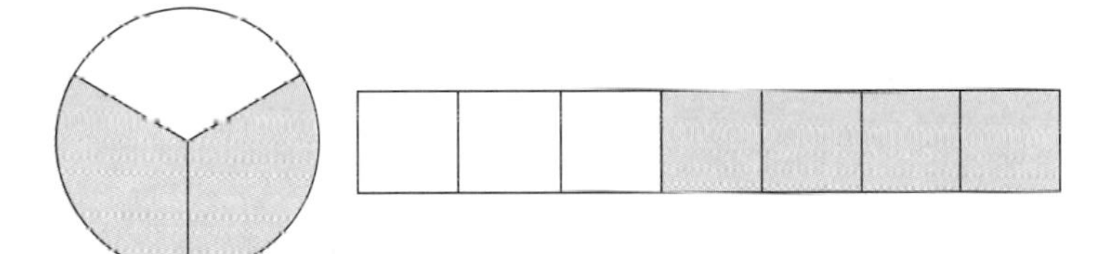

SOLUTION

(a) The figure on the left has 3 equal parts. The 2 shaded parts are represented by the fraction $\frac{2}{3}$.

(b) The 4 shaded parts of the 7-part figure on the right are represented by the fraction $\frac{4}{7}$. ■

WORK PROBLEM 1 AT THE SIDE.

Fractions can be used to show more than one whole object.

OBJECTIVES

1 Use a fraction to show which part of a figure is shaded.

2 Identify the numerator and denominator.

3 Identify proper and improper fractions.

1. Write fractions for the shaded portions.

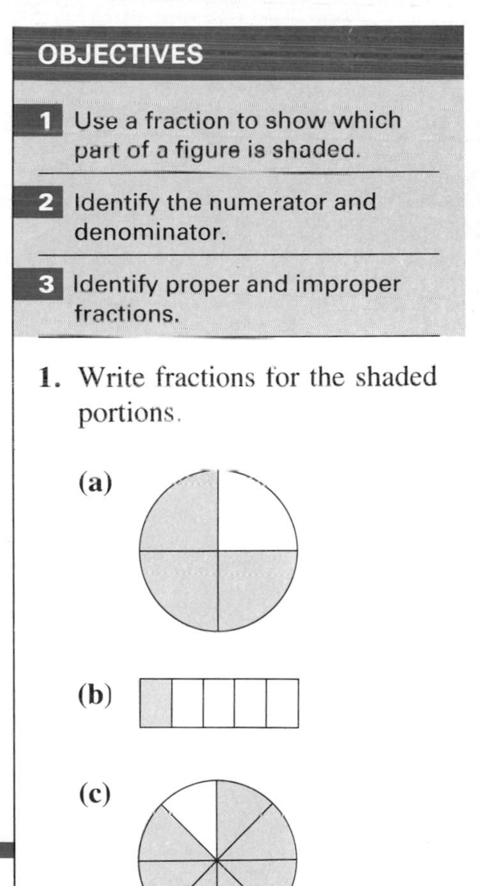

ANSWERS
1. (a) $\frac{3}{4}$ (b) $\frac{1}{5}$ (c) $\frac{7}{8}$

2. Write fractions for the shaded portions.

(a) 1

(b) 1

3. Identify the numerator and the denominator.

(a) $\frac{3}{4}$

(b) $\frac{1}{2}$

(c) $\frac{9}{7}$

(d) $\frac{115}{8}$

ANSWERS

2. (a) $\frac{8}{7}$ (b) $\frac{7}{4}$

3. (a) N: 3; D: 4 (b) N: 1; D: 2
(c) N: 9; D: 7 (d) N: 115; D: 8

EXAMPLE 2 Representing Fractions Greater Than One

Use a fraction to represent the shaded part.

(a) $\frac{1}{4}$ whole object

(b) whole object $\frac{1}{3}$

SOLUTION

(a) An area equal to 5 of the 4 parts is shaded. Write this as $\frac{5}{4}$.

(b) An area equal to 5 of the 3 parts is shaded, so $\frac{5}{3}$ is shaded. ■

WORK PROBLEM 2 AT THE SIDE.

2 In the fraction $\frac{2}{3}$, the number 2 is the **numerator,** and 3 is the **denominator.** The bar between the numerator and the denominator is the **fraction bar.**

$$\text{fraction bar} \rightarrow \frac{2 \leftarrow \text{numerator}}{3 \leftarrow \text{denominator}}$$

The denominator of a fraction shows the number of equivalent parts in the whole, and the numerator shows how many parts are being considered.

Since division by 0 is meaningless, a fraction with a denominator of 0 is meaningless.

EXAMPLE 3 Understanding Numerator and Denominator

Identify the numerator and denominator in each fraction.

(a) $\frac{5}{9}$ **(b)** $\frac{11}{7}$

SOLUTION

(a) $\frac{5 \leftarrow \text{numerator}}{9 \leftarrow \text{denominator}}$ **(b)** $\frac{11 \leftarrow \text{numerator}}{7 \leftarrow \text{denominator}}$

WORK PROBLEM 3 AT THE SIDE.

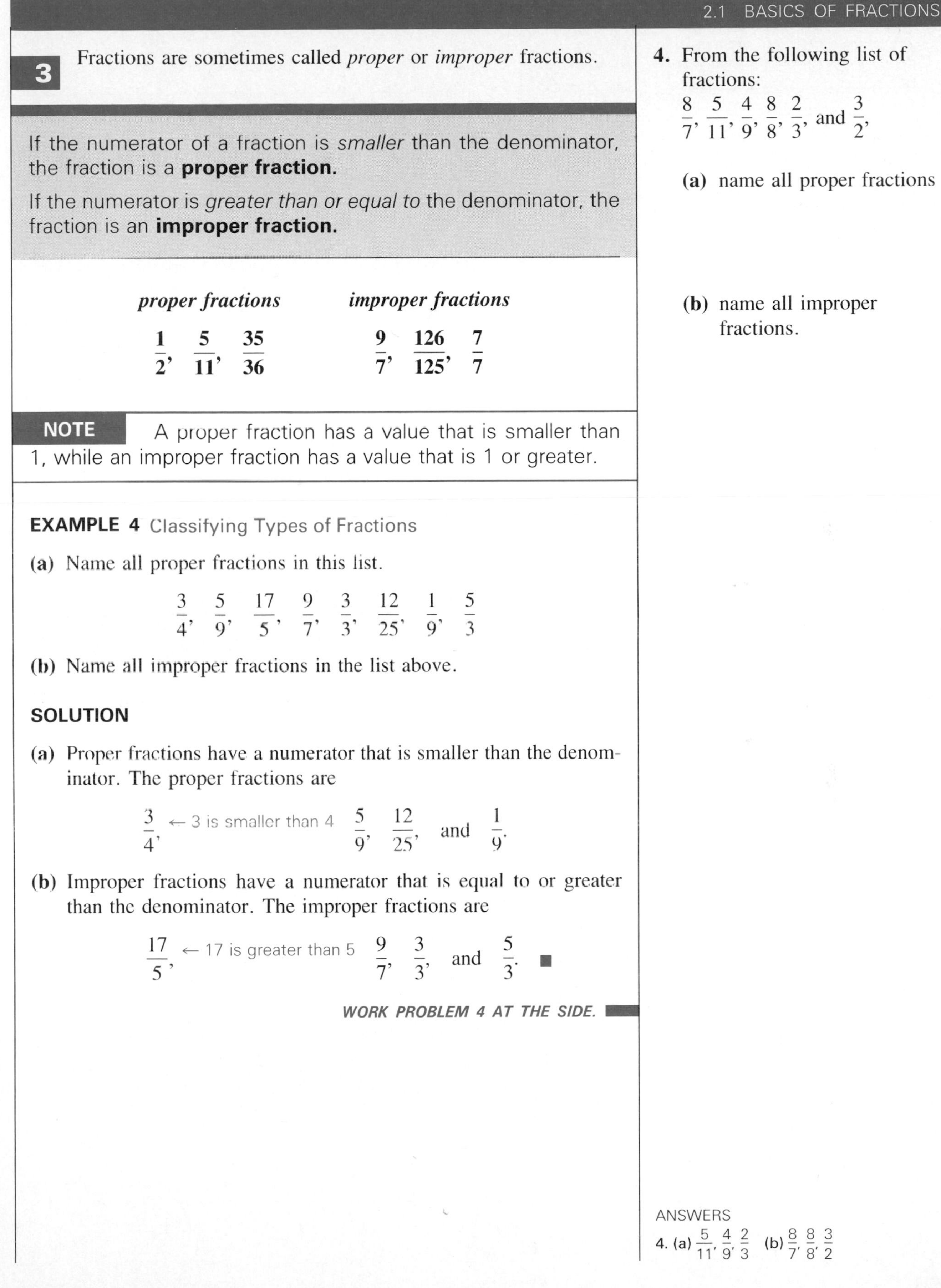

3 Fractions are sometimes called *proper* or *improper* fractions.

If the numerator of a fraction is *smaller* than the denominator, the fraction is a **proper fraction.**

If the numerator is *greater than or equal to* the denominator, the fraction is an **improper fraction.**

proper fractions	***improper fractions***
$\frac{1}{2}, \frac{5}{11}, \frac{35}{36}$	$\frac{9}{7}, \frac{126}{125}, \frac{7}{7}$

NOTE A proper fraction has a value that is smaller than 1, while an improper fraction has a value that is 1 or greater.

EXAMPLE 4 Classifying Types of Fractions

(a) Name all proper fractions in this list.

$$\frac{3}{4}, \frac{5}{9}, \frac{17}{5}, \frac{9}{7}, \frac{3}{3}, \frac{12}{25}, \frac{1}{9}, \frac{5}{3}$$

(b) Name all improper fractions in the list above.

SOLUTION

(a) Proper fractions have a numerator that is smaller than the denominator. The proper fractions are

$$\frac{3}{4}, \leftarrow \text{3 is smaller than 4} \quad \frac{5}{9}, \frac{12}{25}, \text{ and } \frac{1}{9}.$$

(b) Improper fractions have a numerator that is equal to or greater than the denominator. The improper fractions are

$$\frac{17}{5}, \leftarrow \text{17 is greater than 5} \quad \frac{9}{7}, \frac{3}{3}, \text{ and } \frac{5}{3}.$$ ■

WORK PROBLEM 4 AT THE SIDE.

4. From the following list of fractions:
$\frac{8}{7}, \frac{5}{11}, \frac{4}{9}, \frac{8}{8}, \frac{2}{3},$ and $\frac{3}{2}$,

(a) name all proper fractions

(b) name all improper fractions.

ANSWERS

4. (a) $\frac{5}{11}, \frac{4}{9}, \frac{2}{3}$ (b) $\frac{8}{7}, \frac{8}{8}, \frac{3}{2}$

name date hour

2.1 EXERCISES

Write the fraction that represents the shaded area.

Example: 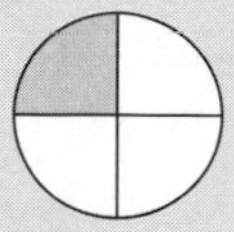 **Solution:**

$\frac{1}{4}$ (There are four parts, and one is shaded.)

1. 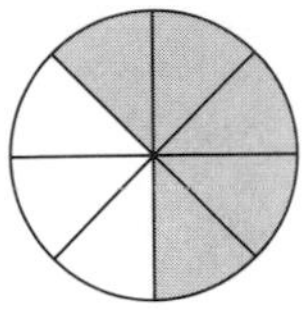 **2.** 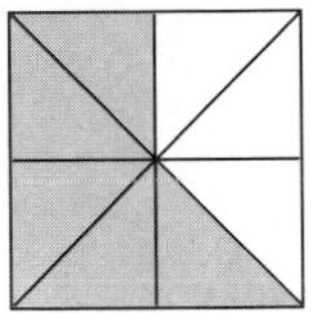 **3.**

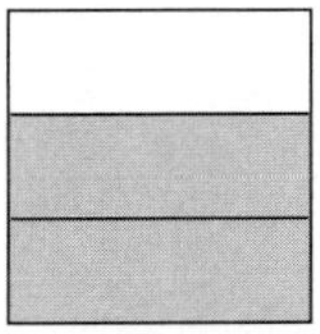

4. 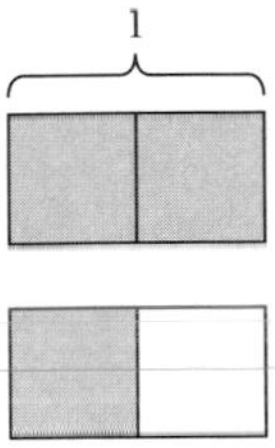 1 **5.** 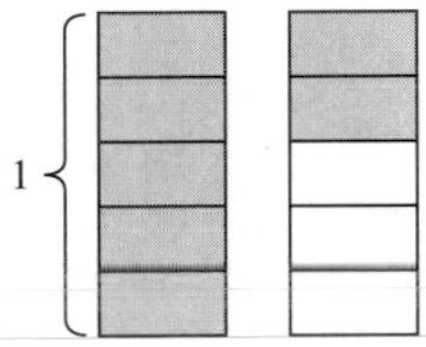 1 **6.** 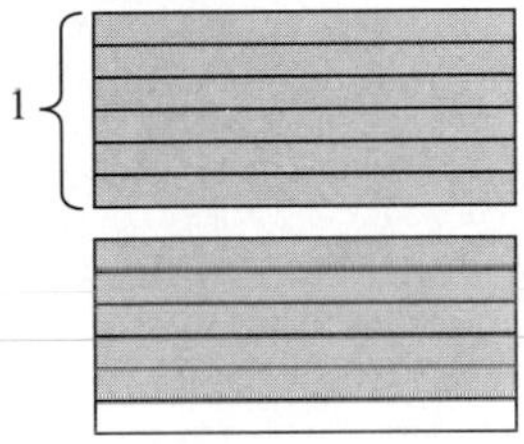 1

7. What fraction of these 11 coins are quarters?

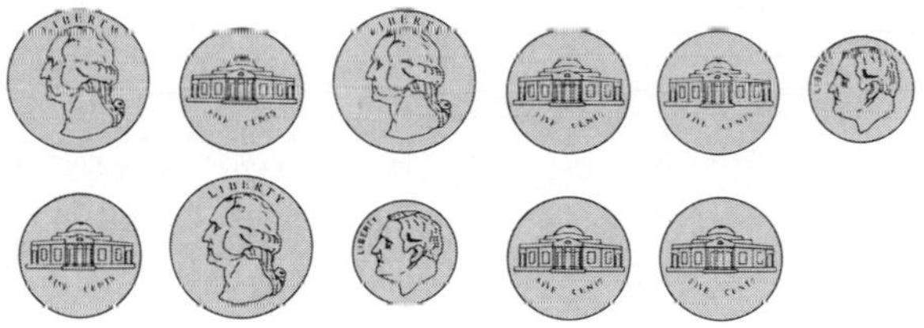

8. What fraction of these 6 stamps are 25¢ stamps?

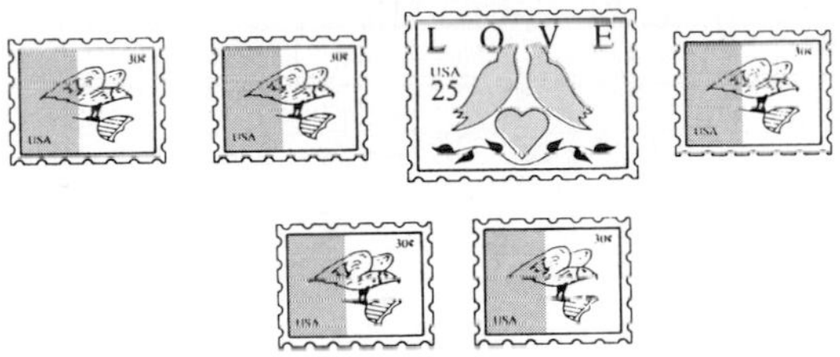

9. In a mathematics class of 25 students, 14 are female. What fraction of the students are female?

10. Of 35 dogs at a dog show, 8 are German shepherds. What fraction of the dogs are *not* German shepherds?

11. Of 71 cars making up a freight train, 58 are boxcars. What fraction of the cars are *not* boxcars?

12. On a basketball team of 10 players, 7 players are sophomores and the rest are freshmen. What fraction of the players are freshmen?

Identify the numerator and denominator.

Example: $\frac{9}{11}$ **Solution:** $\frac{9}{11}$ 9 ← *numerator* (on top), 11 ← *denominator* (on bottom)

	numerator	*denominator*
13. $\frac{5}{7}$	______	______
14. $\frac{11}{14}$	______	______

	numerator	*denominator*		*numerator*	*denominator*
15. $\frac{9}{8}$	________	________	**16.** $\frac{8}{3}$	________	________

Name the proper and improper fractions in each of the following lists.

Example: $\frac{3}{8}, \frac{7}{4}, \frac{5}{6}, \frac{2}{3}, \frac{9}{4}$

Solution: ***proper*** $\frac{3}{8}, \frac{5}{6}, \frac{2}{3}$ ***improper*** $\frac{7}{4}, \frac{9}{4}$

	proper	***improper***
17. $\frac{9}{7}, \frac{1}{4}, \frac{3}{8}, \frac{7}{12}, \frac{11}{4}$	________	________
18. $\frac{3}{17}, \frac{5}{12}, \frac{11}{7}, \frac{14}{13}, \frac{1}{2}$	________	________
19. $\frac{3}{4}, \frac{3}{2}, \frac{9}{11}, \frac{7}{15}, \frac{19}{18}$	________	________
20. $\frac{15}{11}, \frac{13}{12}, \frac{11}{8}, \frac{17}{17}, \frac{19}{12}$	________	________

Complete the following sentences.

21. The fraction $\frac{9}{17}$ represents of the equal parts into which a whole is divided.

22. The fraction $\frac{13}{16}$ represents of the equal parts into which a whole is divided.

REVIEW EXERCISES

Almost every exercise set in the rest of the book ends with a brief set of review exercises. These exercises are designed to help you review ideas needed for the next few sections in the chapter. If you need help with these review exercises, look in the chapter sections indicated.

Multiply each of the following. (For help, see ***Section 1.4.****)*

23. $2 \times 3 \times 3$ **24.** $3 \cdot 5 \cdot 2$ **25.** $5 \cdot 3 \cdot 7$ **26.** $3 \cdot 2 \cdot 7$

Divide each of the following. (For help, see ***Section 1.5.****)*

27. $21 \div 3$ **28.** $56 \div 8$ **29.** $72 \div 9$ **30.** $45 \div 9$

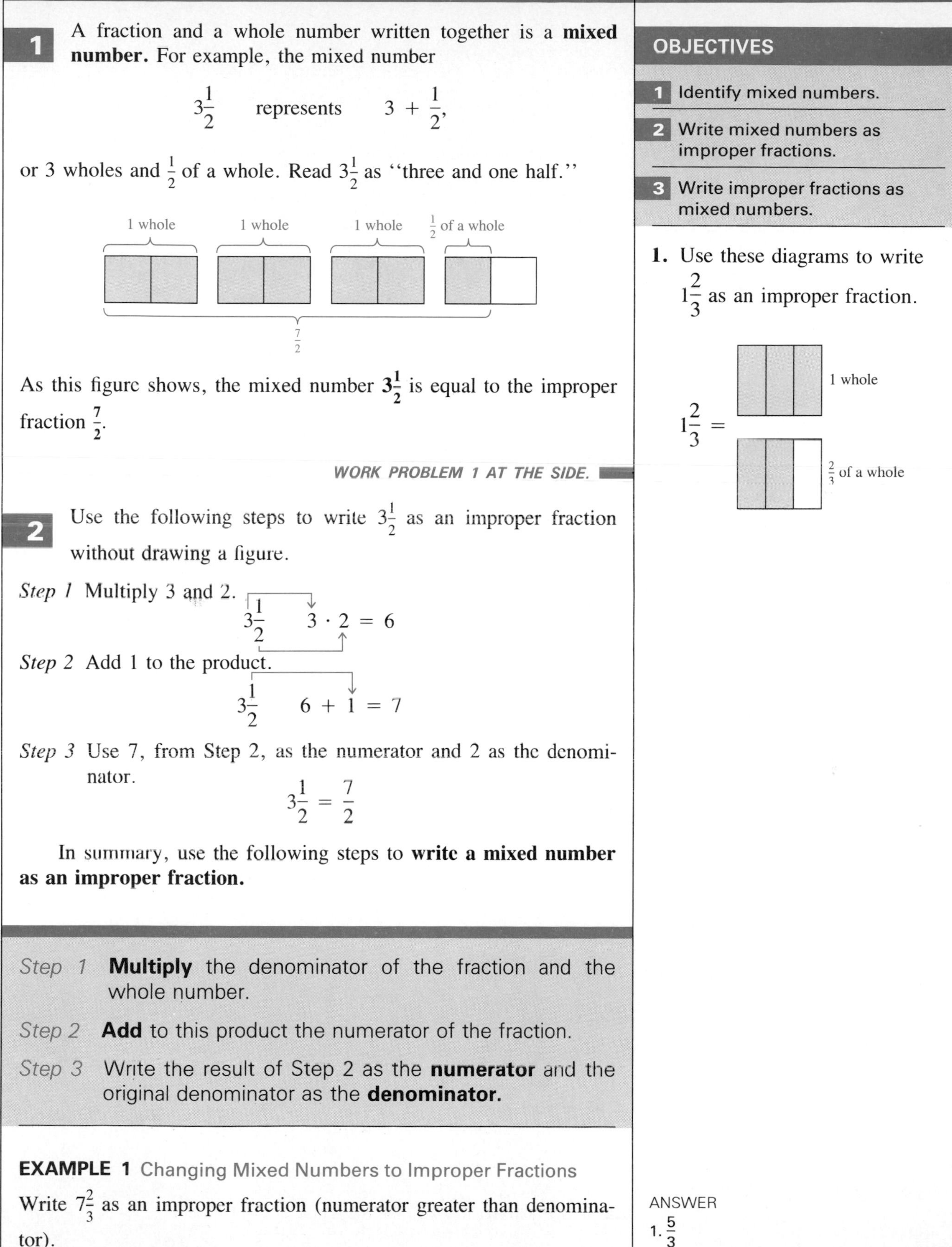

2.2 MIXED NUMBERS

OBJECTIVES

1. Identify mixed numbers.
2. Write mixed numbers as improper fractions.
3. Write improper fractions as mixed numbers.

1 A fraction and a whole number written together is a **mixed number.** For example, the mixed number

$$3\frac{1}{2} \quad \text{represents} \quad 3 + \frac{1}{2},$$

or 3 wholes and $\frac{1}{2}$ of a whole. Read $3\frac{1}{2}$ as "three and one half."

As this figure shows, the mixed number $\mathbf{3\frac{1}{2}}$ is equal to the improper fraction $\frac{7}{2}$.

WORK PROBLEM 1 AT THE SIDE.

1. Use these diagrams to write $1\frac{2}{3}$ as an improper fraction.

$1\frac{2}{3} =$

2 Use the following steps to write $3\frac{1}{2}$ as an improper fraction without drawing a figure.

Step 1 Multiply 3 and 2.

$$3\frac{1}{2} \qquad 3 \cdot 2 = 6$$

Step 2 Add 1 to the product.

$$3\frac{1}{2} \qquad 6 + 1 = 7$$

Step 3 Use 7, from Step 2, as the numerator and 2 as the denominator.

$$3\frac{1}{2} = \frac{7}{2}$$

In summary, use the following steps to **write a mixed number as an improper fraction.**

Step 1 **Multiply** the denominator of the fraction and the whole number.

Step 2 **Add** to this product the numerator of the fraction.

Step 3 Write the result of Step 2 as the **numerator** and the original denominator as the **denominator.**

EXAMPLE 1 Changing Mixed Numbers to Improper Fractions

Write $7\frac{2}{3}$ as an improper fraction (numerator greater than denominator).

ANSWER
1. $\frac{5}{3}$

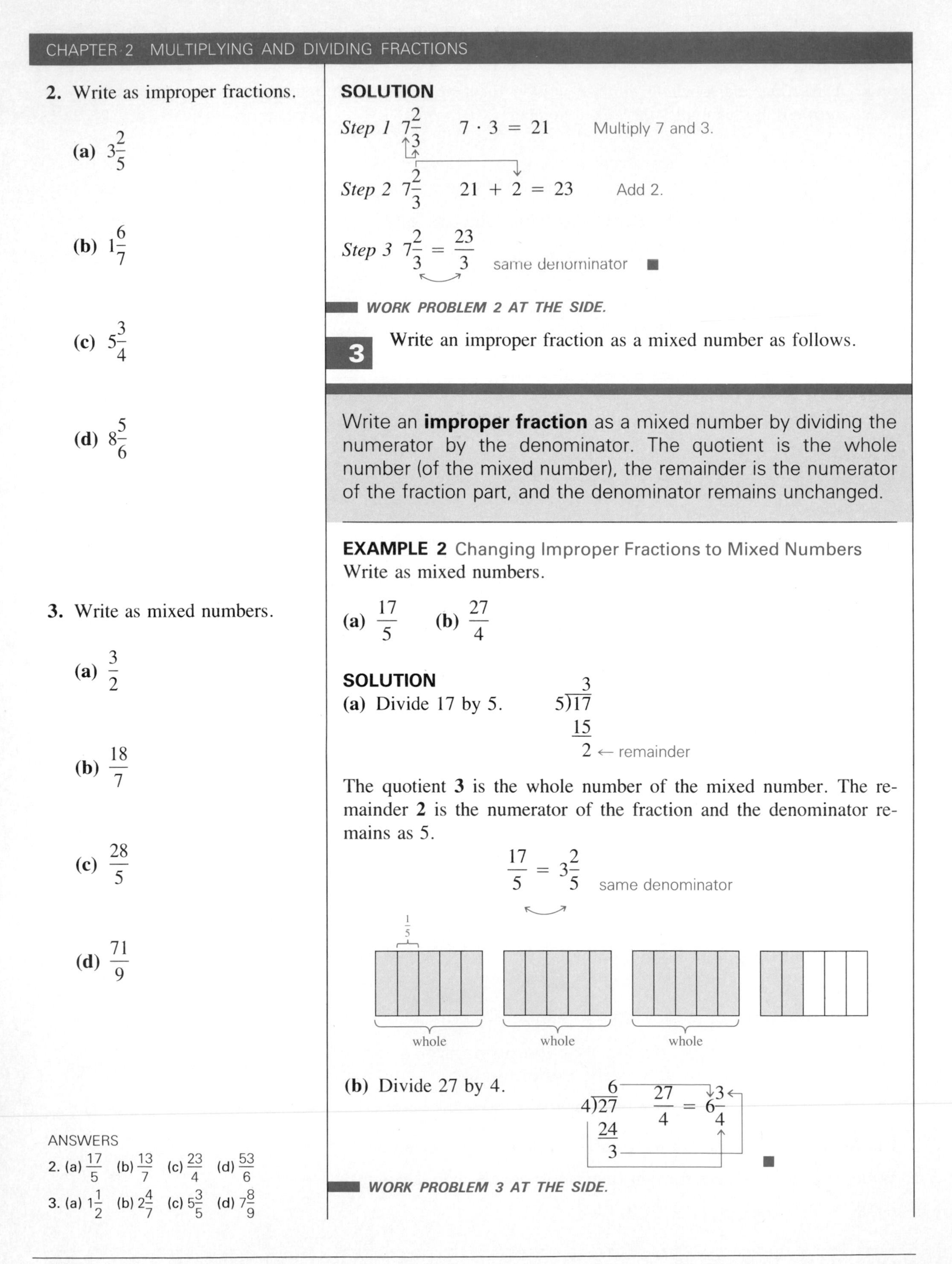

2. Write as improper fractions.

(a) $3\frac{2}{5}$

(b) $1\frac{6}{7}$

(c) $5\frac{3}{4}$

(d) $8\frac{5}{6}$

3. Write as mixed numbers.

(a) $\frac{3}{2}$

(b) $\frac{18}{7}$

(c) $\frac{28}{5}$

(d) $\frac{71}{9}$

ANSWERS

2. (a) $\frac{17}{5}$ (b) $\frac{13}{7}$ (c) $\frac{23}{4}$ (d) $\frac{53}{6}$

3. (a) $1\frac{1}{2}$ (b) $2\frac{4}{7}$ (c) $5\frac{3}{5}$ (d) $7\frac{8}{9}$

SOLUTION

Step 1 $7\frac{2}{3}$ $7 \cdot 3 = 21$ Multiply 7 and 3.

Step 2 $7\frac{2}{3}$ $21 + 2 = 23$ Add 2.

Step 3 $7\frac{2}{3} = \frac{23}{3}$ same denominator ■

WORK PROBLEM 2 AT THE SIDE.

3 Write an improper fraction as a mixed number as follows.

Write an **improper fraction** as a mixed number by dividing the numerator by the denominator. The quotient is the whole number (of the mixed number), the remainder is the numerator of the fraction part, and the denominator remains unchanged.

EXAMPLE 2 Changing Improper Fractions to Mixed Numbers

Write as mixed numbers.

(a) $\frac{17}{5}$ **(b)** $\frac{27}{4}$

SOLUTION

(a) Divide 17 by 5.

$$5\overline{)17} = 3,\quad 15,\quad 2 \leftarrow \text{remainder}$$

The quotient **3** is the whole number of the mixed number. The remainder **2** is the numerator of the fraction and the denominator remains as 5.

$$\frac{17}{5} = 3\frac{2}{5}$$

same denominator

$\frac{1}{5}$

whole whole whole

(b) Divide 27 by 4.

$$4\overline{)27} = 6,\quad 24,\quad 3 \qquad \frac{27}{4} = 6\frac{3}{4}$$

■

WORK PROBLEM 3 AT THE SIDE.

name date hour

2.2 EXERCISES

Write each mixed number as an improper fraction.

Example: $6\frac{3}{7} = \mathbf{\frac{45}{7}}$ **Solution:** $6 \cdot 7 = 42$ $42 + 3 = 45$ $6\frac{3}{7} = \frac{45}{7}$

1. $1\frac{7}{8}$ **2.** $2\frac{5}{6}$ **3.** $3\frac{4}{5}$

4. $1\frac{1}{2}$ **5.** $2\frac{3}{4}$ **6.** $5\frac{1}{4}$

7. $3\frac{4}{9}$ **8.** $7\frac{1}{2}$ **9.** $1\frac{7}{11}$

10. $5\frac{4}{7}$ **11.** $6\frac{1}{3}$ **12.** $8\frac{2}{3}$

13. $11\frac{1}{2}$ **14.** $12\frac{2}{3}$ **15.** $10\frac{3}{4}$

16. $6\frac{2}{3}$ **17.** $3\frac{3}{8}$ **18.** $2\frac{8}{9}$

19. $8\frac{4}{5}$ **20.** $3\frac{4}{7}$ **21.** $4\frac{10}{11}$

22. $13\frac{5}{9}$ **23.** $22\frac{8}{9}$ **24.** $12\frac{9}{10}$

25. $17\frac{12}{13}$ **26.** $19\frac{8}{11}$ **27.** $17\frac{14}{15}$

28. $5\frac{17}{24}$ **29.** $7\frac{13}{19}$ **30.** $8\frac{4}{13}$

Write each improper fraction as a mixed number.

Example: $\frac{13}{5} = 2\frac{3}{5}$ **Solution:** $5\overline{)13}$ with quotient 2, 10, remainder 3; $\frac{13}{5} = 2\frac{3}{5}$

31. $\frac{9}{2}$ **32.** $\frac{7}{5}$ **33.** $\frac{9}{5}$ **34.** $\frac{33}{10}$ **35.** $\frac{14}{11}$

36. $\frac{19}{7}$ **37.** $\frac{28}{9}$ **38.** $\frac{27}{7}$ **39.** $\frac{22}{5}$ **40.** $\frac{40}{9}$

41. $\frac{17}{3}$ **42.** $\frac{29}{3}$ **43.** $\frac{58}{5}$ **44.** $\frac{19}{5}$ **45.** $\frac{47}{9}$

46. $\frac{55}{8}$ **47.** $\frac{50}{7}$ **48.** $\frac{39}{5}$ **49.** $\frac{84}{5}$ **50.** $\frac{92}{3}$

51. $\frac{106}{5}$ **52.** $\frac{113}{4}$ **53.** $\frac{183}{7}$ **54.** $\frac{212}{11}$

CHALLENGE EXERCISES

Write each mixed number as an improper fraction.

55. $255\frac{1}{8}$ **56.** $110\frac{1}{3}$ **57.** $333\frac{1}{3}$

Write each improper fraction as a mixed number.

58. $\frac{1982}{13}$ **59.** $\frac{2573}{15}$ **60.** $\frac{3022}{19}$

REVIEW EXERCISES

Work each of the following problems. (For help, see **Section 1.8.**)

61. $2^2 + 3^2$ **62.** $2^2 \cdot 3^2$ **63.** $15 \cdot 4 + 6$

64. $12 \cdot 8 + 4$ **65.** $5 \cdot 2^2 - 4$ **66.** $3 \cdot 4^2 - 8$

2.3 FACTORS

OBJECTIVES

1 Find factors of a number.

2 Identify primes.

3 Find prime factorizations.

1 Recall that numbers multiplied to give a product are called **factors.** Since $2 \cdot 5 = 10$, both 2 and 5 are factors of 10. The numbers 1 and 10 are also factors of 10, since

$$1 \cdot 10 = 10$$

The various tests for divisibility show 1, 2, 5, and 10 are the only whole-number factors of 10. The products $2 \cdot 5$ and $1 \cdot 10$ are called **factorizations** of 10.

EXAMPLE 1 Using Factors
Find all factorizations of each number that have two factors.

(a) 12 **(b)** 60

SOLUTION

(a) $1 \cdot 12$ $2 \cdot 6$ $3 \cdot 4$

The factors of 12 are 1, 2, 3, 4, 6, and 12.

(b) $1 \cdot 60$ $2 \cdot 30$
$3 \cdot 20$ $4 \cdot 15$
$5 \cdot 12$ $6 \cdot 10$

The factors of 60 are 1, 2, 3, 4, 5, 6, 10, 12, 15, 20, 30, and 60. ■

WORK PROBLEM 1 AT THE SIDE.

1. Find the factors of the following numbers.

(a) 9

(b) 15

(c) 36

(d) 80

A number with a factor other than itself or 1 is called **composite.**

EXAMPLE 2 Identifying Composite Numbers
Which of the following numbers are composite?

(a) 16

Since 16 has a factor of 8, a number other than 16 or 1, the number 16 is composite.

(b) 17

The number 17 has only two factors—17 and 1. It is not composite.

(c) 25

A factor of 25 is 5, so 25 is composite. ■

WORK PROBLEM 2 AT THE SIDE.

2. Which of these numbers are composite?

2, 4, 5, 6, 8, 10, 11, 13, 19, 21, 27, 33, 42

ANSWERS
1. (a) 1, 3, 9 **(b)** 1, 3, 5, 15 **(c)** 1, 2, 3, 4, 6, 9, 12, 18, 36 **(d)** 1, 2, 4, 5, 8, 10, 16, 20, 40, 80
2. 4, 6, 8, 10, 21, 27, 33, 42

3. Which of the following are prime?

3, 4, 7, 9, 13, 19, 29

2 Whole numbers that are not composite are called **prime numbers,** except 0 and 1, which are neither prime nor composite.

A prime number is a whole number that has exactly two different factors, itself and 1.

The number 3 is a prime number, since it can be divided with a remainder of 0 by only itself and 1. The number 6 is not a prime number (it is composite), since 6 can be divided evenly by 2 and 3, as well as by itself and 1.

EXAMPLE 3 Finding Prime Numbers
Which of the following numbers are prime?

2 5 8 11 15

The number 8 can be divided by 4 and 2, so it is not prime. Also, since 15 can be divided by 5 and 3, 15 is not prime. All the other numbers in the list are divisible by only themselves and 1, and are prime. ■

WORK PROBLEM 3 AT THE SIDE.

3 For reference, here are the primes smaller than 100.

2, 3, 5, 7, 11,
13, 17, 19, 23, 29,
31, 37, 41, 43, 47,
53, 59, 61, 67, 71,
73, 79, 83, 91, 97

The *prime factorization* of a number is especially useful.

A **prime factorization** of a number is a factorization in which every factor is a prime number.

ANSWERS
3. 3, 7, 13, 19, 29

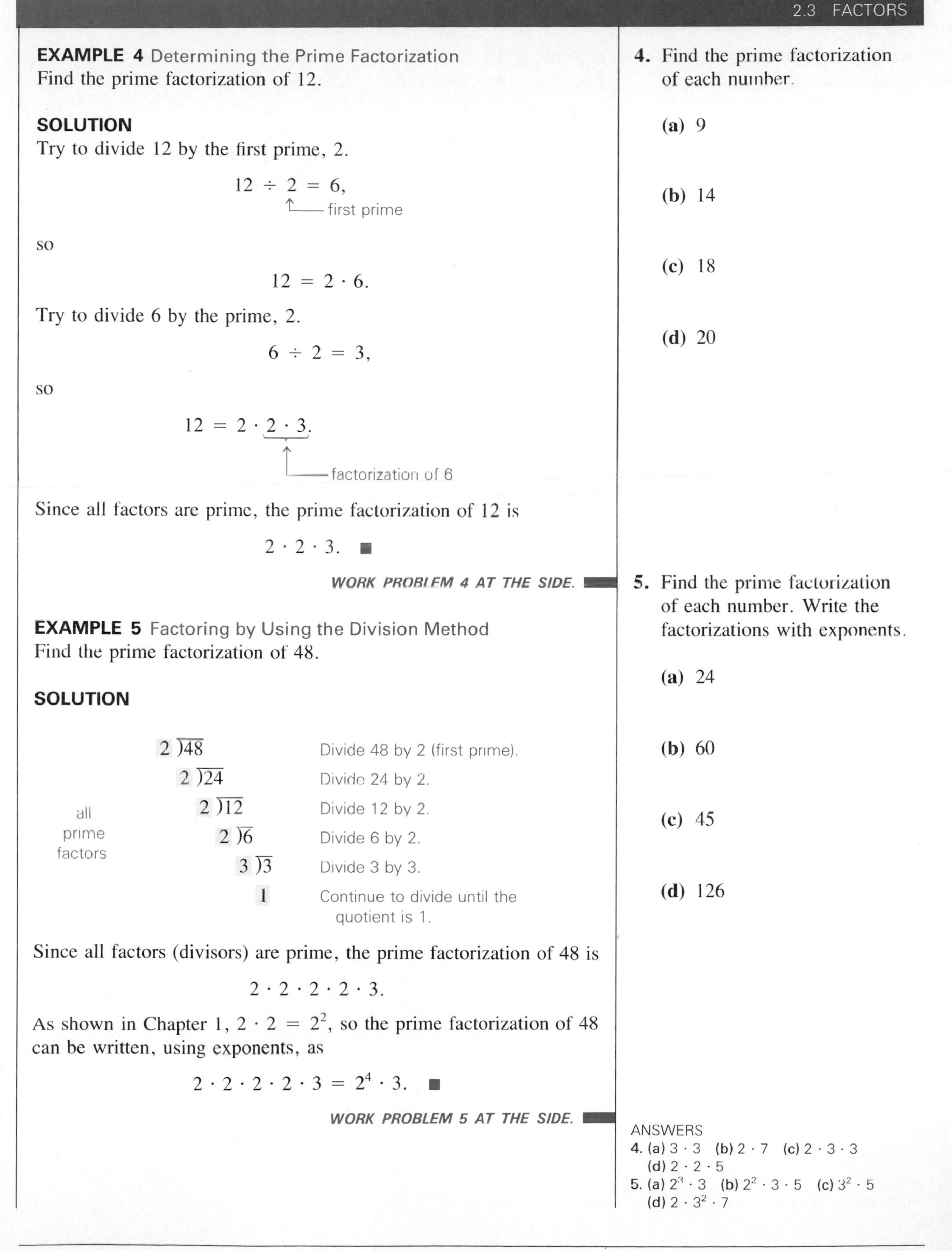

EXAMPLE 4 Determining the Prime Factorization
Find the prime factorization of 12.

SOLUTION
Try to divide 12 by the first prime, 2.

$$12 \div 2 = 6,$$
↑ first prime

so

$$12 = 2 \cdot 6.$$

Try to divide 6 by the prime, 2.

$$6 \div 2 = 3,$$

so

$$12 = 2 \cdot \underbrace{2 \cdot 3}.$$
↑ factorization of 6

Since all factors are prime, the prime factorization of 12 is

$$2 \cdot 2 \cdot 3. \quad ■$$

WORK PROBLEM 4 AT THE SIDE.

EXAMPLE 5 Factoring by Using the Division Method
Find the prime factorization of 48.

SOLUTION

all prime factors	2)48	Divide 48 by 2 (first prime).
	2)24	Divide 24 by 2.
	2)12	Divide 12 by 2.
	2)6	Divide 6 by 2.
	3)3	Divide 3 by 3.
	1	Continue to divide until the quotient is 1.

Since all factors (divisors) are prime, the prime factorization of 48 is

$$2 \cdot 2 \cdot 2 \cdot 2 \cdot 3.$$

As shown in Chapter 1, $2 \cdot 2 = 2^2$, so the prime factorization of 48 can be written, using exponents, as

$$2 \cdot 2 \cdot 2 \cdot 2 \cdot 3 = 2^4 \cdot 3. \quad ■$$

WORK PROBLEM 5 AT THE SIDE.

4. Find the prime factorization of each number.

(a) 9

(b) 14

(c) 18

(d) 20

5. Find the prime factorization of each number. Write the factorizations with exponents.

(a) 24

(b) 60

(c) 45

(d) 126

ANSWERS
4. (a) $3 \cdot 3$ (b) $2 \cdot 7$ (c) $2 \cdot 3 \cdot 3$ (d) $2 \cdot 2 \cdot 5$
5. (a) $2^3 \cdot 3$ (b) $2^2 \cdot 3 \cdot 5$ (c) $3^2 \cdot 5$ (d) $2 \cdot 3^2 \cdot 7$

6. Write the prime factorization of each number by using exponents.

(a) 60

(b) 88

(c) 92

(d) 150

(e) 280

ANSWERS
6. (a) $2^2 \cdot 3 \cdot 5$ (b) $2^3 \cdot 11$ (c) $2^2 \cdot 23$ (d) $2 \cdot 3 \cdot 5^2$ (e) $2^3 \cdot 5 \cdot 7$

EXAMPLE 6 Using Exponents with Prime Factorization
Find the prime factorization of 225.

SOLUTION

all prime factors

$3\overline{)225}$ 225 is not divisible by 2; use 3.
$3\overline{)75}$ Divide 75 by 3.
$5\overline{)25}$ 25 is not divisible by 3; use 5.
$5\overline{)5}$ Divide by 5.
1

Write the prime factorization,

$$3 \cdot 3 \cdot 5 \cdot 5$$

with exponents, as

$$3^2 \cdot 5^2$$ ■

WORK PROBLEM 6 AT THE SIDE.

Another method of factoring uses a diagram that forms the shape of a tree.

EXAMPLE 7 Factoring by Using a Factor Tree
Find the prime factorization of each number.

(a) 30 **(b)** 24 **(c)** 45

SOLUTION

(a) Try to divide by the first prime, 2. Write the factors under the 30. Circle the 2, since it is a prime.

30
↙ ↘
(2) 15

Since 15 cannot be divided by 2 (with remainder of 0), try the next prime, 3.

30
↙ ↘
(2) 15
↙ ↘
(3) (5) ← Circle, since they are primes.

No uncircled factors remain, so the prime factorization (the circled factors) has been found.

$$2 \cdot 3 \cdot 5$$

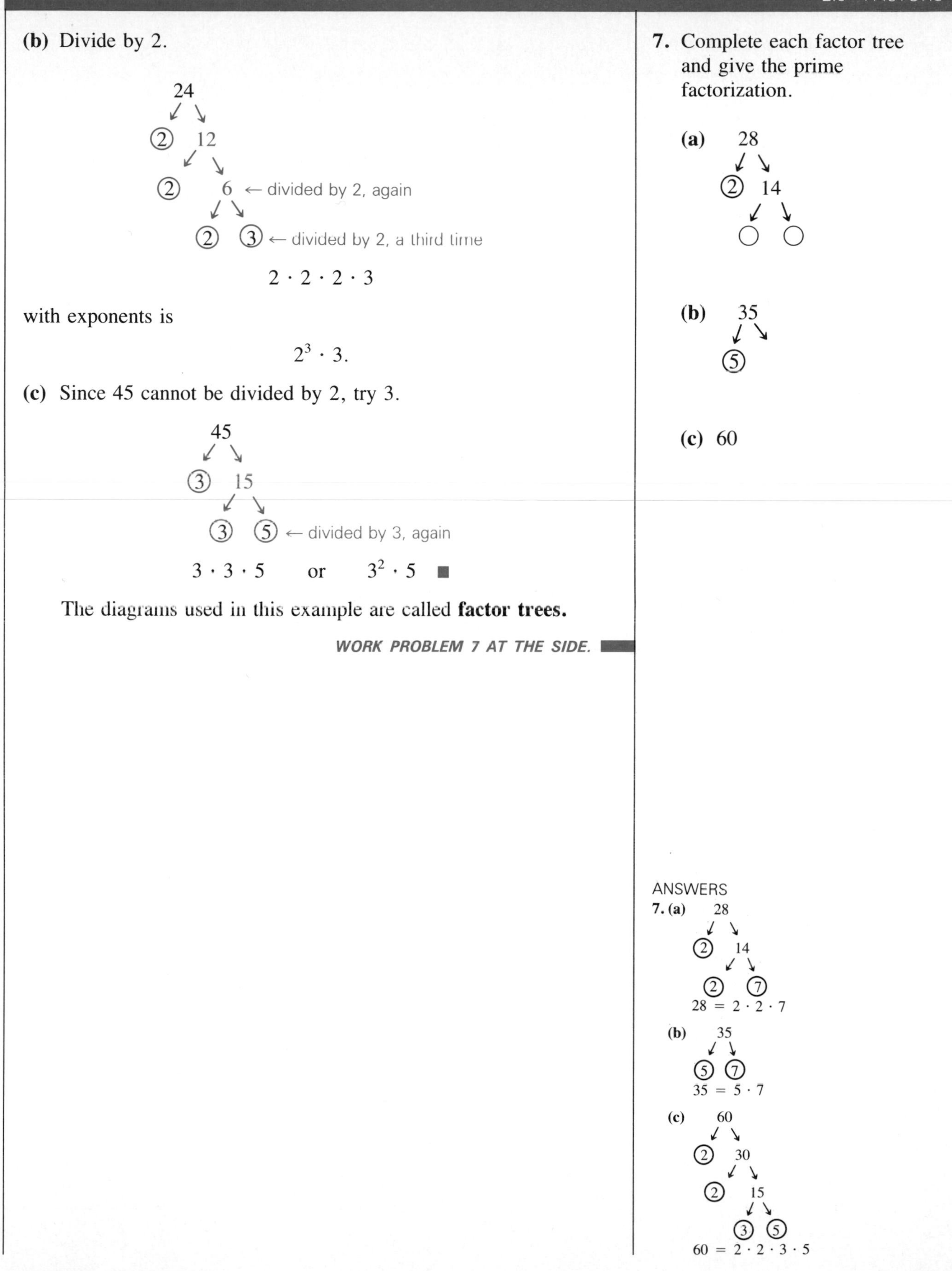
(b) Divide by 2.
24
2
12
2
6 ← divided by 2, again
2
3 ← divided by 2, a third time
2 · 2 · 2 · 3
with exponents is
$2^3 \cdot 3$.
(c) Since 45 cannot be divided by 2, try 3.
45
3
15
3
5 ← divided by 3, again
3 · 3 · 5 or $3^2 \cdot 5$ ■
The diagrams used in this example are called factor trees.
WORK PROBLEM 7 AT THE SIDE.
7. Complete each factor tree and give the prime factorization.
(a) 28
2
14
(b) 35
5
(c) 60
ANSWERS
7. (a) 28
2
14
2
7
28 = 2 · 2 · 7
(b) 35
5
7
35 = 5 · 7
(c) 60
2
30
2
15
3
5
60 = 2 · 2 · 3 · 5

name date hour

2.3 EXERCISES

Find all the factors of each number.

Example: 18

Solution:
Write all factorizations of 18 that have two factors.

$1 \cdot 18 \qquad 2 \cdot 9 \qquad 3 \cdot 6$

The factors of 18 are **1, 2, 3, 6, 9,** and **18.**

1. 6

2. 10

3. 8

4. 24

5. 21

6. 30

7. 18

8. 20

9. 40

10. 55

11. 64

12. 80

Which numbers are prime and which are composite?

Example: 5, 12

Solution:
Because it can be divided by only itself and 1, **5 is prime.** Because it can be divided by 2 and 3, **12 is composite.**

13. 8

14. 6

15. 3

16. 2

17. 9 **18.** 13 **19.** 17 **20.** 18

21. 19 **22.** 23 **23.** 25 **24.** 26

25. 34 **26.** 43 **27.** 45 **28.** 47

Find the prime factorization of the following numbers. Write answers with exponents when repeated factors appear.

Example: 40

Solution:

Division method

$2\overline{)40}$
$2\overline{)20}$
$2\overline{)10}$
$5\overline{)5}$
1

$2 \cdot 2 \cdot 2 \cdot 5$

$\mathbf{2^3 \cdot 5}$ ← 3 factors of 2

Factor tree

40 ← Divide by 2; circle 2, since it is prime.
↙ ↘
(2) 20 ← Divide by 2, again.
↙ ↘
(2) 10 ← Divide by 2, a third time.
↙ ↘
(2) (5)

29. 6 **30.** 10 **31.** 16

32. 30 **33.** 21 **34.** 18

name date hour

35. 32

36. 52

37. 39

38. 45

39. 88

40. 62

41. 75

42. 80

43. 100

44. 120

45. 150

46. 180

47. 225

48. 300

49. 320

50. 340

51. 360

52. 400

As a review, solve each of the following.

Examples: 2^4, 4^3, $2^2 \cdot 3^4$

$2^4 = 2 \cdot 2 \cdot 2 \cdot 2 \leftarrow$ 4 factors of 2
$\mathbf{= 16}$

$4^3 = 4 \cdot 4 \cdot 4 \leftarrow$ 3 factors of 4
$\mathbf{= 64}$

$2^2 \cdot 3^4 = 2 \cdot 2 \cdot 3 \cdot 3 \cdot 3 \cdot 3$
$= 4 \cdot 81$
$\mathbf{= 324}$

53. 5^3 **54.** 3^3 **55.** 8^3 **56.** 6^3

57. 6^4 **58.** 7^4 **59.** $3^3 \cdot 4^2$ **60.** $2^4 \cdot 3^2$

61. $5^3 \cdot 3^2$ **62.** $2^4 \cdot 5^2$ **63.** $6^2 \cdot 4^2$ **64.** $3^5 \cdot 2^2$

CHALLENGE EXERCISES

Find the prime factorization of each number. Write answers by using exponents.

65. 450 **66.** 720

67. 960 **68.** 1125

69. 1600 **70.** 1575

REVIEW EXERCISES

Multiply each of the following. (For help, see ***Section 1.4****.)*

71. $1 \cdot 1 \cdot 5$ **72.** $3 \cdot 2 \cdot 1$ **73.** $7 \cdot 5 \cdot 2$ **74.** $3 \cdot 5 \cdot 7$

Divide each of the following. (For help, see ***Section 1.5****.)*

75. $36 \div 3$ **76.** $21 \div 7$ **77.** $45 \div 5$ **78.** $72 \div 12$

2.4 WRITING A FRACTION IN LOWEST TERMS

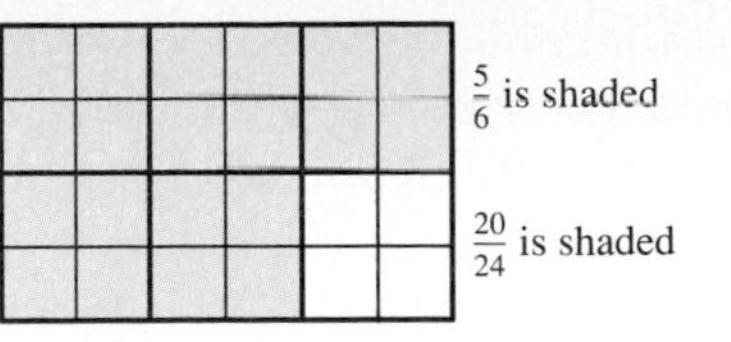

The figure shows areas that are $\frac{5}{6}$ shaded and $\frac{20}{24}$ shaded. Since the shaded areas are equivalent, the fractions $\frac{5}{6}$ and $\frac{20}{24}$ are **equivalent fractions.**

$$\frac{5}{6} = \frac{20}{24}$$

Since the numbers 20 and 24 both have 4 as a factor, 4 is called a **common factor** of the numbers. Other common factors of 20 and 24 are 1 and 2.

WORK PROBLEM 1 AT THE SIDE.

1 The fraction $\frac{5}{6}$ is in lowest terms because the numerator and denominator have no common factor other than 1; however, the fraction $\frac{20}{24}$ is *not* in lowest terms because its numerator and denominator have a common factor of 4.

A fraction is written in *lowest terms* when the numerator and the denominator have no common factor other than 1.

EXAMPLE 1 Understanding Lowest Terms
Are the following fractions in lowest terms?

(a) $\frac{3}{8}$

The numerator and denominator have no common factor other than 1, so the fraction is in lowest terms.

(b) $\frac{21}{36}$

The numerator and denominator have a common factor of 3, so the fraction is not in lowest terms. ■

WORK PROBLEM 2 AT THE SIDE.

OBJECTIVES

1. Tell whether a fraction is written in lowest terms.
2. Write a fraction in lowest terms using common factors.
3. Write a fraction in lowest terms using prime factors.
4. Tell whether two fractions are equivalent.

1. Decide whether the given factor is a common factor of the given numbers.

(a) 12, 18; 6

(b) 32, 48; 16

(c) 9, 12; 4

(d) 56, 73; 1

2. Are the following fractions in lowest terms?

(a) $\frac{1}{2}$

(b) $\frac{3}{12}$

(c) $\frac{18}{54}$

(d) $\frac{12}{17}$

ANSWERS
1. (a) yes (b) yes (c) no (d) yes
2. (a) yes (b) no (c) no (d) yes

2 There are two common methods for writing a fraction in lowest terms. These methods are shown in the next examples. The first method works best when the numerator and denominator are small numbers.

EXAMPLE 2 Changing to Lowest Terms

Write each fraction in lowest terms.

(a) $\frac{20}{24}$ **(b)** $\frac{30}{50}$ **(c)** $\frac{24}{42}$ **(d)** $\frac{60}{72}$

SOLUTION

(a) The largest common factor of 20 and 24 is 4. Divide both numerator and denominator by **4.**

$$\frac{20}{24} = \frac{20}{24} = \frac{5}{6}$$

(b) The largest common factor of 30 and 50 is 10.

$$\frac{30}{50} = \frac{30 \div 10}{50 \div 10} \quad \text{Divide both numerator and denominator by 10.}$$

$$= \frac{3}{5}$$

(c) $\frac{24}{42} = \frac{24 \div 6}{42 \div 6} = \frac{4}{7}$

(d) Suppose we made an error and thought 4 was the largest common factor of 60 and 72. Dividing by 4 would give

$$\frac{60}{72} = \frac{60 \div 4}{72 \div 4} = \frac{15}{18}.$$

But $\frac{15}{18}$ is not in lowest terms, since 15 and 18 have a common factor of 3. Divide by 3.

$$\frac{15}{18} = \frac{15 \div 3}{18 \div 3} = \frac{5}{6}$$

The fraction $\frac{60}{72}$ could have been written in lowest terms in one step by dividing by 12, the largest common factor of 60 and 72. (If necessary, review the rules for divisibility on **page 42.**)

$$\frac{60}{72} = \frac{60 \div 12}{72 \div 12} = \frac{5}{6}$$ ■

WORK PROBLEM 3 AT THE SIDE.

3. Write in lowest terms.

(a) $\frac{3}{6}$

(b) $\frac{6}{8}$

(c) $\frac{22}{33}$

(d) $\frac{12}{40}$

(e) $\frac{32}{80}$

ANSWERS

3. (a) $\frac{1}{2}$ (b) $\frac{3}{4}$ (c) $\frac{2}{3}$ (d) $\frac{3}{10}$ (e) $\frac{2}{5}$

3 The method of writing a fraction in lowest terms by division works well for fractions with small numerators and denominators. For larger numbers, it is better to use the method of **prime factors,** which is shown in the next example.

EXAMPLE 3 Using Prime Factors

Write each of the following in lowest terms.

(a) $\frac{24}{42}$ **(b)** $\frac{180}{54}$ **(c)** $\frac{54}{90}$

SOLUTION

(a) Write the prime factorization of both numerator and denominator. Use the method in **Section 2.3.**

$$\frac{24}{42} = \frac{2 \cdot 2 \cdot 2 \cdot 3}{2 \cdot 3 \cdot 7}$$

Just as with the other method, divide numerator and denominator by any common factors. Use a shortcut called **cancellation** to show this division. Place a **1** by each factor that is canceled.

$$\frac{24}{42} = \frac{\overset{1}{\cancel{2}} \cdot 2 \cdot 2 \cdot \overset{1}{\cancel{3}}}{\underset{1}{\cancel{2}} \cdot \underset{1}{\cancel{3}} \cdot 7}$$

Multiply the remaining factors in both numerator and denominator.

$$\frac{24}{42} = \frac{1 \cdot 2 \cdot 2 \cdot 1}{1 \cdot 1 \cdot 7} = \frac{4}{7}$$

Finally, $\frac{24}{42}$, written in lowest terms, is $\frac{4}{7}$.

(b) Write the prime factorization of both numerator and denominator.

$$\frac{180}{54} = \frac{2 \cdot 2 \cdot 3 \cdot 3 \cdot 5}{2 \cdot 3 \cdot 3 \cdot 3}$$

Now cancel the common factors. **Do not forget the 1's.**

$$\frac{180}{54} = \frac{\overset{1}{\cancel{2}} \cdot 2 \cdot \overset{1}{\cancel{3}} \cdot \overset{1}{\cancel{3}} \cdot 5}{\underset{1}{\cancel{2}} \cdot \underset{1}{\cancel{3}} \cdot \underset{1}{\cancel{3}} \cdot 3}$$

$$= \frac{1 \cdot 2 \cdot 1 \cdot 1 \cdot 5}{1 \cdot 1 \cdot 1 \cdot 3} = \frac{10}{3}$$

(c) $$\frac{54}{90} = \frac{\overset{1}{\cancel{2}} \cdot \overset{1}{\cancel{3}} \cdot \overset{1}{\cancel{3}} \cdot 3}{\underset{1}{\cancel{2}} \cdot \underset{1}{\cancel{3}} \cdot \underset{1}{\cancel{3}} \cdot 5} = \frac{1 \cdot 1 \cdot 1 \cdot 3}{1 \cdot 1 \cdot 1 \cdot 5} = \frac{3}{5}$$ ■

WORK PROBLEM 4 AT THE SIDE.

4. Use the method of prime factors to write each fraction in lowest terms.

(a) $\frac{12}{36}$

(b) $\frac{32}{60}$

(c) $\frac{88}{132}$

(d) $\frac{124}{340}$

ANSWERS

4. (a) $\frac{1}{3}$ (b) $\frac{8}{15}$ (c) $\frac{2}{3}$ (d) $\frac{31}{85}$

5. Are the following fractions equivalent?

(a) $\frac{2}{3}$ and $\frac{3}{4}$

(b) $\frac{10}{18}$ and $\frac{15}{27}$

(c) $\frac{14}{16}$ and $\frac{35}{40}$

(d) $\frac{12}{22}$ and $\frac{18}{32}$

ANSWERS
5. (a) not equivalent (b) equivalent (c) equivalent (d) not equivalent

This method of writing a fraction in lowest terms is summarized below.

Step 1 Write the **prime factorization** of both numerator and denominator.

Step 2 Use **cancellation** to divide numerator and denominator by any common factors.

Step 3 **Multiply** the remaining factors in numerator and denominator.

4 The next example shows how to use the **equivalency test** to tell whether two fractions are equivalent.

EXAMPLE 4 Using the Equivalency Test ("Cross Multiplication")
Are the following fractions equivalent?

(a) $\frac{6}{15}$ and $\frac{8}{20}$ **(b)** $\frac{8}{12}$ and $\frac{21}{30}$

SOLUTION
(a) Find each *cross product.*

$$\frac{6}{15} \quad \frac{8}{20}$$

$15 \cdot 8 = 120$
$6 \cdot 20 = 120$
equivalent

Since the cross products are equivalent, the fractions are equivalent.

(b) Find the cross products.

$$\frac{8}{12} \quad \frac{21}{30}$$

$12 \cdot 21 = 252$
$8 \cdot 30 = 240$
not equivalent

The cross products are not equivalent, so the fractions are not equivalent. ■

NOTE Caution. Cross multiply to determine whether two fractions are equivalent. This is *not* used to multiply or divide fractions.

WORK PROBLEM 5 AT THE SIDE.

name date hour

2.4 EXERCISES

Write each fraction in lowest terms.

Example: $\frac{12}{15} = \mathbf{\frac{4}{5}}$ **Solution:** $\frac{12}{15} = \frac{12 \div 3}{15 \div 3} = \frac{4}{5}$

1. $\frac{6}{8}$ **2.** $\frac{9}{12}$ **3.** $\frac{20}{40}$ **4.** $\frac{12}{36}$

5. $\frac{25}{40}$ **6.** $\frac{16}{24}$ **7.** $\frac{36}{42}$ **8.** $\frac{40}{75}$

9. $\frac{63}{70}$ **10.** $\frac{27}{45}$ **11.** $\frac{180}{210}$ **12.** $\frac{72}{80}$

13. $\frac{36}{63}$ **14.** $\frac{73}{146}$ **15.** $\frac{12}{600}$

16. $\frac{96}{132}$ **17.** $\frac{165}{180}$ **18.** $\frac{60}{108}$

Write the numerator and denominator of each fraction as a product of prime factors. Then write in lowest terms.

Example: $\frac{24}{36}$ **Solution:** $\frac{24}{36} = \frac{2 \cdot 2 \cdot 2 \cdot 3}{2 \cdot 2 \cdot 3 \cdot 3} = \mathbf{\frac{2}{3}}$

19. $\frac{10}{16}$ **20.** $\frac{15}{20}$ **21.** $\frac{25}{45}$

22. $\frac{35}{42}$ **23.** $\frac{60}{150}$ **24.** $\frac{40}{64}$

25. $\frac{36}{12}$ **26.** $\frac{192}{48}$

27. $\frac{77}{264}$ **28.** $\frac{65}{234}$

Decide whether the following pairs of fractions are equivalent or not equivalent.

Example:

$\frac{16}{20}$ and $\frac{36}{45}$ Find cross products.

The fractions are equivalent.

Solution:

$\frac{16}{20}\ \frac{36}{45}$

$20 \cdot 36 = 720$

$16 \cdot 45 = 720$

equivalent

29. $\frac{3}{4}$ and $\frac{18}{24}$

30. $\frac{9}{10}$ and $\frac{27}{30}$

31. $\frac{12}{15}$ and $\frac{35}{45}$

32. $\frac{11}{16}$ and $\frac{32}{48}$

33. $\frac{5}{8}$ and $\frac{3}{4}$

34. $\frac{7}{11}$ and $\frac{9}{12}$

35. $\frac{10}{25}$ and $\frac{16}{40}$

36. $\frac{9}{30}$ and $\frac{12}{40}$

37. $\frac{7}{52}$ and $\frac{9}{40}$

38. $\frac{21}{18}$ and $\frac{7}{5}$

39. $\frac{6}{50}$ and $\frac{9}{75}$

40. $\frac{24}{72}$ and $\frac{30}{90}$

CHALLENGE EXERCISES

Write each fraction in lowest terms.

41. $\frac{105}{252}$

42. $\frac{363}{528}$

43. $\frac{232}{725}$

44. $\frac{492}{1025}$

REVIEW EXERCISES

Find all the factors of each number. (For help, see ***Section 2.3****.)*

45. 8

46. 24

47. 64

48. 55

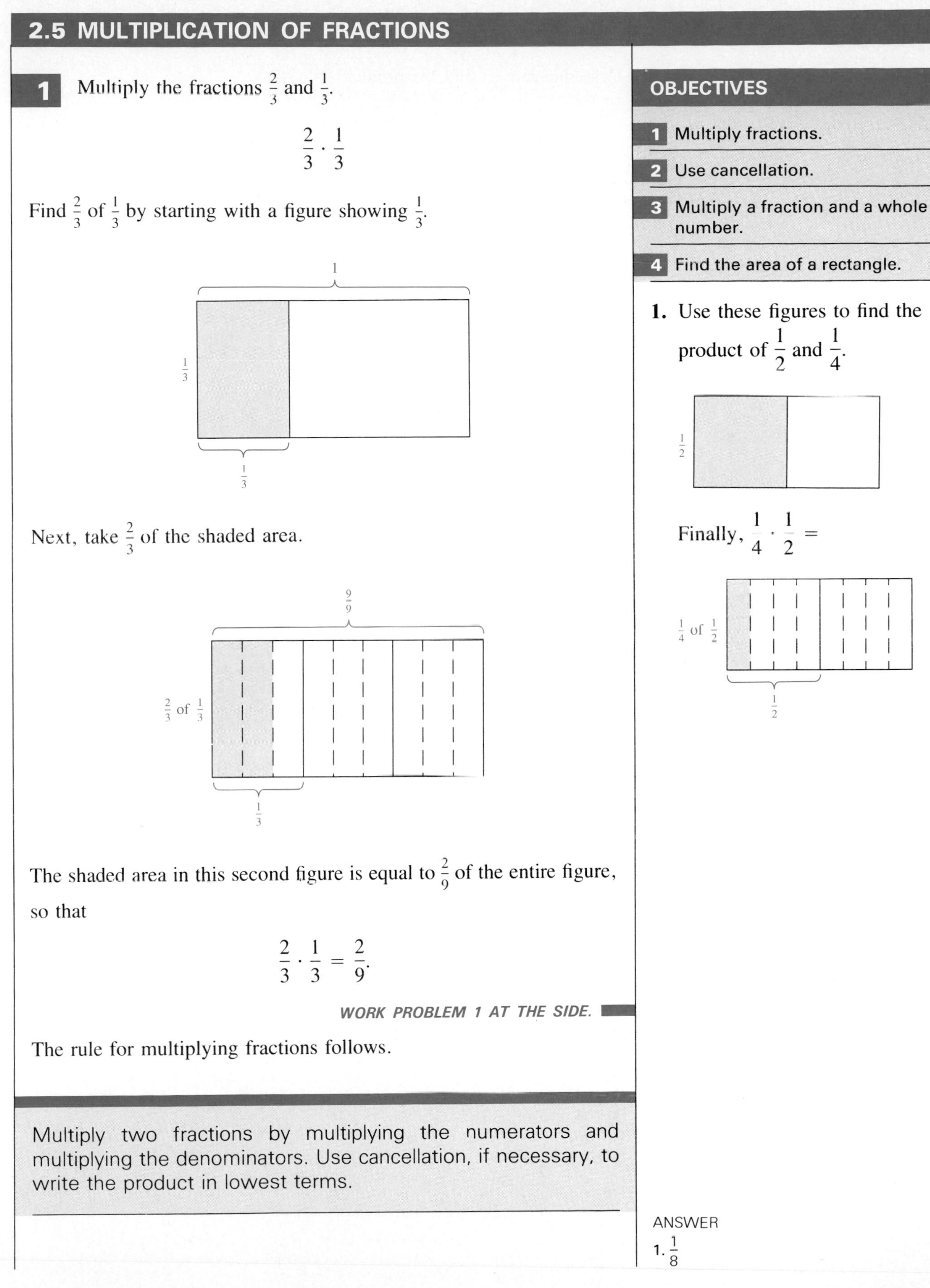

2.5 MULTIPLICATION OF FRACTIONS

OBJECTIVES

1. Multiply fractions.
2. Use cancellation.
3. Multiply a fraction and a whole number.
4. Find the area of a rectangle.

1 Multiply the fractions $\frac{2}{3}$ and $\frac{1}{3}$.

$$\frac{2}{3} \cdot \frac{1}{3}$$

Find $\frac{2}{3}$ of $\frac{1}{3}$ by starting with a figure showing $\frac{1}{3}$.

Next, take $\frac{2}{3}$ of the shaded area.

The shaded area in this second figure is equal to $\frac{2}{9}$ of the entire figure, so that

$$\frac{2}{3} \cdot \frac{1}{3} = \frac{2}{9}.$$

WORK PROBLEM 1 AT THE SIDE.

The rule for multiplying fractions follows.

Multiply two fractions by multiplying the numerators and multiplying the denominators. Use cancellation, if necessary, to write the product in lowest terms.

1. Use these figures to find the product of $\frac{1}{2}$ and $\frac{1}{4}$.

Finally, $\frac{1}{4} \cdot \frac{1}{2} =$

ANSWER

1. $\frac{1}{8}$

2. Multiply. Write answers in lowest terms.

(a) $\frac{3}{8} \cdot \frac{1}{10}$

(b) $\frac{5}{6} \cdot \frac{7}{9}$

(c) $\frac{3}{4} \cdot \frac{5}{9} \cdot \frac{1}{2}$

(d) $\frac{2}{3} \cdot \frac{7}{8} \cdot \frac{3}{4}$

ANSWERS

2. (a) $\frac{3}{80}$ (b) $\frac{35}{54}$ (c) $\frac{5}{24}$ (d) $\frac{7}{16}$

Use this rule to find the product of $\frac{2}{3}$ and $\frac{1}{3}$.

$$\frac{2}{3} \cdot \frac{1}{3} = \frac{2 \cdot 1}{3 \cdot 3}$$

$$= \frac{2}{9} \quad \begin{array}{l}\text{Multiply numerators.}\\ \text{Multiply denominators.}\end{array}$$

Since $\frac{2}{9}$ is in lowest terms,

$$\frac{2}{3} \cdot \frac{1}{3} = \frac{2 \cdot 1}{3 \cdot 3} = \frac{2}{9}. \quad \begin{array}{l}\leftarrow 2 \cdot 1 = 2\\ \leftarrow 3 \cdot 3 = 9\end{array}$$

EXAMPLE 1 Multiplying Fractions

Multiply.

(a) $\frac{5}{8} \cdot \frac{3}{4}$

(b) $\frac{4}{7} \cdot \frac{2}{5}$

(c) $\frac{5}{8} \cdot \frac{3}{4} \cdot \frac{1}{2}$

SOLUTION

(a) Multiply the numerators and multiply the denominators.

$$\frac{5}{8} \cdot \frac{3}{4} = \frac{5 \cdot 3}{8 \cdot 4} = \frac{15}{32}$$

There are no common factors other than 1 for 15 and 32, so the answer is in lowest terms.

(b) $\frac{4}{7} \cdot \frac{2}{5} = \frac{4 \cdot 2}{7 \cdot 5} = \frac{8}{35}$

(c) $\frac{5}{8} \cdot \frac{3}{4} \cdot \frac{1}{2} = \frac{5 \cdot 3 \cdot 1}{8 \cdot 4 \cdot 2} = \frac{15}{64}$ ■

WORK PROBLEM 2 AT THE SIDE.

2 It is often easier to cancel before multiplying, as shown in Example 2.

EXAMPLE 2 Understanding Cancellation

Multiply $\frac{5}{6}$ and $\frac{9}{10}$.

SOLUTION

$$\frac{5}{6} \cdot \frac{9}{10} = \frac{5 \cdot 9}{6 \cdot 10}$$

Since the numerator and denominator have a common factor other than 1, write the prime factorization of each number.

$$\frac{5}{6} \cdot \frac{9}{10} = \frac{5 \cdot 9}{6 \cdot 10} = \frac{5 \cdot 3 \cdot 3}{2 \cdot 3 \cdot 2 \cdot 5}$$

Next, use cancellation.

$$\frac{5}{6} \cdot \frac{9}{10} = \frac{5 \cdot 9}{6 \cdot 10} = \frac{\overset{1}{\cancel{5}} \cdot \overset{1}{\cancel{3}} \cdot 3}{2 \cdot \underset{1}{\cancel{3}} \cdot 2 \cdot \underset{1}{\cancel{5}}}$$

Finally, multiply the remaining factors in the numerator and in the denominator.

$$\frac{5}{6} \cdot \frac{9}{10} = \frac{1 \cdot 1 \cdot 3}{2 \cdot 1 \cdot 2 \cdot 1} = \frac{3}{4}$$ (lowest terms) ■

As a shortcut, instead of writing the prime factorization of each number, find the product of $\frac{5}{6}$ and $\frac{9}{10}$ as follows.

First, divide both 5 and 10 by 5. $\quad \frac{\overset{1}{\cancel{5}}}{6} \cdot \frac{9}{\underset{2}{\cancel{10}}}$

Next, divide both 6 and 9 by 3. $\quad \frac{\overset{1}{\cancel{5}}}{\underset{2}{\cancel{6}}} \cdot \frac{\overset{3}{\cancel{9}}}{\underset{2}{\cancel{10}}}$

Next, multiply. $\quad \frac{1 \cdot 3}{2 \cdot 2} = \frac{3}{4}$

3. Use cancellation to find each of the following.

(a) $\frac{3}{4} \cdot \frac{8}{9}$

(b) $\frac{6}{11} \cdot \frac{33}{21}$

(c) $\frac{25}{4} \cdot \frac{3}{50} \cdot \frac{1}{3}$

(d) $\frac{18}{17} \cdot \frac{1}{36} \cdot \frac{2}{3}$

EXAMPLE 3 Using Cancellation

Use cancellation to multiply. Write in lowest terms.

(a) $\frac{6}{11} \cdot \frac{7}{8}$

(b) $\frac{7}{10} \cdot \frac{20}{21}$

(c) $\frac{35}{12} \cdot \frac{32}{25}$

(d) $\frac{2}{3} \cdot \frac{8}{15} \cdot \frac{3}{4}$

SOLUTION

(a) Divide both 6 and 8 by 2. Next, multiply.

$$\frac{\overset{3}{\cancel{6}}}{11} \cdot \frac{7}{\underset{4}{\cancel{8}}} = \frac{3 \cdot 7}{11 \cdot 4} = \frac{21}{44}$$

(b) Divide 7 and 21 by 7, and divide 10 and 20 by 10.

$$\frac{\overset{1}{\cancel{7}}}{\underset{1}{\cancel{10}}} \cdot \frac{\overset{2}{\cancel{20}}}{\underset{3}{\cancel{21}}} = \frac{1 \cdot 2}{1 \cdot 3} = \frac{2}{3}$$

(c) $\frac{\overset{7}{\cancel{35}}}{\underset{3}{\cancel{12}}} \cdot \frac{\overset{8}{\cancel{32}}}{\underset{5}{\cancel{25}}} = \frac{7 \cdot 8}{3 \cdot 5} = \frac{56}{15}$ or $3\frac{11}{15}$

(d) $\frac{\overset{1}{\cancel{2}}}{\underset{1}{\cancel{3}}} \cdot \frac{\overset{4}{\cancel{8}}}{15} \cdot \frac{\overset{1}{\cancel{3}}}{\underset{\underset{1}{\cancel{2}}}{\cancel{4}}} = \frac{1 \cdot 4 \cdot 1}{1 \cdot 15 \cdot 1} = \frac{4}{15}$ ■

Cancellation is usually most helpful when the fractions involve large numbers.

NOTE There is no specific order that must be used in cancellation.

WORK PROBLEM 3 AT THE SIDE.

ANSWERS

3. (a) $\frac{\overset{1}{\cancel{3}}}{\underset{1}{\cancel{4}}} \cdot \frac{\overset{2}{\cancel{8}}}{\underset{3}{\cancel{9}}} = \frac{2}{3}$

(b) $\frac{\overset{2}{\cancel{6}}}{\underset{1}{\cancel{11}}} \cdot \frac{\overset{3}{\cancel{33}}}{\underset{7}{\cancel{21}}} = \frac{6}{7}$

(c) $\frac{\overset{1}{\cancel{25}}}{4} \cdot \frac{\overset{1}{\cancel{3}}}{\underset{2}{\cancel{50}}} \cdot \frac{1}{\underset{1}{\cancel{3}}} = \frac{1}{8}$

(d) $\frac{\overset{1}{\cancel{18}}}{17} \cdot \frac{1}{\underset{2}{\cancel{36}}} \cdot \frac{2}{3} = \frac{2}{102} = \frac{1}{51}$

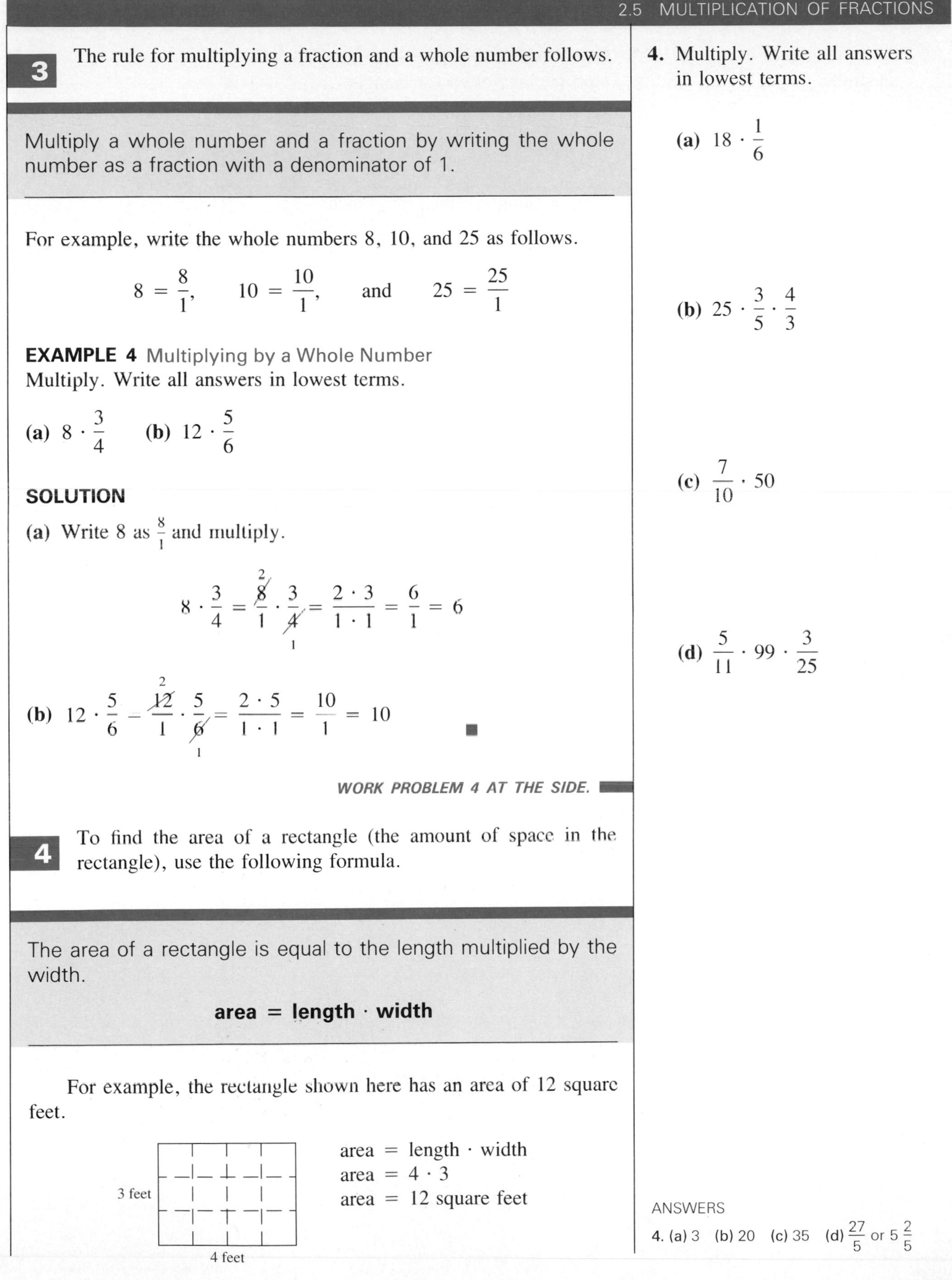

3 The rule for multiplying a fraction and a whole number follows.

Multiply a whole number and a fraction by writing the whole number as a fraction with a denominator of 1.

For example, write the whole numbers 8, 10, and 25 as follows.

$$8 = \frac{8}{1}, \quad 10 = \frac{10}{1}, \quad \text{and} \quad 25 = \frac{25}{1}$$

EXAMPLE 4 Multiplying by a Whole Number
Multiply. Write all answers in lowest terms.

(a) $8 \cdot \frac{3}{4}$ **(b)** $12 \cdot \frac{5}{6}$

SOLUTION

(a) Write 8 as $\frac{8}{1}$ and multiply.

$$8 \cdot \frac{3}{4} = \frac{\overset{2}{\cancel{8}}}{1} \cdot \frac{3}{\underset{1}{\cancel{4}}} = \frac{2 \cdot 3}{1 \cdot 1} = \frac{6}{1} = 6$$

(b) $12 \cdot \frac{5}{6} = \frac{\overset{2}{\cancel{12}}}{1} \cdot \frac{5}{\underset{1}{\cancel{6}}} = \frac{2 \cdot 5}{1 \cdot 1} = \frac{10}{1} = 10$ ■

WORK PROBLEM 4 AT THE SIDE.

4 To find the area of a rectangle (the amount of space in the rectangle), use the following formula.

The area of a rectangle is equal to the length multiplied by the width.

area = length · width

For example, the rectangle shown here has an area of 12 square feet.

area = length · width
area = 4 · 3
area = 12 square feet

4. Multiply. Write all answers in lowest terms.

(a) $18 \cdot \frac{1}{6}$

(b) $25 \cdot \frac{3}{5} \cdot \frac{4}{3}$

(c) $\frac{7}{10} \cdot 50$

(d) $\frac{5}{11} \cdot 99 \cdot \frac{3}{25}$

ANSWERS
4. (a) 3 (b) 20 (c) 35 (d) $\frac{27}{5}$ or $5\frac{2}{5}$

5. Find the area of each rectangle.

(a) $\frac{1}{3}$ foot; $\frac{9}{11}$ foot

(b) $\frac{1}{10}$ inch; $\frac{7}{8}$ inch

(c) a rectangle, $\frac{3}{2}$ yard by $\frac{5}{12}$ yard

ANSWERS

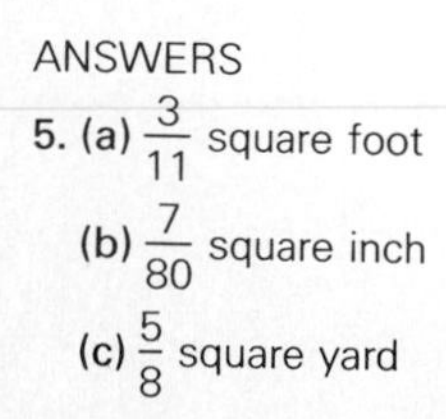

5. (a) $\frac{3}{11}$ square foot

(b) $\frac{7}{80}$ square inch

(c) $\frac{5}{8}$ square yard

EXAMPLE 5 Applying Fraction Skills

Find the area of each rectangle.

(a) $\frac{3}{8}$ foot; $\frac{11}{12}$ foot

(b) a rectangle, $\frac{7}{9}$ inch by $\frac{3}{14}$ inch

SOLUTION

(a) area = length · width

$$\text{area} = \frac{11}{12} \cdot \frac{3}{8}$$

$$= \frac{11}{\underset{4}{\cancel{12}}} \cdot \frac{\overset{1}{\cancel{3}}}{8} \quad \text{Cancel.}$$

$$= \frac{11}{32} \text{ square foot}$$

(b) Multiply the length and width.

$$\text{area} = \frac{7}{9} \cdot \frac{3}{14}$$

$$= \frac{\overset{1}{\cancel{7}}}{\underset{3}{\cancel{9}}} \cdot \frac{\overset{1}{\cancel{3}}}{\underset{2}{\cancel{14}}} \quad \text{Cancel.}$$

$$= \frac{1}{6} \text{ square inch}$$ ■

WORK PROBLEM 5 AT THE SIDE.

name date hour

2.5 EXERCISES

Multiply. Write all answers in lowest terms.

Example: $\frac{3}{8} \cdot \frac{4}{9} = \mathbf{\frac{1}{6}}$ **Solution:** $\frac{\overset{1}{\cancel{3}}}{\underset{2}{\cancel{8}}} \cdot \frac{\overset{1}{\cancel{4}}}{\underset{3}{\cancel{9}}} = \frac{1 \cdot 1}{2 \cdot 3} = \frac{1}{6}$

1. $\frac{1}{4} \cdot \frac{3}{5}$

2. $\frac{1}{7} \cdot \frac{5}{8}$

3. $\frac{1}{9} \cdot \frac{1}{6}$

4. $\frac{1}{4} \cdot \frac{1}{5}$

5. $\frac{3}{8} \cdot \frac{12}{5}$

6. $\frac{4}{9} \cdot \frac{12}{7}$

7. $\frac{5}{6} \cdot \frac{12}{25} \cdot \frac{3}{4}$

8. $\frac{7}{8} \cdot \frac{16}{21} \cdot \frac{1}{2}$

9. $\frac{3}{4} \cdot \frac{5}{6} \cdot \frac{2}{3}$

10. $\frac{3}{8} \cdot \frac{2}{3}$

11. $\frac{9}{10} \cdot \frac{11}{16}$

12. $\frac{5}{9} \cdot \frac{7}{10}$

13. $\frac{21}{30} \cdot \frac{5}{7}$

14. $\frac{6}{11} \cdot \frac{22}{15}$

15. $\frac{15}{7} \cdot \frac{8}{25}$

16. $\frac{32}{15} \cdot \frac{27}{64} \cdot \frac{35}{72}$

17. $\frac{16}{25} \cdot \frac{35}{32} \cdot \frac{15}{64}$

18. $\frac{39}{42} \cdot \frac{7}{13} \cdot \frac{7}{24}$

Multiply. Write all answers in lowest terms.

Example: $27 \cdot \frac{5}{9} = \mathbf{15}$ **Solution:** $27 \cdot \frac{5}{9} = \frac{\overset{3}{\cancel{27}}}{1} \cdot \frac{5}{\underset{1}{\cancel{9}}} = \frac{3 \cdot 5}{1 \cdot 1} = \frac{15}{1} = 15$

19. $10 \cdot \frac{3}{5}$

20. $40 \cdot \frac{7}{10}$

21. $72 \cdot \frac{5}{9}$

22. $38 \cdot \frac{2}{19}$

23. $30 \cdot \frac{1}{6}$

24. $50 \cdot \frac{1}{10}$

25. $21 \cdot \frac{5}{7} \cdot \frac{7}{10}$

26. $35 \cdot \frac{3}{5} \cdot \frac{2}{3}$

27. $100 \cdot \frac{21}{50} \cdot \frac{5}{14}$

28. $300 \cdot \frac{5}{6}$

29. $\frac{3}{4} \cdot 500$

30. $\frac{7}{16} \cdot 800$

31. $\frac{15}{32} \cdot 160$

32. $\frac{12}{25} \cdot 475$

33. $\frac{27}{32} \cdot 640$

34. $\frac{21}{13} \cdot 520 \cdot \frac{7}{20}$

35. $\frac{54}{38} \cdot 684 \cdot \frac{5}{6}$

36. $\frac{76}{43} \cdot 473 \cdot \frac{5}{19}$

name date hour

Find the area of each rectangle.

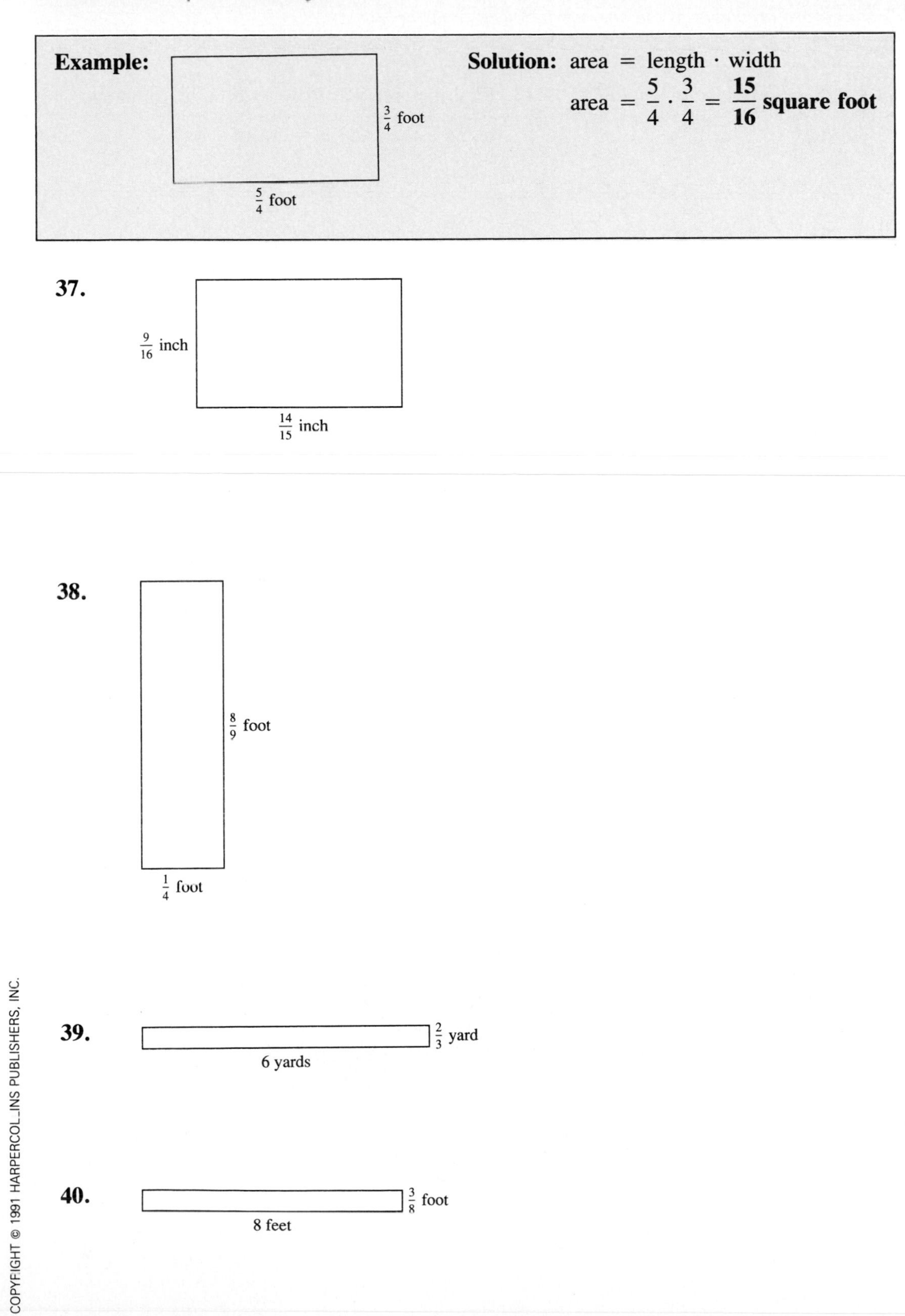

CHALLENGE EXERCISES

Solve each of the following word problems. Write answers in lowest terms or mixed numbers.

41. Find the area of a rectangle having a length of 16 inches and a width of $\frac{1}{4}$ inch.

42. Find the area of a rectangle having a length of 28 inches and a width of $\frac{5}{7}$ inch.

43. Find the floor area of a rabbit cage having a length of 2 yards and a width of $\frac{2}{3}$ yard.

44. Find the floor area of a bird cage having a length of 3 feet and a width of $\frac{11}{12}$ feet.

REVIEW EXERCISES

Solve each of the following word problems. (For help, see ***Section 1.9.****)*

45. If 230 rafts are rented each day, how many rafts are rented in 9 days.

46. There are 18 injections per vial. Find the number of injections in 74 vials.

2.6 APPLICATIONS OF MULTIPLICATION

1 Many word problems are solved by multiplying fractions. Use the following indicator words for multiplication.

product

double

triple

times

of

Look for these indicator words in the following examples.

EXAMPLE 1 Applying Indicator Words

Tom Hudspeth gives $\frac{1}{10}$ of his income to the church. One month he earned \$1980. How much did he give to the church that month?

APPROACH

To find the amount given to the church, the fraction $\frac{1}{10}$ must be multiplied by the monthly earnings (\$1980).

SOLUTION

The indicator word is *of:* Hudspeth gave $\frac{1}{10}$ *of* his income. The word *of* indicates multiplication, so find the amount given to the church by multiplying $\frac{1}{10}$ and \$1980.

$$\begin{aligned} \text{amount} &= \frac{1}{10} \cdot 1980 \\ &= 198 \end{aligned}$$

Hudspeth gave \$198 to the church that month. ■

WORK PROBLEM 1 AT THE SIDE.

EXAMPLE 2 Solving a Fraction Word Problem

Of the 42 students in a biology class, $\frac{2}{3}$ go on a field trip. How many go on the trip?

APPROACH

Find the number of students going on the field trip by multiplying the fraction $\frac{2}{3}$ by the number of students in the class (42).

OBJECTIVE

1 Solve word problems using multiplication.

1. (a) In one state, only $\frac{1}{4}$ of the value of a house is taxed. How much of the value of a \$62,000 house would be taxed?

(b) Suppose $\frac{3}{8}$ of the value of a house is taxed. How much of the value of a \$68,000 house is taxed?

ANSWERS
1. (a) \$15,500 (b) \$25,500

2. At one pharmacy, $\frac{3}{16}$ of the prescriptions are for tranquilizers. Out of 320 prescriptions, how many are for tranquilizers?

SOLUTION
Reword the problem to read

$\frac{2}{3}$ of the students go.
↑
indicator word

Find the number who go by multiplying $\frac{2}{3}$ and 42.

$$\text{number who go} = \frac{2}{3} \cdot 42$$
$$= \frac{2}{\cancel{3}_{1}} \cdot \cancel{42}^{14} = 28$$

28 students go on the trip. ■

WORK PROBLEM 2 AT THE SIDE.

EXAMPLE 3 Finding a Fraction of a Fraction

(a) In her will, a woman divides her estate into 6 equal parts. 5 of the 6 parts are given to relatives. Of the sixth part, $\frac{1}{3}$ goes to the Salvation Army. What fraction of her total estate goes to the Salvation Army?

APPROACH
To find the fraction of the estate going to the Salvation Army, the part not going to relatives $\left(\frac{1}{6}\right)$ is multiplied by the fractional part going to the Salvation Army $\left(\frac{1}{3}\right)$.

SOLUTION
The Salvation Army gets $\frac{1}{3}$ of $\frac{1}{6}$.
↑
indicator word

To find the fraction that the Salvation Army is to receive, multiply $\frac{1}{3}$ and $\frac{1}{6}$.

$$\text{fraction to Salvation Army} = \frac{1}{3} \cdot \frac{1}{6}$$
$$= \frac{1}{18}$$

The Salvation Army gets $\frac{1}{18}$ of the total estate.

ANSWERS
2. 60 prescriptions

(b) $\frac{2}{3}$ of the sixth part goes to the college the woman attended. What fraction of the total estate does the college get?

APPROACH

To find the fraction going to the college, multiply the part *not* going to the Salvation Army $\left(\frac{2}{3}\right)$ by the part not going to the relatives $\left(\frac{1}{6}\right)$.

SOLUTION

The college gets $\frac{2}{3}$ of $\frac{1}{6}$ of the estate.

$$\text{fraction to college} = \frac{2}{3} \cdot \frac{1}{6}$$

$$= \frac{\overset{1}{\cancel{2}}}{3} \cdot \frac{1}{\underset{3}{\cancel{6}}}$$

$$= \frac{1}{9}$$

The college gets $\frac{1}{9}$ of the estate. ■

WORK PROBLEM 3 AT THE SIDE.

3. A company divides its profit-sharing fund into 10 equal parts. 9 of these parts are given to production workers.

(a) Of the tenth part, $\frac{1}{4}$ goes to executives. What fraction of the total fund goes to executives?

(b) Of the tenth part, $\frac{3}{4}$ goes to the office staff. What fraction of the total fund goes to the office staff?

ANSWERS

3. (a) $\frac{1}{40}$ (b) $\frac{3}{40}$

name date hour

2.6 EXERCISES

Solve each of the following word problems. Look for indicator words.

Example: Of the 96 new cars at Andrews Motors, $\frac{5}{8}$ are small cars. How many small cars are there?

Approach: To find the number of small cars, multiply the total number of new cards (96) by the fraction that are small cars $\left(\frac{5}{8}\right)$.

Solution: number of small cars $= \frac{5}{8} \cdot 96$

$$= \frac{5}{\cancel{8}_1} \cdot \cancel{96}^{12}$$

$$= 60$$

There are **60 small cars.**

1. A desk is $\frac{2}{3}$ yard by $\frac{5}{4}$ yard. Find its area.

2. A wading pool is $\frac{7}{8}$ yard by $\frac{10}{9}$ yards. Find its area.

3. Here is a square, $\frac{1}{2}$ inch on a side. Find its area.

4. Is the area of the square in Exercise 3 more or less than $\frac{1}{2}$ square inch?

5. At a certain store, $\frac{1}{5}$ of the items sold are taxable. The store sells 1900 items. How many are taxable?

6. Another store sells 2500 items, of which $\frac{3}{25}$ are classified as junk food. How many of the items are junk food?

7. Lupe needs \$3600 to go to school for one year. She earns $\frac{3}{4}$ of this amount in the summer. How much money does she earn in the summer?

8. John paid \$115 for textbooks this term. Of this amount, the bookstore kept $\frac{1}{5}$. How much money did the bookstore keep?

9. Of the 560 employees of Michigan Tire Service, $\frac{9}{14}$ have given to the United Fund. How many have given to the United Fund?

10. A school gives scholarships to $\frac{5}{24}$ of its 1800 freshmen. How many students received scholarships?

11. At the local hospital, $\frac{5}{11}$ of the volunteers are men. If there are 165 volunteers, how many are men?

12. A hotel has 204 rooms. Of these rooms, $\frac{6}{17}$ have king-sized beds. How many rooms have king-sized beds?

name date hour

The Gomes family earned $32,000 last year. Use this fact to solve Exercises 13–16.

13. They used $\frac{1}{4}$ of their income for taxes. How much were their taxes?

14. They spent $\frac{5}{16}$ on food. How much money did they spend for food?

15. The family saved $\frac{1}{16}$ of their income. How much did they save?

16. They spent $\frac{1}{8}$ of their income on clothing. How much did they spend on clothing?

CHALLENGE EXERCISES

Solve each of the following word problems.

17. Kerry earned $120 in one 8-hour day. How much money did she earn in 5 hours?

18. Jim ran 24 miles in 6 hours. How far did he run in 5 hours?

19. Patty Wilson is running for city council. She needs to get $\frac{5}{8}$ of her votes from the south side of town. Ms. Wilson will need 2400 votes to win. How many votes does she need from the south side?

20. Of all the garage sales in Smallville, $\frac{9}{16}$ have broken televisions. If there are 144 garage sales, how many have broken televisions?

21. A will states that $\frac{7}{8}$ of the estate is to be divided among relatives. Of the remaining $\frac{1}{8}$, $\frac{1}{4}$ goes to the American Cancer Society. What fraction of the estate goes to the American Cancer Society?

22. A couple has invested $\frac{1}{5}$ of their total investment in stocks. Of the $\frac{1}{5}$ invested in stocks, $\frac{1}{8}$ is invested in General Motors. What fraction of the total investment is invested in General Motors?

REVIEW EXERCISES

Solve each of the following word problems. (For help, see ***Section 1.8.****)*

23. There are 18 test kits per carton. Find the number of cartons needed to supply 1332 test kits.

24. Jennifer Quinn drives 396 miles on 12 gallons of gas. Find the number of miles that she gets per gallon.

2.7 DIVIDING FRACTIONS

OBJECTIVES

1. Divide fractions.
2. Solve word problems in which fractions are divided.

As shown in Chapter 1, the division problem $12 \div 3$ asks how many 3's are in 12. In the same way, the division problem $\frac{2}{3} \div \frac{1}{6}$ asks how many $\frac{1}{6}$'s are in $\frac{2}{3}$. Look at the figure.

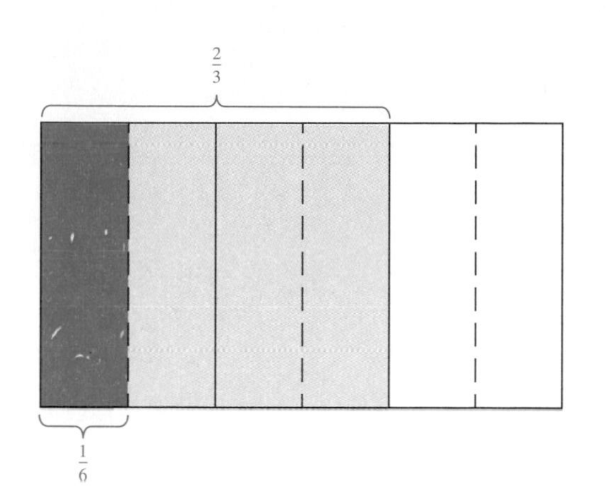

The figure shows that there are 4 of the $\frac{1}{6}$'s in $\frac{2}{3}$, or

$$\frac{2}{3} \div \frac{1}{6} = 4.$$

1 Compare: $\frac{2}{3} \div \frac{1}{6} = 4$ *and* $\frac{2}{3} \cdot \frac{6}{1} = \frac{2}{\cancel{3}_1} \cdot \frac{\cancel{6}^2}{1} = 4$

Invert $\frac{1}{6}$ to get $\frac{6}{1}$.

Divide two fractions by inverting the second fraction (divisor) and multiplying.

EXAMPLE 1 Dividing One Fraction by Another

Divide. Write answers in lowest terms.

(a) $\frac{7}{8} \div \frac{15}{16}$

(b) $\dfrac{\frac{3}{5}}{\frac{7}{10}}$

1. Divide. Write all answers in lowest terms.

(a) $\frac{1}{4} \div \frac{1}{2}$

(b) $\frac{3}{8} \div \frac{2}{3}$

(c) $\dfrac{\frac{9}{10}}{\frac{6}{5}}$

(d) $\dfrac{\frac{5}{9}}{\frac{25}{18}}$

ANSWERS

1. (a) $\frac{1}{2}$ (b) $\frac{9}{16}$ (c) $\frac{3}{4}$ (d) $\frac{2}{5}$

SOLUTION

(a) $\frac{7}{8} \div \frac{15}{16} = \frac{7}{8} \cdot \frac{16}{15}$ ← Invert. (Multiply.)

$$= \frac{7}{\cancel{8}_1} \cdot \frac{\cancel{16}^2}{15}$$

$$= \frac{7 \cdot 2}{1 \cdot 15}$$

$$= \frac{14}{15}$$

(b) $\dfrac{\frac{3}{5}}{\frac{7}{10}} = \frac{3}{5} \div \frac{7}{10}$ Rewrite by using this division symbol.

$$= \frac{3}{\cancel{5}_1} \cdot \frac{\cancel{10}^2}{7}$$ Invert and multiply.

$$= \frac{3 \cdot 2}{1 \cdot 7}$$

$$= \frac{6}{7}$$ ■

WORK PROBLEM 1 AT THE SIDE.

EXAMPLE 2 Dividing with a Whole Number

Divide. Write all answers in lowest terms.

(a) $5 \div \frac{1}{4}$ **(b)** $\frac{2}{3} \div 6$

SOLUTION

(a) Change 5 to $\frac{5}{1}$. Next, invert $\frac{1}{4}$ and multiply.

$$5 \div \frac{1}{4} = \frac{5}{1} \cdot \frac{4}{1}$$

$$= \frac{5 \cdot 4}{1 \cdot 1}$$

$$= \frac{20}{1}$$

$$= 20$$

(b) $\frac{2}{3} \div 6 = \frac{2}{3} \div 6 = \frac{2}{3} \div \frac{6}{1} = \frac{2}{3} \cdot \frac{1}{6}$

Cancel and finish the problem.

$$\frac{\overset{1}{\cancel{2}}}{3} \cdot \frac{1}{\underset{3}{\cancel{6}}} = \frac{1 \cdot 1}{3 \cdot 3}$$

$$= \frac{1}{9} \quad ■$$

WORK PROBLEM 2 AT THE SIDE.

2 Many word problem require division of fractions. Recall typical indicator words for division such as *quotient, divide, divided by,* or *divided into.*

EXAMPLE 3 Applying Fraction Skills

Mary must fill a 12-gallon barrel with a chemical. She has only a $\frac{2}{3}$-gallon container to use. How many times must she fill the $\frac{2}{3}$-gallon container and empty it into the 12-gallon barrel?

APPROACH

To find the number of containers Mary needs to fill the larger barrel, divide the size of the barrel (12 gallons) by the size of the container $\left(\frac{2}{3} \text{ gallon}\right)$.

SOLUTION

This problem can be solved by finding the number of times 12 can be divided by $\frac{2}{3}$.

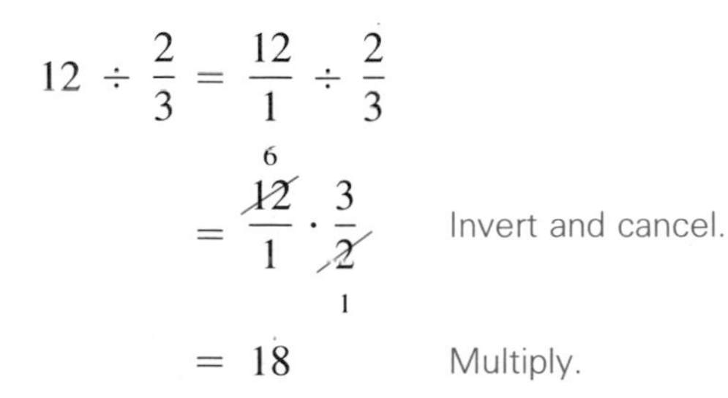

$$12 \div \frac{2}{3} = \frac{12}{1} \div \frac{2}{3}$$

$$= \frac{\overset{6}{\cancel{12}}}{1} \cdot \frac{3}{\underset{1}{\cancel{2}}} \qquad \text{Invert and cancel.}$$

$$= 18 \qquad \text{Multiply.}$$

The bucket must be filled 18 times. ■

WORK PROBLEM 3 AT THE SIDE.

2. Divide. Write all answers in lowest terms.

(a) $4 \div \frac{4}{9}$

(b) $7 \div \frac{1}{3}$

(c) $\frac{5}{8} \div 2$

(d) $\frac{7}{10} \div 3$

3. (a) How many times must a $\frac{3}{4}$-quart jar be filled in order to fill a 15-quart can?

(b) How many $\frac{3}{4}$-quart beverage cans may be filled from 24 quarts of beverage?

ANSWERS

2. (a) 9 (b) 21 (c) $\frac{5}{16}$ (d) $\frac{7}{30}$

3. (a) 20 times (b) 32 cans

name date hour

2.7 EXERCISES

Divide. Write all answers in lowest terms.

Example: $\frac{3}{4} \div \frac{1}{2} = 1\frac{1}{2}$ **Solution:** $\frac{3}{\not{4}_2} \cdot \frac{\not{2}^1}{1} = \frac{3}{2} = 1\frac{1}{2}$

1. $\frac{2}{3} \div \frac{3}{4}$

2. $\frac{5}{6} \div \frac{6}{7}$

3. $\frac{5}{8} \div \frac{4}{3}$

4. $\frac{1}{5} \div \frac{3}{4}$

5. $\frac{5}{9} \div \frac{5}{4}$

6. $\frac{3}{8} \div \frac{3}{7}$

7. $\frac{7}{12} \div \frac{5}{18}$

8. $\frac{13}{20} \div \frac{4}{5}$

9. $\frac{7}{9} \div \frac{12}{5}$

10. $\dfrac{\frac{15}{32}}{\frac{5}{64}}$

11. $\dfrac{\frac{36}{35}}{\frac{15}{14}}$

12. $\dfrac{\frac{28}{15}}{\frac{21}{5}}$

13. $9 \div \frac{1}{2}$

14. $8 \div \frac{2}{5}$

15. $15 \div \frac{3}{4}$

16. $\dfrac{\frac{5}{8}}{6}$

17. $\dfrac{\frac{4}{7}}{8}$

18. $\dfrac{\frac{11}{5}}{3}$

Solve each of the following word problems by using division.

19. Ms. Fullmer has a piece of property with an area that is $\frac{8}{9}$ acre. She wishes to divide it into 4 equal parts for her children. How many acres of land will each child get?

20. Linda wants to make children's dresses to sell at a crafts fair. Each dress requires $\frac{1}{3}$ yard of material. She had 15 yards of material. Find the number of dresses she can make.

21. It takes $\frac{4}{5}$ pound of salt to fill a large salt shaker. How many salt shakers can be filled with 28 pounds of salt?

22. Barbara has 5 quarts of lemonade. If each of her Brownies gets $\frac{1}{3}$ quart of lemonade, how many Brownies does she have?

23. How many $\frac{1}{8}$-ounce vials of medicine can be filled with 7 ounces of medicine?

24. Each guest at a party will eat $\frac{5}{16}$ pound of peanuts. How many guests may be served with 10 pounds of peanuts?

25. Pam had a small pickup truck that would carry $\frac{2}{3}$-cord of firewood. Find the number of trips needed to deliver 40 cords of wood.

26. Anna has a reel of steel cable 600 yards long. Find the number of cable sections $\frac{3}{4}$ yard in length that may be cut from the reel.

27. After reading 320 pages, Tom had read $\frac{5}{8}$ of a book. How many pages long is the entire book?

28. Find the number of $\frac{2}{3}$-quart cans of fruit that can be filled from a vat holding 82 quarts of fruit.

CHALLENGE EXERCISES

29. Walt has driven $\frac{6}{7}$ of the way to a vacation resort. He has driven 216 miles so far. How many *more miles* must he drive to get to the resort?

30. Sheila has been working on a job for 63 hours. The job is $\frac{7}{9}$ finished. How many *more hours* must she work to finish the job?

31. The Bridge Lighting Committee has raised $\frac{7}{8}$ of the funds necessary for their lighting project. If this amounts to $840,000, how much additional money must be raised?

32. A mountain guide has used pack animals for $\frac{14}{15}$ of a trip and must finish the trip on foot. The distance covered with pack animals is 98 miles. Find the number of miles to be completed on foot.

REVIEW EXERCISES

Write each mixed number as an improper fraction. (For help, see ***Section 2.2.****)*

33. $2\frac{5}{6}$

34. $1\frac{7}{11}$

35. $12\frac{2}{3}$

36. $22\frac{8}{9}$

37. $4\frac{10}{11}$

38. $19\frac{8}{11}$

2.8 MULTIPLICATION AND DIVISION OF MIXED NUMBERS

1 Multiply or divide mixed numbers by using the following rule.

Change each mixed number to an improper fraction, and then multiply or divide. Write the answer in lowest terms and change the answer to a mixed number if desired.

EXAMPLE 1 Multiplying Mixed Numbers
Multiply. Write all answers in lowest terms.

(a) $2\frac{1}{2} \cdot 3\frac{1}{5}$

(b) $3\frac{5}{8} \cdot 4\frac{4}{5}$

(c) $1\frac{3}{5} \cdot 3\frac{1}{3}$

SOLUTION

(a) Change each mixed number to an improper fraction.

$$2\frac{1}{2} = \frac{5}{2} \quad \text{and} \quad 3\frac{1}{5} = \frac{16}{5}$$

Next, multiply.

$$2\frac{1}{2} \cdot 3\frac{1}{5} = \frac{5}{2} \cdot \frac{16}{5} = \frac{\overset{1}{\cancel{5}}}{\underset{1}{\cancel{2}}} \cdot \frac{\overset{8}{\cancel{16}}}{\underset{1}{\cancel{5}}} = \frac{1 \cdot 8}{1 \cdot 1} = \frac{8}{1} = 8$$

(b) $3\frac{5}{8} \cdot 4\frac{4}{5} = \frac{29}{8} \cdot \frac{24}{5} = \frac{29}{\underset{1}{\cancel{8}}} \cdot \frac{\overset{3}{\cancel{24}}}{5} = \frac{29 \cdot 3}{1 \cdot 5} = \frac{87}{5}$

As a mixed number,

$$\frac{87}{5} = 17\frac{2}{5}.$$

(c) $1\frac{3}{5} \cdot 3\frac{1}{3} = \frac{8}{\underset{1}{\cancel{5}}} \cdot \frac{\overset{2}{\cancel{10}}}{3} = \frac{8 \cdot 2}{1 \cdot 3} = \frac{16}{3} = 5\frac{1}{3}$ ■

WORK PROBLEM 1 AT THE SIDE.

OBJECTIVES

1. Multiply mixed numbers.
2. Divide mixed numbers.
3. Solve word problems with mixed numbers.

1. Multiply. Write answers in lowest terms.

(a) $2\frac{1}{4} \cdot 7\frac{1}{3}$

(b) $6\frac{1}{2} \cdot 1\frac{3}{5}$

(c) $3\frac{3}{5} \cdot 4\frac{4}{9}$

(d) $5\frac{5}{8} \cdot 4\frac{1}{6}$

ANSWERS
1. (a) $16\frac{1}{2}$ (b) $10\frac{2}{5}$ (c) 16 (d) $23\frac{7}{16}$

2. Divide. Write answer in lowest terms.

(a) $1\frac{1}{4} \div 3\frac{1}{3}$

(b) $3\frac{3}{8} \div 2\frac{4}{7}$

(c) $6 \div 6\frac{2}{3}$

(d) $5\frac{1}{4} \div 3$

2 The next example shows division of mixed numbers.

EXAMPLE 2 Dividing Mixed Numbers

Divide. Write answers in lowest terms.

(a) $2\frac{2}{5} \div 1\frac{1}{2}$ **(b)** $3\frac{1}{3} \div 2\frac{1}{2}$

(c) $8 \div 3\frac{3}{5}$ **(d)** $4\frac{3}{8} \div 5$

SOLUTION

(a) First, change each mixed number to an improper fraction.

$$2\frac{2}{5} \div 1\frac{1}{2} = \frac{12}{5} \div \frac{3}{2}$$

Next, invert the second fraction and multiply.

$$\frac{12}{5} \div \frac{3}{2} = \frac{\overset{4}{\cancel{12}}}{5} \cdot \frac{2}{\underset{1}{\cancel{3}}} = \frac{4 \cdot 2}{5 \cdot 1} = \frac{8}{5} = 1\frac{3}{5}$$

(inverted)

(b) $$3\frac{1}{3} \div 2\frac{1}{2} = \frac{10}{3} \div \frac{5}{2} = \frac{\overset{2}{\cancel{10}}}{3} \cdot \frac{2}{\underset{1}{\cancel{5}}} = \frac{2 \cdot 2}{3 \cdot 1} = \frac{4}{3} = 1\frac{1}{3}$$

(inverted)

(c) $$8 \div 3\frac{3}{5} = \frac{8}{1} \div \frac{18}{5} = \frac{\overset{4}{\cancel{8}}}{1} \cdot \frac{5}{\underset{9}{\cancel{18}}} = \frac{20}{9} = 2\frac{2}{9}$$

(inverted)

Write 8 as $\frac{8}{1}$.

(d) $$4\frac{3}{8} \div 5 = \frac{35}{8} \div \frac{5}{1} = \frac{\overset{7}{\cancel{35}}}{8} \cdot \frac{1}{\underset{1}{\cancel{5}}} = \frac{7}{8}$$ ■

(inverted)

Write 5 as $\frac{5}{1}$.

WORK PROBLEM 2 AT THE SIDE.

ANSWERS

2. (a) $\frac{3}{8}$ (b) $1\frac{5}{16}$ (c) $\frac{9}{10}$ (d) $1\frac{3}{4}$

3 The next two examples show how to solve word problems involving mixed numbers.

EXAMPLE 3 Applying Multiplication Skills

Suppose 11 people each give $3\frac{1}{4}$ pounds of grain to a food drive. How much grain will be collected?

APPROACH

Since several people each give the same amount of grain, multiply to get the total amount collected.

SOLUTION

Multiply the **number** of people and the **amount** of grain that each gives.

$$11 \cdot 3\frac{1}{4} = 11 \cdot \frac{13}{4}$$

$$= \frac{143}{4} = 35\frac{3}{4}$$

The food drive will collect $35\frac{3}{4}$ pounds of grain. ■

WORK PROBLEM 3 AT THE SIDE.

EXAMPLE 4 Applying Division Skills

One tent requires $7\frac{1}{4}$ yards of nylon cloth. How many tents can be made from $65\frac{1}{4}$ yards of the cloth?

APPROACH

Division must be used to find the number of times one number is contained in another.

SOLUTION

Divide the **number** of yards of cloth by the **number** of yards needed for one tent.

$$65\frac{1}{4} \div 7\frac{1}{4} = \frac{261}{4} \div \frac{29}{4}$$

$$= \frac{261}{4} \cdot \frac{4}{29} = \frac{\overset{9}{\cancel{261}}}{\underset{1}{\cancel{4}}} \cdot \frac{\overset{1}{\cancel{4}}}{\underset{1}{\cancel{29}}} = 9$$

9 tents can be made from $65\frac{1}{4}$ yards of cloth. ■

WORK PROBLEM 4 AT THE SIDE.

3. (a) Suppose a dress requires $2\frac{3}{4}$ yards of material. How much material would be needed for 7 dresses?

(b) Tom earns $\$8\frac{1}{4}$ per hour. How much would he earn in $3\frac{1}{2}$ hours? Write the answer as a mixed number.

4. (a) An airplane needs $2\frac{3}{8}$ pounds of a special metal. How many airplanes could be built from $28\frac{1}{2}$ pounds of the metal?

(b) A gem sells for $\$12\frac{2}{3}$ per carat. How many carats could be bought for $152?

ANSWERS

3. (a) $19\frac{1}{4}$ yards (b) $\$28\frac{7}{8}$

4. (a) 12 airplanes (b) 12 carats

name date hour

2.8 EXERCISES

Multiply. Write answers as mixed numbers or whole numbers.

1. $7\frac{1}{2} \cdot 3\frac{1}{3}$

2. $1\frac{1}{2} \cdot 3\frac{3}{4}$

3. $1\frac{2}{3} \cdot 2\frac{7}{10}$

4. $6\frac{1}{4} \cdot 3\frac{1}{5}$

5. $3\frac{1}{9} \cdot 1\frac{2}{7}$

6. $1\frac{1}{4} \cdot 2\frac{1}{2}$

7. $10 \cdot 7\frac{1}{4}$

8. $6 \cdot 2\frac{1}{3}$

9. $9\frac{1}{5} \cdot 15$

10. $6 \cdot 2\frac{2}{3} \cdot 1\frac{1}{2}$

11. $9 \cdot 3\frac{1}{4} \cdot \frac{8}{3}$

12. $\frac{2}{3} \cdot 3\frac{2}{3} \cdot \frac{6}{11}$

Divide. Write answers as mixed numbers or whole numbers.

13. $3\frac{1}{4} \div 2\frac{5}{8}$

14. $2\frac{1}{4} \div 1\frac{1}{8}$

15. $2\frac{1}{2} \div 3\frac{3}{4}$

16. $5\frac{1}{2} \div 4\frac{2}{5}$

17. $1\frac{1}{3} \div 5\frac{1}{2}$

18. $1\frac{1}{2} \div 6\frac{1}{4}$

19. $3 \div 1\frac{1}{4}$

20. $\frac{1}{4} \div 2\frac{1}{2}$

21. $\frac{3}{8} \div 1\frac{1}{4}$

22. $7\frac{1}{2} \div \frac{2}{3}$

23. $5\frac{2}{3} \div \frac{1}{6}$

24. $5\frac{3}{4} \div 2$

name date hour

Solve each of the following word problems by using multiplication or division.

25. Lupe wants to make 16 dresses to sell at a picnic. Each dress needs $3\frac{1}{4}$ yards of material. How many yards does she need?

26. Tom worked $38\frac{1}{4}$ hours at $6 per hour. How much money did he make?

27. Each home in an area needs $36\frac{1}{2}$ yards of rain gutter. How many homes can be fitted with rain gutter if there are $328\frac{1}{2}$ yards of gutter available?

28. For 1 acre of a crop, $7\frac{1}{2}$ gallons of fertilizer must be applied. How many acres can be fertilized with 1200 gallons of fertilizer?

29. Insect spray is mixed $1\frac{3}{4}$ ounces of chemical per gallon of water. How many ounces of chemical are needed for $12\frac{1}{2}$ gallons of water?

30. Each home requires $37\frac{3}{4}$ pounds of roofing nails. How many pounds of roofing nails are needed for 36 homes?

31. A dictionary requires $2\frac{3}{8}$ pounds of paper. How many can be published with 11,875 pounds of paper?

32. A home requires $403\frac{1}{2}$ square feet of glass. Find the number of homes that can be fitted with glass if 29,052 square feet of glass are available.

CHALLENGE EXERCISES

33. A photographer uses $4\frac{1}{4}$ rolls of film at a wedding and $2\frac{3}{8}$ rolls of film at a retirement party. Find the total number of rolls needed for 28 weddings and 16 retirement parties.

34. One necklace can be completed in $6\frac{1}{2}$ minutes, while a bracelet takes $3\frac{1}{8}$ minutes. Find the total time that it takes to complete 36 necklaces and 22 bracelets.

REVIEW EXERCISES

Write each fraction in lowest terms. (For help, see ***Section 2.4****.)*

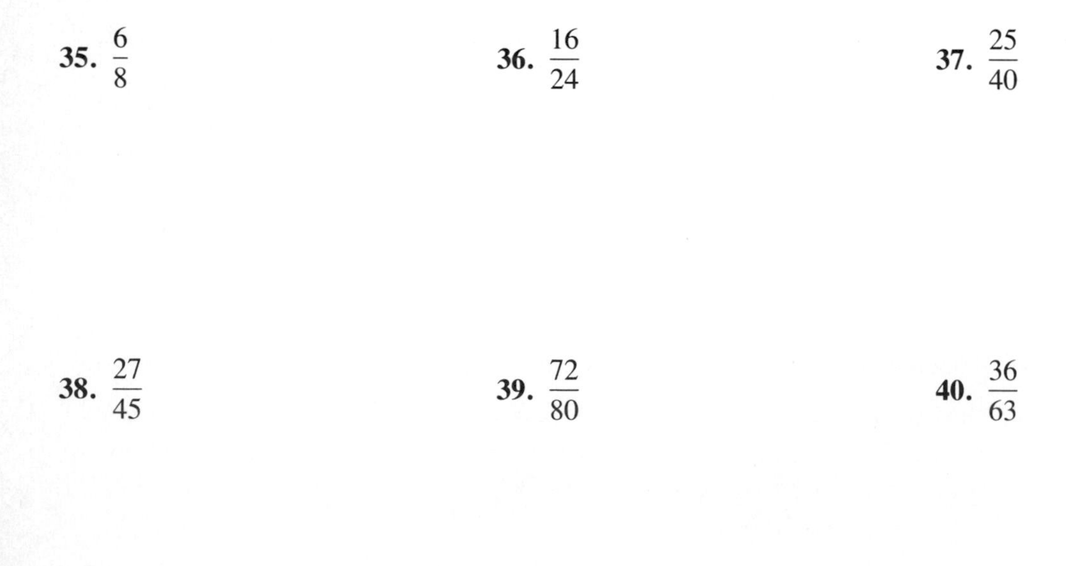

35. $\frac{6}{8}$

36. $\frac{16}{24}$

37. $\frac{25}{40}$

38. $\frac{27}{45}$

39. $\frac{72}{80}$

40. $\frac{36}{63}$

CHAPTER 2 SUMMARY

KEY TERMS

2.1	**numerator**	The number above the division bar in a fraction is called the numerator.
	denominator	The number below the division bar in a fraction is called the denominator.
	proper fraction	In a proper fraction, the numerator is smaller than the denominator.
	improper fraction	In an improper fraction, the numerator is greater than or equal to the denominator.
2.2	**mixed number**	A mixed number includes a fraction and a whole number written together as one.
2.3	**factors**	Numbers that are multiplied to give a product are factors.
	composite number	A composite number has at least one factor other than itself and 1.
	prime number	Any whole number that is not composite, except 0 and 1, is called a prime number.
	prime factorization	A factorization of a number in which every factor is prime is a prime factorization.
2.4	**common factor**	A common factor is a number that can be divided into both the numerator and denominator to reduce the fraction.
	lowest terms	A fraction is written in lowest terms when its numerator and denominator have no common factor other than 1.
2.5	**cancellation**	When multiplying or dividing fractions, the process of dividing a numerator and denominator by a common factor is called cancellation.

QUICK REVIEW

Section Number and Topic	Approach	Example
2.1 Types of Fractions	**Proper:** Numerator smaller than denominator. **Improper:** Numerator equal to or greater than denominator.	$\frac{2}{3}, \frac{3}{4}, \frac{15}{16}, \frac{1}{8}$ $\frac{17}{8}, \frac{19}{12}, \frac{11}{2}, \frac{5}{3}, \frac{7}{7}$
2.2 Converting Fractions	**Mixed to improper:** Multiply denominator by whole number and add numerator. **Improper to mixed:** Divide numerator by denominator and place remainder over denominator.	$7\frac{2}{3} = \frac{23}{3}$ $\frac{17}{5} = 3\frac{2}{5}$

Section Number and Topic	Approach	Example
2.3 Prime Numbers	Determine whether a whole number other than 1 is divisible (with a remainder of 0) only by itself and 1.	Prime numbers below 100 are 2, 3, 5, 7, 11, 13, 17, 19, 23, 29, 31, 37, 41, 43, 47, 53, 59, 61, 67, 71, 73, 79, 83, 91, and 97
2.3 Find the Prime Factorization of a Number	Divide each factor by a prime number by using a diagram that forms the shape of a tree.	Find the prime factorization of 30. Use a factor tree. 30 (2) 15 (3) (5) Prime factors are circled. $2 \cdot 3 \cdot 5$
2.4 Writing Fractions in Lowest Terms	Divide the numerator and denominator by the same number.	$\frac{30}{42} = \frac{30 \div 6}{42 \div 6} = \frac{5}{7}$
2.5 Multiplying Fractions	1. Multiply numerators and denominators. 2. Reduce answers to lowest terms if canceling was not done.	$\frac{6}{11} \cdot \frac{7}{8} = \frac{\overset{3}{\cancel{6}}}{11} \cdot \frac{7}{\underset{4}{\cancel{8}}} = \frac{21}{44}$
2.7 Dividing Fractions	Invert the divisor and multiply as fractions.	$\frac{25}{36} \div \frac{15}{18} = \frac{\overset{5}{\cancel{25}}}{\underset{2}{\cancel{36}}} \cdot \frac{\overset{1}{\cancel{18}}}{\underset{3}{\cancel{15}}}$ $= \frac{5}{2} \cdot \frac{1}{3} = \frac{5}{6}$
2.8 Multiplying Mixed Numbers	1. Change mixed numbers to improper fractions. 2. Cancel if possible. 3. Multiply as numerators and denominators.	$1\frac{3}{5} \cdot 3\frac{1}{3} = \frac{8}{\underset{1}{\cancel{5}}} \cdot \frac{\overset{2}{\cancel{10}}}{3}$ $= \frac{8}{1} \cdot \frac{2}{3}$ $= \frac{16}{3} = 5\frac{1}{3}$
2.8 Dividing Mixed Numbers	Change mixed numbers to improper fractions. Invert the divisor, cancel if possible, and multiply numerators and denominators.	$3\frac{5}{9} \div 2\frac{2}{5} = \frac{32}{9} \div \frac{12}{5}$ $= \frac{\overset{8}{\cancel{32}}}{9} \cdot \frac{5}{\underset{3}{\cancel{12}}} = \frac{40}{27} = 1\frac{13}{27}$

name date hour

CHAPTER 2 REVIEW EXERCISES

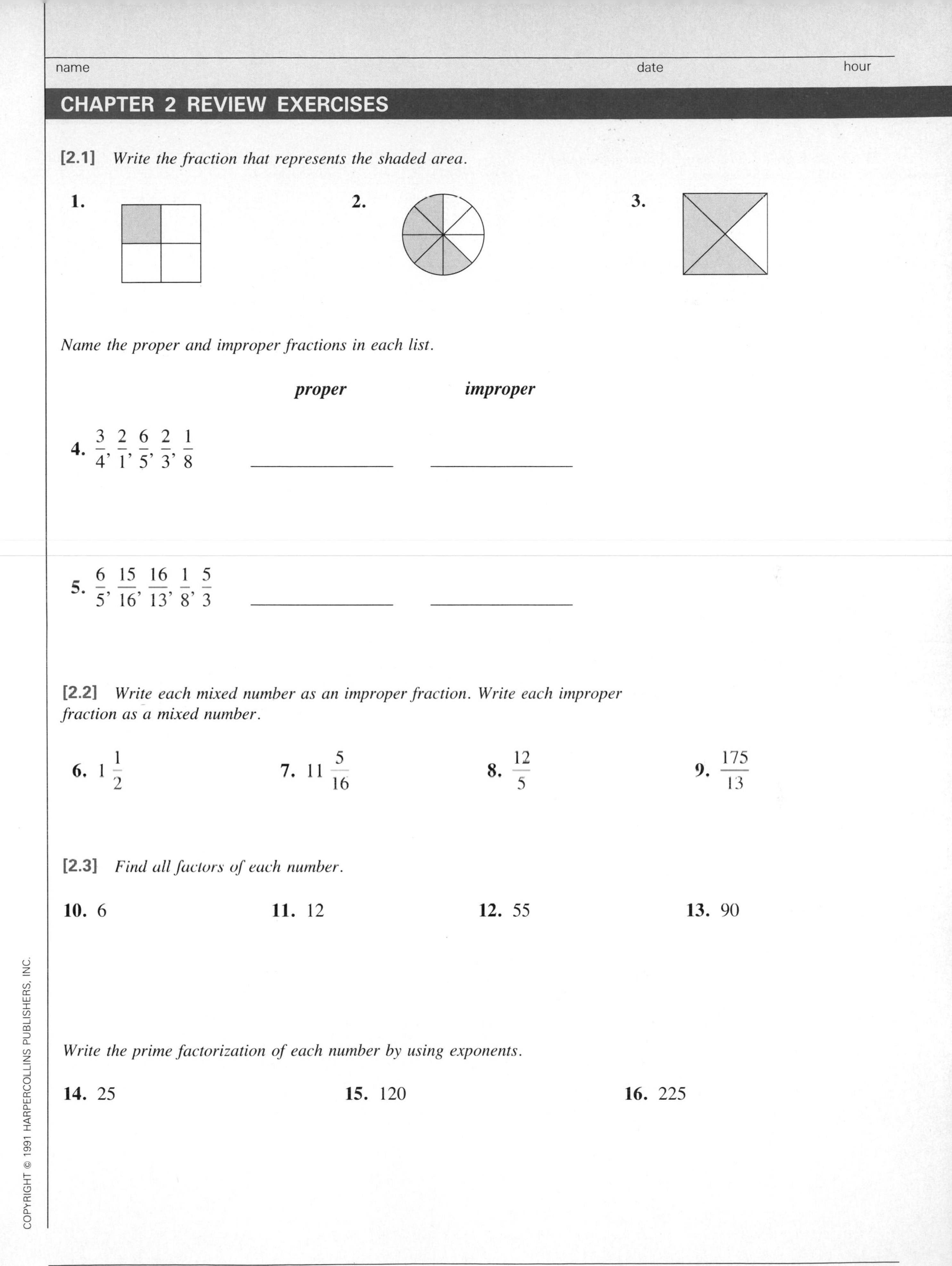

[2.1] *Write the fraction that represents the shaded area.*

1. **2.** **3.**

Name the proper and improper fractions in each list.

	proper	*improper*
4. $\frac{3}{4}, \frac{2}{1}, \frac{6}{5}, \frac{2}{3}, \frac{1}{8}$	______	______
5. $\frac{6}{5}, \frac{15}{16}, \frac{16}{13}, \frac{1}{8}, \frac{5}{3}$	______	______

[2.2] *Write each mixed number as an improper fraction. Write each improper fraction as a mixed number.*

6. $1\frac{1}{2}$ **7.** $11\frac{5}{16}$ **8.** $\frac{12}{5}$ **9.** $\frac{175}{13}$

[2.3] *Find all factors of each number.*

10. 6 **11.** 12 **12.** 55 **13.** 90

Write the prime factorization of each number by using exponents.

14. 25 **15.** 120 **16.** 225

Write each of the following without exponents.

17. 5^2 **18.** $2^2 \cdot 3^2$ **19.** $10^2 \cdot 3^3$ **20.** $3^5 \cdot 2^4$

[2.4] *Write each fraciton in lowest terms.*

21. $\frac{10}{15}$ **22.** $\frac{25}{35}$ **23.** $\frac{175}{190}$

Write the numerator and denominator of each fraction as a product of prime factors. Next, write the fraction in lowest terms.

24. $\frac{25}{60}$ **25.** $\frac{356}{480}$

Decide whether the following pairs of fractions are equivalent or not equivalent.

26. $\frac{3}{5}$ and $\frac{15}{25}$ **27.** $\frac{3}{8}$ and $\frac{15}{40}$

[2.5–2.8] *Multiply. Write all answers in lowest terms.*

28. $\frac{1}{5} \cdot \frac{4}{5}$ **29.** $\frac{3}{5} \cdot \frac{3}{8}$ **30.** $\frac{70}{175} \cdot \frac{5}{14}$

31. $22 \cdot \frac{3}{11}$ **32.** $\frac{5}{16} \cdot 48$ **33.** $\frac{5}{8} \cdot 1000$

Divide. Write all answers in lowest terms.

34. $\frac{2}{3} \div \frac{3}{4}$ **35.** $\frac{5}{6} \div \frac{1}{2}$ **36.** $\dfrac{\frac{15}{18}}{\frac{10}{30}}$

name date hour

37. $\frac{17}{20} \div \frac{34}{80}$

38. $10 \div \frac{5}{8}$

39. $8 \div \frac{6}{7}$

40. $1 \div \frac{7}{8}$

41. $\frac{2}{3} \div 5$

42. $\frac{\frac{12}{13}}{3}$

Find the area of each rectangle.

43. **44.**

45. Find the area of a rectangle having a length of 12 yards and a width of $\frac{1}{3}$ yard.

46. Find the area of a rectangle having a length of 40 feet and a width of $\frac{5}{8}$ foot.

Multiply. Write answers as mixed numbers.

47. $3\frac{1}{2} \cdot 1\frac{1}{8}$

48. $2\frac{1}{4} \cdot 7\frac{1}{8} \cdot 1\frac{1}{3}$

Divide. Write answers in lowest terms.

49. $12\frac{1}{3} \div 2$

50. $3\frac{1}{8} \div 5\frac{5}{7}$

Solve each of the following word problems by using multiplication or division.

51. How many $\frac{3}{4}$-pound peanut cans can be filled with 12 pounds of peanuts?

52. An estate is divided so that 5 children each receive equal shares of $\frac{2}{3}$ of the estate. What fraction of the total estate will each receive?

53. How many horse blankets can be made from $78\frac{3}{4}$ yards of material if each blanket requires $4\frac{3}{8}$ yards?

54. Mike Johnson worked $16\frac{1}{4}$ hours at $6 per hour. How much money did he earn?

55. Joanna Wilson purchases 100 pounds of detergent. After selling $\frac{1}{2}$ of this to her neighbor, she gives $\frac{2}{5}$ of the remaining detergent to her parents. How many pounds of detergent does she have remaining?

56. Jennifer Toms ran $\frac{2}{5}$ of the $\frac{3}{4}$-mile race. how many miles did she run?

57. A company will share $\frac{5}{8}$ of its profits with 4 employees. What fraction of the total profits will each employee receive?

name date hour

MIXED REVIEW EXERCISES

Multiply or divide as indicated. Write all answers in lowest terms.

58. $\frac{1}{2} \cdot \frac{2}{3}$

59. $\frac{7}{8} \div \frac{3}{4}$

60. $12\frac{1}{2} \cdot 1\frac{2}{3}$

61. $\frac{\frac{7}{8}}{6}$

62. $\frac{42}{56} \cdot \frac{71}{84}$

63. $\frac{25}{31} \div \frac{50}{93}$

64. $\frac{15}{31} \cdot 62$

65. $3\frac{1}{4} \div 1\frac{1}{2}$

Write each mixed number as an improper fraction. Write each improper fraction as a mixed number.

66. $\frac{5}{2}$

67. $1\frac{6}{7}$

68. $\frac{198}{4}$

69. $12\frac{19}{21}$

Write the numerator and denominator of each fraction as a product of prime factors, and then write the fraction in lowest terms.

70. $\frac{8}{12}$

71. $\frac{108}{210}$

Write each fraction in lowest terms.

72. $\frac{10}{30}$

73. $\frac{75}{200}$

74. $\frac{29}{87}$

Solve each of the following word problems.

75. The directions on a can of fabric glue say to apply $3\frac{1}{2}$ ounces of glue per square yard. How many ounces are needed for $43\frac{5}{9}$ square yards?

76. A water tank contains 35 gallons when it is $\frac{5}{8}$ full. How much water will it hold when it is full?

77. Each office requires $\frac{7}{8}$ roll of carpet. How many offices can be carpeted with 84 rolls of carpet?

78. A postage stamp is $\frac{2}{3}$ inch by $\frac{3}{4}$ inch. Find its area.

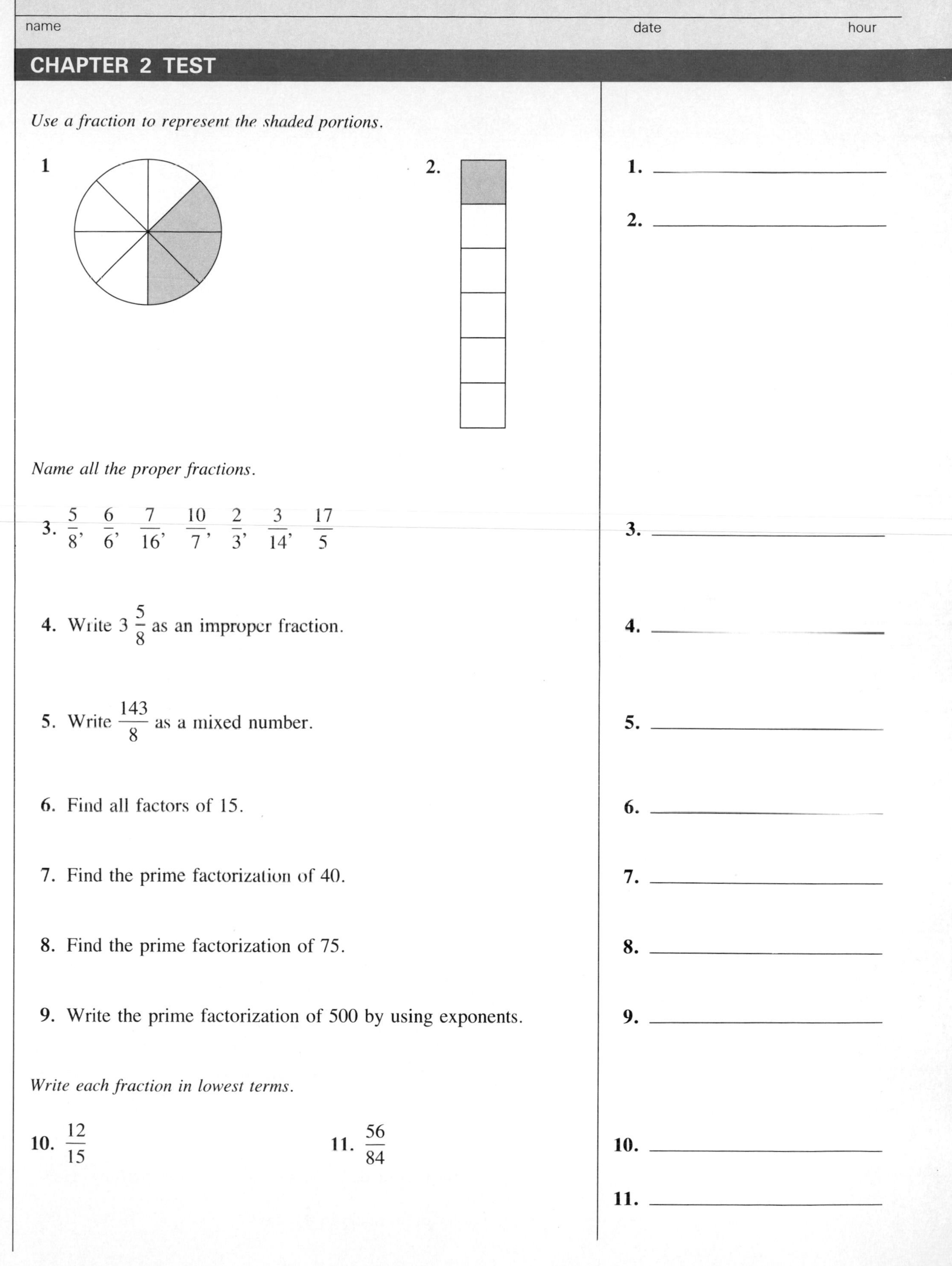

name date hour

CHAPTER 2 TEST

Use a fraction to represent the shaded portions.

1.

2.

1. ______

2. ______

Name all the proper fractions.

3. $\frac{5}{8}, \frac{6}{6}, \frac{7}{16}, \frac{10}{7}, \frac{2}{3}, \frac{3}{14}, \frac{17}{5}$

3. ______

4. Write $3\frac{5}{8}$ as an improper fraction.

4. ______

5. Write $\frac{143}{8}$ as a mixed number.

5. ______

6. Find all factors of 15.

6. ______

7. Find the prime factorization of 40.

7. ______

8. Find the prime factorization of 75.

8. ______

9. Write the prime factorization of 500 by using exponents.

9. ______

Write each fraction in lowest terms.

10. $\frac{12}{15}$

11. $\frac{56}{84}$

10. ______

11. ______

12. ____________

13. ____________

14. ____________

15. ____________

16. ____________

17. ____________

18. ____________

19. ____________

20. ____________

21. ____________

22. ____________

23. ____________

24. ____________

25. ____________

Multiply. Write all answers in lowest terms.

12. $\frac{5}{6} \cdot \frac{1}{3}$

13. $\frac{6}{7} \cdot \frac{2}{9}$

14. $6 \cdot \frac{4}{5}$

15. $24 \cdot \frac{3}{4}$

16. Find the area of a rectangle measuring $\frac{4}{5}$ inch by $\frac{5}{16}$ inch.

17. There are 230 students in the cafeteria. If $\frac{7}{10}$ of them are drinking coffee, how many students are drinking coffee?

Divide. Write all answers in lowest terms.

18. $\frac{3}{5} \div \frac{8}{10}$

19. $\dfrac{7}{\frac{4}{9}}$

20. There are 75 drums of grease on the maintenance platform. If each maintenance truck has a container that holds $\frac{3}{5}$ of a drum of grease, how many maintenance trucks can be filled?

Multiply. Write answers in lowest terms.

21. $5\frac{1}{4} \cdot 3\frac{3}{8}$

22. $1\frac{5}{6} \cdot 4\frac{1}{3}$

Divide. Write answers in lowest terms.

23. $4\frac{4}{5} \div 1\frac{1}{8}$

24. $\dfrac{9\frac{1}{2}}{3\frac{1}{3}}$

25. A rare chemical is made at the rate of $1\frac{1}{4}$ grams per day. How many grams can be made in $8\frac{1}{2}$ days?

ADDING AND SUBTRACTING FRACTIONS

3.1 ADDING AND SUBTRACTING LIKE FRACTIONS

OBJECTIVES

1. Define like and unlike fractions.
2. Add like fractions.
3. Subtract like fractions.

1 Fractions with the same denominators are **like fractions.** Fractions with different denominators are **unlike fractions.**

EXAMPLE 1 Identifying Like and Unlike Fractions

(a) $\frac{3}{4}, \frac{1}{4}, \frac{5}{4}, \frac{6}{4},$ and $\frac{4}{4}$ are like fractions.

All denominators are the same.

(b) $\frac{7}{12}$ and $\frac{12}{7}$ are unlike fractions.

different denominators ■

WORK PROBLEM 1 AT THE SIDE.

1. Write *like* or *unlike* for each pair of fractions.

(a) $\frac{9}{8}$ $\frac{1}{8}$

(b) $\frac{6}{5}$ $\frac{4}{10}$

(c) $\frac{11}{12}$ $\frac{9}{12}$

(d) $\frac{8}{3}$ $\frac{7}{4}$

2 The following figures show how to add the fractions $\frac{2}{7}$ and $\frac{4}{7}$.

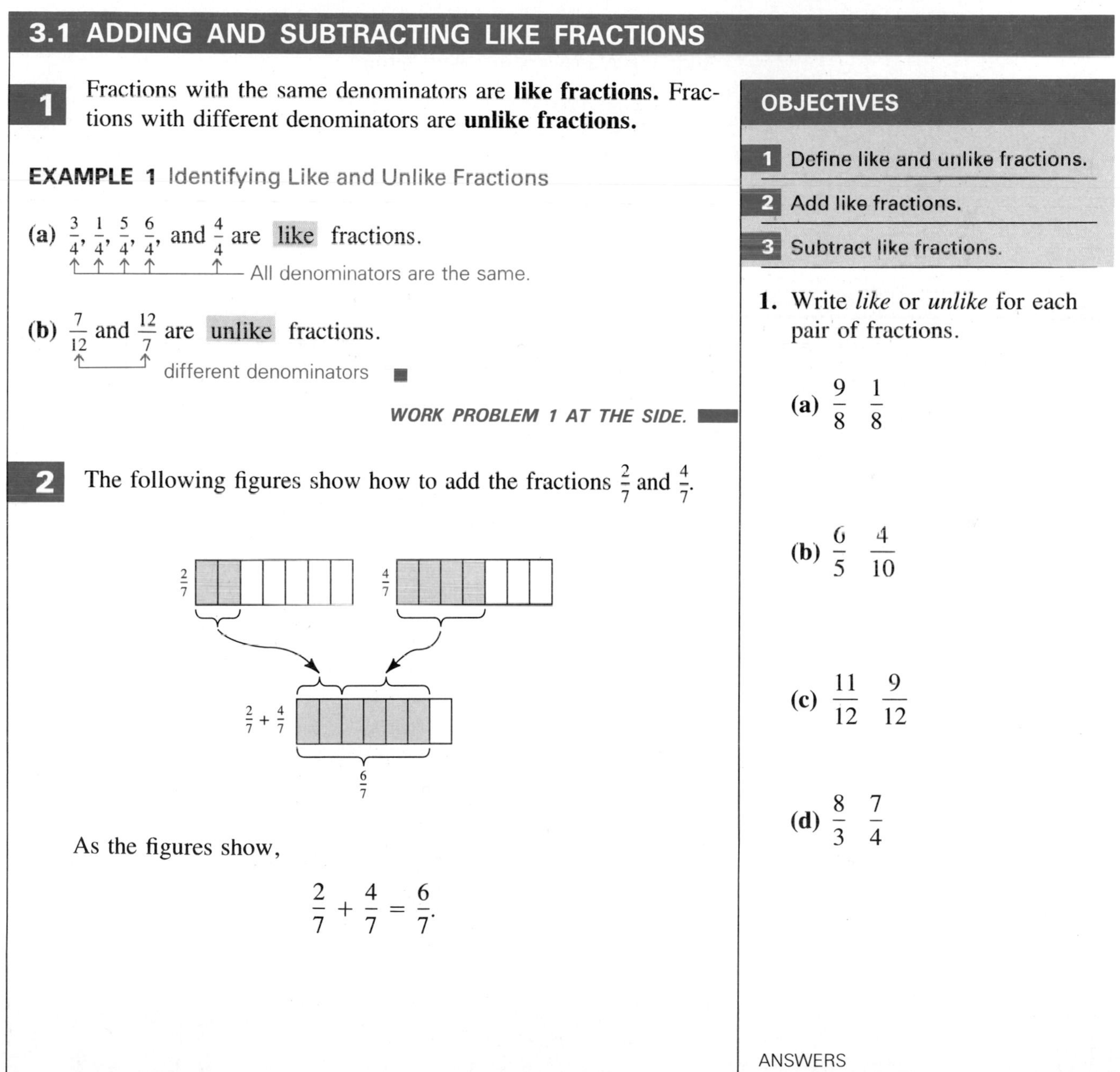

As the figures show,

$$\frac{2}{7} + \frac{4}{7} = \frac{6}{7}.$$

ANSWERS
1. (a) like (b) unlike (c) like (d) unlike

2. Add. Write answers in lowest terms.

(a) $\frac{1}{4} + \frac{2}{4}$

(b) $\frac{5}{9} + \frac{2}{9}$

(c) $\frac{3}{8} + \frac{1}{8}$

(d) $\frac{1}{9} + \frac{2}{9} + \frac{3}{9}$

ANSWERS

2. (a) $\frac{3}{4}$ (b) $\frac{7}{9}$ (c) $\frac{1}{2}$ (d) $\frac{2}{3}$

Add like fractions as follows.

The numerator of the answer (the **sum**) is found by adding the numerators.

The denominator of the sum is the denominator of the like fractions.

Always write the answer in lowest terms.

EXAMPLE 2 Adding Like Fractions

Add. Write answers in lowest terms.

(a) $\frac{1}{5} + \frac{2}{5}$

(b) $\frac{1}{12} + \frac{7}{12} + \frac{1}{12}$

SOLUTION

(a) $\frac{1}{5} + \frac{2}{5} = \frac{1+2}{5} = \frac{3}{5}$ ← Add numerators. ← same denominator

(b) $\frac{1}{12} + \frac{7}{12} + \frac{1}{12} = \frac{1+7+1}{12}$

$= \frac{9}{12} = \frac{3}{4}$ (in lowest terms) ■

WORK PROBLEM 2 AT THE SIDE.

3 The figures show $\frac{7}{8}$ broken into $\frac{4}{8}$ and $\frac{3}{8}$.

$\frac{7}{8}$

$\frac{3}{8}$

$\frac{4}{8}$

Subtracting $\frac{3}{8}$ from $\frac{7}{8}$ gives the answer $\frac{4}{8}$, or

$$\frac{7}{8} - \frac{3}{8} = \frac{4}{8}.$$

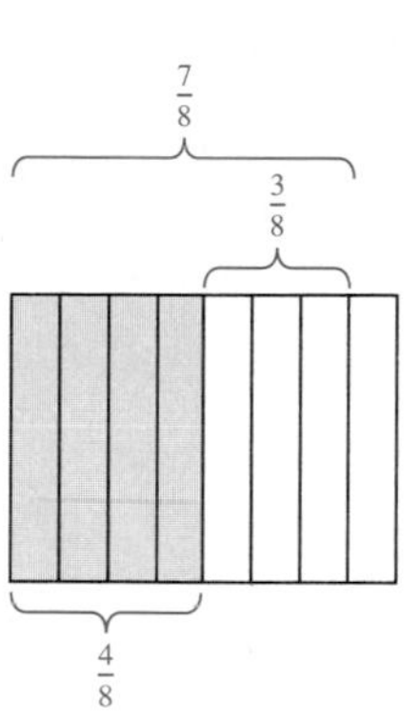

Write in lowest terms.

$$\frac{7}{8} - \frac{3}{8} = \frac{4}{8} = \frac{1}{2}$$

The rule for subtracting like fractions is very similar to that for addition.

The numerator of the answer (the **difference**) is found by subtracting the numerators.

The denominator of the difference is the denominator of the like fractions.

Always write the answer in lowest terms.

EXAMPLE 3 Subtracting Like Fractions
Subtract. Write answers in lowest terms.

$$\frac{11}{12} - \frac{7}{12}$$

3. Subtract. Write answers in lowest terms.

(a) $\frac{11}{15} - \frac{4}{15}$

(b) $\begin{array}{r} \frac{7}{8} \\ -\ \frac{5}{8} \\ \hline \end{array}$

(c) $\frac{21}{3} - \frac{19}{3}$

(d) $\begin{array}{r} \frac{101}{145} \\ -\ \frac{17}{145} \\ \hline \end{array}$

ANSWERS

3. (a) $\frac{7}{15}$ (b) $\frac{1}{4}$ (c) $\frac{2}{3}$ (d) $\frac{84}{145}$

SOLUTION

$$\frac{11}{12} - \frac{7}{12} = \frac{11 - 7}{12} \quad \begin{array}{l} \leftarrow \text{Subtract numerators.} \\ \leftarrow \text{denominator of like fractions} \end{array}$$

$$= \frac{4}{12}$$

Write in lowest terms.

$$\frac{11}{12} - \frac{7}{12} = \frac{4}{12} = \frac{1}{3} \quad ■$$

WORK PROBLEM 3 AT THE SIDE.

name date hour

3.1 EXERCISES

Add. Write answers in lowest terms.

Example: $\frac{2}{9} + \frac{4}{9}$ **Solution:** Add numerators. $\frac{2}{9} + \frac{4}{9} = \frac{2 + 4}{9} = \frac{6}{9}$

Write in lowest terms. $\frac{6}{9} = \mathbf{\frac{2}{3}}$

1. $\frac{1}{3} + \frac{1}{3}$

2. $\frac{3}{8} + \frac{4}{8}$

3. $\frac{1}{10} + \frac{6}{10}$

4. $\frac{9}{11} + \frac{1}{11}$

5. $\frac{1}{4} + \frac{1}{4}$

6. $\frac{1}{8} + \frac{1}{8}$

7. $\begin{array}{r} \frac{1}{9} \\ + \frac{2}{9} \\ \hline \end{array}$

8. $\begin{array}{r} \frac{4}{10} \\ + \frac{1}{10} \\ \hline \end{array}$

9. $\begin{array}{r} \frac{7}{12} \\ + \frac{3}{12} \\ \hline \end{array}$

10. $\frac{8}{15} + \frac{4}{15}$

11. $\frac{6}{20} + \frac{4}{20} + \frac{3}{20}$

12. $\frac{1}{7} + \frac{2}{7} + \frac{3}{7}$

13. $\frac{3}{17} + \frac{2}{17} + \frac{5}{17}$

14. $\frac{5}{11} + \frac{1}{11} + \frac{4}{11}$

15. $\frac{3}{8} + \frac{1}{8} + \frac{2}{8}$

16. $\frac{4}{9} + \frac{1}{9} + \frac{1}{9}$

17. $\frac{2}{54} + \frac{8}{54} + \frac{12}{54}$

18. $\frac{7}{120} + \frac{9}{120} + \frac{18}{120}$

Subtract. Write answers in lowest terms.

Example: $\frac{9}{12} - \frac{5}{12}$ **Solution:** Subtract numerators. $\frac{9}{12} - \frac{5}{12} = \frac{9 - 5}{12} = \frac{4}{12}$

Write in lowest terms. $\frac{4}{12} = \mathbf{\frac{1}{3}}$

19. $\frac{8}{11} - \frac{3}{11}$

20. $\frac{5}{17} - \frac{1}{17}$

21. $\frac{16}{21} - \frac{8}{21}$

22. $\frac{28}{32} - \frac{19}{32}$

23. $\frac{9}{10} - \frac{3}{10}$

24. $\frac{7}{8} - \frac{5}{8}$

25. $\dfrac{14}{15} - \dfrac{4}{15}$

26. $\dfrac{8}{25} - \dfrac{3}{25}$

27. $\dfrac{27}{40} - \dfrac{19}{40}$

28. $\dfrac{38}{55} - \dfrac{16}{55}$

29. $\dfrac{43}{72} - \dfrac{25}{72}$

30. $\dfrac{71}{100} - \dfrac{31}{100}$

31. $\dfrac{87}{144} - \dfrac{71}{144}$

32. $\dfrac{356}{220} - \dfrac{235}{220}$

33. $\dfrac{746}{400} - \dfrac{506}{400}$

Solve each word problem. Write answers in lowest terms.

34. Last month the Jeffersons paid $\frac{2}{12}$ of a debt. This month they paid an additional $\frac{5}{12}$ of the same debt. What fraction of the debt has been paid?

35. Margaret walked $\frac{5}{12}$ mile downhill and then $\frac{1}{12}$ mile along a creek. How far did she walk altogether?

36. The Thompsons owe $\frac{11}{16}$ of a debt. If they pay $\frac{5}{16}$ of it this month, what fraction of the debt will they still owe?

37. Sam must walk $\frac{11}{12}$ of a mile. He has already walked $\frac{5}{12}$ of a mile. How much farther must he walk?

CHALLENGE EXERCISES

38. Matt purchased $\frac{3}{8}$ acre of land one year and $\frac{7}{8}$ acre the next year. He then sold $\frac{5}{8}$ acre of land. How much land does he now have?

39. A forester planted $\frac{5}{12}$ acre in seedlings in the morning and $\frac{11}{12}$ acre in the afternoon. If $\frac{7}{12}$ acre of seedlings were destroyed by frost, how many acres remained?

REVIEW EXERCISES

Find the prime factorization. Do not use exponents. (For help, see Section 2.3.)

40. 6 **41.** 10 **42.** 30

43. 100 **44.** 45 **45.** 75

3.2 LEAST COMMON MULTIPLES

Only *like* fractions can be added or subtracted. Because of this, *unlike* fractions must be rewritten as like fractions before adding or subtracting.

Do this with the *least common multiple* of the denominators.

The **least common multiple (LCM)** of two whole numbers is the smallest whole number divisible by both those numbers.

EXAMPLE 1 Finding the Multiples of a Number

The list shows multiples of 6.

6, 12, 18, 24, 30, . . .

(The three dots show the list continues in the same pattern without stopping.) The next list shows multiples of 9.

9, 18, 27, 36, 45, . . .

The smallest number in both lists is 18, so 18 is the least common multiple of 6 and 9; the number 18 is the smallest whole number divisible by both 6 and 9.

multiples of 6 — 6, 12, 18, 24, 30, . . .
multiples of 9 — 9, 18, 27, 36, 45, . . .

18 is the smallest number in both lists. ■

WORK PROBLEM 1 AT THE SIDE.

2 Example 1 shows how to find the least common multiple by making a list of the common multiples of each number. Although this method works, it is usually easier to find the least common multiple by using prime factorization, as shown in the next example.

EXAMPLE 2 Applying Prime Factorization Knowledge

Use prime factorization to find the least common multiple of 18 and 60.

SOLUTION

Start by finding the prime factorization of each number.

$$18 = 2 \cdot 3 \cdot 3 \qquad 60 = 2 \cdot 2 \cdot 3 \cdot 5$$

Place the factorizations in a table, as shown below.

prime	2	3	5
18 =	$2 \cdot$	$3 \cdot 3$	
60 =	$2 \cdot 2 \cdot$	$3 \cdot$	5

OBJECTIVES

1. Find the least common multiple.
2. Find the least common multiple by using prime factorization and a table.
3. Find the least common multiple by using an alternative method.
4. Write a fraction with an indicated denominator.

1. (a) List the multiples of 10.
10, ____, ____, ____, ____, ____

(b) List the multiples of 12.
12, ____, ____, ____, ____

(c) Find the least common multiple of 10 and 12.

ANSWERS
1. (a) 20, 30, 40, 50, 60, . . . (b) 24, 36, 48, 60, . . . (c) 60

4. Find the least common multiple for each set of numbers.

(a) 7, 21

(b) 8, 9, 12

(c) 18, 20, 30

(d) 12, 15, 18, 45

ANSWERS
4. (a) 21 (b) 72 (c) 180 (d) 180

Circle the largest product in each column.

prime	2	3	5	7
20 =	2 · 2 ·		(5)	
24 =	(2 · 2 · 2) ·	3		
42 =	2 ·	(3) ·		(7)
LCM =	(2 · 2 · 2)	(3)	(5)	(7)

least common multiple (LCM) = **2 · 2 · 2 · 3 · 5 · 7 = 840** ■

WORK PROBLEM 4 AT THE SIDE.

3 Some people like the following *alternative method* for finding the least common multiple. Try both methods, and use the one you prefer. As a review, a list of the first few primes follows.

2, 3, 5, 7, 11, 13, 17

EXAMPLE 5 Alternative Method for Finding the Least Common Multiple

Find the least common multiple of each set of numbers.

(a) 14 and 21 **(b)** 6, 15, 18

SOLUTION

(a) Start by trying to divide 14 and 21 by the prime numbers listed previously. Use the following shortcut.

Divide by 2, the first prime.

```
2 | 14  21̸
    7   21
```

Since 21 cannot be divided evenly by 2, cross 21 out and bring it down.

Divide by 3, the second prime.

```
2 | 14  21̸
3 |  7̸  21
     7   7
```

Since 7 cannot be divided evenly by 5, the third prime, skip 5, and divide by the next prime, 7.

Divide by 7, the fourth prime.

```
2 | 14  21̸
3 |  7̸  21
7 |  7   7
     1   1
```

All quotients are 1.

3.2 LEAST COMMON MULTIPLES

Only *like* fractions can be added or subtracted. Because of this, *unlike* fractions must be rewritten as like fractions before adding or subtracting.

1 Do this with the *least common multiple* of the denominators.

The **least common multiple (LCM)** of two whole numbers is the smallest whole number divisible by both those numbers.

EXAMPLE 1 Finding the Multiples of a Number
The list shows multiples of 6.

6, 12, 18, 24, 30, . . .

(The three dots show the list continues in the same pattern without stopping.) The next list shows multiples of 9.

9, 18, 27, 36, 45, . . .

The smallest number in both lists is 18, so 18 is the least common multiple of 6 and 9; the number 18 is the smallest whole number divisible by both 6 and 9.

multiples of 6 6, 12, 18, 24, 30, . . .
multiples of 9 9, 18, 27, 36, 45, . . .

18 is the smallest number in both lists. ■

WORK PROBLEM 1 AT THE SIDE.

2 Example 1 shows how to find the least common multiple by making a list of the common multiples of each number. Although this method works, it is usually easier to find the least common multiple by using prime factorization, as shown in the next example.

EXAMPLE 2 Applying Prime Factorization Knowledge
Use prime factorization to find the least common multiple of 18 and 60.

SOLUTION
Start by finding the prime factorization of each number.

$$18 = 2 \cdot 3 \cdot 3 \qquad 60 = 2 \cdot 2 \cdot 3 \cdot 5$$

Place the factorizations in a table, as shown below.

prime	2	3	5
18 =	$2 \cdot$	$3 \cdot 3$	
60 =	$2 \cdot 2 \cdot$	$3 \cdot$	5

OBJECTIVES

1. Find the least common multiple.
2. Find the least common multiple by using prime factorization and a table.
3. Find the least common multiple by using an alternative method.
4. Write a fraction with an indicated denominator.

1. (a) List the multiples of 10.
10, ____, ____, ____, ____, ____

(b) List the multiples of 12.
12, ____, ____, ____, ____

(c) Find the least common multiple of 10 and 12.

ANSWERS
1. (a) 20, 30, 40, 50, 60, . . . (b) 24, 36, 48, 60, . . . (c) 60

Circle the largest product in each column.

prime	2	3	5
18 =	2 ·	3 · 3	
60 =	2 · 2 ·	3 ·	5
LCM =	2 · 2	3 · 3	5

Now multiply the circled products to find the least common multiple.

least common multiple (LCM) = **2 · 2 · 3 · 3 · 5 = 180**

The smallest whole number divisible by both 18 and 60 is 180. ■

WORK PROBLEM 2 AT THE SIDE.

EXAMPLE 3 Using Prime Factorization

Find the least common multiple of 12, 18, and 40.

SOLUTION

Write each prime factorization.

12 = 2 · 2 · 3 18 = 2 · 3 · 3 40 = 2 · 2 · 2 · 5

Prepare the following table.

prime	2	3	5
12 =	2 · 2 ·	3	
18 =	2 ·	3 · 3	
40 =	2 · 2 · 2 ·		5

Circle the largest product in each column.

prime	2	3	5
12 =	2 · 2 ·	3	
18 =	2 ·	3 · 3	
40 =	2 · 2 · 2 ·		5
LCM =	2 · 2 · 2	3 · 3	5

(If two products in a column are equal, circle either one.)

Now multiply the circled products.

least common multiple (LCM) = **2 · 2 · 2 · 3 · 3 · 5 = 360**

The smallest whole number divisible by 12, 18, and 40 is 360. ■

WORK PROBLEM 3 AT THE SIDE.

2. Find the least common multiple of 36 and 54.

(a) Find the prime factorization of each number.

(b) Complete this table.

prime	2	3
36 =		
54 =		

(c) Identify the largest product in each column.

prime	2	3
36 =	2 · 2 ·	3 · 3
54 =	2 ·	3 · 3 · 3

(d) Find the least common multiple.

2 · 2 · ____ · ____ · ____ =

3. Find the least common multiple of the denominators in these fractions.

(a) $\frac{1}{2}$ and $\frac{1}{4}$

(b) $\frac{3}{10}$ and $\frac{6}{5}$

(c) $\frac{5}{9}$ and $\frac{1}{6}$

(d) $\frac{2}{15}$ and $\frac{3}{10}$

ANSWERS

2. (a) 36 = 2 · 2 · 3 · 3
54 = 2 · 3 · 3 · 3
(b)

2	3
2 · 2 · 3 · 3	
2	3 · 3 · 3

(c) 2 · 2; 3 · 3 · 3 (d) 3, 3, 3; 108

3. (a) 4 (b) 10 (c) 18 (d) 30

EXAMPLE 4 Finding the Least Common Multiple
Find the least common multiple for each set of numbers.

(a) 5 and 35 **(b)** 20, 24, 42

SOLUTION
(a) Write each prime factorization.

$$5 = 5 \qquad 35 = 5 \cdot 7$$

Prepare the table.

prime	5	7
5 =	5	
35 =	5 ·	7

Circle the largest product in each column, in this case, the only prime number in the column.

prime	5	7
5 =	5	
35 =	(5) ·	(7)
LCM =	(5)	(7)

least common multiple (LCM) = **5 · 7 = 35**

(b) Write each prime factorization

$$20 = 2 \cdot 2 \cdot 5 \qquad 24 = 2 \cdot 2 \cdot 2 \cdot 3 \qquad 42 = 2 \cdot 3 \cdot 7$$

Prepare the table.

prime	2	3	5	7
20 =	2 · 2 ·		5	
24 =	2 · 2 · 2 ·	3		
42 =	2 ·	3 ·		7

4. Find the least common multiple for each set of numbers.

(a) 7, 21

(b) 8, 9, 12

(c) 18, 20, 30

(d) 12, 15, 18, 45

Circle the largest product in each column.

prime	2	3	5	7
20 =	2 · 2 ·		⑤	
24 =	(2 · 2 · 2) ·	3		
42 =	2 ·	③ ·		⑦
LCM =	(2 · 2 · 2)	③	⑤	⑦

least common multiple (LCM) = **2 · 2 · 2 · 3 · 5 · 7 = 840** ■

WORK PROBLEM 4 AT THE SIDE.

3 Some people like the following *alternative method* for finding the least common multiple. Try both methods, and use the one you prefer. As a review, a list of the first few primes follows.

2, 3, 5, 7, 11, 13, 17

EXAMPLE 5 Alternative Method for Finding the Least Common Multiple

Find the least common multiple of each set of numbers.

(a) 14 and 21 **(b)** 6, 15, 18

SOLUTION

(a) Start by trying to divide 14 and 21 by the prime numbers listed previously. Use the following shortcut.

Divide by 2, the first prime.

$$\begin{array}{r|rr} 2 & 14 & \cancel{21} \\ & 7 & 21 \end{array}$$

Since 21 cannot be divided evenly by 2, cross 21 out and bring it down.

Divide by 3, the second prime.

$$\begin{array}{r|rr} 2 & 14 & \cancel{21} \\ 3 & \cancel{7} & 21 \\ & 7 & 7 \end{array}$$

Since 7 cannot be divided evenly by 5, the third prime, skip 5, and divide by the next prime, 7.

Divide by 7, the fourth prime.

$$\begin{array}{r|rr} 2 & 14 & \cancel{21} \\ 3 & \cancel{7} & 21 \\ 7 & 7 & 7 \\ & 1 & 1 \end{array}$$

All quotients are 1.

ANSWERS
4. (a) 21 (b) 72 (c) 180 (d) 180

When all quotients are 1, multiply the prime numbers on the left side.

least common multiple = $\mathbf{2 \cdot 3 \cdot 7 = 42}$

The least common multiple of 14 and 21 is 42.

(b) Divide by 2.

$$\begin{array}{r|rrr} 2 & 6 & \cancel{15} & 18 \\ \hline & 3 & 15 & 9 \end{array}$$

Cross out 15 and bring it down.

Divide by 3.

$$\begin{array}{r|rrr} 2 & 6 & \cancel{15} & 18 \\ \hline 3 & 3 & 15 & 9 \\ \hline & 1 & 5 & 3 \end{array}$$

Divide by 3, again.

$$\begin{array}{r|rrr} 2 & 6 & \cancel{15} & 18 \\ \hline 3 & 3 & 15 & 9 \\ \hline 3 & \cancel{1} & \cancel{5} & 3 \\ \hline & 1 & 5 & 1 \end{array}$$

Finally, divide by 5.

$$\begin{array}{r|rrr} 2 & 6 & \cancel{15} & 18 \\ \hline 3 & 3 & 15 & 9 \\ \hline 3 & \cancel{1} & \cancel{5} & 3 \\ \hline 5 & \cancel{1} & 5 & \cancel{1} \\ \hline & 1 & 1 & 1 \end{array}$$

All quotients are 1.

Multiply the prime numbers on the side.

$2 \cdot 3 \cdot 3 \cdot 5 = 90 \leftarrow$ least common multiple ■

WORK PROBLEMS 5 AND 6 AT THE SIDE.

4 The least common multiple often must be used as the denominator for a list of fractions.

EXAMPLE 5 Writing a Fraction with an Indicated Denominator

Write the fraction $\frac{2}{3}$ by using a denominator of 15.

SOLUTION

Find a numerator, so that

$$\frac{2}{3} = \frac{\quad}{15}.$$

To find the new numerator, first divide 15 by 3.

$$\frac{2}{3} = \frac{\quad}{15} \qquad 15 \div 3 = 5$$

5. In the problems below, the divisions have already been worked out. Multiply the prime numbers on the left to find the least common multiple.

(a)

$$\begin{array}{r|rr} 2 & 6 & 15 \\ \hline 3 & 3 & 15 \\ \hline 5 & 1 & 5 \\ \hline & 1 & 1 \end{array}$$

(b)

$$\begin{array}{r|rr} 2 & 30 & 25 \\ \hline 3 & 15 & 25 \\ \hline 5 & 5 & 25 \\ \hline 5 & 1 & 5 \\ \hline & 1 & 1 \end{array}$$

6. Find the least common multiple of each set of numbers.

(a) 12 and 15

(b) 30 and 45

(c) 4, 8, and 12

(d) 25, 20, 35

ANSWERS
5. (a) 30 (b) 150
6. (a) 60 (b) 90 (c) 24 (d) 700

7. Write each fraction by using the indicated denominator.

(a) $\frac{1}{2} = \frac{\quad}{12}$

(b) $\frac{7}{9} = \frac{\quad}{27}$

(c) $\frac{4}{5} = \frac{\quad}{50}$

(d) $\frac{6}{11} = \frac{\quad}{55}$

ANSWERS

7. (a) $\frac{6}{12}$ (b) $\frac{21}{27}$ (c) $\frac{40}{50}$ (d) $\frac{30}{55}$

Multiply both numerator and denominator of the fraction $\frac{2}{3}$ by 5.

$$\frac{2}{3} = \frac{2 \cdot 5}{3 \cdot 5} = \frac{10}{15} \quad ■$$

This process is just the opposite of writing a fraction in lowest terms. Check the answer by writing $\frac{10}{15}$ in lowest terms; you should get $\frac{2}{3}$.

EXAMPLE 6 Changing to a New Denominator

Write each fraction with the indicated denominator.

(a) $\frac{3}{8} = \frac{\quad}{48}$ **(b)** $\frac{5}{6} = \frac{\quad}{42}$

SOLUTION

(a) Divide 48 by 8, getting 6. Now multiply both numerator and denominator of $\frac{3}{8}$ by 6.

$$\frac{3}{8} = \frac{3 \cdot 6}{8 \cdot 6} = \frac{18}{48}$$

Multiply numerator and denominator by 6.

That is, $\frac{3}{8} = \frac{18}{48}$. As a check, write $\frac{18}{48}$ in lowest terms.

(b) Divide 42 by 6, getting 7. Next, multiply numerator and denominator of $\frac{5}{6}$ by 7.

$$\frac{5}{6} = \frac{5 \cdot 7}{6 \cdot 7} = \frac{35}{42}$$

Multiply numerator and denominator by 7. ■

WORK PROBLEM 7 AT THE SIDE.

name date hour

3.2 EXERCISES

Find the least common multiple of each set of numbers.

Example: 18, 30

Solution:
Complete a table.

prime	2	3	5
18 = 30 =	2 · 2 ·	3 · 3 3 ·	 5

Identify the largest product in each column.

prime	2	3	5
18 = 30 =	2 · (2)·	(3 · 3) 3 ·	 (5)
LCM =	(2)	(3 · 3)	(5)

The least common multiple (LCM) = **2 · 3 · 3 · 5 = 90.**

Alternative Method:

Divide by 2	2	18	30
Divide by 3	3	9	15
Divide by 3	3	3	~~5~~
Divide by 5	5	~~1~~	5
		1	1

Multiply the prime numbers on the side.

2 · 3 · 3 · 5 = 90 = least common multiple

1. 6, 12

2. 5, 10

3. 12, 15

4. 15, 18

5. 18, 24

6. 30, 40

7. 25, 40

8. 15, 35

9. 36, 45

10. 42, 60

11. 12, 18, 20

12. 9, 15, 20

13. 15, 24, 30

14. 12, 20, 24

15. 18, 20, 24

16. 20, 24, 30

17. 6, 8, 10, 12

18. 8, 9, 12, 18

19. 10, 15, 20, 25

20. 12, 15, 18, 20

21. 12, 18, 24, 36

22. 15, 20, 30, 40

23. 10, 30, 50, 60

24. 5, 18, 25, 30

Rewrite each of the following fractions, so that it has a denominator of 36.

Example: $\frac{7}{12} = \frac{21}{36}$

Solution: Divide 36 by 12, getting 3. Next, multiply both numerator and denominator by 3.

$$\frac{7}{12} = \frac{7 \cdot 3}{12 \cdot 3} = \frac{21}{36}$$

25. $\frac{1}{2} = \frac{\quad}{36}$

26. $\frac{5}{9} = \frac{\quad}{36}$

27. $\frac{3}{4} = \frac{\quad}{36}$

28. $\frac{5}{12} = \frac{\quad}{36}$

29. $\frac{3}{9} = \frac{\quad}{36}$

30. $\frac{1}{4} = \frac{\quad}{36}$

name date hour

Rewrite each of the following fractions with the indicated denominators.

31. $\frac{3}{8} = \frac{}{24}$

32. $\frac{5}{16} = \frac{}{32}$

33. $\frac{9}{10} = \frac{}{40}$

34. $\frac{7}{11} = \frac{}{44}$

35. $\frac{7}{8} = \frac{}{56}$

36. $\frac{5}{12} = \frac{}{48}$

37. $\frac{3}{10} = \frac{}{70}$

38. $\frac{7}{8} = \frac{}{96}$

39. $\frac{15}{19} = \frac{}{76}$

40. $\frac{6}{5} = \frac{}{40}$

41. $\frac{9}{7} = \frac{}{56}$

42. $\frac{3}{2} = \frac{}{64}$

43. $\frac{8}{3} = \frac{}{51}$

44. $\frac{9}{5} = \frac{}{120}$

45. $\frac{8}{11} = \frac{}{132}$

46. $\frac{7}{15} = \frac{}{210}$

CHALLENGE EXERCISES

Find the least common multiple of the denominators of the following fractions.

47. $\frac{17}{800}, \frac{23}{3600}$

48. $\frac{53}{288}, \frac{115}{1568}$

49. $\frac{109}{1512}, \frac{47}{392}$

50. $\frac{61}{810}, \frac{37}{1170}$

REVIEW EXERCISES

Write each improper fraction as a mixed number. (For help, see ***Section 2.2****.)*

51. $\frac{7}{5}$

52. $\frac{9}{2}$

53. $\frac{9}{5}$

54. $\frac{14}{11}$

55. $\frac{27}{7}$

56. $\frac{28}{9}$

3.3 ADDING AND SUBTRACTING UNLIKE FRACTIONS

OBJECTIVES

1. Add unlike fractions.
2. Add fractions vertically.
3. Subtract unlike fractions.

1 Unlike fractions can be added by first changing them to like fractions. For example, the diagrams show $\frac{3}{8}$ and $\frac{1}{4}$.

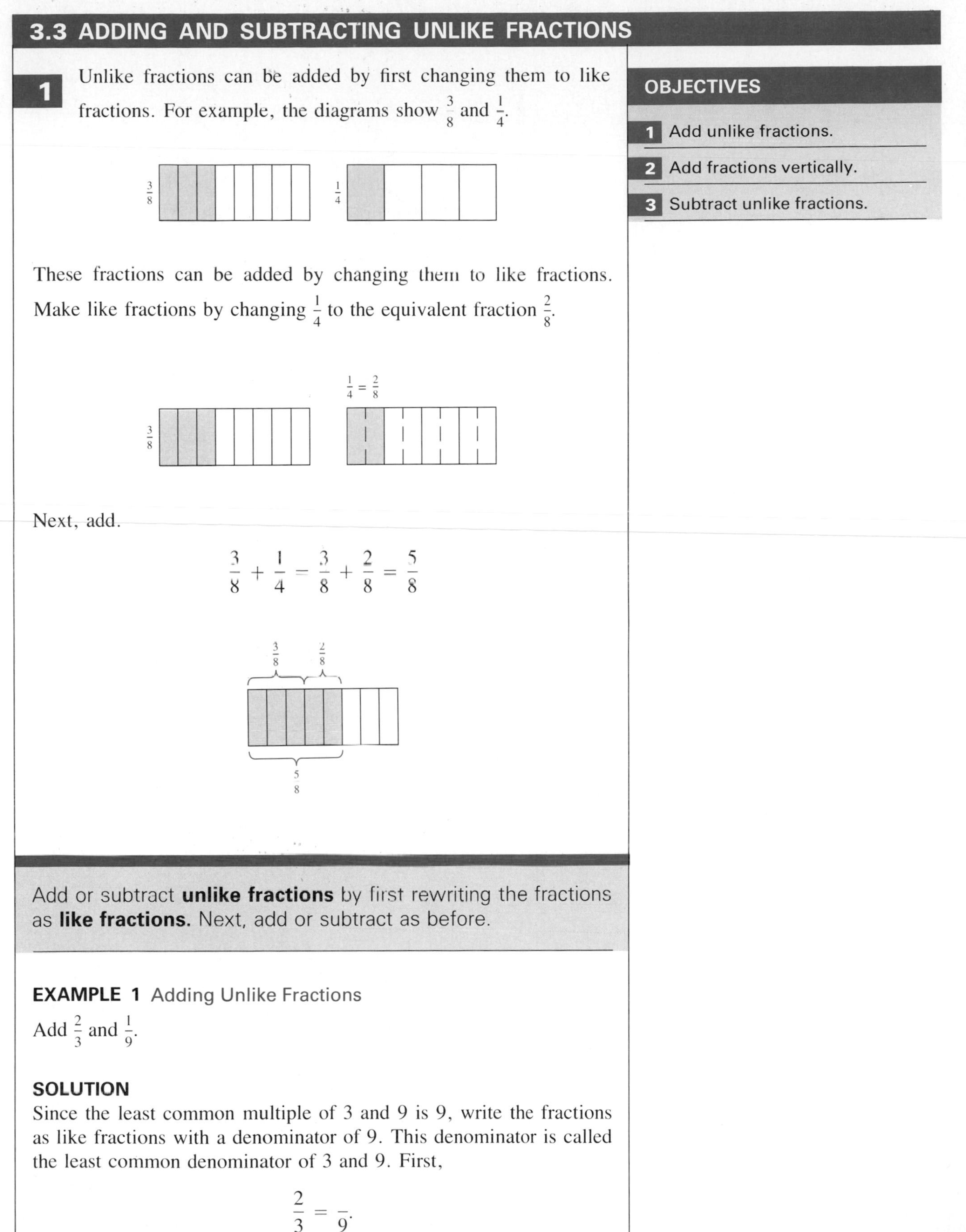

These fractions can be added by changing them to like fractions. Make like fractions by changing $\frac{1}{4}$ to the equivalent fraction $\frac{2}{8}$.

Next, add.

$$\frac{3}{8} + \frac{1}{4} = \frac{3}{8} + \frac{2}{8} = \frac{5}{8}$$

Add or subtract **unlike fractions** by first rewriting the fractions as **like fractions.** Next, add or subtract as before.

EXAMPLE 1 Adding Unlike Fractions

Add $\frac{2}{3}$ and $\frac{1}{9}$.

SOLUTION

Since the least common multiple of 3 and 9 is 9, write the fractions as like fractions with a denominator of 9. This denominator is called the least common denominator of 3 and 9. First,

$$\frac{2}{3} = \frac{}{9}.$$

Divide 9 by 3, getting 3. Next, multiply numerator and denominator by 3.

$$\frac{2}{3} = \frac{2 \cdot 3}{3 \cdot 3} = \frac{6}{9}$$

Next, add the like fractions $\frac{6}{9}$ and $\frac{1}{9}$.

$$\frac{2}{3} + \frac{1}{9} = \frac{6}{9} + \frac{1}{9} = \frac{6+1}{9} = \frac{7}{9} \quad ■$$

WORK PROBLEM 1 AT THE SIDE.

EXAMPLE 2 Adding Fractions

Add the following fractions. Write all answers in lowest terms.

(a) $\frac{1}{4} + \frac{1}{6}$ **(b)** $\frac{6}{15} + \frac{3}{10}$

SOLUTION

(a) The least common multiple of 4 and 6 is 12. Write both fractions as fractions with a least common denominator of 12.

$$\frac{1}{4} = \frac{3}{12} \quad \text{and} \quad \frac{1}{6} = \frac{2}{12}$$

Now add.

Add numerators.

$$\frac{1}{4} + \frac{1}{6} = \frac{3}{12} + \frac{2}{12} = \frac{3+2}{12} = \frac{5}{12}$$

(b) The least common multiple of 15 and 10 is 30, so write both fractions with a least common denominator of 30.

$$\frac{6}{15} + \frac{3}{10} = \frac{12}{30} + \frac{9}{30} = \frac{21}{30}$$

Write in lowest terms. $\frac{21}{30} = \frac{7}{10}$ ■

WORK PROBLEM 2 AT THE SIDE.

1. Add.

(a) $\frac{1}{4} + \frac{1}{2}$

(b) $\frac{3}{8} + \frac{1}{2}$

(c) $\frac{3}{10} + \frac{2}{5}$

(d) $\frac{1}{12} + \frac{5}{6}$

2. Add. Write answers in lowest terms.

(a) $\frac{5}{12} + \frac{1}{4}$

(b) $\frac{2}{15} + \frac{1}{6}$

(c) $\frac{1}{10} + \frac{1}{3} + \frac{1}{6}$

ANSWERS

1. (a) $\frac{3}{4}$ (b) $\frac{7}{8}$ (c) $\frac{7}{10}$ (d) $\frac{11}{12}$

2. (a) $\frac{2}{3}$ (b) $\frac{3}{10}$ (c) $\frac{3}{5}$

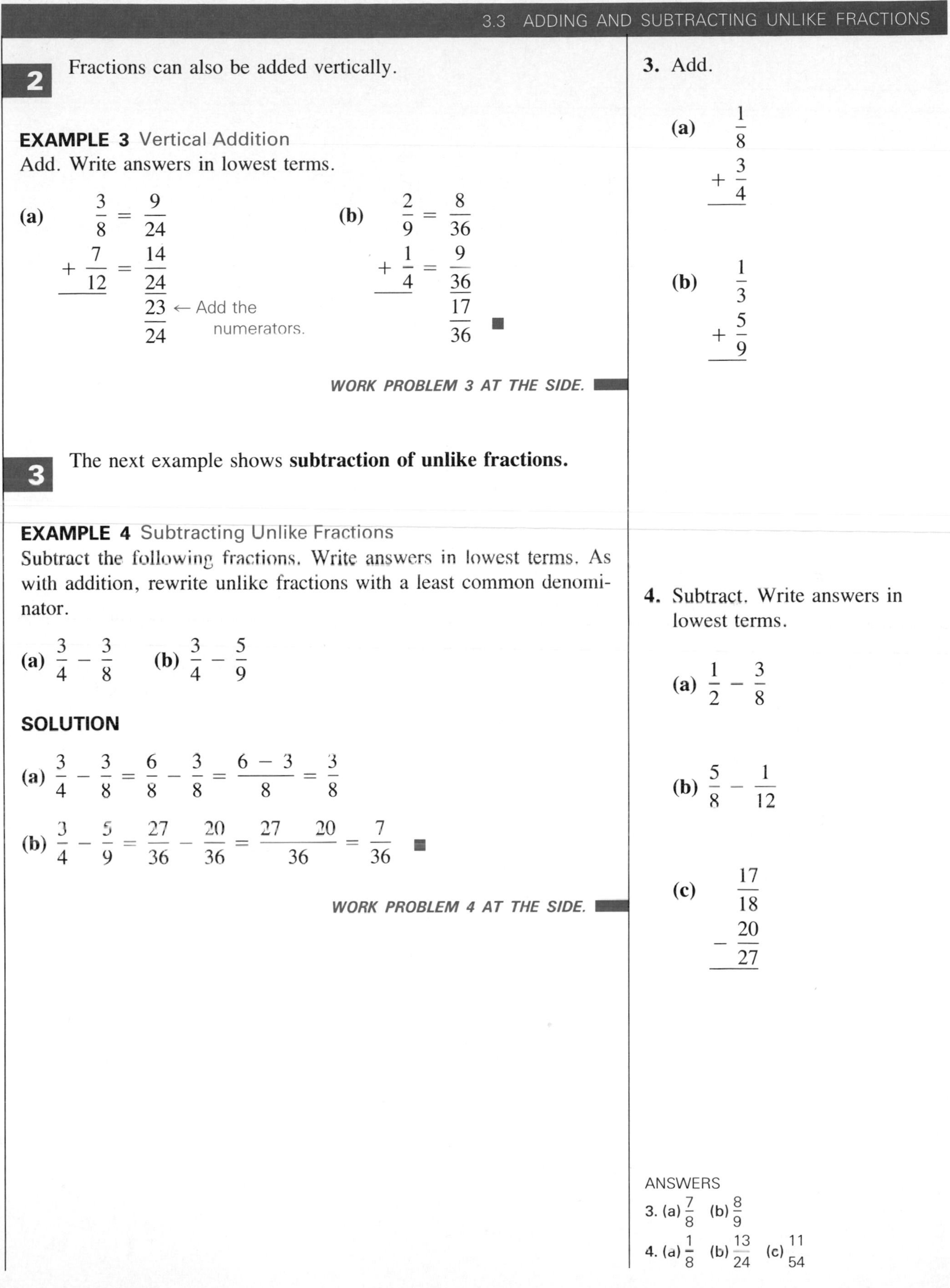

2 Fractions can also be added vertically.

EXAMPLE 3 Vertical Addition
Add. Write answers in lowest terms.

(a)
$$\begin{array}{rcl} \frac{3}{8} & = & \frac{9}{24} \\ + \frac{7}{12} & = & \frac{14}{24} \\ \hline & & \frac{23}{24} \end{array}$$
← Add the numerators.

(b)
$$\begin{array}{rcl} \frac{2}{9} & = & \frac{8}{36} \\ + \frac{1}{4} & = & \frac{9}{36} \\ \hline & & \frac{17}{36} \end{array}$$
■

WORK PROBLEM 3 AT THE SIDE.

3 The next example shows **subtraction of unlike fractions.**

EXAMPLE 4 Subtracting Unlike Fractions
Subtract the following fractions. Write answers in lowest terms. As with addition, rewrite unlike fractions with a least common denominator.

(a) $\frac{3}{4} - \frac{3}{8}$ **(b)** $\frac{3}{4} - \frac{5}{9}$

SOLUTION

(a) $\frac{3}{4} - \frac{3}{8} = \frac{6}{8} - \frac{3}{8} = \frac{6-3}{8} = \frac{3}{8}$

(b) $\frac{3}{4} - \frac{5}{9} = \frac{27}{36} - \frac{20}{36} = \frac{27 - 20}{36} = \frac{7}{36}$ ■

WORK PROBLEM 4 AT THE SIDE.

3. Add.

(a)
$$\begin{array}{r} \frac{1}{8} \\ + \frac{3}{4} \\ \hline \end{array}$$

(b)
$$\begin{array}{r} \frac{1}{3} \\ + \frac{5}{9} \\ \hline \end{array}$$

4. Subtract. Write answers in lowest terms.

(a) $\frac{1}{2} - \frac{3}{8}$

(b) $\frac{5}{8} - \frac{1}{12}$

(c)
$$\begin{array}{r} \frac{17}{18} \\ - \frac{20}{27} \\ \hline \end{array}$$

ANSWERS
3. (a) $\frac{7}{8}$ (b) $\frac{8}{9}$
4. (a) $\frac{1}{8}$ (b) $\frac{13}{24}$ (c) $\frac{11}{54}$

name date hour

3.3 EXERCISES

Add the following fractions. Write answers in lowest terms.

Example: $\frac{2}{3} + \frac{1}{6} = \frac{5}{6}$ **Solution:** $\frac{2}{3} + \frac{1}{6} = \frac{4}{6} + \frac{1}{6} = \frac{4+1}{6} = \frac{5}{6}$

Least common multiple is 6.

1. $\frac{3}{5} + \frac{1}{5}$ **2.** $\frac{2}{7} + \frac{4}{7}$ **3.** $\frac{9}{16} + \frac{3}{16}$

4. $\frac{5}{9} + \frac{1}{9}$ **5.** $\frac{7}{20} + \frac{3}{10}$ **6.** $\frac{5}{8} + \frac{1}{4}$

7. $\frac{9}{11} + \frac{1}{22}$ **8.** $\frac{5}{7} + \frac{3}{14}$ **9.** $\frac{1}{12} + \frac{5}{9}$

10. $\frac{3}{8} + \frac{5}{12}$ **11.** $\frac{1}{6} + \frac{5}{9}$ **12.** $\frac{3}{10} + \frac{5}{8}$

13. $\frac{1}{4} + \frac{2}{9} + \frac{1}{3}$ **14.** $\frac{1}{4} + \frac{1}{8} + \frac{5}{12}$ **15.** $\frac{3}{10} + \frac{2}{5} + \frac{3}{20}$

16. $\frac{1}{6} + \frac{1}{4} + \frac{3}{8}$ **17.** $\frac{7}{10} + \frac{1}{15} + \frac{1}{6}$ **18.** $\frac{1}{2} + \frac{3}{8} + \frac{1}{12}$

19. $\frac{1}{4} + \frac{2}{3}$ **20.** $\frac{7}{12} + \frac{3}{8}$ **21.** $\frac{8}{15} + \frac{3}{10}$ **22.** $\frac{1}{6} + \frac{5}{9}$

Subtract the following fractions. Write answers in lowest terms.

Example: $\frac{3}{5} - \frac{1}{2} = \mathbf{\frac{1}{10}}$ **Solution:** $\frac{3}{5} - \frac{1}{2} = \frac{6}{10} - \frac{5}{10} = \frac{6-5}{10} = \frac{1}{10}$

23. $\frac{5}{6} - \frac{1}{6}$

24. $\frac{11}{12} - \frac{5}{12}$

25. $\frac{2}{3} - \frac{1}{6}$

26. $\frac{7}{8} - \frac{1}{2}$

27. $\frac{5}{12} - \frac{1}{4}$

28. $\frac{5}{6} - \frac{7}{9}$

29. $\frac{3}{4} - \frac{5}{12}$

30. $\frac{8}{15} - \frac{1}{3}$

31. $\frac{8}{9} - \frac{7}{15}$

32. $\frac{7}{8} - \frac{2}{3}$

33. $\frac{5}{9} - \frac{5}{12}$

34. $\frac{5}{8} - \frac{1}{3}$

35. $\frac{2}{3} - \frac{3}{5}$

36. $\frac{7}{10} - \frac{1}{4}$

Solve each of the following word problems.

Example: The county painted a white line on $\frac{1}{6}$ of the road on Monday and on $\frac{1}{4}$ of the road on Tuesday. What fraction of the white line had been painted by the end of Tuesday?

Approach: To find the fraction painted during the two days, add the fraction painted Monday $\left(\frac{1}{6}\right)$ to the fraction painted Tuesday $\left(\frac{1}{4}\right)$.

Solution: The indicator word *and* shows that the fractions should be added.

$\frac{1}{6}$ on Monday and $\frac{1}{4}$ on Tuesday
↑ indicator word

$$\frac{1}{6} + \frac{1}{4} = \frac{2}{12} + \frac{3}{12} = \frac{5}{12}$$

By the end of Tuesday, $\frac{5}{12}$ of the white line had been painted.

37. A buyer for a grain company bought $\frac{3}{8}$ ton of wheat, $\frac{1}{4}$ ton of rice, and $\frac{1}{3}$ ton of barley. How many tons of grain were bought?

38. Fred Thompson paid $\frac{1}{8}$ of a debt in January, $\frac{1}{3}$ in February, $\frac{1}{4}$ in March, and $\frac{1}{12}$ in April. What fraction of the debt was paid in these 4 months?

39. A company has $\frac{3}{4}$ acre of land. They sell $\frac{1}{6}$ acre. How much land is left?

40. Lori has completed $\frac{2}{5}$ of her savings goal. If she saves another $\frac{1}{8}$ of her goal and then $\frac{1}{6}$ more, find the total portion of her goal she has saved.

CHALLENGE EXERCISES

41. Sherri Josephs has completed $\frac{5}{16}$ of the units needed for a degree. She will complete another $\frac{1}{8}$ of the units this term. What fraction of the units will she still need to complete?

42. At one kennel, $\frac{11}{18}$ of the dogs are small, $\frac{1}{9}$ are medium size, and the rest are large. What fraction of the dogs are large?

REVIEW EXERCISES

Multiply or divide as indicated. Write answers as mixed numbers. (For help, see ***Section 2.8.****)*

43. $7\frac{1}{2} \cdot 3\frac{1}{3}$

44. $6\frac{1}{4} \cdot 3\frac{1}{5}$

45. $1\frac{2}{3} \cdot 2\frac{7}{10}$

46. $3\frac{1}{4} \div 2\frac{5}{8}$

47. $1\frac{1}{2} \div 3\frac{3}{4}$

48. $5\frac{3}{4} \div 2$

3.4 ADDING AND SUBTRACTING MIXED NUMBERS

Recall that a mixed number is the sum of a whole number and a fraction. For example,

$$3\tfrac{2}{5} \quad \text{means} \quad 3 + \tfrac{2}{5}.$$

1 Add or subtract mixed numbers by adding or subtracting the whole number parts and the fractions parts.

WORK PROBLEM 1 AT THE SIDE.

EXAMPLE 1 Adding and Subtracting Mixed Numbers

Add or subtract.

(a)

$$\begin{array}{r} 16\tfrac{1}{8} \\ +\ 5\tfrac{5}{8} \\ \hline 21\tfrac{6}{8} \end{array}$$

$21\tfrac{6}{8}$ ← sum of fractions; 21 ← sum of whole numbers

Since $\frac{6}{8}$ in lowest terms is $\frac{3}{4}$, the final answer is $21\frac{3}{4}$.

(b)

$$\begin{array}{rcr} 8\tfrac{5}{8} & = & 8\tfrac{15}{24} \\ 3\tfrac{1}{12} & = & 3\tfrac{2}{24} \\ \hline & & 5\tfrac{13}{24} \end{array}$$

$5\tfrac{13}{24}$ ← Subtract fractions.; 5 ← Subtract whole numbers.

Just as before, check by adding $5\frac{13}{24}$ and $3\frac{1}{12}$; the sum should be $8\frac{5}{8}$. ■

WORK PROBLEM 2 AT THE SIDE.

When you add the fractions of mixed numbers, the answer may be greater than 1. If this happens, **carry** from the fraction to the whole number.

EXAMPLE 2 Carrying When Adding Mixed Numbers

$$\begin{array}{r} 9\tfrac{5}{8} \\ +\ 13\tfrac{7}{8} \\ \hline 22\tfrac{12}{8} \end{array}$$

$22\tfrac{12}{8}$ ← sum of fractions; 22 ← sum of whole numbers

OBJECTIVES

1 Add or subtract mixed numbers.

2 Subtract by using borrowing.

1. Review mixed numbers; convert mixed numbers to improper fractions and improper fractions to mixed numbers.

(a) $\frac{7}{3}$

(b) $\frac{13}{4}$

(c) $5\frac{1}{3}$

(d) $3\frac{2}{7}$

2. Add or subtract.

(a) $6\frac{1}{4} + 3\frac{3}{8}$

(b) $25\frac{3}{5} + 12\frac{3}{10}$

(c) $5\frac{4}{9} - 3\frac{1}{3}$

ANSWERS

1. (a) $2\frac{1}{3}$ (b) $3\frac{1}{4}$ (c) $\frac{16}{3}$ (d) $\frac{23}{7}$

2. (a) $9\frac{5}{8}$ (b) $37\frac{9}{10}$ (c) $2\frac{1}{9}$

3. Add.

(a) $9\frac{3}{4} + 7\frac{1}{2}$

(b) $45\frac{5}{8} + 36\frac{3}{4} + 51\frac{1}{2}$

The improper fraction $\frac{12}{8}$ can be written in lowest terms as $\frac{3}{2}$. Since $\frac{3}{2} = 1\frac{1}{2}$, the sum

$$22\frac{12}{8} = 22 + \frac{12}{8} = 22 + 1\frac{1}{2} = 23\frac{1}{2}. \quad \blacksquare$$

WORK PROBLEM 3 AT THE SIDE.

2 Borrowing is sometimes necessary when subtracting mixed numbers.

EXAMPLE 3 Borrowing When Subtracting Mixed Numbers

Subtract.

(a) $\begin{array}{r} 7 \\ -\ 2\frac{5}{6} \\ \hline \end{array}$ **(b)** $8\frac{1}{3} - 4\frac{3}{5}$

SOLUTION

(a) $\begin{array}{rcl} 7 & = & 7\frac{0}{6} \\ -\ 2\frac{5}{6} & = & 2\frac{5}{6} \\ \hline \end{array}$ There is no fraction $\frac{0}{6}$.

It is not possible to subtract $\frac{5}{6}$ from $\frac{0}{6}$, so borrow from the whole number **7.**

$$7 = 6 + 1 \quad \text{(Borrow 1.)}$$

$$1 = \frac{6}{6}$$

$$= 6 + \frac{6}{6}$$

$$= 6\frac{6}{6}$$

Next, subtract.

$$\begin{array}{rcl} 7 & = & 6\frac{6}{6} \\ -\ 2\frac{5}{6} & = & 2\frac{5}{6} \\ \hline & & 4\frac{1}{6} \end{array}$$

ANSWERS

3. (a) $17\frac{1}{4}$ (b) $133\frac{7}{8}$

(b) $8\frac{1}{3} - 4\frac{3}{5} = 8\frac{5}{15} - 4\frac{9}{15}$ ← least common denominator

It is not possible to subtract $\frac{9}{15}$ from $\frac{5}{15}$, so borrow from the whole number **8.**

$$8\frac{1}{3} = 8 + \frac{1}{3} = 7 + 1 + \frac{5}{15} \quad \text{(Borrow 1.)}$$

$$= 7 + \frac{15}{15} + \frac{5}{15} \quad \left(1 \text{ is } \frac{15}{15}\right)$$

$$= 7 + \frac{20}{15} \quad \leftarrow \frac{15}{15} + \frac{5}{15}$$

$$= 7\frac{20}{15}$$

Next, subtract.

$$8\frac{1}{3} - 4\frac{3}{5} = 7\frac{20}{15} - 4\frac{9}{15} = 3\frac{11}{15}$$

In lowest terms, the answer is $3\frac{11}{15}$. ■

WORK PROBLEM 4 AT THE SIDE.

4. Subtract.

(a) $8\frac{8}{9} - 3\frac{5}{9}$

(b) $2\frac{5}{8} - 1\frac{15}{16}$

(c) $25\frac{1}{6} - 18\frac{11}{15}$

ANSWERS

4. (a) $5\frac{1}{3}$ (b) $\frac{11}{16}$ (c) $6\frac{13}{30}$

name date hour

3.4 EXERCISES

Add. Write all answers in lowest terms.

Examples: $\begin{array}{r} 8\frac{3}{4} \\ +\ 2\frac{1}{8} \\ \hline \mathbf{10\frac{7}{8}} \end{array}$

Solution: $\begin{array}{rcl} 8\frac{3}{4} & = & 8\frac{6}{8} \\ +\ 2\frac{1}{8} & = & 2\frac{1}{8} \\ \hline & & 10\frac{7}{8} \end{array}$ ← Add the fraction parts.
↑ Add the whole number parts.

$\begin{array}{r} 2\frac{3}{5} \\ +\ 9\frac{2}{3} \\ \hline \mathbf{12\frac{4}{15}} \end{array}$

Solution: $\begin{array}{rcl} 2\frac{3}{5} & = & 2\frac{9}{15} \\ +\ 9\frac{2}{3} & = & 9\frac{10}{15} \\ \hline & & 11\frac{19}{15} \end{array}$ $\frac{19}{15} = 1\frac{4}{15}$, so

$11\frac{19}{15} = 11 + 1\frac{4}{15} = 12\frac{4}{15}$

1. $\begin{array}{r} 21\frac{1}{7} \\ +\ 49\frac{3}{7} \\ \hline \end{array}$

2. $\begin{array}{r} 28\frac{1}{9} \\ +\ 32\frac{7}{9} \\ \hline \end{array}$

3. $\begin{array}{r} 51\frac{1}{4} \\ +\ 29\frac{1}{2} \\ \hline \end{array}$

4. $\begin{array}{r} 25\frac{1}{6} \\ +\ 46\frac{2}{3} \\ \hline \end{array}$

5. $\begin{array}{r} 46\frac{3}{8} \\ +\ 15\frac{1}{4} \\ \hline \end{array}$

6. $\begin{array}{r} 26\frac{5}{8} \\ +\ 8\frac{1}{12} \\ \hline \end{array}$

7. $\begin{array}{r} 82\frac{3}{5} \\ +\ 15\frac{4}{5} \\ \hline \end{array}$

8. $\begin{array}{r} 24\frac{5}{6} \\ +\ 18\frac{5}{6} \\ \hline \end{array}$

9. $\begin{array}{r} 126\frac{4}{5} \\ +\ 25\frac{9}{10} \\ \hline \end{array}$

10. $\begin{array}{r} 58\frac{1}{2} \\ +\ 9\frac{7}{8} \\ \hline \end{array}$

11. $\begin{array}{r} 268\frac{9}{10} \\ +\ 35\frac{3}{8} \\ \hline \end{array}$

12. $\begin{array}{r} 59\frac{7}{8} \\ +\ 24\frac{5}{6} \\ \hline \end{array}$

13. $7\frac{1}{4}$, $25\frac{3}{8}$, $+\ 9\frac{1}{2}$

14. $18\frac{3}{5}$, $47\frac{7}{10}$, $+\ 25\frac{8}{15}$

15. $28\frac{1}{4}$, $23\frac{3}{5}$, $+\ 19\frac{9}{10}$

Subtract. Write all answers in lowest terms.

Examples: $5\frac{3}{5} - 2\frac{1}{10} = \mathbf{3\frac{1}{2}}$

Solution:

$$5\frac{3}{5} = 5\frac{6}{10}$$
$$-\ 2\frac{1}{10} = 2\frac{1}{10}$$
$$3\frac{5}{10} = 3\frac{1}{2} \quad \text{(lowest terms)}$$

$6\frac{1}{4} - 4\frac{7}{8} = \mathbf{1\frac{3}{8}}$

Solution:

$$6\frac{1}{4} = 6\frac{2}{8}$$
$$-\ 4\frac{7}{8} = 4\frac{7}{8}$$

Borrow. $6\frac{2}{8} = 5 + 1 + \frac{2}{8}$

$$= 5 + \frac{8}{8} + \frac{2}{8}$$
$$= 5 + \frac{10}{8}$$
$$= 5\frac{10}{8}$$

Subtract.

$$6\frac{1}{4} = 5\frac{10}{8}$$
$$-\ 4\frac{7}{8} = 4\frac{7}{8}$$
$$1\frac{3}{8}$$

16. $7\frac{5}{8}$, $-\ 2\frac{1}{8}$

17. $9\frac{3}{4}$, $-\ 6\frac{1}{4}$

18. $6\frac{7}{12}$, $-\ 2\frac{1}{3}$

19. $11\frac{9}{20}$, $-\ 4\frac{3}{5}$

20. $28\frac{3}{10}$, $-\ 6\frac{1}{15}$

21. $14\frac{11}{16}$, $-\ 8\frac{5}{12}$

22. 19, $-\ 8\frac{7}{8}$

23. $37\frac{1}{2}$, $-\ 24\frac{5}{8}$

24. $68\frac{3}{8}$, $-\ 16\frac{4}{5}$

25. $47\frac{3}{8} - 26\frac{7}{12}$

26. $15\frac{13}{24} - 8\frac{15}{16}$

27. $26\frac{5}{18} - 12\frac{11}{24}$

28. $157\frac{2}{3} - 86\frac{14}{15}$

29. $374\frac{2}{8} - 211\frac{5}{6}$

30. $589\frac{10}{11} - 68\frac{21}{22}$

31. $746\frac{3}{8} - 423\frac{11}{12}$

32. $15 - 8\frac{3}{7}$

33. $21 - 5\frac{7}{8}$

34. $415 - 198\frac{3}{4}$

35. $115 - 62\frac{15}{16}$

36. $316 - 101\frac{12}{13}$

Solve each word problem by using addition or subtraction.

37. Russell worked $15\frac{1}{8}$ hours over the weekend. He worked $6\frac{1}{2}$ hours on Saturday. How many hours did he work on Sunday?

38. A wading pool contains $17\frac{1}{6}$ gallons of water. If $9\frac{2}{3}$ gallons are drained out, how many gallons still remain in the pool?

39. A carpenter has a $12\frac{1}{2}$-foot-long board and a $7\frac{2}{3}$-foot-long board. How many feet of wood does she have in all?

40. On Monday, $5\frac{3}{4}$ tons of cans were recycled, and $9\frac{3}{5}$ tons were recycled on Tuesday. How many tons were recycled in total on these two days?

41. Mike Kane worked $6\frac{3}{8}$ hours on Monday, $7\frac{1}{2}$ hours on Tuesday, $8\frac{3}{4}$ hours on Wednesday, $7\frac{3}{8}$ hours on Thursday, and 8 hours on Friday. How many hours did he work altogether?

42. Hernando Ramirez drove for $5\frac{1}{2}$ hours on the first day of his vacation, $6\frac{1}{4}$ hours on the second day, $3\frac{3}{4}$ hours on the third day, and 7 hours on the fourth day. How many hours did he drive altogether?

43. A concrete truck is loaded with $9\frac{5}{8}$ cubic yards of concrete. The driver unloads $1\frac{1}{2}$ cubic yards at the first stop and $2\frac{3}{4}$ cubic yards at the second stop. The customer at the third stop gets 3 cubic yards. How much concrete is left in the truck?

44. Marv Levenson bought 15 yards of material at a sale. He made a shirt with $3\frac{3}{4}$ yards of the material, a suit for his wife with $4\frac{1}{8}$ yards, and a jacket with $3\frac{7}{8}$ yards. How many yards of material were left over?

45. Stacey Kaufmann worked 40 hours during a certain week. She worked $8\frac{1}{4}$ hours on Monday, $6\frac{3}{8}$ hours on Tuesday, $7\frac{3}{4}$ hours on Wednesday, and $8\frac{3}{4}$ hours on Thursday. How many hours did she work on Friday?

46. Three sides of a parking lot are $108\frac{1}{4}$ feet, $162\frac{3}{8}$ feet, and $143\frac{1}{2}$ feet. If the distance around the lot is $518\frac{3}{4}$ feet, find the length of the fourth side.

47. The Eastside Wholesale Vegetable Market sold $3\frac{1}{4}$ tons of broccoli, $2\frac{3}{8}$ tons of spinach, $7\frac{1}{2}$ tons of corn, and $1\frac{5}{6}$ tons of turnip last month. Find the total number of tons of these vegetables sold by the market last month.

48. Johnson's Texaco Service sold $5\frac{1}{8}$ cases of cheap oil last week, $8\frac{3}{4}$ cases of medium-priced oil, and $12\frac{5}{8}$ cases of expensive oil. Find the total number of cases of oil that Johnson's sold during the week.

CHALLENGE EXERCISES

Find x *in the following figures.*

49.

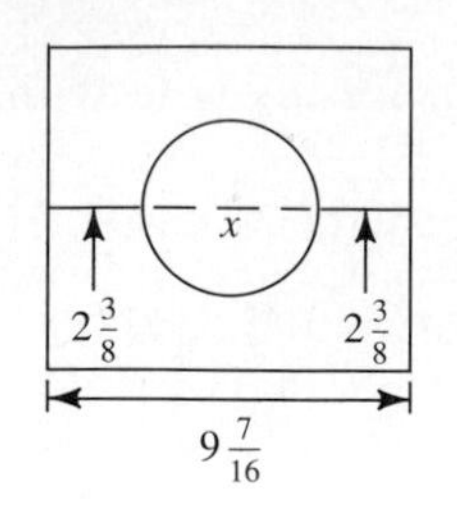

50.

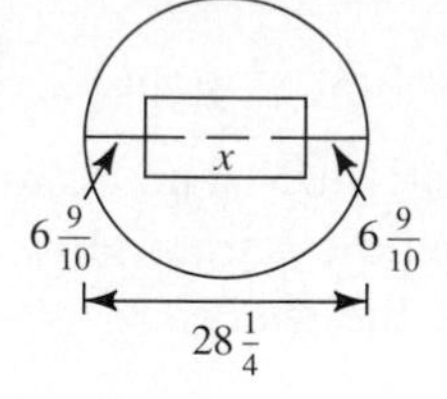

REVIEW EXERCISES

Work each problem, using the order of operations. (For help, see ***Section 1.8.****)*

51. $9^2 + 5 - 2$

52. $2 \cdot 7 - 4$

53. $4 \cdot 1 + 8 \cdot 7 + 3$

54. $8 + 9 \div 3 + 6 \cdot 2$

55. $3^2 \cdot (5 - 2)$

56. $8 \cdot 4 - (15 - 8)$

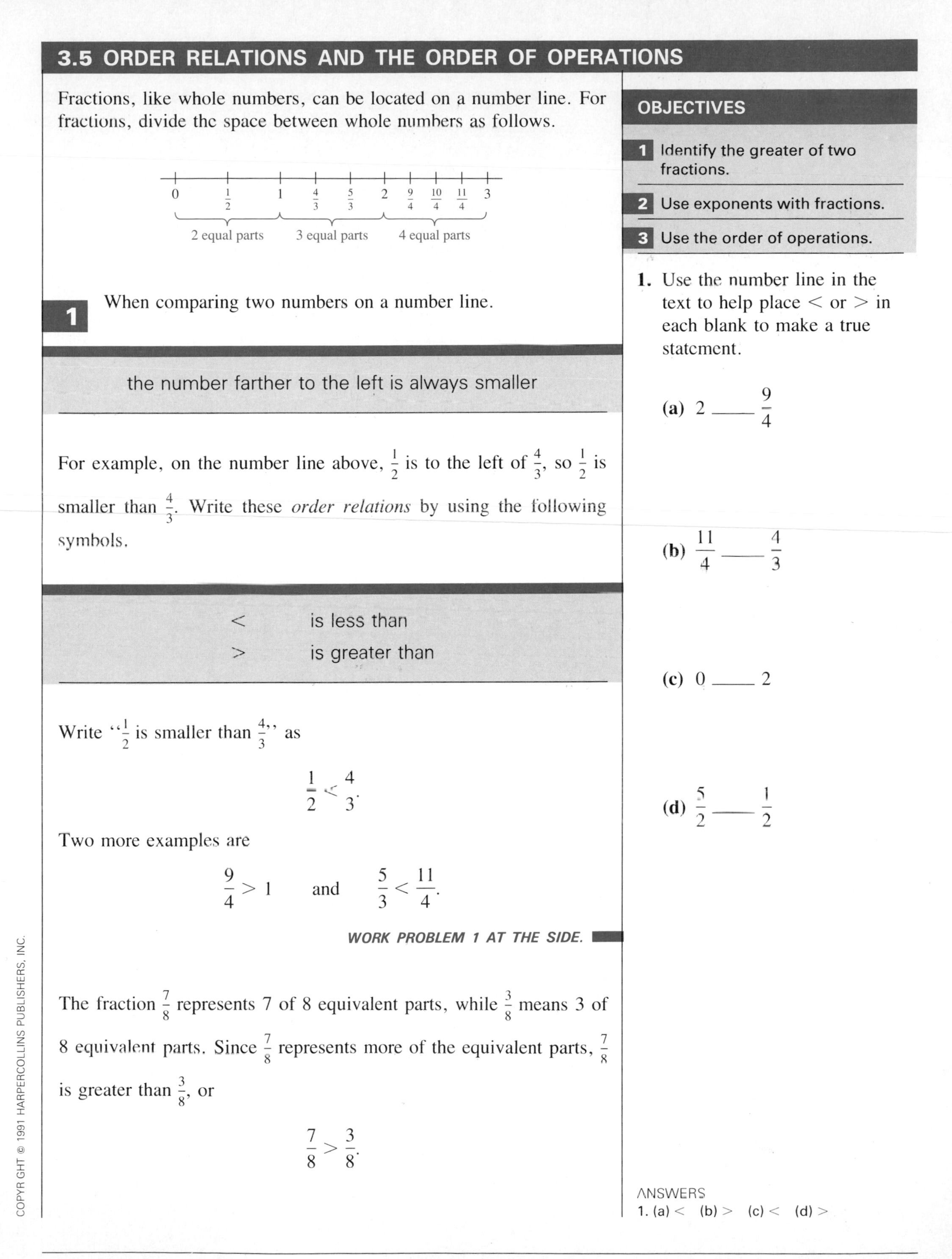

3.5 ORDER RELATIONS AND THE ORDER OF OPERATIONS

OBJECTIVES

1. Identify the greater of two fractions.
2. Use exponents with fractions.
3. Use the order of operations.

Fractions, like whole numbers, can be located on a number line. For fractions, divide the space between whole numbers as follows.

1 When comparing two numbers on a number line.

the number farther to the left is always smaller

For example, on the number line above, $\frac{1}{2}$ is to the left of $\frac{4}{3}$, so $\frac{1}{2}$ is smaller than $\frac{4}{3}$. Write these *order relations* by using the following symbols.

$<$	is less than
$>$	is greater than

Write "$\frac{1}{2}$ is smaller than $\frac{4}{3}$," as

$$\frac{1}{2} < \frac{4}{3}.$$

Two more examples are

$$\frac{9}{4} > 1 \quad \text{and} \quad \frac{5}{3} < \frac{11}{4}.$$

WORK PROBLEM 1 AT THE SIDE.

The fraction $\frac{7}{8}$ represents 7 of 8 equivalent parts, while $\frac{3}{8}$ means 3 of 8 equivalent parts. Since $\frac{7}{8}$ represents more of the equivalent parts, $\frac{7}{8}$ is greater than $\frac{3}{8}$, or

$$\frac{7}{8} > \frac{3}{8}.$$

1. Use the number line in the text to help place $<$ or $>$ in each blank to make a true statement.

(a) $2 \_\_\_ \frac{9}{4}$

(b) $\frac{11}{4} \_\_\_ \frac{4}{3}$

(c) $0 \_\_\_ 2$

(d) $\frac{5}{2} \_\_\_ \frac{1}{2}$

ANSWERS
1. (a) $<$ (b) $>$ (c) $<$ (d) $>$

2. Place $<$ or $>$ in each blank to make a true statement.

(a) $\frac{3}{8}$ ___ $\frac{7}{12}$

(b) $\frac{11}{18}$ ___ $\frac{5}{9}$

(c) $\frac{13}{15}$ ___ $\frac{8}{9}$

(d) $\frac{17}{24}$ ___ $\frac{5}{6}$

Of two like fractions, the fraction with the greater numerator is the greater fraction.

EXAMPLE 1 Identifying the Greater Fraction

Decide which fraction in each pair is greater.

(a) $\frac{7}{8}, \frac{9}{10}$ **(b)** $\frac{8}{5}, \frac{23}{15}$

SOLUTION

(a) Write the fractions as like fractions to tell which fraction is greater. The least common multiple for 8 and 10 is 40, so

$$\frac{7}{8} = \frac{7 \cdot 5}{8 \cdot 5} = \frac{35}{40} \quad \text{and} \quad \frac{9}{10} = \frac{9 \cdot 4}{10 \cdot 4} = \frac{36}{40}.$$

Since $\frac{36}{40}$ is greater than $\frac{35}{40}$,

$$\frac{9}{10} > \frac{7}{8} \quad \text{or} \quad \frac{7}{8} < \frac{9}{10}.$$

The greater fraction is $\frac{9}{10}$.

(b) The least common multiple of 5 and 15 is 15.

$$\frac{8}{5} = \frac{8 \cdot 3}{5 \cdot 3} = \frac{24}{15} \quad \text{and} \quad \frac{23}{15} = \frac{23}{15}$$

This shows that $\frac{8}{5}$ is greater than $\frac{23}{15}$, or

$$\frac{8}{5} > \frac{23}{15}. \quad ■$$

WORK PROBLEM 2 AT THE SIDE.

2 Exponents were used in Chapter 1 to write repeated products. For example,

$$3^2 = 3 \cdot 3 = 9 \text{ and } 5^3 = 5 \cdot 5 \cdot 5 = 125.$$

(exponent: the 2 in 3^2; $3 \cdot 3$: two factors of 3)

The next example shows exponents used with fractions.

ANSWERS

2. (a) $<$ (b) $>$ (c) $<$ (d) $<$

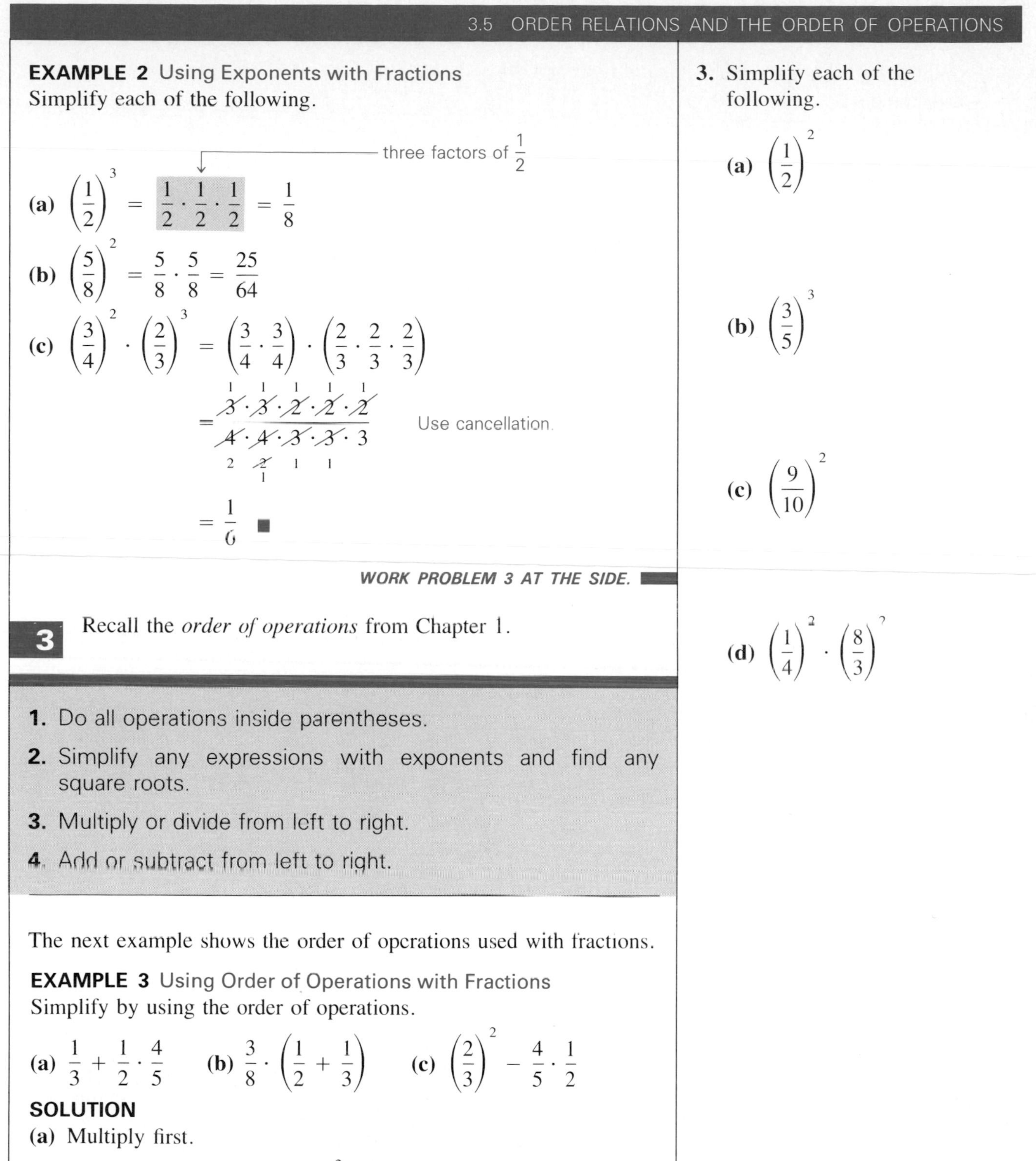

EXAMPLE 2 Using Exponents with Fractions

Simplify each of the following.

(a) $\left(\frac{1}{2}\right)^3 = \frac{1}{2} \cdot \frac{1}{2} \cdot \frac{1}{2} = \frac{1}{8}$ (three factors of $\frac{1}{2}$)

(b) $\left(\frac{5}{8}\right)^2 = \frac{5}{8} \cdot \frac{5}{8} = \frac{25}{64}$

(c) $\left(\frac{3}{4}\right)^2 \cdot \left(\frac{2}{3}\right)^3 = \left(\frac{3}{4} \cdot \frac{3}{4}\right) \cdot \left(\frac{2}{3} \cdot \frac{2}{3} \cdot \frac{2}{3}\right)$

$$= \frac{\cancel{3} \cdot \cancel{3} \cdot \cancel{2} \cdot \cancel{2} \cdot \cancel{2}}{\cancel{4} \cdot \cancel{4} \cdot \cancel{3} \cdot \cancel{3} \cdot 3}$$ Use cancellation.

$$= \frac{1}{6}$$ ■

WORK PROBLEM 3 AT THE SIDE.

3. Simplify each of the following.

(a) $\left(\frac{1}{2}\right)^2$

(b) $\left(\frac{3}{5}\right)^3$

(c) $\left(\frac{9}{10}\right)^2$

(d) $\left(\frac{1}{4}\right)^2 \cdot \left(\frac{8}{3}\right)^2$

3 Recall the *order of operations* from Chapter 1.

1. Do all operations inside parentheses.
2. Simplify any expressions with exponents and find any square roots.
3. Multiply or divide from left to right.
4. Add or subtract from left to right.

The next example shows the order of operations used with fractions.

EXAMPLE 3 Using Order of Operations with Fractions

Simplify by using the order of operations.

(a) $\frac{1}{3} + \frac{1}{2} \cdot \frac{4}{5}$ (b) $\frac{3}{8} \cdot \left(\frac{1}{2} + \frac{1}{3}\right)$ (c) $\left(\frac{2}{3}\right)^2 - \frac{4}{5} \cdot \frac{1}{2}$

SOLUTION

(a) Multiply first.

$$\frac{1}{3} + \frac{1}{\cancel{2}} \cdot \frac{\cancel{4}}{5} = \frac{1}{3} + \frac{2}{5}$$

Next, add. The least common denominator of 3 and 5 is 15.

$$\frac{1}{3} + \frac{2}{5} = \frac{5}{15} + \frac{6}{15} = \frac{11}{15}$$

ANSWERS

3. (a) $\frac{1}{4}$ (b) $\frac{27}{125}$ (c) $\frac{81}{100}$ (d) $\frac{4}{9}$

4. Simplify by using the order of operations.

(a) $\frac{7}{8} - \frac{1}{2} \cdot \frac{4}{3}$

(b) $\frac{1}{4} \cdot \left(\frac{2}{3} - \frac{1}{2}\right)$

(c) $\dfrac{\left(\frac{3}{5}\right)^2}{\frac{3}{2}}$

(d) $\frac{1}{2} \cdot \frac{4}{5} - \left(\frac{1}{3}\right)^2$

ANSWERS

(a) $\frac{5}{24}$ (b) $\frac{1}{24}$ (c) $\frac{6}{25}$ (d) $\frac{13}{45}$

(b) $\frac{3}{8} \cdot \left(\frac{1}{2} + \frac{1}{3}\right) = \frac{3}{8} \cdot \underbrace{\left(\frac{3}{6} + \frac{2}{6}\right)}_{\text{Work in parentheses first.}}$

$= \frac{3}{8} \cdot \frac{5}{6}$

$= \frac{\overset{1}{\cancel{3}} \cdot 5}{8 \cdot \underset{2}{\cancel{6}}}$ Use cancellation.

$= \frac{5}{16}$ Multiply.

(c) $\left(\frac{2}{3}\right)^2 - \frac{4}{5} \cdot \frac{1}{2} = \frac{4}{9} - \frac{4}{5} \cdot \frac{1}{2}$ Evaluate exponential expression first.

$= \frac{4}{9} - \frac{\overset{2}{\cancel{4}}}{5} \cdot \frac{1}{\underset{1}{\cancel{2}}}$ Next, multiply.

$= \frac{4}{9} - \frac{2}{5}$

$= \frac{20}{45} - \frac{18}{45}$ Subtract.

$= \frac{2}{45}$ ■

WORK PROBLEM 4 AT THE SIDE.

name date hour

3.5 EXERCISES

Locate each fraction on the number line.

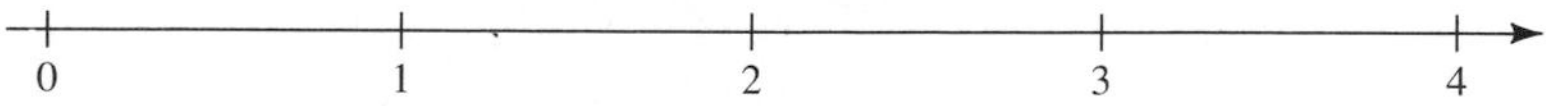

1. $\frac{1}{4}$ **2.** $\frac{3}{8}$ **3.** $\frac{7}{9}$ **4.** $\frac{5}{4}$

5. $\frac{7}{3}$ **6.** $\frac{11}{4}$ **7.** $\frac{13}{6}$ **8.** $\frac{19}{5}$

Write < or > to make a true statement.

Example: $\frac{7}{4}$ $\frac{13}{6}$

Solution:
The least common multiple of 4 and 6 is 12.

$$\frac{7}{4} = \frac{21}{12} \qquad \frac{13}{6} = \frac{26}{12}$$

From this, $\frac{21}{12}$ or $\frac{7}{4}$ is smaller, so write <.

9. $\frac{1}{2}$ $\frac{3}{4}$ **10.** $\frac{2}{3}$ $\frac{5}{6}$ **11.** $\frac{5}{8}$ $\frac{11}{16}$ **12.** $\frac{9}{5}$ $\frac{23}{15}$

13. $\frac{3}{8}$ $\frac{5}{12}$ **14.** $\frac{7}{15}$ $\frac{9}{20}$ **15.** $\frac{7}{12}$ $\frac{11}{18}$ **16.** $\frac{19}{24}$ $\frac{17}{36}$

17. $\frac{11}{24}$ $\frac{19}{36}$ **18.** $\frac{21}{40}$ $\frac{17}{30}$ **19.** $\frac{37}{50}$ $\frac{13}{20}$ **20.** $\frac{5}{12}$ $\frac{11}{27}$

Evaluate each of the following.

Example: $\left(\frac{2}{3}\right)^4 = \frac{2}{3}\cdot\frac{2}{3}\cdot\frac{2}{3}\cdot\frac{2}{3} = \mathbf{\frac{16}{81}}$

21. $\left(\frac{4}{5}\right)^2$ **22.** $\left(\frac{3}{8}\right)^2$ **23.** $\left(\frac{7}{15}\right)^2$ **24.** $\left(\frac{9}{11}\right)^2$

25. $\left(\frac{2}{3}\right)^3$ **26.** $\left(\frac{3}{5}\right)^3$ **27.** $\left(\frac{2}{9}\right)^3$ **28.** $\left(\frac{5}{8}\right)^3$

29. $\left(\frac{3}{2}\right)^4$ **30.** $\left(\frac{4}{3}\right)^4$ **31.** $\left(\frac{1}{2}\right)^5$ **32.** $\left(\frac{2}{3}\right)^5$

name date hour

Use the order of operations to simplify each of the following.

Example: $\left(\frac{2}{3}\right)^2 \cdot \left(\frac{1}{2} \cdot \frac{1}{4}\right)$

Solution: $= \left(\frac{2}{3}\right)^2 \cdot \left(\frac{3}{4}\right)$ Work in parentheses first.

$= \frac{4}{9} \cdot \frac{3}{4}$ Evaluate exponential expression.

$= \mathbf{\frac{1}{3}}$ Multiply.

33. $2^2 + 3 - 2$

34. $3 \cdot 2 + 7 \cdot 4$

35. $8 \cdot 3^2 - \frac{10}{2}$

36. $3^2 \cdot 2^2 + (10 - 6)$

37. $\left(\frac{1}{2}\right)^2 \cdot 4$

38. $6 \cdot \left(\frac{2}{3}\right)^2$

39. $\left(\frac{3}{4}\right)^2 \cdot \left(\frac{1}{3}\right)$

40. $\left(\frac{2}{3}\right)^3 \cdot \left(\frac{1}{2}\right)$

41. $\left(\frac{3}{4}\right)^2 \cdot \left(\frac{2}{3}\right)^2$

42. $\left(\frac{5}{8}\right)^2 \cdot \left(\frac{4}{25}\right)^2$

43. $6 \cdot \left(\frac{2}{3}\right)^2 \cdot \left(\frac{1}{2}\right)^3$

44. $9 \cdot \left(\frac{1}{3}\right)^3 \cdot \left(\frac{4}{3}\right)^2$

45. $\frac{1}{2} \cdot \frac{4}{5} + \frac{2}{3} \cdot \frac{9}{5}$

46. $\frac{3}{4} \cdot \frac{2}{5} + \frac{1}{3} \cdot \frac{3}{5}$

47. $\frac{1}{2} + \left(\frac{1}{2}\right)^2 - \frac{3}{8}$

48. $\frac{2}{3} + \left(\frac{1}{3}\right)^2 - \frac{5}{9}$

49. $\left(\frac{1}{3} + \frac{1}{6}\right) \cdot \frac{1}{2}$

50. $\left(\frac{3}{8} - \frac{1}{4}\right) \cdot \frac{3}{2}$

51. $\frac{9}{8} \div \left(\frac{2}{3} + \frac{1}{12}\right)$

52. $\frac{6}{5} \div \left(\frac{3}{5} - \frac{3}{10}\right)$

53. $\left(\frac{5}{6} - \frac{1}{12}\right) \div \frac{3}{2}$

54. $\left(\frac{8}{5} - \frac{7}{10}\right) \div \frac{3}{5}$

55. $\frac{3}{8} \cdot \left(\frac{1}{4} + \frac{1}{2}\right) \cdot \frac{32}{3}$

56. $\frac{1}{6} \cdot \left(\frac{3}{5} - \frac{1}{10}\right) \cdot \frac{3}{2}$

57. $\left(\frac{3}{4}\right)^2 - \left(\frac{3}{4} - \frac{1}{8}\right) \div \frac{7}{4}$

58. $\left(\frac{5}{6} - \frac{7}{12}\right) - \left(\frac{1}{3}\right)^2 \cdot \frac{3}{4}$

CHALLENGE EXERCISES

59. $\left(\frac{3}{5}\right)^2 \cdot \left(\frac{1}{3} + \frac{2}{9}\right) - \frac{1}{2} \cdot \frac{1}{5}$

60. $\left(\frac{2}{3}\right)^2 \cdot \left(\frac{1}{2} - \frac{1}{8}\right) - \frac{2}{3} \cdot \frac{1}{8}$

REVIEW EXERCISES

Rewrite the following numbers in words. (For help, see ***Section 1.1.****)*

61. 8436

62. 625,115

63. 4,071,280

64. 220,518,315

CHAPTER 3 SUMMARY

KEY TERMS

3.1	**like fractions**	Fractions with the same denominator are called like fractions.
	unlike fractions	Fractions with different denominators are called unlike fractions.
3.2	**least common multiple**	Given two or more whole numbers, the least common multiple is the smallest whole number that is a multiple of these.
3.4	**carrying**	The method used when the sum of the fractions of mixed numbers is greater than 1 is called carrying. Carry from the fraction to the whole number.

QUICK REVIEW

Section Number and Topic	Approach	Example
3.1 Adding Like Fractions	Add numerators and write in lowest terms.	$\frac{3}{4} + \frac{1}{4} + \frac{5}{4} = \frac{3+1+5}{4} = \frac{9}{4} = 2\frac{1}{4}$
3.1 Subtracting Like Fractions	Subtract numerators and write in lowest terms.	$\frac{7}{8} - \frac{5}{8} = \frac{7-5}{8} = \frac{2}{8} = \frac{1}{4}$
3.2 Finding the Least Common Multiple	Method of prime numbers: Use prime numbers to find the least common multiple.	$\frac{1}{3} + \frac{1}{4} + \frac{1}{10}$ (see table below) least common multiple (LCM) $= 2 \cdot 2 \cdot 3 \cdot 5 = 60$
3.3 Adding Unlike Fractions	1. Find the least common multiple. 2. Rewrite fractions with the least common multiple. 3. Add numerators, placing the answer over the least common multiple.	$\frac{1}{3} + \frac{1}{4} + \frac{1}{10}$ LCM = 60; $\frac{1}{3} = \frac{20}{60}$, $\frac{1}{4} = \frac{15}{60}$, $\frac{1}{10} = \frac{6}{60}$; $\frac{20}{60} + \frac{15}{60} + \frac{6}{60} = \frac{41}{60}$

Prime	2	3	5
3 =		(3)	
4 =	(2 · 2)		
10 =	2		(5)
LCM =	(2 · 2)	(3)	(5)

Section Number and Topic	Approach	Example
3.3 Subtracting Unlike Fractions	1. Find the least common multiple. 2. Subtract the numerator of subtrahend. 3. Write the difference over the least common multiple.	$\frac{5}{8} - \frac{1}{3} = \frac{15}{24} - \frac{8}{24} = \frac{7}{24}$
3.4 Adding Mixed Numbers	1. Add fractions. 2. Add whole numbers. 3. Combine the sums of whole numbers and fractions. Write the answer in lowest terms.	$9\frac{2}{3} = 9\frac{8}{12}$ $+\ 6\frac{3}{4} = 6\frac{9}{12}$ $15\frac{17}{12} = 16\frac{5}{12}$
3.4 Subtracting Mixed Numbers	1. Subtract fractions by using borrowing if necessary. 2. Subtract whole numbers. 3. Combine the differences of whole numbers and fractions.	$8\frac{5}{8} = 8\frac{15}{24}$ $-\ 3\frac{1}{12} = 3\frac{2}{24}$ $5\frac{13}{24}$
3.5 Identifying the Larger of Two Fractions	With unlike fractions, change to like fractions first. The fraction with the greater numerator is the greater fraction. $<$ is less than $>$ is greater than	Identify the larger fraction. $\frac{7}{8}, \frac{9}{10}$ $\frac{7}{8} = \frac{7 \cdot 5}{8 \cdot 5} = \frac{35}{40}$ $\frac{9}{10} = \frac{9 \cdot 4}{10 \cdot 4} = \frac{36}{40}$ $\frac{35}{40}$ is smaller than $\frac{36}{40}$, so $\frac{7}{8} < \frac{9}{10}$. or $\frac{9}{10} > \frac{7}{8}$ $\frac{9}{10}$ is greater.
3.5 Using the Order of Operations with Fractions	Follow the order of operations. 1. Do all operations inside parentheses. 2. Simplify any expressions with exponents and find any square roots. 3. Multiply from left to right. 4. Add or subtract from left to right.	Simplify by using the order of operations. $\frac{1}{2} \cdot \frac{2}{3} - \left(\frac{1}{4}\right)^2$ $= \frac{1}{2} \cdot \frac{2}{3} - \frac{1}{16}$ Simplify exponents. $= \frac{2}{6} - \frac{1}{16}$ Next, multiply. $= \frac{16}{48} - \frac{3}{48}$ Change to common denominator. $= \frac{13}{48}$ Subtract.

name date hour

CHAPTER 3 REVIEW EXERCISES

[3.1] *Add or subtract. Write answers in lowest terms.*

1. $\frac{1}{2} + \frac{1}{2}$

2. $\frac{3}{8} + \frac{4}{8}$

3. $\frac{1}{12} + \frac{2}{12} + \frac{1}{12}$

4. $\frac{8}{14} - \frac{3}{14}$

5. $\frac{3}{10} - \frac{1}{10}$

6. $\frac{8}{32} - \frac{4}{32}$

7. $\frac{36}{62} - \frac{10}{62}$

8. $\frac{208}{360} - \frac{170}{360}$

Solve each word problem. Write answers in lowest terms.

9. John traveled $\frac{3}{8}$ of his journey on the first day and $\frac{2}{8}$ of his journey on the second day. What fraction of his travel was completed on these two days?

10. Diane did $\frac{3}{7}$ of her reading in the morning and $\frac{2}{7}$ of her reading in the afternoon. How much less reading did she do in the afternoon than in the morning?

[3.2] *Find the least common multiple of each set of numbers.*

11. 10, 8

12. 5, 12

13. 10, 12, 20

14. 9, 20, 15

15. 6, 8, 5, 15

16. 25, 16, 5, 18

Rewrite each of the following fractions by using the indicated denominators.

17. $\frac{2}{5} = \frac{\quad}{25}$

18. $\frac{7}{12} = \frac{\quad}{48}$

19. $\frac{5}{6} = \frac{\quad}{102}$

20. $\frac{5}{9} = \frac{\quad}{81}$

21. $\frac{7}{16} = \frac{\quad}{144}$

22. $\frac{3}{22} = \frac{\quad}{88}$

[3.1–3.3] *Add or subtract. Write answers in lowest terms.*

23. $\frac{1}{7} + \frac{4}{7}$

24. $\frac{1}{5} + \frac{4}{15}$

25. $\frac{3}{10} + \frac{1}{2} + \frac{1}{5}$

26. $\frac{1}{2} + \frac{3}{8} + \frac{1}{16}$

27. $\frac{1}{4} + \frac{2}{3}$

28. $\frac{5}{9} + \frac{1}{12}$

29. $\frac{9}{16} + \frac{1}{12}$

30. $\frac{4}{9} - \frac{2}{9}$

31. $\frac{7}{8} - \frac{7}{16}$

32. $\frac{7}{16} - \frac{1}{4}$

33. $\frac{5}{8} - \frac{1}{3}$

34. $\frac{8}{15} - \frac{3}{10}$

Solve each of the following word problems.

35. A dump truck contains $\frac{1}{4}$ cubic yard of fine gravel, $\frac{1}{3}$ cubic yard of pea gravel, and $\frac{3}{8}$ cubic yard of coarse gravel. How many cubic yards of gravel are on the truck?

36. Dick has previously installed $\frac{3}{8}$ of his thermostats. He now installs another $\frac{1}{3}$ of his thermostats and then another $\frac{1}{5}$ of them. Find the total portion of his thermostats he has installed.

[3.4] *Add or subtract. Write answers in lowest terms.*

37. $8\frac{1}{4} + 9\frac{3}{4}$

38. $25\frac{3}{4} + 16\frac{3}{8}$

39. $78\frac{3}{7} + 17\frac{6}{7}$

40. $12\frac{3}{5} + 8\frac{5}{8} + 10\frac{5}{16}$

41. $6\frac{2}{3} - 1\frac{1}{2}$

42. $17\frac{1}{2} - 11\frac{1}{3}$

43. $73\frac{1}{2} - 55\frac{2}{3}$

44. $238\frac{1}{8} - 152\frac{3}{8}$

Solve each word problem.

45. The lab has $14\frac{1}{3}$ gallons of distilled water. If $5\frac{1}{2}$ gallons are used, how many gallons remain?

46. The Scouts collected $6\frac{4}{5}$ tons of newspaper on Saturday and $9\frac{2}{3}$ tons on Sunday. Find the total amount of newspaper collected.

name date hour

47. At birth, the Bolton triplets weigh $5\frac{3}{4}$ pounds, $4\frac{7}{8}$ pounds, and $5\frac{1}{3}$ pounds. Find their total weight.

48. A developer wants to build a shopping center. She bought two parcels of land, one, $1\frac{11}{16}$ acres, and the other, $2\frac{3}{4}$ acres. If she needs a total of $8\frac{1}{2}$ acres for the center, how much additional land does she need to buy?

[3.5] *Locate each fraction on the number line.*

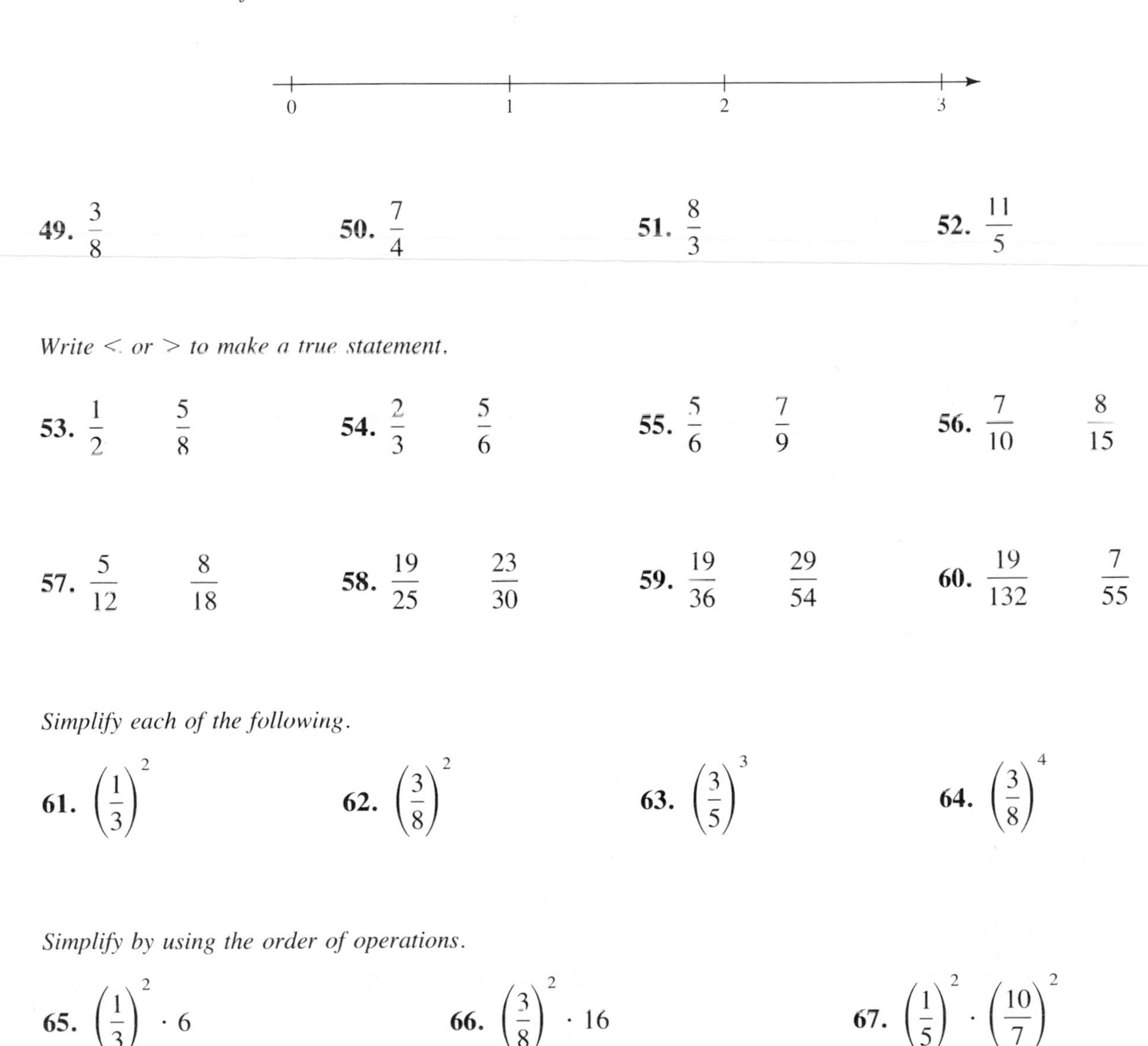

49. $\frac{3}{8}$ **50.** $\frac{7}{4}$ **51.** $\frac{8}{3}$ **52.** $\frac{11}{5}$

Write < or > to make a true statement.

53. $\frac{1}{2}$ $\frac{5}{8}$ **54.** $\frac{2}{3}$ $\frac{5}{6}$ **55.** $\frac{5}{6}$ $\frac{7}{9}$ **56.** $\frac{7}{10}$ $\frac{8}{15}$

57. $\frac{5}{12}$ $\frac{8}{18}$ **58.** $\frac{19}{25}$ $\frac{23}{30}$ **59.** $\frac{19}{36}$ $\frac{29}{54}$ **60.** $\frac{19}{132}$ $\frac{7}{55}$

Simplify each of the following.

61. $\left(\frac{1}{3}\right)^2$ **62.** $\left(\frac{3}{8}\right)^2$ **63.** $\left(\frac{3}{5}\right)^3$ **64.** $\left(\frac{3}{8}\right)^4$

Simplify by using the order of operations.

65. $\left(\frac{1}{3}\right)^2 \cdot 6$ **66.** $\left(\frac{3}{8}\right)^2 \cdot 16$ **67.** $\left(\frac{1}{5}\right)^2 \cdot \left(\frac{10}{7}\right)^2$

68. $\frac{3}{5} \div \left(\frac{1}{10} + \frac{1}{5}\right)$ **69.** $\left(\frac{1}{2}\right)^2 \cdot \left(\frac{1}{4} + \frac{1}{2}\right)$ **70.** $\left(\frac{1}{2}\right)^3 + \left(\frac{1}{4} + \frac{1}{12}\right) \div \frac{1}{9} \cdot \frac{3}{4}$

name date hour

MIXED REVIEW EXERCISES

Solve by using the order of operations as necessary. Write answers in lowest terms.

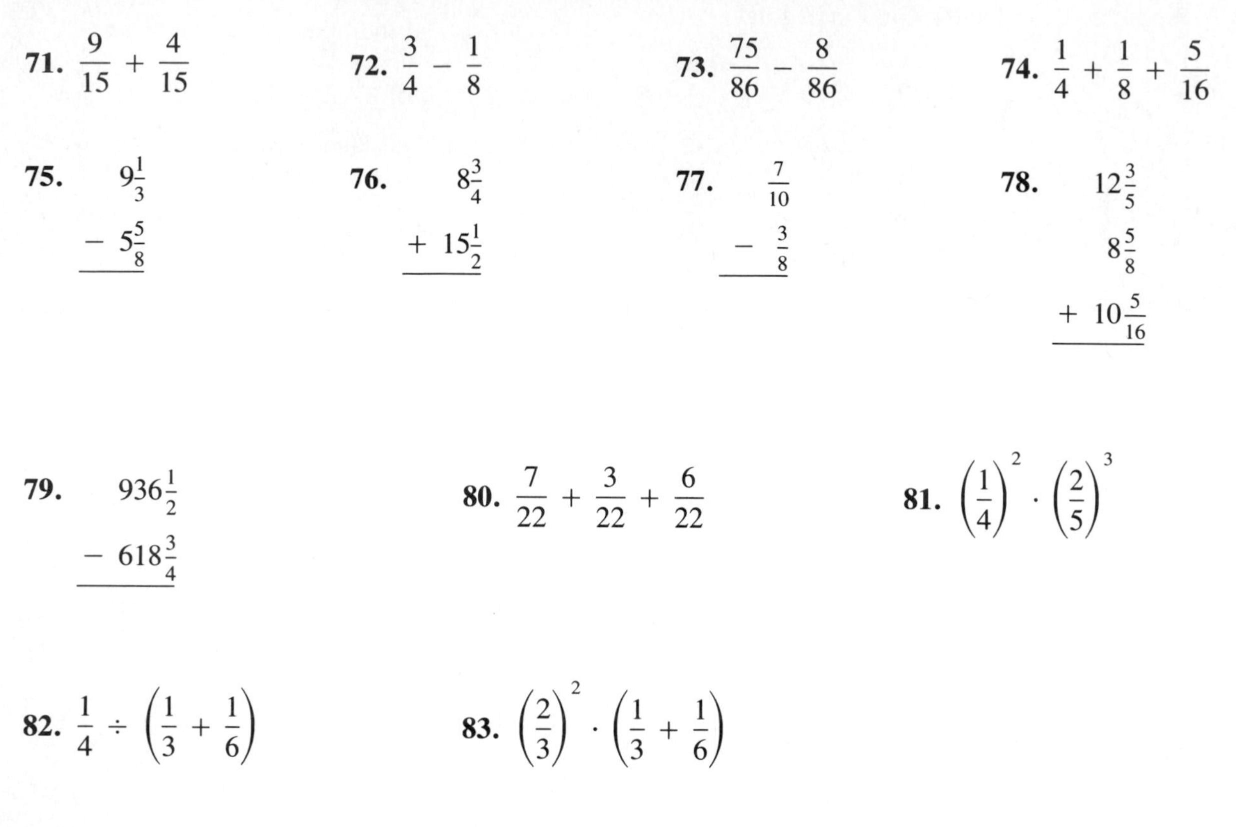

71. $\frac{9}{15} + \frac{4}{15}$

72. $\frac{3}{4} - \frac{1}{8}$

73. $\frac{75}{86} - \frac{8}{86}$

74. $\frac{1}{4} + \frac{1}{8} + \frac{5}{16}$

75. $9\frac{1}{3} - 5\frac{5}{8}$

76. $8\frac{3}{4} + 15\frac{1}{2}$

77. $\frac{7}{10} - \frac{3}{8}$

78. $12\frac{3}{5} + 8\frac{5}{8} + 10\frac{5}{16}$

79. $936\frac{1}{2} - 618\frac{3}{4}$

80. $\frac{7}{22} + \frac{3}{22} + \frac{6}{22}$

81. $\left(\frac{1}{4}\right)^2 \cdot \left(\frac{2}{5}\right)^3$

82. $\frac{1}{4} \div \left(\frac{1}{3} + \frac{1}{6}\right)$

83. $\left(\frac{2}{3}\right)^2 \cdot \left(\frac{1}{3} + \frac{1}{6}\right)$

Write $<$ or $>$ to make a true statement.

84. $\frac{11}{9}$ $\frac{11}{6}$

85. $\frac{10}{11}$ $\frac{32}{33}$

86. $\frac{19}{40}$ $\frac{29}{60}$

87. $\frac{17}{12}$ $\frac{25}{54}$

Find the least common multiple of each set of numbers.

88. 12, 22

89. 2, 16, 36, 42

Rewrite each of the following fractions by using the indicated denominators.

90. $\frac{3}{7} = \frac{\quad}{560}$

91. $\frac{9}{12} = \frac{\quad}{144}$

Solve each word problem.

92. A plumber needs two pieces of pipe. One piece must be $21\frac{5}{16}$ inches long and the other, $7\frac{3}{8}$ inches long. Find the total length needed.

93. The business department has $1\frac{5}{8}$ positions for student help and the science department has $4\frac{5}{6}$ positions. If the college wishes to fill 10 positions, find the number that remain.

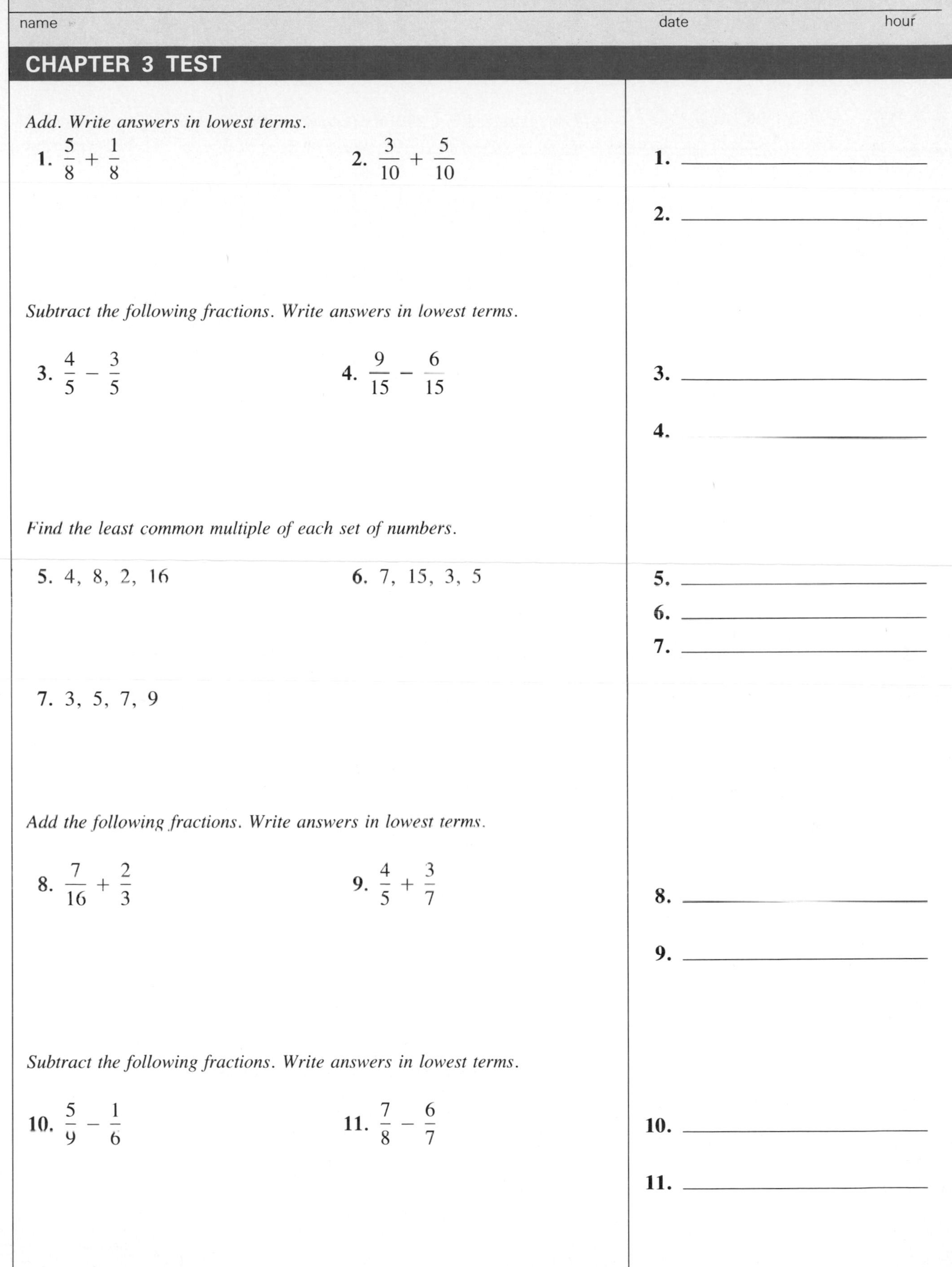

name date hour

CHAPTER 3 TEST

Add. Write answers in lowest terms.

1. $\frac{5}{8} + \frac{1}{8}$

2. $\frac{3}{10} + \frac{5}{10}$

Subtract the following fractions. Write answers in lowest terms.

3. $\frac{4}{5} - \frac{3}{5}$

4. $\frac{9}{15} - \frac{6}{15}$

Find the least common multiple of each set of numbers.

5. 4, 8, 2, 16

6. 7, 15, 3, 5

7. 3, 5, 7, 9

Add the following fractions. Write answers in lowest terms.

8. $\frac{7}{16} + \frac{2}{3}$

9. $\frac{4}{5} + \frac{3}{7}$

Subtract the following fractions. Write answers in lowest terms.

10. $\frac{5}{9} - \frac{1}{6}$

11. $\frac{7}{8} - \frac{6}{7}$

1. ______________
2. ______________
3. ______________
4. ______________
5. ______________
6. ______________
7. ______________
8. ______________
9. ______________
10. ______________
11. ______________

Add or subtract. Write answers in lowest terms.

12. $1\frac{1}{2} + 3\frac{1}{4}$

13. $5\frac{7}{8} + 2\frac{3}{4}$

14. $9\frac{3}{4} - 4\frac{3}{10}$

15. $7\frac{2}{3} - 4\frac{11}{12}$

16. $18\frac{3}{4} + 9\frac{2}{5} + 12\frac{1}{3}$

17. $276\frac{1}{4} - 127\frac{5}{8}$

Solve the following word problems.

18. Howard studied $3\frac{1}{4}$ hours on Monday, $4\frac{1}{6}$ hours on Tuesday, $2\frac{1}{3}$ hours on Wednesday, $3\frac{5}{6}$ hours on Thursday, and $4\frac{2}{3}$ hours on Friday. Find the total number of hours that he studied.

19. The Ecology Club has collected $6\frac{2}{3}$ tons of recyclable glass. How much more glass is needed to fill their $12\frac{1}{4}$-ton dumpster?

Write $<$ or $>$ to make a true statement.

20. $\frac{3}{5}$ $\qquad$ $\frac{13}{20}$

21. $\frac{11}{18}$ $\qquad$ $\frac{17}{24}$

Simplify each of the following. Use the order of operations as needed.

22. $\left(\frac{1}{2}\right)^2 \cdot 2$

23. $\left(\frac{5}{8}\right)^2 \cdot \left(\frac{2}{3}\right)^2$

24. $\left(\frac{5}{6} - \frac{5}{12}\right) \cdot 3$

25. $\frac{2}{3} + \frac{5}{8} \cdot \frac{4}{3}$

12. ______
13. ______
14. ______
15. ______
16. ______
17. ______
18. ______
19. ______
20. ______
21. ______
22. ______
23. ______
24. ______
25. ______

name date hour

CUMULATIVE REVIEW CHAPTERS 1–3

[1.1] *Name the digit that has the given place value in each of the following problems.*

1. 946
hundreds
ones

2. 8,354,917
millions
thousands

[1.2] *Add the following.*

3. 7 + 6 + 4 + 9

4. 15 + 28 + 38 + 42

5. 51,506 + 9 834 + 279 + 15,702

6. 375,899 + 521,742 + 357,968

[1.3] *Subtract.*

7. 722 − 546

8. 3246 − 2983

9. 12,509 − 8 765

10. 3,896,502 − 1,094,807

[1.7] *Round each of the following to the nearest ten, nearest hundred, and nearest thousand.*

	ten	***hundred***	***thousand***
11. 2847	________	________	________
12. 59,803	________	________	________

[1.4] *Multiply.*

13. 3 × 9 × 7

14. 2 × 8 × 5

15. 9 × 4 × 6

16. 79 × 8

17. 58 × 37

18. 845 × 325

19. 1258 × 420

20. 530 × 8

21. 290 × 50

22. 389 × 600

Solve each word problem.

23. Each lawn in an area needs 10 boxes of grass seed. How many boxes would be needed for 80 lawns?

24. A fan blade makes 1800 revolutions in one minute. How many revolutions would it make in 50 minutes?

[1.5] *Divide.*

25. $\frac{56}{7}$

26. $9\overline{)1422}$

27. $26{,}927 \div 3$

[1.6] *Divide.*

28. $17\overline{)9894}$

29. $25\overline{)117{,}750}$

30. $286\overline{)16{,}058}$

31. $506\overline{)16{,}358}$

Solve each word problem.

32. A couple has $11,725 worth of rare coins, which they divide equally among five children. Find the value of the coins received by each child.

33. How many 16-ounce cans of beverage can be filled from a vat holding 9280 ounces of the beverage?

[2.3] *Find the prime factorization of each number. Write answers by using exponents.*

34. 30

35. 100

36. 250

Solve each of the following.

37. $3^2 \cdot 2^4$

38. $4^2 \cdot 5^2$

39. $3^3 \cdot 5^2$

Find each square root.

40. $\sqrt{9}$

41. $\sqrt{64}$

42. $\sqrt{225}$

name date hour

[1.8] *Simplify each problem by using the order of operations.*

43. $8^2 - 8 \cdot 2$

44. $\sqrt{25} + 5 \cdot 9 - 6$

[2.1] *Write* proper *or* improper *for each fraction.*

45. $\frac{5}{6}$

46. $\frac{3}{8}$

47. $\frac{7}{4}$

[2.4] *Write each fraction in lowest terms.*

48. $\frac{25}{40}$

49. $\frac{38}{50}$

50. $\frac{105}{300}$

[2.5–2.8] *Multiply. Write all answers in lowest terms.*

51. $\frac{5}{8} \cdot \frac{4}{3}$

52. $\frac{9}{11} \cdot \frac{5}{18}$

53. $25 \cdot \frac{3}{5}$

Solve each word problem.

54. A rectangle is $\frac{3}{8}$ inch by $\frac{4}{5}$ inch. Find its area.

55. A woman has an estate of $10,000. She leaves $\frac{2}{5}$ to a charity. Of the remainder, $\frac{2}{3}$ goes to her son. How much money does her son get?

Divide. Write all answers in lowest terms.

56. $\frac{5}{8} \div \frac{1}{4}$

57. $\frac{25}{40} \div \frac{10}{35}$

58. $6 \div \frac{3}{4}$

[3.1] *Add or subtract. Write all answers in lowest terms.*

59. $\frac{1}{4} + \frac{1}{4}$

60. $\frac{15}{75} - \frac{10}{75}$

61. $\frac{1}{10} + \frac{2}{10} + \frac{3}{10}$

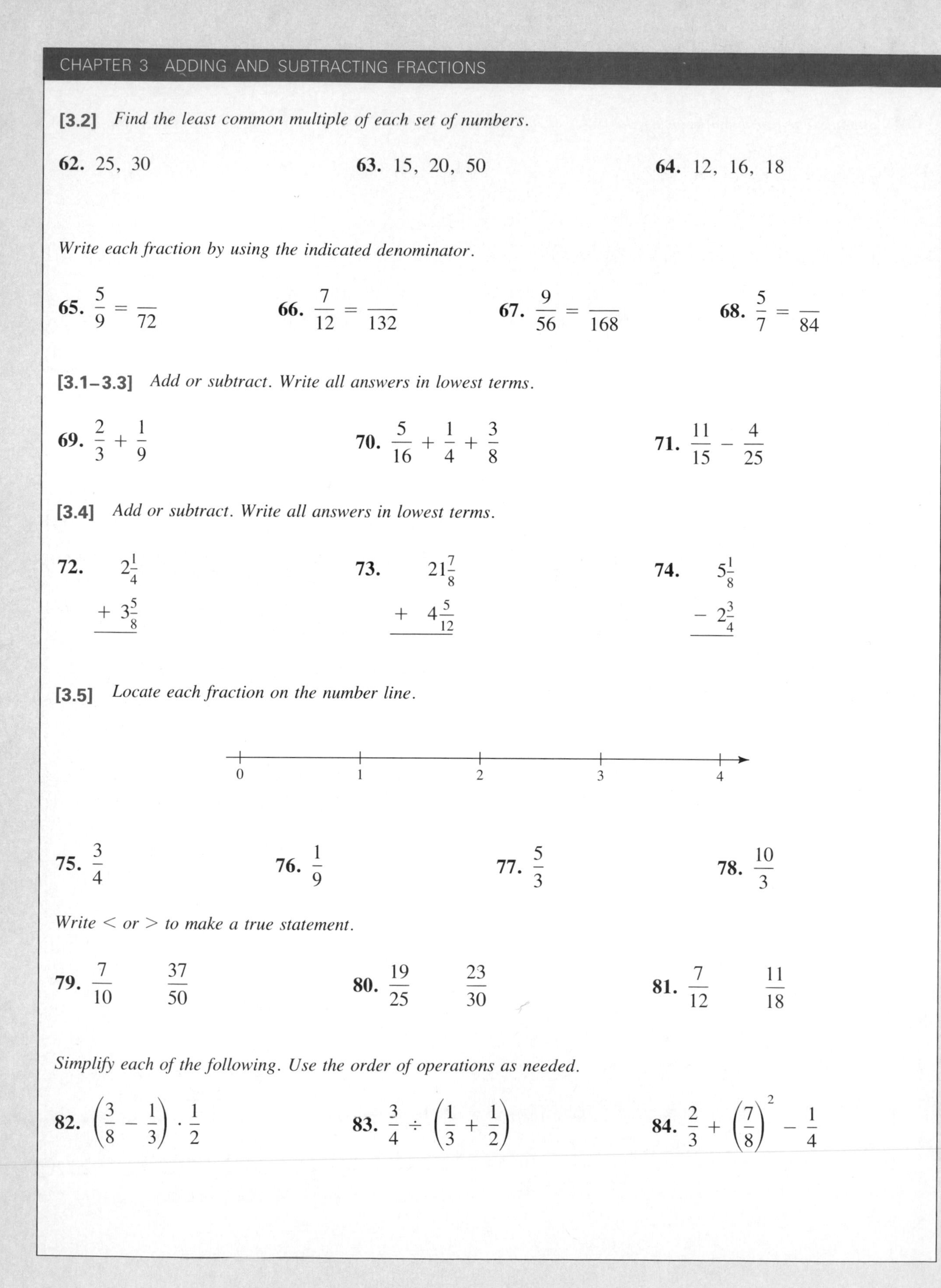

[3.2] *Find the least common multiple of each set of numbers.*

62. 25, 30

63. 15, 20, 50

64. 12, 16, 18

Write each fraction by using the indicated denominator.

65. $\frac{5}{9} = \frac{}{72}$

66. $\frac{7}{12} = \frac{}{132}$

67. $\frac{9}{56} = \frac{}{168}$

68. $\frac{5}{7} = \frac{}{84}$

[3.1–3.3] *Add or subtract. Write all answers in lowest terms.*

69. $\frac{2}{3} + \frac{1}{9}$

70. $\frac{5}{16} + \frac{1}{4} + \frac{3}{8}$

71. $\frac{11}{15} - \frac{4}{25}$

[3.4] *Add or subtract. Write all answers in lowest terms.*

72. $2\frac{1}{4} + 3\frac{5}{8}$

73. $21\frac{7}{8} + 4\frac{5}{12}$

74. $5\frac{1}{8} - 2\frac{3}{4}$

[3.5] *Locate each fraction on the number line.*

75. $\frac{3}{4}$

76. $\frac{1}{9}$

77. $\frac{5}{3}$

78. $\frac{10}{3}$

Write < *or* > *to make a true statement.*

79. $\frac{7}{10} \quad \frac{37}{50}$

80. $\frac{19}{25} \quad \frac{23}{30}$

81. $\frac{7}{12} \quad \frac{11}{18}$

Simplify each of the following. Use the order of operations as needed.

82. $\left(\frac{3}{8} - \frac{1}{3}\right) \cdot \frac{1}{2}$

83. $\frac{3}{4} \div \left(\frac{1}{3} + \frac{1}{2}\right)$

84. $\frac{2}{3} + \left(\frac{7}{8}\right)^2 - \frac{1}{4}$

DECIMALS

Fractions are used to represent parts of a whole. In this chapter, decimals will be used as another way to show parts of a whole. For example, our money system is based on decimals. One dollar is divided into 100 equivalent parts. One cent is one of these parts, and a quarter is 25 of the equal parts.

WORK PROBLEM 1 AT THE SIDE.

4.1 READING AND WRITING DECIMALS

OBJECTIVES

1. Write parts of a whole as decimals.
2. Find the place value of a digit.
3. Read decimals.
4. Change decimals to fractions.

1 Decimals are used when a whole is divided into 10 equivalent parts or into equivalent parts that are powers of 10 (such as 100 or 1000). For example, the square below is cut into 10 equivalent parts. As a fraction, each equivalent part is expressed as

$$\frac{1}{10} \text{ of the whole.}$$

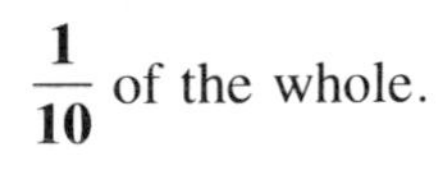

As a decimal, each of the 10 equivalent parts is expressed as

.1 of the whole.

Read .1 as **"one tenth."** The period in .1 is called the **decimal point.**

. 1
↑ decimal point

This square has 7 of its 10 parts shaded. In fractions,

$$\frac{7}{10} \text{ of the parts are shaded.}$$

In decimals,

.7 of the parts are shaded.

1. How many quarters are represented by the shaded portion of each dollar?

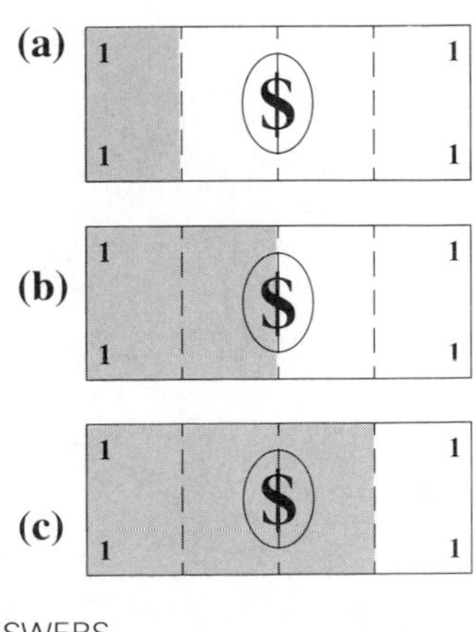

ANSWERS
1. (a) 1 (b) 2 (c) 3

2. Write the portion of each square that is shaded. Use both fractions and decimals.

(a)

(b)

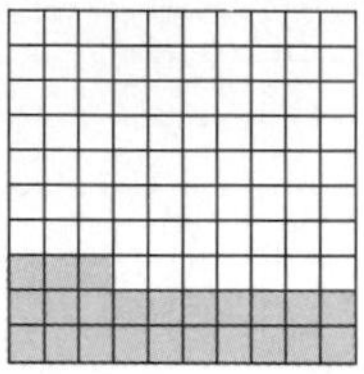

3. Give the fraction form of a decimal.

(a) .7

(b) .9

(c) .13

(d) .87

(e) .119

The next square is cut into 100 equivalent parts. In fractions, each part is

$$\frac{1}{100} \text{ of the whole.}$$

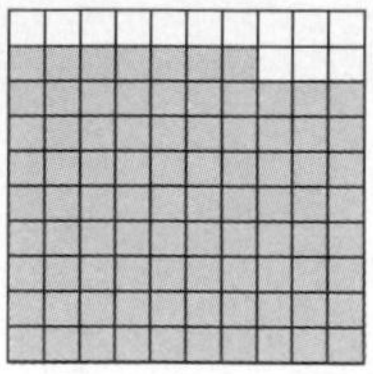

In decimals, each part is

.01 (Read "one hundredth.")

of the whole.

This square has 87 parts shaded. In fractions,

$$\frac{87}{100} \text{ of the total area is shaded.}$$

In decimals,

.87 of the total area is shaded.

WORK PROBLEM 2 AT THE SIDE.

The example below shows several numbers written as both fractions and decimals.

EXAMPLE 1 Using the Decimal Forms of Fractions

	decimal	*fraction*
(a)	.3	$\frac{3}{10}$
(b)	.09	$\frac{9}{100}$
(c)	.71	$\frac{71}{100}$
(d)	.832	$\frac{832}{1000}$ ■

WORK PROBLEM 3 AT THE SIDE.

ANSWERS

2. (a) $\frac{3}{10}$; .3 (b) $\frac{23}{100}$; .23

3. (a) $\frac{7}{10}$ (b) $\frac{9}{10}$ (c) $\frac{13}{100}$ (d) $\frac{87}{100}$

2 The decimal point is used to separate the whole-number part from the fractional part of a number. The most common **place values** used in naming decimals are shown below.

hundred thousands	ten thousands	thousands	hundreds	tens	units, or ones	.	tenths	hundredths	thousandths	ten-thousandths	hundred-thousandths	millionths
100,000	10,000	1000	100	10	1	↑ decimal point	$\frac{1}{10}$	$\frac{1}{100}$	$\frac{1}{1000}$	$\frac{1}{10,000}$	$\frac{1}{100,000}$	$\frac{1}{1,000,000}$

NOTE Each place value is 10 times the place value to its right. If a number appears with **no decimal point,** the number is a whole number and the decimal point, if placed, would be at the far right.

EXAMPLE 2 Identifying the Place Value of a Digit
Give the place values of the digits in each decimal.

(a) 78.36 **(b)** .0093

SOLUTION

(a)

tens	ones		tenths	hundredths
7	8	.	3	6

(b)

	tenths	hundredths	thousandths	ten-thousandths
.	0	0	9	3

■

WORK PROBLEM 4 AT THE SIDE.

3 As mentioned earlier, the decimal .9 can be written as the fraction $\frac{9}{10}$. For this reason, .9 is read "nine tenths."

Also, .67 means

$$\frac{67}{100}$$

so .67 is read "sixty-seven hundredths."

4. Identify the place values of the digits in the following decimals.

(a) 192.61

(b) 3.758

(c) 6

(d) .083

ANSWERS

4. (a) 1 hundreds, 9 tens, 2 ones, . , 6 tenths, 1 hundredths
(b) 3 ones, . , 7 tenths, 5 hundredths, 8 thousandths
(c) .6 tenths
(d) . , 0 tenths, 8 hundredths, 3 thousandths

5. Write the name of each decimal.

(a) .3

(b) .39

(c) .07

(d) .409

(e) .0003

(f) .0703

ANSWERS
5. (a) three tenths
(b) thirty-nine hundredths
(c) seven hundredths
(d) four hundred nine thousandths
(e) three ten-thousandths
(f) seven hundred three ten-thousandths

EXAMPLE 3 Naming a Decimal Number
Write the name of each decimal.

(a) .3 **(b)** .49 **(c)** .82
(d) .918 **(e)** .0106

SOLUTION

(a) Since $.3 = \frac{3}{10}$, the decimal is three tenths.

(b) .49 is forty-nine hundredths.

(c) .82 is eighty-two hundredths.

(d) .918 is nine hundred eighteen thousandths.

(e) .0106 is one hundred six ten-thousandths. ■

WORK PROBLEM 5 AT THE SIDE.

Read a decimal as follows.

Step 1 Read any number to the left of the decimal point as you would any whole number.

Step 2 Next, read the decimal point as **"and."**

Step 3 Finally, read the numbers to the right of the decimal point as an ordinary whole number, followed by the place value of the right-most digit.

EXAMPLE 4 Reading a Decimal
Read each decimal.

(a) 16.9
16 → sixteen; decimal point → and; 9 → nine tenths

The digit 9 is in the tenths place.

(b) 482.35
482 → four hundred eighty-two; decimal point → and; 35 → thirty-five hundredths ← 5 is in the hundredths place.

(c) 5.063 is five and sixty-three thousandths.

(d) 11.1085 is eleven and one thousand eighty-five ten-thousandths. ■

WORK PROBLEM 6 AT THE SIDE.

4 The name of a decimal can be used to change the decimal to a fraction. Use the following rule.

Step 1 Write a decimal with no whole-number part as a fraction by writing the digits after the decimal point as the numerator of the fraction.

Step 2 The denominator is **1** followed by as many zeros as there are digits to the right of the decimal point.

Step 3 If the decimal has a whole-number part, write the decimal as a mixed number.

EXAMPLE 5 Changing a Decimal to a Fraction or Mixed Number
Change each decimal to a fraction.

(a) .19 **(b)** .863 **(c)** 4.2199

SOLUTION
There are two digits to the right of the decimal point, so put 1 followed by two zeros in the denominator of the fraction.

(a) $.19 = \frac{19}{100}$ ← 1 is followed by two 0's.

(b) $.863 = \frac{863}{1000}$ ← 1 is followed by three 0's.

The whole number is included in the mixed number.

(c) $4.2199 = 4\frac{2199}{10{,}000}$ ← 1 is followed by four 0's. ■

WORK PROBLEM 7 AT THE SIDE.

NOTE When a decimal is written as a fraction, the resulting fraction may not be in lowest terms.

6. Write the name of each decimal.

(a) 5.6

(b) 17.9

(c) 80.72

(d) 64.329

7. Change each decimal to a fraction or mixed number.

(a) .7

(b) 7.39

(c) .101

(d) .897

(e) 1.3717

ANSWERS
6. (a) five and six tenths
(b) seventeen and nine tenths
(c) eighty and seventy-two hundredths
(d) sixty-four and three hundred twenty-nine thousandths

7. (a) $\frac{7}{10}$ (b) $7\frac{39}{100}$ (c) $\frac{101}{1000}$ (d) $\frac{897}{1000}$
(e) $1\frac{3717}{10{,}000}$

8. Write each decimal as a fraction or mixed number in lowest terms.

(a) .2

(b) 28.8

(c) .35

(d) 3.05

(e) .505

(f) 420.0802

ANSWERS

8. (a) $\frac{1}{5}$ (b) $28\frac{4}{5}$ (c) $\frac{7}{20}$ (d) $3\frac{1}{20}$ (e) $\frac{101}{200}$ (f) $420\frac{401}{5000}$

EXAMPLE 6 Writing a Decimal as a Fraction or Mixed Number
Write each decimal as a fraction. Write in lowest terms.

(a) $.4 = \frac{4}{10}$ ← 1 is followed by one 0.

Write $\frac{4}{10}$ in lowest terms. $\frac{4}{10} = \frac{2}{5}$

(b) $.75 = \frac{75}{100} = \frac{3}{4}$ (lowest terms)

(c) $18.105 = 18\frac{105}{1000} = 18\frac{21}{200}$

(d) $42.8085 = 42\frac{8085}{10{,}000} = 42\frac{1617}{2000}$ ■

WORK PROBLEM 8 AT THE SIDE.

name date hour

4.1 EXERCISES

Name the digit that has the given place value in each of the following.

Example: 75.638
tenths **6**
thousandths **8**

1. 62.407
tenths
hundredths

2. .3852
tenths
hundredths

3. .591
hundredths
thousandths

4. 0.769
hundredths
thousandths

5. .51472
thousandths
ten-thousandths

6. .89146
tenths
ten-thousandths

7. 27.658
tens
tenths

8. 51.325
tens
tenths

9. 149.0832
hundreds
hundredths

10. 3458.712
hundreds
hundredths

11. 6285.712
thousands
thousandths

12. 5417.6832
thousands
thousandths

Give the place value of each digit in the following decimals.

Example: .309 0 **hundredths** 9 **thousandths**

13. .62 6 2

14. .81 8 1

15. .965 9 6 5

16. .173	1	7	3
17. 47.691	4	7	
	6	9	1
18. 89.325	8	9	
	3	2	5

Write each decimal as a fraction in lowest terms.

Example: $.68 = \frac{17}{25}$

Solution: $.68 = \frac{68}{100} = \frac{17}{25}$ (lowest terms)

19. .7 **20.** .1 **21.** .4 **22.** .8 **23.** .45

24. .85 **25.** .88 **26.** .33 **27.** .41 **28.** .97

29. .02 **30.** .08 **31.** .405 **32.** .805

33. .919 **34.** .179 **35.** .686 **36.** .492

name date hour

4.1 EXERCISES

Write each of the following decimals in words.

Example: 16.28

Solution: **sixteen and twenty-eight hundredths**

37. .7

38. .9

39. .64

40. .82

41. .165

42. .753

43. 12.4

44. 86.9

45. 1.72

46. 4.98

Use digits to write each of the following decimals.

47. three and seven tenths

48. eight and eleven hundredths

49. twenty-seven and thirty-two hundredths

50. fifty-nine and one hundred eleven thousandths

51. four hundred twenty and three hundred eight thousandths

52. two hundred and two hundred twenty-four thousandths

53. seven hundred sixty and nine hundred twenty-one thousandths

54. three thousand, two hundred eighteen, and forty-seven hundredths

CHALLENGE EXERCISES

55. Write the word name for 4322.0531.

56. Write the word name for 22,625.00508.

57. Write 625.4284 as a fraction in lowest terms.

58. Write 714.1372 as a fraction in lowest terms.

REVIEW EXERCISES

Round each of the following to the nearest ten, nearest hundred, and nearest thousand. (For help, see ***Section 1.7.****)*

	ten	***hundred***	***thousand***
59. 8235	________	________	________
60. 3565	________	________	________
61. 46,805	________	________	________
62. 78,634	________	________	________

4.2 ROUNDING DECIMALS

Section 1.7 showed how to round whole numbers. For example, 89 rounded to the nearest ten is 90, and 8512 rounded to the nearest hundred is 8500.

OBJECTIVES

1 Learn the rules for rounding decimals.

2 Round decimals to any given place.

3 Round a number to two or more different places.

It is also important to be able to round decimals. For example, suppose a store sells an item at 2 for 25¢, but a customer wants to buy only one item. The price of each item should be $12\frac{1}{2}$¢, but a customer cannot pay $\frac{1}{2}$¢, so the store *rounds* the price to 13¢ each. Round decimals as follows.

Step 1 Locate the **place** to which the rounding is being done.

Step 2 Look at the next **digit to the right** of the place to which the rounding is being done.

Step 3A If this digit is **less than 5,** drop all digits to the right of the place to which the rounding is being done. Do **not change** the digit in the place to which the rounding is being done.

Step 3B If this digit is **5 or greater,** drop all digits to the right of the place to which the rounding is being done. **Add one** to the digit in the place to which the rounding is being done.

NOTE All digits to the left of the digit being rounded remain unchanged.

The next examples show these rules in use.

EXAMPLE 1 Rounding a Decimal Number

Round 14.39656 to the nearest thousandth.

SOLUTION

Step 1 Use an arrow to locate the place to which the rounding is being done.

14.39 6 56

↑

rounding to the nearest thousandth

Step 2 Check to see if the first digit to the right of the arrow is 5 or greater.

14.396 5 6 — Digit to right of arrow is 5, which is 5 or greater.

1. Round to the nearest thousandth.

(a) .65437

(b) 89.07254

(c) 112.376123

(d) .00913

Step 3 If the digit to the right of the arrow is 5 or greater, increase by one the digit to which the arrow is pointing. Drop all digits to the right of the arrow.

14.396 56 Drop.
↓ increased by 1
14.39 7

14.39656 rounded to the nearest thousandth is 14.397. ■

NOTE The decimal point is **never** moved when rounding.

WORK PROBLEM 1 AT THE SIDE.

EXAMPLE 2 Rounding Decimals to Different Places
Round as shown.

(a) 14.39656 to the nearest tenth

(b) .69413 to the nearest hundredth

(c) .01834 to the nearest thousandth

(d) 57.976 to the nearest tenth

SOLUTION

(a) *Step 1* 14.39656
↑
Use an arrow to locate the place to which the rounding is being done.

Step 2 The first digit to the right of the arrow is 9.

14.3 9 656 5 or greater
↑

Step 3 14.3 9656 Drop all these digits.
↓ increased by 1
14. 4

14.39656 rounded to the nearest tenth is 14.4.

(b) *Step 1* .6 9 413
↑ hundredths place

Step 2 .69 4 13 less than 5
↑

Step 3 .69 413 Drop.
↓ not changed
.6 9

ANSWERS
1. (a) .654 (b) 89.073 (c) 112.376 (d) .009

(c) .01 8 34 — Digit to the right of arrow is 3, which is less than 5.

.018 34 — Drop.

↓ not changed

.01 8

(d) 57. 9 76 — Digit to right is 5 or greater, so add 1.

tenths place

57. 9 76

9 + 1 = 10; so carry 1 to the 7 in the ones place.

7 + 1 = 8

57.976 rounded to the nearest tenth is 58.0. ■

WORK PROBLEM 2 AT THE SIDE.

3 Sometimes the same number must be rounded to two different places.

Round the second time by *going back to the original number.*

EXAMPLE 3 Rounding the Same Number to Different Places
Round each of the following numbers to the nearest hundredth and the nearest tenth.

(a) 36.749 **(b)** .853

SOLUTION

(a) First, round 36.749 to the nearest hundredth as 36.75. Next, going back to the original number, round 36.749 to the nearest tenth as 36.7.

(b) Round .853 to the nearest hundredth as .85. Round .853 to the nearest tenth as .9. ■

WORK PROBLEM 3 AT THE SIDE.

With money, it is common to round to the nearest dollar. (This can be done with federal income tax, for example.)

2. Round each of the following as shown.

(a) 14.69 to the nearest tenth

(b) 5.8163 to the nearest hundredth

(c) .0954 to the nearest thousandth

(d) .8988 to the nearest hundredth

3. Round each of the following numbers to the nearest hundredth and the nearest tenth.

(a) 14.595

(b) 578.0653

(c) .849

(d) 1546.149

ANSWERS
2. (a) 14.7 (b) 5.82 (c) .095 (d) .90
3. (a) 14.60; 14.6 (b) 578.07; 578.1
(c) .85; .8 (d) 1546.15; 1546.1

4. Round to the nearest dollar.

(a) \$74.10

(b) \$136.49

(c) \$510.78

(d) \$5947.88

(e) \$19.83

(f) \$.55

(g) \$1.08

ANSWERS
4. (a) \$74 (b) \$136 (c) \$511
(d) \$5948 (e) \$20 (f) \$1
(g) \$1

EXAMPLE 4 Rounding to the Nearest Dollar
Round to the nearest dollar.

(a) \$48.69 **(b)** \$594.36 **(c)** \$349.88
(d) \$2689.50 **(e)** \$.61

SOLUTION

(a) \$48. 6 9

↑
Tenths digit is 5 or greater, so add 1 to the digit to the left.

\$48.69
↓
8 + 1 = 9

\$48.69 rounded to the nearest dollar is \$49. (Write the answer as \$49 instead of \$49.00 to emphasize that the rounding is to the *nearest dollar*.)

(b) \$594. 3 6

↑
Tenths digit is less than 5, so do not change digit to the left.

\$594.36 rounded to the nearest dollar is \$594.

(c) \$349. 8 8

↑
5 or greater, so add 1 to the digit to left.

\$34 9 .88

↑
9 + 1 = 10; so carry 1 to the 4 in the tens place.

49 + 1 = 50

\$349.88 rounded to the nearest dollar is \$350.

(d) \$2689.50 rounded is \$2690.

(e) \$.61 rounded is \$1. ■

WORK PROBLEM 4 AT THE SIDE.

name date hour

4.2 EXERCISES

Round each of the following to the place shown.

1. 16.8974 to the nearest tenth

2. 389.74 to the nearest tenth

3. 965.4983 to the nearest thousandth

4. 96.81584 to the nearest ten-thousandth

5. 42.399 to the nearest hundredth

6. 61.488 to the nearest tenth

7. 27.90561 to the nearest thousandth

8. 7.3449 to the nearest hundredth

9. 899.498 to the nearest tenth

10. 476.1196 to the nearest thousandth

11. .09864 to the nearest ten-thousandth

12. 176.894 to the nearest ones

Round each of the following to the nearest dollar.

13. \$69.13

14. \$58.41

15. \$139.86

16. \$3958.50

17. \$11,562.59

18. \$14,869.37

Round each of the following to the nearest thousandth, to the nearest hundredth, and to the nearest tenth. Remember to always round from the original number.

	nearest thousandth	***nearest hundredth***	***nearest tenth***
19. 78.414	________	________	________
20. 3689.537	________	________	________
21. .7837	________	________	________
22. 2.548	________	________	________
23. .0875	________	________	________
24. 125.149	________	________	________

As she gets ready to do her income tax return, Ms. Johnson gathers all the figures she will need. She must round each of these figures to the nearest dollar. Round each of the following to the nearest dollar.

25. Income from job, $18,765.48

26. Income from interest on bank account, $67.58

27. Stock dividends, $109.08

28. Amount withheld from check, $2146.49

29. Donations to charity, $379.82

30. Sick pay, $209.74

31. Moving expenses, $705.48

32. Credit union dividend, $115.51

CHALLENGE EXERCISES

Round each of the following to the nearest millionth, to the nearest hundred-thousandth, and to the nearest ten-thousandth.

33. 35.6150479

34. .0056310

35. 614.7899153

36. 22.6719837

37. 1000.0050028

38. 2002.6109952

REVIEW EXERCISES

Add the following numbers, carrying as necessary. Use mental carrying as much as possible. (For help, see ***Section 1.2****.)*

39.
$$\begin{array}{r} 7929 \\ 6076 \\ +\ 8218 \\ \hline \end{array}$$

40.
$$\begin{array}{r} 2078 \\ 5791 \\ 83 \\ 231 \\ +\ 7209 \\ \hline \end{array}$$

41.
$$\begin{array}{r} 2 \\ 816 \\ 43 \\ 7591 \\ +\ 26{,}308 \\ \hline \end{array}$$

42.
$$\begin{array}{r} 96{,}485 \\ 26 \\ 1 \\ 503 \\ +\ \ \ 3879 \\ \hline \end{array}$$

43. 81,976 + 8 + 785 + 20,076 + 7208

44. 17 + 4 + 18,763 + 918 + 32,102

4.3 ADDITION OF DECIMALS

OBJECTIVES

1 Add decimals.

2 Estimate the answer.

1 Decimals can be added by changing each decimal to a fraction and then adding the fractions. However, it is easier to add decimals by using the following rule.

Step 1 Line up the decimal points.

Step 2 Next, add as with whole numbers.

Step 3 The decimal point in the answer appears directly below the decimal points in the problem.

EXAMPLE 1 Adding Decimal Numbers
Add.

(a) 16.92 and 48.34 **(b)** 5.897 + 4.632 + 12.174

SOLUTION
(a) Write the numbers vertically, with decimal points lined up.

$$\begin{array}{r} 16.92 \\ +\ 48.34 \\ \hline \end{array}$$

Decimal points are lined up.

Add as with whole numbers, and place the decimal point in the answer under the decimal points in the problem.

$$\begin{array}{r} {}^{1\,1} \\ 16.92 \\ +\ 48.34 \\ \hline 65.26 \end{array}$$

Decimal point in answer is under decimal points in problem.

(b) Write the numbers vertically with decimal points lined up. Next, add.

$$\begin{array}{r} {}^{1\,1\;\,2\,1} \\ 5.897 \\ 4.632 \\ +\ 12.174 \\ \hline 22.703 \end{array}$$

Decimal points are lined up. ■

WORK PROBLEM 1 AT THE SIDE.

The numbers being added above all had the same number of digits after the decimal point. If the numbers do not all have the same number of digits, using the following rule helps line up the digits to the right of the decimal point and makes addition easier.

Step 1 Find the number with the most digits after the decimal point.

Step 2 Attach zeros to the right of the other numbers (to keep the places lined up), so that they all have the same number of digits after the decimal point.

Step 3 Next, add.

1. Find each sum.

(a) 4.98 + 2.17

(b) 13.761 + 8.325

(c)
$$\begin{array}{r} 3.842 \\ 7.968 \\ +\ 42.329 \\ \hline \end{array}$$

(d)
$$\begin{array}{r} 276.9 \\ 43.8 \\ +\ 1574.2 \\ \hline \end{array}$$

ANSWERS
1. (a) 7.15 (b) 22.086 (c) 54.139
(d) 1894.9

2. Find each of the following sums.

(a) 6.54 + 9.8

(b) 17.921 + 111.1 + 10

(c)
```
    8.64
    9.115
+ 3.0076
```

(d)
```
   14
   29.823
+ 45.7
```

3. Estimate the answer and then add.

(a)
```
   12.940
    6.083
+ 74.100
```

(b)
```
    398.81
     47.658
+ 4158.7
```

(c)
```
  3217.6
   895.41
+   37.288
```

ANSWERS
2. (a) 16.34 (b) 139.021 (c) 20.7626 (d) 89.523
3. (a) 93; 93.123 (b) 4606; 4605.168 (c) 4150; 4150.298

NOTE Attaching zeros to the right of a decimal does not change the value of the number.

EXAMPLE 2 Attaching Zeros as Placeholders Before Adding

Add.

(a) 7.5 + 9.83 (b) 6.42 + 9 + 2.576

SOLUTION

There are two digits after the decimal point in 9.83, so attach one zero to the right in 7.5.

(a)
```
  7.5 0  ← One 0 is attached.
+ 9.8 3
 17.3 3
```

(b)
```
  6.4 2 0  ← One 0 is attached.
  9.0 0 0  ← 9 is a whole number; decimal point and three 0's are attached.
+ 2.5 7 6  ← No 0's are attached.
 17.9 9 6
```
■

Notice in Example 2(b) that the 9 is really 9., with the decimal point at the right. If no decimal point appears in a whole number, place one at the far right.

WORK PROBLEM 2 AT THE SIDE.

2 A common error in working decimal problems is to misplace the decimal point in the answer. Avoid this error by estimating the answer by rounding each number to the nearest whole number and adding. Compare your actual answer with your estimate.

EXAMPLE 3 Estimating a Decimal Answer

Estimate the answer in each of the following and then add.

(a) 16.985 + 1.58 + 12.738 (b) 65.2 + 174.08 + 16.825

SOLUTION

(a)

problem	*estimate*	*actual answer*
16.985	17	16.985
1.58	2	1.580 ← One 0 is attached.
+ 12.738	+ 13	+ 12.738
	32	31.303

Since the actual answer is close to the estimate, the decimal point in the answer is probably in the right place.

(b)

problem	*estimate*	*actual answer*
65.2	65	65.200
174.08	174	174.080
+ 16.825	+ 17	+ 16.825
	256	256.105

Do you think the decimal point is in the right place? ■

WORK PROBLEM 3 AT THE SIDE.

name date hour

4.3 EXERCISES

Find each of the following sums.

Example:

```
   826.28
    53.6
+ 138.152
```

Solution:

```
   826.280   Line up decimal points.
    53.600   Use zeros as placeholders.
+ 138.152
  1018.032
```

1.
```
   48.96
   37.42
+  99.71
```

2.
```
  472.15
   81.73
+  64.31
```

3.
```
  769.08
  406.15
+  83.91
```

4.
```
   25.65
   18.92
+  46.35
```

5.
```
   28.76
   14.1
+  39.25
```

6.
```
   76.5
   89.39
+  42.56
```

7.
```
   34.72
   19.812
+   4.6
```

8.
```
   23.77
   19.812
+  74
```

9.
```
    9.71
    4.8
    3.6
    5.2
+  19.52
```

10.
```
    2.98
    6.43
    7.12
   14.2
+  18.9
```

11.
```
   39.765
  182
    4.719
    8.31
+   5.9
```

12.
```
  489.76
   38
   19.3
    8.5
+ 762.0
```

13. 14.23 + 28 + 74.63 + 18.715 + 64.286

14. 197.4 + 83.72 + 17.43 + 25 + 1.4

15. 27.65 + 18.714 + 9.749 + 3.21

16. 58.546 + 19.2 + 8.735 + 14.58

First estimate the answer, and then add.

Example:	**Solution:**		
problem	***estimate***	***actual answer***	
56.9	57	56.90	← Attach one 0.
3.82	4	3.82	
+ 15.06	+ 15	+ 15.06	
	76	**75.78**	

	problem	***estimate***		***problem***	***estimate***
17.	37.25 18.9 + 7.5		**18.**	24.83 19.7 + 46.19	
19.	482.7 16.92 + 43.87		**20.**	497.62 18.509 + 246.1	
21.	382.504 591.089 + 612.715		**22.**	159.76 382.54 + 179.18	
23.	62.81 539.9 + 5.629		**24.**	332.607 12.5 + 823.39	

First estimate the answer, and then solve each word problem.

25. Joann spent $17.14 for books, $19.36 for a blouse, and $5.14 for lunch. How much did she spend?

26. Tom Rodriquez made $254.19 at the regular rate of pay and $76.49 at the overtime rate. How much money did he make?

27. Chris Howard worked at Blockblaster Video 4.5 days one week, 6.25 days another week, and 3.74 days a third week. How many days did he work altogether?

28. Marge Lial used 7.65 yards of material one week, 8.4 a second week, and 11.23 a third week. How much material did she use altogether?

29. At a bakery, Sue Chee bought $7.42 worth of muffins, $10.09 worth of croissants, and $17.19 worth of cookies for a staff party. How much money did she spend altogether?

30. Jeff McGee wrote checks for $172.15, $89.06, $122.43, and $19.25. Find the total of the checks.

31. At the beginning of a trip to Visalia, a car odometer read 7542.1 miles. It is 186.4 miles to Visalia. What should the odometer read after driving to Visalia *and back?*

32. Ben Whitney sells industrial rubber goods. On one trip he started from Atlanta and drove 226.6 miles to Charlotte, then 153.8 miles to Roanoke, and finally, 341.3 miles back to Atlanta. Find the total length of his trip.

33. Chuck drove on a five-day vacation trip. He drove 8.6 hours the first day, 3.7 hours the second day, 11.3 hours the third day, 2.9 hours the fourth day, and 14.6 hours the fifth day. How many hours did he drive?

34. The accountant at a lumber yard found that one week the payroll was $979.12, utilities were $108.11, advertising was $253.79, and payments to lumber mills were $6985.46. Find the amount spent during the week.

CHALLENGE EXERCISES

Find the distance around these geometric figures.

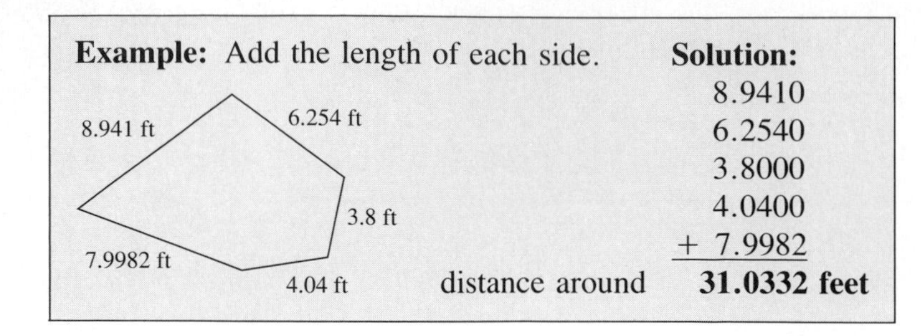

Example: Add the length of each side.

Solution:

$$\begin{array}{r} 8.9410 \\ 6.2540 \\ 3.8000 \\ 4.0400 \\ +\ 7.9982 \\ \hline \end{array}$$

distance around **31.0332 feet**

35.

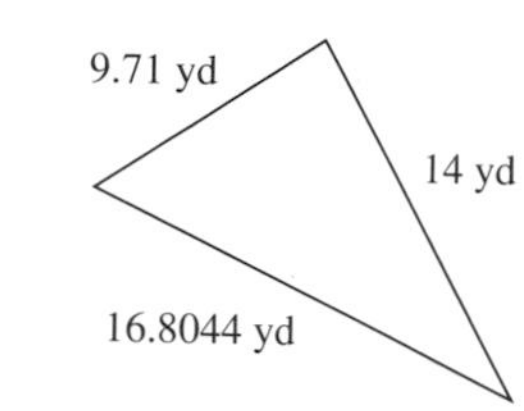

36.

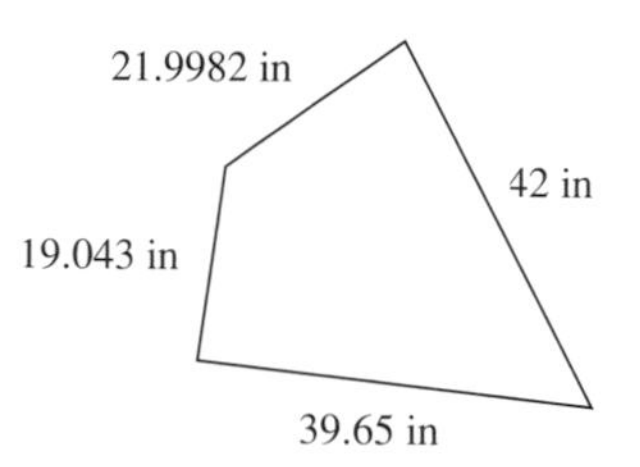

37.

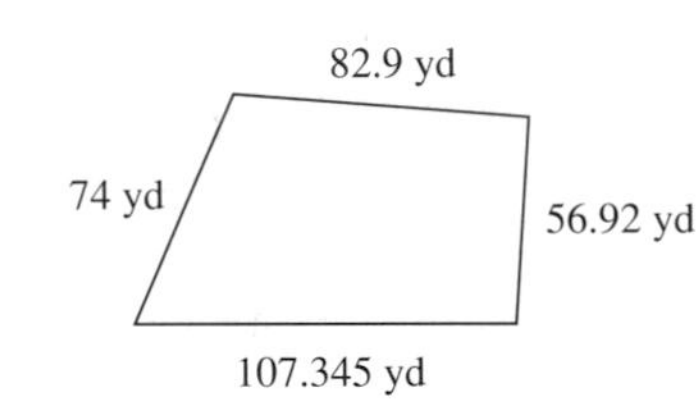

38.

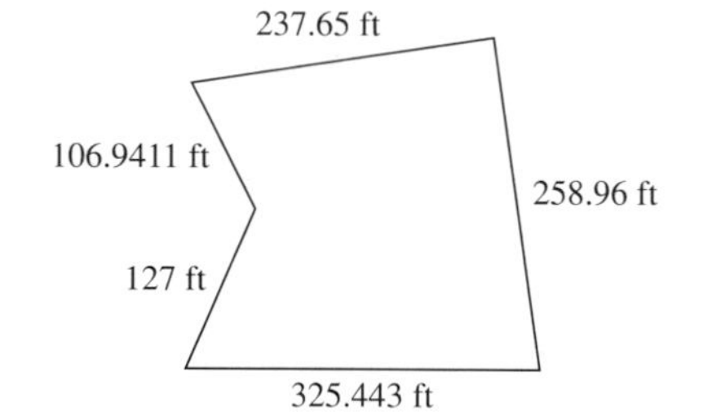

39.

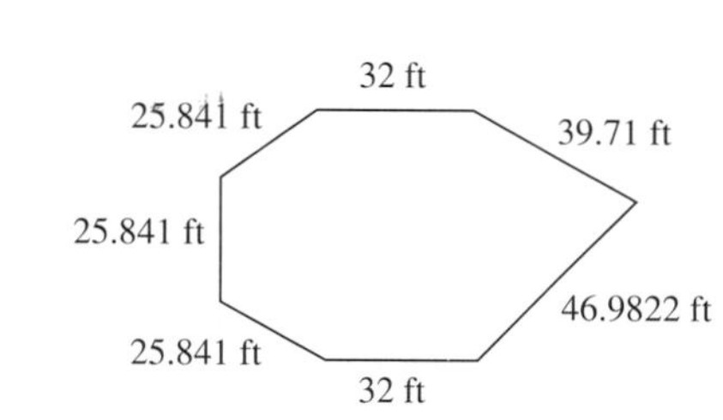

40.

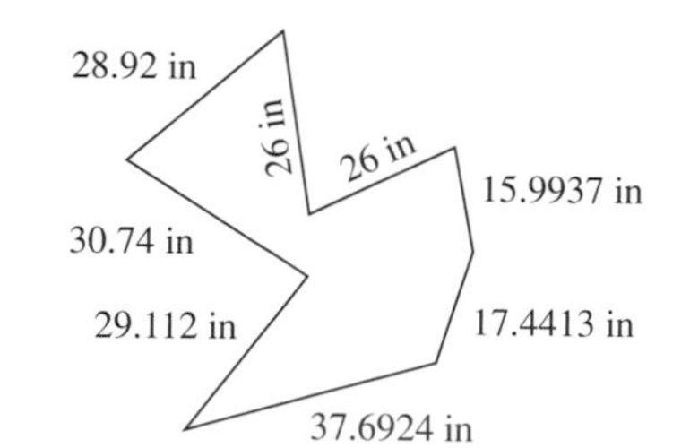

REVIEW EXERCISES

Solve the following problems. (For help, see **Section 1.3**.*)*

41. $\begin{array}{r} 315 \\ -\ 104 \\ \hline \end{array}$

42. $\begin{array}{r} 553 \\ -\ 386 \\ \hline \end{array}$

43. $\begin{array}{r} 6708 \\ -\ 139 \\ \hline \end{array}$

44. $\begin{array}{r} 74{,}000 \\ -\ 5\ 896 \\ \hline \end{array}$

4.4 SUBTRACTION OF DECIMALS

1 Subtraction of decimals is done in much the same way as subtraction of whole numbers. Use the following steps.

Step 1 Write the problem vertically with decimal points lined up in a column.

Step 2 Bring the decimal point straight down.

Step 3 Subtract the numbers as if they were whole numbers. It may be necessary to use zeros as placeholders in one of the numbers.

EXAMPLE 1 Subtracting Decimal Numbers
Subtract each of the following.

(a) 15.82 from 28.93 **(b)** 58.98 from 146.35.

SOLUTION

(a) *Step 1*

```
  28.93 ← number subtracted from (minuend)
- 15.82 ← number being subtracted (subtrahend)
```

Step 2

```
  28.93
- 15.82
     .  ← Bring decimal point down.
```

Step 3

```
  28.93
- 15.82
  13.11 ← Subtract as whole number (difference).
```

(b) Borrowing is needed here.

```
  0 13 15  12 15
  1  4  6 . 3  5
-    5  8 . 9  8
     8  7 . 3  7
```
■

Check the answer by adding 87.37 and 58.98. The sum should be 146.35.

WORK PROBLEM 1 AT THE SIDE.

EXAMPLE 2 Attaching Zeros Before Subtracting
Subtract each of the following.

(a) 16.5 from 28.362 **(b)** 59.7 − 38.914 **(c)** 12 − 5.83

SOLUTION
Use the same steps as above, remembering to attach zeros.

(a)

```
  28.3 62
- 16.5 00 ← Attach two 0's.
  11.8 62 ← Next, subtract as usual.
```

OBJECTIVES

1. Subtract decimals.
2. Estimate the answer.

1. Subtract.

(a) 36.7 from 58.9

(b) 6.425 from 11.813

(c) 19.37 from 35.15

ANSWERS
1. (a) 22.2 (b) 5.388 (c) 15.78

2. Subtract.

(a) 18.651 from 25.3

(b) 5.816 − 4.98

(c) 27 − 19.82

(d) 1 − .32

3. Estimate the answer and then subtract.

(a) 29.76 − 11.985

(b) 157.2 − 98.516

(c) 17.28 − 1.53

(d) 46.091 − 12.837

ANSWERS
2. (a) 6.649 (b) .836 (c) 7.18 (d) .68
3. (a) 18; 17.775 (b) 58; 58.684 (c) 15; 15.75 (d) 33; 33.254

(b)

```
  59.7 00 ← Attach two 0's.
− 38.9 14
  20.7 86 ← Subtract as usual.
```

(c)

```
  12. 00 ← Attach a decimal point and two 0's.
−  5. 83
   6. 17 ← Subtract as usual. ■
```

WORK PROBLEM 2 AT THE SIDE.

2 *Estimate* the answer to make sure the problem has been set up properly and the decimal point has been properly placed in the answer.

EXAMPLE 3 Estimating a Decimal Answer
Estimate the answer and subtract.

(a) 69.42 − 13.78 **(b)** 79.82 − 15.614

SOLUTION
Estimate the answer by rounding each number to the nearest whole number.

(a)

problem	*estimate*	*actual answer*
69.42	69	69.42
− 13.78	− 14	− 13.78
	55 ⟷	55.64

Answers are close, so the problem is probably set up correctly.

(b)

problem	*estimate*	*actual answer*
79.82	80	79.82 0 ← Attach one 0.
− 15.614	− 16	− 15.61 4
	64 ⟷	64.20 6

Answers are close, so the problem is probably set up correctly. ■

WORK PROBLEM 3 AT THE SIDE.

name date hour

4.4 EXERCISES

Subtract.

Example:

$$\begin{array}{r} 71.65 \\ -\ 14.352 \\ \hline \end{array}$$

Solution:

$$\begin{array}{r} 71.650 \\ -\ 14.352 \\ \hline \mathbf{57.298} \end{array}$$

Line up decimal points and attach zeros.

1. $\begin{array}{r} 73.5 \\ -\ 19.2 \\ \hline \end{array}$ **2.** $\begin{array}{r} 47.8 \\ -\ 36.5 \\ \hline \end{array}$ **3.** $\begin{array}{r} 112.2 \\ -\ 96.5 \\ \hline \end{array}$ **4.** $\begin{array}{r} 381.8 \\ -\ 87.9 \\ \hline \end{array}$ **5.** $\begin{array}{r} 283.54 \\ -\ 18.77 \\ \hline \end{array}$

6. $\begin{array}{r} 49.253 \\ -\ 8.714 \\ \hline \end{array}$ **7.** $\begin{array}{r} 58.413 \\ -\ 25.847 \\ \hline \end{array}$ **8.** $\begin{array}{r} 27.905 \\ -\ 18.176 \\ \hline \end{array}$ **9.** $\begin{array}{r} 58.254 \\ -\ 19.7 \\ \hline \end{array}$ **10.** $\begin{array}{r} 47.658 \\ -\ 20.9 \\ \hline \end{array}$

11. $\begin{array}{r} 85.721 \\ -\ 37.68 \\ \hline \end{array}$ **12.** $\begin{array}{r} 472.712 \\ -\ 11.935 \\ \hline \end{array}$ **13.** $\begin{array}{r} 15.7 \\ -\ 2.852 \\ \hline \end{array}$ **14.** $\begin{array}{r} 36.9 \\ -\ 14.582 \\ \hline \end{array}$ **15.** $\begin{array}{r} 72.89 \\ -\ 27.654 \\ \hline \end{array}$

16. $\begin{array}{r} 38.9 \\ -\ 27.807 \\ \hline \end{array}$ **17.** $\begin{array}{r} 479.3 \\ -\ 85.793 \\ \hline \end{array}$ **18.** $\begin{array}{r} 285.4 \\ -\ 56.932 \\ \hline \end{array}$ **19.** $\begin{array}{r} 18 \\ -\ .896 \\ \hline \end{array}$ **20.** $\begin{array}{r} 54 \\ -\ .183 \\ \hline \end{array}$

21. 39.8 − 27.42 **22.** 596.8 − 14.389 **23.** 578.49 − 69.8 **24.** 372.5 − 189.461

Estimate the answer in each of the following and then subtract.

Example: **Solution:**

problem	***estimate***	***actual answer***
$\begin{array}{r} 7.3 \\ -\ 4.962 \\ \hline \end{array}$	$\begin{array}{r} 7 \\ -\ 5 \\ \hline \mathbf{2} \end{array}$	$\begin{array}{r} 7.300 \\ -\ 4.962 \\ \hline \mathbf{2.338} \end{array}$

	problem	***estimate***		***problem***	***estimate***
25.	$\begin{array}{r} 19.74 \\ -\ 6.58 \\ \hline \end{array}$		**26.**	$\begin{array}{r} 27.96 \\ -\ 8.39 \\ \hline \end{array}$	

	problem	estimate		problem	estimate
27.	8.6 − 3.751		**28.**	28 − 13.582	
29.	2 − 1.981		**30.**	31.7 − 4.271	
31.	473.675 − 89.06		**32.**	786.1 − 23.607	

Subtract.

33. 59.8 − 11.7251

34. 20 − 1.37009

35. 114.809 − 52.7172

36. 803.25 − .69815

Solve each word problem. Estimate each answer first.

37. Tom has agreed to work 42.5 hours at a certain job. He has already worked 16.35 hours. How many more hours must he work?

38. Cathy Eastes bought $167.12 worth of clothes. She returned a dress worth $37.95. Find the value of the clothes she kept.

39. A customer gives a clerk a $20 bill to pay for $9.12 in purchases. How much change should the customer get?

40. A man buys $31.09 worth of sporting goods and pays with a $50 bill. How much change should he get?

41. A man receives a bill for $72.18 from Exxon. Of vthis amount, $39.76 is for a tune-up. The rest is for gas. How much is for gas?

42. At one gasoline stop, a car odometer read 39,974.4 miles. At the next stop, it read 38,083.5 miles. How far did the car travel between stops?

43. At the beginning of March, the odometer of Maria DeRisi's company car read 29,086.1 miles. At the end of March, it read 31,561.9 miles. How many miles did Ms. DeRisi drive during the month?

44. Refer to Exercise 43. Suppose that in March, Ms. DeRisi drove the car 897.4 miles on personal business. How many miles was the car driven on company business?

CHALLENGE EXERCISES

Subtract.

45. 386.021 − 179.68231

46. 2.5006 − .005318

47. 221.04 − 218.528683

48. 128.3506 − 97.009398

Work each problem.

49. Mr. Albers had a checking account balance of $129.86 on September 1. During the month, he deposited an additional $1749.82 to the account, and wrote checks totaling $1802.15. The bank charged him with a $2 service charge. Find the amount in the account at the end of the month.

50. On February 1, Lynn Fiorentino had $1009.24 in her checking account. During the month she deposited a tax refund check of $704.42 and her paycheck of $1258.94. She wrote checks totaling $1389.54 and had $200 transferred to her savings account. Find her checking account balance at the end of the month.

CHALLENGE EXERCISES

Find the missing measurement in each figure.

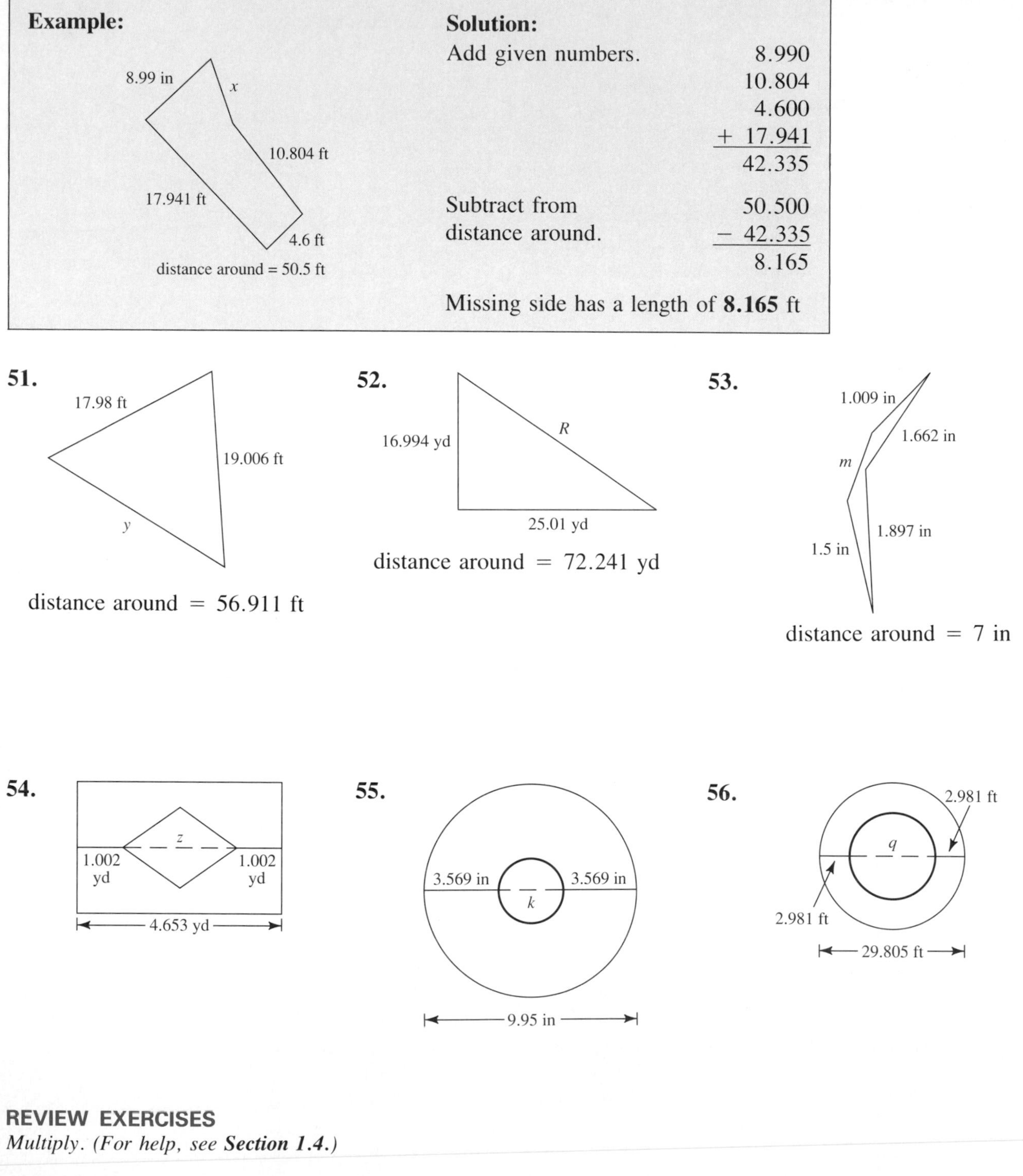

REVIEW EXERCISES

Multiply. (For help, see ***Section 1.4****.)*

57. 837×708

58. 6310×3078

59. 3789×2205

60. 6381×7009

4.5 MULTIPLICATION OF DECIMALS

The decimals .3 and .07 can be multiplied by first writing each decimal as a fraction and then multiplying the fractions.

WORK PROBLEM 1 AT THE SIDE.

Since the fraction $\frac{21}{1000}$ from part (b) at the side is the same as the decimal .021,

$$.3 \cdot .07 = .021.$$

1 To multiply decimals, it is not necessary to rewrite the decimals as fractions. Use the following steps to multiply decimals.

Step 1 Multiply the numbers (factors) as if they were whole numbers. (It is not necessary to line up decimal points.)

Step 2 Find the total number of digits to the right of the decimal points in the numbers being multiplied.

Step 3 Position the decimal point in the answer by counting from right to left the number of decimal places found above. It may be necessary to attach zeros to the left of the digits in the answer.

EXAMPLE 1 Multiplying Decimal Numbers

Multiply 8.34 and 4.2.

SOLUTION

Step 1 Multiply the numbers as if they were whole numbers.

$$\begin{array}{r} 8.34 \\ \times \quad 4.2 \\ \hline 1668 \\ 3336 \\ \hline 35028 \end{array}$$

Step 2 Count the number of digits to the right of the decimal points.

$$\begin{array}{rl} 8.34 & \leftarrow \text{2 decimal places} \\ \times \quad 4.2 & \leftarrow \underline{1} \text{ decimal place} \\ \hline 1668 & \quad 3 \leftarrow \text{total decimal places} \\ 3336 & \\ \hline 35028 & \end{array}$$

Step 3 Count from the right in the answer over 3 places to position the decimal point.

$$\begin{array}{r} 8.34 \\ \times \quad 4.2 \\ \hline 1\ 668 \\ 33\ 36 \\ \hline 35.028 \end{array}$$

Count over 3 places to position decimal point. ■

OBJECTIVES

1 Multiply decimals.

2 Estimate the answer.

1. (a) Write .3 and .07 as fractions.

(b) Multiply the results of part (a).

ANSWERS

1. (a) $\frac{3}{10}, \frac{7}{100}$ (b) $\frac{3}{10} \cdot \frac{7}{100} = \frac{21}{1000}$

2. First estimate the answer, and then multiply to find the exact answer.

(a)
$$\begin{array}{r} 7.9 \\ \times\ 2.6 \\ \hline \end{array}$$

(b)
$$\begin{array}{r} 146.8 \\ \times\ \ \ 3.4 \\ \hline \end{array}$$

(c)
$$\begin{array}{r} 8.51 \\ \times\ 2.3 \\ \hline \end{array}$$

(d)
$$\begin{array}{r} 209.6 \\ \times\ \ \ 5.2 \\ \hline \end{array}$$

3. First estimate, and then multiply to find the exact answer.

(a) 11.62 · 4.01

(b) 5.986 · .53

(c) 17.91 · .89

(d) 58.63 · 17.4

4. Multiply.

(a) .04 · .09

(b) .3 · .005

(c) .063 · .04

(d) .0081 · .007

ANSWERS
2. (a) 24; 20.54 (b) 441; 499.12 (c) 18; 19.573 (d) 1050; 1089.92
3. (a) 48; 46.5962 (b) 6; 3.17258 (c) 18; 15.9399 (d) 1003; 1020.162
4. (a) .0036 (b) .0015 (c) .00252 (d) .0000567

2 It is usually a good idea to **estimate** the answer to make sure the decimal point is in the right place. Estimate the answer to Example 1 as follows.

problem	*estimate*
$\begin{array}{r} 8.34 \\ \times\ \ 4.2 \\ \hline 1\ 668 \\ 33\ 36\ \ \\ \hline 35.028 \end{array}$	$\begin{array}{r} 8 \\ \times\ 4 \\ \hline 32 \end{array}$

The estimate and answer are fairly close, so the decimal point in 35.028 is probably in the correct place.

WORK PROBLEM 2 AT THE SIDE.

EXAMPLE 2 Estimating Before Multiplying

Estimate 76.34 · 12.5. Then multiply to find the exact answer.

SOLUTION

problem	*estimate*	*actual answer*
$\begin{array}{r} 76.34 \\ \times\ \ 12.5 \\ \hline \end{array}$	$\begin{array}{r} 76 \\ \times\ 13 \\ \hline 228 \\ 76\ \\ \hline 988 \end{array}$	$\begin{array}{r} 76.34 \\ \times\ \ 12.5 \\ \hline 38\ 170 \\ 152\ 68\ \ \\ 763\ 4\ \ \ \ \\ \hline 954.250 \end{array}$

76.34 ← 2 decimal places
12.5 ← 1 decimal place
3 decimal places are in answer. ■

WORK PROBLEM 3 AT THE SIDE.

EXAMPLE 3 Attaching Zeros to Place the Decimal Point

Multiply .042 by .03.

SOLUTION

Start by multiplying as above.

$$\begin{array}{r} .042 \\ .03 \\ \hline 126 \end{array}$$

.042 ← 3 decimal places
.03 ← 2 decimal places
126 5 decimal places are in answer.

The answer has only three digits, but five are needed. So attach two zeros at the left.

126
00 126 ← two 0's at left
.00 126 ← decimal point is at left. ■

In problems such as the one in Example 3, estimating is not very helpful because it is very difficult to determine the position of the decimal point.

WORK PROBLEM 4 AT THE SIDE.

name date hour

4.5 EXERCISES

First estimate, and then find the exact answer.

Example: **Solution:**

problem	*estimate*	*actual answer*
83.68	84	83.68
× 7.4	× 7	× 7.4
	588	33472
		58576
		619.232

problem	*estimate*	*problem*	*estimate*
1. 39.6 × 4.8		**2.** 18.7 × 2.3	
3. 47.62 × 2.61		**4.** 21.43 × 12.15	
5. 65.3 × 4.6		**6.** 7.51 × 8.2	
7. 280.9 × 6.85		**8.** 73.52 × 22.34	

Multiply.

Example: **Solution:**

Example	Solution
.093	.093
× 5.6	× 5.6
	558
	465
	.5208

9. .042 × 3.2

10. .571 × 2.9

11. 21.7 × 6.1

12. 85.4 × 3.5

13. $\begin{array}{r} 56.5 \\ \times\ .013 \\ \hline \end{array}$

14. $\begin{array}{r} 896 \\ \times\ .081 \\ \hline \end{array}$

15. $\begin{array}{r} 51.81 \\ \times\ .021 \\ \hline \end{array}$

16. $\begin{array}{r} 36.75 \\ \times\ .088 \\ \hline \end{array}$

17. $\begin{array}{r} 25.94 \\ \times\ 1.1 \\ \hline \end{array}$

18. $\begin{array}{r} 48.05 \\ \times\ .85 \\ \hline \end{array}$

19. $\begin{array}{r} .0892 \\ \times\ .036 \\ \hline \end{array}$

20. $\begin{array}{r} .0907 \\ \times\ .045 \\ \hline \end{array}$

21. $\begin{array}{r} 325.6 \\ \times\ .031 \\ \hline \end{array}$

22. $\begin{array}{r} 5.296 \\ \times\ 21.8 \\ \hline \end{array}$

23. $\begin{array}{r} .0398 \\ \times\ .056 \\ \hline \end{array}$

24. $\begin{array}{r} .0319 \\ \times\ .068 \\ \hline \end{array}$

25. $\begin{array}{r} 7.189 \\ \times\ .0062 \\ \hline \end{array}$

26. $\begin{array}{r} 32.34 \\ \times\ .0075 \\ \hline \end{array}$

27. $\begin{array}{r} 60.98 \\ \times\ .0018 \\ \hline \end{array}$

28. $\begin{array}{r} 3.514 \\ \times\ .0031 \\ \hline \end{array}$

29. $(4.2)(8.7)$

30. $(9.3)(7.2)$

31. $.6 \cdot 8.49$

32. $.3 \cdot 12.5$

33. $(.0063)(.004)$

34. $(.0052)(.009)$

35. $.003 \cdot .002$

36. $.0079 \cdot .006$

Find the amount of money earned on a job by multiplying the number of hours worked and the pay per hour. Find the amount earned in each of the following. Round to the nearest cent.

Example:

25.5 hours at $4.85 per hour

Solution:

$$\begin{array}{r} 25.5 \\ \times\ 4.85 \\ \hline 1275 \\ 2040 \\ 1020 \\ \hline 123.675 \end{array} = \textbf{\$123.68} \text{ rounded}$$

37. 30 hours at $6.04 per hour

38. 45 hours at $4.72 per hour

39. 41.5 hours at $4.51 per hour

40. 32.4 hours at $5.09 per hour

41. 38.4 hours at $5.45 per hour

42. 28.6 hours at $8.25 per hour

43. 60.5 hours at $7.35 per hour

44. 42.2 hours at $6.15 per hour

Find the cost of each of the following.

Example:

6 fasteners at $.18 each

Solution:

$$\begin{array}{r} .18 \\ \times \quad 6 \\ \hline \mathbf{\$1.08} \end{array} \leftarrow \text{cost of 6 fasteners}$$

45. 12 hamburgers at $.85 each

46. 8 quarts of oil at $1.05 each

47. 15 rolls of film at $1.72 each

48. 110 pieces of candy at $.06 each

49. 5100 sheets of poster paper at $.017 each

50. 380 clips at $1.08 each

51. 6400 ball point pens at $.022 each

52. 145 pairs of gloves at $.78 per pair

53. 7540 washers at $.005 each

54. 2182 chips at $1.35 each

Solve each of the following word problems. Estimate each answer first.

55. Judy Lewis has a $28.96-per-month television payment. How much will she pay over 24 months?

56. Chuck's car payment is $180.27 per month for 36 months. How much will he pay altogether?

57. The Dawkins family's state income tax is found by multiplying the family income of $22,906.15 by the decimal .054 and rounding the result to the nearest dollar. Find their tax.

58. A recycling center pays $.142 per pound of plastic. How much would be paid for 196.8 pounds? Round to the nearest cent.

CHALLENGE EXERCISES

59. A fertilizer must be used at the rate of 3.52 gallons per acre. If 158.25 acres are to be fertilized, find the amount of fertilizer remaining in a tank originally containing 600 gallons.

60. Valerie Johnson bought 7.8 yards of a Hawaiian print fabric at $5.62 per yard. If the fabric is paid for with three $20 bills, find the amount of change received. (Ignore sales tax.)

61. Hertz charges $29.95 a day for a certain car rental, plus $.29 per mile. Find the cost of a four-day trip of 926 miles.

62. A motor home rents for $375 per week plus $.35 per mile. Find the rental cost for a three-week trip of 2650 miles.

REVIEW EXERCISES

Divide by using long division and an **R** *to express a remainder. (For help, see* ***Section 1.6.****)*

63. $77\overline{)1650}$

64. $23\overline{)7065}$

65. $26\overline{)62{,}583}$

66. $86\overline{)10{,}327}$

67. $38\overline{)24{,}328}$

68. $21\overline{)149{,}826}$

4.6 DIVISION OF DECIMALS

There are two kinds of decimal division problems; those in which a decimal is divided by a whole number, and those in which a decimal is divided by a decimal.

OBJECTIVES

1. Divide a decimal by a whole number.
2. Divide a decimal by a decimal.
3. Use the order of operations with decimals.

1 Divide a decimal by a whole number with the following rule.

Divide a decimal by a whole number by placing the decimal point in the quotient (answer) directly above the decimal point in the dividend. Then divide as if both numbers were whole numbers.

EXAMPLE 1 Dividing a Decimal by a Whole Number

Divide.

(a) 21.93 by 3 **(b)** $9\overline{)470.7}$

SOLUTION

(a) Write the division problem. $3\overline{)21.93}$

Place the decimal point directly above the decimal point in the dividend.

$3\overline{)21.93}$ directly above

Divide as if the numbers were whole numbers.

$$\begin{array}{r} 7.31 \\ 3\overline{)21.93} \end{array}$$

Check by multiplying the divisor 3 and the quotient 7.31. The product should be equal to the dividend 21.93.

(b) Place the decimal point in the answer above the one in the dividend. Next, divide as if the numbers were whole numbers.

$$\begin{array}{r} 52.3 \\ 9\overline{)470.7} \end{array}$$ directly above ■

WORK PROBLEM 1 AT THE SIDE.

NOTE Attach zeros (extra places) to the dividend until you reach a remainder of 0 or a repeating remainder. This does not change the value of the dividend.

1. Divide.

(a) $4\overline{)93.6}$

(b) $7\overline{)799.4}$

(c) $11\overline{)278.3}$

(d) $129.05 \div 5$

(e) $213.45 \div 15$

ANSWERS

1. (a) 23.4 (b) 114.2 (c) 25.3 (d) 25.81 (e) 14.23

2. Divide. Round answers to the nearest thousandth.

(a) $\frac{34.8}{6}$

(b) $\frac{172.2}{14}$

(c) $\frac{259.5}{30}$

(d) $\frac{403.65}{23}$

ANSWERS
2. (a) 5.8 (b) 12.3 (c) 8.65 (d) 17.55

EXAMPLE 2 Attaching Zeros to Divide
Divide 1.5 by 8.

SOLUTION

$$\begin{array}{r} .1 \\ 8\overline{)1.5} \\ \underline{8} \\ 1. \end{array}$$

Start as above.

Attach as many zeros after the 5 as necessary to make the quotient come out even, or until the desired accuracy is obtained.

$$\begin{array}{r} .1\,875 \\ 8\overline{)1.5\,000} \\ \underline{8} \\ 7\,0 \\ \underline{6\,4} \\ 60 \\ \underline{56} \\ 40 \\ \underline{40} \\ 0 \end{array}$$

← Three 0's are attached.

0 ← comes out even (no remainder) ■

WORK PROBLEM 2 AT THE SIDE.

The next example shows a quotient that must be rounded, since a remainder of 0 does not occur.

EXAMPLE 3 Rounding After a Decimal Division
Divide 4.7 by 3. Round to the nearest thousandth.

SOLUTION
Attach zeros and divide.

$$\begin{array}{r} 1.5\,666 \\ 3\overline{)4.7\,000} \end{array}$$

Attach zeros.

Notice that the digits repeat and will continue to do so. The remainder will never be zero. The answer is

$$1.5666\ldots$$

The dots indicate a repeating pattern that is endless. The decimal does not terminate. These dots are often replaced by a raised bar to indicate that the digits repeat. For example,

$$1.5666\ldots = 1.5\overline{6}.$$

Both indicate that the 6 is a **repeating digit.**

When repeating digits occur, round according to the directions given for the problem.

$$4.7 \div 3 = 1.567 \quad \text{(rounded to the nearest thousandth)}$$

Check this answer by multiplying 3 and 1.567.

$$3 \cdot 1.567 = 4.701 \quad \text{(a little inaccurate because of rounding)}$$ ■

Round a quotient to a given place by carrying out the division to one decimal place past the given place and then rounding to the given place.

WORK PROBLEM 3 AT THE SIDE.

2 Find the rule for dividing a decimal by a decimal by thinking of the quotient as a fraction. For example, think of

$$.3\overline{)27.69} \quad \text{as} \quad \frac{27.69}{.3}.$$

Convert the denominator into a whole number by multiplying numerator and denominator by 10.

$$\frac{27.69}{.3} = \frac{27.69 \times 10}{.3 \times 10} = \frac{276.9}{3}$$

Next, divide as shown above. $\begin{array}{r} 92.3 \\ 3\overline{)276.9} \end{array}$

Check by multiplying 92.3 and 3.

Divide a decimal by another decimal with the following rule.

Step 1 To divide a decimal by another decimal, move the decimal point in the divisor to the right of the last digit. Place a caret (‸) at the right of that digit.

Step 2 Move the decimal point in the dividend as many places to the right as you moved the one in the divisor. Place another ‸ there.

Step 3 The decimal point in the quotient goes directly above the ‸ in the dividend. Then divide as usual.

The next example shows these rules used with the division problem worked above.

EXAMPLE 4 Dividing a Decimal by a Decimal

Divide $.3\overline{)27.69}$.

SOLUTION

$.3_{\wedge}\overline{)27.6_{\wedge}9}$ Move the decimal points in the divisor and dividend one place to the right.

$\begin{array}{r} 92.3 \\ 3\overline{)276_{\wedge}9} \end{array}$ Place the decimal point in the quotient and divide.

Check by using the original numbers.

$$.3 \cdot 92.3 = 27.69$$ ■

3. Divide. Round answers to the nearest thousandth if necessary.

(a) $12\overline{)137.04}$

(b) $6\overline{)20.5}$

(c) 1028.52 ÷ 9

(d) 59.21 ÷ 3

(e) 116.3 ÷ 7

ANSWERS

3. (a) 11.42 (b) 3.417 (c) 114.28 (d) 19.737 (e) 16.614

4. Divide. If the quotient does not come out even, round to the nearest thousandth.

(a) $.2\overline{)28.84}$

(b) $.8\overline{)9.0064}$

(c) $.005\overline{)32}$

(d) $10.0004 \div 8$

(e) $70 \div 2.8$

(f) $5.3091 \div 6.2$

5. Simplify by using the order of operations.

(a) $1.8^2 + 3.5 - 2.6$

(b) $3.64 \div 1.3 \cdot 3.6$

(c) $5.2 + (9.7 - 7.9) \cdot 4.5$

(d) $10.85 - 2.3 \cdot 5.2 \div 3.2$

ANSWERS
4. (a) 144.2 (b) 11.258 (c) 6400
(d) 1.25005 (e) 25 (f) .856
5. (a) 4.14 (b) 10.08 (c) 13.3
(d) 7.1125

EXAMPLE 5 Dividing a Whole Number by a Decimal
Divide 5 by 4.2. Round to the nearest thousandth if necessary.

SOLUTION

$$\begin{array}{r} 1.1904 \\ 4.2\overline{)5.0\,0000} \\ \underline{4\,2} \\ 8\,0 \\ \underline{4\,2} \\ 3\,80 \\ \underline{3\,78} \\ 200 \\ \underline{168} \\ 32 \end{array}$$

Since the remainder is not 0, round the quotient to 1.190. ■

WORK PROBLEM 4 AT THE SIDE.

3 Use the order of operations when a decimal problem involves more than one operation.

ORDER OF OPERATIONS

1. Do all operations inside parentheses.
2. Simplify any expressions with exponents and find any square roots.
3. Multiply or divide from left to right.
4. Add or subtract from left to right.

EXAMPLE 6 Using the Order of Operations
Simplify by using the order of operations.

(a) $2.5^2 + 6.3 + 9.62 = 6.25 + 6.3 + 9.62$ Use the exponent.
$= 12.55 + 9.62$ Add from left to right.
$= 22.17$

(b) $1.82 + (6.7 - 5.2) \cdot 5.8 = 1.82 + 1.5 \cdot 5.8$ Work inside parentheses first.
$= 1.82 + 8.7$ Multiply next.
$= 10.52$ Next, add.

(c) $2.2^2 + 1.8 \times 5.1 \div 1.5 = 4.84 + 1.8 \times 5.1 \div 1.5$ Use the exponent.
$= 4.84 + 6.12$ Multiply and divide.
$= 10.96$ Next, add. ■

WORK PROBLEM 5 AT THE SIDE.

name date hour

4.6 EXERCISES

Divide. Round to the nearest thousandth if necessary.

Example: $8\overline{)39.65}$

Solution:

$$\begin{array}{r} 4.9562 \\ 8\overline{)39.6500} \\ \underline{32} \\ 7\,6 \\ \underline{7\,2} \\ 45 \\ \underline{40} \\ 50 \\ \underline{48} \\ 20 \end{array} \quad = \mathbf{4.956} \text{ (rounded)}$$

1. $6\overline{)24.84}$ **2.** $4\overline{)32.84}$ **3.** $5\overline{)20.01}$ **4.** $15\overline{)30.06}$

5. $4\overline{)14.7389}$ **6.** $7\overline{)38.2465}$ **7.** $4\overline{).008}$ **8.** $6\overline{)32.71}$

9. $9\overline{)81.0009}$ **10.** $8\overline{).032}$ **11.** $8\overline{)93.52}$ **12.** $4\overline{)62.19}$

13. $22\overline{)66.836}$ **14.** $14\overline{)152.86}$ **15.** $16\overline{).064}$ **16.** $12\overline{)144.006}$

17. $35\overline{)43.605}$ **18.** $29\overline{)583.225}$ **19.** $78\overline{)66.059}$ **20.** $54\overline{)77.113}$

21. $3\overline{).027}$ **22.** $.05\overline{)80}$ **23.** $.009\overline{)27}$ **24.** $4.6\overline{)116.38}$

25. $7.5\overline{)16.2}$ **26.** $.8\overline{).002}$ **27.** $2.6\overline{)4.987}$ **28.** $.71\overline{)6.72}$

29. $.004\overline{)16}$ **30.** $.008\overline{)40}$ **31.** $.25\overline{).001}$ **32.** $.89\overline{)27.3214}$

33. $2.43\overline{)9.6153}$ **34.** $.006\overline{).36}$ **35.** $.025\overline{).0759}$ **36.** $.003\overline{)11.628}$

37. $.004\overline{).0016}$

38. $9.15\overline{)6.003}$

39. $.001\overline{).056}$

40. $428.17 \div .034$

41. $921.47 \div 25$

42. $.034 \div 17$

43. $375.429 \div 12$

44. $70.032 \div 1.9$

45. $12.988 \div 24$

46. $764.358 \div 5.36$

47. $.00102 \div .034$

48. $1748.4 \div .043$

Find the cost of each item. Round to the nearest cent (hundredth).

Example:	**Solution:**	
3 rolls of film for $1.89	$\begin{array}{r} \$\ .63 \\ 3\overline{)\ 1.89} \\ \underline{1\ 8} \\ 09 \end{array}$	cost per roll

49. 6 pairs of socks for $5.98

50. 7 tacos for $4.00

51. 4 notepads for $1.69

52. 20 bundles of shingles for $447.20

53. 480 bricks for $187.20

54. 24 cans of beverage for $4.59

55. 500 pencils for $25.50

56. 3 dozen pens (36) for $3.98

Solve each word problem.

57. Kim Hutchinson drove 346.2 miles on the 16.3 gallons of gas in the tank of her Ford Mustang. How many miles per gallon did she get? Round to the nearest tenth.

58. Michael Anderson bought 7.4 yards of fabric, paying a total of $26.27. Find the cost per yard.

59. Adrian Webb bought 619 bricks to build a barbecue, paying $185.70. Find the cost per brick.

60. Lupe Wilson is a newspaper distributor. Last week she paid the newspaper $130.51 for 842 copies. Find the cost per copy. Round to the nearest cent.

61. Darren Jackson earned \$235.50 for 50 hours of work. Find his earnings per hour.

62. At a record manufacturing company, 400 records cost \$289. Find the cost per record. Round to the nearest cent.

63. Aimee Coulter pays \$19.46 per month on a charge account. It will take her 21 months to pay off the account. Find the amount that she owes.

64. Suppose Chris Rodriquez pays \$53.19 per month to Household Finance. How many months will it take to pay off a loan, if \$1436.13 is owed?

*A **gross** of an item contains 144 items. Find the price for one of each of the following items. Do not round.*

65. A gross of pencils for \$12.24

66. A gross of pens for \$133.20

67. 2 gross of large paper clips for \$27.36

68. 2 gross of clothespins for \$61.92

Simplify by using the order of operations.

Example: $5.2^2 + 7.9 - 6.3$	$= 27.04 + 7.9 - 6.3$	Do exponent first.
	$= 34.94 - 6.3$	Add.
	$= \mathbf{28.64}$	Subtract.

69. $3.5^2 + 5.2 - 7.2$

70. $4.3^2 + 6.2 - 9.72$

71. $9.152 \div 4.16 \cdot 1.5$

72. $12.6 \div 8.4 \cdot 6.2$

73. $38.6 + (13.4 - 10.4) \cdot 11.6$

74. $1.06 + (4.85 - 3.95) \cdot 2.25$

75. $8.68 - 4.6 \cdot 10.4 \div 6.4$

76. $25.1 - 11.4 \cdot 7.5 \div 3.75$

CHALLENGE EXERCISES

77. A store offers a personal computer for $1250 with $350 as a down payment. The balance is to be paid in eight equal monthly payments. Find the amount of each payment.

78. Stockdale Marine offers a sailboat for $10,296 with $1800 as a down payment and monthly payments of $236. Find the number of payments needed to pay off the boat.

79. The annual premium for an auto insurance policy is $938. This premium may be paid quarterly by adding an additional $2.75 to each payment. Find the amount of each quarterly payment.

80. Lock and Store charges rent of $936 per year for 200 square feet of storage space. To pay the rent monthly, $1.25 must be added to each payment. Find the amount of each monthly payment.

REVIEW EXERCISES

Write < or > to make a true statement. (For help, see ***Section 3.5****.)*

(For help, see Section 3.5.)

81. $\frac{1}{4} \quad \frac{2}{3}$

82. $\frac{5}{8} \quad \frac{11}{16}$

83. $\frac{9}{5} \quad \frac{23}{15}$

84. $\frac{7}{8} \quad \frac{11}{12}$

85. $\frac{12}{24} \quad \frac{23}{36}$

86. $\frac{21}{40} \quad \frac{17}{30}$

4.7 WRITING FRACTIONS AS DECIMALS

A fraction must often be written as a decimal to perform mathematical calculations or to compare the size of two numbers.

OBJECTIVES

1. Change a fraction to a decimal.
2. Compare the size of fractions and decimals.

1 The fraction $\frac{a}{b}$ means $a \div b$, so $\frac{a}{b}$ can be written as a decimal by dividing a by b.

Write a fraction as a decimal by dividing the numerator of the fraction by the denominator.

EXAMPLE 1 Writing a Fraction as a Decimal

Write $\frac{1}{8}$ as a decimal. (Recall that $\frac{1}{8}$ means $1 \div 8$.)

SOLUTION

Divide 1 by 8. $8\overline{)1}$

The decimal point for any whole number is to the right of the number.

$8\overline{)1.}$ ← Decimal point is placed here.

Attach zeros and divide.

Decimal point comes straight up.

```
   .125
8)1.000  ← Three 0's are attached.
   8
   20
   16
    40
    40
     0
```

Therefore, $\frac{1}{8} = .125$. ■

WORK PROBLEM 1 AT THE SIDE.

EXAMPLE 2 Changing to a Decimal and Rounding

Write $\frac{2}{3}$ as a decimal and round to the nearest thousandth.

SOLUTION

Divide 2 by 3.

$3\overline{)2}$

1. Write each fraction or mixed number as a decimal.

(a) $\frac{1}{4}$

(b) $2\frac{1}{2}$

(c) $\frac{3}{8}$

(d) $7\frac{2}{5}$

(e) $\frac{7}{8}$

ANSWERS
1. (a) .25 (b) 2.5 (c) .375 (d) 7.4 (e) .875

2. Write as decimals. Round to the nearest thousandth.

(a) $\frac{1}{3}$

(b) $\frac{5}{9}$

(c) $\frac{7}{11}$

(d) $\frac{3}{7}$

(e) $\frac{5}{6}$

Attach four zeros for rounding to the nearest thousandth.

$$\begin{array}{r} .6666 \\ 3\overline{)2.0000} \\ \underline{1\ 8} \\ 20 \\ \underline{18} \\ 20 \\ \underline{18} \\ 20 \\ \underline{18} \\ 2 \end{array}$$

← Four 0's are attached for rounding.

To the nearest thousandth, $\frac{2}{3} = .667$. ■

WORK PROBLEM 2 AT THE SIDE.

The following table lists some of the most common fractions, written as decimals, from smallest to largest. Some of the divisions have been carried out completely, and others have been rounded to the nearest ten-thousandth.

$\frac{1}{16} = .0625$	$\frac{3}{16} = .1875$	$\frac{3}{8} = .375$	$\frac{11}{16} = .6875$
$\frac{1}{9} = .1111$	$\frac{1}{5} = .2$	$\frac{1}{2} = .5$	$\frac{3}{4} = .75$
$\frac{1}{8} = .125$	$\frac{1}{4} = .25$	$\frac{9}{16} = .5625$	$\frac{13}{16} = .8125$
$\frac{1}{7} = .1429$	$\frac{5}{16} = .3125$	$\frac{5}{8} = .625$	$\frac{5}{6} = .8333$
$\frac{1}{6} = .1667$	$\frac{1}{3} = .3333$	$\frac{2}{3} = .6667$	$\frac{7}{8} = .875$

2 Fractions can be compared to see which is greater by first writing each as a decimal. The decimals can then be compared by writing each with the same number of decimal places.

EXAMPLE 3 Arranging Numbers in Order

Write the following numbers in order, from smallest to greatest.

(a)	.49	.487	.4903
(b)	$\frac{5}{8}$	.62	.6182

ANSWERS

2. (a) .333 (b) .556 (c) .636 (d) .429 (e) .833

SOLUTION

(a) The decimal .4903 has four decimal places, so write each decimal with four places.

.4900	.4870	.4903
↑↑ Two 0's are attached.	↑ One 0 is attached.	

Mentally ignore the decimal points, and arrange the numbers in order.

4870	4900	4903
smallest		greatest

Next, write the original decimals in the correct order.

.487 .49 .4903

(b) The fraction $\frac{5}{8}$ written as a decimal is .625.

Since the decimal .6182 has four decimal places, write each decimal with four places.

.6250 .6200 .6182

Ignoring the decimal point, arrange the numbers in order.

6182	6200	6250
smallest		greatest

Next, write the original numbers in the correct order.

.6182 .62 $\frac{5}{8}$ ■

WORK PROBLEM 3 AT THE SIDE.

3. Arrange in order, from smallest to greatest.

(a) .7, .703, .7029

(b) 2.59, 2.507, 2.613, 2.6

(c) $\frac{3}{4}$, $\frac{11}{12}$, .82

(d) $\frac{1}{4}$, $\frac{2}{5}$, $\frac{3}{7}$, .428

ANSWERS

3. (a) .7, .7029, .703 (b) 2.507, 2.59, 2.6, 2.613 (c) $\frac{3}{4}$, .82, $\frac{11}{12}$ (d) $\frac{1}{4}$, $\frac{2}{5}$, .428, $\frac{3}{7}$

name date hour

4.7 EXERCISES

Change each of the following fractions to decimals. Round to the nearest thousandth if necessary.

Example: $\frac{9}{16} = .563$ **Solution:**

$$\begin{array}{r} .5625 \\ 16\overline{)9.0000} \\ \underline{8\ 0} \\ 1\ 00 \\ \underline{96} \\ 40 \\ \underline{32} \\ 80 \\ \underline{80} \\ 0 \end{array}$$

1. $\frac{1}{2}$ **2.** $\frac{1}{4}$ **3.** $\frac{1}{3}$ **4.** $\frac{1}{5}$

5. $\frac{1}{6}$ **6.** $\frac{1}{7}$ **7.** $5\frac{1}{8}$ **8.** $\frac{1}{9}$

9. $\frac{1}{10}$ **10.** $10\frac{3}{4}$ **11.** $\frac{2}{3}$ **12.** $\frac{3}{8}$

13. $\frac{4}{5}$ **14.** $\frac{17}{25}$ **15.** $3\frac{22}{25}$ **16.** $14\frac{19}{20}$

17. $\frac{11}{16}$

18. $\frac{13}{16}$

19. $\frac{6}{7}$

20. $22\frac{3}{11}$

21. $\frac{13}{18}$

22. $\frac{7}{18}$

23. $15\frac{17}{36}$

24. $\frac{29}{36}$

Determine the decimal or fraction equivalent for each of the following. Round decimals to the nearest thousandth and write fractions in lowest terms.

Example:

fraction	*decimal*
$\frac{3}{8}$	________

Solution:

$$\begin{array}{r} .375 \\ 8\overline{)3.000} \\ \underline{2\,4} \\ 60 \\ \underline{56} \\ 40 \\ \underline{40} \\ 0 \end{array}$$

$\frac{3}{8}$ = **.375**

	fraction	*decimal*		*fraction*	*decimal*
25.	$\frac{1}{4}$	________	**26.**	$\frac{1}{8}$	________

name ______ date ______ hour ______

	fraction	*decimal*		*fraction*	*decimal*
27.	______	.4	**28.**	______	.75
29.	$\frac{7}{8}$	______	**30.**	$\frac{2}{3}$	______
31.	______	.875	**32.**	______	.111
33.	______	.35	**34.**	______	.7
35.	$\frac{7}{20}$	______	**36.**	$\frac{1}{40}$	______

	fraction	*decimal*		*fraction*	*decimal*
37.	________	.65	**38.**	________	.05
39.	$\frac{5}{6}$	________	**40.**	$\frac{1}{10}$	________
41.	________	.1	**42.**	________	.01
43.	________	.15	**44.**	________	.85
45.	$\frac{3}{5}$	________	**46.**	$\frac{1}{6}$	________

name date hour

Find the smallest of the two numbers. Write $<$ or $>$ to make a true statement.

Example:

$\frac{7}{8}$ ____ .87

Solution:

Change $\frac{7}{8}$ to a decimal.

$$\begin{array}{r} .875 \\ 8\overline{)7.000} \\ \underline{6\,4} \\ 60 \\ \underline{56} \\ 40 \\ \underline{40} \\ 0 \end{array}$$

$\frac{7}{8} = .875$

Write .87 as .870 to see that .870 is smaller than .875.

Therefore, $\frac{7}{8} > .87$

47. $\frac{3}{8}$ ____ .38

48. $\frac{4}{5}$ ____ .75

49. $\frac{1}{4}$ ____ .28

50. $\frac{1}{3}$ ____ .35

51. $\frac{5}{8}$ ____ .60

52. $\frac{5}{6}$ ____ .75

53. $\frac{7}{8}$ ____ .90

54. $\frac{1}{9}$ ____ .12

55. $\frac{2}{3}$ ____ .64

56. $\frac{1}{8}$ ____ .12

57. $\frac{1}{6}$ ____ .18

58. $\frac{1}{20}$ ____ .04

Circle the smallest of the given numbers.

Example:

$\frac{4}{9}$, .44, .451

Solution:

Change $\frac{4}{9}$ to a decimal.

$$\begin{array}{r} .4444 \\ 9\overline{)4.0000} \\ \underline{3\,6} \\ 40 \\ \underline{36} \\ 40 \\ \underline{36} \\ 40 \\ \underline{36} \\ 4 \end{array}$$

$\frac{4}{9}$ = .444 (rounded)

Write .44 as .440 to see that .440 is smaller than .444 or .451.

Therefore, .44 is smaller than $\frac{4}{9}$ and .451. Circle **(.44)**

59. $\frac{3}{5}$, .062, .55

60. $\frac{5}{6}$, .83, $\frac{7}{8}$

61. $\frac{1}{2}$, $\frac{5}{8}$, .506

62. $\frac{3}{11}$, $\frac{1}{3}$, .28

63. .0909, .091, $\frac{1}{11}$

64. .8061, $\frac{4}{5}$, .80609

65. .084, $\frac{1}{12}$, .08395

66. $\frac{3}{8}$, .3759, $\frac{1}{15}$

67. .17, $\frac{1}{6}$, $\frac{2}{13}$

68. .7962, $\frac{11}{14}$, .7909

69. .8925, .893, $\frac{7}{8}$

70. .4598, $\frac{7}{15}$, $\frac{9}{19}$

Arrange in order from smallest to greatest.

Example: .863, $\frac{7}{8}$, .88, .8632

Solution: .8630, .8750, .8800, .8632

(smallest) **.863** **.8632** $\mathbf{\frac{7}{8}}$ **.88** (greatest)

71. .54, .5455, .5399

72. .72, .7114, .7, .7006

73. 5.8, 5.79, 5.4443, 5.804

74. 12.99, 12.5, 13.0001, 12.77

75. .628, .62812, .609, $\frac{5}{8}$

76. .27, .281, $\frac{2}{7}$, .296

77. .8751, .876, $\frac{7}{8}$, .8902

78. .38, $\frac{3}{8}$, $\frac{2}{5}$, $\frac{3}{7}$

79. .043, .051, .506, $\frac{1}{20}$

80. .729, .7305, .709, $\frac{7}{3}$

81. $\frac{3}{4}$, $\frac{12}{15}$, .762, .7781

82. $\frac{1}{7}$, .15, $\frac{2}{15}$, .1501

CHALLENGE EXERCISES

Arrange in order from smallest to greatest.

83. $\frac{6}{11}$, $\frac{5}{9}$, $\frac{4}{7}$, .571

84. $\frac{8}{13}$, $\frac{10}{17}$, .615, $\frac{11}{19}$

85. $\frac{3}{11}$, $\frac{4}{15}$, .25, $\frac{1}{3}$

86. .223, $\frac{2}{11}$, $\frac{2}{9}$, $\frac{1}{4}$

87. $\frac{3}{16}$, $\frac{1}{4}$, $\frac{1}{5}$, .188

88. $\frac{7}{20}$, $\frac{1}{3}$, $\frac{3}{8}$, .361

REVIEW EXERCISES

Write each fraction in lowest terms. (For help, see **Section 2.4.**)

89. $\frac{9}{12}$

90. $\frac{30}{60}$

91. $\frac{60}{80}$

92. $\frac{40}{75}$

93. $\frac{96}{132}$

94. $\frac{26}{98}$

CHAPTER 4 SUMMARY

KEY TERMS

4.1	**decimals**	In addition to fractions, decimals are another way to show parts of a whole.
	decimal point	The starting point in the decimal system. The point or period that is used to separate the whole-number part from the fractional part of a number.
	place value	The value assigned to each place to the right or left of the decimal point—for example, ones, tenths, hundredths. Whole numbers are to the left of the decimal point and decimal numbers are to the right.
4.2	**rounding**	The reduction of a number with more decimals to a number with fewer decimals. The rounded number is less accurate than the original number.
4.5	**estimating**	The process of approximating an answer to make sure the decimal is in the correct place.
4.6	**repeating decimal**	A repeating decimal has a digit that repeats, such as the 6 in .1666. . . . The dots indicate that the decimal does not terminate but continues to repeat.

QUICK REVIEW

Section Number and Topic	Approach	Example
4.1 Reading and Writing Decimals	. decimal point ("and") 0 tenths 7 hundredths 3 thousandths 2 ten-thousandths 6 hundred-thousandths 5 millionths	Read and write the following decimals. 5.2 — five and two tenths 211.25 — two hundred eleven and twenty-five hundredths
4.2 Rounding Decimals	Find the digit in the position to which the number is being rounded. If the digit to the right is 5 or greater, add 1; if the digit to the right is 4 or less, leave as is, dropping all digits to the right of the digit being rounded.	Round .073265 to the nearest ten-thousandth. .073 2 65 — ten-thousandth position Since the digit to the right is 6, increase the ten-thousandth digit by 1. .073265 rounds to .0733.
4.3 "and" **4.4** Addition and Subtraction of Decimals	Decimal points must be in a column. Attach zeros to keep digits in their correct columns.	Add 5.68 + 785.3 + .007 + 10.1062. Line up decimal points. 5.6800 785.3000 .0070 + 10.1062 801.0932 Attach zeros.

Section Number and Topic	Approach	Example
4.5 Multiplication of Decimals	Multiply as if decimals are whole numbers. Place decimal point in the product as follows: 1. Count digits to the right of decimal points in both the multiplicand and the multiplier. 2. In the product, count from right to left the same number of places as in step 1. Zeros may be attached on the left if necessary.	Multiply .169 × .21. .169 — 3 decimal places × .21 — 2 decimal places 169 — 5 decimal places 338 — are in answer. .03549 ↑ Attach one zero.
4.6 Dividing by a Decimal	1. Move the decimal point in the divisor all the way to the right. 2. Move the decimal point in the dividend the same number of places to the right. 3. Place a decimal point in the answer position directly above the dividend decimal point. 4. Divide as with whole numbers.	Divide 52.8 by .75. 70.4 .75)52.80 0 — Move decimal points two places to the right. 52 5 30 00 300 300 0 Check. 70.4 × .75 3520 4928 52.800
4.7 Writing Fractions as Decimals	Divide the numerator by the denominator. Round if necessary.	Write $\frac{1}{8}$ as a decimal. .125 8)1.000 8 20 16 40 40 0 $\frac{1}{8} = .125$
4.7 Write Decimals as Fractions	Think of the decimal as being written in words; next, write in fraction form. Reduce to lowest terms.	Write .47 as a fraction. Think of .47 as "forty-seven hundredths"; next, write as $\frac{47}{100}$.

name date hour

CHAPTER 4 REVIEW EXERCISES

[4.1] *Name the digit that has the given place value in each of the following.*

1. 6.58
tenths
hundredths

2. .7853
tenths
ten-thousandths

3. 435.621
tenths
thousandths

4. 896.503
tenths
hundredths

5. 620.738
tenths
thousandths

Write each of the following decimals as fractions in lowest terms.

6. .9

7. .75

8. .03

9. .875

10. .6158

11. .8895

Write the word name of each of the following decimals.

12. .8

13. .45

14. 12.87

15. 335.708

16. 42.105

[4.2] *Round each of the following to the place shown.*

17. 275.635 to the nearest tenth

18. 72.789 to the nearest hundredth

19. 896.253 to the nearest hundredth

20. .0235 to the nearest thousandth

21. 87.4798 to the nearest one

Round each of the following to the nearest dollar.

22. $15.83

23. $81.51

24. $17,625.79

Round each of the income and expense items to the nearest dollar.

25. Income from washers and dryers is $78.58.

26. Lawn care costs $37.28.

27. Water and garbage bill is $19.20.

[4.3] *First estimate the answer, then add.*

28.
$$\begin{array}{r} 78.56 \\ 22.15 \\ +\ 39.68 \\ \hline \end{array}$$

29.
$$\begin{array}{r} 5.8 \\ 23.96 \\ +\ 15.09 \\ \hline \end{array}$$

30.
$$\begin{array}{r} 3.58 \\ 7.9 \\ 5.7 \\ 8.65 \\ +\ 20.4 \\ \hline \end{array}$$

31.
$$\begin{array}{r} 75.6 \\ 1.63 \\ 22.045 \\ 1.88 \\ +\ 33.7 \\ \hline \end{array}$$

32. $45.6 + 7.09 + 5.63 + 78.09$

33. $11.206 + 50.3 + 77.8 + 9.05 + 11.7$

[4.4] *First estimate the answer and next, subtract.*

34.
$$\begin{array}{r} 28.6 \\ -\ 17.4 \\ \hline \end{array}$$

35.
$$\begin{array}{r} 28.5 \\ -\ 17.8 \\ \hline \end{array}$$

36.
$$\begin{array}{r} 36.356 \\ -\ 25.348 \\ \hline \end{array}$$

37.
$$\begin{array}{r} 8.731 \\ -\ 7.8 \\ \hline \end{array}$$

[4.3–4.4] *Solve the following word problems.*

38. Tim agreed to donate 30 hours of work to the day school. If he has already worked 19.6 hours, how many more hours does he have to work?

39. A shopper cashed a paycheck for $215.53 to pay for $43.89 worth of groceries. Find his change.

40. Joey spent $1.18 for toothpaste, $5.83 for a gift, and $15.94 for a humidifier. How much did he spend?

41. Roseanne jogged 2.36 miles on Monday, 3.58 miles on Tuesday, 1.78 miles on Wednesday, 3.9 miles on Thursday, and 5.35 miles on Friday. How many miles did she jog altogether?

[4.5] *Multiply.*

42. $\begin{array}{r} .312 \\ \times\ 5.6 \\ \hline \end{array}$

43. $\begin{array}{r} 6.138 \\ \times\ .037 \\ \hline \end{array}$

44. $\begin{array}{r} 42.09 \\ \times\ .023 \\ \hline \end{array}$

45. (5.6)(.02)

Find the cost of each of the following.

46. 41 pounds of rabbit food at $2.46 per pound

47. 48 packs of gum at $.088 per pack

[4.2–4.5] *Solve the following word problems.*

48. Adriene worked 36 hours at $3.57 an hour. Find her total earnings.

49. The 18 monthly payments on a microwave oven are $26.48. Find the total amount of the payments.

[4.6] *Divide. Round to the nearest thousandth.*

50. $3\overline{)43.5}$

51. $7.56\overline{)35.648}$

52. $.05\overline{)775}$

53. $.00048 \div .0012$

Solve the following word problems.

54. Stock in the Beaver Corporation sells for \$33.75 per share. How many shares of stock can be purchased for \$2970?

55. Irma Reynolds earned \$286.40 for 44.75 hours of work. Find her earnings per hour.

Simplify by using the order of operations. Round to the nearest thousandth.

56. $3.5^2 + 8.7 - 1.95$

57. $3.16 \div 3.95 - .33$

[4.7] *Write each fraction as a decimal. Round to the nearest thousandth.*

58. $\frac{4}{5}$

59. $\frac{16}{25}$

60. $\frac{14}{21}$

61. $\frac{1}{9}$

Arrange in order from smallest to greatest.

62. .68, .6821, .67295

63. .215, .22, .209, .2102

64. .17, $\frac{3}{20}$, $\frac{1}{6}$, .159

name date hour

MIXED REVIEW EXERCISES

Solve the following problems. Round to the nearest thousandth if necessary.

65. $\begin{array}{r} 72.11 \\ 5.06 \\ 31.673 \\ 4.08 \\ +\ 9.7 \\ \hline \end{array}$

66. $\begin{array}{r} 72.8 \\ \times\ 3.5 \\ \hline \end{array}$

67. $1648.3 \div .46$

68. $\begin{array}{r} 70.2 \\ -\ 35.668 \\ \hline \end{array}$

69. $\begin{array}{r} 34.28 \\ \times\ .08 \\ \hline \end{array}$

70. $.387\overline{).047}$

71. $\begin{array}{r} 72.105 \\ 8.2 \\ +\ 95.37 \\ \hline \end{array}$

72. $9\overline{)81.36}$

73. $\begin{array}{r} 21.059 \\ -\ 3.8 \\ \hline \end{array}$

74. $(.07)(.38)$

75. $\begin{array}{r} 1.60 \\ \times\ .508 \\ \hline \end{array}$

76. $.218\overline{)17.63}$

77. $(5.6 - 1.22) \cdot 4.8 \div 3.15$

78. $2^3 + 1.3 - 5.6$

Write each fraction as a decimal. Round to the nearest thousandth.

79. $\frac{2}{5}$

80. $\frac{11}{20}$

81. $\frac{3}{10}$

82. $\frac{1}{7}$

Arrange in order from smallest to greatest.

83. $\frac{3}{8}$, .381, .3749, $\frac{2}{5}$

84. .348, $\frac{7}{20}$, $\frac{8}{23}$, .375

Solve the following word problems.

85. Ed drove 285.6 miles on 18.4 gallons of gas. How many miles per gallon did he get. Round to the nearest tenth of a mile.

86. The Willis family pays $1.09 per gallon of heating oil. Find the cost of 412.7 gallons of heating oil.

87. Find the number of months needed to repay a loan of $3136.50, if each monthly payment is $174.25.

88. A bank charge card bill includes credits of $38.52 and $14.24, and charges of $9.36, $17.29, $85.82, $23.59, $16.73, and $46.71. If the balance due on the account was originally zero, find the balance now due.

89. Toni Shellan spent $89.14 for books, $14.37 for supplies, $.75 for parking, and $3.68 for lunch. How much money remains if she started with $125.50?

90. Econo Car Rental charges $19.95 per day plus $7.80 per day for insurance. In addition, there is a mileage charge of $.15 per mile. In 4 days, 652 miles were traveled. Find the cost of the rental including insurance.

name date hour

CHAPTER 4 TEST

Write each decimal as a fraction.

1. .937 **2.** .3053

Write each decimal as a fraction in lowest terms.

3. .875 **4.** .0075

Round as shown.

5. 725.60895 to the nearest thousandth

6. .705149 to the nearest ten-thousandth

Round to the nearest dollar.

7. $611.49 **8.** $7859.51

Add.

9. 7.6059 + 82.0128 + 32.59

10. 53.182 + 4.631 + 782.052 + .031

Subtract.

11. 23.602 from 79.1 **12.** 667.815 − .996

Multiply.

13. 45.79 · .03 **14.** .0069 · .007

Divide.

15. $\frac{96}{.015}$ **16.** $.15\overline{)90.7545}$

1. ______
2. ______
3. ______
4. ______
5. ______
6. ______
7. ______
8. ______
9. ______
10. ______
11. ______
12. ______
13. ______
14. ______
15. ______
16. ______

Arrange in order from smallest to greatest.

17. ______________

17. .508, .516, .5108

18. ______________

18. .44, .451, $\frac{9}{20}$, .4606

19. ______________

19. .602, $\frac{3}{5}$, .5983, $\frac{2}{3}$

Use the order of operations to simplify.

20. ______________

20. $6.3^2 - 5.9 + 3.4 \cdot .5$

Solve the following word problems.

21. ______________

21. Jennifer had $271.15 in her checking account. In one week, she wrote checks for $5.73, $17.92, $29.56, and $153.18. Find her checking account balance after writing these checks.

22. ______________

22. How many gallons of chlorine are in 10.6 drums containing 38.25 gallons each?

23. ______________

23. A new car traveled 352 miles on 12.7 gallons of gas. Find the number of miles traveled per gallon. Round to the nearest tenth of a mile.

24. ______________

24. Alfafa seeds cost $34.75 for a 12.5-pound sack. Find the cost per pound.

25. ______________

25. A utility bill shows daily usage of 4.6 therms of gas. Find the number of therms used in 32 days.

5

RATIO AND PROPORTION

Many quantities in everyday life are measured in *units,* such as 10 *dollars,* 5 *feet,* or 17 *days*. Two quantities measured by the same units can be compared by forming a *ratio*.

5.1 RATIOS

OBJECTIVES

1. Write ratios as fractions.
2. Solve ratio problems involving mixed numbers.
3. Solve ratio problems after converting units.

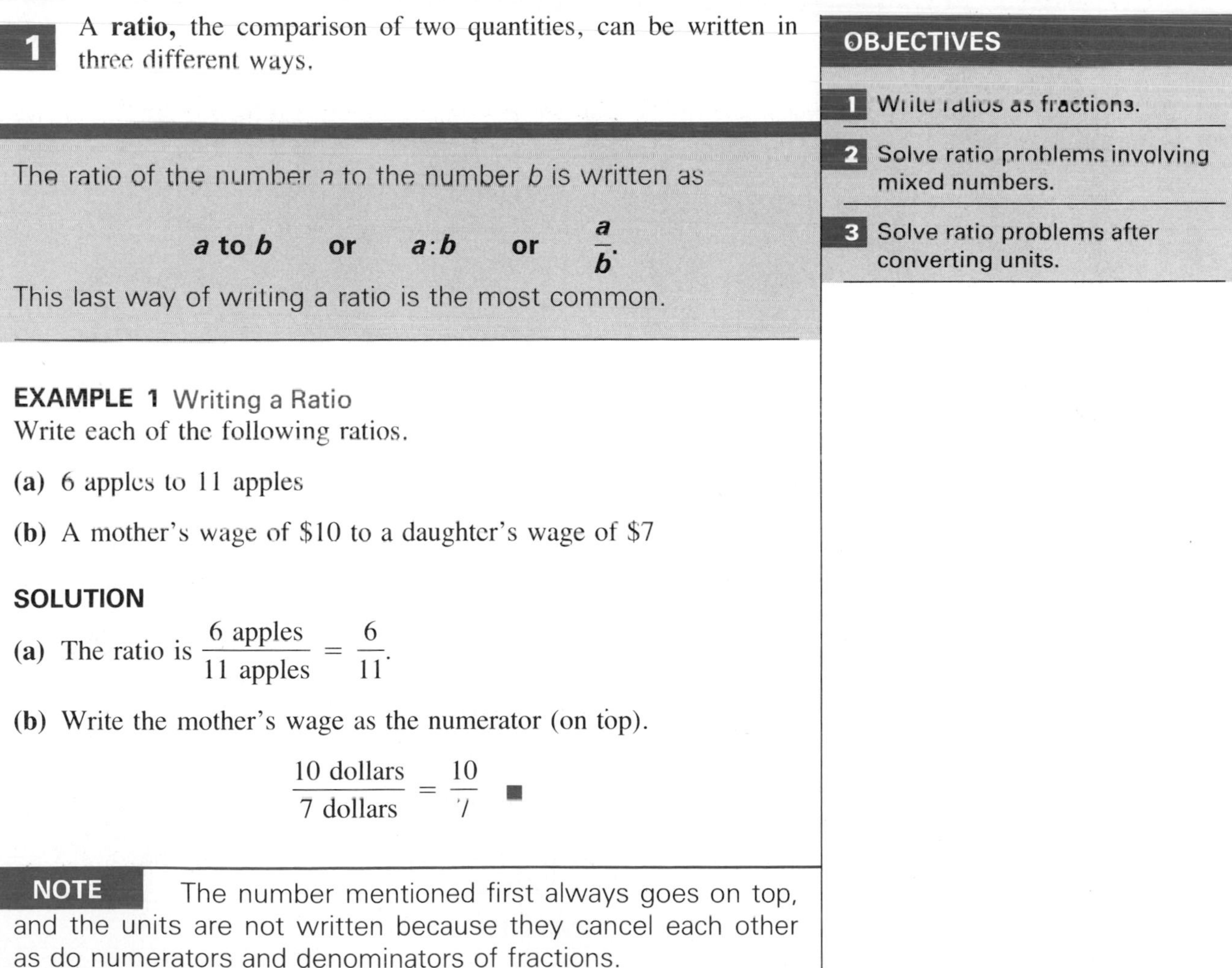

1 A **ratio,** the comparison of two quantities, can be written in three different ways.

The ratio of the number a to the number b is written as

$$a \text{ to } b \quad \text{or} \quad a:b \quad \text{or} \quad \frac{a}{b}.$$

This last way of writing a ratio is the most common.

EXAMPLE 1 Writing a Ratio
Write each of the following ratios.

(a) 6 apples to 11 apples

(b) A mother's wage of \$10 to a daughter's wage of \$7

SOLUTION

(a) The ratio is $\frac{6 \text{ apples}}{11 \text{ apples}} = \frac{6}{11}$.

(b) Write the mother's wage as the numerator (on top).

$$\frac{10 \text{ dollars}}{7 \text{ dollars}} = \frac{10}{7} \quad ■$$

NOTE The number mentioned first always goes on top, and the units are not written because they cancel each other as do numerators and denominators of fractions.

1. Write each ratio.

(a) 8 basketball goals to 15 basketball goals

(b) $85 from one job to $18 from another job

(c) 12 men to 17 men

(d) 17 women to 12 women

2. Write each ratio in lowest terms.

(a) 9 quarts to 15 quarts

(b) 16 tons to 12 tons

(c) Write the ratio of width to length for this rectangle.

7 ft

4 ft

ANSWERS

1. (a) $\frac{8}{15}$ (b) $\frac{85}{18}$ (c) $\frac{12}{17}$ (d) $\frac{17}{12}$

2. (a) $\frac{3}{5}$ (b) $\frac{4}{3}$ (c) $\frac{4}{7}$

Remember: the ratio of a to b is written in this way.

$\frac{a}{b}$ a ← **mentioned first**, b ← **mentioned second**

WORK PROBLEM 1 AT THE SIDE.

Since a ratio is a fraction, a ratio can be written in lowest terms just as a fraction is reduced to lowest terms.

EXAMPLE 2 Writing a Ratio in Lowest Terms

Write each of the following as ratios in lowest terms.

(a) 60 days to 80 days

(b) 10 ounces of medicine to 120 ounces of medicine.

(c) A large van holds 18 people; a small van holds 8.

SOLUTION

(a) The ratio is $\frac{60}{80}$. Write this ratio in lowest terms by dividing numerator and denominator by 20.

$$\frac{60}{80} = \frac{60 \div 20}{80 \div 20} = \frac{3}{4}$$

(b) The ratio is $\frac{10}{120}$. Divide numerator and denominator by 10.

$$\frac{10}{120} = \frac{10 \div 10}{120 \div 10} = \frac{1}{12}$$

(c) The ratio is $\frac{18}{8} = \frac{18 \div 2}{8 \div 2} = \frac{9}{4}$. ■

NOTE Although $\frac{9}{4} = 2\frac{1}{4}$, ratios are not written as mixed numbers. However, the ratio $\frac{9}{4}$ does mean the large van holds $2\frac{1}{4}$ times as many people as the small van.

WORK PROBLEM 2 AT THE SIDE.

EXAMPLE 3 Application of a Ratio

The price of a quart of juice increased from $.98 to $1.24. Find the ratio of the increase in price to the original price.

APPROACH
The increase in price must be found by subtracting the original price from the new price ($1.24 − $.98). This increase is then written as the numerator over the fraction bar, and the original price as the denominator, to form a ratio.

SOLUTION
The increase in price is

$$\$1.24 - \$.98 = \$.26.$$

The ratio of the increase in price to the original price is

$$\frac{\$.26}{\$.98}. \begin{array}{l} \leftarrow \text{increase} \\ \leftarrow \text{original price} \end{array}$$

Simplify this ratio by multiplying numerator and denominator by 100. Multiplying by 100 $\left(\frac{100}{100} = 1\right)$ eliminates the decimals in the ratio and results in whole numbers.

$$\frac{.26}{.98} = \frac{.26 \times 100}{.98 \times 100} = \frac{26}{98}$$

Write the ratio in lowest terms as $\frac{13}{49}$. ■

WORK PROBLEM 3 AT THE SIDE.

2 The next example shows a mixed number used in ratios.

EXAMPLE 4 Using a Mixed Number in a Ratio
Write each ratio in lowest terms.

(a) 2 days to $2\frac{1}{4}$ days **(b)** $3\frac{1}{4}$ to $1\frac{1}{2}$

SOLUTION
(a) Write the ratio as follows.

$$\frac{2 \text{ days}}{2\frac{1}{4} \text{ days}} = \frac{2}{2\frac{1}{4}}$$

Next, write $2\frac{1}{4}$ as the improper fraction $\frac{9}{4}$.

$$\frac{2}{2\frac{1}{4}} = \frac{2}{\frac{9}{4}}$$

3. Give both ratios in lowest terms.

(a) A price increased from $1.28 to $1.50. Find the ratio of the increase in price to the original price.

(b) A price decreased from $12 to $9. Find the ratio of the decrease in price to the original price.

ANSWERS
3. (a) $\frac{11}{64}$ (b) $\frac{1}{4}$

4. Write each ratio in lowest terms.

(a) $3\frac{1}{2}$ to 4

(b) $5\frac{5}{8}$ pounds to $3\frac{3}{4}$ pounds

(c) $2\frac{2}{3}$ days to $4\frac{5}{12}$ days

ANSWERS

4. (a) $\frac{7}{8}$ (b) $\frac{3}{2}$ (c) $\frac{32}{53}$

Invert and multiply. (Remember: $2 = \frac{2}{1}$.)

$$\frac{2}{\frac{9}{4}} = \frac{2}{1} \div \frac{9}{4} = \frac{2}{1} \cdot \frac{4}{9} = \frac{8}{9}$$

The ratio, in lowest terms, is $\frac{8}{9}$.

(b) Write the ratio as $\frac{3\frac{1}{4}}{1\frac{1}{2}}$. Write this ratio in lowest terms by first writing $3\frac{1}{4}$ and $1\frac{1}{2}$ as improper fractions.

$$3\frac{1}{4} = \frac{13}{4} \quad \text{and} \quad 1\frac{1}{2} = \frac{3}{2}$$

The ratio is

$$\frac{3\frac{1}{4}}{1\frac{1}{2}} = \frac{\frac{13}{4}}{\frac{3}{2}}.$$

Write as a division problem.

$$= \frac{13}{4} \div \frac{3}{2}$$

Invert and multiply.

$$= \frac{13}{4} \cdot \frac{2}{3}$$

$$= \frac{26}{12} = \frac{13}{6} \quad ■$$

WORK PROBLEM 4 AT THE SIDE.

3 When ratios compare measurements, the measurements must be in the *same* unit. For example, *feet* must be compared to *feet, hours* to *hours, pints* to *pints,* and *inches* to *inches*.

EXAMPLE 5 Application Using Measurement

Write the ratio of the lengths of these boards.

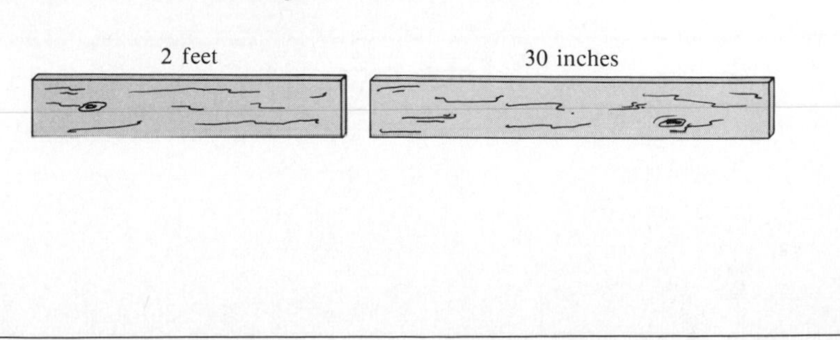

SOLUTION
First, express the lengths in the same unit, such as inches. Since

$$1 \text{ foot} = 12 \text{ inches},$$

2 feet is

$$2 \cdot 12 \text{ inches} = 24 \text{ inches}.$$

The length of the board on the left is 24 inches, so the ratio of the lengths is

$$\frac{24 \text{ inches}}{30 \text{ inches}} = \frac{24}{30}.$$

Write the ratio in lowest terms.

$$\frac{24}{30} = \frac{4}{5}$$

The shorter board is $\frac{4}{5}$ the length of the longer board. ■

WORK PROBLEM 5 AT THE SIDE.

The tables below provide a list of time relationships and a list of relationships among common measures.

Length	**Capacity (Volume)**
12 inches = 1 foot	2 cups = 1 pint
3 feet = 1 yard	2 pints = 1 quart
5280 feet = 1 mile	4 quarts = 1 gallon

Weight	**Time**
16 ounces = 1 pound	1 week = 7 days
2000 pounds = 1 ton	1 day = 24 hours
	1 hours = 60 minutes
	1 minute = 60 seconds

EXAMPLE 6 Application Using Common Measures
Write each ratio in lowest terms.

(a) 9 hours to 2 days **(b)** 5 yards to 8 feet

(c) 9 quarts to 24 pints **(d)** $7\frac{1}{2}$ inches to 1 foot

5. Write each ratio in lowest terms.

(a) 9 quarts to 15 quarts

(b) 16 tons to 12 tons

(c) 35 oranges to 20 oranges

ANSWERS
5. (a) $\frac{3}{5}$ (b) $\frac{4}{3}$ (c) $\frac{7}{4}$

6. Write each ratio in lowest terms.

(a) 9 inches to 6 feet

(b) 5 hours to 3 days

(c) 7 yards to 14 feet

(d) 8 pints to 9 quarts

(e) 25 minutes to 2 hours

(f) $2\frac{1}{2}$ hours to 2 days

ANSWERS
6. (a) $\frac{1}{8}$ (b) $\frac{5}{72}$ (c) $\frac{3}{2}$ (d) $\frac{4}{9}$
(e) $\frac{5}{24}$ (f) $\frac{5}{96}$

SOLUTION

(a) Since there are 24 hours in one day, there are 24 · 2 (= 48 hours) in 2 days. The ratio is

$$\frac{9 \text{ hours}}{48 \text{ hours}} = \frac{9}{48}. \quad \text{after canceling units (hours)}$$

Write the ratio in lowest terms.

$$\frac{9}{48} = \frac{3}{16}$$

(b) Since 1 yard = 3 feet, 5 yards = 5 · 3 = 15 feet. The ratio is

$$\frac{15 \text{ feet}}{8 \text{ feet}} = \frac{15}{8}.$$

(c) 1 quart = 2 pints, so 9 quarts = 9 · 2 = 18 pints. The ratio is

$$\frac{18 \text{ pints}}{24 \text{ pints}} = \frac{18}{24} = \frac{3}{4}. \quad \text{lowest terms}$$

(d) Since 1 foot = 12 inches, the ratio is

$$\frac{7\frac{1}{2} \text{ inches}}{12 \text{ inches}} = \frac{\frac{15}{2}}{12} = \frac{\frac{15}{2}}{\frac{12}{1}} = \frac{15}{2} \div \frac{12}{1} = \frac{15}{2} \cdot \frac{1}{12} = \frac{5}{8}. \quad \blacksquare$$

WORK PROBLEM 6 AT THE SIDE.

name date hour

5.1 EXERCISES

Write each ratio in lowest terms.

1. 8 to 9

2. 11 to 15

3. 75 to 100

4. 20 to 72

5. 90 cents to 120 cents

6. 150 cents to 165 cents

7. 80 miles to 50 miles

8. 300 miles to 450 miles

9. \$30 to \$70

10. \$95 to \$125

11. $6\frac{1}{2}$ to 4

12. $3\frac{1}{4}$ to 6

13. 3 to $2\frac{1}{2}$

14. 5 to $1\frac{1}{4}$

15. $1\frac{1}{4}$ to $1\frac{1}{2}$

16. $2\frac{1}{3}$ to $2\frac{2}{3}$

Write each ratio in lowest terms. Be sure to make all necessary conversions. (See the lists of relationships among measures in Example 6.)

17. 4 feet to 30 inches

18. 6 yards to 12 feet

19. 5 minutes to 1 hour

20. 8 quarts to 5 pints

21. 15 hours to 2 days

22. 80¢ to \$3

23. $1\frac{1}{2}$ days to 1 week

24. $3\frac{1}{2}$ days to $1\frac{1}{2}$ weeks

Solve each word problem. Write each ratio in lowest terms.

25. Mr. Wilkins is 35 years old, and his daughter is 10. Find the ratio of his age to hers.

26. The Empire State Building is 1100 feet tall, and the Sears Tower in Chicago is 1300 feet tall. Write the ratio of the height of the Empire State Building to the height of the Sears Tower.

27. One refrigerator holds $2\frac{2}{3}$ cubic feet of food, and another holds 6 cubic feet. Find the ratio of the amounts of storage of the small refrigerator to the larger refrigerator.

28. One car has an $11\frac{1}{2}$-gallon tank, and another has a 15-gallon tank. Find the ratios of the amounts the tanks hold.

29. A building is $24\frac{1}{2}$ feet tall. It casts a shadow $8\frac{1}{4}$ feet long. Find the ratio of the height of the building to the length of its shadow.

30. The price of a car increased from $9000 to $12,000. Find the ratio of the increase in price to the original price.

name date hour

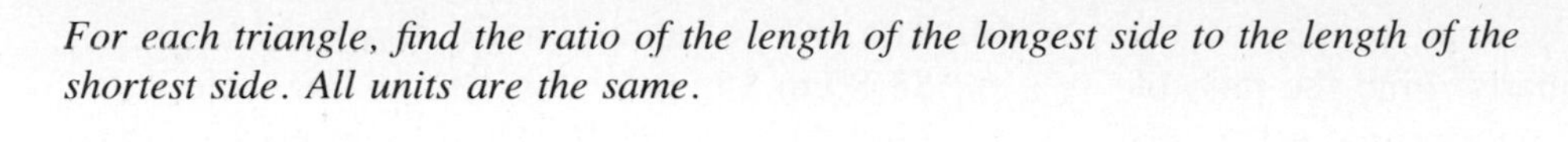

For each triangle, find the ratio of the length of the longest side to the length of the shortest side. All units are the same.

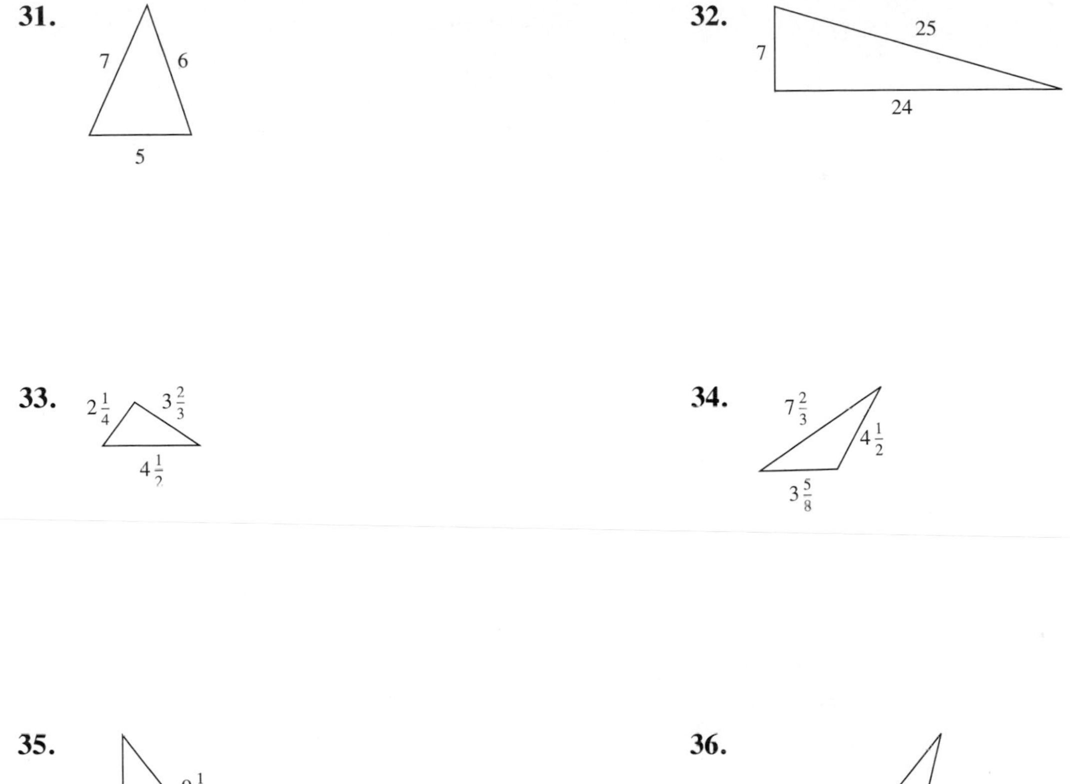

35.

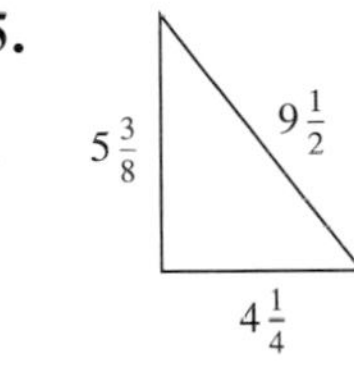

36.

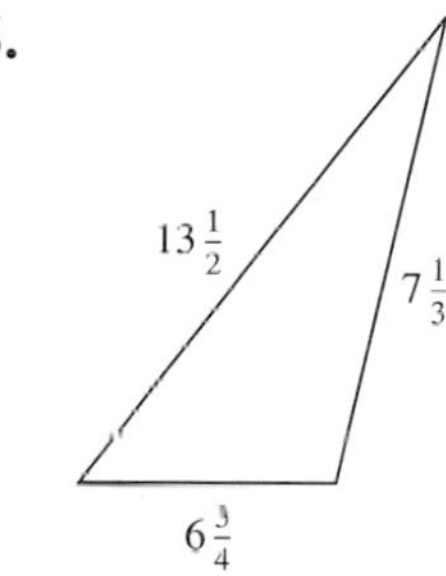

CHALLENGE EXERCISES

37. What is the ratio of $59\frac{1}{2}$ days to $8\frac{3}{4}$ weeks?

38. Find the ratio of $18\frac{1}{2}$ inches to $4\frac{1}{2}$ feet?

39. The price of oil recently went from \$6.60 to \$9.90 per case of 12 quarts. Find the ratio of the increase in price to the original price.

40. The price of an antibiotic decreased from \$8.80 to \$5.60 for a bottle of 100 tablets. Find the ratio of the decrease in price to the original price.

REVIEW EXERCISES

Multiply or divide as indicated. Round to the nearest thousandth if necessary. (For help, see ***Sections 4.5 and 4.6.****)*

41. $\begin{array}{r} 18.7 \\ \times\ 2.3 \\ \hline \end{array}$

42. $\begin{array}{r} 48.51 \\ \times\ 3.62 \\ \hline \end{array}$

43. $\begin{array}{r} 32.81 \\ \times\ 16.83 \\ \hline \end{array}$

44. $.95\overline{).829}$

45. $.71\overline{)6.72}$

46. $4.6\overline{)116.38}$

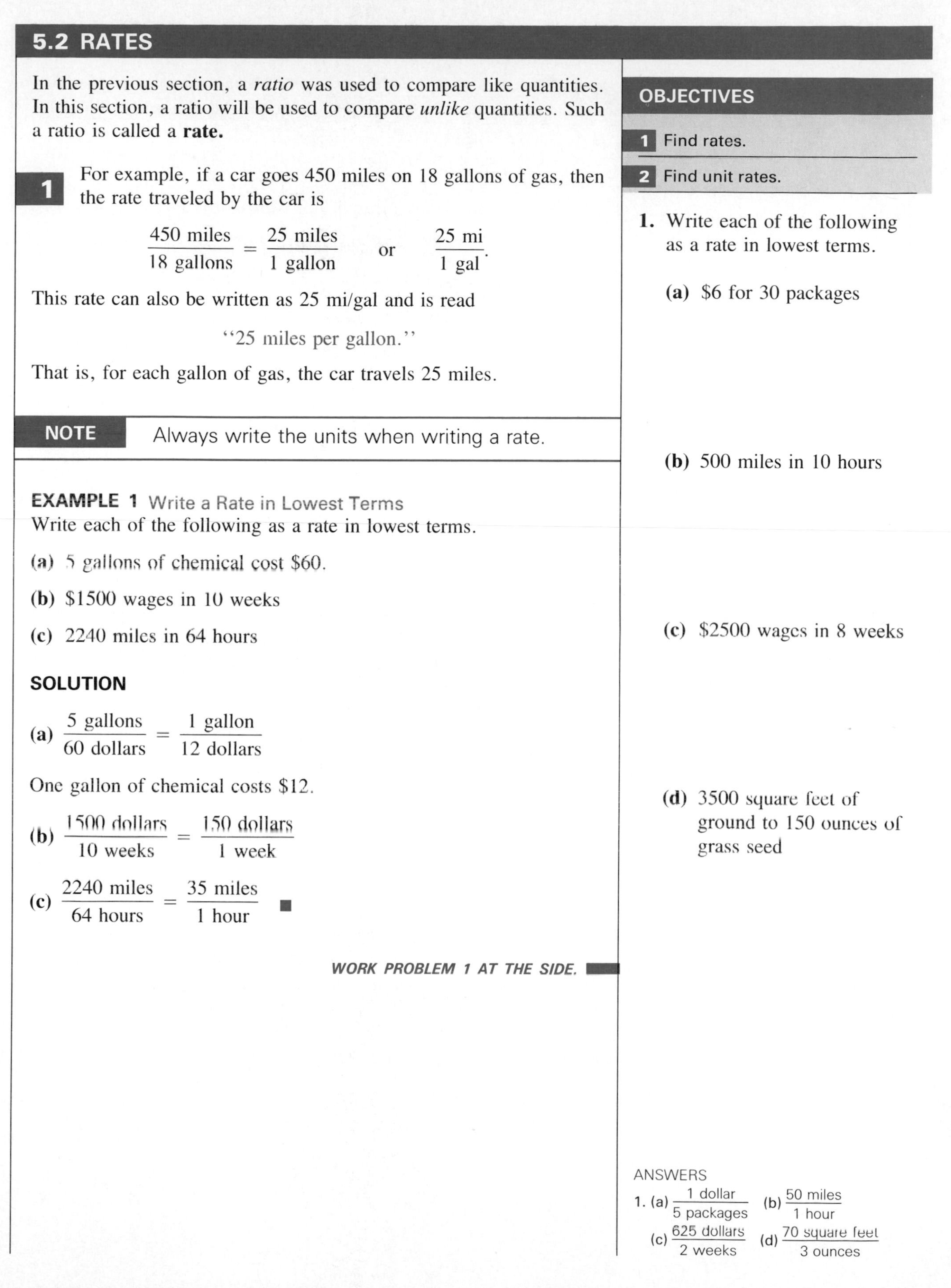

5.2 RATES

In the previous section, a *ratio* was used to compare like quantities. In this section, a ratio will be used to compare *unlike* quantities. Such a ratio is called a **rate.**

OBJECTIVES

1. Find rates.
2. Find unit rates.

1 For example, if a car goes 450 miles on 18 gallons of gas, then the rate traveled by the car is

$$\frac{450 \text{ miles}}{18 \text{ gallons}} = \frac{25 \text{ miles}}{1 \text{ gallon}} \quad \text{or} \quad \frac{25 \text{ mi}}{1 \text{ gal}}.$$

This rate can also be written as 25 mi/gal and is read

"25 miles per gallon."

That is, for each gallon of gas, the car travels 25 miles.

NOTE Always write the units when writing a rate.

EXAMPLE 1 Write a Rate in Lowest Terms

Write each of the following as a rate in lowest terms.

(a) 5 gallons of chemical cost \$60.

(b) \$1500 wages in 10 weeks

(c) 2240 miles in 64 hours

SOLUTION

(a) $\frac{5 \text{ gallons}}{60 \text{ dollars}} = \frac{1 \text{ gallon}}{12 \text{ dollars}}$

One gallon of chemical costs \$12.

(b) $\frac{1500 \text{ dollars}}{10 \text{ weeks}} = \frac{150 \text{ dollars}}{1 \text{ week}}$

(c) $\frac{2240 \text{ miles}}{64 \text{ hours}} = \frac{35 \text{ miles}}{1 \text{ hour}}$ ■

WORK PROBLEM 1 AT THE SIDE.

1. Write each of the following as a rate in lowest terms.

(a) \$6 for 30 packages

(b) 500 miles in 10 hours

(c) \$2500 wages in 8 weeks

(d) 3500 square feet of ground to 150 ounces of grass seed

ANSWERS

1. (a) $\frac{1 \text{ dollar}}{5 \text{ packages}}$ (b) $\frac{50 \text{ miles}}{1 \text{ hour}}$ (c) $\frac{625 \text{ dollars}}{2 \text{ weeks}}$ (d) $\frac{70 \text{ square feet}}{3 \text{ ounces}}$

2. Write each of the following as a unit rate.

(a) 510 miles on 20 gallons of gas

(b) 28 gallons in 7 days

(c) \$850 in 5 days

(d) 24-pound turkey for 8 people

ANSWERS
2. (a) 25.5 miles/gallon
(b) 4 gallons/day
(c) 170 dollars/day
(d) 3 pounds/person

2 The rate with 1 unit in the denominator is one of the most useful rates. An example of this rate, called a **unit rate,** is

$$\frac{\$1.25}{1 \text{ gallon}} = \$1.25/\text{gal} \quad \text{or} \quad \$1.25 \text{ per gallon.}$$

For unit rates, look for indicator words such as

of, *in*, *for*, *on*, and *per*.

EXAMPLE 2 Finding a Unit Rate
Write each of the following as a unit rate.

(a) 337.5 miles on 13.5 gallons of gas

(b) 510 miles in 17 hours

(c) \$810 in 6 days

SOLUTION

(a) The indicator word is *on*.

337.5 miles on 13.5 gallons of gas
↑ indicator word

The unit rate gives the number of miles traveled on *one gallon of gas*.

$$\frac{337.5 \text{ miles}}{13.5 \text{ gallons}} \quad \text{Write the rate.}$$

Divide by 13.5 to find the unit rate.

$$13.5\overline{)337.5} = 25.$$

$$\frac{337.5 \text{ miles} \div 13.5}{13.5 \text{ gallons} \div 13.5} = \frac{25 \text{ miles}}{1 \text{ gallon}}$$

The unit rate is 25 mi/gal.

(b) $\frac{510 \text{ miles}}{17 \text{ hours}} = \frac{30 \text{ miles}}{1 \text{ hour}}$ Write the rate and divide.

$= 30$ mi/hr unit rate

(c) $\frac{810 \text{ dollars}}{6 \text{ days}} = \frac{135 \text{ dollars}}{1 \text{ day}} = \$135/\text{day}$ ■

WORK PROBLEM 2 AT THE SIDE.

A unit rate that gives a *cost per item* is called a *unit price.* For example, $1.25 per gallon is a unit price, as is $47 per day or $125 per week.

EXAMPLE 3 Determining the Best Buy

The local store charges the following prices for Mrs. Butterworth's pancake syrup. Find the best buy (lowest cost per unit).

size	*price*
12 oz	$1.28
24 oz	$1.79
36 oz	$2.73

APPROACH

The best buy is determined by finding the cost per unit—here, per ounce. Divide the price by the number of ounces in the container.

SOLUTION

Find the best buy by dividing the price by the quantity to get the price per ounce.

size	***unit cost*** *(dollars per ounce)*
12 oz	$\frac{1.28}{12} = .107$
24 oz	$\frac{1.79}{24} = .075$ ← best buy
36 oz	$\frac{2.73}{36} = .076$

The best buy is the 24-ounce bottle. ■

WORK PROBLEM 3 AT THE SIDE.

3. Find the best buy (lowest cost per unit) for the following.

10 pounds costing $25
15 pounds costing $38.70
22 pounds costing $49.50
31 pounds costing $70.68

ANSWERS
3. 22 pounds (at $2.25 per pound)

name date hour

5.2 EXERCISES

Write each of the following as a rate in lowest terms.

1. 54 miles in 27 minutes

2. 80 feet in 20 seconds

3. 250 yards in 50 seconds

4. 925 miles in 30 hours

5. 14 people for 28 dresses

6. 12 teams for 60 horses

7. 35 gallons in 5 hours

8. 68 pills for 17 people

9. 11 chapters for 132 pages

10. 25 doctors for 150 patients

11. 72 miles on 4 gallons

12. 132 miles on 8 gallons

Write each of the following as a unit rate.

13. \$60 in 5 hours

14. \$2500 in 20 days

15. \$1260 in 14 days

16. \$57.75 in 7 hours

17. \$101.25 for 45 pounds

18. \$238.56 for 21 gallons

19. 311.08 miles on 15.4 gallons

20. 175.68 miles on 9.6 gallons

21. 7.5 pounds for 5 people

22. 8.75 pounds for 7 people

23. \$413.20 for 4 days

24. \$742.50 for 9 days

Find the best buy (based on the price per ounce) for each of the following.

25. black pepper
4 oz for $.89
8 oz for $2.13

26. kosher dills
32 oz for $1.30
46 oz for $1.97

27. cereal
10 oz for $1.22
15 oz for $1.66
20 oz for $2.11

28. spaghetti sauce
$15\frac{1}{2}$ oz for $1.07
32 oz for $1.59
48 oz for $2.51

29. chunky peanut butter
12 oz for $1.09
18 oz for $1.43
28 oz for $2.29
40 oz for $3.19

30. pork and beans
8 oz for $.37
16 oz for $.40
21 oz for $.59
31 oz for $.85
53 oz for $1.38

Solve each of the following word problems.

31. A 20-pound turkey will serve 15 people. Give the rate in pounds per person.

32. One recipe calls for 15 heads of lettuce to make salad for 35 people. Give the rate in heads per person.

33. Russ makes $85.82 in 7 hours. What is his rate per hour?

34. Find the cost of 1 gallon of gas if 18 gallons cost $28.08.

35. Ms. Johnson bought 150 shares of stock for \$1725. Find the cost of one share.

36. A company pays \$6450 in dividends for the 2500 shares of its stock. Find the dividend per share.

37. John can pack six crates of berries in 18 minutes. Give his rate in crates per minute and in minutes per crate.

38. Sandra can plow 5 acres in 10 hours. Give her rate in acres per hour and in hours per acre.

CHALLENGE EXERCISES

39. The 4.6 yards of fabric needed for a dress coat costs \$51.75. Find the cost of 1 yard of fabric.

40. The cost to lay 42.4 square yards of carpet is \$691.12. Find the cost of 1 square yard of carpet.

REVIEW EXERCISES

Multiply the following and express fractional answers as mixed numbers. (For help, see ***Section 2.8.****)*

41. $10 \cdot 7\frac{1}{4}$

42. $12 \cdot 5\frac{2}{3}$

43. $18\frac{2}{5} \cdot 30$

44. $3\frac{5}{8} \cdot 5$

45. $1\frac{1}{6} \cdot 3$

46. $26 \cdot 2\frac{5}{8}$

5.3 PROPORTIONS

1 A **proportion** states that two ratios are equivalent. For example,

$$\frac{3}{4} = \frac{15}{20}$$

is a proportion that says the ratios $\frac{3}{4}$ and $\frac{15}{20}$ are equivalent.

This proportion is read in either of two ways:

three is to four as fifteen is to twenty

or

three fourths equals fifteen twentieths.

EXAMPLE 1 Writing a Proportion
Write each of the following proportions.

(a) 6 is to 11 as 18 is to 33. **(b)** 9 is to 6 as 3 is to 2.

SOLUTION

(a) $\frac{6}{11} = \frac{18}{33}$ **(b)** $\frac{9}{6} = \frac{3}{2}$ ■

WORK PROBLEM 1 AT THE SIDE.

2 There are two ways to see whether a proportion is true. One is to *write both of the ratios in lowest terms.*

EXAMPLE 2 Writing Both Ratios in Lowest Terms
Are the following proportions true?

(a) $\frac{5}{9} = \frac{18}{27}$ **(b)** $\frac{16}{12} = \frac{28}{21}$

SOLUTION

(a) Write each fraction in lowest terms. $\frac{5}{9}$ is already in lowest terms, and

$$\frac{18}{27} = \frac{2}{3}. \quad \text{lowest terms}$$

Since $\frac{2}{3}$ is not equivalent to $\frac{5}{9}$, the proportion is false.

(b) Write both fractions in lowest terms.

$$\frac{16}{12} = \frac{4}{3} \quad \text{and} \quad \frac{28}{21} = \frac{4}{3}$$

Both fractions are equivalent to $\frac{4}{3}$, so the proportion is true. ■

WORK PROBLEM 2 AT THE SIDE.

OBJECTIVES

1 Write proportions.

2 Decide whether proportions are true.

3 Find cross products.

1. Write each proportion.

(a) 5 is to 2 as 20 is to 8.

(b) 6 is to 35 as 12 is to 70.

(c) 5 is to 7 as 35 is to 49.

(d) 11 is to 28 as 66 is to 168.

2. Are these proportions true?

(a) $\frac{6}{12} = \frac{15}{30}$

(b) $\frac{15}{18} = \frac{2}{3}$

(c) $\frac{25}{40} = \frac{30}{48}$

(d) $\frac{45}{99} = \frac{60}{132}$

(e) $\frac{21}{45} = \frac{56}{120}$

ANSWERS
1. (a) $\frac{5}{2} = \frac{20}{8}$ (b) $\frac{6}{35} = \frac{12}{70}$ (c) $\frac{5}{7} = \frac{35}{49}$ (d) $\frac{11}{28} = \frac{66}{168}$
2. (a) true (b) false (c) true (d) true (e) true

3. Cross multiply to see whether the following proportions are true.

(a) $\frac{5}{9} = \frac{10}{18}$

(b) $\frac{32}{15} = \frac{16}{8}$

(c) $\frac{10}{17} = \frac{20}{34}$

(d) $\frac{2\frac{1}{2}}{6} = \frac{5}{12}$

(e) $\frac{3}{4\frac{1}{4}} = \frac{24}{34}$

(f) $\frac{5\frac{1}{2}}{6\frac{1}{4}} = \frac{24}{25}$

3 Another way to see whether a proportion is true is called the **cross products method,** or **cross multiplication.**

To see whether a proportion is true, first cross multiply one way, then cross multiply the other way, as shown here.

$$\frac{a}{b} = \frac{c}{d} \qquad bc \qquad ad$$

If the two answers are the same, the proportion is true. In summary: In a true proportion, the cross products are equivalent.

EXAMPLE 3 Using Cross Products

Use the cross multiplication method to see whether the following proportions are true.

(a) $\frac{3}{5} = \frac{12}{20}$ **(b)** $\frac{2\frac{1}{4}}{3\frac{1}{2}} = \frac{9}{16}$

SOLUTION

(a) Cross multiply one way and then the other way.

$$\frac{3}{5} = \frac{12}{20} \qquad 5 \cdot 12 = 60 \qquad 3 \cdot 20 = 60$$

same

Since both answers are the same, the proportion is *true*.

(b) Cross multiply.

changed to improper fraction

$$\frac{2\frac{1}{4}}{3\frac{1}{2}} = \frac{9}{16}$$

$$3\frac{1}{2} \cdot 9 = \frac{7}{2} \cdot \frac{9}{1} = \frac{63}{2}$$

$$2\frac{1}{4} \cdot 16 = \frac{9}{4} \cdot \frac{16}{1} = 36$$

different

This proportion is *false*. ■

NOTE The numbers in a proportion do not have to be whole numbers.

WORK PROBLEM 3 AT THE SIDE.

ANSWERS
3. (a) true (b) false (c) true (d) true (e) true (f) false

name date hour

5.3 EXERCISES

Write each of the following proportions.

1. 13 is to 15 as 26 is to 30.

2. 40 is to 8 as 25 is to 5.

3. 20 is to 45 as 4 is to 9.

4. 50 is to 70 as 15 is to 21.

5. 120 is to 150 as 8 is to 10.

6. 6 is to 9 as 10 is to 15.

7. 32 is to 48 as 22 is to 33.

8. 25 is to 30 as 40 is to 48.

9. $1\frac{1}{2}$ is to 8 as 6 is to 32.

10. 6 is to $3\frac{1}{4}$ as 24 is to 13.

Use the method of writing in lowest terms to decide whether the following proportions are true or false.

11. $\frac{6}{10} = \frac{3}{5}$

12. $\frac{15}{20} = \frac{3}{4}$

13. $\frac{5}{8} = \frac{25}{40}$

14. $\frac{12}{18} = \frac{2}{3}$

15. $\frac{7}{10} = \frac{20}{30}$

16. $\frac{21}{35} = \frac{2}{3}$

17. $\frac{42}{15} = \frac{28}{10}$

18. $\frac{18}{16} = \frac{36}{32}$

19. $\frac{32}{18} = \frac{48}{27}$

20. $\frac{15}{48} = \frac{10}{24}$

21. $\frac{7}{6} = \frac{54}{48}$

22. $\frac{20}{16} = \frac{48}{36}$

Use the cross products method to decide whether the following proportions are true or false. Circle the correct answer.

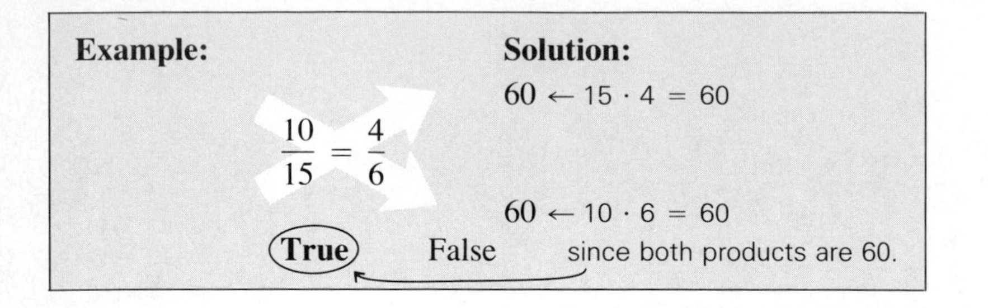

23. $\frac{4}{7} = \frac{12}{21}$

True False

24. $\frac{6}{8} = \frac{15}{20}$

True False

25. $\frac{7}{10} = \frac{82}{120}$

True False

26. $\frac{16}{40} = \frac{22}{55}$

True False

27. $\frac{110}{18} = \frac{160}{27}$

True False

28. $\frac{420}{600} = \frac{14}{20}$

True False

name date hour

29. $\dfrac{3\frac{1}{2}}{4} = \dfrac{7}{8}$

True False

30. $\dfrac{5\frac{3}{4}}{9} = \dfrac{23}{36}$

True False

31. $\dfrac{9}{16} = \dfrac{2\frac{7}{8}}{5}$

True False

32. $\dfrac{17}{24} = \dfrac{6\frac{1}{3}}{8}$

True False

33. $\dfrac{6}{3\frac{2}{3}} = \dfrac{18}{11}$

True False

34. $\dfrac{16}{13} = \dfrac{2}{1\frac{5}{8}}$

True False

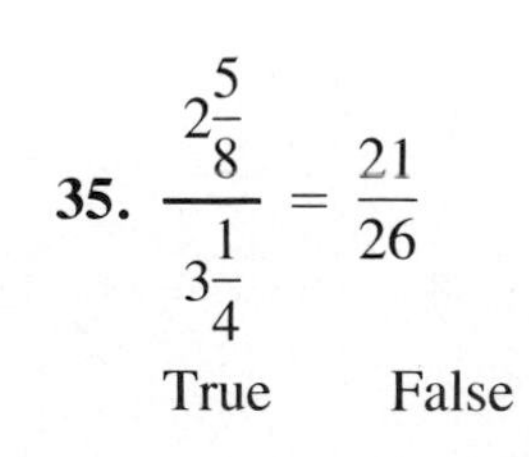

35. $\dfrac{2\frac{5}{8}}{3\frac{1}{4}} = \dfrac{21}{26}$

True False

36. $\dfrac{28}{17} = \dfrac{9\frac{1}{3}}{5\frac{2}{3}}$

True False

CHALLENGE EXERCISES

37. $\frac{8.15}{2.03} = \frac{6.09}{24.45}$
True False

38. $\frac{423.8}{17.11} = \frac{51.33}{1271.4}$
True False

39. $\frac{6.12}{24.48} = \frac{62.7}{250.8}$
True False

40. $\frac{36.72}{146.88} = \frac{376.2}{1504.8}$
True False

REVIEW EXERCISES

Write each fraction in lowest terms. Write in proper fractions or as mixed numbers. (For help, see ***Sections 2.2 and 2.4****.)*

41. $\frac{16}{24}$

42. $\frac{24}{40}$

43. $\frac{60}{48}$

44. $\frac{20}{15}$

45. $\frac{36}{63}$

46. $\frac{73}{146}$

47. $\frac{55}{8}$

48. $\frac{19}{7}$

5.4 SOLVING PROPORTIONS

1 Four numbers are used in a proportion. If three of these numbers are known, the fourth can be found. For example, find the value of the missing number in the proportion

$$\frac{3}{5} = \frac{x}{40}.$$

Here x is used as a placeholder for the unknown number. In a proportion, the cross products are equivalent.

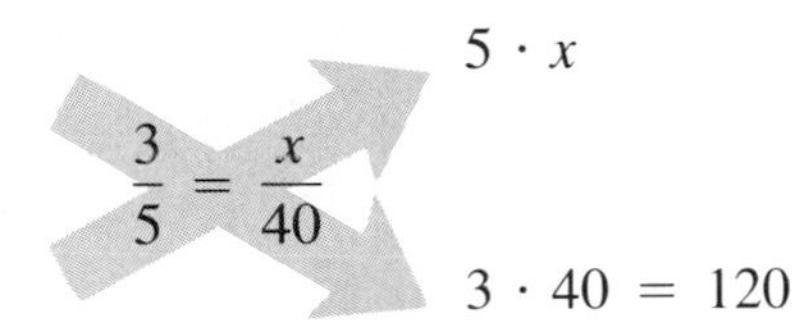

Since the cross products are equivalent,

$$5 \cdot x = 120.$$

The equal sign says that $5 \cdot x$ and 120 are the same. If both $5 \cdot x$ and 120 are divided by 5, the results should still be equivalent.

$$\frac{5 \cdot x}{5} = \frac{120}{5} \quad \text{Divide both sides by 5.}$$

Cancel 5's on the left and divide 120 by 5 to get

$$1 \cdot x = 24.$$

Since $1 \cdot x$ is the same as x,

$$x = 24.$$

The missing number in the proportion is 24, so the complete proportion is

$$\frac{3}{5} = \frac{24}{40}. \quad \leftarrow x \text{ is 24.}$$

Check by finding the cross products.

WORK PROBLEM 1 AT THE SIDE.

Solve a proportion for a missing number with the following steps.

Step 1 Write all ratios in lowest terms.

Step 2 Find the two cross products.

Step 3 Show the cross products are equivalent.

Step 4 Divide both products by the number multiplying (next to) the letter.

OBJECTIVES

1 Find the missing number in a proportion.

2 Find the missing number with mixed numbers.

1. Is this proportion true?

$$\frac{3}{5} = \frac{24}{40}$$

ANSWER
1. yes

EXAMPLE 1 Solving for a Missing Number
Find the missing number in each proportion.

(a) $\frac{6}{9} = \frac{p}{21}$ **(b)** $\frac{16}{k} = \frac{32}{20}$ **(c)** $\frac{7}{12} = \frac{14}{y}$

SOLUTION

(a) *Step 1* Write $\frac{6}{9}$ in lowest terms to get the proportion

$$\frac{2}{3} = \frac{p}{21}.$$

Step 2 Find the cross products.

$$3 \cdot p$$

$$2 \cdot 21 = 42$$

Step 3 Show the cross products are equivalent.

$$3 \cdot p = 42$$

Step 4 Divide both sides by the number multiplying p.

$$\frac{3 \cdot p}{3} = \frac{42}{3}$$

Cancel on the left, and divide 42 by 3.

$$p = 14$$

The complete proportion is

$$\frac{2}{3} = \frac{14}{21}.$$ Replace p with 14.

Check that 14 is the correct answer by finding cross products.

(b) Write $\frac{32}{20}$ in lowest terms to get $\frac{16}{k} = \frac{8}{5}$.

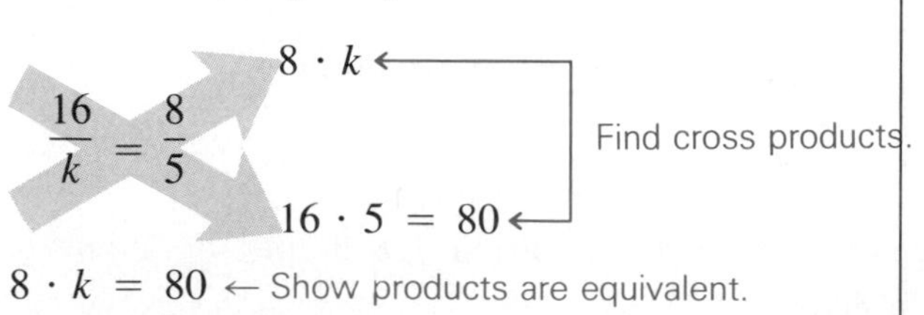

$8 \cdot k = 80$ ← Show products are equivalent.

$\frac{8 \cdot k}{8} = \frac{80}{8}$ ← Divide both sides by 8.

$k = 10$ ← Find k.

(c) Multiply. $\frac{7}{12} = \frac{14}{y}$ $12 \cdot 14 = 168$ $7 \cdot y$

Cross products are equivalent.

$$7 \cdot y = 168$$

Divide by 7. $$\frac{7 \cdot y}{7} = \frac{168}{7}$$

$$y = 24$$ ■

WORK PROBLEM 2 AT THE SIDE.

2 The following examples show how the numbers in a proportion can be mixed numbers.

EXAMPLE 2 Solving for a Mixed Number in a Proportion
Find the missing number in each of the following proportions.

(a) $\frac{1}{2} = \frac{m}{9}$ **(b)** $\frac{3}{10} = \frac{5}{y}$

SOLUTION

(a) Multiply. $\frac{1}{2} = \frac{m}{9}$ $2 \cdot m$ $1 \cdot 9 = 9$

Cross products are equivalent. $$2 \cdot m = 9$$

Divide both sides by 2. $$\frac{2 \cdot m}{2} = \frac{9}{2}$$

$$m = \frac{9}{2} = 4\frac{1}{2}$$

With the mixed number, the complete proportion is

$$\frac{1}{2} = \frac{4\frac{1}{2}}{9}.$$

Check by finding cross products.

2. Find the value of the missing number.

(a) $\frac{1}{2} = \frac{x}{12}$

(b) $\frac{9}{10} = \frac{y}{80}$

(c) $\frac{5}{9} = \frac{15}{m}$

(d) $\frac{12}{p} = \frac{6}{8}$

(e) $\frac{15}{20} = \frac{30}{a}$

ANSWERS
2. (a) $x = 6$ (b) $y = 72$ (c) $m = 27$
(d) $p = 16$ (e) $a = 40$

3. Find the missing numbers. Write answers as mixed numbers.

(a) $\frac{5}{7} = \frac{2}{r}$

(b) $\frac{3}{4} = \frac{a}{6}$

(c) $\frac{z}{10} = \frac{1}{6}$

(d) $\frac{9}{x} = \frac{6}{5}$

ANSWERS

3. (a) $r = 2\frac{4}{5}$ (b) $a = 4\frac{1}{2}$ (c) $z = 1\frac{2}{3}$ (d) $x = 7\frac{1}{2}$

(b) Multiply.

$$\frac{3}{10} = \frac{5}{y} \qquad 10 \cdot 5 = 50 \qquad 3 \cdot y$$

$$3 \cdot y = 50$$

Divide by 3. $\frac{3 \cdot y}{3} = \frac{50}{3}$

$$y = \frac{50}{3} = 16\frac{2}{3} \quad ■$$

WORK PROBLEM 3 AT THE SIDE.

EXAMPLE 3 Using Mixed Numbers and Decimals

Find the missing number in each proportion.

(a) $\frac{2\frac{1}{5}}{6} = \frac{x}{10}$ **(b)** $\frac{4}{z} = \frac{3}{1\frac{1}{4}}$ **(c)** $\frac{.6}{1.5} = \frac{m}{2}$

SOLUTION

(a) Cross multiply.

$$\frac{2\frac{1}{5}}{6} = \frac{x}{10} \qquad 6 \cdot x \qquad 2\frac{1}{5} \cdot 10$$

Find $2\frac{1}{5} \cdot 10$.

$$2\frac{1}{5} \cdot 10 = \frac{11}{5} \cdot 10 = 22$$

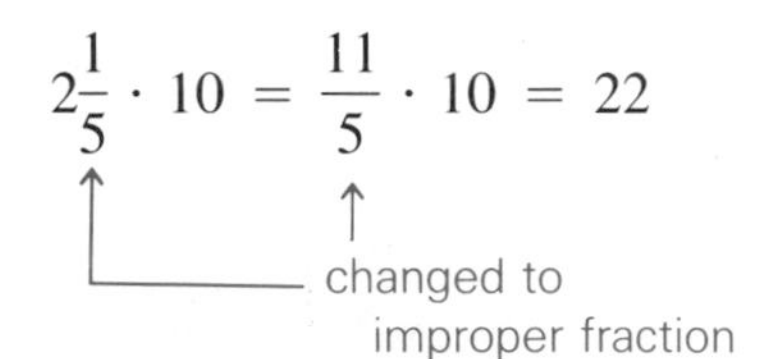

Show the cross products are equivalent.

$$6 \cdot x = 22$$

Divide by 6.

$$\frac{6 \cdot x}{6} = \frac{22}{6}$$

$$x = \frac{22}{6} = \frac{11}{3} = 3\frac{2}{3}$$

(b) Cross multiply.

$$4 \cdot 1\frac{1}{4} = 3 \cdot z$$

$$4 \cdot \frac{5}{4} = 3 \cdot z$$

$$5 = 3 \cdot z$$

Divide by 3.

$$\frac{5}{3} = \frac{3 \cdot z}{3}$$

$$\frac{5}{3} = z$$

$$1\frac{2}{3} = z$$

(c) Cross multiply.

$$.6 \cdot 2 = 1.5 \cdot m$$

$$1.2 = 1.5 \cdot m$$

Divide by 1.5.

$$\frac{1.2}{1.5} = \frac{1.5 \cdot m}{1.5}$$

$$\frac{1.2}{1.5} = m$$

Multiply numerator and denominator of $\frac{1.2}{1.5}$ by 10. Multiplying by 10 $\left(\frac{10}{10} = 1\right)$ eliminates the decimals in the ratio and results in whole numbers.

$$\frac{1.2}{1.5} = \frac{1.2 \cdot 10}{1.5 \cdot 10} = \frac{12}{15} = \frac{4}{5} \quad ■$$

WORK PROBLEM 4 AT THE SIDE.

4. Find the missing numbers. Write answers as mixed numbers.

(a) $\frac{3\frac{1}{4}}{2} = \frac{m}{8}$

(b) $\frac{k}{3} = \frac{1\frac{2}{3}}{5}$

(c) $\frac{8}{5\frac{1}{4}} = \frac{6}{p}$

(d) $\frac{9}{a} = \frac{2}{1\frac{1}{4}}$

(e) $\frac{k}{6} - \frac{.5}{1.2}$

(f) $\frac{0}{2} = \frac{x}{5\frac{1}{3}}$

ANSWERS

4. (a) $m = 13$ (b) $k = 1$ (c) $p = 3\frac{15}{16}$ (d) $a = 5\frac{5}{8}$ (e) $k = 2\frac{1}{2}$ (f) $x = 0$

name date hour

5.4 EXERCISES

Find the missing number in each of the following proportions.

Example: $\frac{6}{15} = \frac{x}{5}$ **Solution:**

$$6 \cdot 5 = 15 \cdot x$$
$$30 = 15 \cdot x$$
$$\frac{30}{15} = \frac{15 \cdot x}{15} \quad \text{Divide by 15.}$$
$$2 = x$$
$$\mathbf{x = 2}$$

1. $\frac{1}{3} = \frac{r}{12}$

2. $\frac{z}{6} = \frac{15}{18}$

3. $\frac{k}{36} = \frac{7}{12}$

4. $\frac{5}{w} = \frac{20}{8}$

5. $\frac{16}{b} = \frac{4}{5}$

6. $\frac{21}{15} = \frac{7}{t}$

7. $\frac{10}{5} = \frac{x}{20}$

8. $\frac{4}{8} = \frac{q}{32}$

9. $\frac{n}{25} = \frac{4}{20}$

10. $\frac{6}{p} = \frac{4}{8}$

11. $\frac{8}{s} = \frac{24}{30}$

12. $\frac{32}{5} = \frac{d}{10}$

13. $\frac{15}{42} = \frac{h}{14}$

14. $\frac{25}{100} = \frac{8}{x}$

15. $\frac{y}{9} = \frac{62}{18}$

16. $\frac{3}{11} = \frac{u}{110}$

17. $\frac{50}{7} = \frac{r}{28}$

18. $\frac{29}{18} = \frac{58}{t}$

Find the missing number in each of the following proportions. Write answers as mixed numbers.

Example: $\frac{z}{12} = \frac{5}{8}$ **Solution:** $8 \cdot z = 5 \cdot 12$

$8 \cdot z = 60$

$\frac{8 \cdot z}{8} = \frac{60}{8}$ Divide by 8.

$z = \frac{60}{8} = \frac{15}{2} = 7\frac{1}{2}$

$\mathbf{z = 7\frac{1}{2}}$

19. $\frac{1}{2} = \frac{k}{7}$

20. $\frac{2}{3} = \frac{5}{x}$

21. $\frac{2}{9} = \frac{p}{12}$

22. $\frac{5}{6} = \frac{a}{10}$

23. $\frac{m}{3} = \frac{7}{8}$

24. $\frac{f}{10} = \frac{2}{7}$

25. $\frac{3}{q} = \frac{8}{9}$

26. $\frac{2}{z} = \frac{9}{7}$

27. $\frac{10}{15} = \frac{y}{8}$

28. $\frac{30}{20} = \frac{5}{u}$

29. $\frac{2\frac{1}{4}}{3} = \frac{b}{8}$

30. $\frac{6}{1\frac{1}{2}} = \frac{4}{r}$

31. $\dfrac{2\frac{1}{3}}{1\frac{1}{2}} = \dfrac{r}{2\frac{1}{4}}$

32. $\dfrac{1\frac{1}{4}}{m} = \dfrac{2\frac{1}{2}}{5}$

33. $\dfrac{3}{p} = \dfrac{.6}{4.2}$

34. $\dfrac{8}{9} = \dfrac{y}{1.5}$

35. $\dfrac{0}{5} = \dfrac{m}{8}$

36. $\dfrac{k}{4} = \dfrac{0}{7}$

CHALLENGE EXERCISES

Write a proportion for each of the following and then find the missing number.

	proportion	*missing number*
37. 6 is to 5 as 18 is to x.	__________	__________
38. 10 is to m as 25 is to 10.	__________	__________
39. 42 is to a as 30 is to 60.	__________	__________

	proportion	*missing number*
40. 25 is to 20 as r is to 40.	________	________
41. k is to 8 as $\frac{4}{3}$ is to 2.	________	________
42. $2\frac{1}{2}$ is to 1 as w is to 3.	________	________

REVIEW EXERCISES

Write each of the following as a proportion. (For help, see **Section 5.3.***)*

43. 10 is to 4 as 40 is to 16.

44. 50 is to 70 as 15 is to 21.

45. 6 is to 9 as 10 is to 15.

46. 6 is to $3\frac{1}{4}$ as 24 is to 13.

5.5 APPLICATIONS OF PROPORTIONS

1 Proportions can be used to solve word problems. Start by using a letter, such as x, for the unknown number. Place units next to each number and arrange the numbers in a proportion, so that corresponding places have the same units.

$$\frac{\text{dollars}}{\text{quarts}} = \frac{\text{dollars}}{\text{quarts}}$$ Correct, since dollars and quarts are in the same corresponding positions.

$$\frac{\text{quarts}}{\text{quarts}} = \frac{\text{dollars}}{\text{dollars}}$$ correct

$$\frac{\text{quarts}}{\text{dollars}} = \frac{\text{dollars}}{\text{quarts}}$$ incorrect

For these word problems, remember the indicator words mentioned earlier:

of, in, for, on, and *per.*

EXAMPLE 1 Applying Proportion to Find Cost

One store sells 12 pencils for 78¢. Find the cost of 18 pencils.

APPROACH

Use a proportion of ratios of pencils to cents to help solve this problem.

SOLUTION

Let x represent the unknown cost of 18 pencils. Think "12 pencils for 78 cents," and "18 pencils for x cents." Write a proportion by using a bar to represent the indicator word *for*.

Bar represents the indicator *for*. → $$\frac{12 \text{ pencils}}{78 \text{ cents}} = \frac{18 \text{ pencils}}{x \text{ cents}}$$

Check to be certain that corresponding units are the same. Since they are, drop the units.

$$\frac{12}{78} = \frac{18}{x}$$

Write $\frac{12}{78}$ in lowest terms to get the proportion

$$\frac{6}{39} = \frac{18}{x}.$$

Find the value of x by first finding the cross products.

$$6 \cdot x = 39 \cdot 18$$

$$6 \cdot x = 702$$

Divide by 6. $$\frac{6 \cdot x}{6} = \frac{702}{6}$$

$$x = 117$$

The cost is 117 cents, or \$1.17, for 18 pencils. ■

WORK PROBLEM 1 AT THE SIDE.

OBJECTIVE

1 Use proportions to solve word problems.

1. (a) 6 tickets cost 15 dollars. Find the cost of 14 tickets.

(b) 12 apples cost \$2.52. Find the cost of 5 apples.

(c) 25 books cost \$212.50. Find the cost of one book.

ANSWERS

1. (a) \$35 (b) \$1.05 (c) \$8.50

2. (a) If 2 pounds of fertilizer will cover 50 square feet of garden, how many pounds would be needed for 225 square feet?

(b) The tax on a $20 item is $1. Find the tax on a $110 item.

(c) If 8 ounces of extract must be mixed with 20 ounces of syrup, how many ounces of extract must be mixed with 50 ounces of syrup?

(d) A certain motorcycle uses 3 gallons of gas to compete in 10 races. How many gallons of gas would be needed for 30 races?

ANSWERS
2. (a) 9 pounds (b) $5.50
(c) 20 ounces (d) 9 gallons

EXAMPLE 2 Applications of Proportion
A car can go 150 miles on 6 gallons of gas. How far can it go on 14 gallons of gas?

APPROACH
A proportion using ratios of miles to gallons of gas is useful in solving this problem.

SOLUTION
Use m to stand for the unknown number of miles the car can go. Set up a proportion, making sure like units are in the same position on both sides.

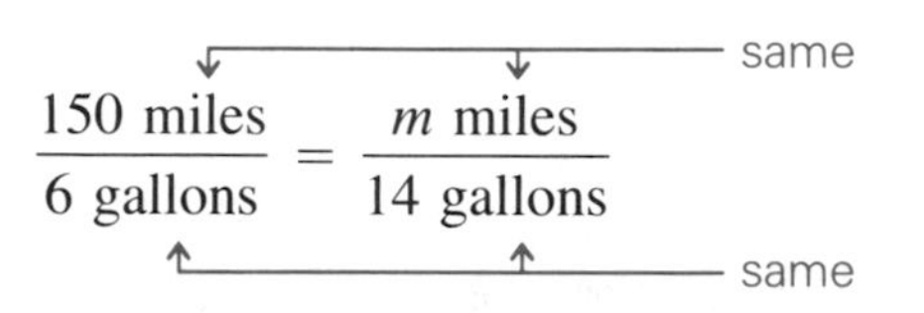

Drop the units.

$$\frac{150}{6} = \frac{m}{14}$$

Write $\frac{150}{6}$ in lowest terms.

$$\frac{25}{1} = \frac{m}{14}$$

Find the cross products.

$$25 \cdot 14 = 1 \cdot m$$
$$350 = 1 \cdot m$$

Since $1 \cdot m = m$,

$$350 = m.$$

The car can go 350 miles on 14 gallons of gas. ■

WORK PROBLEM 2 AT THE SIDE.

name date hour

5.5 EXERCISES

Set up proportions for each of the following problems and then solve them.

Example: 8 pounds of vegetables cost \$5. Find the cost of 24 pounds.
Solution: Let x represent the cost of 24 pounds.

$$\frac{8 \text{ pounds}}{5 \text{ dollars}} = \frac{24 \text{ pounds}}{x \text{ dollars}} \qquad \frac{8}{5} = \frac{24}{x}$$ Drop the units.

$$8 \cdot x = 5 \cdot 24$$ Find cross products.

$$8 \cdot x = 120$$

$$\frac{8 \cdot x}{8} = \frac{120}{8}$$ Divide by 8.

$$x = 15$$ **\$15**

1. A gardening service charges \$30 to install 50 square feet of sod. Find the charge to install 125 square feet.

2. On a road map, a length of 3 inches represents 8 miles. How many inches represent a distance of 24 miles?

3. 6 magazines cost \$15. Find the cost of 14 magazines.

4. 22 ties cost \$176. Find the cost of 12 ties.

5. Five pounds of grass seed cover 3500 square feet of ground. How many pounds are needed for 4900 square feet?

6. Anna earns \$280.84 in 7 days. How much does she earn in 10 days?

7. Tom makes \$255.75 in 5 days. How much does he make in 3 days?

8. If 5 ounces of a medicine must be mixed with 11 ounces of water, how many ounces of medicine would be mixed with 99 ounces of water?

9. The distance between two cities on a road map is 5 inches. The two cities are really 600 miles apart. The map distance between two other cities is 9 inches. How many miles apart are these cities?

10. The distance between two cities is 700 miles. On a map, the cities are 10 inches apart. Two other cities are 1030 miles apart. How many inches apart are they on the map?

11. If 4 days of parking in the city cost $80, find the cost of parking for 11 days.

12. An 8-minute phone call cost 96¢. Find the cost of a 6-minute call.

13. A carpenter charges $150.50 to install a deck railing 10 feet long. How much would she charge to install a deck railing 18 feet long?

14. Stan paid $240,000 for a 5-unit apartment house. Find the cost of a 12-unit apartment house.

15. John plants his seeds early in the year. To keep them from freezing, he covers the ground with black plastic. A piece with an area of 80 square feet costs $12. Find the cost of a piece with an area of 200 square feet.

16. A taxi ride of 3 miles costs $6.90. Find the cost of a ride of 5 miles.

CHALLENGE EXERCISES

17. To make vitamin C tablets, 248 pounds of ascorbic acid crystals are combined with 575 pounds of binding and filler. How many pounds of ascorbic acid crystals are needed for 2875 pounds of binder and filler?

18. To make battery acid, Ralph mixes $4\frac{1}{2}$ gallons of pure acid with $12\frac{1}{2}$ gallons of water. How much acid would be needed for $37\frac{1}{2}$ gallons of water?

19. Tax on a $5220 computer system is $339.30. Find the tax on a $7400 computer system.

20. If $8\frac{1}{2}$ yards of material are needed for 5 dresses, how much material is needed for 28 dresses?

REVIEW EXERCISES

Multiply or divide as indicated. (For help, see ***Sections 1.4 and 1.6.****)*

21. $.12 \times 100$

22. 6.17×100

23. 2.87×100

24. $25.8 \div 100$

25. $1.93 \div 100$

26. $5 \div 100$

CHAPTER 5 SUMMARY

KEY TERMS

5.1 ratio — A ratio is a comparison of two quantities. For example, the ratio of 6 apples to 11 apples is written as $\frac{6}{11}$.

5.2 rate — The type of ratio used to compare unlike quantities is a rate. An example is 22 miles per gallon to express mileage.

unit rate — A unit rate is a rate with 1 unit in the denominator.

unit price — The unit price is the unit rate that gives a cost per item or quantity.

5.3 proportion — A proportion states that two ratios are equivalent.

cross products — The method of cross products, or cross multiplication, is used to determine whether a proportion is true.

QUICK REVIEW

Section Number and Topic	Approach	Example
5.1 Writing a Ratio	A ratio is a comparison of two quantities. Ratios are usually written in the form $\frac{a}{b}$, with the number that is mentioned first written as the numerator. Units are not written.	Write the following as a ratio in lowest terms. 60 ounces of chemical to 160 ounces of solvent $\frac{60}{160} = \frac{60 \div 20}{160 \div 20} = \frac{3}{8}$
5.1 Using a Mixed Number in a Ratio	First write the ratio with mixed numbers. Next, change mixed numbers to improper fractions, then invert and multiply. Finally, change to lowest terms.	Write as a ratio in lowest terms. $2\frac{1}{2}$ to $3\frac{3}{4}$ $\frac{2\frac{1}{2}}{3\frac{3}{4}}$ ratio in mixed numbers $= \frac{\frac{5}{2}}{\frac{15}{4}}$ ratio in improper fractions $= \frac{5}{2} \div \frac{15}{4} = \frac{5}{2} \cdot \frac{4}{15}$ Invert and multiply. $= \frac{\overset{1}{\cancel{5}}}{\underset{1}{\cancel{2}}} \cdot \frac{\overset{2}{\cancel{4}}}{\underset{3}{\cancel{15}}} = \frac{2}{3}$ ratio in lowest terms

Section Number and Topic	Approach	Example
5.1 Ratios that Compare Measurements	The measurements must always be in the same units.	Write as a ratio in lowest terms. 8 inches to 6 feet Since there are 12 inches in 1 foot, there are $12 \cdot 6 = 72$ inches in 6 feet. The ratio is $\frac{8}{72} = \frac{8 \div 8}{72 \div 8} = \frac{1}{9}.$
5.2 Write a Rate	A rate is a ratio used to compare unlike quantities.	Write as a rate in lowest terms. 450 miles in 10 hours $\frac{450 \text{ miles}}{10 \text{ hours}} = \frac{45 \text{ miles}}{1 \text{ hour}}$
5.2 Finding a Unit Rate	A unit rate has a 1 in the denominator. A unit rate that gives a cost per item is called a unit price.	Write as a unit rate. \$1278 in 9 days $\frac{1278 \text{ dollars}}{9 \text{ days}} = \frac{142 \text{ dollars}}{1 \text{ day}}$ $= \frac{\$142}{\text{day}}$
5.3 Writing a Proportion	A proportion states that two ratios are equivalent. This proportion, $\frac{5}{6} = \frac{25}{30},$ is read as "5 is to 6 as 25 is to 30," or as "five sixths equals twenty-five thirtieths."	Write as a proportion. 8 is to 40 as 32 is to 160 $\frac{8}{40} = \frac{32}{160}$

Section Number and Topic	Approach	Example
5.3 Using Cross Products	To see whether a proportion is true, cross multiply one way, then cross multiply the other way. $\frac{a}{b} = \frac{c}{d}$ → bc, ad. If the two answers are the same, the proportion is true.	Cross multiply to see whether the following proportion is true. $\frac{6}{8\frac{1}{2}} = \frac{24}{34}$ Cross multiply. $8\frac{1}{2} \cdot 24 = \frac{17}{\not{2}_1} \cdot \frac{\not{24}^{12}}{1} = 204$; $\frac{6}{8\frac{1}{2}} = \frac{24}{34}$; $6 \cdot 34 = 204$ (same). The proportion is true.
5.4 Solving a Proportion for a Missing Number	Solve for a missing number by using these steps. *Step 1* Write all ratios in lowest terms. *Step 2* Find the two cross products. *Step 3* Show the products are equivalent. *Step 4* Divide both the products by the number multiplying the letter.	Find the value of the missing number. $\frac{12}{p} = \frac{6}{8}$ *Step 1* $\frac{12}{p} = \frac{3}{4}$ *Step 2* $\frac{12}{p} = \frac{3}{4}$; $3 \cdot p$; $12 \cdot 4 = 48$ *Step 3* $3p = 48$ *Step 4* $\frac{3p}{3} = \frac{48}{3}$; $p = 16$

Section Number and Topic	Approach	Example
5.5 Applications of Proportions	Use a letter such as x for the unknown in a word problem. Place units next to corresponding units in the proportion and solve for the missing number.	Use a proportion to solve. The cost of 150 pine seedlings is $12. Find the cost of 1250 seedlings. $\frac{150}{12} = \frac{1250}{x}$ $\frac{25}{2} = \frac{1250}{x}$ $25 \cdot x = 2 \cdot 1250$ $25x = 2500$ $\frac{25x}{25} = \frac{2500}{25}$ $x = \$100$

name date hour

CHAPTER 5 REVIEW EXERCISES

[5.1–5.2] *Write each of the following ratios in lowest terms. Change to the same units when necessary.*

1. 6 apples to 11 apples

2. 12 tablets to 8 tablets

3. 25 miles to 2 gallons

4. 50 feet to 90 feet

5. 38 hits in 76 times at bat

6. 20 victories in 30 games

7. 5 hours to 100 minutes

8. 550 passengers to 11 buses

9. $100 to 20 hours

10. $45 to 300 miles

11. $2\frac{1}{2}$ to 5

12. $6\frac{1}{4}$ to $12\frac{1}{2}$

When writing the following ratios, be sure to make all necessary conversions to the same units by using the list of relationships in Section 5.1.

13. 3 inches to 1 foot

14. 4 feet to 10 inches

15. 3 hours to 30 minutes

16. 3 quarts to 1 gallon

17. 20 hours to 3 days

18. 6 quarts to 5 pints

19. $2\frac{1}{2}$ days to 1 week

20. $3\frac{1}{2}$ yards to 4 feet

Solve each word problem. Write each ratio in lowest terms.

21. John has sales of $35 in products and Marcia has sales of $50 in products. Find the ratio of his sales to hers.

22. The domestic car gets 35 miles per gallon, while the import gets 30 miles per gallon. Find the ratio of the domestic car's mileage to the import car's.

Find the best buy for each of the following.

23. spice
16 oz for $2.80
8 oz for $1.45
3 oz for $1.15

24. canned peaches
30 oz for $1.50
15 oz for $.79
8 oz for $.45

[5.3] *Write each of the following proportions.*

25. 5 is to 10 as 20 is to 40.

26. 7 is to 2 as 35 is to 10.

27. $1\frac{1}{4}$ is to 5 as 3 is to 12.

Use the method of writing in lowest terms to decide whether the following proportions are true or false.

28. $\frac{3}{5} = \frac{9}{15}$

29. $\frac{16}{48} = \frac{9}{36}$

30. $\frac{47}{10} = \frac{98}{20}$

31. $\frac{64}{36} = \frac{96}{54}$

32. $\frac{32}{72} = \frac{24}{54}$

33. $\frac{110}{140} = \frac{22}{26}$

name date hour

Use the cross products method to decide whether the following proportions are true or false. Circle the correct answer.

34. $\frac{6}{10} = \frac{12}{20}$
True False

35. $\frac{6}{8} = \frac{21}{28}$
True False

36. $\frac{32}{40} = \frac{44}{55}$
True False

[5.4] *Find the missing number in each of the following proportions.*

37. $\frac{1}{4} = \frac{p}{8}$

38. $\frac{16}{b} = \frac{12}{15}$

39. $\frac{100}{14} = \frac{p}{56}$

Find the missing number in each of the following proportions. Write answers as mixed numbers.

40. $\frac{5}{8} = \frac{a}{20}$

41. $\frac{p}{24} = \frac{11}{18}$

42. $\frac{f}{10} = \frac{2\frac{1}{2}}{8}$

Write a proportion for each of the following and then find the missing number.

	proportion	***missing number***
43. 5 is to 8 as x is to 16.	____________	____________
44. 7 is to 10 as 14 is to y.	____________	____________

[5.5] *Set up proportions for each of the following and then solve them.*

45. 16 files are pulled in 5 minutes. Find the number that can be pulled in 20 minutes.

46. A cyclist travels 25 miles in 2 hours. How far can she travel in $5\frac{1}{2}$ hours?

47. 6 gallons of chemicals cover 4500 square feet. How many gallons are needed to cover 13,500 square feet?

48. The interest on a \$2000 loan is \$380. Find the interest on a \$3000 loan.

49. A mechanic charges \$53 for a 2-hour job. How much will he charge for a 5-hour job?

50. A box of 12 floor tiles costs \$10.50. Find the cost of 180 floor tiles.

CHALLENGE EXERCISES

Write each of the following proportions.

51. 110 is to 140 as 22 is to 28.

52. 8 is to $3\frac{1}{2}$ as 32 is to 14.

53. 15 is to 5 as 21 is to 7.

Write a proportion for each of the following and then find the missing number.

	proportion	***missing number***
54. p is to 6 as 8 is to 15.	__________	__________
55. 4 is to w as 15 is to 2.	__________	__________

name date hour

Find the missing number in each of the following proportions.

56. $\frac{b}{45} = \frac{70}{30}$

57. $\frac{k}{52} = \frac{0}{20}$

58. $\frac{64}{10} = \frac{x}{20}$

Use the cross products method to decide whether the following proportions are true. Circle the correct answer.

59. $\frac{55}{18} = \frac{80}{27}$
True False

60. $\frac{56}{60} = \frac{180}{194}$
True False

61. $\frac{66}{8} = \frac{165}{20}$
True False

Find the missing number in each of the following proportions. Write fractional answers as mixed numbers.

62. $\frac{2}{7} = \frac{7}{y}$

63. $\frac{7}{1\frac{1}{2}} = \frac{e}{6}$

64. $\frac{3}{5} = \frac{z}{4\frac{1}{2}}$

Write each of the following ratios in lowest terms. Change to the same units when necessary.

65. 4 dollars to 10 quarters

66. $4\frac{1}{8}$ to 10

67. 10 yards to 8 feet

68. 3 pints to 4 quarts

69. 15 minutes to 1 hour

70. $1\frac{3}{10}$ to $4\frac{1}{2}$

Set up a proportion for each of the following and then solve it.

71. If 8 ounces of a medicine must be mixed with 20 ounces of water, how many ounces of medicine must be mixed with 50 ounces of water?

72. A certain lawn mower uses 3 gallons of gas to cut 10 acres of lawn. How many gallons of gas would be needed for 30 acres?

73. A Hershey bar contains 200 calories. How many bars would you need to eat to get 500 calories?

74. Jose can assemble 12 car parts in 40 minutes. How many minutes would he need to assemble 15 parts?

name date hour

CHAPTER 5 TEST

Write each of the following ratios in lowest terms.

1. 9 ships to 11 ships — 1. ____________

2. 128 feet of fencing will cost $225. — 2. ____________

3. Write 12 instructions for 8 patients. — 3. ____________

4. The little theater holds 340 people. The auditorium holds 1120 people. — 4. ____________

5. Find the best buy on cornflakes. — 5. ____________
   28 oz for $2.15
   18 oz for $1.67
   10 oz for $1.29

Write each of the following ratios in lowest terms. Refer to the lists of relationships among measures in Section 5.1.

6. 3 quarts to 180 gallons — 6. ____________

7. 9 inches to 15 feet — 7. ____________

Write each of the following proportions.

8. 7 is to 15 as 21 is to 45. — 8. ____________

9. 12 is to 7 as 36 is to 21. — 9. ____________

10. ______________________

11. ______________________

12. ______________________

13. ______________________

14. ______________________

15. ______________________

16. ______________________

17. ______________________

18. ______________________

19. ______________________

20. ______________________

Decide whether the following proportions are true or false.

10. $\frac{6}{15} = \frac{18}{45}$

11. $\frac{7}{20} = \frac{28}{75}$

12. $\frac{1\frac{1}{4}}{2} = \frac{6\frac{1}{4}}{10}$

13. $\frac{4\frac{3}{8}}{7\frac{1}{2}} = \frac{13\frac{1}{8}}{22\frac{1}{2}}$

Find the missing number in each proportion.

14. $\frac{5}{9} = \frac{f}{45}$

15. $\frac{1}{3} = \frac{h}{8}$

16. $\frac{6\frac{1}{2}}{4} = \frac{g}{20}$

Solve the following word problems.

17. A student types 240 words in 5 minutes. How many words can he type in 12 minutes?

18. A boat travels 82 miles in 4 hours. How far can it go in 9 hours?

19. A tree grows at the rate of 3 yards in 2 years. How much will it grow in 15 years?

20. Two points on a map are $3\frac{1}{2}$ inches apart. If this represents a distance of 7 miles, how many miles are represented by $6\frac{3}{4}$ inches?

6

PERCENT

6.1 BASICS OF PERCENT

OBJECTIVES

1. Learn the meaning of percent.
2. Write percents as decimals.
3. Write decimals as percents.

The figure here has one hundred equivalent squares. Eleven of these are shaded. The shaded portion is $\frac{11}{100}$, or .11, of the total.

The shaded portion is also 11% of the total, or "eleven parts of 100 parts." Read **11%** as "eleven percent."

1 As shown above, percent is a ratio with a denominator of 100.

Percent means **per one hundred.**

EXAMPLE 1 Understanding Percent

(a) If 43 of 100 students are men, then **43%** of the students are men.

(b) If a person pays a tax of $7 per $100, then the tax rate is **7%.** ■

WORK PROBLEM 1 AT THE SIDE.

1. Write as percents.

(a) In a group of people, 56 out of 100 are driving small cars.

(b) The tax is $14 per $100.

(c) Out of 100 gallons of gas, 18 were unleaded.

ANSWERS
1. (a) 56% (b) 14% (c) 18%

2. Write as a decimal.

(a) 34%

(b) 82%

(c) 56%

2 Since $p\%$ means p parts of 100 parts, then

$$\frac{p}{100} = p \div 100 = p \cdot \frac{1}{100} = p \cdot (.01)$$

and

$$p\% = \underbrace{\frac{p}{100}}_{\text{as a fraction}} \quad \text{or} \quad p\% = p \cdot \underbrace{(.01)}_{\text{as a decimal}}.$$

EXAMPLE 2 Writing Percent as a Decimal
Write each percent as a decimal.

(a) 47% **(b)** 76%

SOLUTION
Since $p\% = p \cdot (.01)$,

(a) $47\% = 47 \cdot (.01) = .47$.

(b) $76\% = 76 \cdot (.01) = .76$ ■

WORK PROBLEM 2 AT THE SIDE.

Example 2 suggests the following rule for writing a percent as a decimal.

Write a percent as a decimal by dropping the percent sign and dividing by 100. This results in moving the decimal point two places to the left.

EXAMPLE 3 Changing to Decimal by Moving the Decimal Point
Write each percent as a decimal.

(a) 17% **(b)** 83.4% **(c)** 4.9% **(d)** .6%

SOLUTION

(a) .17 ← Percent sign is dropped.
↑ Decimal point is moved two places to the left.
17% = .17

(b) 83.4% = .834

(c) .04 9 0 is attached and decimal point is moved two places to the left.
4.9% = .049

(d) .6% = .006 ■

ANSWERS
2. (a) .34 (b) .82 (c) .56

NOTE Since 1% is equivalent to .01, any fraction of a percent smaller than 1% is less than .01.

WORK PROBLEM 3 AT THE SIDE.

3 Decimals can also be written as percents. For example, the decimal .78 is the same as the fraction

$$\frac{78}{100}.$$

This fraction means 78 of 100 parts, or 78%. The following rule gives the same result.

Write a decimal as a percent by multiplying by 100. This results in moving the decimal point two places to the right. Next, attach a percent sign.

EXAMPLE 4 Changing to Percent by Moving the Decimal Point
Write each decimal as a percent.

(a) .21 **(b)** .529 **(c)** 1.92

SOLUTION

(a) .21% ← Percent sign is attached.
↑ Decimal point is moved two places to the right.
.21 = 21%

(b) .529 = 52.9%

(c) 1.92 = 192% ■

NOTE Since the number 1 is equivalent to 100%, all whole numbers will be 100% or larger.

WORK PROBLEM 4 AT THE SIDE.

3. Write each percent as a decimal.

(a) 61%

(b) 4%

(c) 18.7%

(d) .8%

4. Write as a percent.

(a) .74

(b) .12

(c) .09

(d) .617

(e) .834

(f) 3.14

(g) 2.87

ANSWERS
3. (a) .61 (b) .04 (c) .187 (d) .008
4. (a) 74% (b) 12% (c) 9% (d) 61.7% (e) 83.4% (f) 314% (g) 287%

name date hour

6.1 EXERCISES

Write each of the following percents as decimals.

Examples: 42% 1.4%

Solutions: .42 ← Percent sign is dropped. 01.4

↑ Decimal point is moved two places to the left. ↑ two places left

42% = **.42** 1.4% = **.014**

1. 50%

2. 70%

3. 25%

4. 87%

5. 65%

6. 32%

7. 140%

8. 250%

9. 7%

10. 9%

11. 14.9%

12. 27.6%

13. .8%

14. .3%

15. .25%

16. .035%

Write each of the following decimals as percents.

Examples: .23 | .8652

Solutions: .23 % ← Percent sign is attached.
↑ Decimal point is moved two places to the right.
.23 = **23%**

.86 52%
.8652 = **86.52%**

17. .60

18. .82

19. .75

20. .31

21. .02

22. .08

23. .125

24. .875

25. .453

26. .324

27. 3.82

28. 4.61

29. 1.5

30. 2.3

31. .0312

32. .0625

33. 3.725

34. .0025

35. .0075

36. .005

In each of the following, write percents as decimals and decimals as percents.

37. In Georgetown, the sales tax is 6%.

38. At Nichols College, 47% of the students are female.

39. In one company, 65% of the salespeople are women.

40. Only 38.6% of those registered actually voted.

41. The sales tax rate in one state is .065.

42. A charity has raised .25 of the needed funds.

43. Sales this year are 2.6 times last year's sales.

44. Enrollment this year is 5.6 times last year's enrollment.

45. Only .005 of the total population has this genetic defect.

46. Defects in production remain at .0073 of total production.

47. The patient's cholesterol was decreased by 47.8%.

48. Success with the diet was 248.7% greater than anticipated.

CHALLENGE EXERCISES

Write a percent for both the shaded and unshaded part of each figure.

49.

50.

51. 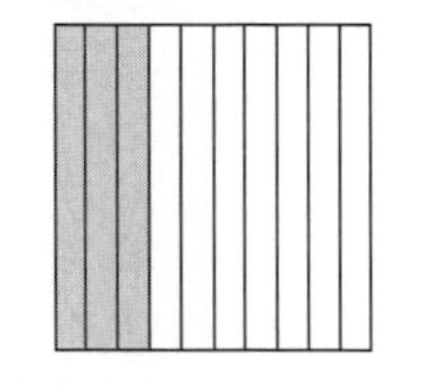

52. 

53.

54.

REVIEW EXERCISES

Change each of the following fractions to decimals. Round to the nearest thousandth if necessary. (For help, see ***Section 4.7****.)*

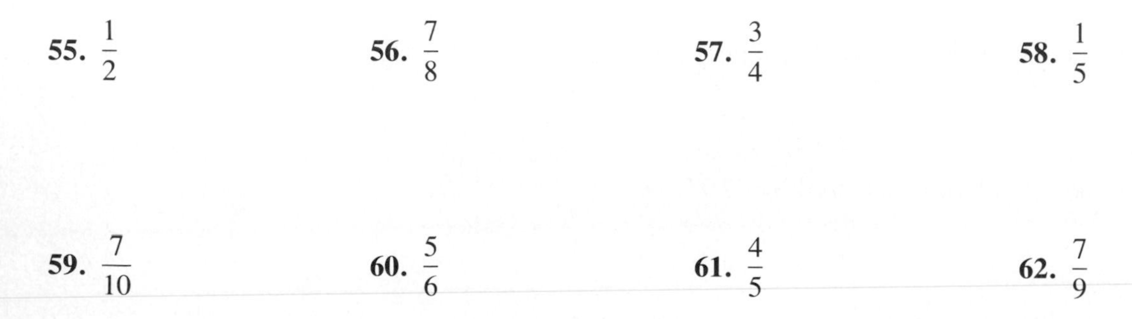

55. $\frac{1}{2}$

56. $\frac{7}{8}$

57. $\frac{3}{4}$

58. $\frac{1}{5}$

59. $\frac{7}{10}$

60. $\frac{5}{6}$

61. $\frac{4}{5}$

62. $\frac{7}{9}$

6.2 PERCENTS AND FRACTIONS

OBJECTIVES

1. Write percents as fractions.
2. Write fractions as percents.
3. Use the table of percent equivalents.

1 Percents can be written as fractions by using the result from the previous section.

$$p\% = \frac{p}{100}, \text{ as a fraction.}$$

EXAMPLE 1 Writing a Percent as a Fraction
Write each percent as a fraction or mixed number in lowest terms.

(a) 45% **(b)** 76%

(c) 150%

SOLUTION

(a) As shown earlier, 45% can be written as a decimal.

$$45\% = 45 \cdot (.01) = .45 \quad \text{(percent sign dropped)}$$

Since .45 means 45 hundredths,

$$.45 = \frac{45}{100} = \frac{9}{20}. \quad \text{(lowest terms)}$$

It is not necessary, however, to write 45% as a decimal first.

$$45\% = \frac{45}{100} \quad \text{from the rule above}$$

$$= \frac{9}{20} \quad \text{(lowest terms)}$$

(b) Write 76% as $\frac{76}{100}$.

The percent becomes the *numerator*.
The *denominator* is always 100.

Write $\frac{76}{100}$ in lowest terms.

$$\frac{76}{100} = \frac{19}{25}$$

(c) $150\% = \frac{150}{100} = \frac{3}{2} = 1\frac{1}{2}$ (mixed number)

WORK PROBLEM 1 AT THE SIDE.

The next example shows how to write decimal percents as fractions.

1. Write each percent as a fraction or mixed number in lowest terms.

(a) 34%

(b) 11%

(c) 54%

(d) 75%

(e) 120%

(f) 210%

ANSWERS

1. (a) $\frac{17}{50}$ (b) $\frac{11}{100}$ (c) $\frac{27}{50}$ (d) $\frac{3}{4}$ (e) $1\frac{1}{5}$ (f) $2\frac{1}{10}$

2. Write as fractions in lowest terms.

(a) 24.5%

(b) 62.5%

(c) 8.5%

(d) $66\frac{2}{3}\%$

(e) $12\frac{1}{3}\%$

(f) $37\frac{1}{2}\%$

ANSWERS

2. (a) $\frac{49}{200}$ (b) $\frac{5}{8}$ (c) $\frac{17}{200}$ (d) $\frac{2}{3}$ (e) $\frac{37}{300}$ (f) $\frac{3}{8}$

EXAMPLE 2 Writing a Decimal Percent as a Fraction

Write each percent as a fraction in lowest terms.

(a) 15.5% **(b)** $33\frac{1}{3}\%$

SOLUTION

(a) Place 15.5 over 100.

$$15.5\% = \frac{15.5}{100}$$

Multiply numerator and denominator by 10 $\left(\frac{10}{10} = 1\right)$. This is done to get a whole number in both the numerator and denominator.

$$\frac{15.5}{100} = \frac{15.5 \cdot 10}{100 \cdot 10} = \frac{155}{1000}$$

Write in lowest terms.

$$\frac{155}{1000} = \frac{31}{200}$$

(b) Place $33\frac{1}{3}$ over 100.

$$33\frac{1}{3}\% = \frac{33\frac{1}{3}}{100}$$

Write $33\frac{1}{3}$ as the improper fraction $\frac{\mathbf{100}}{\mathbf{3}}$**.** Now,

$$\frac{33\frac{1}{3}}{100} = \frac{\frac{100}{3}}{100}.$$

Divide these fractions by inverting the bottom fraction and multiplying.

$$\frac{\frac{100}{3}}{100} = \frac{\frac{100}{3}}{\frac{100}{1}} = \frac{100}{3} \div \frac{100}{1} = \frac{\overset{1}{\cancel{100}}}{3} \cdot \frac{1}{\underset{1}{\cancel{100}}} = \frac{1}{3}$$ ■

WORK PROBLEM 2 AT THE SIDE.

2 Use the result given at the beginning of this section to write fractions as percents.

$$p\% = \frac{p}{100}$$

EXAMPLE 3 Writing a Fraction as a Percent
Write each fraction as a percent. Round to the nearest tenth if necessary.

(a) $\frac{3}{5}$ **(b)** $\frac{7}{8}$ **(c)** $\frac{5}{6}$

SOLUTION
(a) Write this fraction as a percent by solving for p in the proportion

$$\frac{3}{5} = \frac{p}{100}.$$

Find cross products.

$$3 \cdot 100 = 5 \cdot p$$
$$300 = 5 \cdot p$$

Divide both sides by 5

$$\frac{300}{5} = \frac{5 \cdot p}{5}$$
$$60 = p$$

This result means that $\frac{3}{5} = \frac{60}{100}$ or 60%.

(b) Write a proportion.

$$\frac{7}{8} = \frac{p}{100}$$

$$7 \cdot 100 = 8 \cdot p \quad \text{Cross multiply.}$$
$$700 = 8 \cdot p$$
$$\frac{700}{8} = p \quad \text{Divide.}$$
$$87\frac{1}{2} = 87.5 = p$$

Finally, $\frac{7}{8} = 87\frac{1}{2}\%$ or 87.5%.

3. Write as percents. Round to the nearest tenth if necessary.

(a) $\frac{3}{10}$

(b) $\frac{7}{25}$

(c) $\frac{1}{4}$

(d) $\frac{3}{8}$

(e) $\frac{1}{12}$

(f) $\frac{2}{9}$

ANSWERS
3. (a) 30% (b) 28% (c) 25%
(d) 37.5% (e) 8.3% (f) 22.2%

(c) Start with a proportion.

$$\frac{5}{6} = \frac{p}{100}$$

$$5 \cdot 100 = 6 \cdot p \qquad \text{Cross multiply.}$$

$$500 = 6 \cdot p$$

$$\frac{500}{6} = p \qquad \text{Divide.}$$

$$83\frac{1}{3} = p$$

or $$83.3 = p \qquad \text{rounded to the nearest tenth}$$

Solving this proportion shows

$$\frac{5}{6} = 83\frac{1}{3}\% \quad \text{or} \quad 83.3\% \quad \text{rounded.}$$ ■

WORK PROBLEM 3 AT THE SIDE.

3 The table on the next page summarizes common fractions and their decimal and percent equivalents. Some of the decimals have been rounded to the nearest thousandth.

EXAMPLE 4 Using a Conversion Table
Read the following from the chart.

(a) $\frac{1}{12}$ as a percent

(b) .375 as a fraction

(c) $\frac{13}{16}$ as a percent

SOLUTION

(a) Find $\frac{1}{12}$ in the "fraction column." The percent is $8\frac{1}{3}\%$.

(b) Look in the "decimal column" for .375. The fraction is $\frac{3}{8}$.

(c) Find $\frac{13}{16}$ in the "fraction column." The percent is $81\frac{1}{4}\%$ or 81.25%. ■

NOTE Remember that the bar over a digit shows that the digit is repeating. In $.08\overline{3}$, the **3** is repeating; in $.16\overline{6}$, the **6** is repeating.

WORK PROBLEM 4 AT THE SIDE.

percent	*decimal*	*fraction*
1%	.01	$\frac{1}{100}$
5%	.05	$\frac{1}{20}$
$6\frac{1}{4}$% or 6.25%	.0625	$\frac{1}{16}$
$8\frac{1}{3}$%	$.08\overline{3}$	$\frac{1}{12}$
10%	.1	$\frac{1}{10}$
$12\frac{1}{2}$% or 12.5%	.125	$\frac{1}{8}$
$16\frac{2}{3}$%	$.16\overline{6}$	$\frac{1}{6}$
$18\frac{3}{4}$% or 18.75%	.1875	$\frac{3}{16}$
20%	.2	$\frac{1}{5}$
25%	.25	$\frac{1}{4}$
30%	.3	$\frac{3}{10}$
$31\frac{1}{4}$% or 31.25%	.3125	$\frac{5}{16}$
$33\frac{1}{3}$%	$.33\overline{3}$	$\frac{1}{3}$
$37\frac{1}{2}$% or 37.5%	.375	$\frac{3}{8}$
40%	.4	$\frac{2}{5}$
$43\frac{3}{4}$% or 43.75%	.4375	$\frac{7}{16}$
50%	.5	$\frac{1}{2}$
$56\frac{1}{4}$% or 56.25%	.5625	$\frac{9}{16}$
60%	.6	$\frac{3}{5}$
$62\frac{1}{2}$% or 62.5%	.625	$\frac{5}{8}$

4. Read the following common fractions, decimals, and percents from the table.

(a) $\frac{1}{4}$ as a percent

(b) $12\frac{1}{2}$% as a fraction

(c) $.33\overline{3}$ as a fraction

(d) 10% as a fraction

(e) $\frac{7}{8}$ as a percent

(f) $\frac{1}{2}$ as a percent

(g) $66\frac{2}{3}$% as a fraction

(h) $\frac{1}{5}$ as a percent

ANSWERS

4. (a) 25% (b) $\frac{1}{8}$ (c) $\frac{1}{3}$ (d) $\frac{1}{10}$ (e) 87.5% (f) 50% (g) $\frac{2}{3}$ (h) 20%

percent	*decimal*	*fraction*
$66\frac{2}{3}\%$	$.66\overline{6}$	$\frac{2}{3}$
$68\frac{3}{4}\%$ or 68.75%	.6875	$\frac{11}{16}$
70%	.7	$\frac{7}{10}$
75%	.75	$\frac{3}{4}$
80%	.8	$\frac{4}{5}$
$81\frac{1}{4}\%$ or 81.25%	.8125	$\frac{13}{16}$
$83\frac{1}{3}\%$	$.83\overline{3}$	$\frac{5}{6}$
$87\frac{1}{2}\%$ or 87.5%	.875	$\frac{7}{8}$
90%	.9	$\frac{9}{10}$
$93\frac{3}{4}\%$ or 93.75%	.9375	$\frac{15}{16}$
100%	1.	1

name date hour

6.2 EXERCISES

Write each percent as a fraction or mixed number in lowest terms.

Example: 20%

Solution: $20\% = \frac{20}{100} = \frac{1}{5}$ ← in lowest terms

↑ numerator (20) — Denominator is always 100.

1. 10% **2.** 50% **3.** 25% **4.** 75%

5. 85% **6.** 15% **7.** 37.5% **8.** 87.5%

9. 6.25% **10.** 43.75% **11.** $16\frac{2}{3}\%$ **12.** $83\frac{1}{3}\%$

13. $6\frac{2}{3}\%$ **14.** $46\frac{2}{3}\%$ **15.** .4% **16.** .9%

17. 120% **18.** 150% **19.** 225% **20.** 320%

Write each fraction as a percent. Round percents to the nearest tenth if necessary.

Examples: $\frac{1}{4}$ $\frac{3}{7}$

Solutions:

$$\frac{1}{4} = \frac{p}{100}$$
$$100 \cdot 1 = 4 \cdot p$$
$$100 = 4 \cdot p$$
$$\frac{100}{4} = p$$
$$25 = p$$
$$\frac{1}{4} = \mathbf{25\%}$$

$$\frac{3}{7} = \frac{p}{100}$$
$$3 \cdot 100 = 7 \cdot p$$
$$300 = 7 \cdot p$$
$$\frac{300}{7} = p$$
$$42.9 = p \quad \text{(rounded)}$$
$$\frac{3}{7} = \mathbf{42.9\%} \quad \text{(rounded)}$$

21. $\frac{3}{4}$ **22.** $\frac{1}{2}$ **23.** $\frac{7}{10}$ **24.** $\frac{1}{10}$

25. $\frac{57}{100}$	26. $\frac{79}{100}$	27. $\frac{1}{5}$	28. $\frac{4}{5}$
29. $\frac{5}{8}$	30. $\frac{1}{8}$	31. $\frac{7}{8}$	32. $\frac{3}{8}$
33. $\frac{12}{25}$	34. $\frac{18}{25}$	35. $\frac{37}{50}$	36. $\frac{41}{50}$
37. $\frac{11}{20}$	38. $\frac{19}{20}$	39. $\frac{5}{6}$	40. $\frac{1}{6}$
41. $\frac{5}{9}$	42. $\frac{7}{9}$	43. $\frac{1}{7}$	44. $\frac{5}{7}$

Complete this chart. Round decimals to the nearest thousandth and percents to the nearest tenth of a percent.

Example:

fraction	*decimal*	*percent*
$\frac{3}{50}$	**.06**	**6%**

	fraction	*decimal*	*percent*
45.	______	.1	______
46.	$\frac{1}{9}$	______	______
47.	______	______	12.5%
48.	$\frac{1}{6}$	______	______
49.	______	.2	______
50.	______	______	25%
51.	$\frac{3}{10}$	______	______

	fraction	decimal	percent
52.	$\frac{1}{3}$	______	______
53.	______	.4	______
54.	______	______	50%
55.	______	______	60%
56.	$\frac{5}{8}$	______	______
57.	$\frac{2}{3}$	______	______
58.	______	.7	______
59.	$\frac{3}{4}$	______	______
60.	$\frac{4}{5}$	______	______
61.	$\frac{5}{6}$	______	______
62.	______	______	87.5%
63.	______	______	90%
64.	______	______	100%
65.	$\frac{1}{200}$	______	______
66.	$\frac{7}{500}$	______	______
67.	______	______	.3%
68.	______	______	.8%
69.	______	2.5	______
70.	______	1.7	______
71.	$3\frac{1}{4}$	______	______
72.	$2\frac{4}{5}$	______	______

CHALLENGE EXERCISES

In the following word problems, write the answer as a fraction, as a decimal, and as percent.

73. A student answered 15 of 20 questions correctly. What portion was correct?

74. Of 25 boards, 13 are redwood. What portion is redwood?

75. The price of a television set was $500. It was reduced $100. By what portion was the price reduced?

76. To pass a French class, Carolyn must study 8 hours a week. So far this week, she has studied 5 hours. What portion of the necessary time must she still study?

77. Of the five women at the office, three are college students. What portion are college students?

78. A hamburger stand has 25 employees. Of these, 14 have no previous experience. What portion have no previous experience?

79. A real estate office has 40 salespeople. If 22 of these salespeople drive Cadillacs and the rest drive Lincolns, what portion drive Lincolns?

80. A zoo has 125 animals, including 100 that are not members of endangered species. What portion is endangered?

81. An antibiotic is used to treat 380 people. If 342 people do not have a side reaction to the antibiotic, find the portion that do have a side reaction.

82. An apple grower's cooperative has 250 members. If 100 of the growers use a certain insecticide, find the portion that do not use the insecticide.

REVIEW EXERCISES

Find the missing number in each of the following proportions. (For help, see ***Section 5.4****.)*

83. $\frac{10}{5} = \frac{x}{20}$

84. $\frac{n}{50} = \frac{8}{20}$

85. $\frac{4}{y} = \frac{12}{15}$

86. $\frac{6}{x} = \frac{4}{8}$

87. $\frac{42}{30} = \frac{14}{b}$

88. $\frac{6}{22} = \frac{a}{220}$

6.3 THE PERCENT PROPORTION

There are two ways to solve percent problems. One method uses ratios and is discussed in this section, while the other method is explained in Section 6.6.

OBJECTIVES

1. Learn the percent proportion.
2. Solve for a missing letter in a proportion.

Recall that a statement of two equivalent ratios is called a proportion.

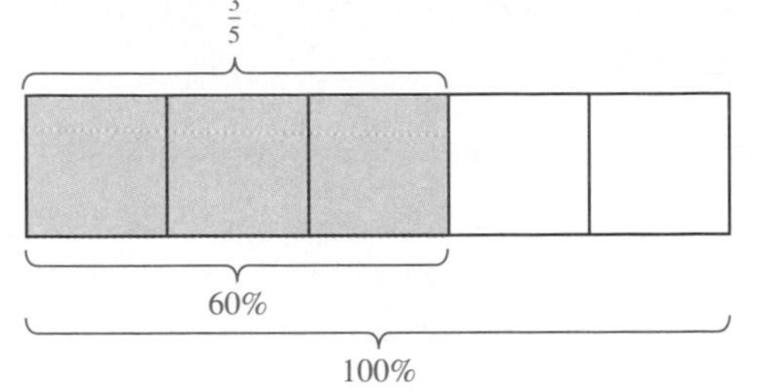

For example, the fraction $\frac{3}{5}$ is the same as the ratio 3 to 5, and 60% is the ratio 60 to 100. As the figure above shows, these two ratios are equivalent and make a proportion.

WORK PROBLEM 1 AT THE SIDE.

Write a proportion to solve a percent problem as follows.

> *Amount* is to *base* as percent is to 100, or
>
> $$\frac{\textbf{amount}}{\textbf{base}} = \frac{\textbf{percent}}{\textbf{100}}. \quad \leftarrow \text{always 100}$$
>
> With letters, the proportion is
>
> $$\frac{a}{b} = \frac{p}{100}. \quad \textit{percent proportion}$$

The final statement in the box is the **percent proportion.** In the figure at the top of the page, the **base** is 5 (the entire quantity), the **amount** is 3 (the part of the total), and the **percent** is 60. Write the percent proportion as follows.

$$\frac{a \rightarrow 3}{b \rightarrow 5} = \frac{60 \leftarrow p}{100 \leftarrow 100}$$

2 As shown in Section 5.3, if any two of the three values in the percent proportion are known, the third can be found by solving the proportion.

EXAMPLE 1 Using the Percent Proportion

Use the percent proportion and solve for the missing number.

(a) $a = 12$, $p = 25$, find b **(b)** $a = 30$, $b = 50$, find p

(c) $b = 150$, $p = 18$, find a

1. As a review of proportions, use the method of cross products to decide whether these proportions are *true* or *false*.

(a) $\frac{3}{5} = \frac{60}{100}$

(b) $\frac{5}{7} = \frac{35}{49}$

(c) $\frac{21}{50} = \frac{168}{400}$

(d) $\frac{104}{37} = \frac{515}{185}$

(e) $\frac{29}{83} = \frac{232}{660}$

ANSWERS
1. (a) true (b) true (c) true (d) false (e) false

2. Find the value of the missing letter.

(a) $a = 40, p = 20$

(b) $a = 52, b = 208$

(c) $b = 350, p = 12$

(d) $b = 5000, p = 27$

(e) $a = 74, b = 185$

SOLUTION

(a) Replace a with 12 and p with 25.

$$\frac{a}{b} = \frac{p}{100} \quad \text{percent proportion}$$

$$\frac{12}{b} = \frac{25}{100} \quad \text{or} \quad \frac{12}{b} = \frac{1}{4} \quad \text{(lowest terms)}$$

Find the cross products to solve this proportion.

$$\frac{12}{b} = \frac{1}{4} \qquad 1 \cdot b \qquad 12 \cdot 4$$

Show the products are equivalent.

$$1 \cdot b = 12 \cdot 4$$
$$b = 48$$

The base is 48.

(b) Use the percent proportion.

$$\frac{30}{50} = \frac{p}{100} \quad \text{percent proportion}$$

$$\frac{3}{5} = \frac{p}{100} \quad \text{(lowest terms)}$$

$$3 \cdot 100 = 5 \cdot p \quad \text{cross products}$$

$$300 = 5 \cdot p$$

$$\frac{300}{5} = \frac{5 \cdot p}{5} \quad \text{Divide by 5.}$$

$$60 = p$$

The percent is 60, written as 60%.

(c) $$\frac{a}{150} = \frac{18}{100} \quad \text{or} \quad \frac{a}{150} = \frac{9}{50} \quad \text{(lowest terms)}$$

$$50 \cdot a = 150 \cdot 9 \quad \text{cross products}$$

$$50 \cdot a = 1350$$

$$\frac{50 \cdot a}{50} = \frac{1350}{50} \quad \text{Divide by 50.}$$

$$a = 27$$

The amount is 27. ■

WORK PROBLEM 2 AT THE SIDE.

ANSWERS
2. (a) $b = 200$ (b) $p = 25$ (c) $a = 42$
(d) $a = 1350$ (e) $p = 40$

name date hour

6.3 EXERCISES

Find the value of the missing letter in the percent proportion $\frac{a}{b} = \frac{p}{100}$. *Round to the nearest tenth if necessary.*

Examples:

$a = 10, p = 50$, find b $\quad$ $a = 80, b = 120$, find p $\quad$ $b = 90, p = 75$, find a

Solutions:

$$\frac{10}{b} = \frac{50}{100}$$
$$\frac{10}{b} = \frac{1}{2}$$
lowest terms
$$10 \cdot 2 = 1 \cdot b$$
$$20 = b$$
$$\mathbf{20} = b$$

$$\frac{80}{120} = \frac{p}{100}$$
$$\frac{2}{3} = \frac{p}{100}$$
lowest terms
$$2 \cdot 100 = 3 \cdot p$$
$$200 = 3 \cdot p$$
$$\frac{200}{3} = \frac{3 \cdot p}{3}$$
$$\mathbf{66.7} = p \quad \text{(rounded)}$$
The percent is 66.7%.

$$\frac{a}{90} = \frac{75}{100}$$
$$\frac{a}{90} = \frac{3}{4}$$
lowest terms
$$4 \cdot a = 90 \cdot 3$$
$$4 \cdot a = 270$$
$$\frac{4 \cdot a}{4} = \frac{270}{4}$$
$$a = \mathbf{67.5}$$

1. $a = 40, p = 25$

2. $a = 90, p = 50$

3. $a = 180, p = 20$

4. $a = 72, p = 75$

5. $a = 8, p = 40$

6. $a = 11, p = 10$

7. $a = 25, p = 6$

8. $a = 61, p = 12$

9. $a = 70, b = 140$

10. $a = 12, b = 48$

11. $a = 105, b = 35$

12. $a = 72, b = 24$

13. $a = 1\frac{1}{2}, b = 4\frac{1}{2}$

14. $a = 9\frac{1}{4}, b = 27\frac{3}{4}$

15. $b = 52, p = 25$

16. $b = 80, p = 50$

17. $b = 72, p = 15$

18. $b = 112, p = 38$

19. $b = 47.2, p = 28$

20. $b = 79.6, p = 13$

Solve each of the following problems. Round answers to the nearest tenth.

21. Find b if a is 89 and p is 25.

22. p is 40 and b is 120. Find a.

23. b is 380 and a is 95. Find p.

24. Suppose a is 15 and b is 2500. Find p.

CHALLENGE EXERCISES

25. Find p if b is 1850 and a is 157.25.

26. What is a, if p is $25\frac{1}{2}$ and b is 2800?

27. b is 8116 and a is 994.21. Find p.

28. Suppose a is 550 and p is $6\frac{1}{4}$. Find b.

REVIEW EXERCISES

Write each fraction as a percent. Round to the nearest tenth of a percent if necessary. (For help, see ***Section 6.2.****)*

29. $\frac{3}{4}$

30. $\frac{3}{10}$

31. $\frac{57}{100}$

32. $\frac{1}{5}$

33. $\frac{7}{8}$

34. $\frac{1}{6}$

35. $\frac{5}{7}$

36. $\frac{17}{25}$

6.4 IDENTIFYING THE PARTS IN A PERCENT PROBLEM

The next section shows how to solve percent problems. As a help in solving these problems, remember:

All the problems in this chapter involve comparisons between parts and a whole.

Solving these problems requires identifying the three parts of a percent problem, a, b, and p.

1 Look for p, percent, first.

The **percent** is the ratio of a part to a whole, with 100 as denominator.

NOTE The percent p appears with the word *percent* or with the symbol %.

EXAMPLE 1 Finding Percent in a Percent Problem

Find p in each of the following.

(a) 32% of the 900 men were too tall.

$\underbrace{32\%}_{p}$

p is 32. The number 32 appears with a %.

(b) 150 is 25 percent of what number?

$\underbrace{25}_{p}$

(c) What percent of the 350 women will go?

$\underbrace{\text{What percent}}_{p}$ (an unknown) ■

WORK PROBLEM 1 AT THE SIDE.

2 Next, look for b, the base.

The **base** is the entire quantity, or the total.

NOTE The b often appears after the word *of.*

OBJECTIVES

Identify the

1 percent;

2 base;

3 amount.

1. Identify p.

(a) Of the 850 people, 96% approve.

(b) Of the $1000 prize, a total of 42% must be paid in taxes.

(c) Find the sales tax by mutliplying $590 and 5 percent.

(d) 600 is 55% of what number?

(e) What percent of the 75 crates will be sold today?

ANSWERS

1. (a) 96 (b) 42 (c) 5 (d) 55 (e) Here, p is unknown.

2. Find b.

(a) Of the 850 people, 96% approve.

(b) Of the $1000 prize, a total of 42% must be paid in taxes.

(c) Find the sales tax by multiplying $590 and 5 percent.

(d) 600 is 55% of what number?

(e) What percent of the 75 crates will be sold today?

3. Find a.

(a) 12.5% of 1500 is 187.5.

(b) 59 is 25% of 236.

(c) 500 is what percent of 4250?

(d) The tax of 5% on an item costing $650 is $32.50.

ANSWERS
2. (a) 850 (b) $1000 (c) $590
(d) what number (an unknown)
(e) 75
3. (a) 187.5 (b) 59 (c) 500 (d) $32.50

EXAMPLE 2 Finding Base in a Percent Problem
Find b in each of the following.

(a) 32% of the 900 men were too tall.
900 → b

b is 900. The number 900 appears after the word *of*.

(b) 150 is 25 percent of what number?
what number → b (an unknown)

(c) 85% of 7000 is what number?
7000 → b ■

WORK PROBLEM 2 AT THE SIDE.

3 Finally, look for a, the amount.

The **amount** is the part being compared with the whole.

NOTE If you have trouble identifying a, find b and p first. The remaining number is a.

EXAMPLE 3 Finding Amount in a Percent Problem
Find a in each of the following.

(a) 54% of 700 is 378.
Find p and b.

54% of 700 is 378
54% → p with % sign; 700 → b whole

54% of 700 is 378
378 → a part

The amount, a, is 378.

(b) 150 is 25% of what number?
25% → p; what number → b (an unknown)

150 is the remaining number, so $a = 150$.

(c) 85% of 7000 is what number?
85% → p; 7000 → b; what number → a (an unknown) ■

WORK PROBLEM 3 AT THE SIDE.

name date hour

6.4 EXERCISES

Identify p, b, and a in each of the following. Do not try to solve for any unknowns.

Example: **60** is **75%** of **what number**?
Solution: **60** → a, **75%** → p, **what number** → b

Example: Of the **592** tomato plants, **75%**, or **444**, are ready.
Solution: **592** → b, **75%** → p, **444** → a

	p	b	a
1. 65% of 1000 is 650.			
2. 15% of 750 is 112.5.			
3. 81% of what number is 748?			
4. 93% of what number is 11.5?			
5. What is 12% of 48?			
6. What is 37% of 950?			
7. 18 is 72% of what number?			
8. 46 is 13% of what number?			

	p	b	a
9. 52 is what percent of 104?	________	________	________
10. 91 is what percent of 273?	________	________	________
11. What percent of 50 is 30?	________	________	________
12. What percent of 190 is 85?	________	________	________
13. 29.81 is what percent of 508?	________	________	________
14. 16.74 is 11.9% of what number?	________	________	________
15. .68% of 487 is what number?	________	________	________
16. What number is 12.42% of 1408.7?	________	________	________

Find p, b, and a in the following word problems. Do not try to solve for any unknowns.

17. A team won 18 of the 24 games it played. What percent of its games did it win?

18. A chemical is 42% pure. Of 800 grams of the chemical, how much is pure?

19. Sales tax of $8 is charged on an item costing $200. What percent sales tax is charged?

20. From Tom's check of $340, 17% is withheld. How much is withheld?

21. Of the baby salmon shipped from a hatchery, 7% do not arrive healthy. If 1500 salmon are shipped, how many do not arrive healthy?

22. There are 590 quarts of grape juice in a vat holding a total of 1700 quarts of fruit juice. What percent is grape juice?

CHALLENGE EXERCISES

Find p, b, and a in the following word problems. Do not try to solve for any unknowns.

23. In one storm, Springbrook got 15% of the season's snowfall. That winter, Springbrook had 30 inches of snow. How many inches fell in the one storm?

24. In one state, the sales tax is 6%. On a purchase, the amount of tax was $48. Find the cost of the item purchased.

25. A medical clinic found that 16.8% of the patients were late for their appointments. The number of patients who were late was 504. Find the total number of patients.

26. 1848 automobiles are tested for exhaust emissions and 231 do not pass the test. Find the percent that do not pass.

REVIEW EXERCISES

Write a proportion for each of the following and then find the missing number. (For help, see ***Section 5.4****.)*

	proportion	*missing number*
27. 21 is to x as 15 is to 30.	________	________
28. 50 is to 40 as b is to 80.	________	________
29. y is to 16 as $\frac{5}{4}$ is to 4.	________	________
30. 12 is to 10 as 36 is to a.	________	________

6.5 USING PROPORTIONS TO SOLVE PERCENT PROBLEMS

The percent proportion given earlier,

$$\frac{\text{amount}}{\text{base}} = \frac{\text{percent}}{100}$$

or

$$\frac{a}{b} = \frac{p}{100},$$

uses three letters, a, b, and p. As shown earlier, if any two of these letters are known, the third can be found.

NOTE Remember that in the percent proportion, b (base) is the entire quantity, a (amount) is part of the total, and p is the percent.

1 The first example shows the percent proportion used to find a, the amount. (Remember: the amount is a part of the whole.)

EXAMPLE 1 Finding a with the Percent Proportion
Find 15% of 160.

SOLUTION
Here p (percent) is 15 and b (base) is 160. (Recall that b often comes after *of*.) Now find a.

$$\frac{a}{b} = \frac{p}{100} \qquad \frac{a}{160} = \frac{15}{100} \qquad \frac{a}{160} = \frac{3}{20} \quad \text{(lowest terms)}$$

Find the cross products in the proportion.

$$20 \cdot a = 160 \cdot 3$$
$$20 \cdot a = 480$$
$$\frac{20 \cdot a}{20} = \frac{480}{20} \quad \text{Divide each side by 20.}$$
$$a = 24$$

24 is 15% of 160. ■

WORK PROBLEM 1 AT THE SIDE.

OBJECTIVES

Solve percent problems for

1. a (amount);
2. b (base);
3. p (percent).

1. Use the percent proportion.

(a) Find 25% of 760.

(b) Find 74% of 1500.

(c) Find 9% of 3250.

(d) Find 86% of 509.

ANSWERS
1. (a) 190 (b) 1110 (c) 292.5 (d) 437.74

2. Use the shortcut way to find a.

(a) Find 57% of 9000.

(b) Find 11% of 82.

(c) Find 130% of 64.

(d) Find .6% of 120.

Just as with the word problems given earlier, the word *of* is an indicator word meaning *multiply*. For example,

$$15\% \text{ of } 160$$

means

$$15\% \cdot 160.$$

Because of this, there is a shortcut way to find the amount, a.

To find a:

Find p. Write the percent as a decimal.

Multiply this decimal and b.

EXAMPLE 2 Finding a by Using the Shortcut

Use the shortcut to find a.

(a) Find 42% of 830.

(b) Find 140% of 60.

(c) Find .4% of 50.

SOLUTION

(a) Here p is 42. Write 42% as the decimal .42. Multiply .42 and b, which is 830.

$$a = .42 \cdot 830$$
$$a = 348.6$$

It is a good idea to estimate the answer, to make sure no mistakes were made with decimal points. Estimate 42% as 40% or .4, and estimate 830 as 800. Next, 40% of 800 is

$$.4 \cdot 800 = 320,$$

so that 348.6 is a reasonable answer.

(b) In this problem, p is 140. Write 140% as the decimal 1.40. Next, multiply 1.40 and 60.

$$a = 1.40 \cdot 60 = 84$$

(c) $a = .004 \cdot 50 = .2$

↑ Write percent as a decimal.

An estimate would not be very useful here. ■

WORK PROBLEM 2 AT THE SIDE.

ANSWERS

2. (a) 5130 (b) 9.02 (c) 83.2 (d) .72

EXAMPLE 3 Solving for *a* By Using the Shortcut
Video Production has 850 employees. Of these employees, 28% are students. How many of the employees are students?

APPROACH
As above, *of* is an indicator word for multiplication.

SOLUTION

28% of the employees are students
↑ indicator word

The total number of employees is 850, so $b = 850$. The percent is 28 ($p = 28$). Find a to find the number of students.

$$a = .28 \cdot 850 = 238$$
↑ Write as a decimal.

Video Productions has 238 student employees. ■

WORK PROBLEM 3 AT THE SIDE.

2 The next example shows how to use the percent proportion to find b, the base.

NOTE Remember, the base is the entire quantity, or the total.

EXAMPLE 4 Finding *b* with the Percent Proportion

(a) 8 is 4% of what number?

(b) 135 is 15% of what number?

SOLUTION
(a) Here $p = 4$, b is unknown, and $a = 8$. Use the percent proportion to find b.

$$\frac{a}{b} = \frac{p}{100} \qquad \frac{8}{b} = \frac{4}{100} \qquad \frac{8}{b} = \frac{1}{25} \quad \text{(lowest terms)}$$

Find cross products.

$$8 \cdot 25 = 1 \cdot b$$
$$200 = b$$

8 is 4% of 200.

3. Use the shortcut way to find a.

(a) In a town of 1560 people, 25% are members of the Grange. How many people in the town belong to the Grange?

(b) Of 7400 students, 52% are female. How many students are female?

ANSWERS
3. (a) 390 (b) 3848

4. Use the percent proportion to find the unknown.

(a) 50 is 25% of what number?

(b) 21 is 10% of what number?

(c) 1548 is 72% of what number?

(d) 97.5 is 12.5% of what number?

5. (a) A farmer lost 15% of her crop to bad weather. The loss was 429 tons. Find the total number of tons in the crop.

(b) A metal alloy contains 8% zinc. The alloy contains 450 pounds of zinc. Find the total weight of the alloy.

ANSWERS
4. (a) 200 (b) 210 (c) 2150 (d) 780
5. (a) 2860 tons (b) 5625 pounds

(b) $p = 15$ and $a = 135$, so

$$\frac{135}{b} = \frac{15}{100}$$

$$\frac{135}{b} = \frac{3}{20} \quad \text{(lowest terms)}$$

$$135 \cdot 20 = 3 \cdot b \quad \text{cross products}$$

$$2700 = 3 \cdot b$$

$$\frac{2700}{3} = \frac{3 \cdot b}{3} \quad \text{Divide each side by 3.}$$

$$900 = b.$$

135 is 15% of 900. ■

WORK PROBLEM 4 AT THE SIDE.

EXAMPLE 5 Applying the Percent Proportion

At a certain company, 78 employees are absent because of illness. If this is 5% of the total number of employees, how many employees does the company have?

APPROACH

From the information in the problem, the percent is 5 ($p = 5$) and the amount, or part of the total number of employees is 78 ($a = 78$). The total number of employees or entire quantity, which is the base (b), must be found.

SOLUTION

Use the percent proportion to find b (the total number of employees).

$$\frac{78}{b} = \frac{5}{100}$$

$$\frac{78}{b} = \frac{1}{20} \quad \text{(lowest terms)}$$

Find cross products.

$$78 \cdot 20 = 1 \cdot b$$

$$1560 = b$$

Estimate the answer. 78 is approximately 80, and 5% is equivalent to the fraction $\frac{1}{20}$. Since 80 is $\frac{1}{20}$ of 1600, or

$$80 \cdot 20 = 1600,$$

1560 is a reasonable answer.
The company has 1560 employees. ■

WORK PROBLEM 5 AT THE SIDE.

3 Finally, if a and b are known, the percent proportion can be used to find p.

EXAMPLE 6 Using the Percent Proportion to Find p

(a) 13 is what percent of 52?

(b) What percent of 500 is 100?

SOLUTION

(a) The base is $b = 52$ (52 follows *of*) and $a = 13$. Next, find p.

$$\frac{a}{b} = \frac{p}{100}$$

$$\frac{13}{52} = \frac{p}{100}$$

$$\frac{1}{4} = \frac{p}{100} \quad \text{(lowest terms)}$$

Find cross products.

$$1 \cdot 100 = 4 \cdot p$$

$$\frac{100}{4} = \frac{4 \cdot p}{4} \quad \text{Divide both sides by 4.}$$

$$25 = p$$

13 is 25% of 52.

(b) $b = 500$, $a = 100$

$$\frac{100}{500} = \frac{p}{100}$$

$$\frac{1}{5} = \frac{p}{100} \quad \text{(lowest terms)}$$

$$1 \cdot 100 = 5 \cdot p \quad \text{cross products}$$

$$100 = 5 \cdot p$$

$$\frac{100}{5} = \frac{5 \cdot p}{5} \quad \text{Divide both sides by 5.}$$

$$20 = p$$

20% of 500 is 100. ■

WORK PROBLEM 6 AT THE SIDE.

6. (a) 15 is what percent of 60?

(b) What percent of 250 is 50?

(c) What percent of 980 is 441?

(d) 48 is what percent of 12?

ANSWERS

6. (a) 25% **(b)** 20% **(c)** 45% **(d)** 400%

7. (a) The Ski Club has received \$289 of the \$425 needed for an annual picnic. What percent of the total has been collected?

(b) An automobile that should get 28 miles per gallon gets only 23.8 miles per gallon. What percent of the possible mileage does the car actually get?

ANSWERS
7. (a) 68% (b) 85%

EXAMPLE 7 Applying the Percent Proportion

A roof is expected to last 20 years before needing replacement. If the roof is now 15 years old, what percent of the roof's life has been used?

APPROACH

The expected life of the roof is the entire quantity or base ($b = 20$). The part of the roof that is already used is the amount ($a = 15$). The percent (p) of the roof's life that is already used must be found.

SOLUTION

Use the percent proportion to find p, the percent of roof life used.

$$\frac{15}{20} = \frac{p}{100} \quad \frac{3}{4} = \frac{p}{100} \qquad \text{(lowest terms)}$$

Find cross products.

$$3 \cdot 100 = 4 \cdot p$$
$$300 = 4 \cdot p$$
$$\frac{300}{4} = \frac{4 \cdot p}{4} \qquad \text{Divide each side by 4.}$$
$$75 = p$$

75% of the roof's life has been used. ■

WORK PROBLEM 7 AT THE SIDE.

name date hour

6.5 EXERCISES

Find the amount.

Example: 47% of 5000 **Solution:** $.47 \cdot 5000 = 2350$

$a = \mathbf{2350}$

1. 25% of 484

2. 20% of 1500

3. 14% of 780

4. 12% of 350

5. 5% of 96

6. 9% of 150

7. 125% of 20

8. 175% of 50

9. 22.5% of 1100

10. 38.2% of 4250

11. 3% of 128

12. 7% of 850

13. 200% of 371

14. 150% of 768

15. 15.5% of 275

16. 46.1% of 843

17. .9% of 2400

18. .3% of 1400

Find the base.

Example: 64% of what number is 1600? **Solution:** $p = 64$ and $a = 1600$, so

$$\frac{1600}{b} = \frac{64}{100}$$

$$\frac{1600}{b} = \frac{16}{25}. \quad \text{(lowest terms)}$$

Solve, to get $b = \mathbf{2500}$.

19. 50 is 10% of what number?

20. 10 is 5% of what number?

21. 35% of what number is 84?

22. 75% of what number is 78?

23. 300 is 25% of what number?

24. 84 is 28% of what number?

25. 748 is 110% of what number?

26. 77 is 140% of what number?

27. $12\frac{1}{2}$% of what number is 350?

28. $4\frac{1}{2}$% of what number is 135?

Find the percent. Round to the nearest tenth if necessary.

> **Example:** What percent of 80 is 16?
>
> **Solution:** $b = 80$ and $a = 16$, so
>
> $$\frac{16}{80} = \frac{p}{100}$$
>
> $$\frac{1}{5} = \frac{p}{100} \quad \text{(lowest terms)}$$
>
> $$p = 20.$$
>
> **20%** of 80 is 16.

29. 35 is what percent of 70?

30. 15 is what percent of 75?

31. 13 is what percent of 25?

32. 650 is what percent of 1000?

33. 8 is what percent of 400?

34. 7 is what percent of 350?

35. 12 is what percent of 800?

36. 40 is what percent of 1600?

37. What percent of 124 is 16?

38. What percent of 492 is 12?

39. What percent of 500 is 46?

40. What percent of 105 is 54?

Solve each word problem. Round to the nearest tenth if necessary.

41. A library has 270 visitors on Saturday, 20% of whom are children. How many are children?

42. Ruth Wren spent 8% of her savings on textbooks. Her savings were $720. Find the amount that she spent on textbooks.

43. A survey at an intersection found that of 2200 drivers, 38% were wearing seat belts. How many drivers in the survey were wearing seat belts?

44. A home valued at $95,000 will gain 6% in value this year. Find the gain in value this year.

45. An Atlantic City casino advertises that it gives a 97.4% payback on slot machines, and the balance is retained by the casino. The amount retained by the casino is $4823. Find the total amount played on the slot machines.

46. A resort hotel states that 35% of its rooms are for nonsmokers. The resort allows smoking in 468 rooms. Find the total number of rooms.

47. This year, there are 550 scholarship applicants, which is 110% of the number of applicants last year. Find the number of applicants last year.

48. At Goldie's Sandwich Shop, 16% of the customers order a dill pickle. 372 dill pickles are sold. Find the total number of customers.

49. Total daily circulation of the *Herald* is 180,000. If complimentary (nonpaid) circulation amounts to 5400 copies per day, what percent of the total circulation is nonpaid circulation?

50. The number of ballots cast in a parish election is 14,907. If the number of registered voters in the parish is 19,856, what percent has voted?

51. Rocky Mountain Water estimates 11,700 gallons of their water will be used in steam irons. If 755,000 gallons are sold, what percent will be used in steam irons?

52. Barbara's Antiquery says that of its 3800 items in stock, 3344 are just plain junk, and the rest are antiques. What percent of the number of items in stock is antiques?

53. This month's sales goal for Easy Writer Pen Company is 2,380,000 ball-point pens. If sales of 1,844,500 pens have been made, what percent of the goal has been reached?

54. If the increase in an automobile's cost is \$808 over last year's price of \$11,600, what is the percent of increase?

55. In one chemistry class, 72% of the students passed. If 108 students passed, how many students were in the class?

56. In the last election, 64% of the eligible people actually voted. If there were 288 voters, how many people were eligible?

CHALLENGE EXERCISES

57. Bill Poole owns 52.5% of a travel agency that specializes in European travel. His ownership is valued at \$446,250. Find the total value of the travel agency.

58. Raw steel production by the nation's steel mills decreased by 2.5% from last week. The decrease amounted to 50,475 tons. Find the steel production last week.

59. A shipment of posters valued at $175 was damaged in transit. The estimated damage was 35%. What was the value of the undamaged posters?

60. A family of four with a monthly income of $2400 spends 90% of its earnings and saves the balance. Find (*a*) the monthly savings and (*b*) the annual savings of this family.

REVIEW EXERCISES

Multiply or divide as necessary. (For help, see ***Sections 4.5 and 4.6.****)*

61. $\begin{array}{r} 48.05 \\ \times \quad .85 \\ \hline \end{array}$

62. $\begin{array}{r} .0892 \\ \times \quad .018 \\ \hline \end{array}$

63. $\begin{array}{r} 325.6 \\ \times \quad .031 \\ \hline \end{array}$

64. $\begin{array}{r} 51.81 \\ \times \quad .021 \\ \hline \end{array}$

65. $306 \div .085$

66. $24 \div .3$

67. $86 \div 344$

68. $120 \div 24$

6.6 THE PERCENT EQUATION

The last section showed how to use a proportion to solve percent problems. This section shows an alternative way to solve these problems by using the *percent equation*. The percent equation is just a rearrangement of the percent proportion.

OBJECTIVES

Use the percent equation to solve for

1 amount;

2 base;

3 percent.

PERCENT EQUATION

percent · base = amount

Be sure to write the percent as a decimal or fraction before using the equation.

Because of the 100 in the percent proportion, it was not necessary to write the percent as a decimal. However, there is no 100 in the percent equation, so it is necessary to first write the percent as a decimal or fraction.

Some of the examples solved earlier will be reworked by using the percent equation. This will show a comparison of the two methods.

1 The first example shows how to find the amount, a.

EXAMPLE 1 Solving for the Amount (a)

(a) Find 15% of 160.

(b) Find 140% of 60.

(c) Find .4% of 250.

SOLUTION

(a) Write 15% as the decimal .15. The base (the whole, which often comes after the word *of*) is 160. Next, use the percent equation.

$$\text{amount} = \text{percent} \cdot \text{base}$$
$$\text{amount} = .15 \cdot 160$$

Multiply .15 and 160 to get

$$\text{amount} = 24.$$

24 is 15% of 160.

(b) Write 140% as the decimal 1.40. The base is 60.

$$\text{amount} = \text{percent} \cdot \text{base}$$
$$= 1.40 \cdot 60$$
$$= 84$$

84 is 140% of 60.

1. Find each of the following.

(a) 28% of 750

(b) 42% of 95

(c) 125% of 84

(d) 143% of 250

(e) .5% of 600

(f) .75% of 80

ANSWERS
1. (a) 210 (b) 39.9 (c) 105 (d) 357.5
(e) 3 (f) .6

(c) Write .4% as the decimal .004. The base is 250.

$$\begin{aligned} \text{amount} &= \text{percent} \cdot \text{base} \\ &= .004 \cdot 250 \\ &= 1 \end{aligned}$$

1 is .4% of 250. ■

WORK PROBLEM 1 AT THE SIDE.

The next example shows the percent equation used to find the base, b.

NOTE Do not forget that *of* is an indicator word for *multiply.*

EXAMPLE 2 Solving for the Base (b)

(a) 8 is 4% of what number? **(b)** 135 is 15% of what number?

(c) $8\frac{1}{2}\%$ of what number is 102?

SOLUTION

(a) The amount is 8 and the percent is 4% or the decimal .04. The base is unknown.

8 is 4% of what number?
↑ indicator word

Next, use the percent equation.

$$\begin{aligned} \text{amount} &= \text{percent} \cdot \text{base} \\ 8 &= .04 \cdot b && b \text{ is the base.} \\ \frac{8}{.04} &= \frac{.04 \cdot b}{.04} && \text{Divide both sides by .04.} \\ 200 &= b = \text{base} \end{aligned}$$

8 is 4% of 200.

(b) Write 15% as .15. The amount is 135. Next, use the percent equation to find the base.

$$\begin{aligned} \text{amount} &= \text{percent} \cdot \text{base} \\ 135 &= .15 \cdot b \end{aligned}$$

Divide each side by .15.

$$\begin{aligned} \frac{135}{.15} &= \frac{.15 \cdot b}{.15} \\ 900 &= b \quad \text{base} \end{aligned}$$

135 is 15% of 900.

(c) Write $8\frac{1}{2}\%$ as 8.5%, or the decimal .085. The amount is 102. Use the percent equation,

$$\begin{aligned} \text{amount} &= \text{percent} \cdot \text{base.} \\ 102 &= .085 \cdot b \end{aligned}$$

Divide each side by .085.

$$\frac{102}{.085} = \frac{.085 \cdot b}{.085}$$

$$1200 = b \qquad \text{base}$$

102 is $8\frac{1}{2}\%$ of 1200. ■

WORK PROBLEM 2 AT THE SIDE.

3 The final example shows how to use the equation to find the percent p.

EXAMPLE 3 Solving For Percent (p)

(a) 13 is what percent of 52?

(b) What percent of 500 is 100?

(c) What percent of 300 is 390?

(d) 6 is what percent of 1200?

SOLUTION

(a) Since 52 follows *of*, the base is 52. The amount of 13, and the percent is unknown. Use the percent equation,

$$\text{amount} = \text{percent} \cdot \text{base}$$

If p is the percent, then

$$13 = p \cdot 52.$$

Divide each side by 52.

$$\frac{13}{52} = \frac{p \cdot 52}{52}$$

$$.25 = p = 25\% \qquad \text{Write the decimal as percent.}$$

13 is 25% of 52.

This problem could also be solved by using *of* as an indicator word for multiplication, and *is* as an indicator word for "equals."

$$\begin{array}{ccccc} 13 & \text{is} & \text{what percent} & \text{of} & 52? \\ \downarrow & \downarrow & \downarrow & \downarrow & \downarrow \\ 13 & = & p & \cdot & 52 \end{array}$$

2. Find the base.

(a) 24 is 30% of what number?

(b) 22.5 is 18% of what number?

(c) 143 is 55% of what number?

(d) $6\frac{3}{4}\%$ of what number is 54?

ANSWERS
2. (a) 80 (b) 125 (c) 260 (d) 800

3. (a) What percent of 120 is 24?

(b) 36 is what percent of 90?

(c) What percent of 140 is 210?

(d) 3 is what percent of 375?

ANSWERS
3. (a) 20% (b) 40% (c) 150%
(d) .8%

(b) The base is 500 and the amount is 100. Let p be the percent.

$$\text{amount} = \text{percent} \cdot \text{base}$$
$$100 = p \cdot 500$$

Divide each side by 500.

$$\frac{100}{500} = \frac{p \cdot 500}{500}$$
$$.20 = p = 20\% \qquad \text{Write as percent.}$$

(c) The base is 300 and the amount is 390. Let p be the percent.

$$\text{amount} = \text{percent} \cdot \text{base}$$
$$390 = p \cdot 300$$

Divide each side by 300.

$$\frac{390}{300} = \frac{p \cdot 300}{300}$$
$$1.3 = p = 130\% \qquad \text{Write decimal as percent.}$$

130% of 300 is 390.

(d) Since 1200 follows *of,* the base is 1200. The amount is 6. Let p be the unknown.

$$\text{amount} = \text{percent} \cdot \text{base}$$
$$6 = p \cdot 1200$$

Divide each side by 1200.

$$\frac{6}{1200} = \frac{p \cdot 1200}{1200}$$
$$.005 = p = .5\% \qquad \text{Write decimal as percent.}$$

6 is .5% of 1200 ■

WORK PROBLEM 3 AT THE SIDE.

name date hour

6.6 EXERCISES

Find the amount.

Example: 14% of 750

Solution:
amount = percent · base
amount = .14 · 750   Write 14% as the decimal .14.
amount = **105**

1. 40% of 620

2. 70% of 850

3. 85% of 900

4. 65% of 420

5. 16% of 520

6. 22% of 860

7. 8% of 140

8. 5% of 960

9. 125% of 72

10. 150% of 90

11. 12.4% of 8300

12. 13.2% of 9400

13. .4% of 750

14. .8% of 520

15. .3% of 480

Find the base.

Example: 48% of what number is 2496?

Solution:
amount = percent · base
2496 = .48 · b Let b = base.

$$\frac{2496}{.48} = \frac{.48 \cdot b}{.48}$$ Divide by .48.

5200 = b = base

16. 40 is 20% of what number?

17. 32 is 40% of what number?

18. 50% of what number is 31?

19. 75% of what number is 375?

20. 270 is 30% of what number?

21. 500 is 40% of what number?

22. $12\frac{1}{2}\%$ of what number is 135?

23. $6\frac{1}{4}\%$ of what number is 25?

24. $1\frac{1}{4}\%$ of what number is 3.75?

25. $2\frac{1}{2}\%$ of what number is 5?

Find the percent.

Example: What percent of 80 is 20?

Solution: amount = percent · base

$$20 = p \cdot 80$$

$$\frac{20}{80} = \frac{p \cdot 80}{80} \quad \text{Divide by 80.}$$

$$.25 = p = 25\% \quad \text{Write the percent.}$$

26. 40 is what percent of 80?

27. 30 is what percent of 75?

28. 19 is what percent of 25?

29. 75 is what percent of 125?

30. What percent of 120 is 162?

31. What percent of 250 is 87.5?

name date hour

32. What percent of 80 is 1.2?

33. What percent of 1200 is 3?

34. 294.4 is what percent of 920?

35. 1224 is what percent of 850?

Solve each word problem.

36. A gardener has 52 clients, 25% of whom are residential. Find the number of residential clients.

37. For a tour of the eastern United States, a travel agent promised a trip of 3300 miles. Exactly 35% of the trip was by air. How many miles would be traveled by air?

38. One family earns $2400 per month and saves $300 per month. What percent of their income is saved?

39. A family drove 135 miles of its 500-mile vacation. What percent of the total number of miles did the family drive?

40. A tank of an industrial chemical is 25% full. The tank now contains 175 gallons. How many gallons will it contain when it is full?

41. Walt has completed 75% of the units needed for a degree. If he has completed 90 units, how many are needed for a degree?

CHALLENGE EXERCISES

42. Steel production this year rose 16% from the 1,702,000 tons produced last year. Find this year's steel production.

43. An ad for steel-belted radial tires promises 15% better mileage. If mileage has been 25.6 miles per gallon in the past, what mileage could be expected after new tires are installed? (Round to the nearest tenth of a mile.)

44. Electrical components originally priced at $14,000 are discounted 15%. Find the price after the discount.

REVIEW EXERCISES

Identify p, b, and a in each of the following. Do not try to solve for any unknowns. (For help, see ***Section 6.4.****)*

	p	b	a
45. 15% of 375 is 56.25	________	________	________
46. What is 8% of 470?	________	________	________
47. 36 is 72% of what number?	________	________	________
48. 6% of 1224 is 73.44	________	________	________
49. What percent of 650 is 146.15?	________	________	________
50. 182 is what percent of 546?	________	________	________

6.7 APPLICATIONS OF PERCENT

Percent has many applications in daily life. This section discusses percent as it applies to sales tax, commissions, discounts, and the percent of change (increase and decrease).

OBJECTIVES

1. Find sales tax.
2. Find commissions.
3. Find the discount and sale price.
4. Find the percent of change.

1 States, counties, and cities often collect taxes on sales to customers. The **sales tax** is a percent of the total sale. Use the following formula to find sales tax.

amount of sales tax = cost of item · rate of tax

EXAMPLE 1 Solving for Sales Tax
Western Supply sold a saddle for \$374. If the sales tax rate is 5%, how much tax was paid?

APPROACH
Use the formula above with the cost of the saddle (\$374) as cost of the item and the tax rate (5%) to find the amount of sales tax.

SOLUTION

$$\begin{aligned}\text{amount of sales tax} &= \$374 \cdot 5\% \\ &= \$374 \cdot .05 \\ &= \$18.70\end{aligned}$$

The tax paid on the saddle is \$18.70. The customer buying the saddle would pay a total cost of \$374 + \$18.70 = \$392.70. ■

WORK PROBLEM 1 AT THE SIDE.

EXAMPLE 2 Find the Sales Tax Rate
The sales tax on a \$10,800 car is \$648. Find the rate (the percent) of the sales tax.

APPROACH
Use the formula for sales tax, sales tax = cost of item · rate of tax

SOLUTION
Solve for the rate of tax. The cost of the car is \$10,800, and the amount of sales tax is \$648. Use r for the rate of tax.

$$\begin{aligned}\text{sales tax} &= \text{cost of item} \cdot \text{rate of tax} \\ 648 &= 10{,}800 \cdot r\end{aligned}$$

Divide both sides by 10,800.

$$\frac{648}{10{,}800} = \frac{10{,}800 \cdot r}{10{,}800}$$

$$.06 = r$$

$$6\% = r \quad \text{Write the decimal as percent.}$$

The sales tax rate is 6%. ■

WORK PROBLEM 2 AT THE SIDE.

1. Suppose the sales tax in a state is 4%. Find the amount of the tax for items costing

   (a) \$100.

   (b) \$476.

   (c) \$1572.

   (d) \$7893.

2. Find the rate of sales tax.

   (a) The tax on a \$700 item is \$35.

   (b) The tax on a \$12 item is \$.84.

   (c) The tax on a \$9000 item is \$270.

ANSWERS
1. (a) \$4 (b) \$19.04 (c) \$62.88 (d) \$315.72
2. (a) 5% (b) 7% (c) 3%

3. Find the commission.

(a) Scott Hanson works on a commission rate of 5% and has sales for the month of $24,200.

(b) The sales are $46,000 with a commission rate of 2%.

4. Find the rate of commission.

(a) A commission of $660 is earned for selling $13,200 worth of beverages.

(b) Sheila Couture earns $1580 for selling $39,500 worth of ink.

ANSWERS
3. (a) $1210 (b) $920
4. (a) 5% (b) 4%

2 In payment by **commission,** a salesperson is paid a percent of the dollar value of the total sales. Use the following formula to find commission.

> amount of commission =
> rate or percent of commission · amount of sales.

EXAMPLE 3 Determining the Amount of Commission
Jill Beauteo sold power drills worth $19,500. If her commission rate is 11%, find the amount of her commission.

APPROACH
Use the formula for the amount of commission where the rate of commission (11%) is multiplied by the amount of sales ($19,500).

SOLUTION

$$\begin{aligned}\text{amount of commission} &= 11\% \cdot 19{,}500\\ &= .11 \cdot 19{,}500\\ &= 2145\end{aligned}$$

She earned $2145 for selling the power drills. ■

WORK PROBLEM 3 AT THE SIDE.

EXAMPLE 4 Finding the Rate of Commission
A salesperson earned $510 for selling $17,000 worth of paper. Find the rate of commission.

APPROACH
Use the percent proportion with $b = 17{,}000$, $a = 510$, and p unknown.

SOLUTION

$$\frac{510}{17{,}000} = \frac{p}{100}$$

Solve this proportion to find that $p = 3$, and the rate of commission is 3%. ■

NOTE The commission formula, a form of the percent equation, could also have been used to solve this problem.

WORK PROBLEM 4 AT THE SIDE.

3 A store may sell some of its items at a savings, or **discount,** to attract additional customers. Use the following formula to find discount.

amount of discount = original price · rate of discount

Subtract to find the sale price.

sale price = original price − amount of discount

EXAMPLE 5 Application of a Sales Discount

A furniture store has a sofa with an original price of $470 on sale at 15% off. Find the sale price of the sofa.

APPROACH

This problem is solved in two steps. First, the amount of the discount is found by multiplying the original price ($470) by the rate of the discount (15%). The second step is to subtract the amount of discount from the original price, resulting in the sale price.

SOLUTION

First find the amount of the discount.

$$\begin{aligned} \text{amount of discount} &= 470 \cdot .15 \\ &= 70.50 \end{aligned}$$

or $70.50. Find the sale price of the sofa by subtracting the amount of the discount from the original price.

$$\begin{aligned} \text{sale price} &= \$470 - \$70.50 \\ &= \$399.50 \end{aligned}$$

During the sale, the sofa can be bought for $399.50. ■

WORK PROBLEM 5 AT THE SIDE.

4 A common occurrence is an increase or decrease in sales, production, population, and many other areas. This type of problem involves finding the percent of change. Use the percent proportion to find the percent of increase.

$$\frac{\textbf{increase}}{\textbf{original}} = \frac{p}{\textbf{100}}$$

5. Find the amount of the discount and the sale price.

(a) The Tannery Leather Shop offers a 20% discount on a hide originally priced at $260.

(b) Eastside Department Store has women's swimsuits on sale at 40% off. One swimsuit was originally priced at $34.

ANSWERS

5. (a) $52; $208 (b) $13.60; $20.40

6. Find the percent of increase.

(a) Production increased from 12,400 units last year to 16,430 this year.

(b) The number of flu cases rose from 496 cases last week to 620 this week.

EXAMPLE 6 Finding the Percent of Increase

Attendance at county parks climbed from 18,300 last month to 56,730 this month. Find the percent of increase.

APPROACH

Subtract the attendance last month (18,300) from the attendance this month (56,730) to find the increase in attendance. Next, use the percent proportion, with $b = 18{,}300$ (last month's attendance), $a = 38{,}430$ (increase in attendance) and p unknown.

SOLUTION

$$56{,}730 - 18{,}300 = 38{,}430 \qquad \text{increase in attendance}$$

$$\frac{38{,}430}{18{,}300} = \frac{p}{100}$$

Solve this proportion to find that $p = 210$ and the percent of increase is 210%. ■

WORK PROBLEM 6 AT THE SIDE.

7. Find the percent of decrease.

(a) The number of students absent this week fell to 558 from 720 last week.

(b) The number of patients admitted fell from 760 last month to 646 this month.

Use percent proportion to find the percent of decrease.

$$\frac{\textbf{decrease}}{\textbf{original}} = \frac{p}{100}$$

EXAMPLE 7 Finding the Percent of Decrease

The number of production employees this week fell to 1406 people from 1480 people last week. Find the percent of decrease.

APPROACH

Subtract the number of employees this week (1406) from the number of employees last week (1480) to find the decrease. Next, use the percent proportion, with $b = 1480$ (last week's employees), $a = 74$ (decrease in employees), and p unknown.

SOLUTION

$$1480 - 1407 = 74 \qquad \text{decrease in number of employees}$$

$$\frac{74}{1480} = \frac{p}{100}$$

Solve this proportion to find that $p = 5$ and the percent of decrease is 5%. ■

NOTE When solving for percent of increase or decrease, the base is always the previous period or value before the change. The increase or decrease is the amount.

WORK PROBLEM 7 AT THE SIDE.

ANSWERS
6. (a) 32.5% (b) 25%
7. (a) 22.5% (b) 15%

name date hour

6.7 EXERCISES

Find the amount of the sales tax and the total cost (amount of sale + amount of tax = total cost). Round to the nearest cent.

	amount of sale	*tax rate*	*amount of tax*	*total cost*
1.	$100	2%	________	________
2.	$200	4%	________	________
3.	$50	5%	________	________
4.	$170	3%	________	________
5.	$215	6%	________	________
6.	$15	5%	________	________
7.	$10	4%	________	________
8.	$78	2%	________	________

Find the commission earned. Round to the nearest cent.

	sales	*rate of commission*	*commission*
9.	$100	10%	________
10.	$500	6%	________
11.	$1000	22%	________
12.	$5783	3%	________

13. $3200 36% ______________

14. $1500 1% ______________

15. $6225 10% ______________

16. $75,000 5% ______________

Find the amount of discount and the amount paid after the discount. Round to the nearest cent.

	original price	*rate of discount*	*amount of discount*	*sale price*
17.	$100	25%	______________	______________
18.	$200	10%	______________	______________
19.	$780	35%	______________	______________
20.	$38	60%	______________	______________
21.	$17.50	25%	______________	______________
22.	$22.50	20%	______________	______________
23.	$12.50	10%	______________	______________
24.	$24.95	40%	______________	______________

Solve the following word problems. Round money to the nearest cent and rates to the nearest tenth of a percent.

25. The sales tax rate is 5% and the sales are $358. Find the amount of sales tax.

26. Julie Stevens had sales of $16,204 in the month of October. Her rate of commission is 11%. Find the amount of commission she earned.

27. Alan's Shoes sells shoes at 30% off the regular price. Find the sale price of a shoe normally priced at $45.

28. A television set sells for $650 plus 4% sales tax. Find the price of the set including sales tax.

29. A gold bracelet costs $1450 with a sales tax of $58. Find the rate of sales tax.

30. Stephen Louis can purchase a new car at 8% below window sticker price. Find the amount he can save on a car with a window sticker price of $13,608.

31. Enrollment in computer science courses increased from 2480 students last semester to 3286 students this semester. Find the percent of increase.

32. The number of industrial accidents this month fell to 989 accidents from 1276 accidents last month. Find the percent of decrease.

33. A "super 35% off sale" begins today. What is the price of a hair dryer normally priced at $20?

34. In one day, Bobbi's Boutique sold $1145 worth of items. The sales tax is 6%. Find the total amount of money received that day.

35. A salesperson at the Sewing Mart was paid $240 in commissions on sales of $4000. Find the rate of commission.

36. What is the sale price of a bedroom set priced at $995, with a discount of 45%?

37. What is the sales tax on a car costing $10,907.58, if the sales tax rate is 6.25%.

38. Easthills Ski Center has just been sold for $1,692,804. The real estate agent selling the Center earned a commission of $23,700. Find the rate of commission, rounding to the nearest tenth of a percent.

39. The average number of hours worked in manufacturing jobs last week fell to 40.9 from 41.1. Find the percent of decrease.

40. Coleco Industries, Inc., makers of Cabbage Patch and Alf dolls, and Trivial Pursuit and Scrabble games, had sales of $504.5 million in one year and $500.7 million the year before. Find the percent of increase.

41. The county charges a property tax of 2% of the cash value of a building. Find the tax on a building worth $78,000.

42. The income tax in one area is 4% of all income. Find the tax on an income of $29,500.

43. A television set normally priced at $960 is on sale for 35% off. Find the discount and the sale price.

44. This week, cars are on sale at 15% off. Find the discount and the sale price of a car originally priced at $14,000.

CHALLENGE EXERCISES

45. A real estate agent sells a house for $129,605. A sales commission of 6% is charged. The agent gets 55% of this commission. How much money does the agent get?

46. What is the total price of a boat with an original price of $13,905, if it is sold at an 18% discount. Sales tax is 4.75%.

47. The local real estate agents' association collects a fee of 2% on all money received by its members. The members charge 6% of the selling price of a property as their fee. How much does the association get, if its members sell property worth a total of $8,680,000?

48. Annual business defaults climbed from $60 million one year to $840 million the next year. Find the percent of increase.

REVIEW EXERCISES

Use the percent equation (amount = percent · base) to find amount, base, or percent. (For help, see **Section 6.6.***)*

49. 20% of 310

50. 12.4% of 4150

51. .3% of 960

52. 540 is 30% of what number?

53. $6\frac{1}{4}$% of what number is 50?

54. 20 is what percent of 160?

55. 147.2 is what percent of 460?

6.8 SIMPLE INTEREST

Interest is a charge for borrowing money. The amount of money borrowed is called the **principal.** The charge for interest is often given as a percent, called the interest rate or **rate of interest.** The rate of interest is assumed to be *per year,* unless stated otherwise.

OBJECTIVES

1. Find the simple interest on a loan.
2. Find the total amount due on a loan.

Use the following formula to find interest.

interest = principal · rate · time

The formula, written in letters, is

$$I = p \cdot r \cdot t.$$

NOTE Simple interest is used for most short-term business loans, most real estate loans, and many automobile and consumer loans.

EXAMPLE 1 Finding Interest for a Year
Find the interest on $2000 at 12% for 1 year.

SOLUTION
The amount borrowed (principal) is $2000. The interest rate is 129 or .12, and the time of the loan is 1 year. Using the formula

$$I = p \cdot r \cdot t,$$

gives $I = 2000 \cdot (.12) \cdot 1$

or $I = 240.$

The interest is $240. ■

WORK PROBLEM 1 AT THE SIDE.

EXAMPLE 2 Finding Interest for More Than a Year
Find the interest on $2200 at 11% for three and a half years.

SOLUTION
The principal is $2200. The rate is 11% or .11, and the time is $3\frac{1}{2}$ or 3.5 years. Use the formula $I = p \cdot r \cdot t$.

$$I = 2200 \cdot (.11) \cdot (3.5)$$
$$I = 847$$

The interest charge is $847. ■

WORK PROBLEM 2 AT THE SIDE.

1. Find the interest.

(a) $800 at 9% for 1 year

(b) $1570 at 11% for 2 years

2. Find the interest.

(a) $75 at 12% for $4\frac{1}{2}$ years

(b) $1900 at 10% for $3\frac{1}{4}$ years

(c) $11,500 at 12% for $2\frac{3}{4}$ years

ANSWERS
1. (a) $72 (b) $345.40
2. (a) $40.50 (b) $617.50 (c) $3795

3. Find the interest.

(a) \$1200 at 12% for 3 months

(b) \$35,000 at 14% for 2 months.

4. Find the total amount due on a loan of

(a) \$480 at 8% for 8 months.

(b) \$7200 at 12% for 4 years.

(c) \$800 at 10% for $2\frac{1}{2}$ years.

ANSWERS
3. (a) \$36 (b) \$816.67
4. (a) \$505.60 (b) \$10,656 (c) \$1000

Interest rates are given *per year*. For loan periods of less than one year, be careful to express time as a fraction of a year.

If time is given in months, for example, use a denominator of 12, since there are 12 months in a year. A loan of 9 months would be for $\frac{9}{12}$ of a year.

EXAMPLE 3 Finding Interest for Less Than 1 Year
Find the interest on \$840 at 8% for 9 months.

SOLUTION

The principal is \$840. The rate is 8% or .08, and the time is $\frac{9}{12}$ of a year. Use the formula $I = p \cdot r \cdot t$.

$$I = 840 \cdot (.08) \cdot \frac{9}{12}$$

$$= 67.2 \cdot \frac{3}{4} \qquad \left(\frac{9}{12} \text{ in lowest terms}\right)$$

$$= \frac{(67.2) \cdot 3}{4}$$

$$= \frac{201.6}{4} = 50.40$$

The interest is \$50.40. ■

WORK PROBLEM 3 AT THE SIDE.

When a loan is repaid, the interest is added to the original principal to find the total amount due.

amount due = principal + interest

EXAMPLE 4 Calculating the Total Amount Due
A loan of \$900 has been made at 10% interest for three months. Find the total amount due.

SOLUTION
First find the interest, and then add the principal and the interest to find the total amount.

$$I = 900 \cdot (.10) \cdot \frac{3}{12}$$

$$I = 22.50$$

The interest is \$22.50. The total

$$\text{amount due} = \text{principal} + \text{interest}$$
$$= \$900 + \$22.50 = \$922.50.$$ ■

WORK PROBLEM 4 AT THE SIDE.

name date hour

6.8 EXERCISES

Find the interest on each of the following.

	principal	rate	time in years	interest
1.	$100	10%	1	________
2.	$300	10%	3	________
3.	$500	12%	4	________
4.	$1000	14%	2	________
5.	$80	10%	1	________
6.	$175	12%	2	________
7.	$1500	13%	6	________
8.	$5280	14%	5	________
9.	$760	10%	$2\frac{1}{2}$	________
10.	$320	12%	$1\frac{1}{2}$	________
11.	$620	15%	$1\frac{1}{4}$	________
12.	$1000	16%	$3\frac{1}{4}$	________

Find the interest on each of the following. Round to the nearest cent if necessary.

	principal	rate	time in months	interest
13.	$200	16%	6	________

	principal	rate	time in months	interest
14.	$400	9%	9	____________
15.	$500	11%	12	____________
16.	$1000	12%	18	____________
17.	$820	8%	24	____________
18.	$92	9%	5	____________
19.	$780	16%	10	____________
20.	$522	8%	6	____________
21.	$650	12%	3	____________
22.	$2480	14%	3	____________
23.	$14,500	8%	7	____________
24.	$10,800	13%	5	____________

Find the total amount due on the following loans.

	principal	rate	time	total amount due
25.	$300	11%	1 year	____________
26.	$500	16%	6 months	____________
27.	$780	12%	3 months	____________
28.	$1020	8%	2 years	____________

	principal	*rate*	*time*	*total amount due*
29.	$1500	10%	18 months	__________
30.	$5000	15%	5 months	__________
31.	$2210	7%	6 months	__________
32.	$5280	12%	1 year	__________
33.	$1780	10%	6 months	__________
34.	$15,400	14%	5 years	__________
35.	$18,200	16%	8 months	__________
36.	$22,400	9%	9 months	__________

Solve the following work problems. Round to the nearest cent.

37. Marie Perino deposits $680 at 16% for 1 year. How much interest will she earn?

38. Bugby Pest Control invests $1500 at 18% for 6 months. What amount of interest will the company earn?

39. Joann Selzy lends $6500 for 18 months at 16%. How much interest will she earn?

40. A consumer borrows $1000 at 10% for 3 months. Find the total amount due.

41. An investor deposits $4000 at 16% for 2 years. How much interest will be earned?

42. A loan of $1350 will be paid back with 12% interest at the end of 9 months. Find the total amount due.

43. A mother lends \$7500 to her daughter for 6 months and charges 9% interest. Find the interest charged on the loan.

44. Silvio Di Loreto deposits \$780 at 11% for 5 months. Find the interest he will earn.

CHALLENGE EXERCISES

45. The employee credit union pays $7\frac{1}{4}\%$ interest. Sarah Wilson deposits \$2100 in her account for $\frac{1}{4}$ year. Find the amount of interest she earns.

46. Ms. Henderson owes \$1900 in taxes. She is charged a penalty of $12\frac{1}{4}\%$ annual interest and pays the taxes and penalty after 6 months. Find the total amount she must pay.

47. A small business has a spare \$4500 that is invested at 11.25% interest for 5 months. How much interest is earned?

48. An auto parts store loans \$16,200 to another business for 9 months at an interest rate of 14.9%. How much interest will be earned on the loan?

REVIEW EXERCISES

Rewrite each of the following fractions with the indicated denominator. (For help, see ***Section 3.2.****)*

49. $\frac{3}{4} = \frac{}{8}$

50. $\frac{2}{3} = \frac{}{9}$

51. $\frac{5}{12} = \frac{}{48}$

52. $\frac{1}{2} = \frac{}{36}$

53. $\frac{3}{4} = \frac{}{60}$

54. $\frac{4}{5} = \frac{}{100}$

55. $\frac{15}{19} = \frac{}{76}$

56. $\frac{7}{15} = \frac{}{210}$

CHAPTER 6 SUMMARY

KEY WORDS

6.1 percent — Percent means per one hundred. A percent is a ratio with a denominator of 100.

6.3–6.4 percent proportion — The proportion $\frac{a}{b} = \frac{p}{100}$ is used to solve percent problems.

base — The base in a percent problem is the entire quantity or the total.

amount — The amount in a percent problem is the part being compared with the whole.

6.6 percent equation — The percent equation is percent · base = amount, and is an alternative way to solve percent problems.

6.7 sales tax — Sales tax is a percent of the total sales charged as a tax.

commission — Commission is a percent of the dollar value of total sales paid to a salesperson.

discount — Discount is often expressed as a percent of the original price; it is then deducted from the original price, resulting in the sale price.

percent of increase or decrease — Percent of increase or decrease is the amount of increase or decrease expressed as a percent of the original amount.

6.8 interest — Interest is a charge for borrowing money.

interest formula — The interest formula is the formula used to calculate interest (interest = principal · rate · time or $I = p \cdot r \cdot t$).

principal — Principal is the amount of money borrowed.

rate of interest — Often referred to as "rate," it is the charge for interest and is given as a percent.

QUICK REVIEW

Section Number and Topic	Approach	Example
6.1 Writing a Percent as a Decimal	Move the decimal point two places to the left and drop the % sign.	50% (. 5 0 %) = .5
6.1 Writing a Decimal as Percent	Move the decimal point two places to the right and attach a % sign.	.75 (. 7 5) = 75%

Section Number and Topic	Approach	Example
6.2 Writing a Fraction as a Percent	Use a proportion and solve for p to change a fraction to percent.	$\frac{2}{5}$ $\frac{2}{5} = \frac{p}{100}$ proportion $2 \cdot 100 = 5 \cdot p$ cross products $200 = 5 \cdot p$ $\frac{200}{5} = \frac{5p}{5}$ Divide by 5. $40 = p$ $\frac{2}{5} = 40\%$
6.3 Learning the Percent Proportion	**Amount** is to **base** as percent is to 100, or $\frac{a}{b} = \frac{p}{100}.$	Use the percent proportion to solve for the missing number. $a = 30$, $b = 50$, find p $\frac{30}{50} = \frac{p}{100}$ percent proportion $\frac{3}{5} = \frac{p}{100}$ lowest terms $3 \cdot 100 = 5 \cdot p$ cross products $300 = 5 \cdot p$ $\frac{300}{5} = \frac{5p}{5}$ Divide by 5. $60 = p$ The percent is 60, 60%.
6.4 Identifying Percent (p), Base (b), and Amount (a) in a Percent Problem	The percent (p) appears with the word **percent** or with the symbol **%.** The base (b) often appears after the word **of.** Base is the entire quantity. The amount (a) is the part of the total. If $\boldsymbol{p}$ and $\boldsymbol{b}$ are found first, the remaining number is $\boldsymbol{a}$.	Find p, b, and a in each of the following. 10% of the 500 pies are apple. (10% = p, 500 = b, apple = a) 20 is 5% of what number? (20 = a, 5% = p, what number = b) What % of 220 is 33? (What % = p, 220 = b, 33 = a)

Section Number and Topic	Approach	Example
6.5 Applying the Percent Proportion	Read the problem and identify p, b, and a. Use the percent proportion.	A tank contains 35% distilled water. 28 gallons of distilled water are in the tank when it is full. Find the volume of the tank. $p = 35$ and $a = 28$ Use the percent proportion to find b. $\frac{a}{b} = \frac{p}{100}$ $\frac{28}{b} = \frac{35}{100}$ $\frac{28}{b} = \frac{7}{20}$ lowest terms $560 = 7b$ cross products $b = 80$ The volume of the tank is 80 gallons.
6.6 Using the Percent Equation	The percent equation is percent · base = amount. Identify p, b, and a and solve for the missing quantity. Always write percent as a decimal before using the equation.	Solve each of the following. **(a)** Find 15% of 160. amount = percent · base $a = .15 \cdot 160$ $a = 24$ **(b)** 8 is 4% of what number? amount = percent · base $8 = .04 \cdot b$ $\frac{8}{.04} = \frac{.04 \cdot b}{.04}$ $200 = b$ **(c)** 13 is what percent of 52% amount = percent · base $13 = p \cdot 52$ $\frac{13}{52} = \frac{p \cdot 52}{52}$ $.25 = p$ $p = 25\%$

Section Number and Topic	Approach	Example
6.7 Solving for Sales Tax	Use the formula amount of sales tax = cost of item · rate of tax.	The cost of an item is \$450, and the sales tax is 6%. Find the sales tax. amount of sales tax = \$450 · 6% = \$450 · .06 = \$27
6.7 Finding Commissions	Use the formula amount of commission = rate or percent of commission · amount of sales.	The sales are \$92,000 with a commission rate of 3%. Find the commission. amount of commission = \$92,000 · 3% = \$92,000 · .03 = \$2760
6.7 Finding the Discount and the Sale Price	Use the formulas amount of discount = original price · rate of discount sale price = original price − amount of discount	A dishwasher originally priced at \$480 is offered at a 25% discount. Find the amount of the discount and the sale price. amount of discount = 480 · .25 = \$120 sale price = \$480 − \$120 = \$360
6.7 Finding the Percent of Change	Calculate the change (increase or decrease), which is the amount (a). Base (b) is the previous period or value before the change.	Enrollment rose from 3820 students to 5157 students. Find the percent of increase. 5157 − 3820 = 1337 increase $\frac{1337}{3820} = \frac{p}{100}$ Solve the proportion to find that p = 35 and the percent of increase is 35%.
6.8 Finding Simple Interest	Use the formula interest = principal · rate · time $(I = p \cdot r \cdot t)$	\$2800 is deposited at 8% for 3 months. Find the amount of interest. $I = p \cdot r \cdot t$ $= 2800 \cdot (.08) \cdot \frac{3}{12}$ $= 224 \cdot \frac{1}{4} = \frac{224 \cdot 1}{4}$ $I = \$56$

name date hour

CHAPTER 6 REVIEW EXERCISES

[6.1] *Write each of the following percents as decimals and decimals as percents.*

1. 50% **2.** 250% **3.** 17.6% **4.** .036%

5. 2.8 **6.** .09 **7.** .375 **8.** .002

[6.2] *Write each percent as a fraction in lowest terms and each fraction as percent. Round to the nearest tenth of a percent if necessary.*

9. 38% **10.** 6.25% **11.** $66\frac{2}{3}\%$ **12.** .025%

13. $\frac{3}{4}$ **14.** $\frac{7}{8}$ **15.** $\frac{7}{10}$ **16.** $\frac{1}{200}$

Complete this chart.

fraction	*decimal*	*percent*
$\frac{1}{8}$	**17.**	**18.**
19.	.05	**20.**
21.	**22.**	87.5%

[6.3] *Find the value of the missing letter in the percent proportion* $\frac{a}{b} = \frac{p}{100}$.

23. $a = 100$, $p = 10$ **24.** $b = 480$, $p = 20$

[6.4] *Identify p, b, and a in each of the following. Do not try to solve.*

25. 40% of 150 is 60. **26.** 73 is what percent of 90?

27. Find 46% of 1040. **28.** 418 is 30% of what number?

29. A golfer lost 3 of his 8 balls. What percent were lost?

30. Only 78% of one type of tulip bulbs bloom. There are 640 bulbs. How many will bloom?

[6.5] *Find the amount.*

31. 10% of 600

32. 92% of 3080

33. .9% of 4800

34. .2% of 1400

Find the base.

35. 50 is 10% of what number?

36. 128 is 5% of what number?

37. 338.8 is 140% of what number?

38. 2.5% of what number is 425?

Find the percent. Round to the nearest tenth of a percent if necessary.

39. 75 is what percent of 150?

40. What percent of 2400 is 30?

41. What percent of 190 is 18?

42. What percent of 720 is 150?

name date hour

[6.1–6.5] *Solve each word problem. Round to the nearest tenth if necessary.*

43. Rick Luttrell pays 27% of his total earnings in state and federal income tax. His earnings are $32,000 per year. Find his tax.

44. A bond selling for $1230 is paying interest of 6.3%. Find the amount of the interest.

[6.6] *Use the percent equation to find each of the following.*

45. 46% of 84

46. 114% of 32

47. .128 is what percent of 32?

48. 75 is what percent of 60?

49. 33.6 is 28% of what number?

50. 46 is 32% of what number?

[6.7] *Find the amount of sales tax and the total cost. Round to the nearest cent.*

	amount of sale	*tax rate*	*amount of tax*	*total cost*
51.	$100	3%	______	______
52.	$57	2%	______	______

Find the commission earned.

	sales	*rate of commission*	*commission*
53.	$360	25%	______
54.	$18,950	6%	______

Find the amount of discount and the sale price. Round to the nearest cent.

	original price	rate of discount	amount of discount	sale price
55.	$100	10%	______________	______________
56.	$585.50	15%	______________	______________

[6.8] *Find the interest on each of the following.*

	principal	rate	time in years	interest
57.	$100	14%	1	______________
58.	$2800	18%	$4\frac{1}{2}$	______________

Find the interest on each of the following.

	principal	rate	time in months	interest
59.	$100	15%	6	______________
60.	$1150	10%	18	______________

Find the total amount due on the following loans.

	principal	rate	time	total amount due
61.	$200	17%	2 years	______________
62.	$1420	18%	2 months	______________

name date hour

MIXED REVIEW EXERCISES

Find the value of the missing letter in the percent proportion $\frac{a}{b} = \frac{p}{100}$.

63. $b = 50, p = 70$

64. $a = 350, p = 25$

Use the percent equation to find each of the following.

65. 24% of 97

66. 195 is what percent of 130?

67. .6% of 85

68. 125 is 40% of what number?

69. 38 is what percent of 95?

70. 107.242 is 43% of what number?

Write the following percents as decimals and the following decimals as percents.

71. 25%

72. 100%

73. .50

74. 6.8

75. 8.5%

76. .719

77. .25%

78. .0006

Write each percent as a fraction in lowest terms and each fraction as a percent. Round to the nearest tenth of a percent if necessary.

79. 45%

80. $\frac{1}{2}$

81. 37.5%

82. $\frac{3}{8}$

83. $14\frac{1}{2}\%$

84. $\frac{1}{5}$

85. .5%

86. $\frac{1}{400}$

Solve each of the following word problems. Round to the nearest tenth of a percent if necessary.

87. Capital Marine invests $5000 at 13% for 18 months. What amount of interest will the company earn?

88. A consumer borrows $1000 at 15% for 3 months. Find the total amount due.

89. After spending $38, a shopper has $342 remaining. What percent of the original amount of money still remains?

90. Continental Airlines traffic rose to 3.28 billion revenue passenger miles from 3.12 billion revenue passenger miles. Find the percent of increase.

91. A real estate agent sold two properties, one for $105,000 and the other for $145,000. The agent receives a commission of $1\frac{1}{2}\%$ percent of total sales. Find the commission that she earned.

92. A mail carrier's route went from 472 residential stops to 436 residential stops. Find the percent of decrease.

93. Overall, weekly hours of employment last week rose to 34.9 from 34.7. Find the percent of increase.

94. Sales tax is 6% and the sales tax collected is $478.20. Find the total sales.

95. A young couple established a budget allowing 25% for rent, 30% for food, 8% for clothing, 20% for travel and recreation, and the remainder for savings. The man takes home $950 per month, and the woman takes home $14,500 per year. How much money will the couple save in a year?

96. In a 1-year period, air traffic at Continental Airlines rose to 16.33 billion revenue passenger miles from 15.22 billion revenue passenger miles. Find the percent of increase.

name date hour

CHAPTER 6 TEST

Write each percent as a decimal.

1. 75% **2.** .05%

1. ______
2. ______

Write each decimal as percent.

3. .25 **4.** .375

3. ______
4. ______

Write as fractions in lowest terms.

5. 37.5% **6.** .5%

5. ______

6. ______

Write each fraction as a percent.

7. $\frac{1}{2}$ **8.** $\frac{5}{8}$

7. ______
8. ______

Find the value of the missing letter in the percent proportion.

9. $a = 81, \quad p = 30$ **10.** $b = 550, \quad p = 25$

9. ______
10. ______

Identify a, b, and p. Do not try to solve.

11. 15% of 920 is 138.

11. ______

Find each of the following.

12. 8% of 200

12. ______

13. 75 is 10% of what number?

13. ______

14. 150 is what percent of 600?

14. ______

15. Erica has saved 72% of the amount needed for a down payment on a home. She has saved $12,096. Find the total down payment needed.

15. ______

16. ______________

____

17. ______________

____

18. ______________

19. ______________

20. ______________

21. ______________

____

22. ______________

____

23. ______________

24. ______________

25. ______________

26. ______________

Find the amount of sales tax and the total cost.

	sales	***rate of sales tax***		***sales***	***rate of sales tax***
16.	$28	4%	**17.**	$73	5%

18. The price of a copy machine is $1875 plus sales tax of $112.50. Find the sales tax rate.

19. One store pays its salespeople a commission of 16% of all sales. Find the commission earned by a person having sales c $2460.

20. Enrollment in mathematics courses increased from 1440 students last semester to 1944 students this semester. Find the percent of increase.

Find the amount of discount and the sale price.

	original price	***rate of discount***		***original price***	***rate of discount***
21.	$48	8%	**22.**	$175	37.5%

Find the interest on each of the following.

	principal	***rate***	***time***
23.	$480	15%	$2\frac{1}{2}$ years
24.	$1750	16%	9 months

25. A parent borrows $2800 to help her child finish college. The loan is for 6 months at 14% interest. Find the amount of interest owed.

26. If Gil Eckern borrows $17,200 with interest at 11% for 1 year, find the total amount due at the end of the year.

name date hour

CUMULATIVE REVIEW CHAPTERS 4–6

[4.1] *Name the digit that has the given place value in each of the following.*

1. 8.64 tenths hundredths

2. 93.4718 tenths ten-thousandths

Write each decimal as a fraction in lowest terms.

3. .75

4. .125

5. .04

6. .875

[4.2] *Round each of the following to the place shown.*

7. 61.628 to the nearest tenth

8. .6596 to the nearest thousandth

Round each of the following to the nearest dollar.

9. $25.49

10. $182.54

11. $4729.87

[4.3–4.6] *Add, subtract, multiply, or divide the following.*

12.
$$\begin{array}{r} 17.63 \\ 8.79 \\ +\ 64.52 \\ \hline \end{array}$$

13.
$$\begin{array}{r} 9.03 \\ 38.62 \\ 835.9 \\ 3.609 \\ +\ 73.17 \\ \hline \end{array}$$

14. $32.6 + 15.92 + 3.7 + 6.04$

15.
$$\begin{array}{r} 26.31 \\ -\ 15.17 \\ \hline \end{array}$$

16.
$$\begin{array}{r} 953.462 \\ -\ 148.638 \\ \hline \end{array}$$

17. $138.16 - 75.52$

18.
$$\begin{array}{r} .836 \\ \times\ 3.2 \\ \hline \end{array}$$

19.
$$\begin{array}{r} 72.71 \\ \times\ .305 \\ \hline \end{array}$$

20. $.006 \cdot .079$

21. $7\overline{)128.52}$

22. $5.26\overline{)64.961}$

23. $2.7625 \div .025$

[4.7] *Write each fraction as a decimal. Round to the nearest thousandth if necessary.*

24. $\frac{2}{5}$

25. $\frac{7}{8}$

26. $\frac{17}{20}$

27. $\frac{12}{14}$

[5.1] *Write each of the following ratios in lowest terms. Be sure to make all necessary conversions.*

28. $3\frac{1}{2}$ to 7

29. $6\frac{1}{4}$ to $12\frac{1}{2}$

30. $1\frac{5}{8}$ to 13

31. 8 minutes to 1 hour

32. 3 quarts to 3 gallons

[5.2] *Use the method of writing in lowest terms to decide whether the following proportions are true.*

33. $\frac{5}{10} = \frac{10}{20}$

34. $\frac{16}{24} = \frac{2}{6}$

35. $\frac{12}{14} = \frac{17}{21}$

36. $\frac{63}{21} = \frac{48}{16}$

name date hour

Use the cross products method to decide whether the following proportions are true. Circle the correct answer.

37. $\frac{8}{20} = \frac{40}{100}$

true **false**

38. $\frac{64}{144} = \frac{48}{108}$

true **false**

[5.3] *Find the missing numbers in each of the following proportions. Write answers as mixed numbers whenever possible.*

39. $\frac{1}{5} = \frac{t}{15}$

40. $\frac{315}{45} = \frac{21}{z}$

41. $\frac{9}{x} = \frac{57}{114}$

42. $\frac{y}{30} = \frac{7\frac{1}{2}}{24}$

[5.5] *Set up proportions for each of the following and then solve.*

43. 7 watches can be cleaned in 3 hours. Find the number of watches that can be cleaned in 12 hours.

44. If $3\frac{1}{2}$ ounces of weed killer is needed to make 6 gallons of spray, how much weed killer is needed for 102 gallons of spray?

[6.1] *Write each of the following percents as decimals. Write each of the following decimals as percents.*

45. 25%

46. 139.7%

47. .025%

48. 2.62%

49. .68

50. 2.71

51. .023

[6.2] *Write each percent as a fraction in lowest terms. Write each fraction as a percent. Round to the nearest thousandth if necessary.*

52. 12.5%

53. $37\frac{1}{2}\%$

54. $\frac{7}{8}$

55. $\frac{1}{200}$

[6.3] *Find the value of the missing letter in the percent proportion* $\frac{a}{b} = \frac{p}{100}$. *Round to the nearest tenth if necessary.*

56. $a = 50$, $p = 5$

57. $b = 240$, $p = 10$

[6.4] *Identify p, b, and a in each of the following. Do not try to solve these.*

58. 25% of 240 is 60.

59. 18 is what percent of 300?

60. The number of people taking the insurance exam is 272. If 204 people pass the exam, what percent pass the exam?

[6.5] *Find the amount.*

61. 25% of 500

62. 5.4% of 1200

63. 150% of 614

Find the base.

64. 78 is 40% of what number?

65. $5\frac{1}{2}\%$ of what number is 88?

name date hour

Find the percent. Round to the nearest tenth of a percent if necessary.

66. What percent of 110 is 55?

67. 36 is what percent of 90?

[6.6]

68. Use the percent equation and find the percent, if the base is 28 and the amount is 7.

[6.7] *Find the amount of sales tax and the total cost. Round to the nearest cent.*

	amount of sale	tax rate	amount of tax	total cost
69.	\$25	2%	________	________
70.	\$196	7%	________	________

Find the commission earned.

	sales	rate of commission	commission
71.	\$14,622	5%	________
72.	\$179,280	1.2%	________

Find the amount of discount and the sale price. Round to the nearest cent if necessary.

	original price	rate of discount	amount of discount	sale price
73.	$76	35%	________	________
74.	$238.50	22.5%	________	________

75. The number of entrants in this year's triathalon fell to 624 athletes from 960 last year. Find the percent of decrease.

76. The number of electronic defects has increased from 660 defects last month to 891 defects this month. Find the percent of increase.

[6.8] *Find the total amount due on the following loans.*

	principal	rate	time	total amount to be repaid
77.	$357	18%	2 years	________
78.	$18,350	11%	9 months	________

7

MEASUREMENT

We measure things almost every day. The distance to school or work, an area to be covered with carpet, the length of yard goods, the weight of vegetables or fruit in the grocery store—these are just a few examples of things we measure.

In the United States, there are two common systems of measurement. For everyday use, the **English system** (American system of units) of quarts, pounds, feet, and miles is usually used. The United States, however, is gradually switching to a different system of measurement for manufacturing, science, medicine, sports, and other fields; this system is called the **metric system.**

7.1 THE ENGLISH SYSTEM

OBJECTIVES

1. Know the basic units in the English system.
2. Use unit fractions.
3. Convert among units.

1 The English system of measurement uses feet to measure length, quarts to measure volume or capacity, and pounds to measure weight, for example. Here are some of the relationships among measures in the English system.

Length	Volume (Capacity)
12 inches = 1 foot	2 cups = 1 pint
3 feet = 1 yard	2 pints = 1 quart
5280 feet = 1 mile	4 quarts = 1 gallon

Weight

16 ounces = 1 pound

2000 pounds = 1 ton

The relationships in this table are ones you should memorize if you do not already know them. Also, memorize the following time relationships.

1 week	= 7 days
1 day	= 24 hours
1 hour	= 60 minutes
1 minute	= 60 seconds

WORK PROBLEM 1 AT THE SIDE.

As you can see, there is no simple or "natural" way to convert among these various measures. Many of these measures came from early royalty and were based on the sizes of parts of the body. For example, one yard was the distance from the tip of a king's nose to his thumb when his arm was outstretched.

EXAMPLE 1 Knowing English Measure
How many are in each of the following?

(a) inches in a foot 12

(b) pints in a quart 2

(c) hours in a day 24 ■

WORK PROBLEM 2 AT THE SIDE.

2 Conversions from one unit of measure to another are often needed. There are two ways of converting measurements. Both ways are shown here. Study each way and use the method you prefer. Some conversions can be done mentally.

EXAMPLE 2 Converting from One Unit of Measure to Another
Convert each measurement.

(a) 7 feet to inches

(b) 4 pounds to ounces

(c) 20 quarts to gallons

SOLUTION

(a) Since *1 foot = 12 inches,*

$$7 \text{ feet} = 7 \cdot 12 = 84 \text{ inches.}$$

(b) From the table, *1 pound = 16 ounces,* so

$$4 \text{ pounds} = 4 \cdot 16 = 64 \text{ ounces.}$$

(c) *4 quarts = 1 gallon,* so

$$20 \text{ quarts} = \frac{20}{4} = 5 \text{ gallons} \quad ■$$

1. After memorizing all the unit conversions, answer these questions.

(a) 24 hours = _______ day

(b) 1 gallon = _______ quarts

(c) 60 seconds = _______ minute

(d) 1 yard = _______ feet

(e) 1 foot = _______ inches

2. How many are in each of the following?

(a) feet in a mile

(b) quarts in a gallon

(c) ounces in a pound

(d) days in a week

(e) minutes in an hour

ANSWERS
1. (a) 1 (b) 4 (c) 1 (d) 3 (e) 12
2. (a) 5280 feet (b) 4 quarts (c) 16 ounces (d) 7 days (e) 60 minutes

NOTE When converting units, multiply when changing from a greater unit to a smaller unit and divide when changing from a smaller unit to a greater unit.

WORK PROBLEM 3 AT THE SIDE.

3 As in Example 2, some conversions are made by multiplying and some are made by dividing. People often have trouble deciding what to do. This method eliminates this problem because it uses **unit fractions,** which are fractions that are equivalent to 1.

For example, since *16 ounces = 1 pound,*

$$\frac{16 \text{ ounces}}{1 \text{ pound}} \quad \text{and} \quad \frac{1 \text{ pound}}{16 \text{ ounces}}$$

are both unit fractions.

Convert 32 ounces to pounds, for example, with a unit fraction that has the desired unit, pounds, on top.

$$32 \text{ ounces} = \frac{32 \text{ ounces}}{1} \cdot \frac{1 \text{ pound}}{16 \text{ ounces}}$$

↑ unit fraction

Cancel units and cancel numbers.

$$32 \text{ ounces} = \frac{\overset{2}{\cancel{32}} \cancel{\text{ounces}}}{1} \cdot \frac{1 \text{ pound}}{\underset{1}{\cancel{16}} \cancel{\text{ounces}}}$$

Cancel units. Cancel numbers.

Multiply.

$$32 \text{ ounces} = 2 \text{ pounds}$$

Use the following rule for *selecting* a unit fraction.

The unit being converted to should be in the numerator.
The unit being eliminated should be in the denominator.

EXAMPLE 3 Understanding Unit Fractions

Convert 60 inches to feet.

SOLUTION

Use a unit fraction with feet (the unit being converted to) in the numerator, and inches (the unit being eliminated) in the denominator. Since *12 inches = 1 foot,* the necessary unit fraction is

$$\frac{1 \text{ foot}}{12 \text{ inches}}.$$

← unit to convert to
← unit to eliminate

Next, multiply 60 inches and this unit fraction. Cancel all units and numbers possible.

3. Convert each measurement.

(a) 12 feet to inches

(b) 8 yards to feet

(c) 9 pounds to ounces

(d) 2 tons to pounds

(e) 36 pints to quarts

(f) 120 minutes to hours

(g) 3 weeks to days

ANSWERS
3. (a) 144 inches (b) 24 feet (c) 144 ounces (d) 4000 pounds (e) 18 quarts (f) 2 hours (g) 21 days

4. Convert the following.

(a) 36 inches to feet

(b) 180 inches to feet

(c) 9 inches to feet

5. Convert the following.

(a) 18 feet to inches

(b) 7 feet to inches

(c) 72 inches to feet

ANSWERS

4. (a) 3 feet (b) 15 feet (c) $\frac{3}{4}$ foot

5. (a) 216 inches (b) 84 inches (c) 6 feet

NOTE If no units will cancel, a mistake was made in choosing the unit fraction.

$$\frac{\overset{5}{\cancel{60 \text{ inches}}}}{1} \cdot \frac{1 \text{ foot}}{\underset{1}{\cancel{12 \text{ inches}}}} = 5 \text{ feet}$$

Inches cancel, leaving only feet. ■

WORK PROBLEM 4 AT THE SIDE.

EXAMPLE 4 Using Unit Fractions

Convert 9 feet to inches.

SOLUTION

$$\frac{9 \text{ feet}}{1} \cdot \frac{12 \text{ inches}}{1 \text{ foot}}$$

unit fraction
12 inches = 1 foot

To eliminate feet, these units must cancel each other.

$$\frac{9 \cancel{\text{ feet}}}{1} \cdot \frac{12 \text{ inches}}{1 \cancel{\text{ foot}}} = \frac{108}{1} = 108 \text{ inches}$$ ■

WORK PROBLEM 5 AT THE SIDE.

EXAMPLE 5 Using Unit Fractions with Volume (Capacity) Measurement

(a) Convert 9 pints to quarts.

(b) Convert $7\frac{1}{2}$ gallons to quarts.

(c) Convert 24 pints to gallons.

SOLUTION

(a) $\frac{9 \text{ pints}}{1} \cdot \frac{1 \text{ quart}}{2 \text{ pints}}$

Convert from pints to quarts by choosing a unit fraction with **pints** in the denominator and **quarts** in the numerator.

$$\frac{1 \text{ quart}}{2 \text{ pints}} \quad \textit{not as} \quad \frac{2 \text{ pints}}{1 \text{ quart}}$$

Next, multiply.

$$\frac{9 \cancel{\text{ pints}}}{1} \cdot \frac{1 \text{ quart}}{2 \cancel{\text{ pints}}} = \frac{9 \text{ quarts}}{2} = 4\frac{1}{2} \text{ quarts}$$

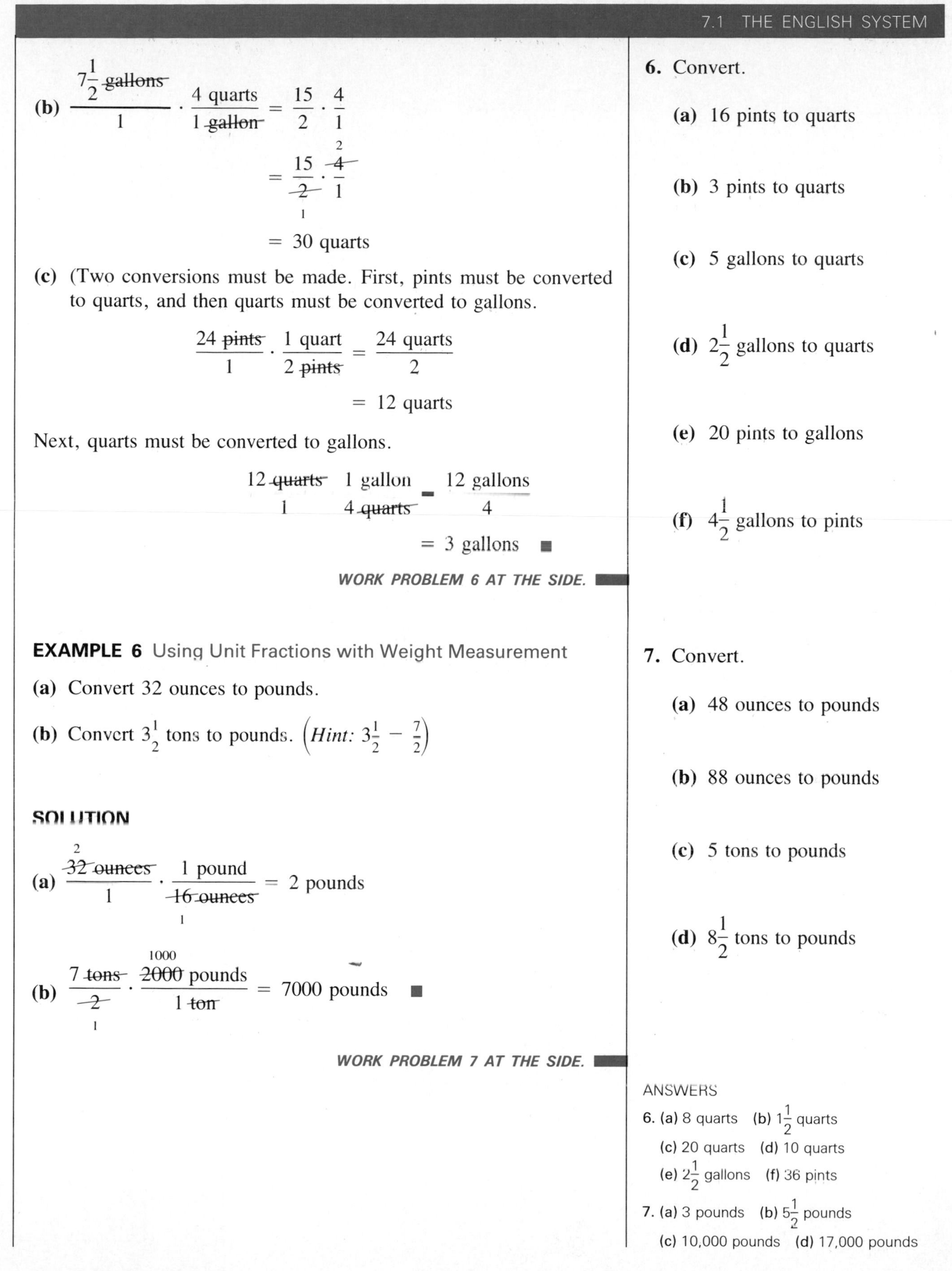

(b) $\dfrac{7\frac{1}{2} \cancel{\text{gallons}}}{1} \cdot \dfrac{4 \text{ quarts}}{1 \cancel{\text{gallon}}} = \dfrac{15}{2} \cdot \dfrac{4}{1}$

$= \dfrac{15}{\cancel{2}_{1}} \cdot \dfrac{\cancel{4}^{2}}{1}$

$= 30 \text{ quarts}$

(c) (Two conversions must be made. First, pints must be converted to quarts, and then quarts must be converted to gallons.

$$\frac{24 \cancel{\text{pints}}}{1} \cdot \frac{1 \text{ quart}}{2 \cancel{\text{pints}}} = \frac{24 \text{ quarts}}{2}$$

$$= 12 \text{ quarts}$$

Next, quarts must be converted to gallons.

$$\frac{12 \cancel{\text{quarts}}}{1} \cdot \frac{1 \text{ gallon}}{4 \cancel{\text{quarts}}} = \frac{12 \text{ gallons}}{4}$$

$$= 3 \text{ gallons}$$ ■

WORK PROBLEM 6 AT THE SIDE.

6. Convert.

(a) 16 pints to quarts

(b) 3 pints to quarts

(c) 5 gallons to quarts

(d) $2\frac{1}{2}$ gallons to quarts

(e) 20 pints to gallons

(f) $4\frac{1}{2}$ gallons to pints

EXAMPLE 6 Using Unit Fractions with Weight Measurement

(a) Convert 32 ounces to pounds.

(b) Convert $3\frac{1}{2}$ tons to pounds. $\left(\textit{Hint: } 3\frac{1}{2} = \frac{7}{2}\right)$

SOLUTION

(a) $\dfrac{\cancel{32}^{2} \cancel{\text{ounces}}}{1} \cdot \dfrac{1 \text{ pound}}{\cancel{16}_{1} \cancel{\text{ounces}}} = 2 \text{ pounds}$

(b) $\dfrac{7 \cancel{\text{tons}}}{\cancel{2}_{1}} \cdot \dfrac{\cancel{2000}^{1000} \text{ pounds}}{1 \cancel{\text{ton}}} = 7000 \text{ pounds}$ ■

WORK PROBLEM 7 AT THE SIDE.

7. Convert.

(a) 48 ounces to pounds

(b) 88 ounces to pounds

(c) 5 tons to pounds

(d) $8\frac{1}{2}$ tons to pounds

ANSWERS

6. (a) 8 quarts (b) $1\frac{1}{2}$ quarts (c) 20 quarts (d) 10 quarts (e) $2\frac{1}{2}$ gallons (f) 36 pints

7. (a) 3 pounds (b) $5\frac{1}{2}$ pounds (c) 10,000 pounds (d) 17,000 pounds

name date hour

7.1 EXERCISES

Convert the following.

Example: 1 hour = minutes

Solution: 1 hour = **60** minutes

1. 1 yard = inches

2. ounces = 1 pound

3. pints = 1 quart

4. 1 ton = pounds

5. 1 mile = feet

6. quarts = 1 gallon

7. pounds = 1 ton

8. 1 yard = feet

Convert the following. Use unit fractions.

Example: 3000 pounds = $1\frac{1}{2}$ tons

Solution: $\frac{3000 \text{ pounds}}{1} \cdot \frac{1 \text{ ton}}{2000 \text{ pounds}} = \frac{3}{2} = 1\frac{1}{2}$ tons

9. 6 feet = yards

10. 36 feet = yards

11. 8 quarts = gallons

12. 24 quarts = gallons

13. 2000 pounds = tons

14. 8000 pounds = tons

15. 18 inches = feet

16. 54 inches = feet

17. 28 pints = quarts = gallons

18. 68 pints = quarts = gallons

19. 136 ounces = pounds

20. 56 ounces = pounds

21. 3 tons = pounds

22. 8 quarts = pints

23. $5\frac{1}{2}$ pounds = ounces

24. $4\frac{1}{4}$ gallons = quarts = pints

25. $15\frac{1}{2}$ feet = inches

26. $1\frac{1}{2}$ yards = feet = inches

27. 2 miles = feet

28. 7 feet = inches

29. 27,720 feet = miles

30. 10 gallons = quarts

31. 17 pints = quarts = gallons

32. 7920 feet = miles

33. 36 cups = pints = quarts

34. 66 inches = feet

35. 21,120 feet = miles

36. $5\frac{1}{2}$ quarts = pints

37. 5 tons = pounds

38. $1\frac{1}{2}$ tons = pounds

39. $7\frac{1}{8}$ tons = pounds

40. $2\frac{1}{2}$ pounds = ounces

41. 3 days = hours

42. 7 hours = minutes

CHALLENGE EXERCISES

Convert the following.

43. $8\frac{1}{4}$ hours = ______ seconds

44. 2880 pints = ______ gallons

45. $1\frac{1}{2}$ miles = ______ inches

46. $3\frac{1}{2}$ tons = ______ ounces

REVIEW EXERCISES

Place $<$ or $>$ in each blank to make a true statement. (For help, see ***Section 3.5****.)*

47. 2 weeks ______ 15 days

48. 72 hours ______ 4 days

49. 4 hours ______ 185 minutes

50. 2 years ______ 28 months

51. 32 days ______ 4 weeks

52. 14 minutes ______ 780 seconds

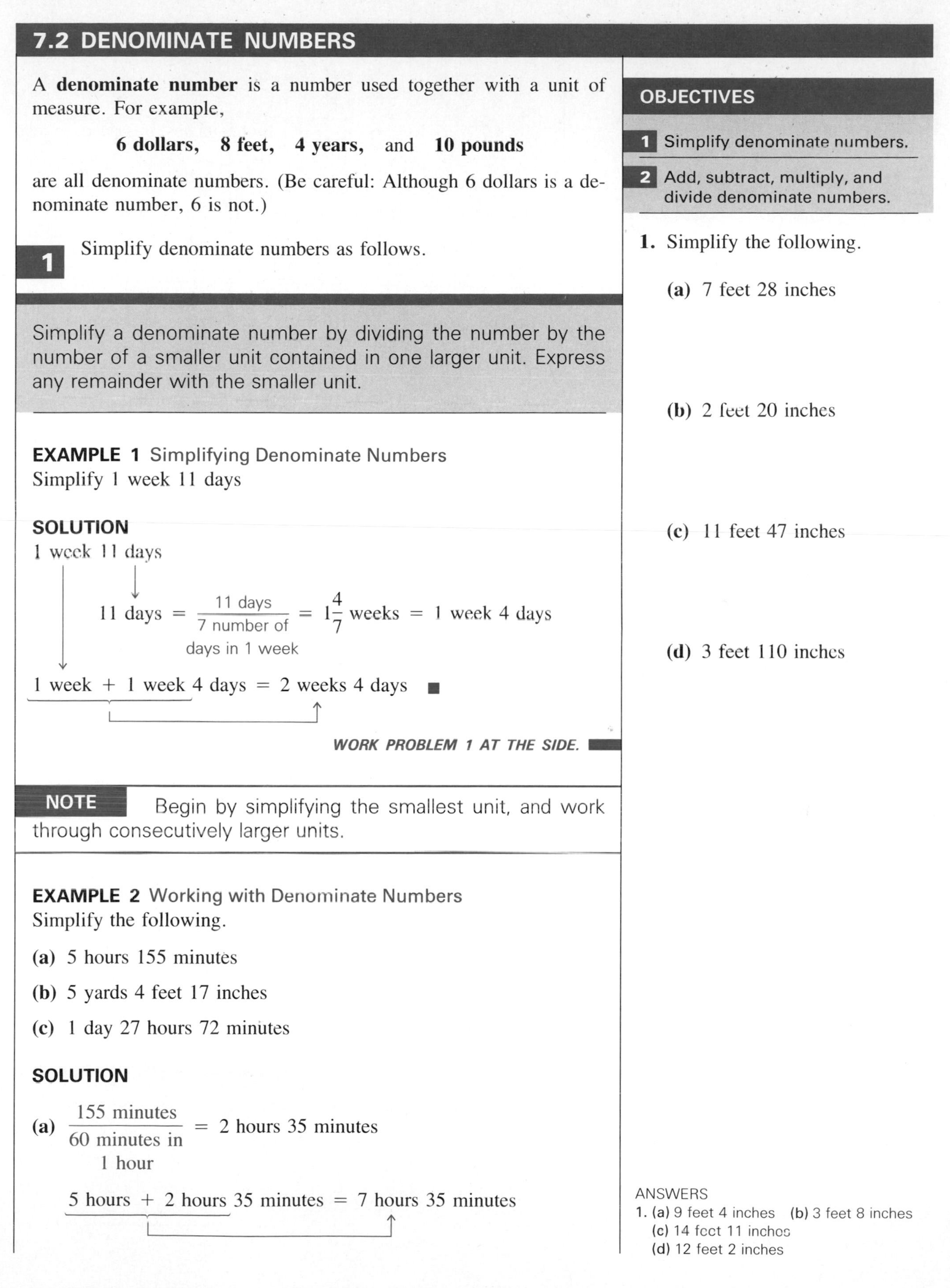

7.2 DENOMINATE NUMBERS

A **denominate number** is a number used together with a unit of measure. For example,

6 dollars, **8 feet,** **4 years,** and **10 pounds**

are all denominate numbers. (Be careful: Although 6 dollars is a denominate number, 6 is not.)

OBJECTIVES

1. Simplify denominate numbers.
2. Add, subtract, multiply, and divide denominate numbers.

1 Simplify denominate numbers as follows.

Simplify a denominate number by dividing the number by the number of a smaller unit contained in one larger unit. Express any remainder with the smaller unit.

EXAMPLE 1 Simplifying Denominate Numbers
Simplify 1 week 11 days

SOLUTION
1 week 11 days

$$11 \text{ days} = \frac{11 \text{ days}}{7 \text{ number of days in 1 week}} = 1\frac{4}{7} \text{ weeks} = 1 \text{ week } 4 \text{ days}$$

1 week + 1 week 4 days = 2 weeks 4 days ■

WORK PROBLEM 1 AT THE SIDE.

NOTE Begin by simplifying the smallest unit, and work through consecutively larger units.

EXAMPLE 2 Working with Denominate Numbers
Simplify the following.

(a) 5 hours 155 minutes

(b) 5 yards 4 feet 17 inches

(c) 1 day 27 hours 72 minutes

SOLUTION

(a) $$\frac{155 \text{ minutes}}{60 \text{ minutes in 1 hour}} = 2 \text{ hours } 35 \text{ minutes}$$

5 hours + 2 hours 35 minutes = 7 hours 35 minutes

1. Simplify the following.

(a) 7 feet 28 inches

(b) 2 feet 20 inches

(c) 11 feet 47 inches

(d) 3 feet 110 inches

ANSWERS
1. (a) 9 feet 4 inches (b) 3 feet 8 inches
(c) 14 feet 11 inches
(d) 12 feet 2 inches

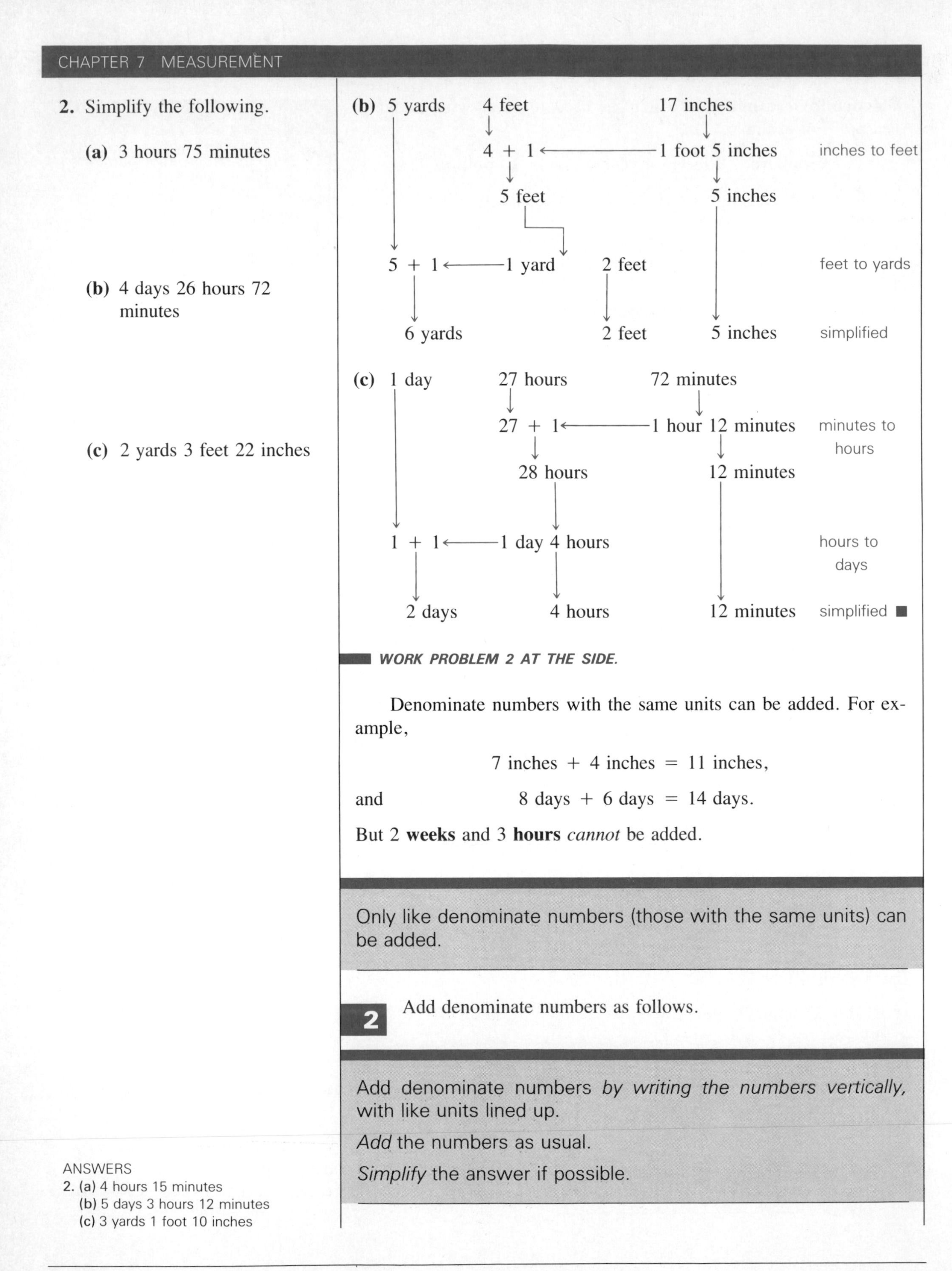

2. Simplify the following.

(a) 3 hours 75 minutes

(b) 4 days 26 hours 72 minutes

(c) 2 yards 3 feet 22 inches

(b) 5 yards 4 feet 17 inches

4 + 1 ← 1 foot 5 inches — inches to feet

5 feet, 5 inches

5 + 1 ← 1 yard, 2 feet — feet to yards

6 yards, 2 feet, 5 inches — simplified

(c) 1 day 27 hours 72 minutes

27 + 1 ← 1 hour 12 minutes — minutes to hours

28 hours, 12 minutes

1 + 1 ← 1 day 4 hours — hours to days

2 days, 4 hours, 12 minutes — simplified ■

WORK PROBLEM 2 AT THE SIDE.

Denominate numbers with the same units can be added. For example,

7 inches + 4 inches = 11 inches,

and 8 days + 6 days = 14 days.

But 2 **weeks** and 3 **hours** *cannot* be added.

Only like denominate numbers (those with the same units) can be added.

2 Add denominate numbers as follows.

Add denominate numbers *by writing the numbers vertically,* with like units lined up.

Add the numbers as usual.

Simplify the answer if possible.

ANSWERS

2. (a) 4 hours 15 minutes
(b) 5 days 3 hours 12 minutes
(c) 3 yards 1 foot 10 inches

EXAMPLE 3 Adding Denominate Numbers
Add 5 gallons 3 quarts and 2 gallons 3 quarts.

SOLUTION

$$\begin{array}{r} 5 \text{ gallons } 3 \text{ quarts} \\ +\ 2 \text{ gallons } 3 \text{ quarts} \\ \hline 7 \text{ gallons } 6 \text{ quarts} \end{array}$$

$$\downarrow$$

$$= \underbrace{7 \text{ gallons} + 1 \text{ gallon}}\ 2 \text{ quarts}$$

$$\downarrow \qquad\qquad \downarrow$$

$$8 \text{ gallons} \qquad 2 \text{ quarts}$$ ■

WORK PROBLEM 3 AT THE SIDE.

Subtract denominate numbers as follows.

Subtract denominate numbers *by writing the numbers vertically,* with like units lined up.

Subtract units from units by borrowing if necessary.

EXAMPLE 4 Subtracting Denominate Numbers
Subtract the following.

(a) 2 pounds 11 ounces from 4 pounds 8 ounces

(b) 5 days 3 hours 15 minutes from 9 days 1 hour 47 minutes

SOLUTION
(a) Write the numbers vertically.

$$\begin{array}{r} 4 \text{ pounds } \ \ 8 \text{ ounces} \\ -\ 2 \text{ pounds } 11 \text{ ounces} \\ \hline \end{array}$$

It is necessary to borrow in the ounces column. Borrow 1 pound from the 4 pounds, and add 16 ounces (1 pound = 16 ounces) to the 8 ounces.

$$\begin{array}{rr} 4 \text{ pounds} & 8 \text{ ounces} \leftarrow \text{original number} \\ -\ 1 \text{ pound} & +\ 16 \text{ ounces} \\ \uparrow & \uparrow \end{array}$$

Subtract 1 pound. Add 16 ounces.

$$3 \text{ pounds } 24 \text{ ounces} \leftarrow \text{result}$$

3. Add the following.

(a) 4 feet 6 inches and 3 feet 8 inches

(b) 1 gallon 3 quarts 2 pints and 2 gallons 2 quarts and 3 pints

ANSWERS
3. (a) 8 feet 2 inches
(b) 4 gallons 3 quarts 1 pint

4. Subtract.

(a) 7 pounds 10 ounces from 12 pounds 3 ounces

(b) 4 gallons 3 pints from 9 gallons 1 pint

(c) 2 weeks 3 days from 8 weeks 1 day

(d) 1 yard 1 foot 9 inches from 3 yards 2 feet 6 inches

ANSWERS
4. (a) 4 pounds 9 ounces
(b) 4 gallons 3 quarts
(c) 5 weeks 5 days
(d) 2 yards 9 inches

Next, subtract.

$$\begin{array}{r} 3 \text{ pounds } 24 \text{ ounces} \\ -\ 2 \text{ pounds } 11 \text{ ounces} \\ \hline 1 \text{ pound } 13 \text{ ounces} \end{array} \leftarrow \text{answer}$$

Here is a shortcut way to show this work.

borrow 1 pound = 16 ounces
16 ounces + 8 ounces = 24

$$\begin{array}{r} \overset{3}{\cancel{4}} \text{ pounds } \overset{24}{\cancel{8}} \text{ ounces} \\ -\ 2 \text{ pounds } 11 \text{ ounces} \\ \hline 1 \text{ pound } 13 \text{ ounces} \end{array}$$

(b)

$$\begin{array}{r} \overset{8}{\cancel{9}} \text{ days } \overset{25}{\cancel{1}} \text{ hour } 47 \text{ minutes} \\ -\ 5 \text{ days } 3 \text{ hours } 15 \text{ minutes} \\ \hline 3 \text{ days } 22 \text{ hours } 32 \text{ minutes} \end{array}$$ ■

WORK PROBLEM 4 AT THE SIDE.

Multiply a denominate number as follows.

Multiply a denominate number *by multiplying each unit* of the denominate number.

Simplify the answer.

EXAMPLE 5 Multiplying a Denominate Number

Multiply the following.

(a) 3 feet 7 inches by 2

(b) 9 days 13 hours by 4

SOLUTION

(a)

$$\begin{array}{r} 3 \text{ feet } 7 \text{ inches} \\ \times \qquad\qquad 2 \\ \hline 6 \text{ feet } 14 \text{ inches} \end{array}$$

$6 + 1 = 6 \text{ feet} + 1 \text{ foot } 2 \text{ inches}$

7 feet 2 inches

(b)

$$\begin{array}{r} 9 \text{ days } 13 \text{ hours} \\ \times \qquad\qquad 4 \\ \hline 36 \text{ days } 52 \text{ hours} \end{array}$$

Since 1 day = 24 hours,

52 hours = 2 days 4 hours,

and 36 days + 2 days 4 hours = 38 days 4 hours. ■

WORK PROBLEM 5 AT THE SIDE.

Use the following rule to divide a denominate number.

> Divide a denominate number *by dividing each part* of the denominate number *separately,* beginning at the left. Any remainder must be added to the next smaller unit. Continue dividing until all parts have been divided.

EXAMPLE 6 Dividing a Denominate Number

Divide the following.

(a) 7 pounds 13 ounces by 5

(b) 4 yards 2 feet 9 inches by 3

SOLUTION

(a)

$$\begin{array}{rll} & 1 \text{ pound} \quad 9 \text{ ounces} & \\ 5\overline{)7 \text{ pounds}} & \overline{13 \text{ ounces}} & \\ \underline{5 \text{ pounds}} & & \\ 2 \text{ pounds} = & 32 \text{ ounces} & \leftarrow \text{Remainder is 2 pounds, which is 32 ounces.} \\ & \overline{45 \text{ ounces}} & \leftarrow \text{Bring down } 13 + 32 = 45 \text{ (ounces).} \\ & \underline{45 \text{ ounces}} & \leftarrow 9 \cdot 5 = 45 \\ & 0 & \leftarrow 45 - 45 = 0 \end{array}$$

Check: 5 times 1 pound 9 ounces is 7 pounds 13 ounces.

(b)

$$\begin{array}{rlll} & 1 \text{ yard} \quad 1 \text{ foot} \quad 11 \text{ inches} & & \\ 3\overline{)4 \text{ yards}} & \overline{2 \text{ feet} \quad 9 \text{ inches}} & & \\ \underline{3} & & & \\ 1 \text{ yard} = & \underline{3 \text{ feet}} & & \\ & 5 \text{ feet} & \leftarrow 2 + 3 = 5 \text{ (feet)} & \\ & \underline{3} & \leftarrow 1 \cdot 3 = 3 & \\ & 2 \text{ feet} = & \underline{24 \text{ inches}} & \leftarrow 5 - 3 = 2 \text{ (feet)}, = 24 \text{ inches} \\ & & 33 \text{ inches} & \leftarrow 9 + 24 = 33 \\ & & \underline{33} & \leftarrow 11 \cdot 3 = 33 \\ & & 0 \; ■ & 33 - 33 = 0 \end{array}$$

divide 5 by 3

divide 33 by 33

WORK PROBLEM 6 AT THE SIDE.

5. Multiply.

(a)

$$\begin{array}{r} 1 \text{ hour } 18 \text{ minutes} \\ \times \qquad\qquad 3 \\ \hline \end{array}$$

(b)

$$\begin{array}{r} 5 \text{ pounds } 6 \text{ ounces} \\ \times \qquad\qquad 4 \\ \hline \end{array}$$

(c)

$$\begin{array}{r} 3 \text{ quarts } 1 \text{ pint} \\ \times \qquad\qquad 5 \\ \hline \end{array}$$

6. Divide.

(a) $2\overline{)3 \text{ gallons } 2 \text{ quarts}}$

(b) $4\overline{)5 \text{ weeks } 3 \text{ days } 8 \text{ hours}}$

ANSWERS
5. (a) 3 hours 54 minutes
(b) 21 pounds 8 ounces
(c) 17 quarts 1 pint
6. (a) 1 gallon 3 quarts
(b) 1 week 2 days 14 hours

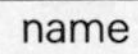

name date hour

7.2 EXERCISES

Simplify the following.

Example: 2 quarts 5 pints

Solution: 2 quarts 5 pints

$\left(5 \text{ pints} = \frac{5}{2} \text{ quarts}\right)$

$\frac{5}{2} = 2 \text{ quarts } 1 \text{ pint}$

$2 + 2 =$ **4 quarts 1 pint**

1. 1 foot 16 inches

2. 1 yard 5 feet

3. 1 pound 20 ounces

4. 5 pounds 30 ounces

5. 1 quart 7 pints

6. 2 quarts 3 pints

7. 1 mile 7250 feet

8. 10 miles 10,000 feet

9. 9 days 36 hours

10. 3 days 42 hours

11. 7 gallons 6 quarts

12. 3 gallons 12 quarts

13. 2 yards 3 feet 18 inches

14. 5 yards 2 feet 22 inches

15. 2 weeks 9 days 60 hours

16. 5 weeks 21 days 54 hours

17. 5 gallons 6 quarts 3 pints

18. 110 gallons 32 quarts 15 pints

Add the following.

Example:

2 feet 8 inches
+ 1 foot 6 inches
4 feet 2 inches

Solution:

2 feet 8 inches
+ 1 foot 6 inches
3 feet 14 inches
↓
1 foot 2 inches
↓
3 + 1 = 4 feet 2 inches

19. 8 feet 3 inches
+ 2 feet 10 inches

20. 6 feet 5 inches
+ 6 feet 8 inches

21. 6 quarts 2 pints
+ 3 quarts 1 pint

22. 3 quarts 1 pint
+ 4 quarts 3 pints

23.
```
  7 pounds 12 ounces
+ 3 pounds 10 ounces
```

24.
```
  12 pounds  7 ounces
+  4 pounds 15 ounces
```

25.
```
  7 weeks 2 days 19 hours
+ 3 weeks 5 days 18 hours
```

26.
```
  4 days 19 hours 57 minutes
+ 3 days 21 hours  9 minutes
```

Subtract the following.

Example:
```
  5 pounds  6 ounces
− 1 pound  10 ounces
  3 pounds 12 ounces
```
(final line in bold)

Solution:
```
                 → 16 ounces + 6 ounces
  4           22
  5̸ pounds    6̸ ounces
− 1 pound    10 ounces
  3 pounds   12 ounces
```

27.
```
  2 gallons 3 quarts
− 1 gallon  2 quarts
```

28.
```
  8 gallons 2 quarts
  5 gallons 1 quart
```

29.
```
  6 tons 800 pounds
− 4 tons 500 pounds
```

30.
```
  3 tons 1200 pounds
− 2 tons  800 pounds
```

31.
```
  3 yards 1 foot
− 2 yards 2 feet
```

32.
```
  10 gallons 1 quart  1 pint
−  8 gallons 2 quarts 1 pint
```

33. 15 days 15 hours 39 minutes
− 8 days 22 hours 42 minutes

34. 11 weeks 3 days 14 hours
− 2 weeks 5 days 23 hours

Multiply the following.

Example:
2 gallons 1 quart
× 4
9 gallons

Solution:
2 gallons 1 quart
× 4
8 gallons 4 quarts
↓
1 gallon
8 + 1 = 9 gallons

35. 6 feet 4 inches
× 2

36. 12 feet 1 inch
× 4

37. 5 tons 800 pounds
× 6

38. 5 days 14 hours
× 7

39. 75 yards 2 feet
× 10

40. 3 gallons 3 quarts
× 4

41. 2 yards 1 foot 6 inches
× 6

42. 9 days 17 hours 11 minutes
× 8

Divide the following.

Example: $\begin{array}{r} \textbf{2 yards 2 feet} \\ 2\overline{)5 \text{ yards } 1 \text{ foot}} \end{array}$

Solution: $\begin{array}{l} \quad\ \ \textbf{2 yards 2 feet} \\ 2\overline{)5 \text{ yards } 1 \text{ foot}} \\ \quad \underline{4} \\ \quad 1 \ = \ \underline{3} \quad 1 \text{ yard} = 3 \text{ feet} \\ \qquad\quad\ \ 4 \\ \qquad\quad\ \ \underline{4} \\ \qquad\quad\ \ 0 \end{array}$

43. $2\overline{)4 \text{ gallons } 2 \text{ quarts}}$

44. $4\overline{)8 \text{ feet } 12 \text{ inches}}$

45. $4\overline{)650 \text{ pounds } 8 \text{ ounces}}$

46. $5\overline{)8 \text{ yards } 1 \text{ foot}}$

47. $2\overline{)6 \text{ yards } 2 \text{ feet}}$

48. $4\overline{)8 \text{ miles } 240 \text{ feet}}$

49. $5\overline{)4 \text{ weeks } 5 \text{ days } 13 \text{ hours}}$

50. $7\overline{)12 \text{ weeks } 3 \text{ days } 19 \text{ hours}}$

CHALLENGE EXERCISES

Solve the following and simplify by using the units given.

51. 3 gallons 3 quarts 1 pint 1 cup
+ 2 gallons 2 quarts 1 pint 1 cup

52. 6 weeks 5 days 23 hours 10 minutes
− 2 weeks 6 days 20 hours 15 minutes

53. 3 weeks 4 days 10 hours 38 minutes
× 3

54. 3)8 gallons 3 quarts 1 pint 2 cups

REVIEW EXERCISES

Name the digit that has the given place value in each of the following. (For help, see ***Section 4.1.****)*

55. .6071
hundredths
thousandths

56. 782.107
hundredths
thousandths

57. 7250.618
thousands
hundredth

58. 1358.6256
ten-thousandths
tenths

7.3 THE METRIC SYSTEM—LENGTH

OBJECTIVES

1 Know the common metric prefixes.

2 Convert metric measurements.

As noted earlier, the United States is gradually switching to the **metric system** of weights and measures for many uses.

The basic unit of length in the metric system is the **meter.** The symbol for meter is **m.** A meter is a little shorter than five of these pages, laid side by side. A meter is a little longer than one yard.

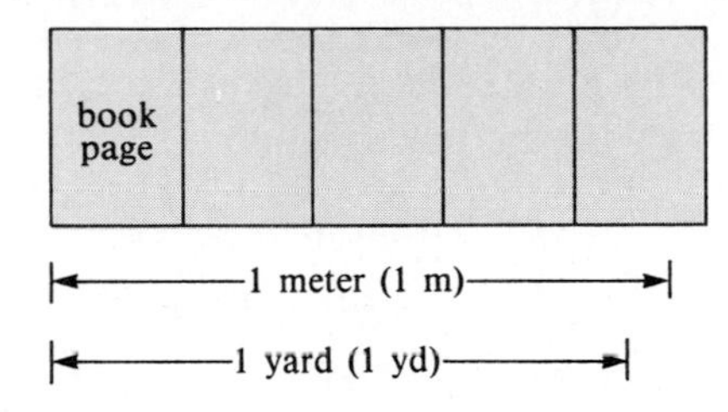

1 One advantage to using the metric system is the ease with which measurements are converted within the system. This is done by multiplying and dividing by 10 and multiples of 10 and in words by using **prefixes.** The common prefixes for the metric system are shown below. The ones most used in daily life are in colored boxes.

prefix	*kilo*	*hecto*	*deka*	*unit*	*deci*	*centi*	*milli*
meaning	1000	100	10	1	$\frac{1}{10}$	$\frac{1}{100}$	$\frac{1}{1000}$
abbreviation	k	h	da		d	c	m

Small items are often measured with **centimeters** and **millimeters.** As the chart shows, *centi-* means one hundredth, and *milli-* means one thousandth. Centimeter is abbreviated **cm,** and millimeter is abbreviated **mm.**

1 centimeter = 1 cm = .01 m
100 cm = 1 m

The figures below show 1 cm and 1 inch.

1 cm   1 inch

A centimeter is a little greater than the width of a paper clip.

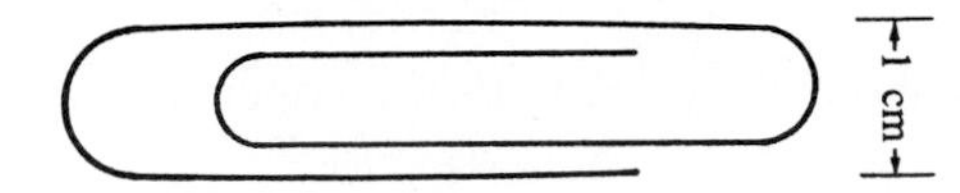

2 The next example shows how to convert from one metric unit to another.

EXAMPLE 1 Converting from One Unit of Measure to Another
Convert as indicated.

(a) 2.5 m to cm **(b)** 18.6 cm to m

1. Convert.

(a) 3.67 m to cm

(b) .92 m to cm

(c) 432.7 cm to m

(d) 34 cm to m

SOLUTION

(a) A meter is a greater unit of length than a centimeter. Thus, 2.5 meters will contain many centimeters. To find out how many, *multiply* 2.5 and the number of cm in 1 m (100).

$$2.5 \text{ m} = 2.5 \cdot 100 = 250 \text{ cm}$$

This same conversion can be made by using unit fractions. Keep cm and eliminate m with the unit fraction

$$\frac{100 \text{ cm}}{1 \text{ m}}.$$

Multiply.

$$\frac{2.5 \text{ m}}{1} \cdot \frac{100 \text{ cm}}{1 \text{ m}} = \frac{2.5 \cancel{\text{m}}}{1} \cdot \frac{100 \text{ cm}}{1 \cancel{\text{m}}}$$
$$= 250 \text{ cm}$$

(b) A centimeter is a smaller unit of length than a meter, so 18.6 cm will be a very small number of meters or part of a meter. Multiply 18.6 by the number of meters in a centimeter (.01).

$$18.6 \text{ cm} = 18.6 \cdot .01 = .186 \text{ m}$$

Alternatively, use unit fractions.

$$18.6 \text{ cm} = \frac{18.6 \cancel{\text{cm}}}{1} \cdot \frac{1 \text{ m}}{100 \cancel{\text{cm}}} = .186 \text{ m} \quad \blacksquare$$

NOTE When converting units, multiply when changing from a greater unit to a smaller unit; divide when changing from a smaller unit to a greater unit.

WORK PROBLEM 1 AT THE SIDE.

Lengths shorter than a centimeter can be measured in millimeters. As mentioned earlier,

$$1 \text{ mm} = .001 \text{ m}$$
$$1000 \text{ mm} = 1 \text{ m}.$$

A millimeter is about the thickness of a paper match.

1 mm

ANSWERS
1. (a) 367 cm (b) 92 cm (c) 4.327 m (d) .34 m

EXAMPLE 2 Converting Units of Measure
Convert the following.

(a) 42 m to mm **(b)** 251 mm to m

SOLUTION

(a) Since a meter is a much greater unit than a millimeter, 42 meters will contain many millimeters. Multiply 42 by the number of millimeters in a meter (1000).

$$42 \text{ m} = 42 \cdot 1000 = 42{,}000 \text{ mm}$$

With unit fractions,

$$42 \text{ m} = \frac{42 \cancel{\text{m}}}{1} \cdot \frac{1000 \text{ mm}}{1 \cancel{\text{m}}} = 42{,}000 \text{ mm}.$$

(b) This conversion is from a smaller to a greater unit of measure, so multiply by .001 (the number of meters in a millimeter).

$$251 \text{ mm} = 251 \cdot .001 = .251 \text{ m}$$

With unit fractions,

$$251 \text{ mm} = \frac{251 \cancel{\text{mm}}}{1} \cdot \frac{1 \text{ m}}{1000 \cancel{\text{mm}}} = .251 \text{ m}. \blacksquare$$

WORK PROBLEM 2 AT THE SIDE.

Long distances are measured with **kilometers.** The prefix *kilo-* means 1000, so with the symbol km for kilometer,

$$1 \text{ km} = 1000 \text{ m}$$
$$.001 \text{ km} = 1 \text{ m}.$$

A kilometer is approximately the length of five city blocks or approximately $\frac{6}{10}$ of a mile. The popular 10-km race is approximately 6 miles long.

EXAMPLE 3 Practicing the Conversion of Units
Convert the following.

(a) 37 km to m **(b)** 583 m to km

SOLUTION

(a) Kilometers are greater than meters, so multiply by 1000, the number of meters in 1 kilometer.

$$37 \text{ km} = 37 \cdot 1000 = 37{,}000 \text{ m}$$

Or with unit fractions,

$$37 \text{ km} = \frac{37 \cancel{\text{km}}}{1} \cdot \frac{1000 \text{ m}}{1 \cancel{\text{km}}} = 37{,}000 \text{ m}.$$

2. Convert.

(a) 23.8 m to mm

(b) .74 m to mm

(c) 856 mm to m

(d) 32 mm to m

ANSWERS
2. (a) 23,800 mm (b) 740 mm
(c) .856 m (d) .032 m

3. Convert.

(a) 68.2 km to m

(b) .47 km to m

(c) 1275 m to km

(d) 763 m to km

(b) Meters are smaller than kilometers, so multiply by .001 to get kilometers.

$$583 \text{ m} = 583 \cdot .001 = .583 \text{ km}$$

Or use unit fractions.

$$\frac{583 \cancel{\text{m}}}{1} \cdot \frac{1 \text{ km}}{1000 \cancel{\text{m}}} = .583 \text{ km}$$ ■

WORK PROBLEM 3 AT THE SIDE.

Conversions among metric units can be made quickly by using a line with all the metric prefixes marked on it. (Common prefixes are in color.)

kilo	hecto	deka	unit	deci	centi	milli
1000	100	10	1	$\frac{1}{10}$	$\frac{1}{100}$	$\frac{1}{1000}$

EXAMPLE 4 Using a Metric Conversion Line
Use the metric conversion line for each of the following conversions.

(a) .0392 km to m

(b) .426 m to cm

(c) 765.4 cm to km

SOLUTION

(a) On the line, starting at *kilo* and moving to the basic *unit* requires moving three places to the right. So move the decimal in .0392 three places to the right.

. 0 3 9 2 — Decimal is moved three places to the right.

.0392 km = 39.2 m

(b) Starting at the basic *unit* and moving to *centi* requires moving two places to the right.

. 4 2 6 — Decimal is moved two places to the right.

.426 m = 42.6 cm

ANSWERS
3. (a) 68,200 m (b) 470 m
(c) 1.275 km (d) .763 km

(c) From *centi* to *kilo* is five places to the *left*.

0 0 7 6 5 . 4   Decimal is moved five places to the left.

765.4 cm = .007654 km ■

NOTE Two zeros must be added as placeholders in (c) above.

WORK PROBLEM 4 AT THE SIDE.

EXAMPLE 5 Practice with the Metric Conversion Line

Use the conversion line to convert each of the following.

(a) .17 m to mm

(b) 352.6 m to km

(c) 48782 mm to km

SOLUTION

(a) Starting at the *unit* position and moving to *milli* requires moving three places to the right.

. 1 7 0   One zero is added to position the decimal three places to the right.

.17 m = 170 mm

(b) From the basic unit to kilo is three places to the left.

3 5 2 . 6   Decimal is moved three places to the left.

352.6 m = .3526 km

(c) From milli to kilo is six places to the left.

0 4 8 7 8 2 .   Decimal is moved six places left, zero is added.

48782 mm = .048782 ■

NOTE When no decimal appears in a number (e.g., 48782 mm), the decimal is assumed to be at the far right.

WORK PROBLEM 5 AT THE SIDE.

4. Convert.

(a) 37.4 km to m

(b) 12.5 m to cm

(c) 46.9 cm to mm

(d) 89.4 mm to m

(e) 354.9 m to km

5. Convert.

(a) .0421 km to m

(b) 3.715 m to km

(c) 37812.1 cm to km

(d) 1.003 km to mm

(e) 5521.6 mm to km

ANSWERS

4. (a) 37,400 m (b) 1250 cm (c) 469 mm (d) .0894 m (e) .3549 km

5. (a) 42.1 m (b) .003715 km (c) .378121 km (d) 1,003,000 mm (e) .0055216 km

name date hour

7.3 EXERCISES

Convert the following.

Example 1 mm = **.001** m   **Solution:** 1 mm = m
1 mm = 1 · .001 m = .001 m

1. 1 m = cm   **2.** 1 m = km   **3.** 1 m = mm

4. 1 km = m   **5.** 1 cm = m   **6.** 1 mm = m

Use the scale below to measure the following.

7. the width of your hand in centimeters

8. the width of your hand in millimeters

9. the width of your pencil or pen in millimeters

10. the width of your pencil or pen in centimeters

Convert each of the following.

Example:

68 cm to m   **.68** m

Solution:

68 cm to m
68 cm = 68 · .01 m = .68 m

1 2 3 4 5 6 7 8 9 10 11 12 13 14 15
METRIC (centimeters)

11. 8 m to mm

12. 14.76 m to cm

13. 1.5 m to mm

14. 78.1 m to cm

15. 52.5 cm to m

16. 918 mm to m

17. 8500 cm to m

18. 250 mm to m

19. 118.6 cm to mm

20. 17.4 cm to mm

21. 5.3 km to m

22. 9.24 km to m

23. 27,500 m to km

24. 14,592 m to km

25. 2.4 km to cm

26. 1.95 km to cm

Solve the following problems.

27. Jim's waist size is 82 cm. Give his waist size in mm.

28. A dog is 598 mm in length. Give its length in cm.

29. A driver is told to turn left after driving .48 km. How many meters is this?

30. A building is 92.6 m tall. How many kilometers is this?

31. Is 82 cm greater than or smaller than 1 m?

32. Is 1022 m greater than or smaller than 1 km?

CHALLENGE EXERCISES

Convert each of the following.

33. .00981 km to cm

34. 53187 cm to km

35. 37031.6 mm to km

36. .00183 km to mm

REVIEW EXERCISES

Write each decimal as a fraction in lowest terms. (For help, see ***Section 4.1.****)*

37. .875

38. .0065

39. .08

40. .075

41. .375

42. .0025

7.4 THE METRIC SYSTEM—VOLUME AND WEIGHT

The liter is the basic unit of volume or capacity in the metric system. A liter is the volume of a box measuring 10 cm on each side.

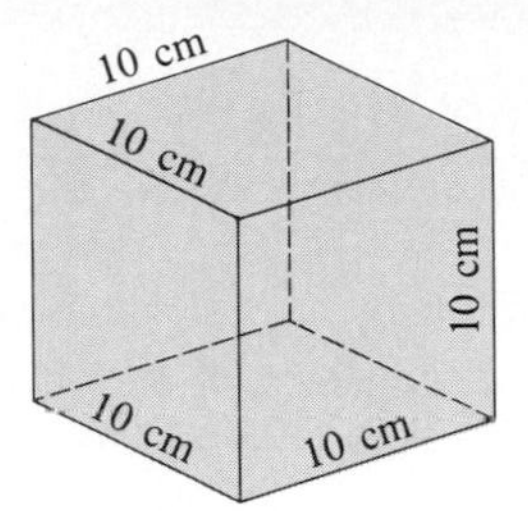

1 A liter is a little more than a quart. A liter is about the volume of $2\frac{3}{4}$ cans of soft drink.

Use the prefixes *milli, centi,* and *kilo* of the previous section with liters.

$$1000 \text{ ml} = 1\ \ell$$
$$100 \text{ cl} = 1\ \ell$$
$$.001 \text{ kl} = 1\ \ell$$

The script letter ℓ is used to symbolize liter—this avoids confusing the letter l with the numeral 1.

Convert among the various prefixes with the metric conversion line from the previous section or with unit fractions.

kilo	hecto	deka	unit	deci	centi	milli
1000	100	10	1	$\frac{1}{10}$	$\frac{1}{100}$	$\frac{1}{1000}$

EXAMPLE 1 Converting Units of Volume

Convert.

(a) 5 ℓ to cl

with the metric conversion line

Go two places to the right.

5 . 0 0. Attach two 0's.

$5\ \ell = 500$ cl

with unit fractions

$$5\ \ell \cdot \frac{100 \text{ cl}}{1\ \ell} = 500 \text{ cl}$$

(b) 20 ml to ℓ

with the metric conversion line

Go three places to the left.

$20 \text{ ml} = .02\ \ell$

with unit fractions

$$20 \text{ ml} \cdot \frac{1\ \ell}{1000 \text{ ml}} = .02\ \ell$$

WORK PROBLEM 1 AT THE SIDE.

OBJECTIVES

1 Know the basic metric units of volume.

2 Know the basic metric units of weight.

1. Convert.

(a) 6.2 ℓ to cl

(b) 300 ml to ℓ

(c) 17.1 kl to ℓ

(d) .96 ℓ to ml

ANSWERS

1. (a) 620 cl (b) .3 ℓ (c) 17,100 ℓ (d) 960 ml

2 The **gram** is the basic unit of weight in the metric system.

A gram is the weight of one cubic centimeter of water at 4°C.

In other words, the water in a box measuring 1 cm on each side weighs 1 gram. (4°C is a measure of the temperature of the water.)

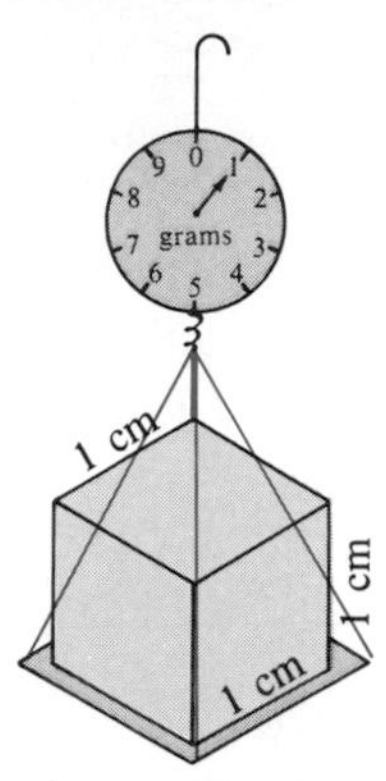

1 gram of water

One gram is about the weight of two paper clips. A nickel weighs about 5 grams. Since a gram is such a small weight, the **kilogram** (1000 grams) is the weight most often used. Use **g** for grams and **kg** for kilograms.

1 kg = 1000 g
.001 kg = 1 g

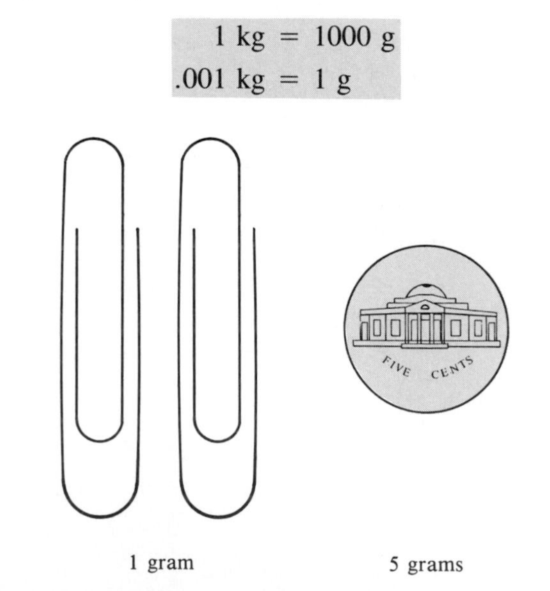

1 gram 5 grams

A kilogram is slightly greater than the weight of this book; 1 kg is about 2.2 pounds. The word **kilo** is a common abbreviation for kilogram.

Extremely small weights can be measured in **milligrams** and **centigrams.**

1000 mg = 1 g
100 cg = 1 g

NOTE These measures are so small that they are used mainly in scientific or medical applications.

EXAMPLE 2 Converting Units of Weight

Convert.

(a) 512 g to kg

(b) 825 cg to g

SOLUTION

(a) Go three places to the left.

$$512 \text{ g} = .512 \text{ kg}$$

Or with unit fractions,

$$512 \cancel{\text{g}} \cdot \frac{1 \text{ kg}}{1000 \cancel{\text{g}}} = .512 \text{ kg}.$$

(b) Go two places to the left.

$$825 \text{ cg} = 8.25 \text{ g}$$

Or with unit fractions,

$$825 \cancel{\text{cg}} \cdot \frac{1 \text{ g}}{100 \cancel{\text{cg}}} = 8.25 \text{ g}.$$ ■

WORK PROBLEM 2 AT THE SIDE.

2. Convert.

(a) 192 g to kg

(b) 25 mg to g

(c) 418 g to cg

(d) 2509 cg to g

ANSWERS

2. (a) .192 kg (b) .025 g (c) 41,800 cg (d) 25.09 g

name date hour

7.4 EXERCISES

Convert the following.

Example:		**Solution:**
378 ml to ℓ	**.378 ℓ**	378 ml to ℓ 378 ml = 378 · .001 ℓ = .378 ℓ

1. 6 ℓ to cl

2. 4.1 ℓ to ml

3. 8.7 ℓ to ml

4. 12.5 ℓ to cl

5. 925 cl to ℓ

6. 412 ml to ℓ

7. 8974 ml to ℓ

8. 5639 cl to ℓ

9. 8.4 ℓ to kl

10. 2.79 ℓ to kl

11. 8.64 kl to ℓ

12. 4.53 kl to ℓ

13. 8000 g to kg

14. 25,000 g to kg

15. 5.2 kg to g

16. 12.42 kg to g

17. 4.2 g to mg

18. 3.89 g to cg

19. 7634 cg to g

20. 598 mg to g

Today, medical measurements are usually given in the metric system. Since we convert among metric units of measure by moving the decimal point, it is possible that mistakes can be made. Examine the following dosages and indicate whether they are reasonable or unreasonable.

21. Drink 2.1 kiloliters of Kaopectate after each meal.

22. Apply a bandage 5 centimeters square as needed.

23. Soak your feet in 3 milligrams of Epsom salts per 4 liters of water.

24. Inject $\frac{1}{2}$ liter of insulin each morning.

25. Take 15 milliliters of cough syrup every four hours.

26. Take 2 kilograms of aspirin three times a day.

27. Administer 94.3 liters of antibiotic every six hours.

28. Take an aspirin weighing .002 gram.

CHALLENGE EXERCISES

Solve the following word problems.

29. One nickel weighs 5 grams. How many nickels are in 1 kilogram of nickels?

30. Sea water contains about 3.5 grams of salt per 1000 milliliters of water. How many grams of salt would be in 1 liter of sea water?

31. Helium weighs about .0002 grams per milliliter. How much would 1 liter of helium weigh?

32. About 1500 grams of sugar can be dissolved in a liter of warm water. How much sugar could be dissolved in 1 milliliter of warm water?

REVIEW EXERCISES

Multiply. (For help, see **Section 4.5.***)*

33. $\begin{array}{r} .454 \\ \times \quad 18 \\ \hline \end{array}$

34. $\begin{array}{r} 38.625 \\ \times \quad 12 \\ \hline \end{array}$

35. $\begin{array}{r} 1.6093 \\ \times \quad 24 \\ \hline \end{array}$

36. 12×1.506

37. 28×23.6078

38. 8×4.04

7.5 METRIC TO ENGLISH CONVERSIONS

1 Until everyone thinks in the metric system as naturally as they do in the English system, it will be necessary to make conversions from one system to the other. Approximate conversions can be made with the help of the following table.

Metric to English			*English to Metric*		
from metric	to English	multiply by	from English	to metric	multiply by
meters	yards	**1.094**	yards	meters	**.9144**
meters	feet	**3.28**	feet	meters	**.305**
meters	inches	**39.37**	inches	meters	**.0254**
kilometers	miles	**.6214**	miles	kilometers	**1.6093**
grams	pounds	**.0022**	pounds	grams	**454**
kilograms	pounds	**2.20**	pounds	kilograms	**.454**
liters	quarts	**1.06**	quarts	liters	**.946**
liters	gallons	**.264**	gallons	liters	**3.785**

EXAMPLE 1 Converting Metric and English (Length)

Convert.

(a) 10 meters to yards

(b) 24 yards to meters

(c) 6 meters to inches

SOLUTION

(a) In the "Metric to English" table, the number for conversion from meters to yards is 1.094. Multiply 10 and 1.094.

$$10 \text{ meters} = 10 \cdot 1.094 = 10.94 \text{ yards}$$

Use a unit fraction as follows.

$$10 \text{ meters} = 10 \text{ meters} \cdot \frac{1.094 \text{ yards}}{\text{meters}} = 10.94 \text{ yards}$$

(b) In the "English to Metric" table, the number for yards to meters is .9144. Multiply 24 and .9144.

$$24 \text{ yards} = 24 \cdot .9144 = 21.9456 \text{ meters}$$

(c) $6 \text{ meters} = 6 \cdot 39.37 = 236.22 \text{ inches}$ ■

WORK PROBLEM 1 AT THE SIDE.

OBJECTIVES

1 Convert from metric to English. Convert from English to metric.

2 Convert temperatures by using the order of operations.

1. Convert.

(a) 23 meters to yards

(b) 12.2 meters to inches

(c) 29.3 meters to feet

(d) 109.3 meters to yards

(e) 35 yards to meters

(f) 14.7 feet to meters

ANSWERS

1. (a) 25.162 yards (b) 480.314 inches (c) 96.104 feet (d) 119.5742 yards (e) 32.004 meters (f) 4.4835 meters

EXAMPLE 2 Converting Metric and English (Weight)
Convert.

(a) 5 kilograms to pounds

(b) 85 pounds to kilograms

(c) .25 pounds to grams

SOLUTION
In the "Metric to English" table, find the number for kilograms to pounds.

(a) 5 kilograms = $5 \cdot 2.20$ = 11 pounds

(b) 85 pounds = $85 \cdot .454$ = 38.59 kilograms

(c) .25 pounds = $.25 \cdot 454$ = 113.5 grams ■

WORK PROBLEM 2 AT THE SIDE.

2. Convert.

(a) 17 kilograms to pounds

(b) 215 pounds to kilograms

(c) 1.92 pounds to grams

EXAMPLE 3 Converting Metric and English (Volume)
Convert the following.

(a) 9 liters to quarts

(b) 25 quarts to liters

(c) 18 gallons to liters

SOLUTION
In the "Metric to English" table, find the number for liters to quarts.

(a) 9 liters = $9 \cdot 1.06$ = 9.54 quarts

(b) 25 quarts = $25 \cdot .946$ = 23.65 liters

(c) 18 gallons = $18 \cdot 3.785$ = 68.13 liters ■

WORK PROBLEM 3 AT THE SIDE.

3. Convert.

(a) 5 liters to quarts

(b) 32 quarts to liters

(c) 50 gallons to liters

2 In the metric system, temperature is measured on the **Celsius scale.** On the Celsius scale, water freezes at 0° and boils at 100°. The thermometer on the next page shows some typical temperatures in both Celsius and Fahrenheit (the system we now use). For example, room temperature is 20°C or 68°F, and body temperature is about 37°C or 98.6°F.

Convert a reading from Fahrenheit to Celsius with the following formula.

$$C = \frac{5(F - 32)}{9}$$

ANSWERS
2. (a) 37.4 pounds (b) 97.61 kilograms (c) 871.68 grams
3. (a) 5.3 quarts (b) 30.272 liters (c) 189.25 liters

By using this formula and the order of operations, a temperature in Fahrenheit can be converted to Celsius. Recall the order of operations from Chapter 1 and Chapter 3.

1. Do all operations inside parentheses.
2. Simplify any expressions with exponents and find any square roots.
3. Multiply or divide from left to right.
4. Add or subtract from left to right.

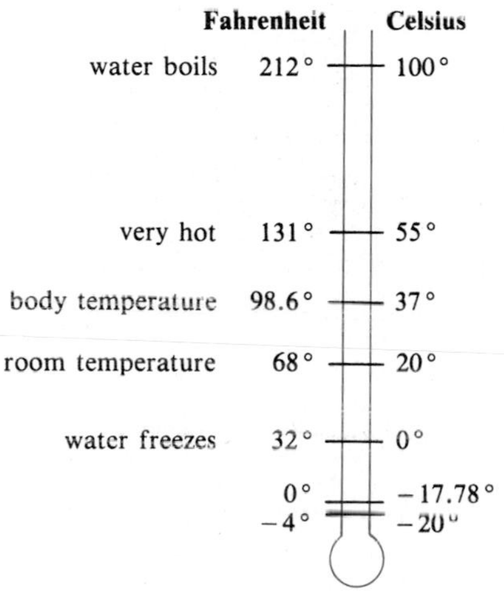

	Fahrenheit	Celsius
water boils	212°	100°
very hot	131°	55°
body temperature	98.6°	37°
room temperature	68°	20°
water freezes	32°	0°
	0°	−17.78°
	−4°	−20°

EXAMPLE 4 Converting Fahrenheit to Celsius
Convert 68°F to Celsius.

SOLUTION
Use the formula and the order of operations.

$$C = \frac{5(F - 32)}{9}$$

$$= \frac{5(68 - 32)}{9}$$

$$= \frac{5(36)}{9} \quad \text{Work in parentheses first.}$$

$$= \frac{5(\cancel{36}^{\,4})}{\cancel{9}_{\,1}} \quad \text{Use cancellation.}$$

$$= 20 \quad \text{Multiply.}$$

Thus, 68°F = 20°C. ■

WORK PROBLEM 4 AT THE SIDE.

4. Convert to Celsius.

(a) 104°F

(b) 176°F

(c) 212°F

(d) 98.6°F

ANSWERS
4. (a) 40°C (b) 80°C (c) 100°C (d) 37°C

5. Convert to Fahrenheit.

(a) 100°C

(b) 25°C

(c) 80°C

Convert from Celsius to Fahrenheit with the following formula.

$$F = \frac{9 \cdot C}{5} + 32$$

EXAMPLE 5 Converting Celsius to Fahrenheit
Convert 15°C to Fahrenheit.

SOLUTION
Use the formula and the order of operations.

$$\begin{aligned} F &= \frac{9 \cdot C}{5} + 32 \\ &= \frac{9 \cdot 15}{5} + 32 \\ &= \frac{9 \cdot \overset{3}{\cancel{15}}}{\underset{1}{\cancel{5}}} + 32 && \text{Use cancellation.} \\ &= 27 + 32 && \text{Multiply.} \\ &= 59 && \text{Add.} \end{aligned}$$

Thus, 15°C = 59°F. ■

WORK PROBLEM 5 AT THE SIDE.

ANSWERS
5. (a) 212°F (b) 77°F (c) 176°F

name date hour

7.5 EXERCISES

Use the table on page 453 to make the following conversions from metric to English and English to metric. Round answers to the nearest tenth of a unit.

Example: 36 meters to yards **Solution:** 36 meters to yards
$36 \cdot 1.094 = 39.384 =$ **39.4 yards**

1. 20 meters to yards

2. 50 yards to meters

3. 80 meters to feet

4. 2.9 meters to inches

5. 16 feet to meters

6. 3.2 yards to meters

7. 982 yards to meters

8. 12.2 kilometers to miles

9. 47.2 pounds to grams

10. 7.68 kilograms to pounds

11. 28.6 liters to quarts

12. 59.4 liters to quarts

13. 28.2 gallons to liters

14. 16 quarts to liters

Write the following measurements in the units indicated.

15. your height in meters

16. your height in centimeters

17. your weight in grams

18. your weight in kilograms

Solve the following word problems.

19. Suppose you decide to put together a do-it-yourself picture frame measuring 22 cm by 31 cm. The wood for the frame costs $1.20 per foot. Find the approximate cost of the wood. (*Hint:* start by finding the number of meters of wood in the frame.)

20. Gasoline sells for $1.40 per gallon. Find the cost of 1 liter.

21. Paint sells for \$9.20 per gallon. Find the cost of 4 liters.

22. A 3-liter bottle of beverage sells for \$2.80. A 1-gallon bottle of the same beverage sells for \$3.50. What is the better value?

Use the conversion formulas on pages 454 and 456 and the order of operations to convert Fahrenheit temperatures to Celsius and Celsius temperatures to Fahrenheit. Round to the nearest degree if necessary.

23. 68°F

24. 86°F

25. 104°F

26. 536°F

27. 98°F

28. 114°F

29. 35°C

30. 0°C

31. 25°C

32. 100°C

33. 15°C

34. 40°C

CHALLENGE EXERCISES

Solve the following word problems. Round to the nearest degree.

35. The highest temperature ever recorded on earth was 136°F at Aziza, Libya. Convert this temperature to Celsius.

36. A recipe for French pastry calls for an oven temperature of 175°C. Convert this to Fahrenheit.

37. A kiln for firing pottery reaches a temperature of 500°C. Convert this to Fahrenheit.

38. The temperature of the water in a lake in December is 34°F. Convert this to Celsius.

REVIEW EXERCISES

Work each problem. (For help, see ***Section 1.8****.)*

39. $2 \cdot 8 + 2 \cdot 8$

40. $2 \cdot 12.2 + 2 \cdot 5.6$

41. 9^2

42. 7^2

43. $(5^2) + (4^2)$

44. $(12^2) + (3^2)$

CHAPTER 7 SUMMARY

KEY TERMS

7.1 English system — The English system of measurement (American system of units) is the most common system of measurement used in the United States. Common units in this system include quarts, pounds, feet, and miles.

metric system — The metric system of measurement is an international system of measurement used in manufacturing, science, medicine, sports, and other fields. The system uses liters, grams, and meters.

7.2 denominate number — A denominate number is a number with a unit of measure. For example, 6 dollars, 8 feet, 4 years, and 10 pounds are denominate numbers.

7.3 metric conversion line — The metric conversion line is a line showing the various metric measurement prefixes and their size relationship to each other. It looks like this.

kilo	hecto	deka	unit	deci	centi	milli
1000	100	10	1	$\frac{1}{10}$	$\frac{1}{100}$	$\frac{1}{1000}$

7.5 Celsius — The Celsius scale is the scale used to measure temperature in the metric system. Water boils at 100°C and freezes at 0°C on the Celsius scale.

Fahrenheit — The Fahrenheit scale is the scale used to measure temperature in the English system. Water boils at 212°F and freezes at 32°F on the Fahrenheit scale.

QUICK REVIEW

Section Number and Topic	Approach	Example
7.1 Knowing and Converting English Measurement	To convert units, multiply when changing from a greater unit to a smaller unit; divide when changing from a smaller unit to a greater unit.	Convert each measurement. **(a)** 5 feet to inches 5 feet = $5 \cdot 12$ = 60 inches **(b)** 3 pounds to ounces 3 pounds = $3 \cdot 16$ = 48 ounces **(c)** 15 quarts to gallons 15 quarts = $\frac{15}{4} = 3\frac{3}{4}$ gallons
7.1 Using Unit Fractions	Conversions are made by multiplying by a unit fraction. The unit being converted should be in the numerator. The unit being eliminated should be in the denominator.	Convert 32 ounces to pounds. 32 ounces $= \frac{32 \text{ ounces}}{1} \cdot \frac{1 \text{ pound}}{16 \text{ ounces}}$ unit fraction $= \frac{\overset{2}{\cancel{32}} \cancel{\text{ounces}}}{1} \cdot \frac{1 \text{ pound}}{\underset{1}{\cancel{16}} \cancel{\text{ounces}}}$ Cancel unit. = 2 pounds multiply

Section Number and Topic	Approach	Example
7.2 Simplifying a Denominate Number	Divide the number by the number of the smaller units contained in one larger unit. Express any remainder with the smaller unit.	Simplify 6 feet 14 inches. 6 feet 14 inches 14 inches = $\frac{\text{14 inches}}{\text{12 number of inches in 1 foot}}$ = 1 foot 2 inches 6 feet + 1 foot 2 inches = 7 feet 2 inches
7.2 Adding Denominate Numbers	Write the numbers vertically with like units lined up. Add number as usual. Simplify the answer if possible.	Add 3 feet 8 inches and 4 feet 6 inches. 3 feet 8 inches 4 feet 6 inches 7 feet 14 inches 7 feet + 1 foot 2 inches 8 feet 2 inches
7.2 Subtracting Denominate Numbers	Write the numbers vertically with like units lined up. Subtract units from units, borrowing if necessary.	Subtract 2 weeks 3 days from 8 weeks 1 day. borrow 1 week = 7 days 7 days + 1 day = 7 8 ~~8~~ weeks ~~1~~ day − 2 weeks 3 days 5 weeks 5 days
7.2 Multiplying a Denominate Number	Multiply each unit of the denominate number. Simplify the answer.	Multiply 5 gallons 3 quarts by 3. 5 gallons 3 quarts × 3 15 gallons 9 quarts Since 4 quarts = 1 gallon, 9 quarts = 2 gallons 1 quart, and 15 gallons + 2 gallons 1 quart = 17 gallons 1 quart.

Section Number and Topic	Approach	Example
7.2 Dividing a Denominate Number	Divide each part of the denominate number separately, beginning at the left. Any remainder must be added to the next smaller unit. Continue dividing until all the parts have been divided.	Divide 7 pounds 13 ounces by 5. 1 pound 9 ounces 5)7 pounds 13 ounces 5 pounds 2 pounds = 32 ounces remainder 45 (13 + 32 = 45) 45 0
7.3 and **7.4** Converting Units Within the Metric System	Use the metric conversion line, moving the decimal point the proper number of places to either the right or left. kilo 1000 · hecto 100 · deka 10 · unit 1 · deci $\frac{1}{10}$ · centi $\frac{1}{100}$ · milli $\frac{1}{1000}$	Convert each of the following. **(a)** 68.2 km to m 6 8 . 2 0 0 Decimal is moved three places to the right. 68.2 km = 68,200 m **(b)** 300 ml to ℓ 3 0 0 . Decimal is moved three places to the left. 300 ml = .3ℓ **(c)** 825 cg to g 8 2 5 . Decimal is moved two places to the left. 825 cg = 8.25 g
7.5 Converting from Metric to English and English to Metric	Multiply the given quantity by the value from the table.	Convert the following. **(a)** 23 meters to yards 23 × 1.094 = 25.162 yards **(b)** 1.92 pounds to grams 1.92 × 454 = 871.68 grams **(c)** 5 liters to quarts 5 × 1.06 = 5.3 quarts

Section Number and Topic	Approach	Example
7.5 Converting Fahrenheit to Celsius	Use the formula $$C = \frac{5(F - 32)}{9}$$ and the order of operations.	Convert 176°F to Celsius. $C = \frac{5(176 - 32)}{9}$ $= \frac{5(\overset{16}{\cancel{144}})}{\underset{1}{\cancel{9}}}$ Cancel. $= 80$ Multiply. 176°F = 80°C
7.5 Converting Celsius to Fahrenheit	Use the formula $$F = \frac{9 \cdot C}{5} + 32$$ and the order of operations.	Convert 80°C to Fahrenheit. $F = \frac{9 \cdot 80}{5} + 32$ $= \frac{9 \cdot \overset{16}{\cancel{80}}}{\underset{1}{\cancel{5}}} + 32$ Cancel. $= 144 + 32$ Multiply. $= 176$ Add. 80°C = 176°F

name date hour

CHAPTER 7 REVIEW EXERCISES

[7.1] *Convert the following.*

1. 1 pound = ______ ounces **2.** ______ feet = 1 yard **3.** 1 ton = ______ pounds

4. ______ quarts = 1 gallon **5.** ______ feet = 1 mile **6.** 1 yard = ______ inches

7. 12 feet = ______ yards **8.** 7 days = ______ hours **9.** 4 miles = ______ feet

Simplify the following.

10. 1 foot 19 inches **11.** 2 miles 6000 feet

12. 2 yards 7 feet 15 inches **13.** 2 quarts 3 pints 1 cup

[7.2] *Solve the following and simplify by using the units given.*

14.
```
   5 feet 7 inches
+  2 feet 8 inches
```

15.
```
   8 pounds 6 ounces
−  3 pounds 8 ounces
```

16.
```
   5 pounds 10 ounces
×                   5
```

17. 2)10 feet 8 inches

[7.3] *Convert each of the following measures as indicated.*

18. 1 m to cm **19.** 1 km to m **20.** 8 m to cm

21. 3781 mm to m

22. .056 m to mm

23. 1.27 km to m

[7.4] *Convert each of the following measures as indicated.*

24. 3 ℓ to cl

25. 680 g to kg

26. 485 cg to mg

27. 17.6 kl to ℓ

28. 5.3 g to cg

29. 5000 g to kg

[7.5] *Use the table to make the following conversions. Round answers to the nearest tenth of a unit.*

30. 10 m to yards

31. 1.4 m to inches

32. 108 km to miles

33. 800 miles to km

34. 23 quarts to ℓ

35. 41.5 ℓ to quarts

Solve the following word problems.

36. The speed limit is 55 miles per hour. How many kilometers per hour is this?

37. A certain fabric sells for $6.95 per yard. Find the selling price per meter.

Use the conversion formulas on page 454 and the order of operations to convert each temperature to Fahrenheit or Celsius. Round to the nearest tenth of a degree if necessary.

38. 77°F

39. 176°F

40. 92°F

41. 159°F

42. 50°C

43. 280°C

44. Water boils at 100°C. Convert this temperature to Fahrenheit.

45. The thermostat on an automobile opens at 180°F. Convert this temperature to Celsius.

MIXED REVIEW EXERCISES

Solve the following and simplify by using the units given.

46.
5 quarts 2 pints
+ 1 quart 1 pint

47.
7 weeks 5 days 3 hours
− 2 weeks 6 days 9 hours

48.
5 weeks 2 days 9 hours
× 4

49. 5)6 gallons 2 quarts 3 pints

Convert each of the following measures as indicated.

50. 2.75 cm to mm

51. 42,885 mm to m

52. 1.3 km to cm

53. 7835 mg to cg

54. 67,500 mg to kg

55. .0894 kl to ℓ

Convert the following.

56. ____ ounces = 5 pounds

57. 90 inches = ____ feet

58. 364 days = ____ weeks

Simplify the following.

59. 2 days 25 hours 150 minutes

60. 2 miles 11,350 feet 110 inches

Use the table in the text to make the following conversions. Round answers to the nearest tenth of a unit.

61. 24 feet to m

62. 5.8 yards to meters

63. 52.3 gallons to ℓ

64. 3.5 pounds to grams

Solve the following word problems.

65. Gasoline sells for $2.00 per gallon. Find the price per liter.

66. A commercial chemical tank is labeled "1000-liter capacity." How many gallons will the tank hold?

67. Find the length of a football field in meters. (*Hint:* a football field is 100 yards long.)

68. A welding torch is advertised as reaching 5000°F. Convert this temperature to Celsius.

name date hour

CHAPTER 7 TEST

Convert the following measurements.

1. 1 gallon = ? quarts

2. 36 feet = ? yards

3. 228 hours = ? days

4. 42 pints = ? gallons

5. 128 ounces = ? pounds

6. 38,016 feet = ? miles

7. Solve.
9 feet 4 inches
+ 7 feet 9 inches

8. Solve.
5 gallons 3 quarts 1 pint
− 2 gallons 2 quarts 2 pints

9. Solve.
5 pounds 13 ounces
× 6

10. Solve. 4)6 yards 2 feet 8 inches

Convert the following measurements as indicated.

11. 250 cm to m

12. 4.6 m to cm

13. .4 km to cm

14. 8412 g to kg

15. 9 ℓ to cl

16. 15.6 cl to ml

1. ______
2. ______
3. ______
4. ______
5. ______
6. ______
7. ______
8. ______
9. ______
10. ______
11. ______
12. ______
13. ______
14. ______
15. ______
16. ______

17. ______________

17. The distance from the library to the bookstore is 198 meters. How many kilometers is this?

18. ______________

18. A chemical sample weighs 2.61 kilograms. Give its weight in grams.

Convert the following measurements as indicated. Round to the nearest tenth of a unit.

19. ______________
20. ______________

19. 18.4 feet to meters

20. 72.1 pounds to kilograms

21. ______________
22. ______________

21. 948 grams to pounds

22. 6.4 miles to kilometers

23. ______________

23. Unleaded gasoline sells for $.33 per liter. Find the cost of 1 gallon.

24. ______________

24. A lap-top computer weighs 7.7 pounds. Find the weight in kilograms.

Use the conversion formulas on page 454 and the order of operations to convert the following temperatures. Round to the nearest tenth of a degree.

25. ______________
26. ______________

25. 76°F to Celsius

26. 172°C to Fahrenheit

8

GEOMETRY

Geometric figures and shapes are all around us—from houses composed of rectangular and square shapes to spherically and cylindrically shaped oil storage tanks. This chapter discusses the basic terms of geometry and the most common geometric shapes.

8.1 BASIC GEOMETRIC TERMS

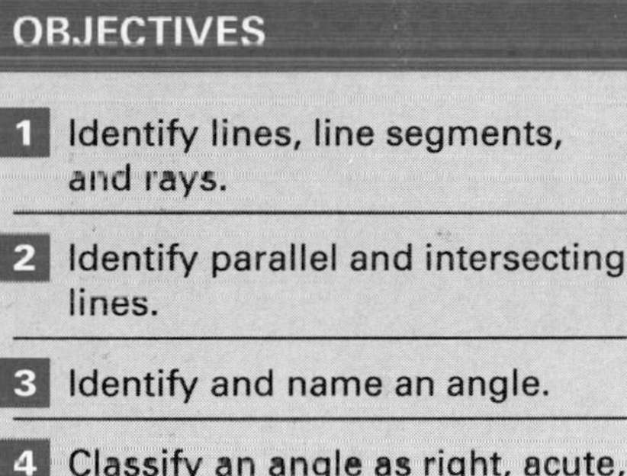
OBJECTIVES

1. Identify lines, line segments, and rays.
2. Identify parallel and intersecting lines.
3. Identify and name an angle.
4. Classify an angle as right, acute, straight, or obtuse.
5. Identify perpendicular lines.

Geometry starts with the idea of a point. A **point** is a location in space. A point is represented by a dot and is named by writing a capital letter next to the dot.

Point P

1 A **line** is a straight collection of points that goes on forever in opposite directions. A line is drawn by using arrowheads to show that it never ends. The line is named by using the letters of any two points on the line.

A part of a line that has two endpoints is called a **line segment.** A line segment is named for its endpoints. The segment with endpoints P and Q is shown below. It can be named $\overline{PQ}$ or $\overline{QP}$.

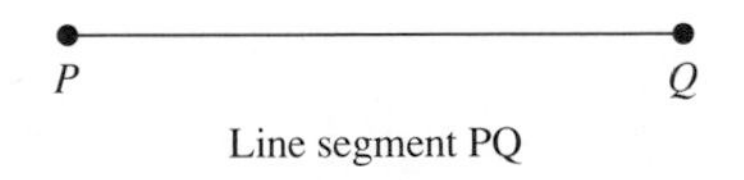

Line segment PQ

A **ray** is a part of a line that has only one endpoint and continues forever in one direction. A ray is named by using the endpoint and some other point on the ray. The endpoint is always mentioned first when naming a ray.

Ray PQ

1. Identify each of the following as a line, line segment, or ray.

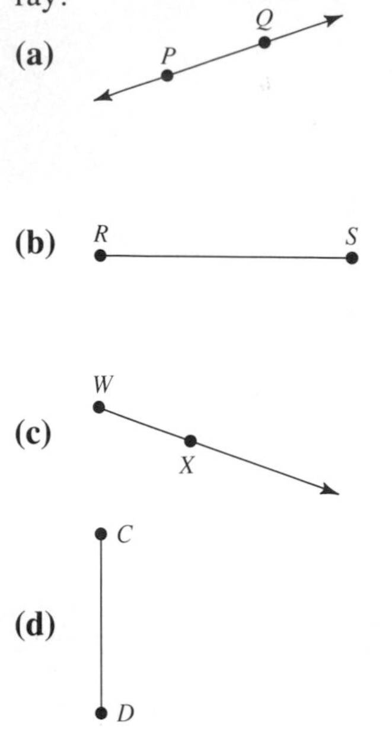

2. Label each pair of lines as parallel or intersecting.

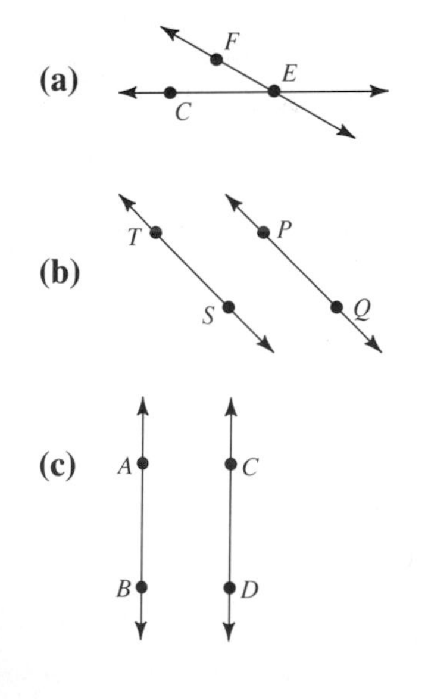

ANSWERS
1. (a) line (b) line segment (c) ray (d) line segment
2. (a) intersecting (b) parallel (c) parallel

EXAMPLE 1 Identifying Lines, Rays, and Line Segments
Identify each of the following as a line, line segment, or ray.

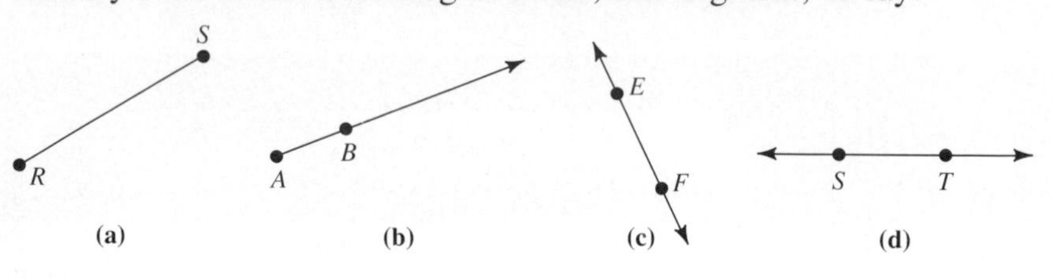

SOLUTION
Both **(c)** and **(d)** continue forever in both directions, so they are lines. **(a)** has two endpoints, so it is a line segment. **(b)** starts at *A* and continues forever. It is a ray. ■

WORK PROBLEM 1 AT THE SIDE.

2 Lines that are in the same plane but that never intersect are called **parallel lines,** while lines that cross or merge are called **intersecting lines.**

EXAMPLE 2 Identifying Parallel and Intersecting Lines
Label each pair of lines as parallel or intersecting.

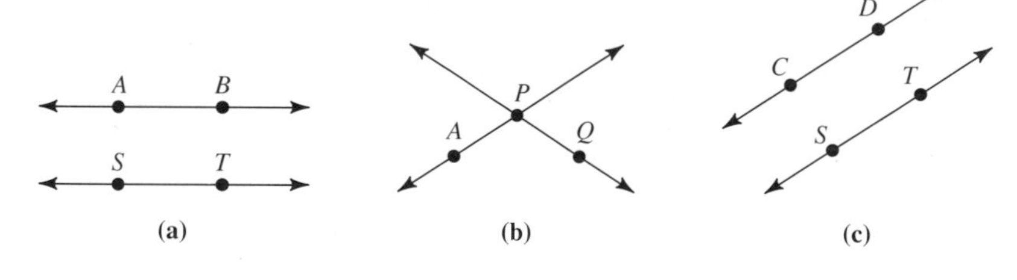

SOLUTION
The lines in **(a)** and **(c)** never intersect. They are parallel lines. The lines in **(b)** cross at *P*, so they are intersecting lines. ■

WORK PROBLEM 2 AT THE SIDE.

3 An **angle** is a geometric figure that consists of two rays that have a common endpoint. This common endpoint is called the **vertex.**

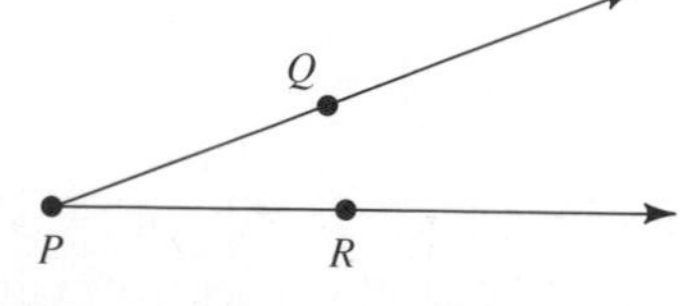

The rays *PQ* and *PR* are called *sides*. The angle can be named

$$\text{angle } QPR, \quad \angle QPR, \quad \angle RPQ, \quad \text{or} \quad \angle P.$$

NOTE The name of the vertex is in the middle or is written alone.

EXAMPLE 3 Identifying and Naming an Angle
Identify the highlighted angle.

The vertex of the angle is P, and the angle can be named as $\angle BPA$ or $\angle APB$ or $\angle P$. ■

WORK PROBLEM 3 AT THE SIDE.

3. Identify the highlighted angle.

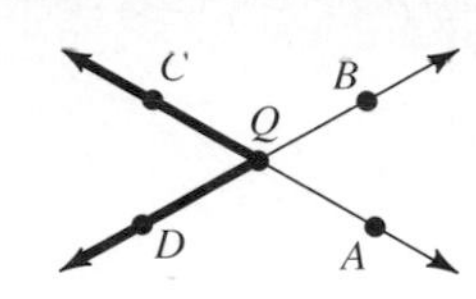

4 Angles are measured by a unit that is called a **degree,** denoted by the symbol °. To determine the size of an angle of 1°, divide a circle into 360 equal parts. Then draw two rays from the center to two adjacent points on the circle. The angle formed measures 1°.

If a circle is divided into four equal segments, each angle measures 360° ÷ 4 or 90°.

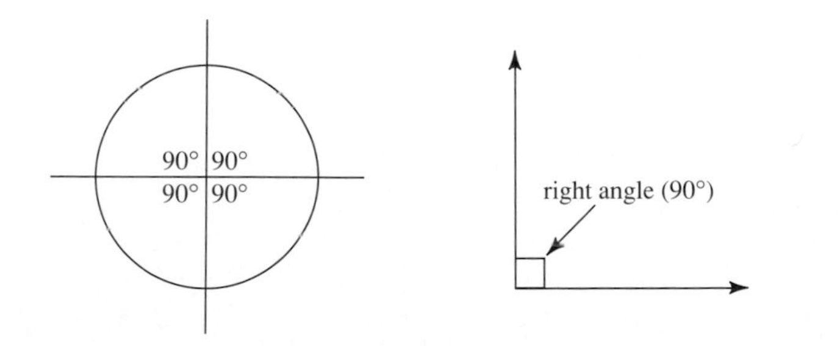

This particular angle has a special name. A **right angle** is an angle that measures 90°. It is indicated by a box drawn between the two sides (or rays of the endpoint).

Some other types of angles include the following:

an **acute angle,** which is an angle that measures between 0° and 90°;

an **obtuse angle,** which is an angle that measures between 90° and 180°;

a **straight angle,** which is an angle that measures 180°.

ANSWERS
3. $\angle CQD$ or $\angle DQC$ or $\angle Q$

4. Label each of the following as acute, right, obtuse, or straight angles.

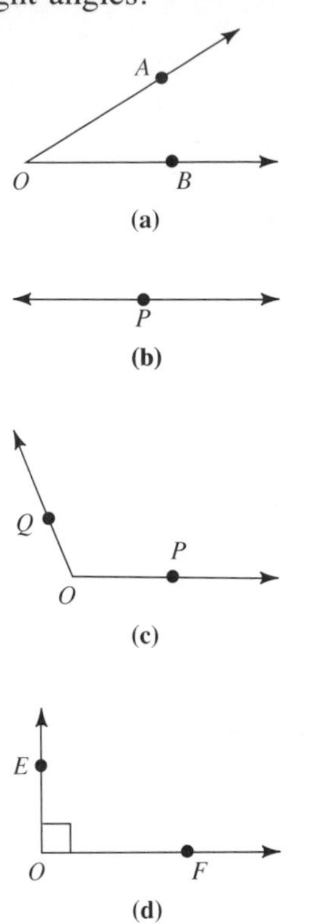

5. Which of the following pairs of lines are perpendicular?

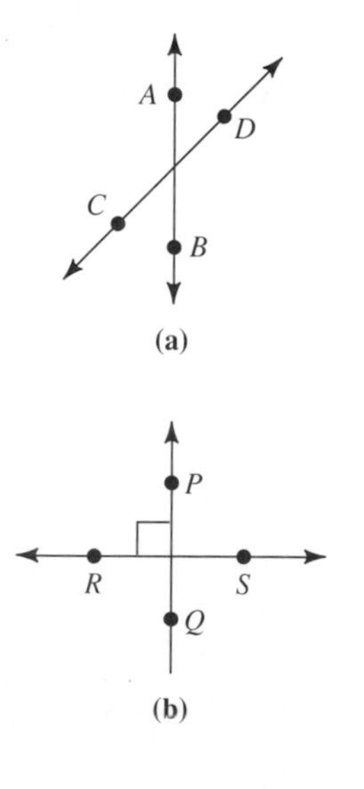

ANSWERS
4. (a) acute (b) straight (c) obtuse (d) right
5. (b)

EXAMPLE 4 Classifying an Angle

Label each of the following angles as acute, right, obtuse, or straight.

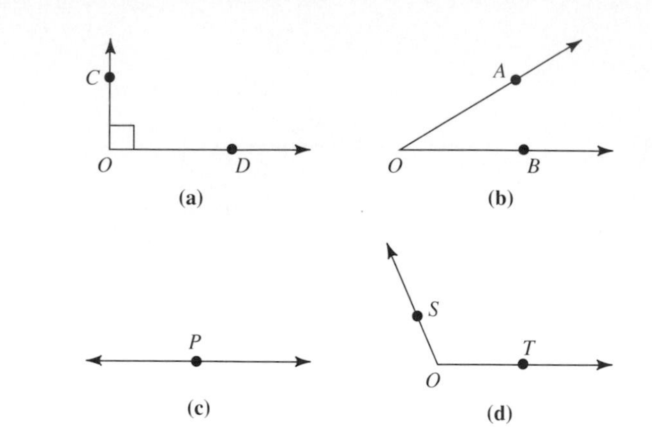

SOLUTION

The angle in **(a)** is a right angle (identified by a box). The angle in **(b)** is an acute angle (between 0° and 90°). The angle in **(c)** is a straight angle (180°), and the angle in **(d)** is an obtuse angle (between 90° and 180°). ■

WORK PROBLEM 4 AT THE SIDE.

Two lines are perpendicular if they intersect to form a right angle.

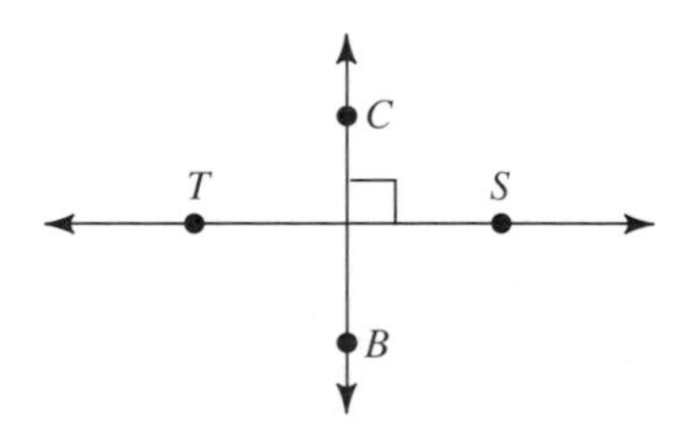

Lines *CB* and *ST* are perpendicular, since they intersect at right angles. This can be written in the following way: $\overleftrightarrow{CB} \perp \overleftrightarrow{ST}$.

EXAMPLE 5 Identifying Perpendicular Lines

Which of the following pairs of lines are perpendicular?

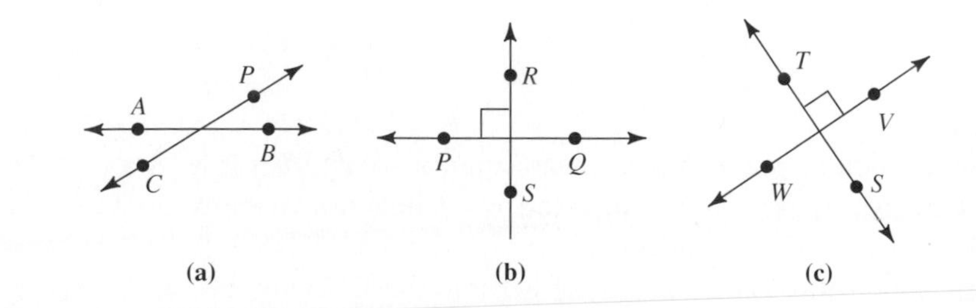

SOLUTION

The lines in **(b)** and the lines in **(c)** are perpendicular to each other, since they intersect at right angles. ■

WORK PROBLEM 5 AT THE SIDE.

name date hour

8.1 EXERCISES

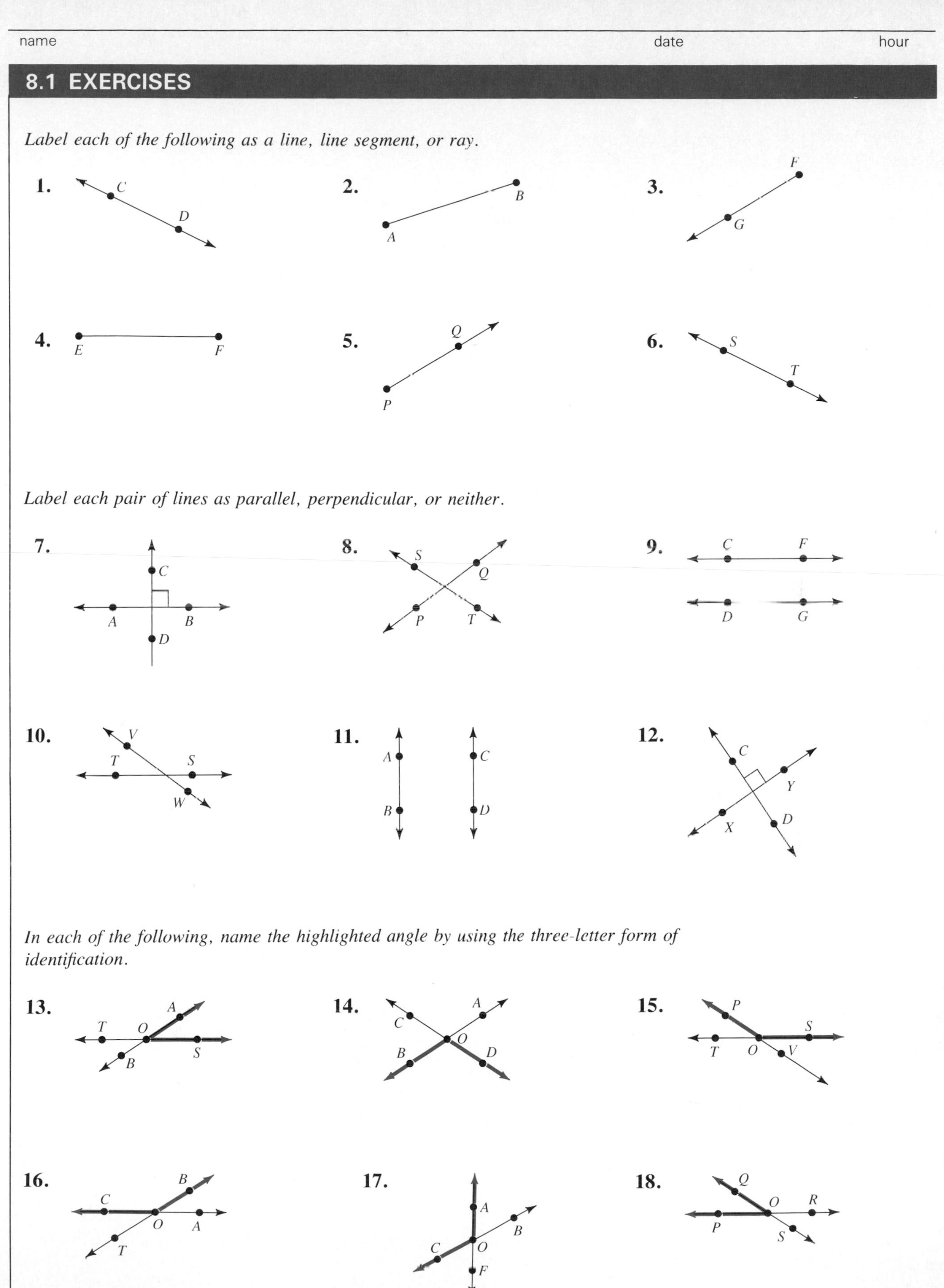

Label each of the following as a line, line segment, or ray.

1.

2.

3.

4.

5.

6.

Label each pair of lines as parallel, perpendicular, or neither.

7.

8.

9.

10.

11.

12.

In each of the following, name the highlighted angle by using the three-letter form of identification.

13.

14.

15.

16.

17.

18.

Label each of the following as acute, right, obtuse, or straight angles.

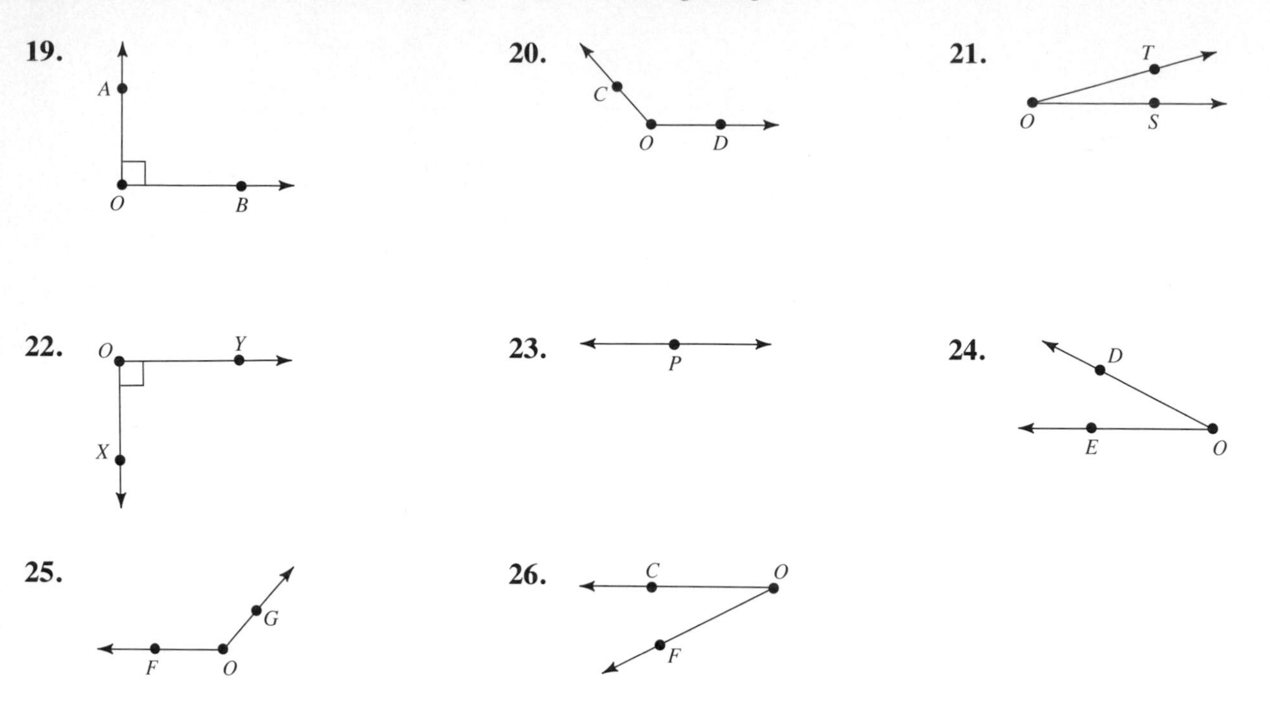

CHALLENGE EXERCISES

Use the diagram given to label the statements that follow as true or false.

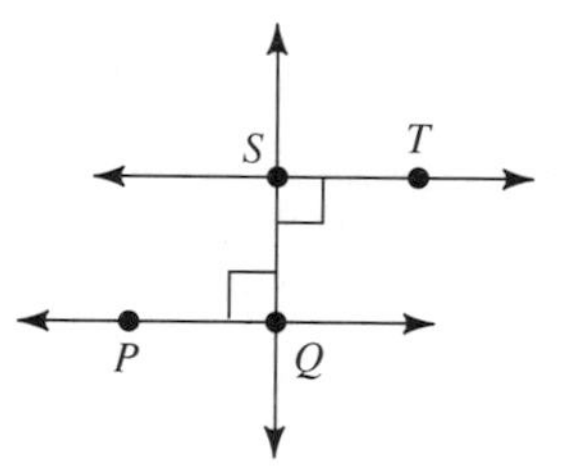

27. $\angle SQP$ is greater than 90°.

28. Lines SQ and PQ are perpendicular.

29. $\angle SQP$ is smaller than $\angle QST$.

30. Lines ST and PQ are parallel.

REVIEW EXERCISES

Evaluate the following expressions. (For help, see ***Sections 1.2 and 1.3****.)*

31. $36 + 45$

32. $83 - 26$

33. $180 - 75 - 15$

34. $115 + (80 - 45)$

35. $63 - 37 + 43$

36. $90 - (37 + 15)$

8.2 ANGLES AND THEIR RELATIONSHIPS

1 Two angles are called **complementary,** if their sum is 90°. If two angles are complementary, either angle is said to be the **complement** of the other.

EXAMPLE 1 Identifying Complementary Angles
Identify each pair of complementary angles.

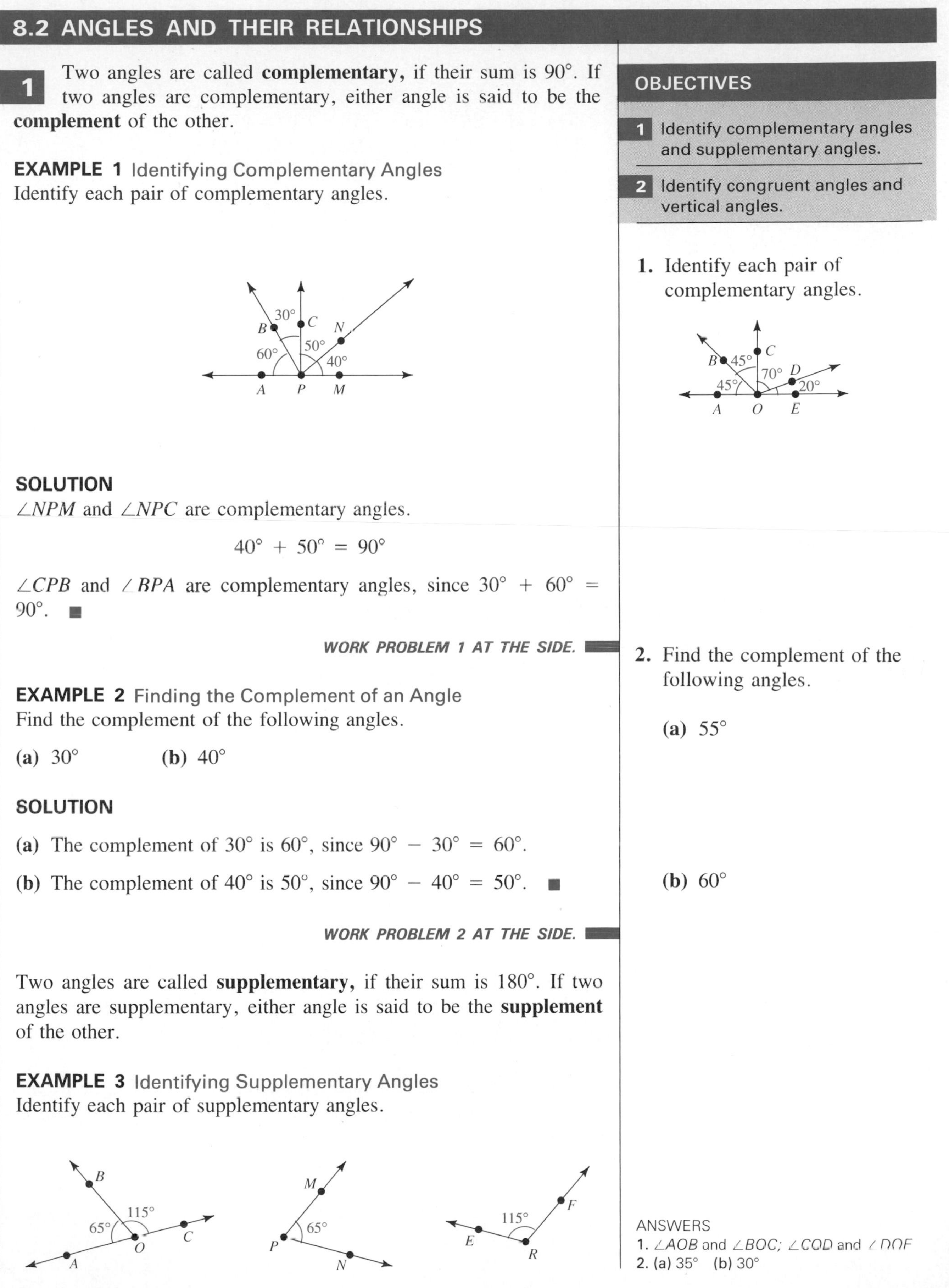

SOLUTION
$\angle NPM$ and $\angle NPC$ are complementary angles.

$$40° + 50° = 90°$$

$\angle CPB$ and $\angle BPA$ are complementary angles, since $30° + 60° = 90°$. ■

WORK PROBLEM 1 AT THE SIDE.

EXAMPLE 2 Finding the Complement of an Angle
Find the complement of the following angles.

(a) 30° **(b)** 40°

SOLUTION

(a) The complement of 30° is 60°, since $90° - 30° = 60°$.

(b) The complement of 40° is 50°, since $90° - 40° = 50°$. ■

WORK PROBLEM 2 AT THE SIDE.

Two angles are called **supplementary,** if their sum is 180°. If two angles are supplementary, either angle is said to be the **supplement** of the other.

EXAMPLE 3 Identifying Supplementary Angles
Identify each pair of supplementary angles.

OBJECTIVES

1 Identify complementary angles and supplementary angles.

2 Identify congruent angles and vertical angles.

1. Identify each pair of complementary angles.

2. Find the complement of the following angles.

(a) 55°

(b) 60°

ANSWERS
1. $\angle AOB$ and $\angle BOC$; $\angle COD$ and $\angle DOE$
2. (a) 35° (b) 30°

3. Identify each pair of supplementary angles.

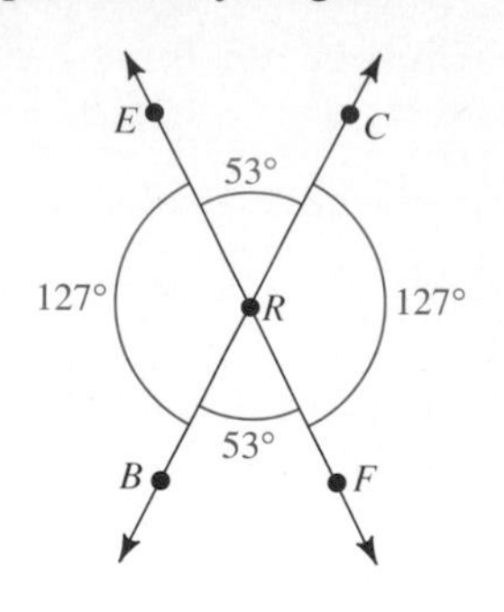

SOLUTION

$\angle BOA$ and $\angle BOC$, since $115° + 65° = 180°$.
$\angle BOA$ and $\angle ERF$, since $115° + 65° = 180°$.
$\angle BOC$ and $\angle MPN$, since $115° + 65° = 180°$.
$\angle MPN$ and $\angle ERF$, since $115° + 65° = 180°$. ■

WORK PROBLEM 3 AT THE SIDE.

EXAMPLE 4 Finding the Supplement of an Angle
Find the supplement of the following angles.

(a) 70° **(b)** 140°

SOLUTION

(a) The supplement of 70° is 110°, since $180° - 70° = 110°$.

(b) The supplement of 140° is 40°, since $180° - 140° = 40°$. ■

4. Find the supplement of each angle.

(a) 120°

(b) 40°

WORK PROBLEM 4 AT THE SIDE.

Two angles are congruent if they are equivalent in size. If two angles are congruent, this is written as

$$\angle A \cong \angle B$$

and read as, "angle A is congruent to angle B."

EXAMPLE 5 Identifying Congruent Angles
Identify the angles that are congruent.

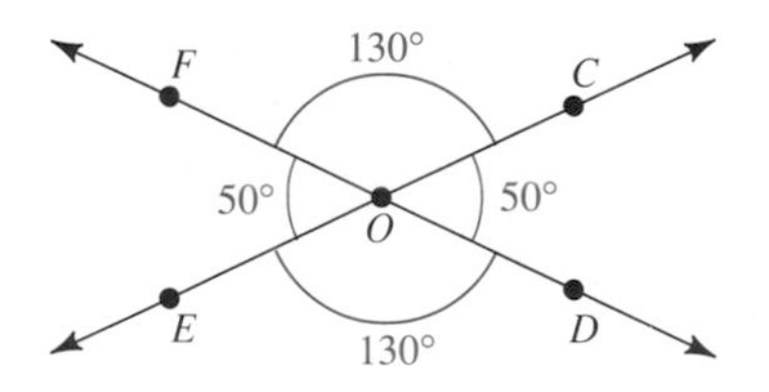

$\angle FOC \cong \angle EOD$ and $\angle COD \cong \angle EOF$ ■

5. Identify the angles that are congruent.

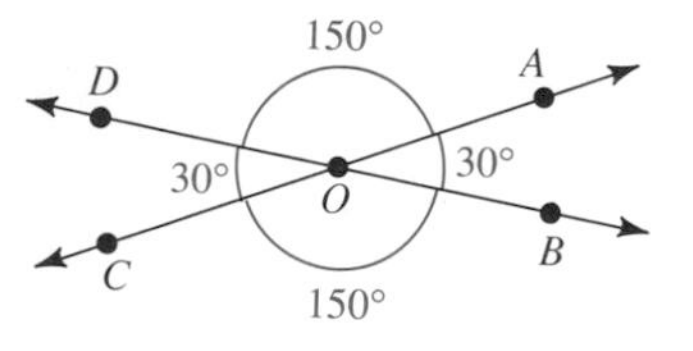

WORK PROBLEM 5 AT THE SIDE.

Angles that do not share a common side are called **nonadjacent** angles. Two nonadjacent angles formed by intersecting lines are called **vertical** angles.

ANSWERS
3. $\angle CRF$ and $\angle BRF$
$\angle CRE$ and $\angle ERB$
$\angle BRF$ and $\angle BRE$
$\angle CRE$ and $\angle CRF$
4. (a) 60° (b) 140°
5. $\angle BOC \cong \angle AOD$, $\angle AOB \cong \angle DOC$

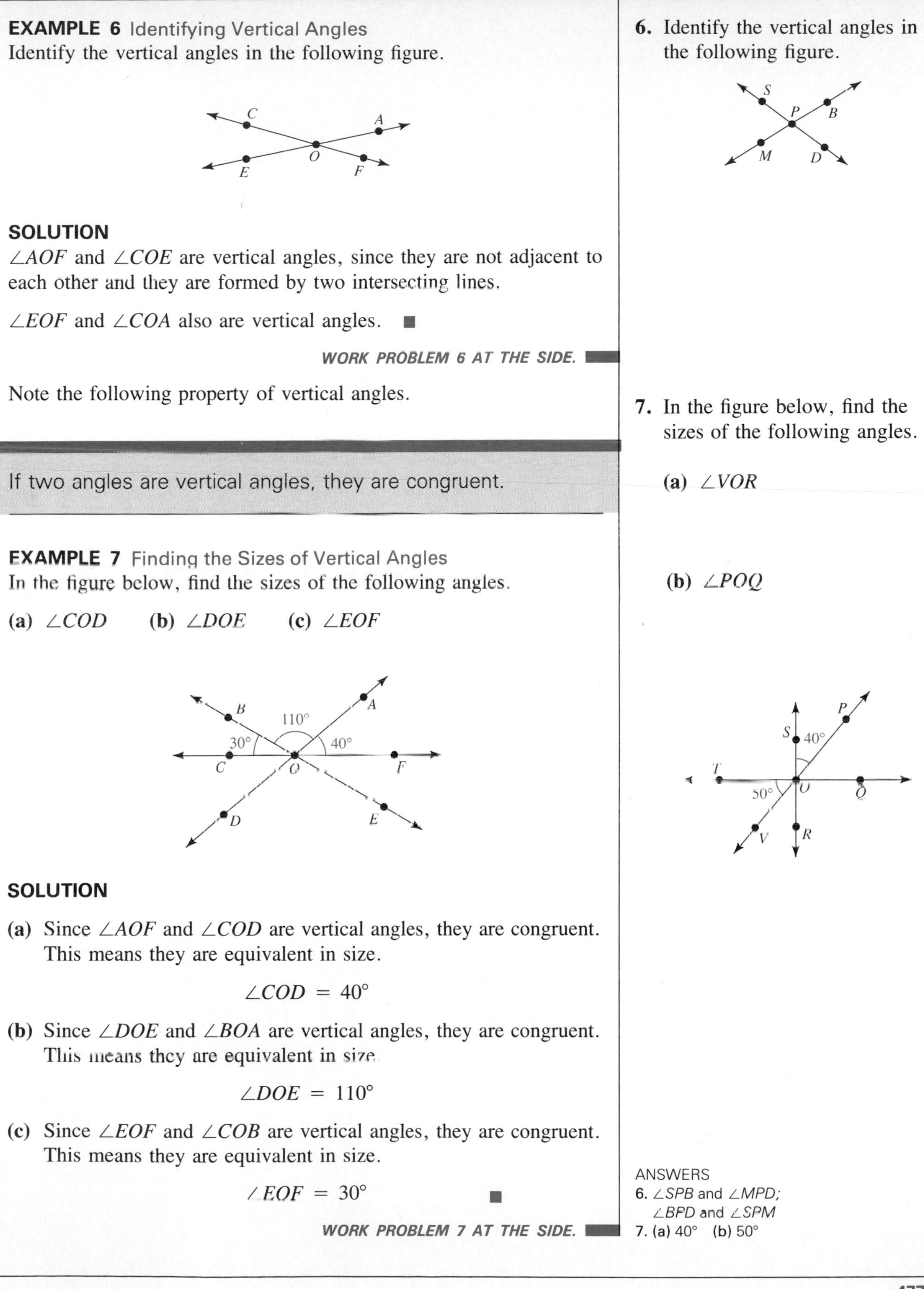

EXAMPLE 6 Identifying Vertical Angles
Identify the vertical angles in the following figure.

SOLUTION

$\angle AOF$ and $\angle COE$ are vertical angles, since they are not adjacent to each other and they are formed by two intersecting lines.

$\angle EOF$ and $\angle COA$ also are vertical angles. ■

WORK PROBLEM 6 AT THE SIDE.

Note the following property of vertical angles.

If two angles are vertical angles, they are congruent.

EXAMPLE 7 Finding the Sizes of Vertical Angles
In the figure below, find the sizes of the following angles.

(a) $\angle COD$ **(b)** $\angle DOE$ **(c)** $\angle EOF$

SOLUTION

(a) Since $\angle AOF$ and $\angle COD$ are vertical angles, they are congruent. This means they are equivalent in size.

$$\angle COD = 40°$$

(b) Since $\angle DOE$ and $\angle BOA$ are vertical angles, they are congruent. This means they are equivalent in size.

$$\angle DOE = 110°$$

(c) Since $\angle EOF$ and $\angle COB$ are vertical angles, they are congruent. This means they are equivalent in size.

$$\angle EOF = 30° \quad ■$$

WORK PROBLEM 7 AT THE SIDE.

6. Identify the vertical angles in the following figure.

7. In the figure below, find the sizes of the following angles.

(a) $\angle VOR$

(b) $\angle POQ$

ANSWERS
6. $\angle SPB$ and $\angle MPD$; $\angle BPD$ and $\angle SPM$
7. (a) 40° (b) 50°

name date hour

8.2 EXERCISES

Find the complement of each angle.

1. 40° **2.** 35° **3.** 65° **4.** 60°

Find the supplement of each angle.

5. 130° **6.** 75° **7.** 30° **8.** 110°

Identify each pair of complementary angles.

9. **10.**

Identify each pair of supplementary angles.

11.

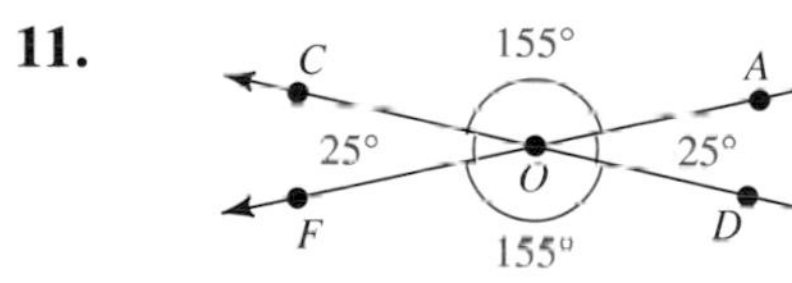

12.

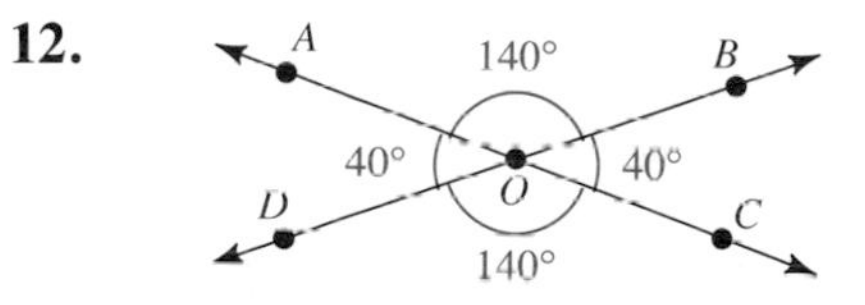

In each of the following, identify the angles that are congruent.

13.

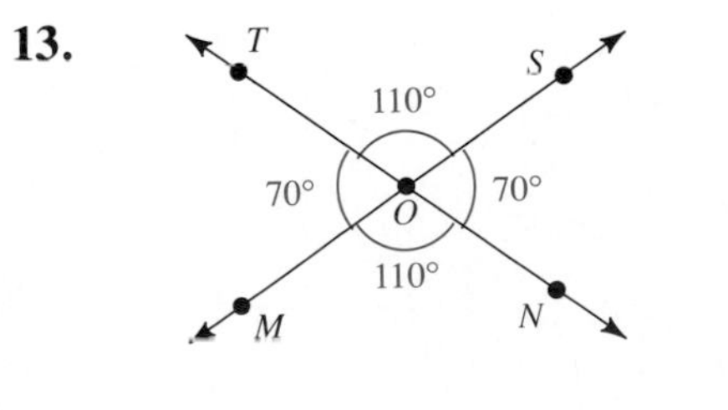

14.

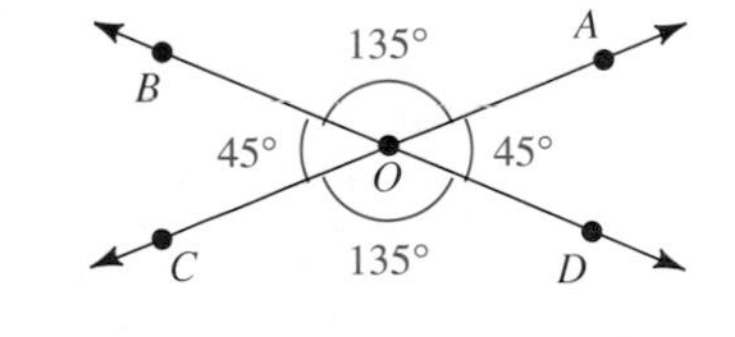

In the following figure, ∠AOH = 37° and ∠COE = 63°. Find the sizes of the following angles.

15. ∠*GOH*

16. ∠*AOC*

17. ∠*EOF*

18. ∠*GOF*

CHALLENGE EXERCISES

In Exercises 19–21, use the following diagram. If ∠BOD is 54°, find the size of each of the following.

19. ∠*AOC*

20. ∠*AOD*

21. ∠*COB*

In each figure, $\overline{AB}$ is parallel to $\overline{CD}$. Identify pairs of congruent angles.

22.

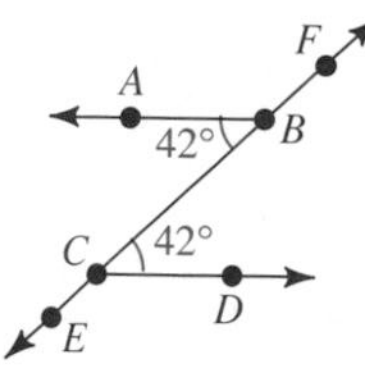

23.

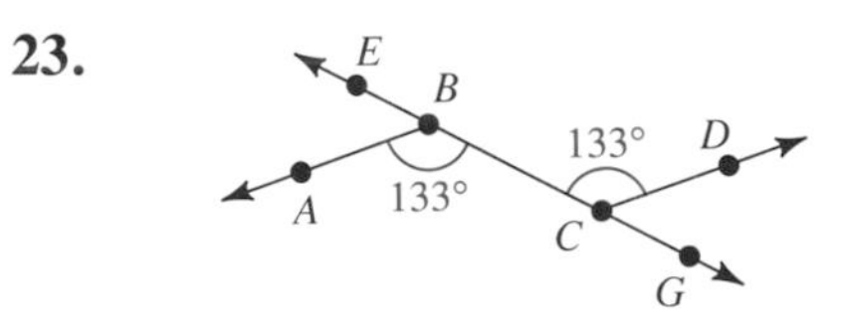

REVIEW EXERCISES

Evaluate the following expressions. (For help, see ***Sections 1.4 and 1.5****.)*

24. $18 \div 9$

25. $16 \cdot 4 \div 2$

26. $36 \cdot 2 \div 4$

27. $81 \div 9 \cdot 4$

28. $144 \div 12 \cdot 4$

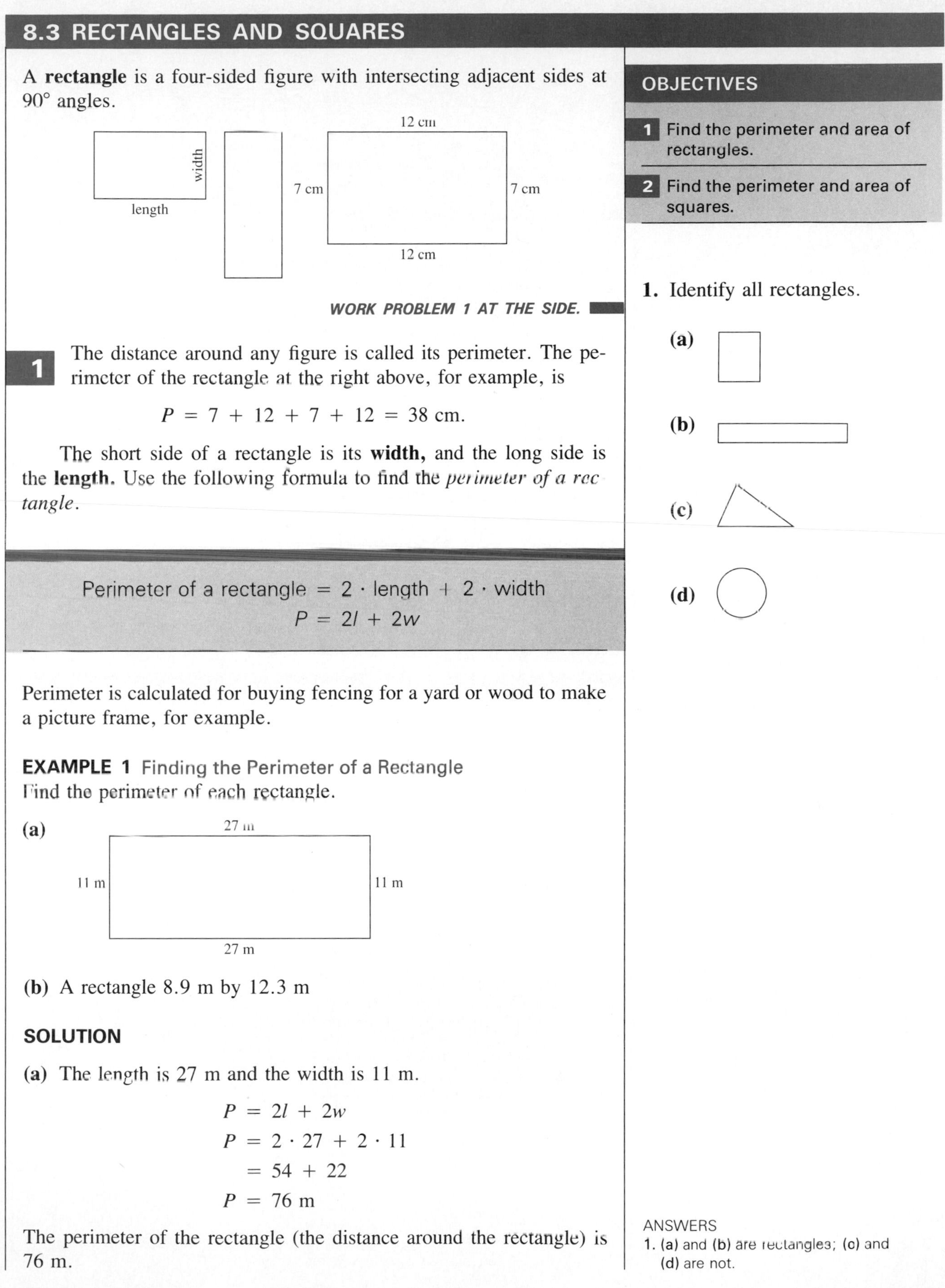

8.3 RECTANGLES AND SQUARES

OBJECTIVES

1. Find the perimeter and area of rectangles.
2. Find the perimeter and area of squares.

A **rectangle** is a four-sided figure with intersecting adjacent sides at 90° angles.

WORK PROBLEM 1 AT THE SIDE.

1 The distance around any figure is called its perimeter. The perimeter of the rectangle at the right above, for example, is

$$P = 7 + 12 + 7 + 12 = 38 \text{ cm}.$$

The short side of a rectangle is its **width,** and the long side is the **length.** Use the following formula to find the *perimeter of a rectangle*.

Perimeter of a rectangle = 2 · length + 2 · width

$$P = 2l + 2w$$

Perimeter is calculated for buying fencing for a yard or wood to make a picture frame, for example.

EXAMPLE 1 Finding the Perimeter of a Rectangle

Find the perimeter of each rectangle.

(a)

(b) A rectangle 8.9 m by 12.3 m

SOLUTION

(a) The length is 27 m and the width is 11 m.

$$P = 2l + 2w$$
$$P = 2 \cdot 27 + 2 \cdot 11$$
$$= 54 + 22$$
$$P = 76 \text{ m}$$

The perimeter of the rectangle (the distance around the rectangle) is 76 m.

1. Identify all rectangles.

(a) **(b)** **(c)** **(d)**

ANSWERS
1. (a) and (b) are rectangles; (c) and (d) are not.

2. Find the perimeter of each rectangle.

(a)

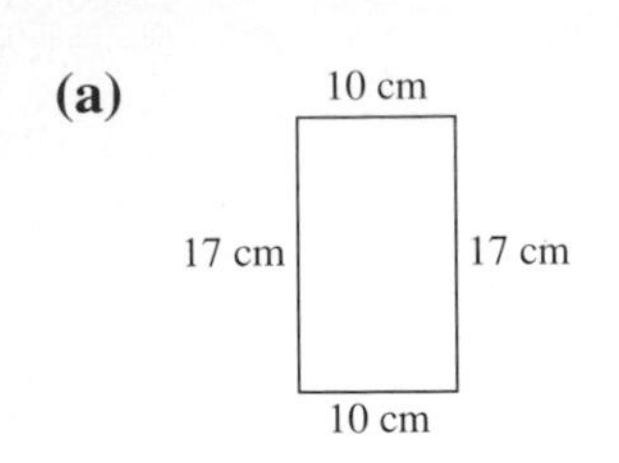

(b)

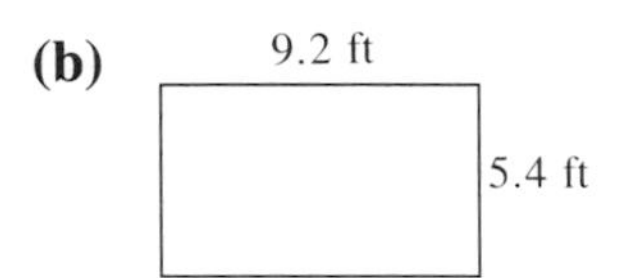

(c) 6 m by 11 m

(d) 4.9 yd by 8.2 yd

ANSWERS
2. (a) 54 cm (b) 29.2 ft (c) 34 m
(d) 26.2 yd

(b)

$$P = 2l + 2w$$
$$P = 2 \cdot 8.9 + 2 \cdot 12.3$$
$$= 17.8 + 24.6$$
$$P = 42.4 \text{ m}$$ ■

WORK PROBLEM 2 AT THE SIDE.

The perimeter of a rectangle is the distance around the rectangle, and the **area** of a rectangle is the space taken up by the rectangle.

Since area is the product of two distances, it is measured in square units. A square inch and a square centimeter are shown here.

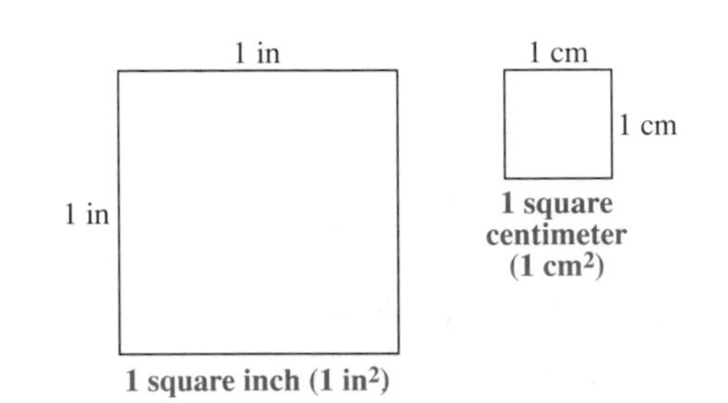

Use the following formula to find the area of a rectangle.

Area of a rectangle = length · width

$$A = lw$$

For example, area is calculated for finding the amount of paint to buy for a wall or the cost of carpet for a room.

EXAMPLE 2 Finding the Area of a Rectangle

Find the area of each rectangle.

(a)

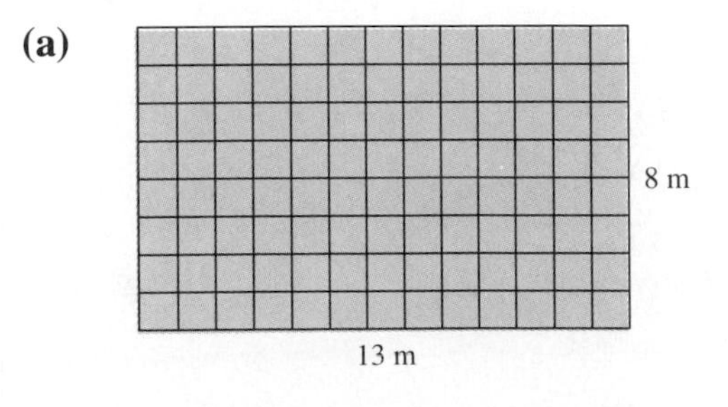

(b) 7 cm by 21 cm

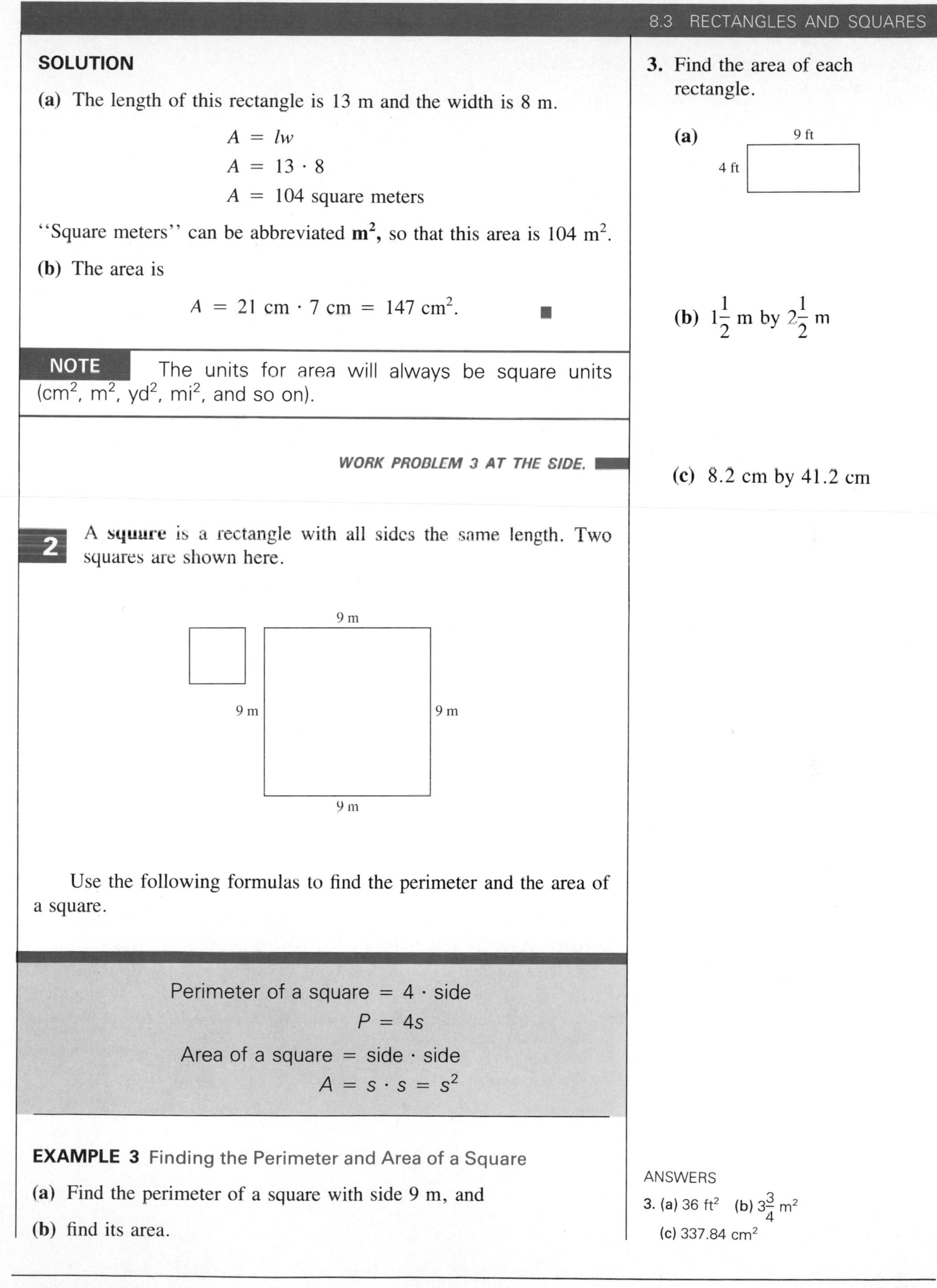

SOLUTION

(a) The length of this rectangle is 13 m and the width is 8 m.

$$A = lw$$
$$A = 13 \cdot 8$$
$$A = 104 \text{ square meters}$$

"Square meters" can be abbreviated $\mathbf{m^2}$, so that this area is 104 m^2.

(b) The area is

$$A = 21 \text{ cm} \cdot 7 \text{ cm} = 147 \text{ cm}^2. \quad ■$$

NOTE The units for area will always be square units (cm^2, m^2, yd^2, mi^2, and so on).

WORK PROBLEM 3 AT THE SIDE.

3. Find the area of each rectangle.

(a)

(b) $1\frac{1}{2}$ m by $2\frac{1}{2}$ m

(c) 8.2 cm by 41.2 cm

2 A **square** is a rectangle with all sides the same length. Two squares are shown here.

Use the following formulas to find the perimeter and the area of a square.

$$\text{Perimeter of a square} = 4 \cdot \text{side}$$
$$P = 4s$$
$$\text{Area of a square} = \text{side} \cdot \text{side}$$
$$A = s \cdot s = s^2$$

EXAMPLE 3 Finding the Perimeter and Area of a Square

(a) Find the perimeter of a square with side 9 m, and

(b) find its area.

ANSWERS

3. (a) 36 ft^2 (b) $3\frac{3}{4}$ m^2 (c) 337.84 cm^2

4. Find the perimeter and area of each square.

(a)

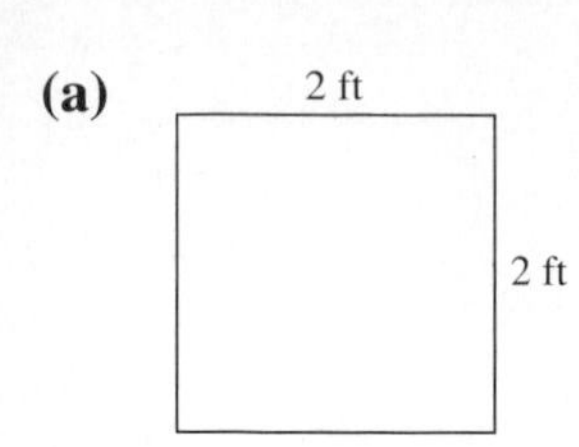

(b) 5.6 m on a side

(c) 28.3 yd on a side

SOLUTION

(a)
$$P = 4s$$
$$P = 4 \cdot 9$$
$$P = 36 \text{ m}$$

(b)
$$A = s^2$$
$$A = 9^2$$
$$= 9 \cdot 9$$
$$A = 81 \text{ m}^2 \quad ■$$

WORK PROBLEM 4 AT THE SIDE.

NOTE Problems that involve irregular shapes can be more easily solved if we first break up the given area into smaller areas of more familiar figures such as squares and rectangles.

EXAMPLE 4 Applying the Concept of Area

A room has the shape shown here.

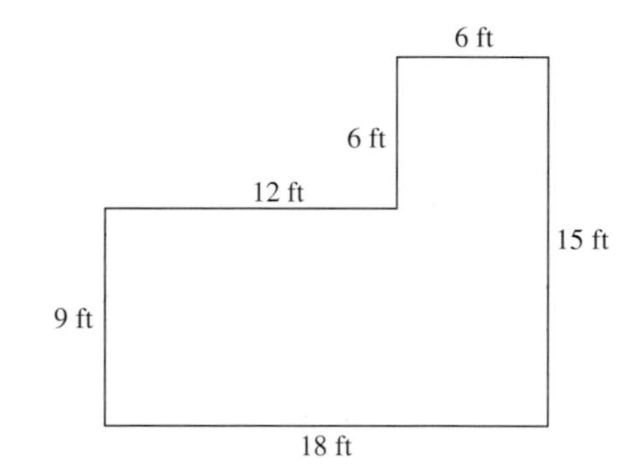

Carpet costs $20.50 per square yard. Find the total cost of the carpet needed for this room.

Carpet covers space, not length, so the *area* of the floor must be found. Break the area into a rectangle and a square, as shown below. Also, change all lengths to yards, since the cost of carpeting is given per square *yard*. (*Recall:* a yd = 3 ft.).

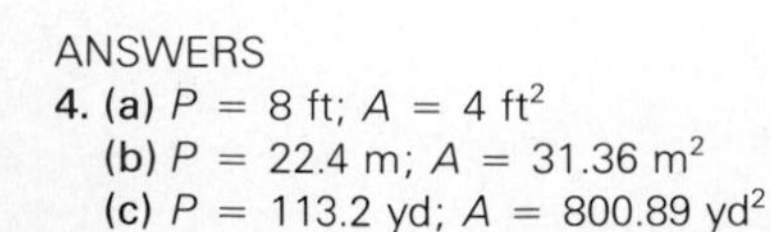
ANSWERS
4. (a) $P = 8$ ft; $A = 4 \text{ ft}^2$
(b) $P = 22.4$ m; $A = 31.36 \text{ m}^2$
(c) $P = 113.2$ yd; $A = 800.89 \text{ yd}^2$

Thus,
$$6 \text{ ft} = \frac{6 \text{ ft}}{3} = 2 \text{ yd}$$
$$12 \text{ ft} = \frac{12 \text{ ft}}{3} = 4 \text{ yd}$$
$$9 \text{ ft} = \frac{9 \text{ ft}}{3} = 3 \text{ yd}$$
$$18 \text{ ft} = \frac{18 \text{ ft}}{3} = 6 \text{ yd}$$
$$15 \text{ ft} = \frac{15 \text{ ft}}{3} = 5 \text{ yd}$$

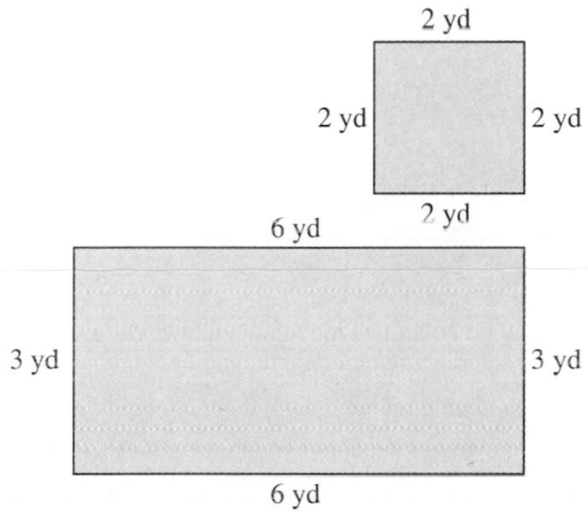

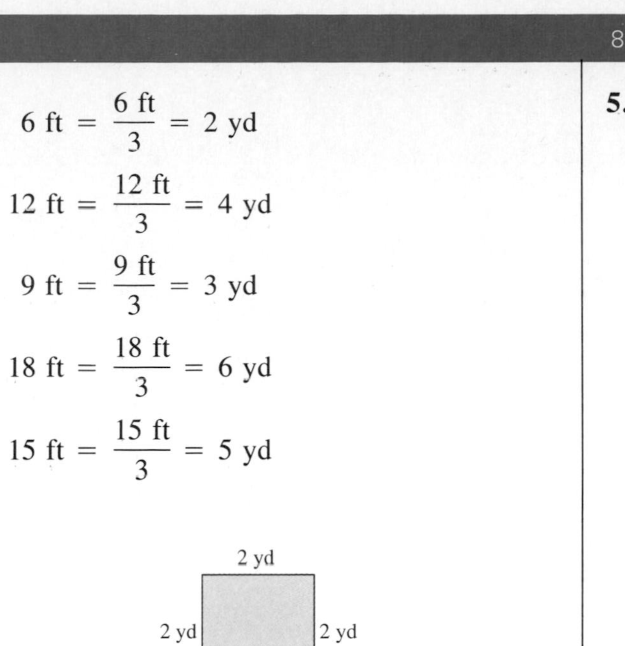

The area of the square is
$$2 \cdot 2 = 4 \text{ yd}^2$$
and the area of the rectangle is
$$3 \cdot 6 = 18 \text{ yd}^2.$$
The total area that needs to be carpeted is the sum of the areas of the rectangle and the square, so the total area is $4 + 18 = 22$ yd^2. Thus, 22 yd^2 of carpet is needed, at \$20.50 per square yard. The cost of the carpet is
$$22 \cdot \$20.50 = \$451. \blacksquare$$

WORK PROBLEM 5 AT THE SIDE.

5. Carpet costs \$19 per square yard. Find the cost of carpeting the following rooms.

(a)

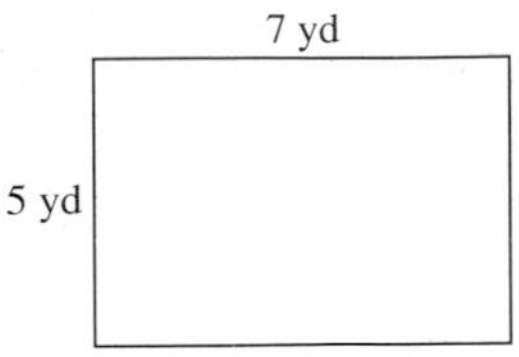

(b)

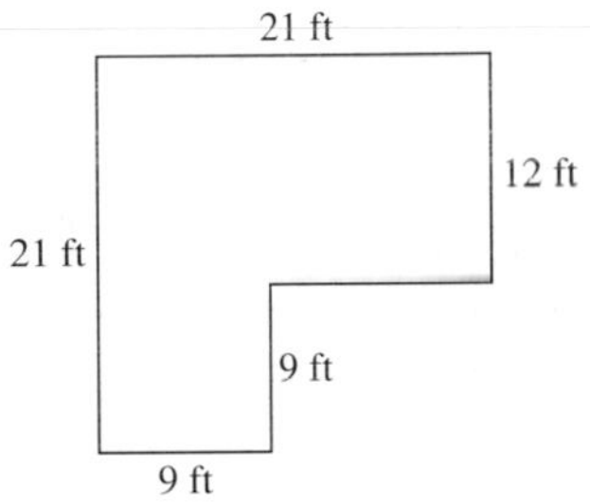

(c) a room 24 ft by 30 ft

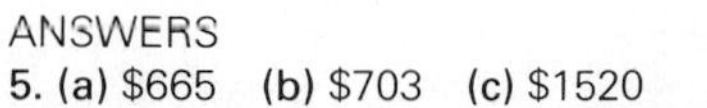

ANSWERS
5. (a) \$665 (b) \$703 (c) \$1520

name date hour

8.3 EXERCISES

Find the perimeter and area of each rectangle or square.

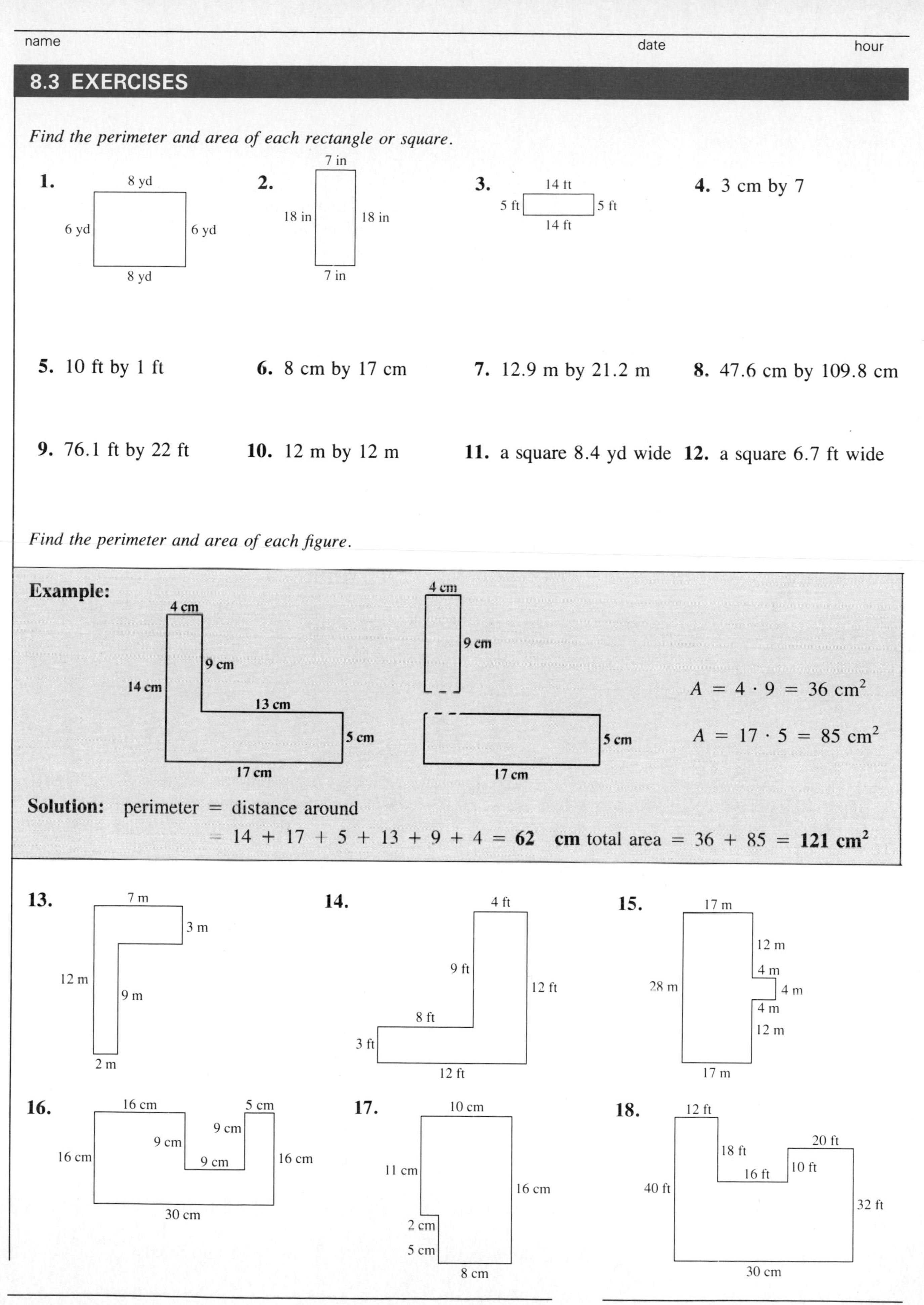

4. 3 cm by 7

5. 10 ft by 1 ft **6.** 8 cm by 17 cm **7.** 12.9 m by 21.2 m **8.** 47.6 cm by 109.8 cm

9. 76.1 ft by 22 ft **10.** 12 m by 12 m **11.** a square 8.4 yd wide **12.** a square 6.7 ft wide

Find the perimeter and area of each figure.

Example:

$A = 4 \cdot 9 = 36 \text{ cm}^2$

$A = 17 \cdot 5 = 85 \text{ cm}^2$

Solution: perimeter = distance around

$= 14 + 17 + 5 + 13 + 9 + 4 = \mathbf{62}$ **cm** total area $= 36 + 85 = \mathbf{121\ cm^2}$

Solve each word problem.

19. A photograph is placed in a frame 24 in by 30 in. Find the area of the photograph.

20. A lot is 124 ft by 172 ft. County rules require that nothing be built on land within 12 ft of any edge of the lot. Find the area that cannot be built on.

21. A room is 18 ft by 24 ft. What is the cost of carpeting this room, if carpet is \$13 per yd^2.

22. A floor is 10 ft by 15 ft. What is the cost of tiling the floor, if tile costs \$10 per ft^2.

CHALLENGE EXERCISES

23. Suppose that Exercise 18 shows the floor plan of a restaurant. Find the cost of covering this area with floor tile at \$.96 per ft^2.

24. A table is 6 ft by 3 ft. Half of the table is covered with material that costs \$2.25 per ft^2. The other half is to be covered with material that costs \$4.50 per ft^2. Find the cost of covering the table.

25. Find the cost of weatherstripping (insulation around the edge) needed for this office window at \$1.92 per foot.

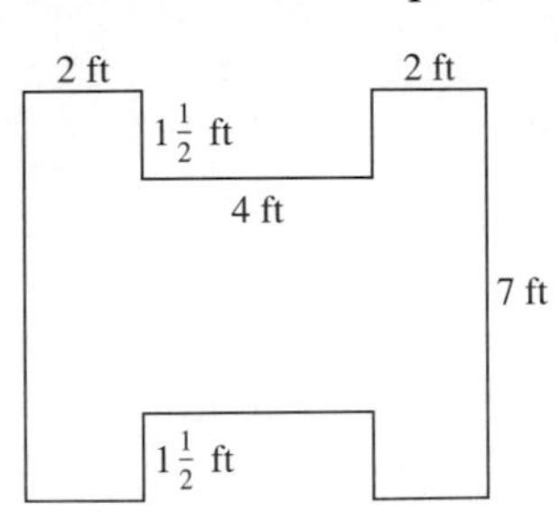

26. Find the cost of fencing needed for this field. Fencing along the highway costs \$4.25 per foot. Fencing for the other three sides costs \$2.75 per foot.

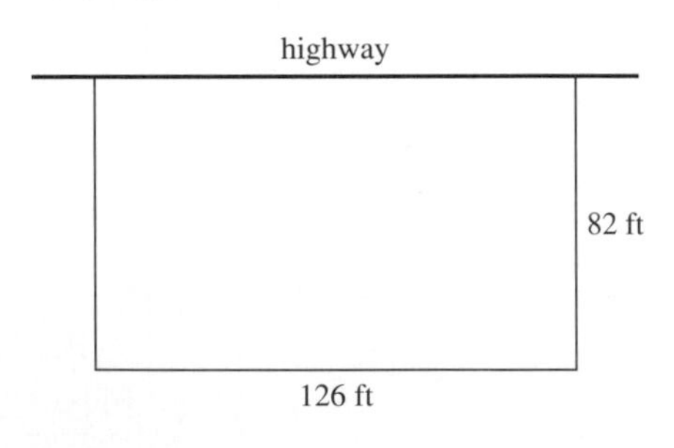

REVIEW EXERCISES

Convert the following measurements. (For help, see ***Sections 7.3 and 7.4****.)*

27. 30 cm to m

28. 40 km to m

29. 50 ft to in

30. 75 ft to yd

31. 2700 in to yd

32. 400 mm to cm

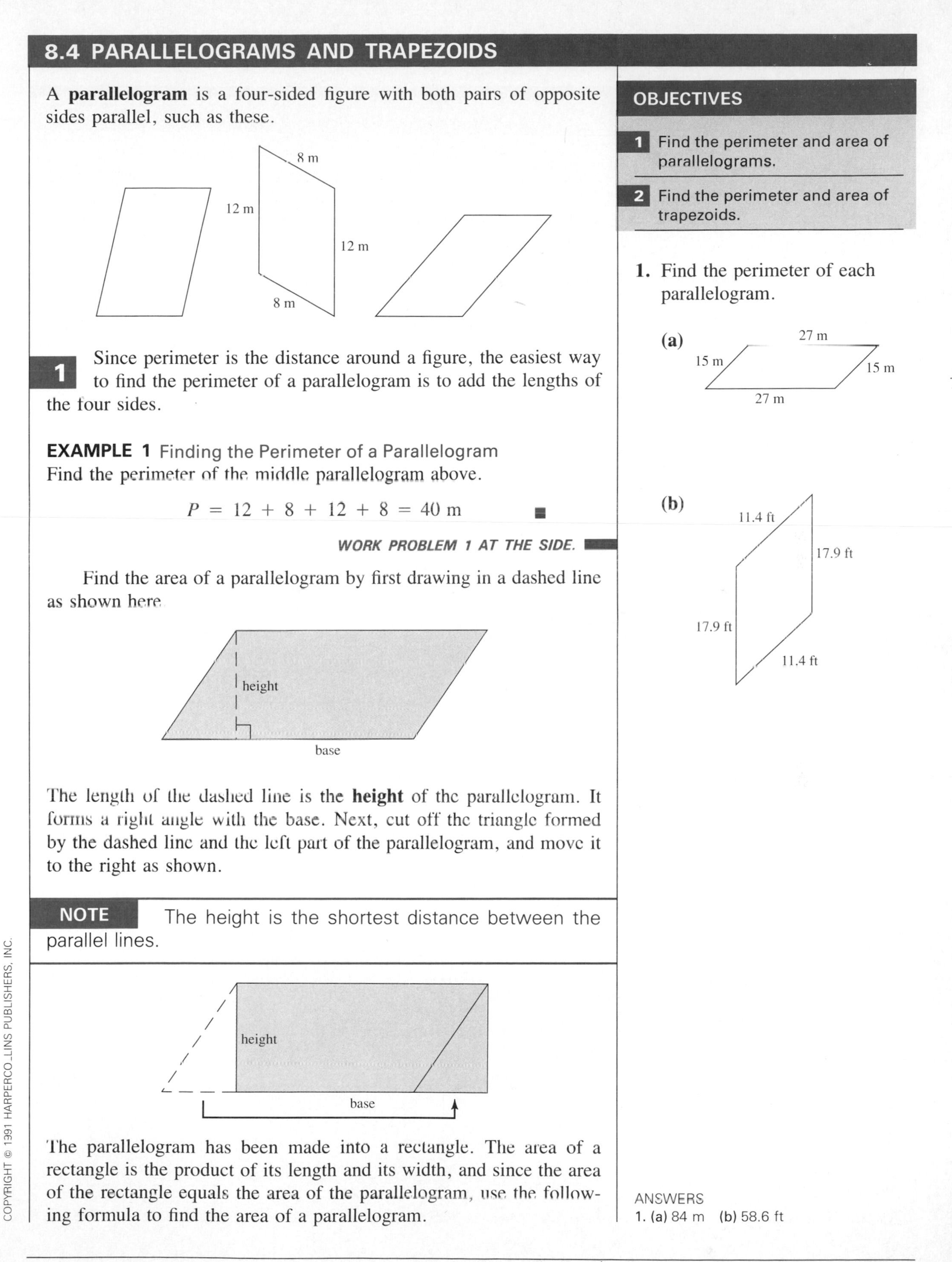

8.4 PARALLELOGRAMS AND TRAPEZOIDS

OBJECTIVES

1. Find the perimeter and area of parallelograms.
2. Find the perimeter and area of trapezoids.

A **parallelogram** is a four-sided figure with both pairs of opposite sides parallel, such as these.

1 Since perimeter is the distance around a figure, the easiest way to find the perimeter of a parallelogram is to add the lengths of the four sides.

EXAMPLE 1 Finding the Perimeter of a Parallelogram

Find the perimeter of the middle parallelogram above.

$$P = 12 + 8 + 12 + 8 = 40 \text{ m}$$

WORK PROBLEM 1 AT THE SIDE.

1. Find the perimeter of each parallelogram.

(a)

(b)

Find the area of a parallelogram by first drawing in a dashed line as shown here.

The length of the dashed line is the **height** of the parallelogram. It forms a right angle with the base. Next, cut off the triangle formed by the dashed line and the left part of the parallelogram, and move it to the right as shown.

NOTE The height is the shortest distance between the parallel lines.

The parallelogram has been made into a rectangle. The area of a rectangle is the product of its length and its width, and since the area of the rectangle equals the area of the parallelogram, use the following formula to find the area of a parallelogram.

ANSWERS

1. (a) 84 m (b) 58.6 ft

2. Find the area of each parallelogram.

(a)

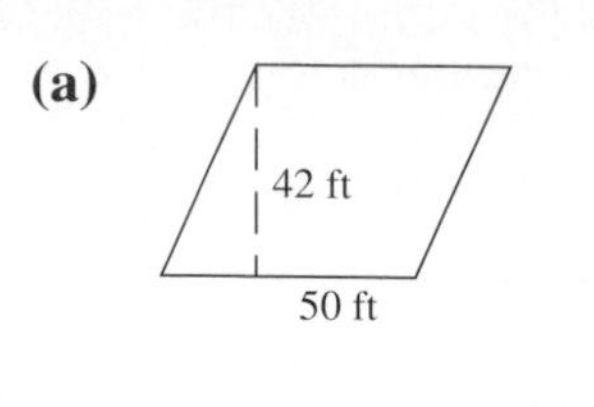

(b)

16.2 cm

25.1 cm

(c) a parallelogram with base $12\frac{1}{2}$ m and height $4\frac{3}{4}$ m

3. Find the perimeter of each trapezoid.

(a)

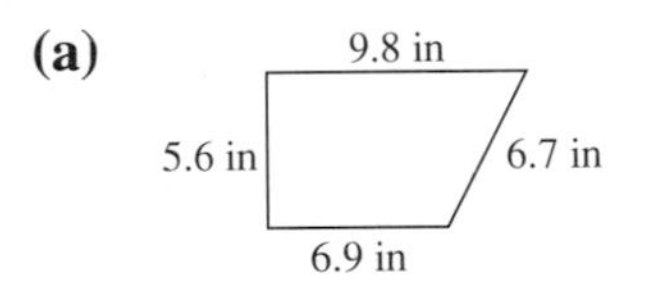

(b)

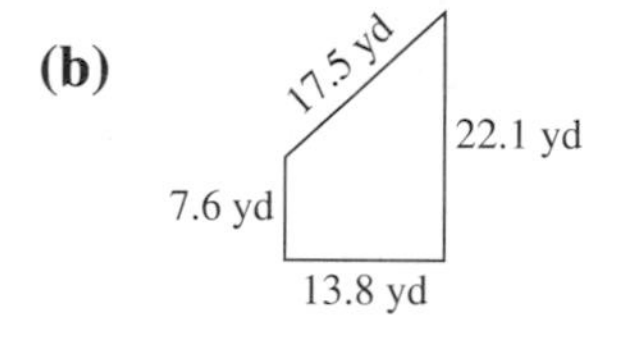

(c) a trapezoid with sides 39.76 cm, 19.24 cm, 74.90 cm, and 26.45 cm

ANSWERS
2. (a) 2100 ft^2 (b) 406.62 cm^2
(c) $59\frac{3}{8}$ m^2 or 59.375 m^2
3. (a) 29 in (b) 61 yd (c) 160.35 cm

Area of parallelogram = base · height

$$A = bh$$

EXAMPLE 2 Finding the Area of a Parallelogram

Find the area of each parallelogram.

(a) 19 cm; 24 cm

(b) 47 m; 30 m; 24 m; 30 m; 47 m

SOLUTION

(a) The base is 24 cm and the height is 19 cm. The area is

$$A = 19 \cdot 24$$
$$A = 456 \text{ cm}^2.$$

(b) $A = 47 \cdot 24 = 1128 \text{ m}^2$

Notice that the 30-m sides are not used in finding the area. ■

WORK PROBLEM 2 AT THE SIDE.

2 A **trapezoid** is a four-sided figure with only one pair of opposite sides parallel. Here are some examples of trapezoids.

32 m; 17 m; 19 m; 46 m

Find the perimeter of a trapezoid by adding the lengths of the four sides.

EXAMPLE 3 Finding the Perimeter of a Trapezoid

Find the perimeter of the middle trapezoid above.

$$P = 17 + 32 + 19 + 46 = 114 \text{ m}.$$ ■

WORK PROBLEM 3 AT THE SIDE.

Use the following formula to find the area of a trapezoid.

$$\text{Area} = \frac{1}{2} \cdot \text{height} \cdot (\text{short base} + \text{long base})$$

$$A = \frac{1}{2}h(b + B) \quad \text{or} \quad A = h\left(\frac{b + B}{2}\right)$$

EXAMPLE 4 Finding the Areas of Trapezoids

(a) Find the area of this trapezoid.

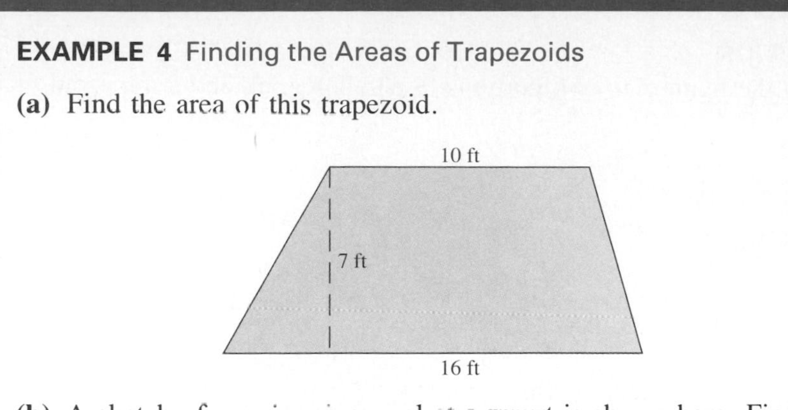

(b) A sketch of a swimming pool at a resort is shown here. Find the area of the surface of the pool.

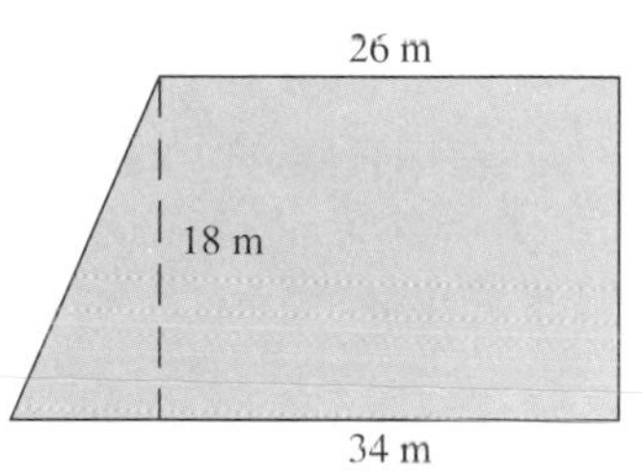

SOLUTION

(a) The height is 7 ft, the short base is 10 ft, and the long base is 16 ft.

$$A = \frac{1}{2}h(b + B)$$

$$A = \frac{1}{2} \cdot 7 \cdot (10 + 16) = \frac{1}{\cancel{2}_1} \cdot 7 \cdot \cancel{26}^{13} = 91 \text{ ft}^2$$

(b)

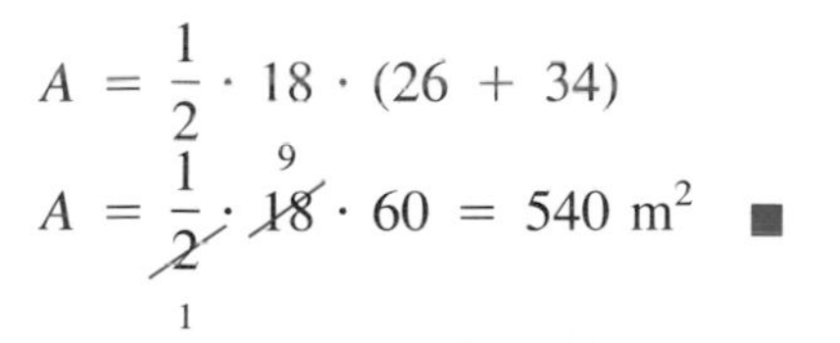

$$A = \frac{1}{2} \cdot 18 \cdot (26 + 34)$$

$$A = \frac{1}{\cancel{2}_1} \cdot \cancel{18}^{9} \cdot 60 = 540 \text{ m}^2 \quad ■$$

WORK PROBLEM 4 AT THE SIDE.

EXAMPLE 5 Applying the Concept of Area

Find the area of this figure.

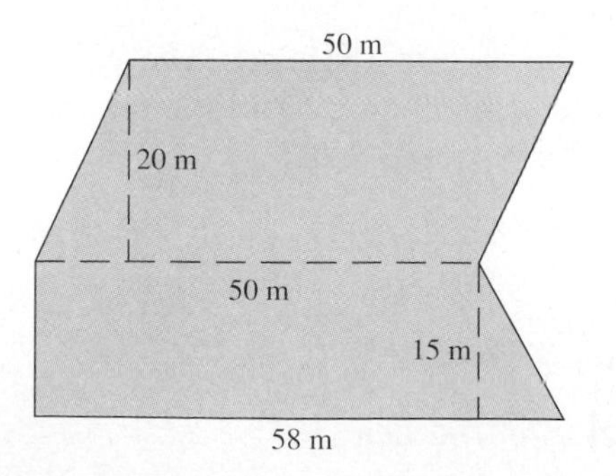

4. Find the area of each trapezoid.

(a)

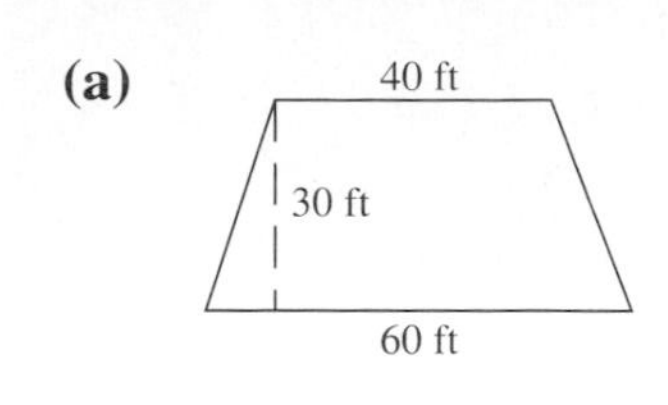

(b)

51 cm

92 cm

64 cm

(c) a trapezoid with height 14.7 m, short base 8.2 m, and long base 10.2 m

ANSWERS

4. (a) 1500 ft² (b) 3978 cm² (c) 135.24 m²

5. Find each area.

(a)

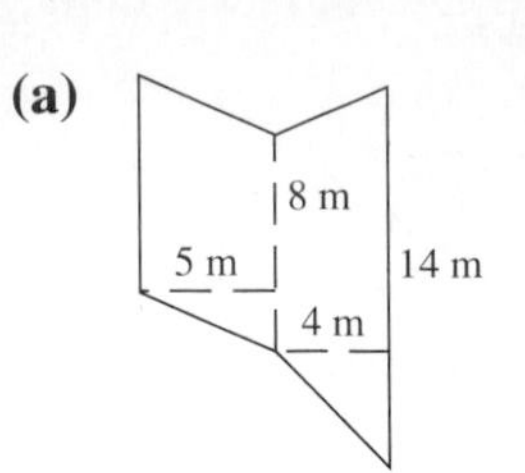

(b)

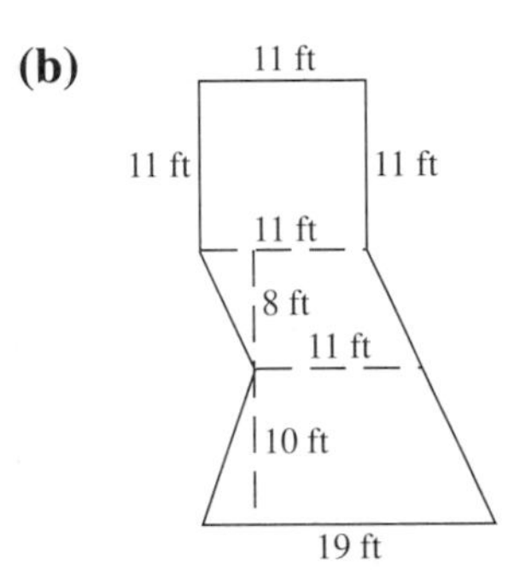

6. Find the cost of carpeting the areas in Problem 5. The cost of carpet is as follows:

(a) Region (a), \$18.50 per square meter.

(b) Region (b), \$1.64 per square foot.

ANSWERS
5. (a) 84 m^2 (b) 359 ft^2
6. (a) \$1554 (b) \$588.76

SOLUTION
Break the figure into two portions, a parallelogram and a trapezoid.

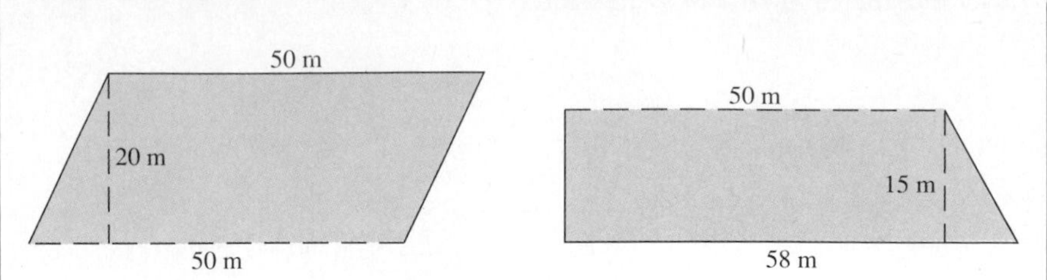

Find the area of each figure separately.

$$\text{area of parallelogram} = b \cdot h = 50 \cdot 20 = 1000 \text{ m}^2$$

$$\text{area of trapezoid} = \frac{1}{2}h(b + B) = \frac{1}{2} \cdot 15 \cdot (50 + 58) = \frac{1}{2} \cdot 15 \cdot 108 = 810 \text{ m}^2$$

The total area is the sum of these areas.

$$\text{Total area} = \text{area of parallelogram} + \text{area of trapezoid} = 1000 \text{ m}^2 + 810 \text{ m}^2 = 1810 \text{ m}^2$$ ■

NOTE The problem was solved by dividing the given figure into smaller figures with areas that are easily found.

WORK PROBLEM 5 AT THE SIDE.

EXAMPLE 6 Applying Knowledge of Area
Suppose the figure in Example 5 represents the floor plan of a hotel lobby. What is the cost of tiling the floor, if tile costs \$5.11 per square meter?

SOLUTION
From Example 5, the area is 1810 m^2. The cost of the floor tile at \$5.11 per square meter is found by multiplying the number of square meters and the cost per square meter.

$$\text{Cost} = 1810 \text{ m}^2 \cdot \$5.11 \text{ per square meter}$$
$$\text{Cost} = \$9249.10$$

The cost of the tile is \$9249.10. ■

WORK PROBLEM 6 AT THE SIDE.

name date hour

8.4 EXERCISES

Find the perimeter of each figure.

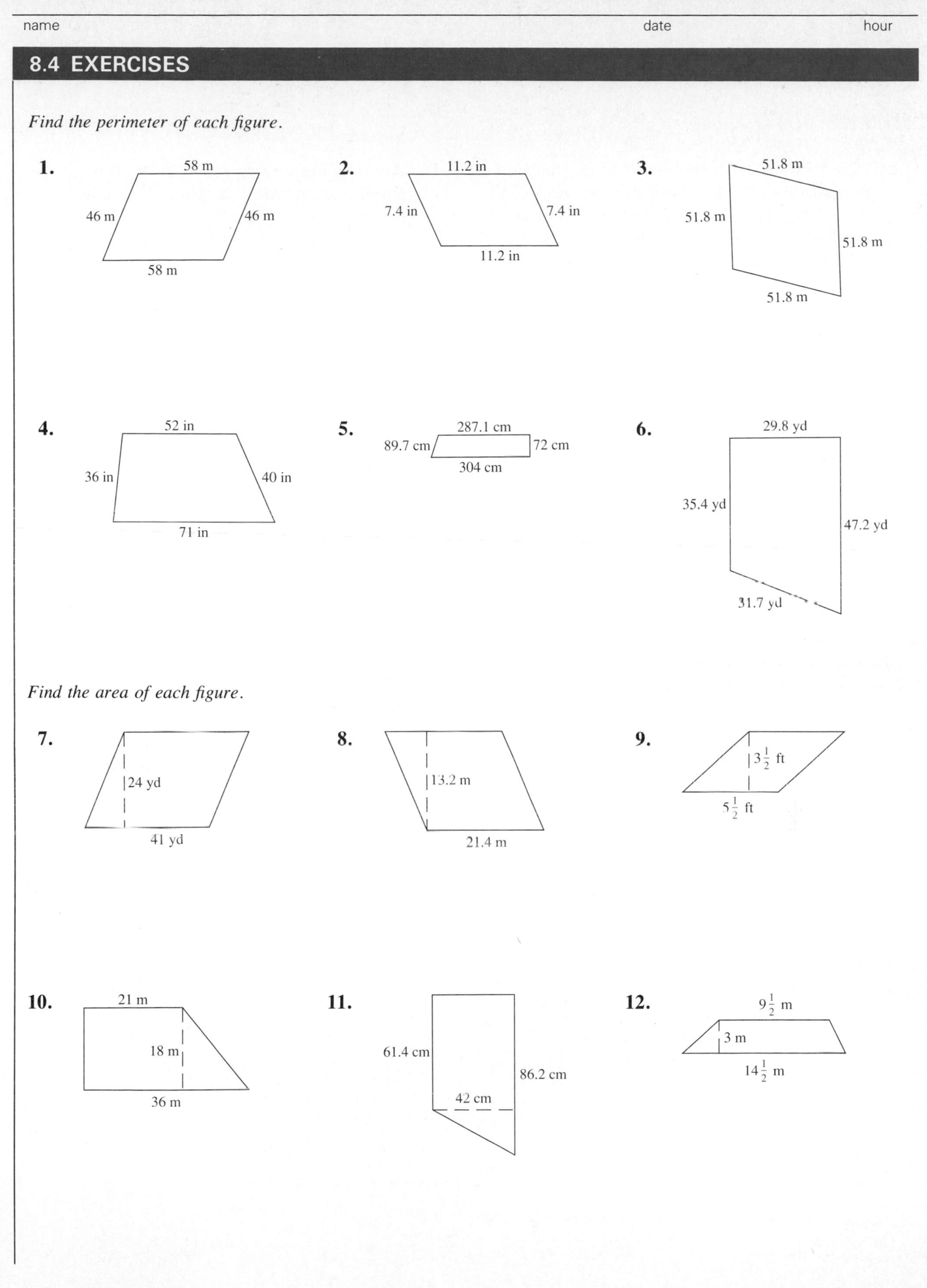

Solve each word problem.

13. The lobby in a resort hotel is in the shape of a trapezoid. The height of the trapezoid is 32 ft and the bases are 47 ft and 59 ft. Carpet costing \$1.75 per ft^2 is to be laid in the lobby. Find the cost of the carpet.

14. A swimming pool is in the shape of a parallelogram with a height of 9.6 m and base of 12.4 m. Find the cost of a solar cover for the pool at a cost of \$4.92 per m^2.

15. The backyard of a new home is shaped like a trapezoid with a height of 45 ft and bases of 80 ft and 110 ft. What is the cost of planting a lawn in the yard, if the landscaper charges \$.22 per ft^2.

16. An auditorium stage has a hardwood floor that is shaped like a parallelogram with a height of 20 ft and a base of 40 ft. If a company charges \$.60 per ft^2 to refinish floors, what is the cost of refinishing the floor.

CHALLENGE EXERCISES

Find the area of each figure.

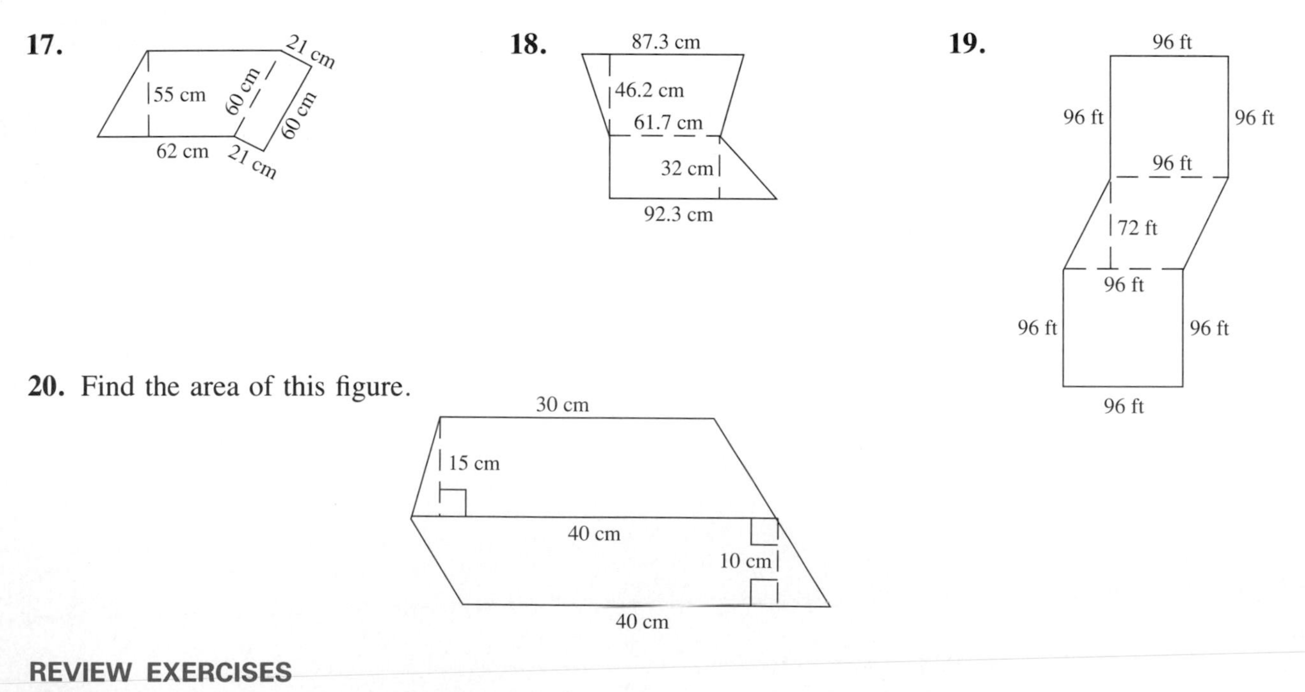

20. Find the area of this figure.

REVIEW EXERCISES

Multiply the following. Write all answers in lowest terms. (For help, see ***Section 2.5****.)*

21. $6\frac{1}{2} \cdot 8$

22. $\frac{33}{4} \cdot \frac{16}{11}$

23. $25 \cdot \frac{50}{7} \cdot \frac{28}{75}$

24. $\frac{15}{34} \cdot \frac{17}{5} \cdot 1\frac{1}{3}$

8.5 TRIANGLES

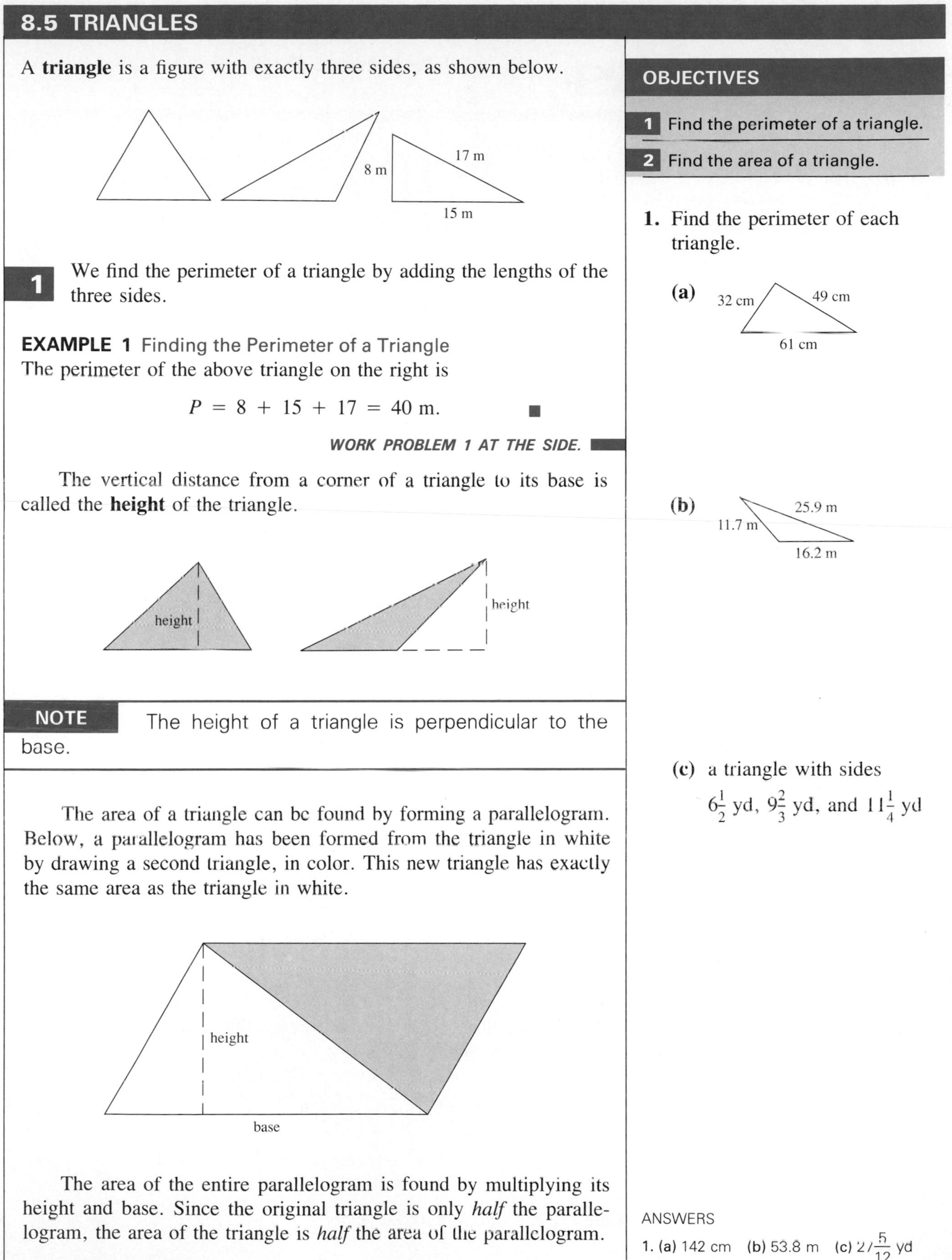

A **triangle** is a figure with exactly three sides, as shown below.

8 m   17 m   15 m

1 We find the perimeter of a triangle by adding the lengths of the three sides.

EXAMPLE 1 Finding the Perimeter of a Triangle
The perimeter of the above triangle on the right is

$$P = 8 + 15 + 17 = 40 \text{ m.}$$ ■

WORK PROBLEM 1 AT THE SIDE.

The vertical distance from a corner of a triangle to its base is called the **height** of the triangle.

height   height

NOTE The height of a triangle is perpendicular to the base.

The area of a triangle can be found by forming a parallelogram. Below, a parallelogram has been formed from the triangle in white by drawing a second triangle, in color. This new triangle has exactly the same area as the triangle in white.

height   base

The area of the entire parallelogram is found by multiplying its height and base. Since the original triangle is only *half* the parallelogram, the area of the triangle is *half* the area of the parallelogram.

OBJECTIVES

1. Find the perimeter of a triangle.
2. Find the area of a triangle.

1. Find the perimeter of each triangle.

(a) 32 cm, 49 cm, 61 cm

(b) 11.7 m, 25.9 m, 16.2 m

(c) a triangle with sides $6\frac{1}{2}$ yd, $9\frac{2}{3}$ yd, and $11\frac{1}{4}$ yd

ANSWERS
1. (a) 142 cm (b) 53.8 m (c) $27\frac{5}{12}$ yd

2. Find the area of each triangle.

(a)

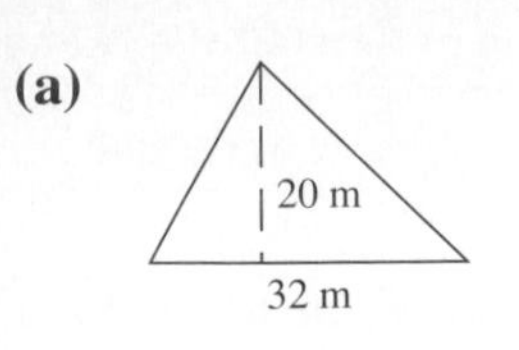

(b)

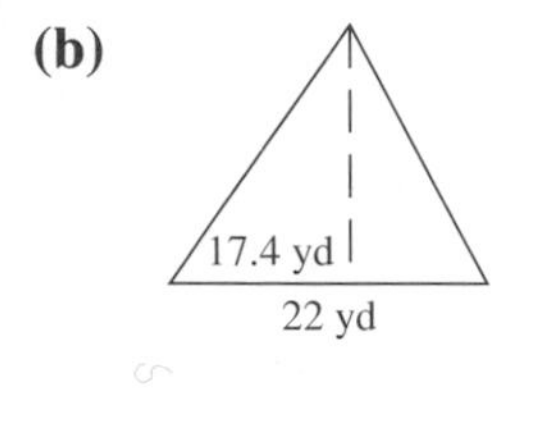

(c)

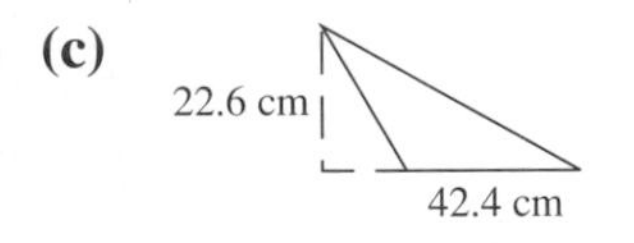

(d)

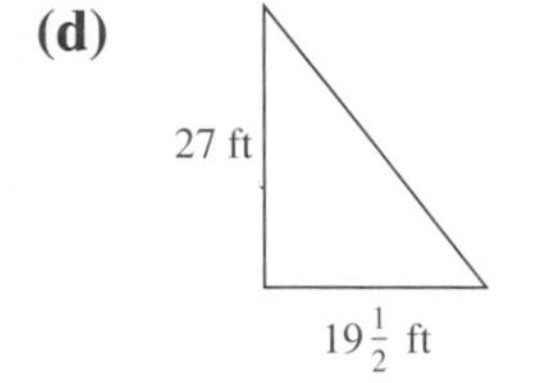

ANSWERS
2. (a) 320 m^2 (b) 191.4 yd^2
(c) 479.12 cm^2
(d) $263\frac{1}{4}$ ft^2 or 263.25 ft^2

2 Use the following formula to find the area of a triangle.

$$\text{area of triangle} = \frac{1}{2} \cdot \text{base} \cdot \text{height}$$

$$A = \frac{1}{2}bh$$

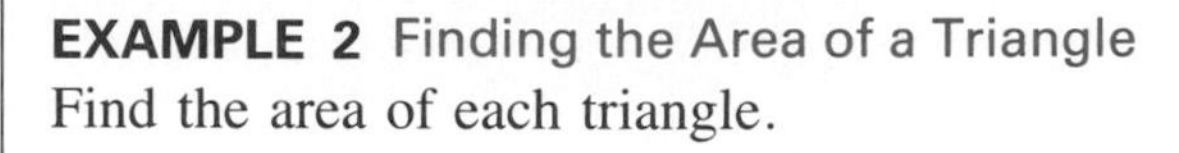

EXAMPLE 2 Finding the Area of a Triangle
Find the area of each triangle.

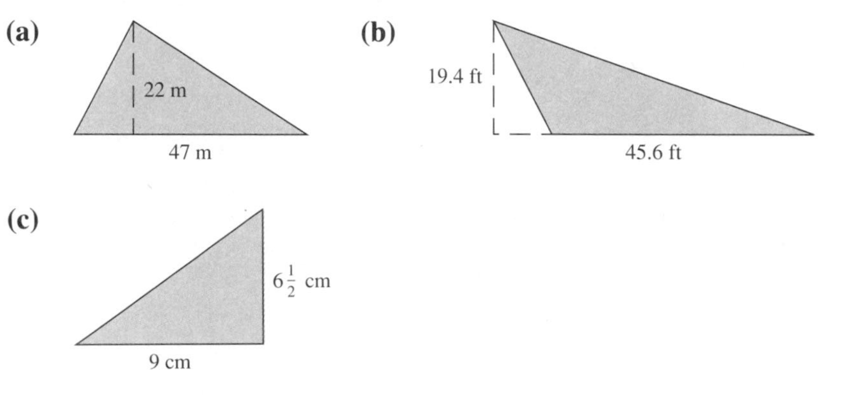

SOLUTION

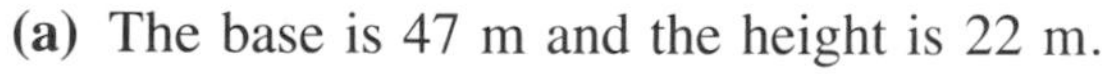

(a) The base is 47 m and the height is 22 m.

$$A = \frac{1}{2}bh$$

$$A = \frac{1}{\cancel{2}_1} \cdot 47 \cdot \cancel{22}^{11}$$

$$A = 517 \text{ m}^2$$

(b) The height is outside the triangle.

$$A = \frac{1}{2} \cdot 45.6 \cdot 19.4$$

$$A = 442.32 \text{ ft}^2$$

(c) The height is the side of the triangle that is perpendicular to the horizontal side.

$$A = \frac{1}{2} \cdot 9 \cdot 6\frac{1}{2}$$

$$A = 29\frac{1}{4} \text{ cm}^2 \text{ or } 29.25 \text{ cm}^2 \quad \blacksquare$$

WORK PROBLEM 2 AT THE SIDE.

EXAMPLE 3 Applying the Concept of Area
The department of transportation designs a triangular sign with a base of 30 centimeters and a height of 40 centimeters. If the material for the sign costs \$1.25 per square centimeter, what is the total cost of material for the sign?

SOLUTION

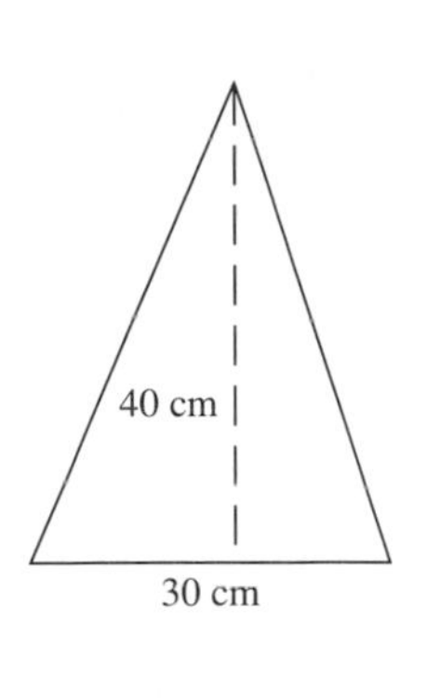

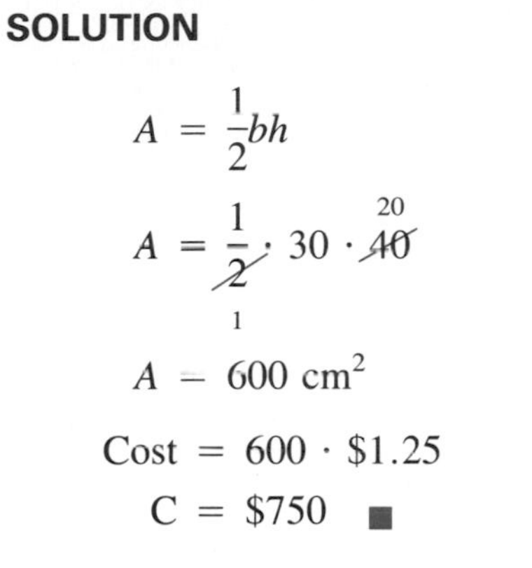

$$A = \frac{1}{2}bh$$

$$A = \frac{1}{\not{2}_1} \cdot 30 \cdot \overset{20}{\not{40}}$$

$$A = 600 \text{ cm}^2$$

$$\text{Cost} = 600 \cdot \$1.25$$

$$C = \$750$$ ■

WORK PROBLEM 3 AT THE SIDE.

3. A triangular sign has a 10-foot base and height of 5 feet. Material for the sign costs \$2.75 per square foot. Find the total cost of material to make the sign.

EXAMPLE 4 Using the Concept of Area
A certain design uses nine pieces of metal for its surface. These pieces are triangular with a base of 6 inches and a height of 8 inches. How much material is needed for the figure?

SOLUTION

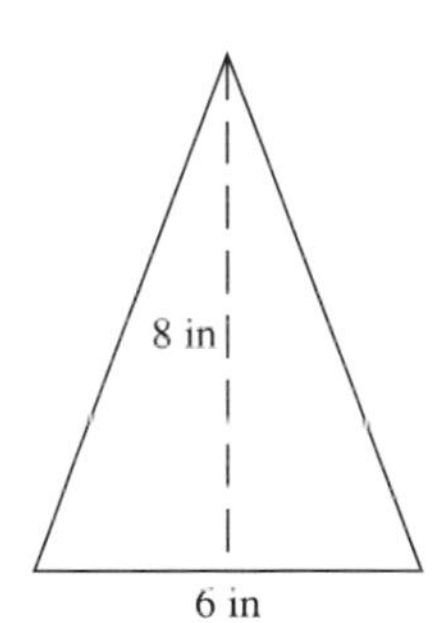

$$A = \frac{1}{2}bh$$

$$A = \frac{1}{\not{2}_1} \cdot \overset{3}{\not{6}} \cdot 8$$

$$A = 24 \text{ in}^2$$

$$\text{total material needed} = 9 \cdot 24 = 216 \text{ in}^2.$$ ■

WORK PROBLEM 4 AT THE SIDE.

4. Wooden triangular braces have a base of 10 inches and are 15 inches high. If 20 braces are used to build a boat, what is the total amount of material needed for the braces?

ANSWERS
3. \$68.75 **4.** 1500 in^2

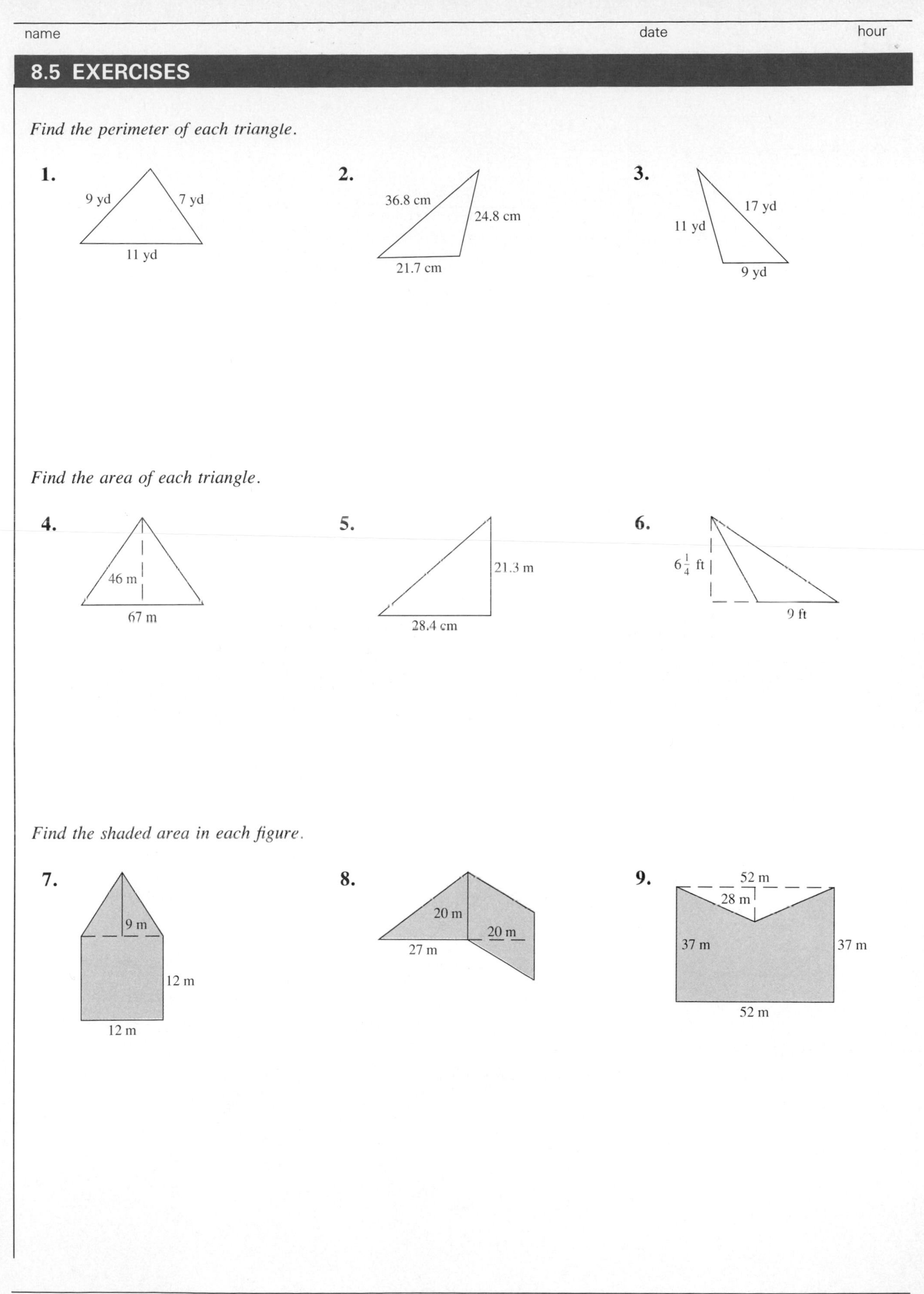

name date hour

8.5 EXERCISES

Find the perimeter of each triangle.

1.

2.

3.

Find the area of each triangle.

4.

5.

6.

Find the shaded area in each figure.

7.

8.

9.

10. **11.** **12.**

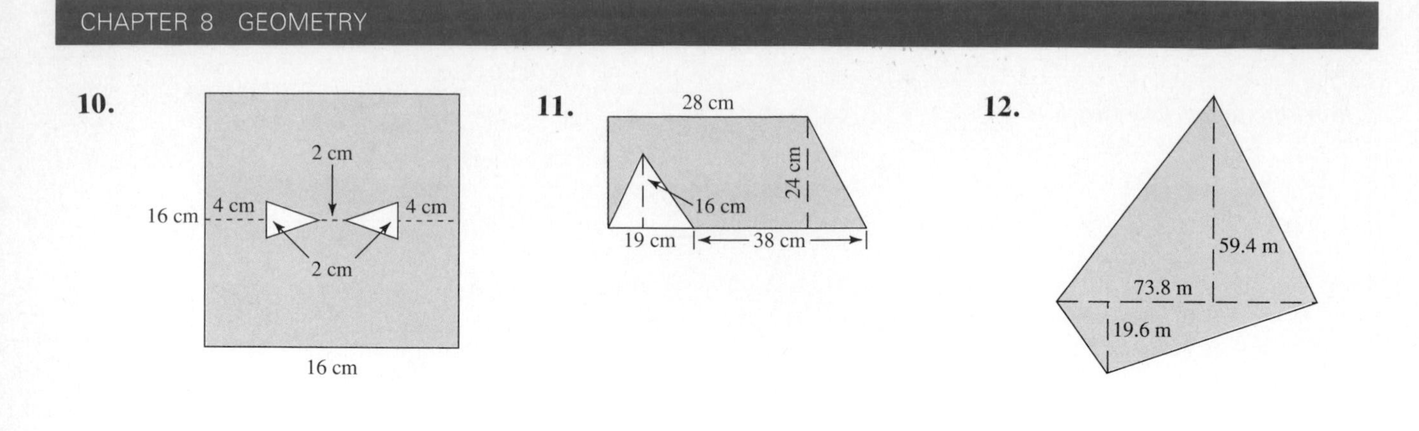

CHALLENGE EXERCISES

Solve each word problem.

13. **(a)** Find the area of one side of the house.
(b) Find the area of one roof section.

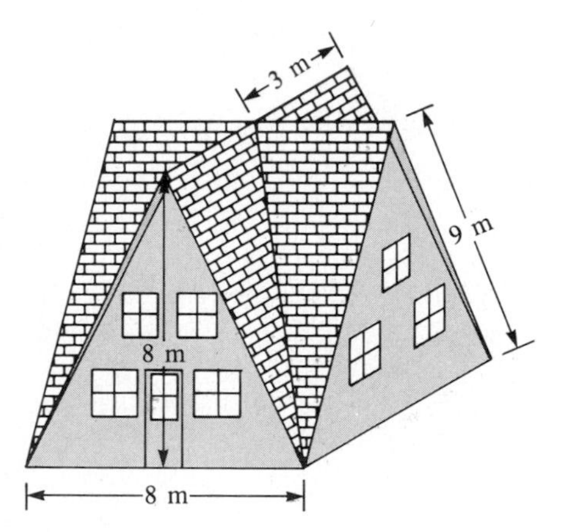

14. The sketch shows the plan for an office building. The shaded area will be a parking lot. What is the cost of building the parking lot, if the contractor charges $28.00 per square yard.

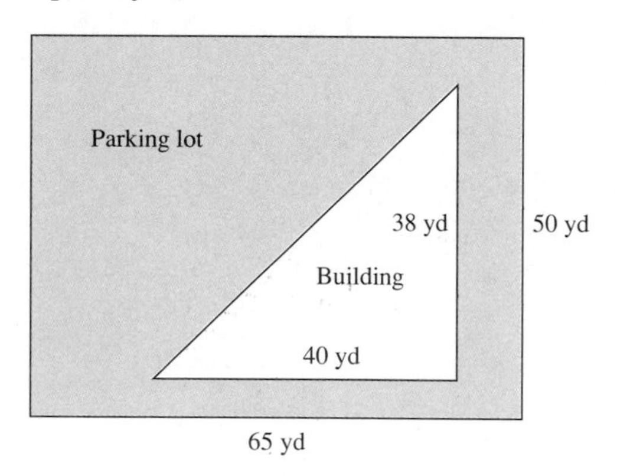

REVIEW EXERCISES

Multiply the following decimals. (For help, see ***Section 4.5.****)*

15. 3.14 · 16

16. 2.13 · 4.65

17. .71 · .34

18. .2 · 1.24

8.6 CIRCLES

1 A **circle** is a figure with all points the same distance from a fixed point within it. This fixed point is called the **center** of the circle. The distance across the circle, passing through the center, is called the **diameter,** d, of the circle. The distance from the center to any point on the circle is called the **radius,** r, of the circle.

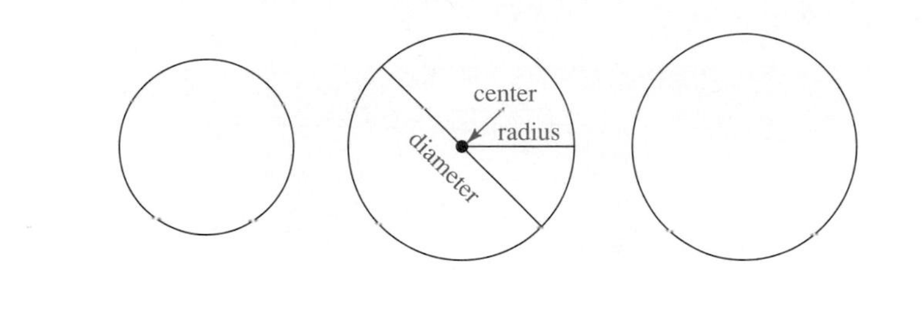

As this figure shows,

diameter = 2 · radius

or $d = 2r$

or $r = \frac{1}{2}d.$

WORK PROBLEM 1 AT THE SIDE.

EXAMPLE 1 Finding the Diameter and Radius of a Circle

Find the missing value in each circle.

(a)

$d = ?$

$r = 9$ cm

(b)

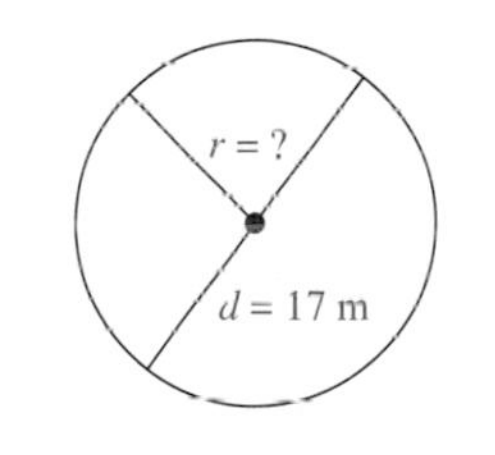

SOLUTION

(a) Since the radius is 9, the diameter is

$$d = 2r$$
$$d = 2 \cdot 9$$
$$d = 18 \text{ cm.}$$

(b) The radius is half the diameter.

$$r = \frac{1}{2}d$$
$$r = \frac{1}{2} \cdot 17$$
$$r = \frac{17}{2} \text{ or } 8\frac{1}{2} \text{ m}$$ ■

WORK PROBLEM 2 AT THE SIDE.

OBJECTIVES

1. Identify the parts of a circle.
2. Find the circumference of a circle.
3. Find the area of a circle.

1. Find the radius of each circle.

(a)

40 ft

(b)

9 m

2. Find the missing value in each circle.

(a)

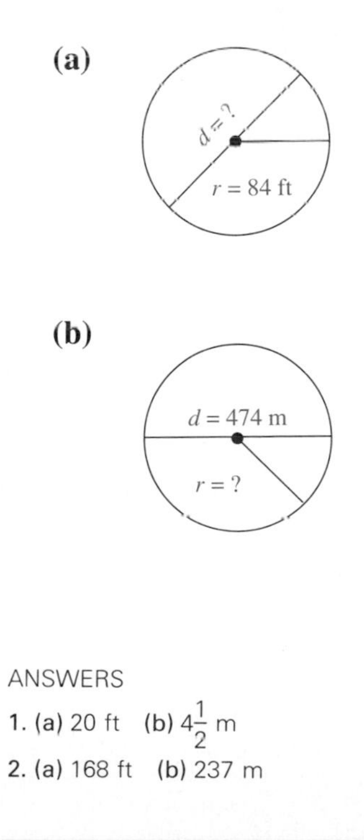

(b)

ANSWERS

1. (a) 20 ft (b) $4\frac{1}{2}$ m
2. (a) 168 ft (b) 237 m

2 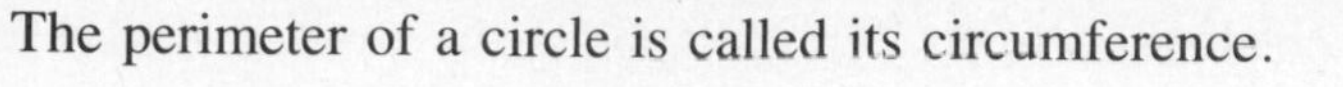
The perimeter of a circle is called its circumference.

NOTE Circumference is the distance around a circle.

The diameter of the can in the photograph is about 10.6 cm, and the circumference of the can is about 33.3 cm. Divide the circumference by the diameter to get

$$\frac{\text{circumference}}{\text{diameter}} = \frac{33.3}{10.6} = 3.14. \qquad \text{(rounded)}$$

The ratio of the circumference of *any* circle to its diameter is always about 3.14. This number is called π (the Greek letter pi). There is no decimal that is exactly equal to π, but approximately,

$$\pi = 3.14159265359.$$

Most problems with π will tell you the value of π to use.

It is common to round π to 3.14.

Use the following formulas to find the circumference of a circle.

$$\text{Circumference} = \pi \cdot \text{diameter}$$
$$C = \pi d$$
$$C = 2\pi r \qquad (\text{since } d = 2r)$$

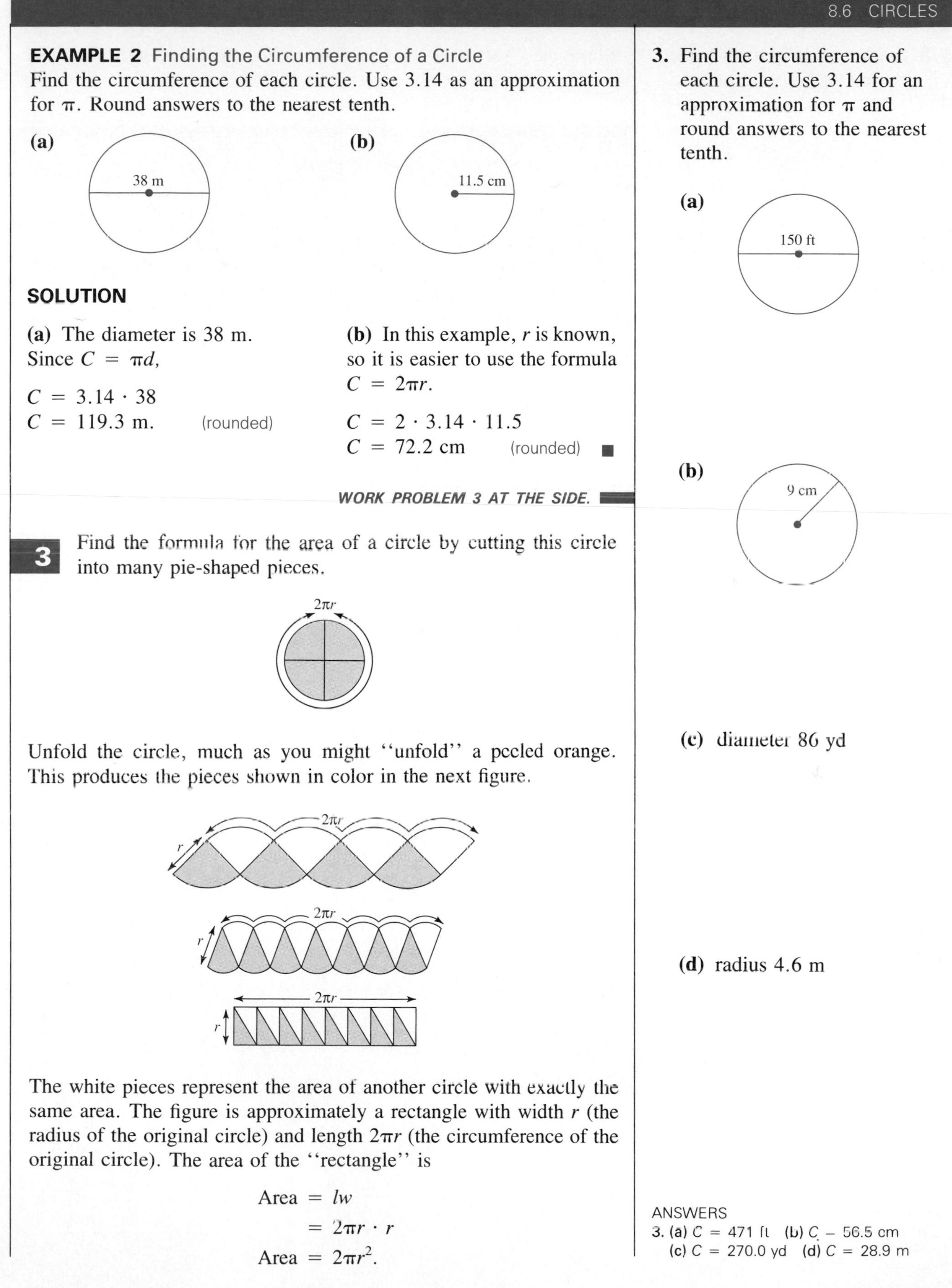

EXAMPLE 2 Finding the Circumference of a Circle

Find the circumference of each circle. Use 3.14 as an approximation for π. Round answers to the nearest tenth.

(a) 38 m

(b) 11.5 cm

SOLUTION

(a) The diameter is 38 m. Since $C = \pi d$,

$C = 3.14 \cdot 38$
$C = 119.3$ m. (rounded)

(b) In this example, r is known, so it is easier to use the formula $C = 2\pi r$.

$C = 2 \cdot 3.14 \cdot 11.5$
$C = 72.2$ cm (rounded) ■

WORK PROBLEM 3 AT THE SIDE.

3. Find the circumference of each circle. Use 3.14 for an approximation for π and round answers to the nearest tenth.

(a) 150 ft

(b) 9 cm

(c) diameter 86 yd

(d) radius 4.6 m

3 Find the formula for the area of a circle by cutting this circle into many pie-shaped pieces.

Unfold the circle, much as you might "unfold" a peeled orange. This produces the pieces shown in color in the next figure.

The white pieces represent the area of another circle with exactly the same area. The figure is approximately a rectangle with width r (the radius of the original circle) and length $2\pi r$ (the circumference of the original circle). The area of the "rectangle" is

$$\text{Area} = lw$$
$$= 2\pi r \cdot r$$
$$\text{Area} = 2\pi r^2.$$

ANSWERS
3. (a) $C = 471$ ft (b) $C = 56.5$ cm (c) $C = 270.0$ yd (d) $C = 28.9$ m

4. Find the area of each circle. Round to the nearest tenth.

(a)

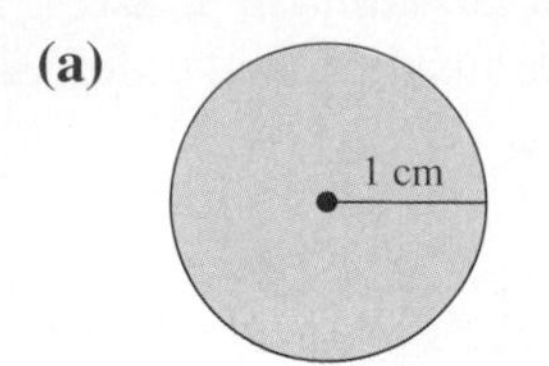

(b)

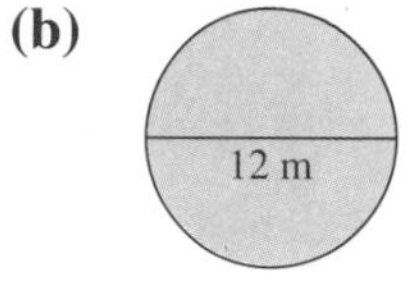

(Hint: r = ________.)

(c)

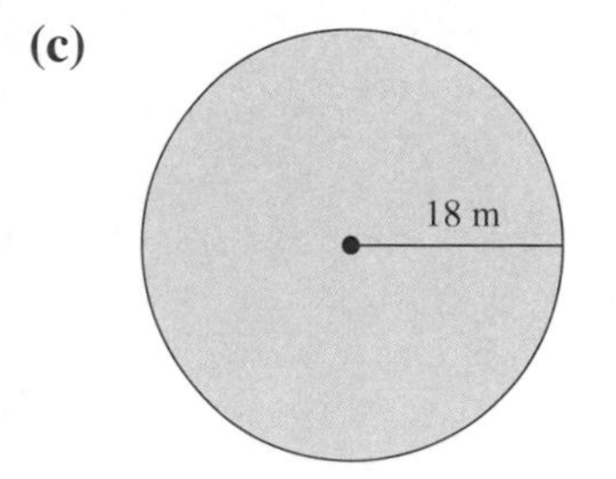

ANSWERS
4. (a) 3.14 cm^2 (b) 113.0 m^2
(c) 1017.4 m^2

Since the "rectangle" has an area twice the area of the original circle,

Area of circle = πr^2.

EXAMPLE 3 Finding the Area of a Circle

Find the area of a circle with radius 8.2 cm. The formula is $A = \pi r^2$. Use 8.2 for r and 3.14 as an approximation for π.

$$A = 3.14 \cdot 8.2 \cdot 8.2$$
$$A = 211.1 \text{ cm}^2 \quad \text{(rounded)}$$ ■

WORK PROBLEM 4 AT THE SIDE.

A half circle is called a **semicircle.** The area of a semicircle is *half* the area of a circle.

EXAMPLE 4 Finding the Area of a Semicircle

Find the area of each semicircle. Use 3.14 as an approximation for π. Round answers to the nearest tenth.

(a)

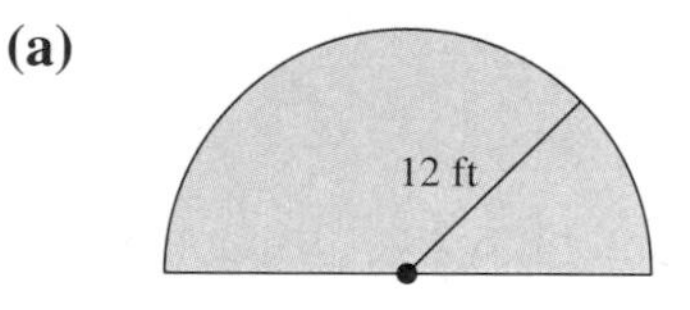

(b) A semicircle with $r = 7.8$ cm.

SOLUTION

(a) First, find the area of an entire circle with a radius of 12 ft.

$$A = \pi r^2$$
$$A = 3.14 \cdot 12 \cdot 12$$
$$A = 452.2 \text{ ft}^2 \quad \text{(rounded)}$$

The area of the semicircle is

$$\frac{1}{2} \cdot 452.2 = 226.1 \text{ ft}^2.$$

(b) The area of the circle is

$$A = 3.14 \cdot 7.8 \cdot 7.8$$
$$A = 191.0 \text{ cm}^2 \quad \text{(rounded)}$$

The area of the semicircle is

$$\frac{1}{2} \cdot 191 = 95.5 \text{ cm}^2. \blacksquare$$

WORK PROBLEM 5 AT THE SIDE.

EXAMPLE 5 Applying the Concept of Circumference

A circular rug is 8 feet in diameter. The cost of fringe for the edge is \$2.25 per foot. What will it cost to add fringe to the rug? Use 3.14 as an approximation for π.

SOLUTION

$$\text{Circumference} = \pi \cdot d$$
$$C = 3.14 \cdot 8$$
$$C = 25.12 \text{ ft}$$

$$\text{Cost} = \text{Cost per foot} \cdot \text{Circumference}$$
$$\text{Cost} = \$2.25 \cdot 25.12$$
$$\text{Cost} = \$56.52 \blacksquare$$

WORK PROBLEM 6 AT THE SIDE.

EXAMPLE 6 Applying the Concept of Area

Find the cost of covering the rug in Example 5 with a plastic cover. The material for the cover costs \$1.50 per square foot. Use 3.14 as an approximation for π.

SOLUTION

The radius, $r = \frac{1}{2} \cdot \text{diameter} = \frac{1}{2} \cdot 8 = 4$ ft.

So,
$$A = \pi r^2$$
$$A = 3.14 \cdot 4 \cdot 4$$
$$A = 50.24 \text{ sq ft.}$$

Thus,
$$\text{Cost} = 50.24 \cdot \$1.50$$
$$\text{Cost} = \$75.36. \blacksquare$$

WORK PROBLEM 7 AT THE SIDE.

5. Find the area of each semicircle. Use 3.14 as an approximation for π and round answers to the nearest tenth.

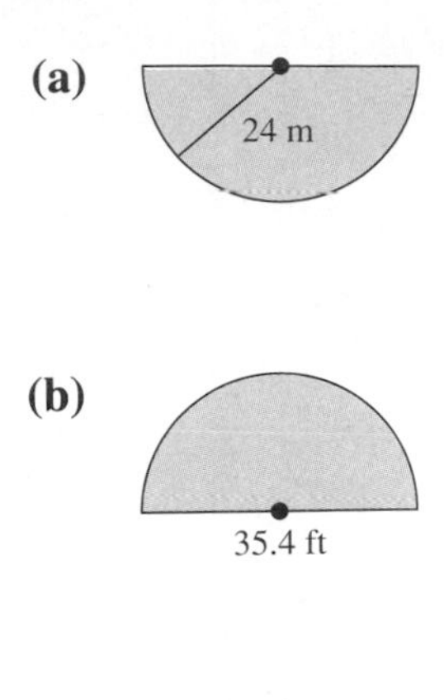

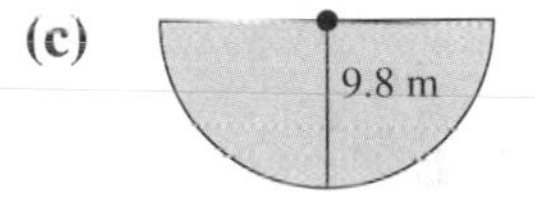

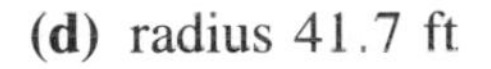

(d) radius 41.7 ft

6. Find the cost of binding around the edge of a circular rug 6 feet in diameter. Material costs \$1.75 per foot. Use 3.14 as an approximation for π.

7. Find the cost of covering the rug in Problem 6 on the previous page with plastic. Plastic costs \$2 per square foot.

ANSWERS
5. (a) 904.3 m^2 (b) 491.9 ft^2 (c) 150.8 m^2 (d) 2730.1 ft^2
6. \$32.97
7. \$56.52

name date hour

8.6 EXERCISES

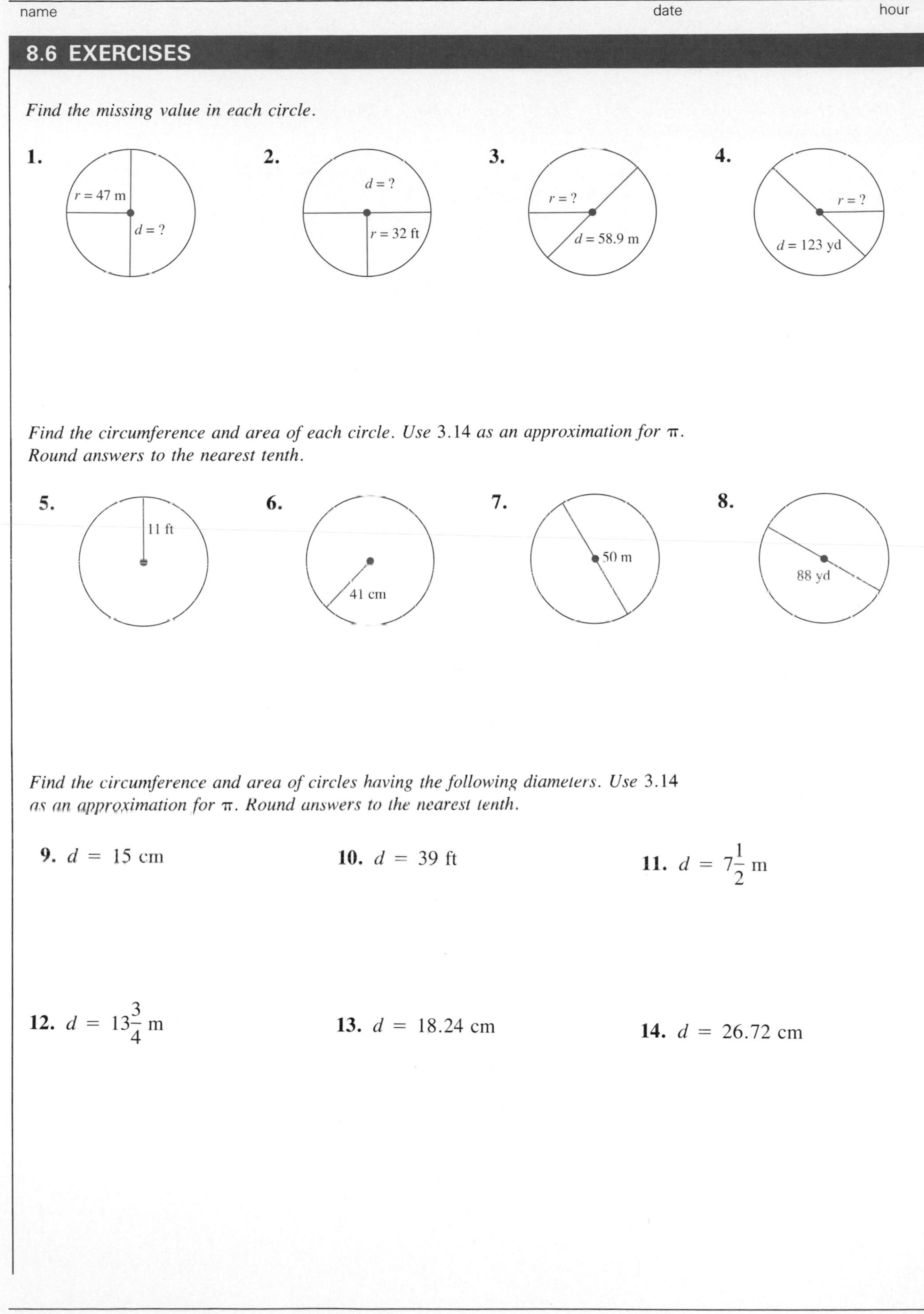

Find the missing value in each circle.

Find the circumference and area of each circle. Use 3.14 *as an approximation for* π*. Round answers to the nearest tenth.*

Find the circumference and area of circles having the following diameters. Use 3.14 *as an approximation for* π*. Round answers to the nearest tenth.*

9. $d = 15$ cm

10. $d = 39$ ft

11. $d = 7\frac{1}{2}$ m

12. $d = 13\frac{3}{4}$ m

13. $d = 18.24$ cm

14. $d = 26.72$ cm

Find each shaded area in Exercises 15–17. Use 3.14 as an approximation for π. Round answers to the nearest tenth.

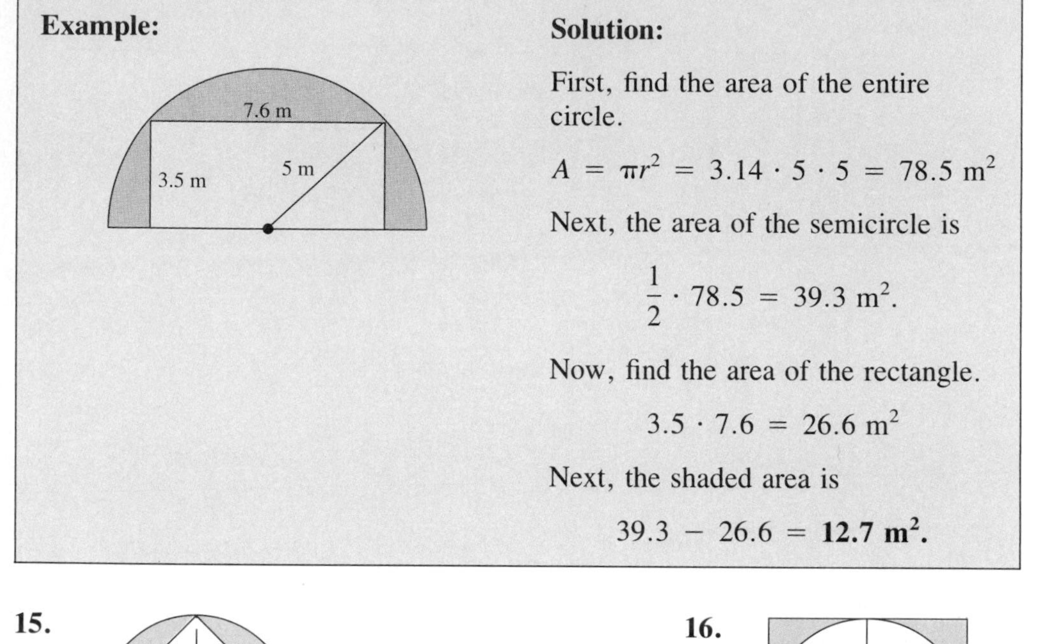

Example:

Solution:

First, find the area of the entire circle.

$A = \pi r^2 = 3.14 \cdot 5 \cdot 5 = 78.5 \text{ m}^2$

Next, the area of the semicircle is

$$\frac{1}{2} \cdot 78.5 = 39.3 \text{ m}^2.$$

Now, find the area of the rectangle.

$$3.5 \cdot 7.6 = 26.6 \text{ m}^2$$

Next, the shaded area is

$$39.3 - 26.6 = \mathbf{12.7 \text{ m}^2}.$$

15.

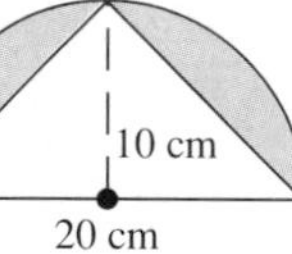

16.

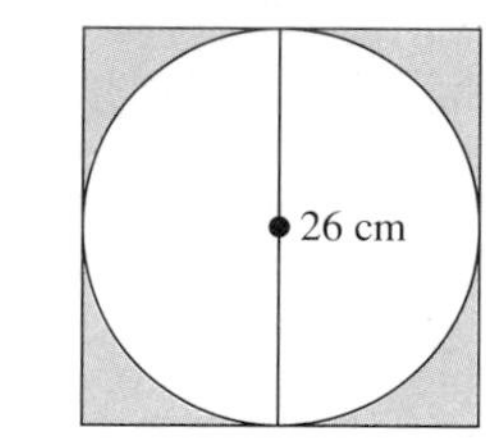

17.

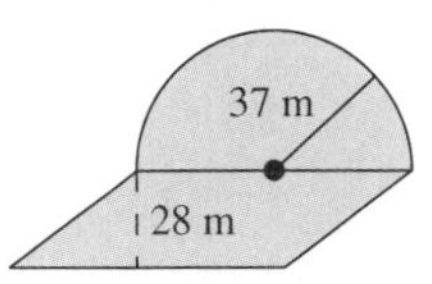

name date hour

Solve each word problem.

18. How far does a point on the tread of a tire move in one turn, if the diameter of the tire is 70 cm?

19. If you swing a ball held at the end of a string 2 m long, how far will the ball travel on each turn?

20. A wave energy extraction device is a huge undersea dome used to harness the power of ocean waves. The dome is 250 ft in diameter. Find its circumference.

21. Find the area of the dome in Exercise 20.

Find the shaded area in each figure. Use 3.14 as an approximation for π.

22. **23.**

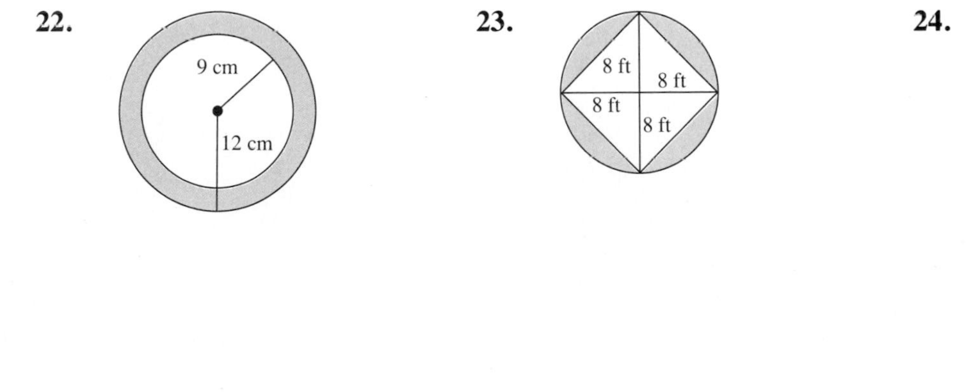

24.

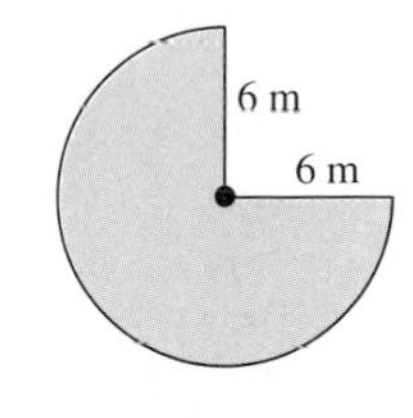

CHALLENGE EXERCISES

Solve each word problem.

25. Find the cost of sod, at \$1.76 per square foot, for the following playing field.

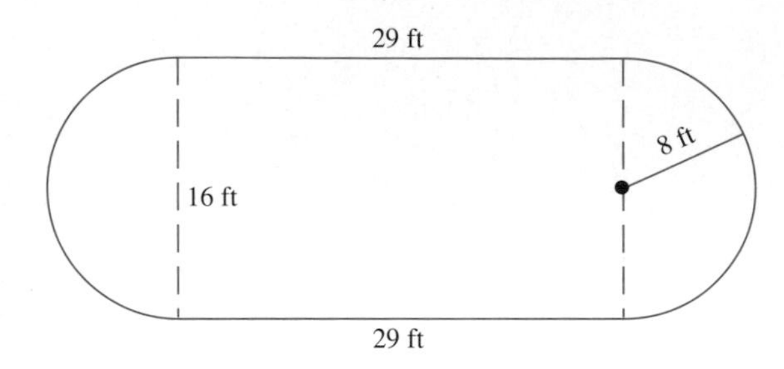

26. Find the area of this skating rink.

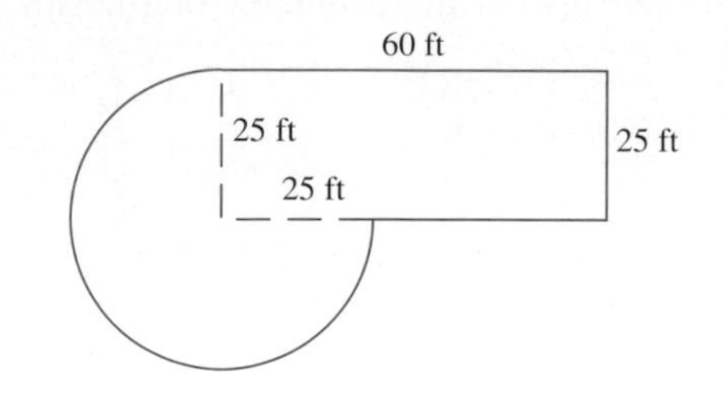

REVIEW EXERCISES

Divide the following fractions. (For help, see ***Section 2.7.****)*

27. $\frac{6}{7} \div \frac{12}{21}$

28. $\frac{2}{5} \div \frac{8}{15}$

29. $\dfrac{\frac{8}{7}}{4}$

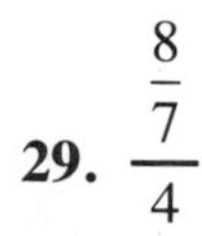

30. $\dfrac{\frac{8}{7}}{\frac{16}{21}}$

8.7 VOLUME

The space inside a solid (three-dimensional) figure is called its **volume.** Volume is measured in **cubic units.** One cubic inch (abbreviated 1 in^3) and one cubic centimeter (1 cc or 1 cm^3) are shown here.

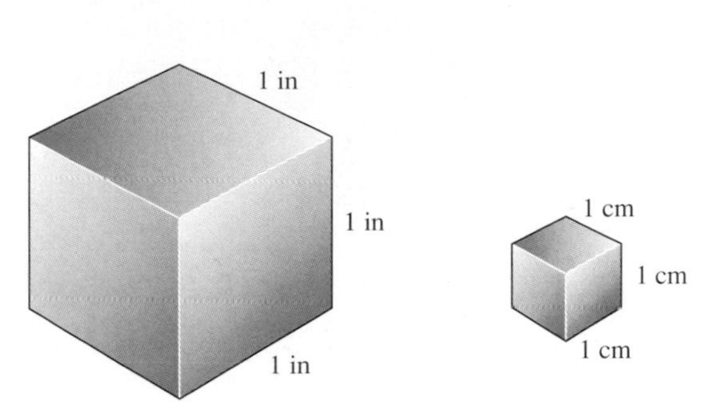

1 Use the following formula to find the volume of a box.

Volume of box = length · width · height

$$V = lwh$$

EXAMPLE 1 Finding the Volume of a Box

Find the volume of each box.

(a) **(b)**

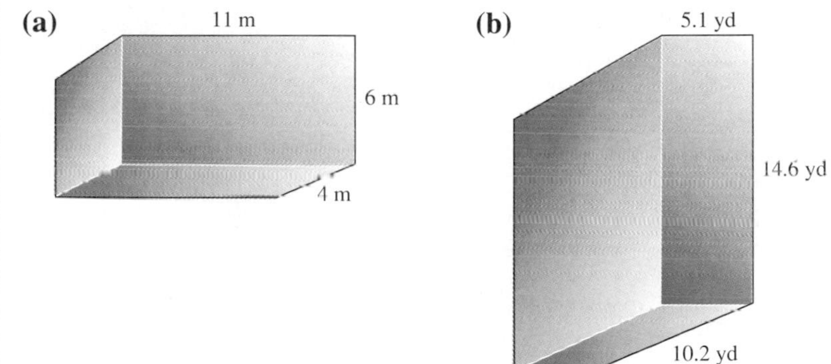

SOLUTION

(a) The length is 11 m, the width is 4 m, and the height is 6 m. The volume is

$$V = lwh$$
$$V = 11 \cdot 6 \cdot 4 = 264 \text{ m}^3.$$

(b) $V = 14.6 \cdot 5.1 \cdot 10.2$

$V = 759.5 \text{ yd}^3$ (rounded to the nearest tenth) ■

WORK PROBLEM 1 AT THE SIDE.

OBJECTIVES

Find the volume of a

1. box;
2. sphere;
3. cylinder;
4. cone and pyramid.

1. Find the volume of each box. Round to the nearest tenth if necessary.

(a) 7 cm, 8 cm, 9 cm

(b) 23.4 cm, 52.3 cm, 15.2 cm

(c) length $6\frac{1}{4}$ ft, width $3\frac{1}{2}$ ft, height 2 ft

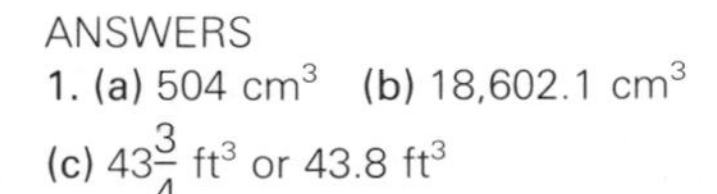

ANSWERS

1. (a) 504 cm^3 (b) 18,602.1 cm^3 (c) $43\frac{3}{4}$ ft^3 or 43.8 ft^3

2. Find the volume of each sphere. Use 3.14 as an approximation for π. Round to the nearest tenth.

(a)

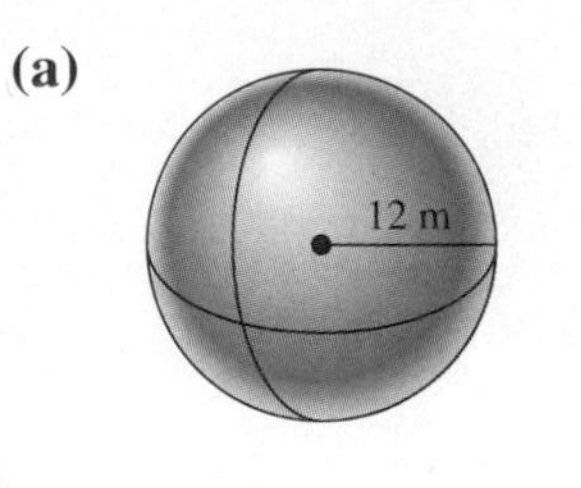

(b)

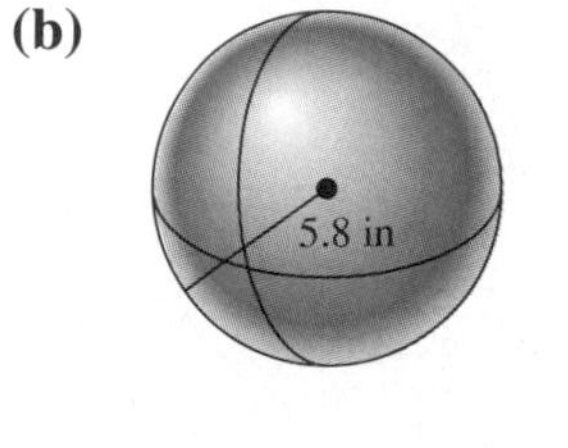

(c) radius 1.6 yd

ANSWERS
(a) 7234.6 m^3 (b) 816.9 in^3
(c) 17.1 yd^3

2 A **sphere** is shown here. Examples of spheres include baseballs and some oil storage tanks.

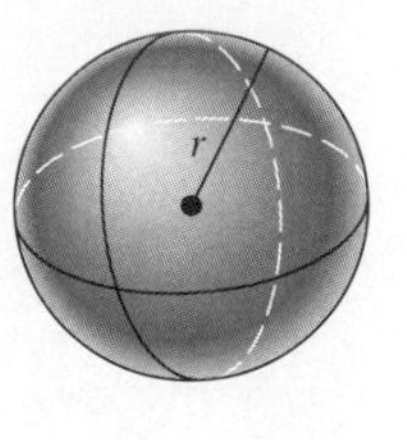

As with circles, the **radius** of a sphere is the distance from the center to the edge of the sphere. Use the following formula to find the volume of a sphere.

$$\text{Volume of sphere} = \frac{4}{3} \cdot \pi \cdot r \cdot r \cdot r$$

$$V = \frac{4}{3}\pi r^3$$

EXAMPLE 2 Finding the Volume of a Sphere
Find the volume of each sphere. Use 3.14 as an approximation for π and round to the nearest tenth.

(a)

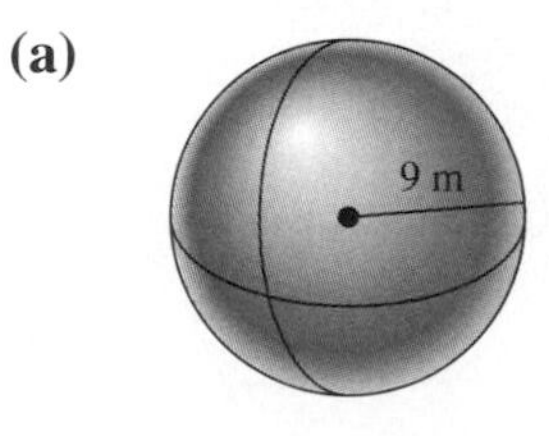

(b)

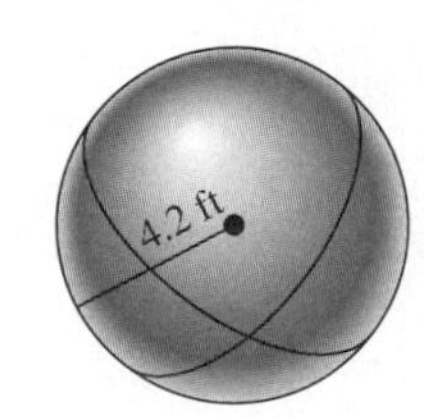

SOLUTION

(a) $V = \frac{4}{3}\pi r^3$

$V = \frac{4}{3} \cdot 3.14 \cdot 9 \cdot 9 \cdot 9$

$V = 3052.1 \text{ m}^3$ (rounded)

(b) $V = \frac{4}{3} \cdot 3.14 \cdot 4.2 \cdot 4.2 \cdot 4.2$

$V = 310.2 \text{ ft}^3$ (rounded) ■

WORK PROBLEM 2 AT THE SIDE.

Half a sphere is called a **hemisphere.** The volume of a hemisphere is half the volume of a sphere. Use the following formula to find the volume of a hemisphere.

$$\text{Volume of hemisphere} = \frac{1}{2} \cdot \frac{4}{3} \cdot \pi \cdot r \cdot r \cdot r$$

$$V = \frac{2}{3}\pi r^3$$

EXAMPLE 3 Finding the Volume of a Hemisphere

Find the volume of each hemisphere. Use 3.14 as an approximation for π and round to the nearest tenth.

(a) 7 m **(b)** 32.9 cm

SOLUTION

(a) $V = \frac{2}{3}\pi r^3$

$V = \frac{2}{3} \cdot 3.14 \cdot 7 \cdot 7 \cdot 7$

$V = 718.0 \text{ m}^3$ (rounded)

(b) $V = \frac{2}{3} \cdot 3.14 \cdot 32.9 \cdot 32.9 \cdot 32.9$

$V = 74{,}546.3 \text{ cm}^3$ (rounded) ■

WORK PROBLEM 3 AT THE SIDE.

3 Several **cylinders** are shown here.

radius

height

These are called right circular cylinders because the top and bottom are circles, and the sides are perpendicular to the top and bottom. A tin can is an example of a right circular cylinder.

3. Find the volume of each hemisphere. Use 3.14 as an approximation for π and round to the nearest tenth.

(a) 15 ft

(b) 28.7 in.

(c) radius 3.7 yd

ANSWERS
3. (a) 7065 ft³ (b) 49,486.2 in³
(c) 106.0 yd³

4. Find the volume of each cylinder. Use 3.14 as an approximation for π. Round to the nearest tenth.

(a)

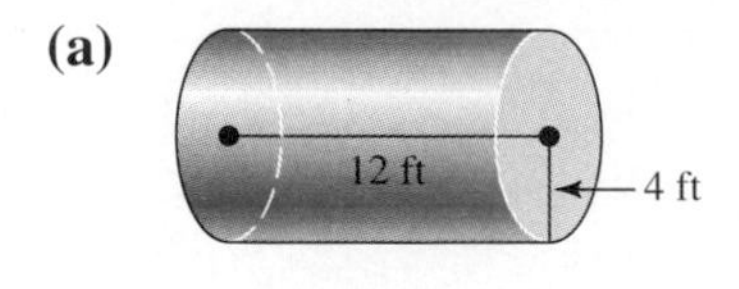

(b)

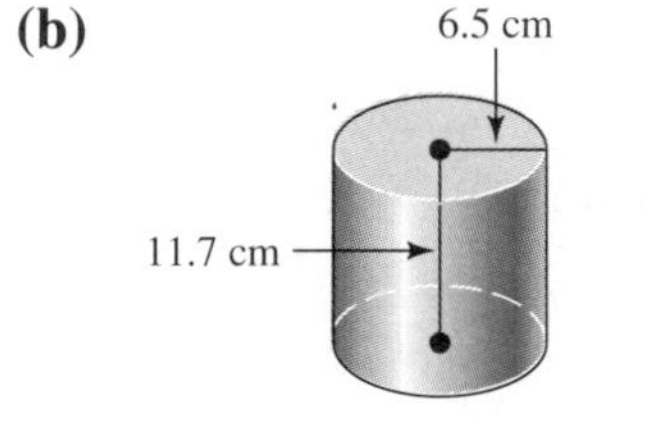

(c) radius 1.9 ft, height 7.6 ft

ANSWERS
4. (a) 602.9 ft^3 (b) 1552.2 cm^3
(c) 86.1 ft^3

Use the following formula to find the volume of a cylinder.

$$\text{volume of cylinder} = \pi \cdot r \cdot r \cdot h$$
$$V = \pi r^2 h$$

EXAMPLE 4 Finding the Volume of a Cylinder
Find the volume of each cylinder. Use 3.14 as an approximation for π and round to the nearest tenth.

(a)

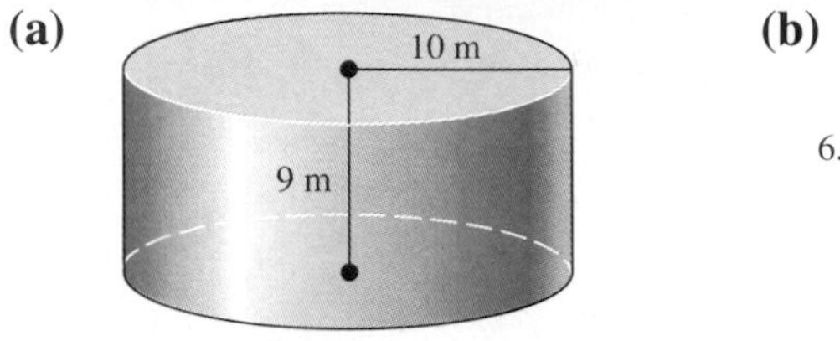

(b)

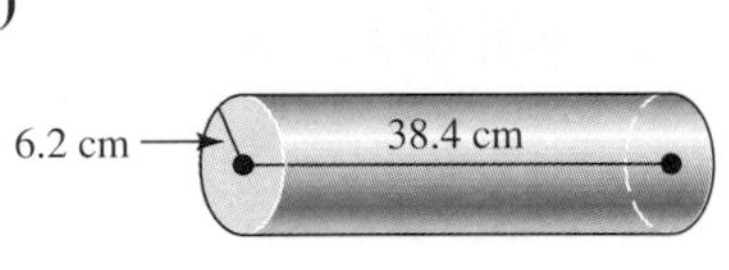

SOLUTION

(a) The radius is 10 m and the height is 9 m. The volume is

$$V = \pi r^2 h$$
$$V = 3.14 \cdot 10 \cdot 10 \cdot 9$$
$$V = 2826 \text{ m}^3.$$

(b) $V = 3.14 \cdot 6.2 \cdot 6.2 \cdot 38.4$
$V = 4634.9 \text{ cm}^3$ (rounded) ■

WORK PROBLEM 4 AT THE SIDE.

4 A cone and a pyramid are shown here.

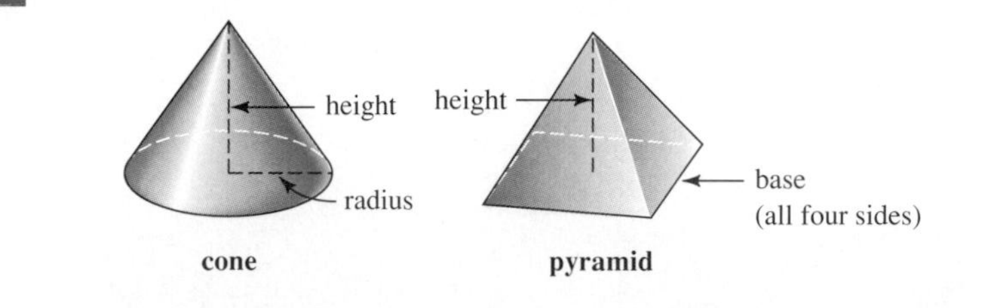

Use the following formula to find the volume of a cone.

$$\text{Volume of a cone} = \frac{1}{3} \cdot \pi \cdot r^2 \cdot h$$
$$V = \frac{1}{3}\pi r^2 h$$

EXAMPLE 5 Finding the Volume of a Cone
Find the volume of the cone. Use 3.14 as an approximation for π and round to the nearest tenth.

9 cm
4 cm

SOLUTION
The radius is 4 cm, and the height is 9 cm.

$$V = \frac{1}{3}\pi r^2 h$$

$$V = \frac{1}{3} \cdot 3.14 \cdot 4 \cdot 4 \cdot 9$$

$$V = 150.7 \text{ cm}^3 \quad \text{(rounded)}$$ ■

WORK PROBLEM 5 AT THE SIDE.

Use the following formula to find the volume of a pyramid.

$$\text{Volume of a pyramid} = \frac{1}{3} \quad b \cdot h$$

$$V = \frac{1}{3}bh$$

where b is the area of the base of the pyramid and h is the height of the pyramid.

EXAMPLE 6 Finding the Volume of a Pyramid
Find the volume of the pyramid. Round answer to the nearest tenth.

$h = 11$ cm
4 cm
5 cm

SOLUTION
The area of the base is

$$4 \cdot 5 = 20$$

So $b = 20$.

Next, find the volume.

$$V = \frac{1}{3}bh$$

$$V = \frac{1}{3} \cdot 20 \cdot 11$$

$$V = 73.3 \text{ cm}^3 \quad \text{(rounded)}$$ ■

WORK PROBLEM 6 AT THE SIDE.

5. Find the volume of a cone with base radius 7 cm and height 12 cm. Use 3.14 as an approximation for π, and round to the nearest tenth.

6. Find the volume of a pyramid with base 4 ft by 7 ft and height 9 ft.

ANSWERS
5. 615.4 cm^3
6. 84 ft^3

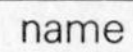

name date hour

8.7 EXERCISES

Find the volume of each figure. Use 3.14 as an approximation for π and round answers to the nearest tenth.

1. 11 cm, 4 cm, 12 cm

2. 2 ft, 10 ft, 9 ft

3. 12 in, 12 in, 12 in

4. 15 in, 3 in, 14 in

5. 22 m

6. 1.53 m

7. 12 in

8. 7.4 in

9. 5 ft, 6 ft

10. 12 m, 15 m

11. 18 ft, 5 ft

12. 12 in, 21 in

Find the volume of each figure. Use 3.14 as an approximation for π and round answers to the nearest tenth.

13.

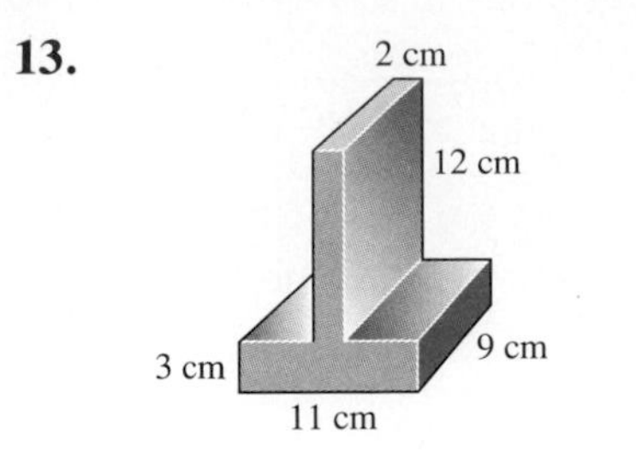

14. Find the volume of the shaded part.

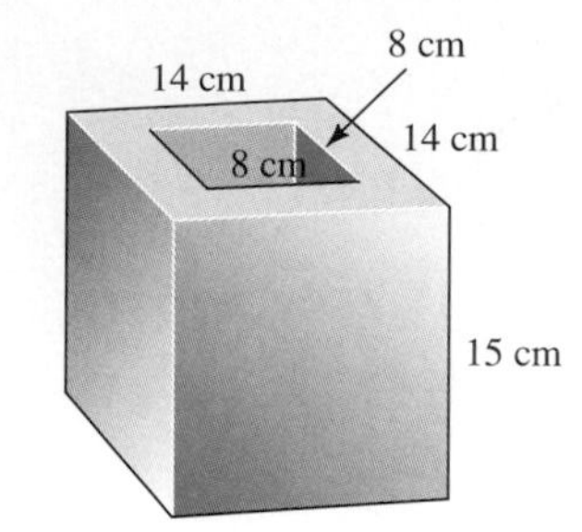

Find each volume.

15.

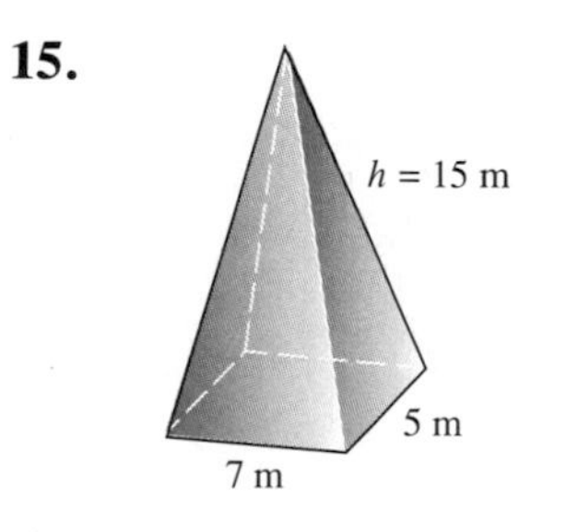

16.

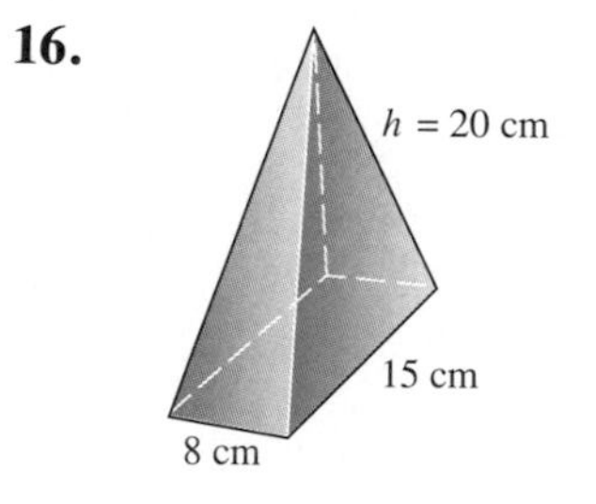

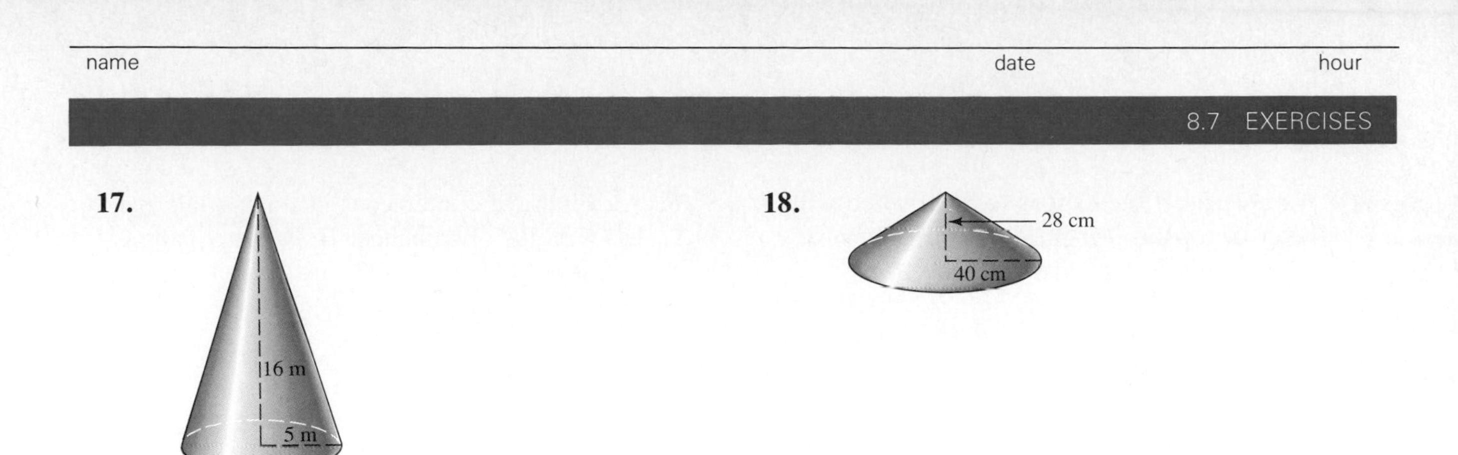

Find the volume of each of the following. Round answers to the nearest tenth. Use 3.14 *as an approximation for* π.

19. a coffee can, radius 6.3 cm and height 15.8 cm

20. a jelly jar, radius 3.2 cm and height 9.5 cm

21. an oil can, diameter 7.2 cm and height 10.5 cm

22. a cardboard mailing tube, diameter 2 cm and height 40 cm

CHALLENGE EXERCISES

23. Find the volume of the shaded part.

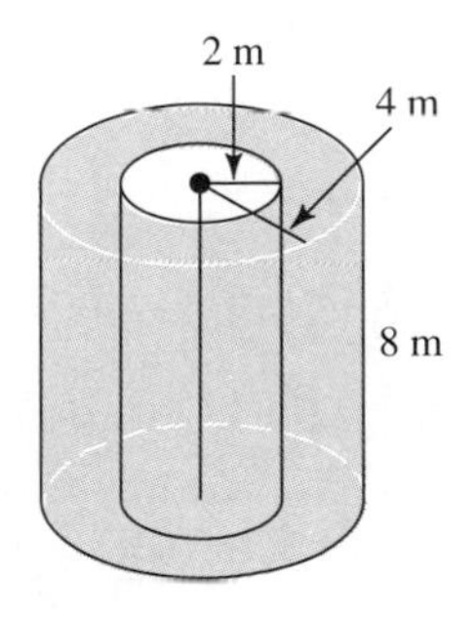

24. Find the volume.

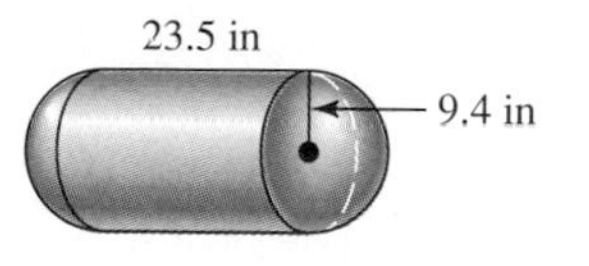

25. An area 54 ft by 18 ft is to be covered with a layer of topsoil 3 inches thick. How many cubic yards of topsoil will be needed?

26. A cylindrical container is 20 inches tall and has a radius of 6 inches. How many gallons will it hold (1 gal = 231 in^3). Round answer to the nearest tenth of a gallon.

REVIEW EXERCISES

Add or subtract the following decimals. (For help, see ***Sections 4.3–4.4.****)*

27. 47.123 + 36.345

28. 36.28 − 15.19

29. 23.165 − 14.235

30. 16.215 + 23.841 − 11.799

8.8 PYTHAGOREAN FORMULA

Recall the formula for area of a square, $A = s^2$. The square on the left has area

$$A = s^2 = 5 \cdot 5 = 25 \text{ cm}^2.$$

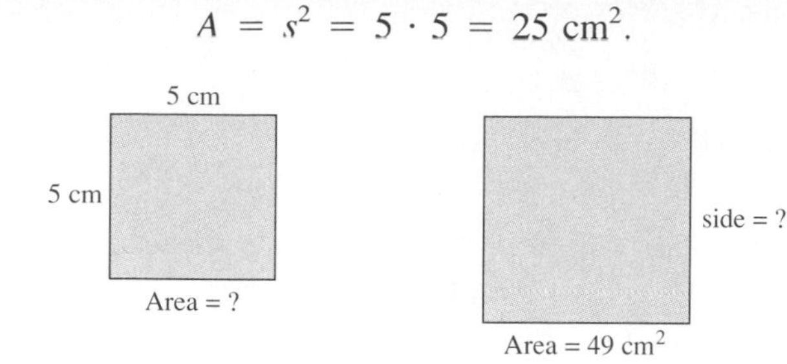

The square on the right has an area of 49 cm^2. To find the length of a side of this square, ask "What number can be multiplied by itself to give 49?" Since $7 \cdot 7 = 49$, the length of each side of the square on the right is 7 cm.

Remember: $7 \cdot 7 = 49$, so 7 is the **square root** of 49, or $\sqrt{49} = 7$. Also, $\sqrt{81} = 9$, since $9 \cdot 9 = 81$.

WORK PROBLEM 1 AT THE SIDE.

A number that has a whole number as its square root is called a *perfect square*. For example, 9 is a perfect square because $\sqrt{9} = 3$, and 3 is a whole number.

The first few perfect squares are listed here.

$\sqrt{1} = 1$	$\sqrt{16} = 4$	$\sqrt{49} = 7$	$\sqrt{100} = 10$
$\sqrt{4} = 2$	$\sqrt{25} = 5$	$\sqrt{64} = 8$	$\sqrt{121} = 11$
$\sqrt{9} = 3$	$\sqrt{36} = 6$	$\sqrt{81} = 9$	$\sqrt{144} = 12$

1 If a number is not a perfect square, then its approximate square root can be found from a square root table, such as the one inside the back cover, or from a calculator with a square root key.

EXAMPLE 1 Finding the Square Root of a Number

Find each square root. Round to the nearest thousandth.

(a) $\sqrt{7}$ **(b)** $\sqrt{35}$ **(c)** $\sqrt{124}$ **(d)** $\sqrt{200}$

SOLUTION

(a) Find $\sqrt{7}$ from the table by locating 7 in the "number column." Next, under the "square root column," find

$$\sqrt{7} = 2.646.$$

The number 2.646 is an *approximate* square root of 7. Check this by multiplying 2.646 and 2.646; the result should be close to 7. A more accurate square root can be found with a calculator that has a square root key.

OBJECTIVES

1 Use the square root table.

2 Find the missing length in a right triangle.

3 Solve problems with right triangles.

1. Find each square root.

(a) $\sqrt{4}$

(b) $\sqrt{25}$

(c) $\sqrt{64}$

(d) $\sqrt{100}$

(e) $\sqrt{144}$

ANSWERS

1. (a) 2 (b) 5 (c) 8 (d) 10 (e) 12

2. Use the table or a calculator with a square root key to find each square root. Round to the nearest thousandth if necessary.

(a) $\sqrt{11}$

(b) $\sqrt{29}$

(c) $\sqrt{56}$

(d) $\sqrt{92}$

(e) $\sqrt{147}$

ANSWERS
2. (a) 3.317 (b) 5.385 (c) 7.483
(d) 9.592 (e) 12.124

(b) $\sqrt{35} = 5.916$

(c) $\sqrt{124} = 11.136$

(d) $\sqrt{200} = 14.142$ ■

WORK PROBLEM 2 AT THE SIDE.

2 Square roots are used with the *Pythagorean formula,* which applies only to **right triangles** (a triangle with a 90° angle). The Pythagorean formula comes to us from the Pythagoreans, who lived in Italy about 2500 years ago. They may have discovered the formula from floor tiles, as shown here.

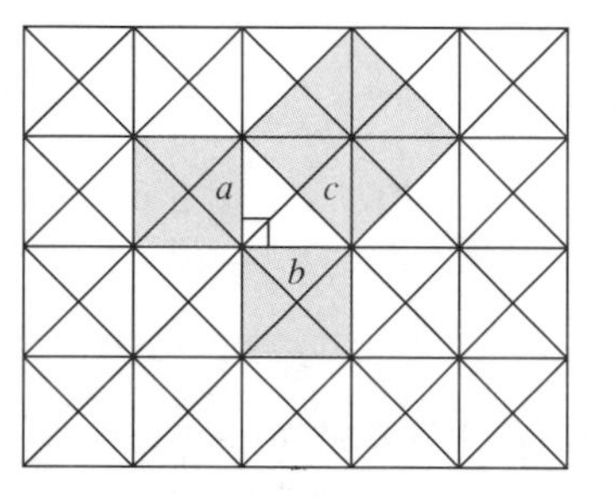

The square drawn on side a contains four triangles. The square on side b contains four triangles. The square on side c contains eight triangles. The number of triangles in the square on side c equals the sum of the number of triangles in the square on sides a and b.

The Pythagorean formula follows and shows how to find the lengths of the legs (the two shorter sides) and the hypotenuse (the side opposite the right angle) of a right angle.

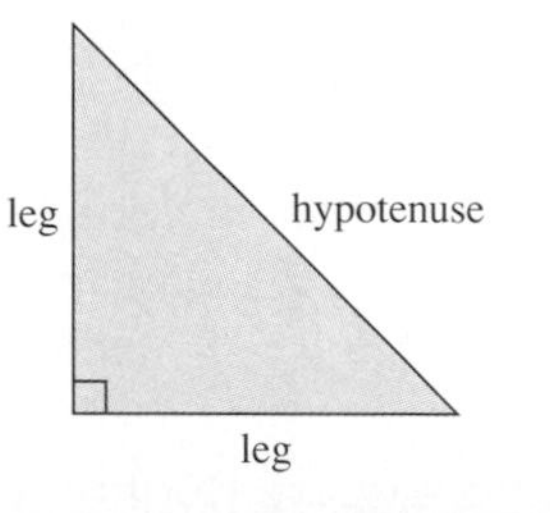

PYTHAGOREAN FORMULA

$$(\textbf{hypotenuse})^2 = (\textbf{leg})^2 + (\textbf{leg})^2$$

To find the hypotenuse, use

$$\textbf{hypotenuse} = \sqrt{(\textbf{leg})^2 + (\textbf{leg})^2}.$$

To find a leg, use

$$\textbf{leg} = \sqrt{(\textbf{hypotenuse})^2 - (\textbf{leg})^2}.$$

Remember: A square drawn in the corner of an angle shows a right angle.

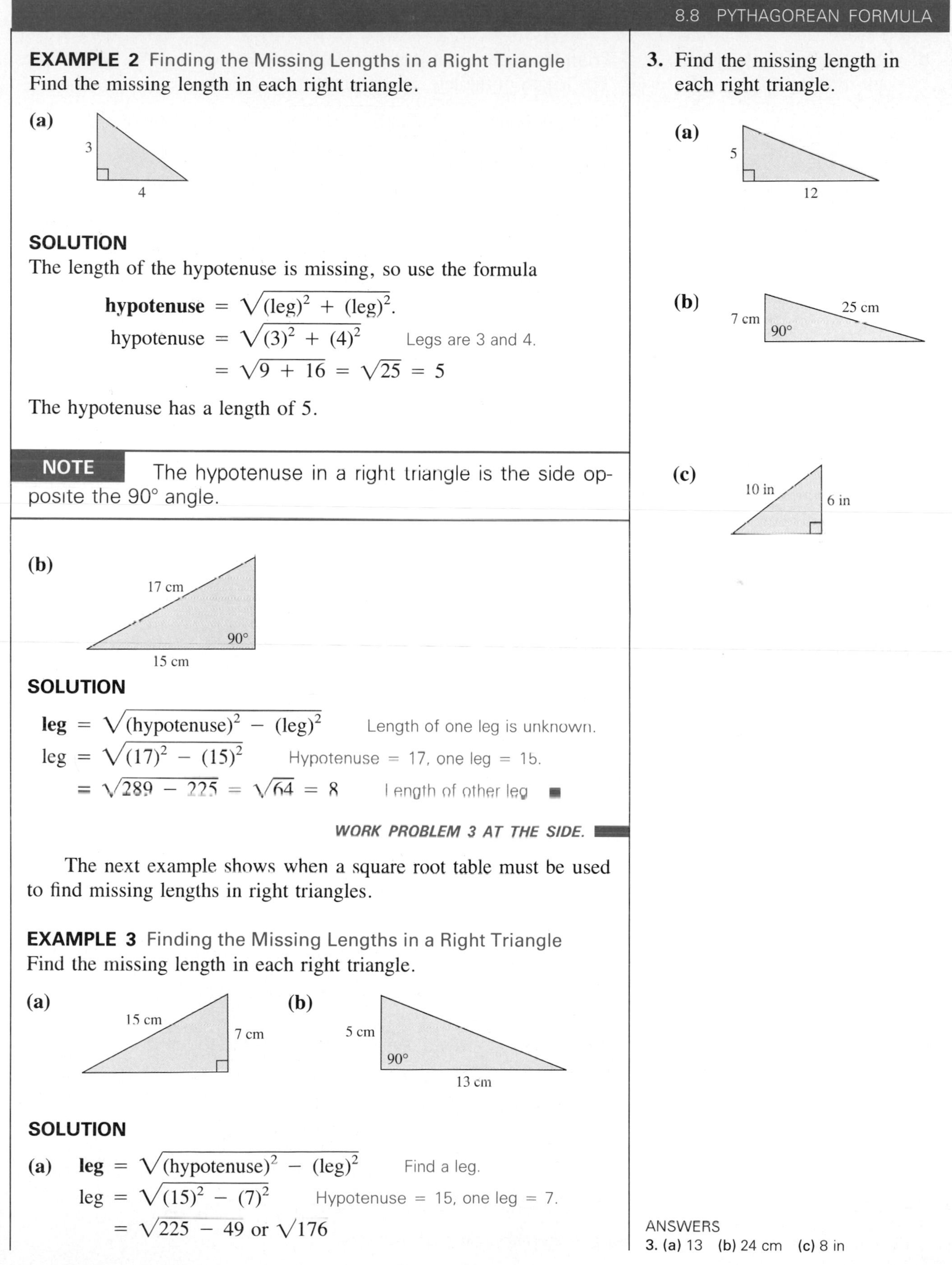

EXAMPLE 2 Finding the Missing Lengths in a Right Triangle
Find the missing length in each right triangle.

(a)

SOLUTION
The length of the hypotenuse is missing, so use the formula

$$\textbf{hypotenuse} = \sqrt{(\text{leg})^2 + (\text{leg})^2}.$$

$$\text{hypotenuse} = \sqrt{(3)^2 + (4)^2} \quad \text{Legs are 3 and 4.}$$

$$= \sqrt{9 + 16} = \sqrt{25} = 5$$

The hypotenuse has a length of 5.

NOTE The hypotenuse in a right triangle is the side opposite the 90° angle.

(b)

SOLUTION

$$\textbf{leg} = \sqrt{(\text{hypotenuse})^2 - (\text{leg})^2} \quad \text{Length of one leg is unknown.}$$

$$\text{leg} = \sqrt{(17)^2 - (15)^2} \quad \text{Hypotenuse} = 17\text{, one leg} = 15.$$

$$= \sqrt{289 - 225} = \sqrt{64} = 8 \quad \text{Length of other leg} \blacksquare$$

WORK PROBLEM 3 AT THE SIDE.

The next example shows when a square root table must be used to find missing lengths in right triangles.

EXAMPLE 3 Finding the Missing Lengths in a Right Triangle
Find the missing length in each right triangle.

(a)

(b)

SOLUTION

(a) $\textbf{leg} = \sqrt{(\text{hypotenuse})^2 - (\text{leg})^2}$ Find a leg.

$\text{leg} = \sqrt{(15)^2 - (7)^2}$ Hypotenuse = 15, one leg = 7.

$= \sqrt{225 - 49}$ or $\sqrt{176}$

3. Find the missing length in each right triangle.

(a)

(b)

(c)

ANSWERS
3. (a) 13 (b) 24 cm (c) 8 in

4. Find the missing length in each triangle.

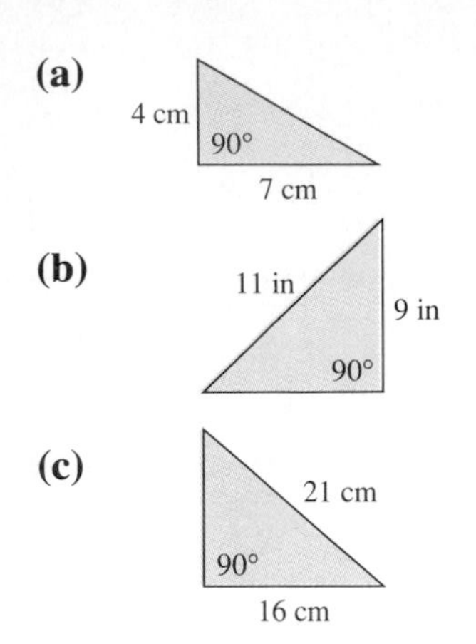

5. These sketches show ladders leaning against buildings. Find the missing lengths. Round to the nearest thousandth of a foot.

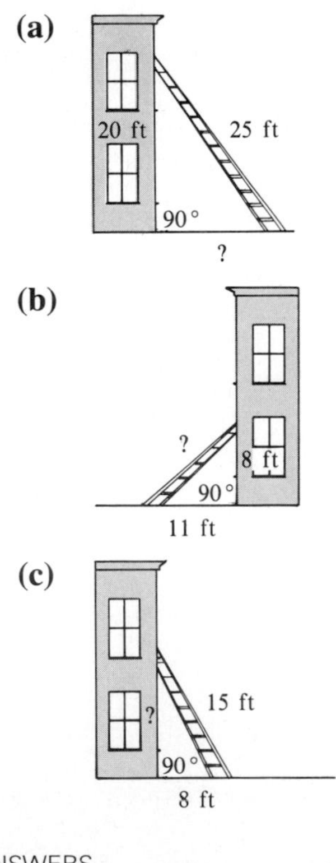

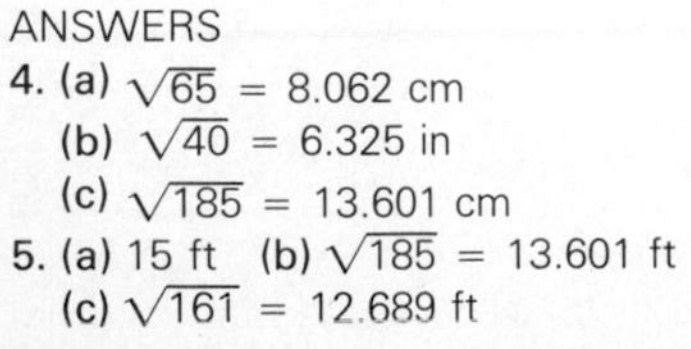

ANSWERS

4. (a) $\sqrt{65} = 8.062$ cm
(b) $\sqrt{40} = 6.325$ in
(c) $\sqrt{185} = 13.601$ cm

5. (a) 15 ft (b) $\sqrt{185} = 13.601$ ft
(c) $\sqrt{161} = 12.689$ ft

From the square root table inside the back cover, $\sqrt{176} = 13.266$. The length of the leg is approximately 13.266 cm.

(b) $\textbf{hypotenuse} = \sqrt{(\text{leg})^2 + (\text{leg})^2}$ Use this formula.

$= \sqrt{(5)^2 + (13)^2}$ Legs are 5 and 13.

$= \sqrt{25 + 169} = \sqrt{194}$

From the table, $\sqrt{194} = 13.928$. Approximate length of hypotenuse ■

WORK PROBLEM 4 AT THE SIDE.

3 The next example shows an application of the Pythagorean formula.

EXAMPLE 4 Using the Pythagorean Formula

A television antenna is on the roof of a house, as shown. Find the length of the support wire shown in the sketch.

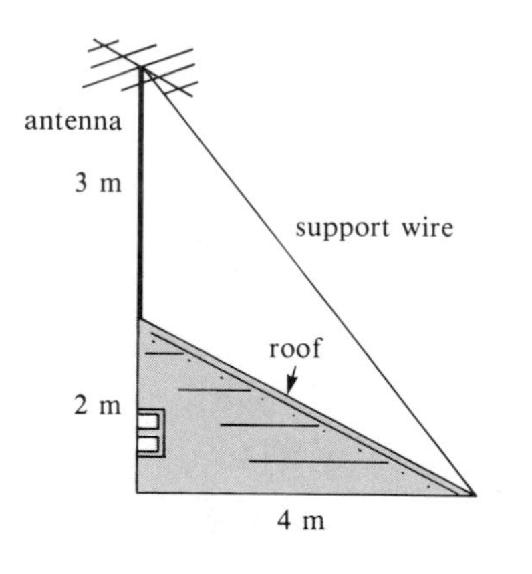

SOLUTION

The total length of the side at the left is 3 m + 2 m = 5 m.

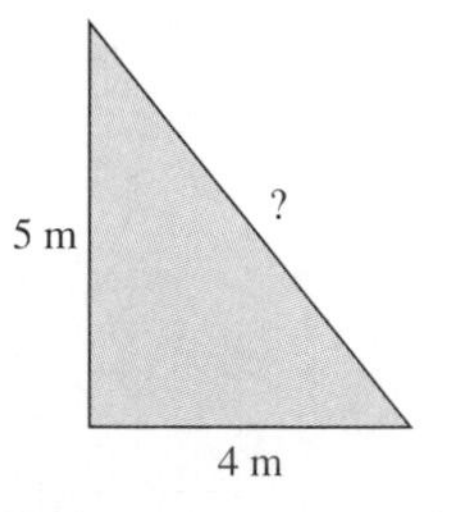

Find the length of the support wire with the formula for the hypotenuse.

$\textbf{hypotenuse} = \sqrt{(\text{leg})^2 + (\text{leg})^2}$

hypotenuse $= \sqrt{(5)^2 + (4)^2}$ Legs are 4 and 5.

$= \sqrt{25 + 16} = \sqrt{41} = 6.403$ From the table

The support wire has a length of about 6.403 m. ■

WORK PROBLEM 5 AT THE SIDE.

name date hour

8.8 EXERCISES

Find each square root. Starting with Exercise 5, use the table inside the back cover.

1. $\sqrt{25}$ **2.** $\sqrt{36}$ **3.** $\sqrt{64}$ **4.** $\sqrt{81}$

5. $\sqrt{11}$ **6.** $\sqrt{23}$ **7.** $\sqrt{45}$ **8.** $\sqrt{52}$

9. $\sqrt{73}$ **10.** $\sqrt{80}$ **11.** $\sqrt{106}$ **12.** $\sqrt{125}$

13. $\sqrt{190}$ **14.** $\sqrt{160}$ **15.** $\sqrt{173}$ **16.** $\sqrt{195}$

Find the areas of the squares on the sides of the right triangles in Exercises 17 and 18. Check to see if the Pythagorean formula holds true.

Find the missing length in each right triangle. Use a calculator and round to the nearest thousandth, or write the answer with a square root sign.

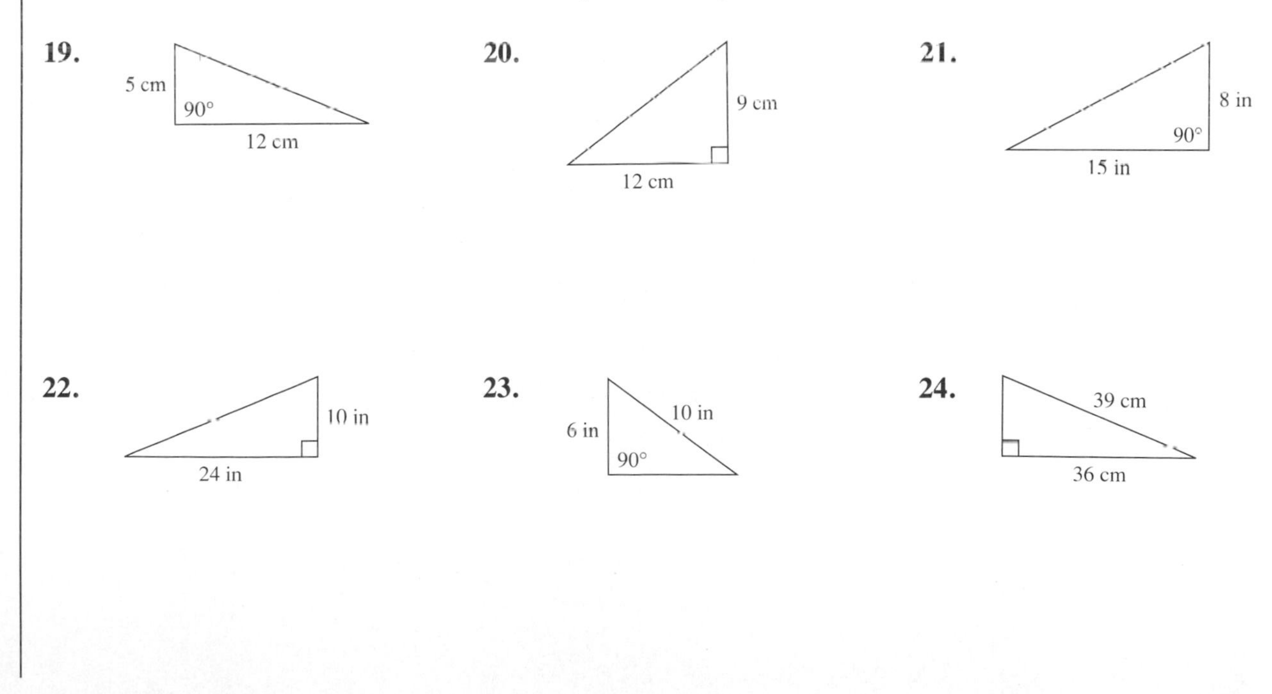

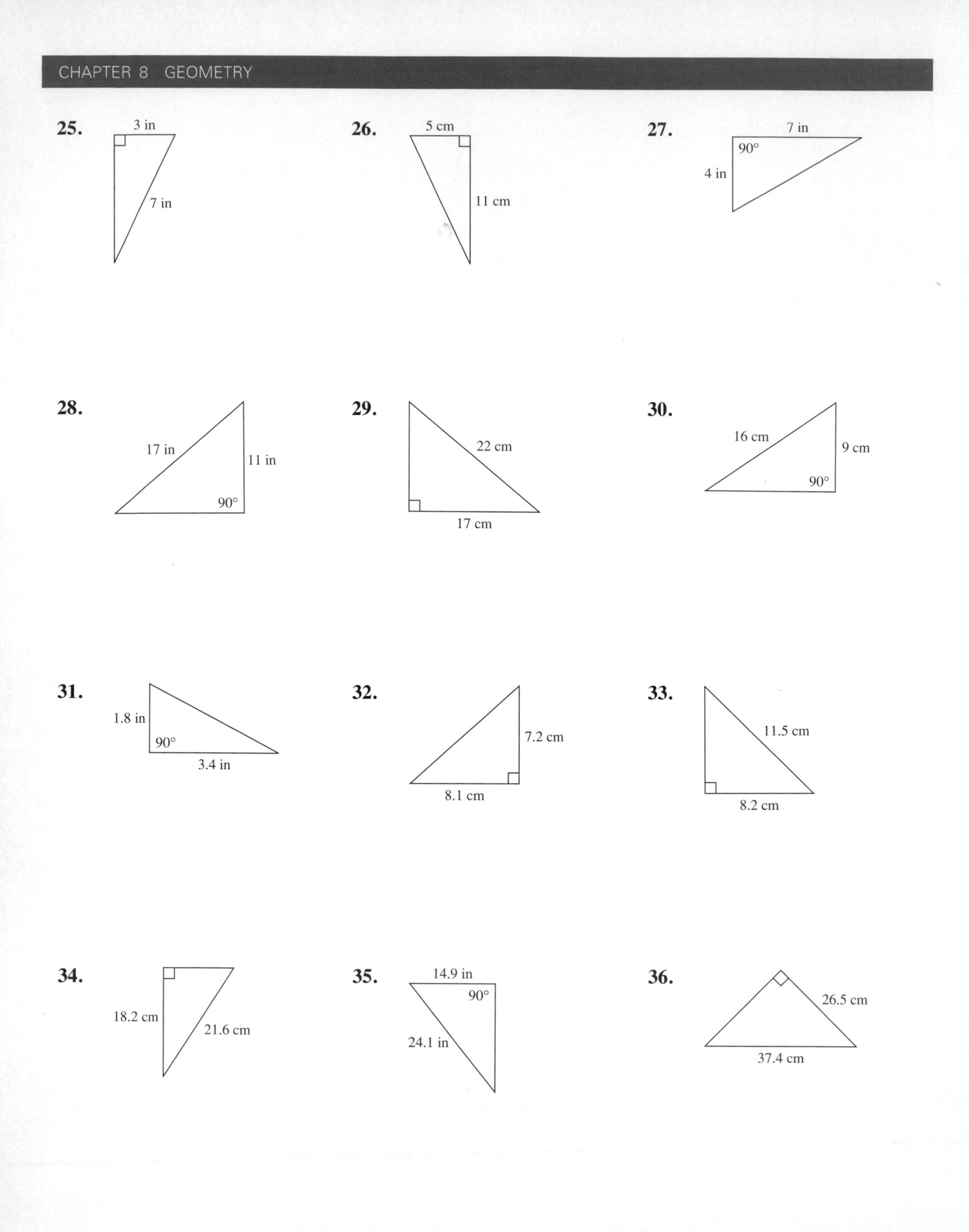
25.
3 in
7 in
26.
5 cm
11 cm
27.
7 in
90°
4 in
28.
17 in
11 in
90°
29.
22 cm
17 cm
30.
16 cm
9 cm
90°
31.
1.8 in
90°
3.4 in
32.
7.2 cm
8.1 cm
33.
11.5 cm
8.2 cm
34.
18.2 cm
21.6 cm
35.
14.9 in
90°
24.1 in
36.
26.5 cm
37.4 cm

Solve each word problem.

37. Find the length of this loading dock.

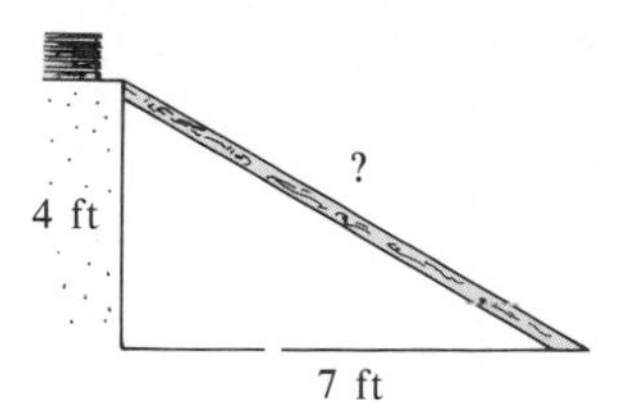

38. Find the marked length in this roof plan.

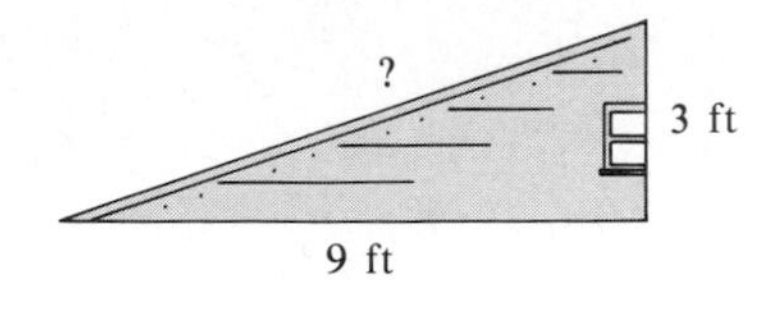

39. A boat goes 11 miles south and then 15 miles east. How far is it, in a straight line, from its starting point?

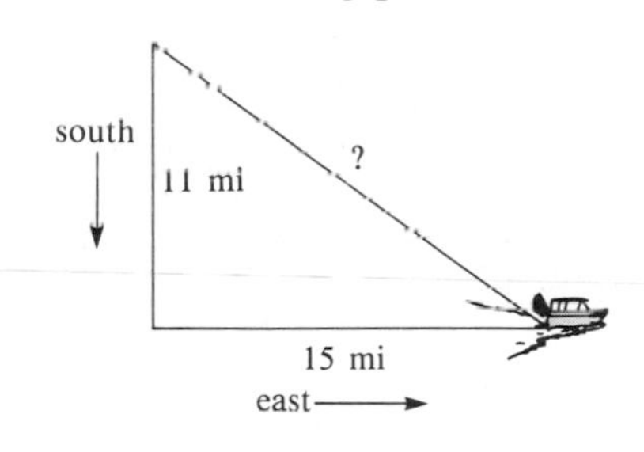

40. Find the length of this telephone pole.

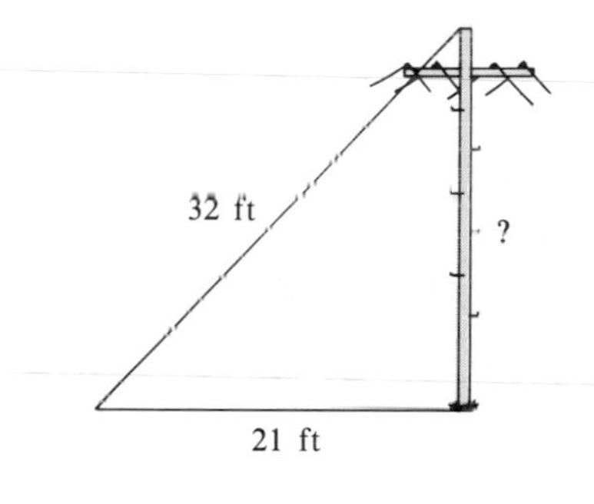

Find the distance between the center of the holes in these metal plates.

41.

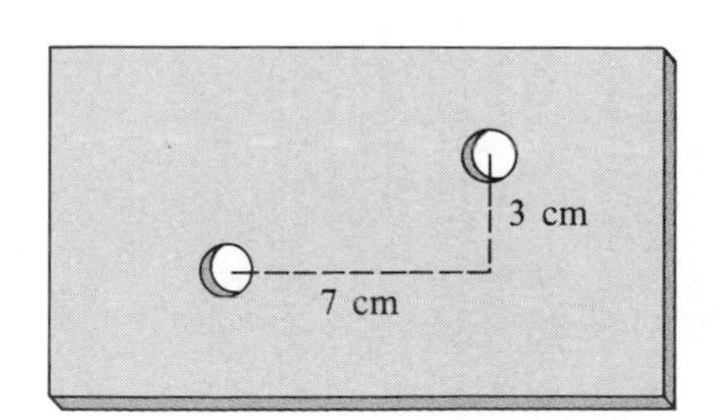

42.

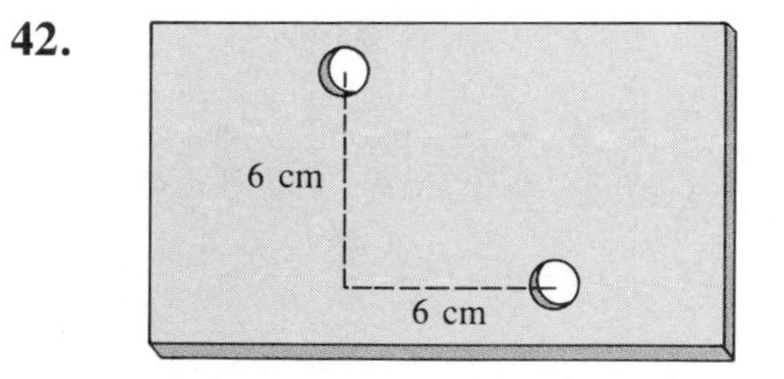

CHALLENGE EXERCISES

43. Find lengths BC and BD in the following triangle.

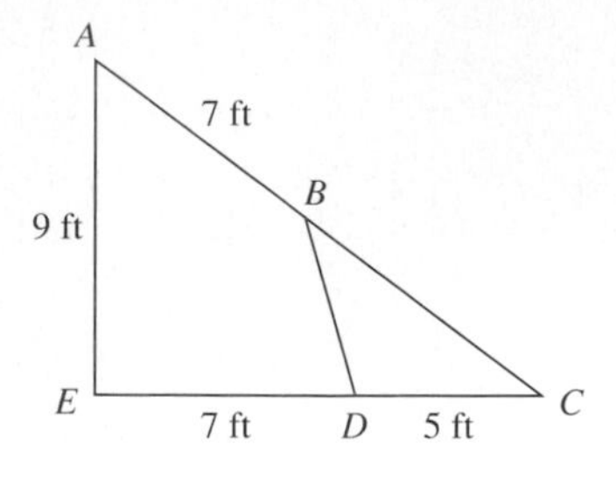

44. Find the length of CD and DB. Round answers to the nearest thousandth.

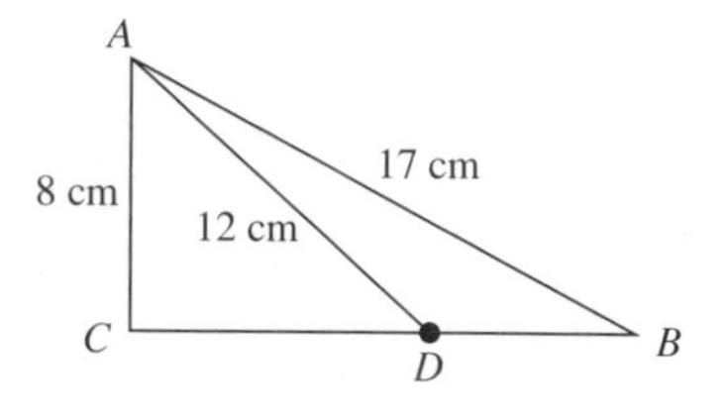

REVIEW EXERCISES

Find the missing number in each of the following proportions. (For help, see ***Section 5.4.****)*

45. $\frac{2}{9} = \frac{x}{36}$

46. $\frac{7}{x} = \frac{21}{24}$

47. $\frac{2\frac{1}{2}}{3\frac{1}{3}} = \frac{x}{1\frac{1}{4}}$

48. $\frac{17}{19} = \frac{51}{x}$

8.9 SIMILAR TRIANGLES

Two triangles with the same shape (but not necessarily the same size) are called **similar triangles.** Three pairs of similar triangles are shown here.

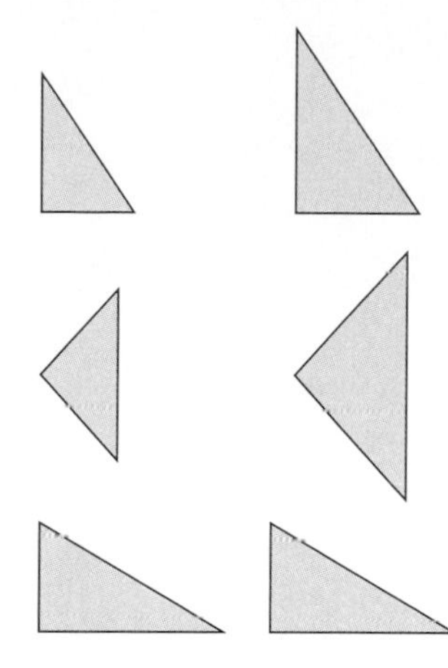

1 The two triangles shown below are the same in shape, so they are similar triangles. Angles A and P are equivalent in degrees and are called **corresponding angles.** Angles B and Q are corresponding angles, as are angles C and R.

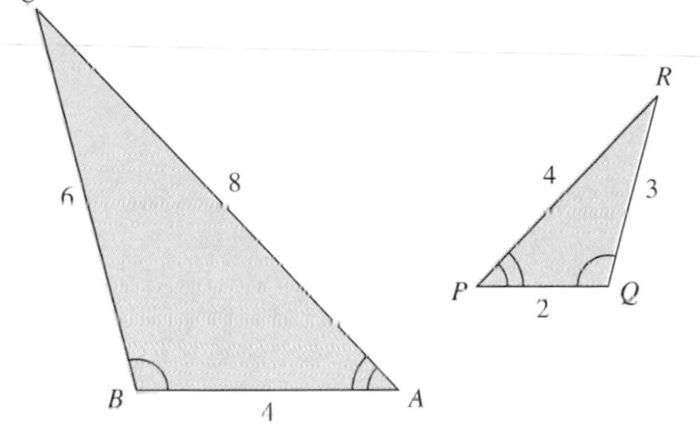

Sides PR and AC are called **corresponding sides,** since they are *opposite* corresponding angles. Also, QR and BC are corresponding sides, as are PQ and AB. Although corresponding angles are equal, corresponding sides do not need to be the same in length. In the triangles here, each side in the smaller triangle is half the length of the corresponding side in the larger triangle.

WORK PROBLEM 1 AT THE SIDE.

2 Similar triangles are useful because of the following property.

In similar triangles, the ratios of the lengths of corresponding sides are equal.

OBJECTIVES

1. Identify corresponding parts in similar triangles.
2. Find missing lengths.
3. Solve problems with similar triangles.

1. Identify corresponding angles and sides in these similar triangles.

P Z N M X Y

Angles:
P and ______
N and ______
M and ______
Sides:
PN and ______
PM and ______
NM and ______

ANSWERS
1. *Z; X; Y; ZX; ZY; XY*

2. Use the same triangles as in Example 1, but find EF.

EXAMPLE 1 Finding the Lengths of Missing Sides in Similar Triangles

Find the length of DF in the smaller triangle. Assume the triangles are similar.

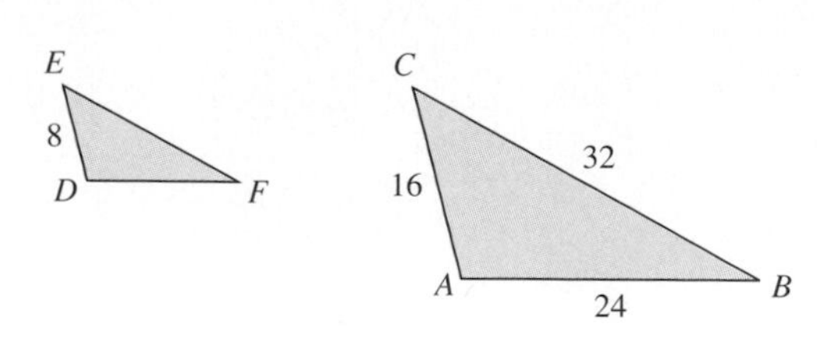

SOLUTION

Sides ED and CA are corresponding sides. The ratio of the lengths of these sides is

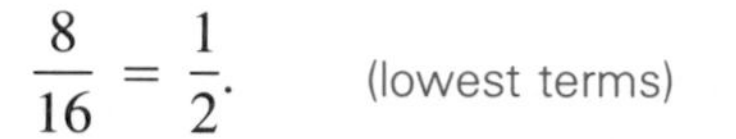

$$\frac{8}{16} = \frac{1}{2}. \quad \text{(lowest terms)}$$

As mentioned above, the ratios of the lengths of corresponding sides are equal. Side DF in the smaller triangle corresponds to side AB in the larger triangle. Since the ratios of corresponding sides are equal,

$$\frac{DF}{24} = \frac{1}{2}.$$

To solve this proportion, first find cross products.

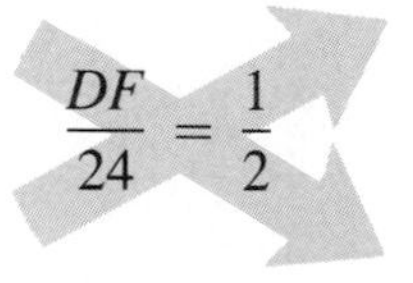

$$24 \cdot 1 = 24$$
$$2 \cdot DF = 2DF$$

Since cross products must be equal,

$$2DF = 24.$$

Find DF by dividing both sides by 2.

$$\frac{2DF}{2} = \frac{24}{2}$$
$$DF = 12$$

Side DF has length 12. ■

WORK PROBLEM 2 AT THE SIDE.

ANSWER
2. 16

EXAMPLE 2 Using a Ratio to Find a Missing Side

Find x in the smaller triangle. Assume the triangles are similar.

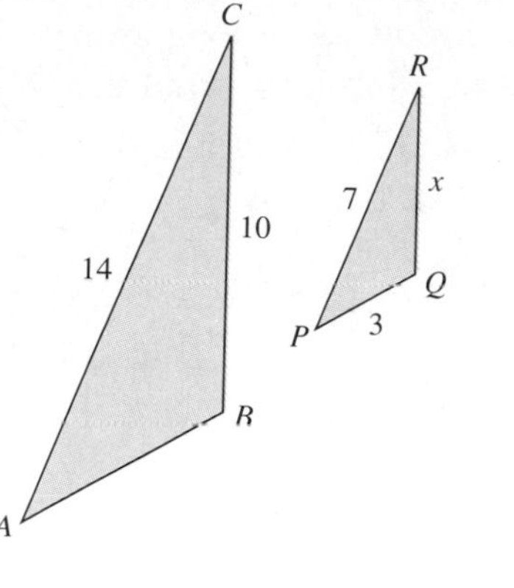

SOLUTION

Sides PR and AC are corresponding sides. The ratio of their lengths is

$$\frac{7}{14} = \frac{1}{2}. \quad \text{Lowest terms}$$

The two triangles are similar, so the ratio of any pair of corresponding sides will also equal $\frac{1}{2}$. Since sides RQ and CB are corresponding sides,

$$\frac{RQ}{CB} = \frac{1}{2}.$$

Replace RQ with x and CB with 10 to get the proportion

$$\frac{x}{10} = \frac{1}{2}.$$

Find cross products.

$$\frac{x}{10} = \frac{1}{2} \qquad \begin{array}{l} 10 \cdot 1 = 10 \\ 2 \cdot x = 2x \end{array}$$

Cross products must be equal, so

$$2x = 10.$$

Divide both sides by 2.

$$\frac{2x}{2} = \frac{10}{2}$$

$$x = 5$$

Side RQ has length 5. ■

WORK PROBLEM 3 AT THE SIDE.

3. (a) Find the length of side AB in Example 2.

(b) Find x and y if the triangles are similar.

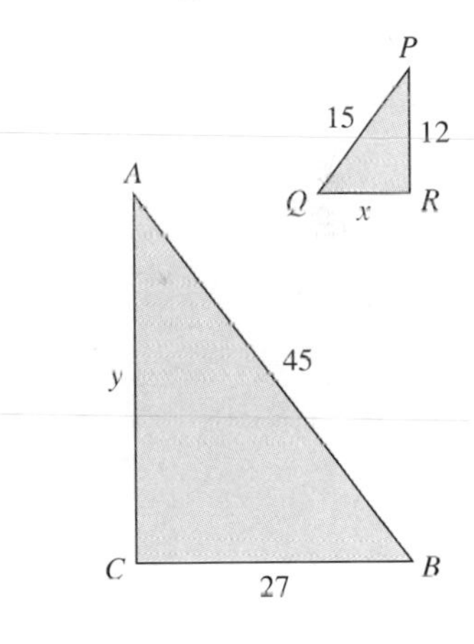

ANSWERS

3. (a) 6 (b) $x = 9$, $y = 36$.

4. Find the height of the flagpole.

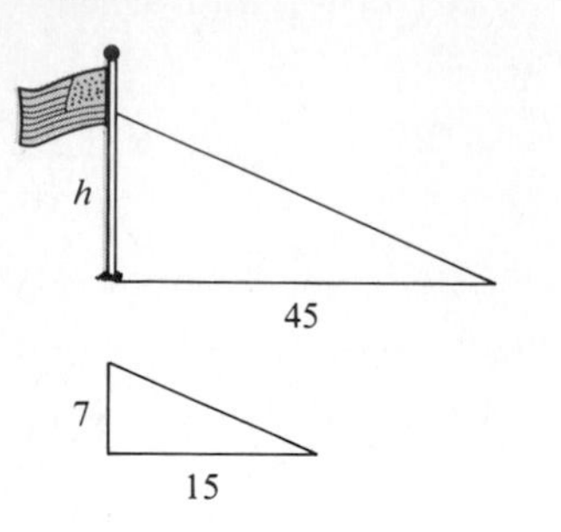

ANSWER
4. 21

3 The next example shows an application of similar triangles.

EXAMPLE 3 Using Similar Triangles in Applications

A flagpole casts a shadow 99 m long at the same time that a pole 10 m tall casts a shadow 18 m long. Find the height of the flagpole.

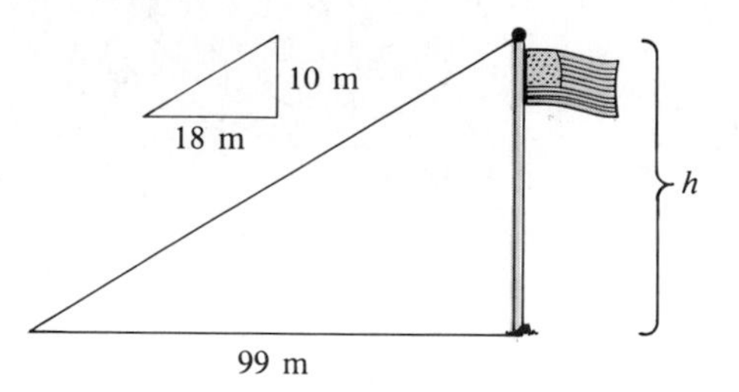

SOLUTION

The triangles shown are similar, so write a proportion to find h.

$$\frac{h}{10} = \frac{99}{18}$$

Find cross products and place them equal to each other.

$$18 \cdot h = 10 \cdot 99$$
$$18h = 990$$

Divide both sides by 18.

$$\frac{18h}{18} = \frac{990}{18}$$
$$h = 55$$

The flagpole is 55 m high. ■

WORK PROBLEM 4 AT THE SIDE.

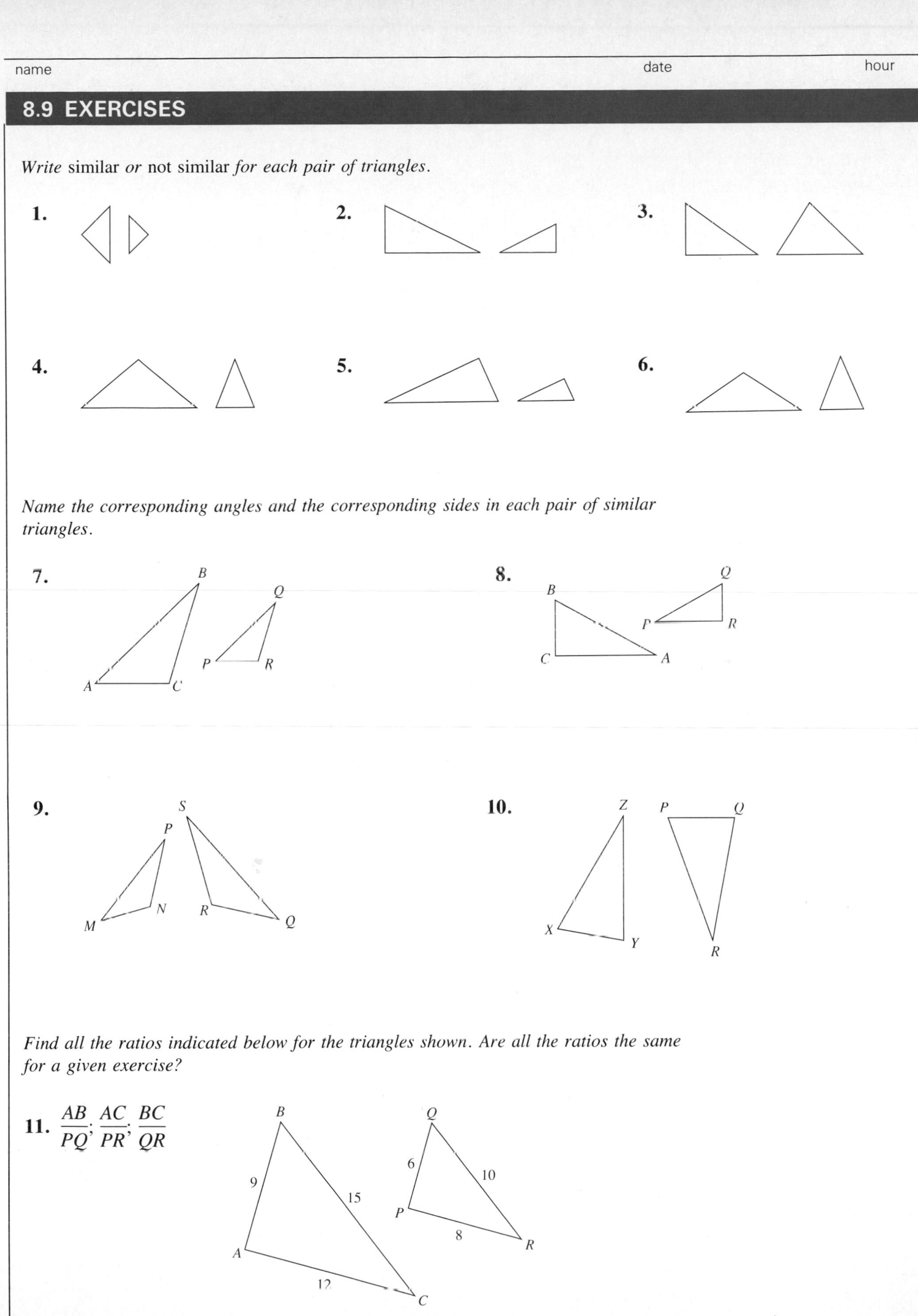

name date hour

8.9 EXERCISES

Write similar *or* not similar *for each pair of triangles.*

1.

2.

3.

4.

5.

6.

Name the corresponding angles and the corresponding sides in each pair of similar triangles.

7. B, Q, P, R, A, C

8. Q, B, P, R, C, A

9. S, P, N, R, M, Q

10. Z, P, Q, X, Y, R

Find all the ratios indicated below for the triangles shown. Are all the ratios the same for a given exercise?

11. $\frac{AB}{PQ}$; $\frac{AC}{PR}$; $\frac{BC}{QR}$

12. $\frac{AB}{PQ}; \frac{AC}{PR}; \frac{BC}{QR}$

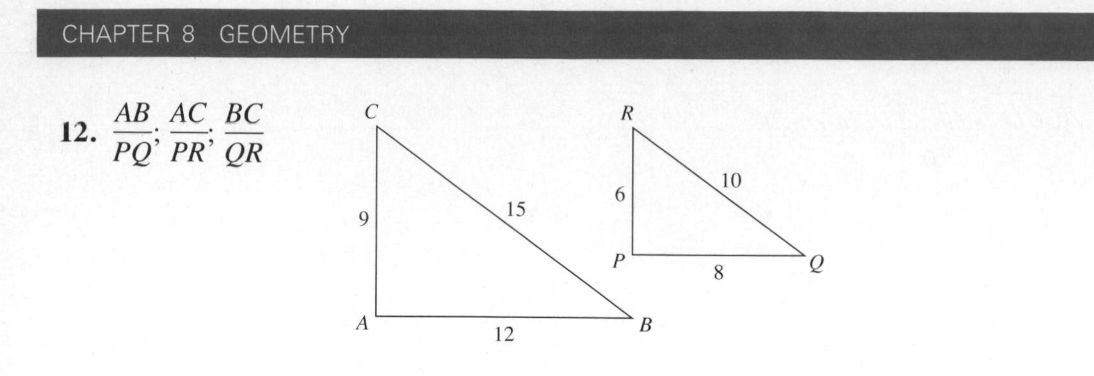

Find the missing lengths in each pair of similar triangles.

13.

14.

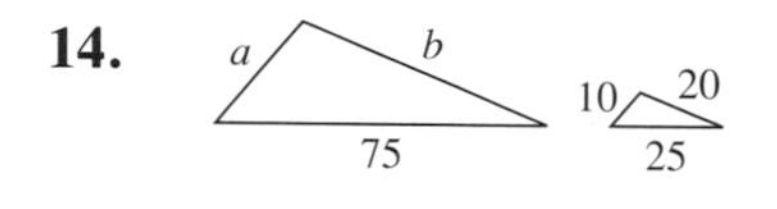

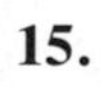

15.

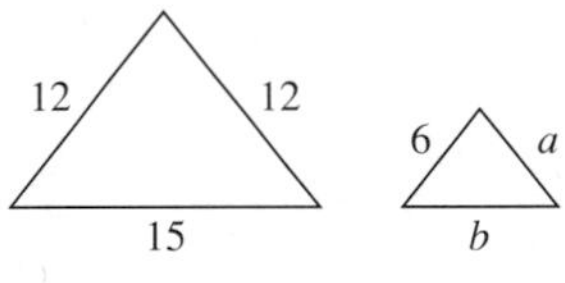

16.

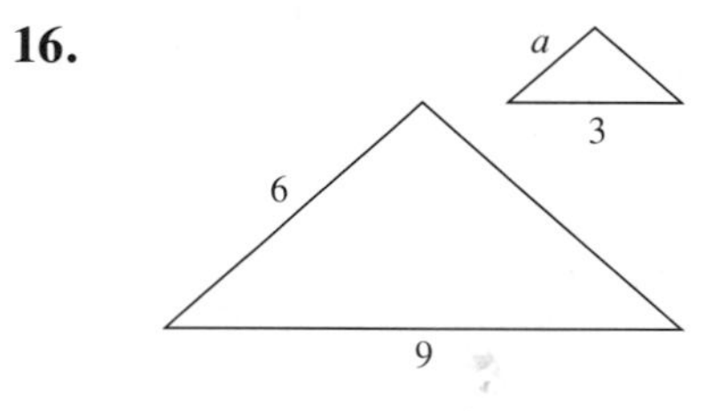

17.

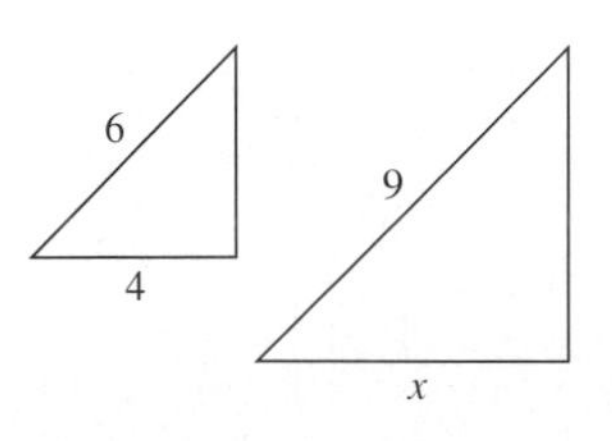

18.

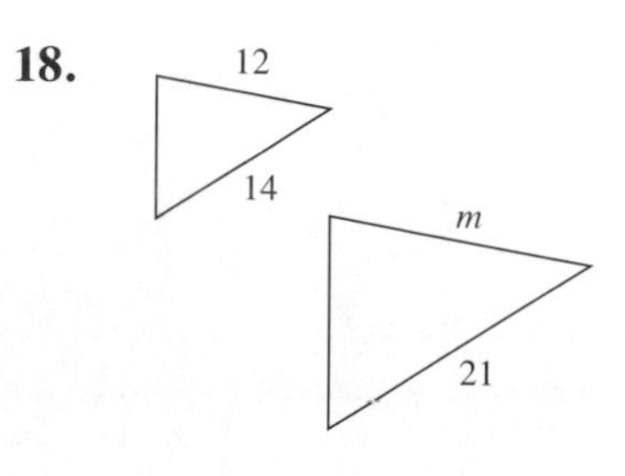

Solve the following problems.

19. The height of the house shown here can be found by using similar triangles and a proportion. Find the height of the house by writing a proportion and solving it.

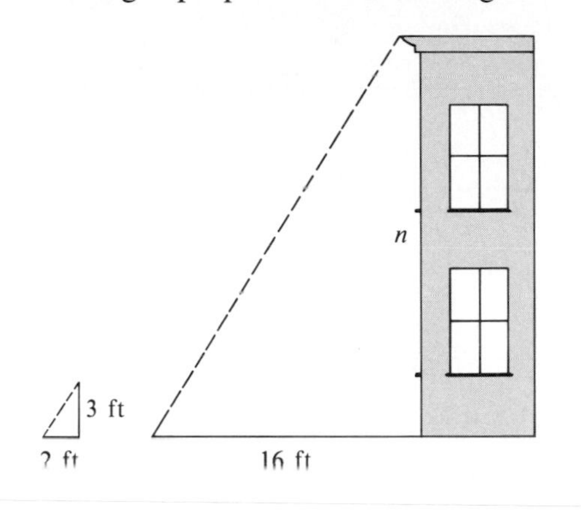

20. A sailor on the USS *Ramapo* saw one of the highest waves ever recorded. He used the height of the ship's mast, the length of the deck, and similar triangles to find the height of the wave. Using the information in the figure, write a proportion and then find the height of the wave.

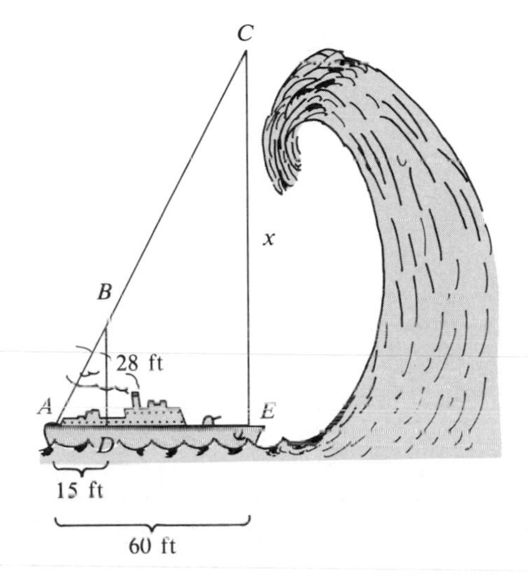

CHALLENGE EXERCISES

21. Use similar triangles and a proportion to find the length of the lake shown here. (*Hint:* The side 100 m long in the smaller triangle corresponds to a side of 100 + 120 = 220 m in the larger triangle.)

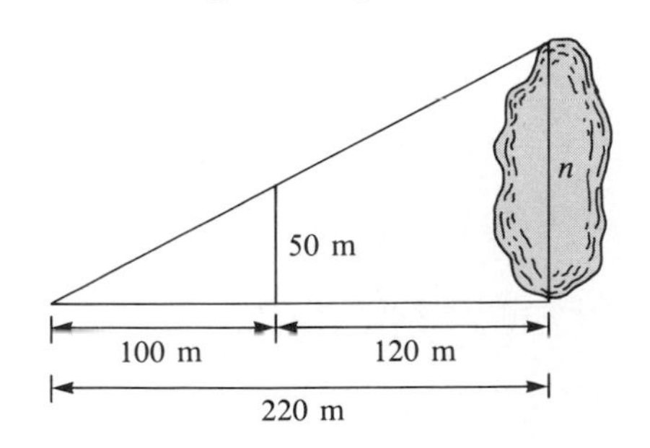

22. To find the height of the tree, find y and then add $5\frac{1}{2}$ feet for the distance from the ground to eye level.

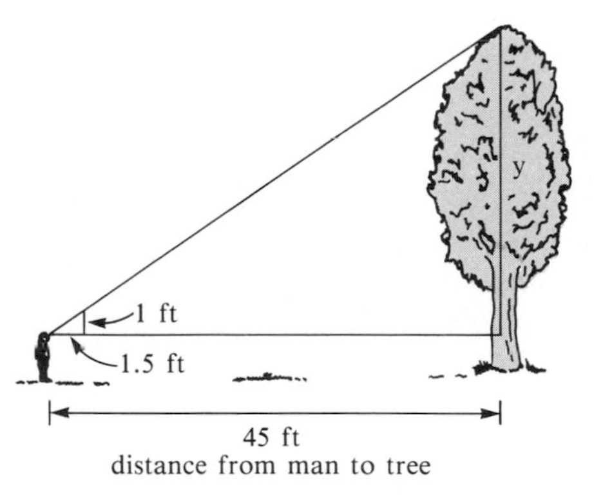

23. One triangle has sides of length 20 m, 30 m, and 35 m. A second, similar triangle, has a shortest side of length 50 m. Find the lengths of the other two sides of this second triangle.

Find the missing length in each of the following. Round answers to the nearest tenth.

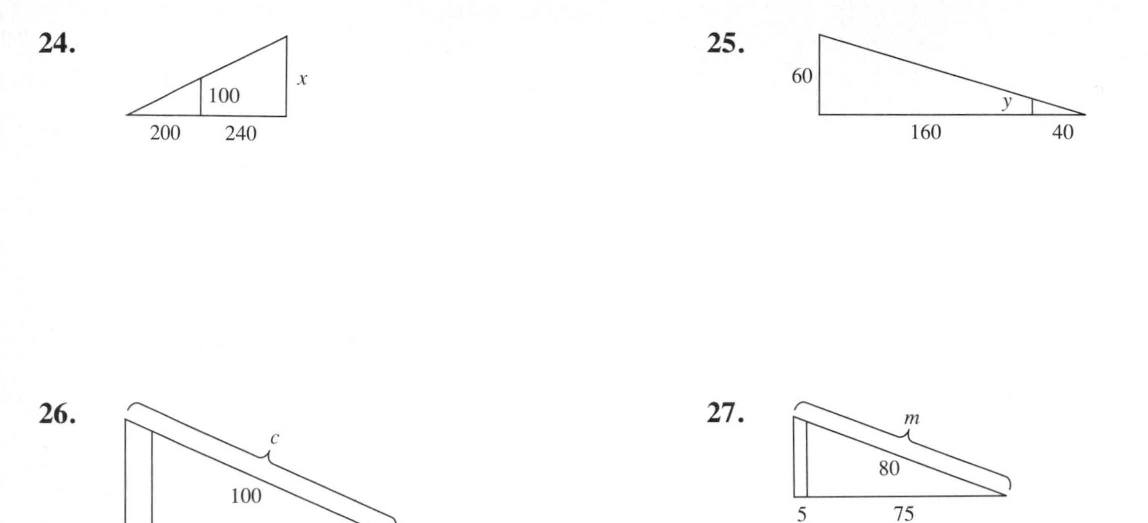

REVIEW EXERCISES

Work each problem by using the order of operations. (For help, see ***Section 1.8.****)*

28. $8 \div 4 + 3 \cdot 2$

29. $16 \div 4 \cdot 2$

30. $13 - 2 \cdot 5 + 7$

31. $18 + 7 \cdot 2 - 33 \div 3$

32. $11 + 56 \div 8 + 9 \cdot 2 - 30$

CHAPTER 8 SUMMARY

KEY TERMS

8.1	**line**	A line is a straight collection of points that goes on forever in opposite directions.
	line segment	A line segment is a part of a line bounded by endpoints.
	ray	A ray is a part of a line that has one endpoint and extends forever in one direction.
8.2	**degree**	A degree is the size of an angle that intercepts $\frac{1}{360}$th of the circumference of a circle.
	right angle	A right angle is an angle that measures 90°.
	acute angle	An acute angle is an angle that measures between 0° and 90°.
	obtuse angle	An obtuse angle is an angle that measures between 90° and 180°.
	straight angle	A straight angle is an angle that measures 180°.
	perpendicular lines	Perpendicular lines are two lines that intersect to form a right angle.
	parallel lines	Parallel lines are two lines that never intersect and are equidistant from each other.
	complementary angles	Complementary angles are two angles with a sum of 90°.
	supplementary angles	Supplementary angles are two angles with a sum of 180°.
	congruent angles	Congruent angles are angles that are equivalent in degrees.
	vertical angles	Vertical angles are two nonadjacent angles formed by intersecting lines.
8.3, 8.4, 8.5, 8.6	**perimeter**	Perimeter is the distance around a figure.
8.3, 8.4, 8.5, 8.6	**area**	Area is the space taken up by a figure.
8.3	**rectangle**	A rectangle is a four-sided figure with all sides meeting at 90° angles.
	square	A square is a rectangle with all four sides of equivalent length.
8.4	**parallelogram**	A parallelogram is a four-sided figure with both pairs of opposite sides parallel.
	trapezoid	A trapezoid is a four-sided figure with only one pair of opposite sides parallel.
8.5	**triangle**	A triangle is a figure with exactly three sides.

8.6	**circle**	A circle is a figure with all points the same distance from a fixed center point.
	diameter	Diameter is the distance across the circle, passing through the center.
	radius	Radius is the distance from the center to any point on the circle.
	circumference	Circumference is the distance around a circle.
	semicircle	A semicircle is a half-circle.
8.8	**volume**	Volume is the space inside a three-dimensional figure.
	square root	A square root is one of two equal factors of a number.
	hypotenuse	The hypotenuse is the side of a right triangle opposite the 90° angle.
8.9	**similar triangle**	Similar triangles are two triangles with the same shape but not necessarily the same size.

QUICK REVIEW

Section Number and Topic	Approach	Example
8.1 Identifying Lines, Line Segments, and Rays	If a line has one endpoint, it is a ray, while if it has two endpoints, it is a line segment.	Identify each of the following as a line, line segment, or ray. O S (a) P Q (b) S T (c) **(a)** is a ray, **(b)** is a line, and **(c)** is a line segment.
8.1 Identifying Parallel and Perpendicular Lines	If two lines intersect at right angles, they are perpendicular. If two lines never intersect, they are parallel.	Label each pair of lines as parallel or perpendicular. (a) (b) **(a)** shows two perpendicular lines (they intersect at 90°). **(b)** shows two parallel lines (they never intersect).
8.2 Identifying Complementary and Supplementary Angles	If the sum of two angles is 90°, they are complementary. If the sum of two angles is 180°, they are supplementary.	Find the complement and supplement of 35°. The complement of 35° is 55°, and the supplement is 145°.

Section Number and Topic	Approach	Example
8.2 Identifying Congruent and Vertical Angles	If two angles are equivalent in degrees, the angles are congruent. Two nonadjacent angles formed by intersecting lines are called vertical angles.	Identify the congruent and vertical angles in the following figure. 1 2 3 4 $\angle 1 \cong \angle 3$ and $\angle 2 \cong \angle 4$ $\angle 1$ and $\angle 3$ are vertical angles. $\angle 2$ and $\angle 4$ are vertical angles.
8.3 Finding Perimeter and Area of a Rectangle	Use the formula perimeter = $2 \cdot$ length $+ 2 \cdot$ width or area = length $\cdot$ width.	Find the perimeter and area of the rectangle. 3 m 2 m $P = 2l + 2w = 2 \cdot 3 + 2 \cdot 2$ $P = 6 + 4 = 10$ m $A = l \cdot w = 3 \cdot 2 = 6\ \text{m}^2$
8.3 Finding Perimeter and Area of a Square	Use the formula $P = 4 \cdot \text{side}$ or $A = (\text{side})^2$.	Find the perimeter and area of a square with a side of 6 meters. $P = 4 \cdot s = 4 \cdot 6 = 24$ m $A = s^2 = (6)^2 = 6 \cdot 6 = 36\ \text{m}^2$
8.4 Finding Perimeter and Area of a Parallelogram	Use the formula P = sum of the lengths of the sides or $A = \text{base} \cdot \text{height}$.	Find the perimeter and area of the parallelogram. 5 cm 6 cm 4 cm 6 cm 5 cm $P = 5 + 6 + 5 + 6 = 22$ cm $A = 5 \cdot 4 = 20\ \text{cm}^2$

Section Number and Topic	Approach	Example
8.4 Find Perimeter and Area of a Trapezoid	Use the formulas P = sum of the lengths of the sides and $A = \frac{1}{2} \cdot \text{height} \cdot \left(\begin{matrix}\text{short}\\ \text{base}\end{matrix} + \begin{matrix}\text{long}\\ \text{base}\end{matrix}\right)$	Find the perimeter and area of the trapezoid. 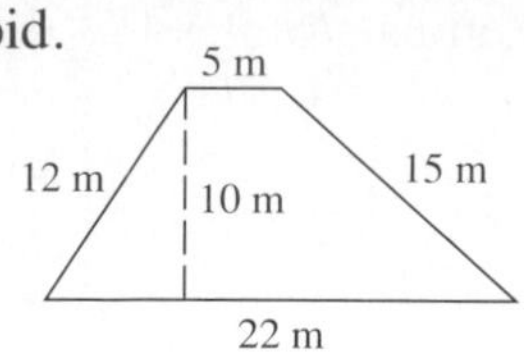 $P = 5 + 15 + 22 + 12 = 54$ m $A = \frac{1}{2} \cdot 10 \cdot (5 + 22)$ $= 5 \cdot (27)$ $= 135 \text{ m}^2$
8.5 Finding Perimeter and Area of a Triangle	Use the formulas P = sum of the lengths of the sides and $A = \frac{1}{2} \cdot \text{base} \cdot \text{height}.$	Find the perimeter and area of the triangle. 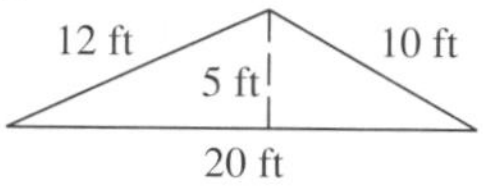 $P = 12 + 10 + 20 = 42$ ft $A = \frac{1}{2} \cdot b \cdot h$ $A = \frac{1}{2} \cdot 20 \cdot 5$ $A = 50 \text{ ft}^2$
8.6 Finding the Diameter of a Circle, Given the Radius	Use the formula diameter = 2 · radius.	Find the diameter of a circle, if the radius is 3 meters. $d = 2 \cdot r = 2 \cdot 3 = 6$ meters
8.6 Finding the Radius of a Circle, Given the Diameter	Use the formula radius $= \frac{1}{2} \cdot$ diameter.	Find the radius of a circle, if the diameter is 4 inches. $r = \frac{1}{2} \cdot d = \frac{1}{2} \cdot 4 = 2$ inches
8.6 Finding the Circumference and Area of a Circle	Use the formulas $C = 2\pi r$ or $C = \pi d$ and $A = \pi r^2.$	Find the circumference and area of a circle with a radius of 3 cm. Circumference $= 2\pi r = 2 \cdot 3.14 \cdot 3$ $C = 18.84$ cm Area $= \pi r^2 = 3.14 \cdot 3 \cdot 3$ $A = 28.26 \text{ cm}^2$

Section Number and Topic	Approach	Example
8.7 Finding the Volume of a Cube	Use the formula volume = length · width · height.	Find the volume of the following cube. (5 cm, 3 cm, 6 cm) $V = l \cdot w \cdot h$ $V = 5 \cdot 3 \cdot 6$ $V = 90 \text{ cm}^3$
8.7 Finding the Volume of a Sphere	Use the formula $\text{volume} = \frac{4}{3}\pi r^3$ where r is the radius of the sphere.	Find the volume of a sphere with radius of 5 m. $V = \frac{4}{3}\pi r^3$ $V = \frac{4}{3} \cdot 3.14 \cdot 5 \cdot 5 \cdot 5$ $V = 523.33 \text{ m}^3$
8.7 Finding the Volume of a Hemisphere	Use the formula $\text{Volume} = \frac{1}{2}\left(\frac{4}{3}\pi r^3\right) = \frac{2}{3}\pi r^3$ where r is the radius of the hemisphere.	Find the volume of a hemisphere with a radius of 20 cm. $V = \frac{2}{3}\pi r^3$ $V = \frac{2}{3} \cdot 3.14 \cdot 20 \cdot 20 \cdot 20$ $V = 16{,}746.67 \text{ cm}^3$
8.7 Finding the Volume of a Cylinder	Use the formula $\text{Volume} = \pi r^2 h$ where r is the radius of the base and h is the height.	Find the volume of a cylinder that is 10 m high with a radius of 4 m. $V = \pi r^2 h$ $V = \pi \cdot 4 \cdot 4 \cdot 10$ $V = 502.4 \text{ m}^3$
8.7 Finding the Volume of a Cone	Use the formula $\text{Volume} = \frac{1}{3}\pi r^2 h$ where r is the radius of the base and h is the height of the cone.	Find the volume of a cone with a height of 9 in and a base of radius 4 in. $V = \frac{1}{3}\pi r^2 h$ $V = \frac{1}{3} \cdot 3.14 \cdot 4 \cdot 4 \cdot 9$ $V = 150.72 \text{ in}^3$

Section Number and Topic	Approach	Example
8.7 Finding the Volume of a Pyramid	Use the formula $$V = \frac{1}{3}bh$$ where b is the area of the base and h is the height.	Find the volume of a pyramid with a base 2 cm by 2 cm and a height of 3 cm. $V = \frac{1}{3}bh$ $V = \frac{1}{3} \cdot 2 \cdot 2 \cdot 3 = 4 \text{ cm}^3$
8.7 Finding the Square Root of a Number	Use the table inside the back cover or a calculator to find the square root.	$\sqrt{43} = 6.557$
8.8 Finding the Hypotenuse in a Right Triangle	To find the hypotenuse, use the formula hypotenuse $= \sqrt{(\text{leg})^2 + (\text{leg})^2}$	Find the hypotenuse of the following triangle. hypotenuse $= \sqrt{(6)^2 + (5)^2}$ $= \sqrt{36 + 25}$ $= \sqrt{61} = 7.81$ 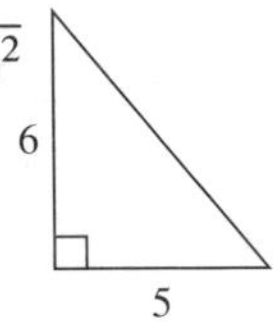
8.9 Finding a Leg of a Right Triangle	To find a leg, use the formula leg $= \sqrt{(\text{hypotenuse})^2 - (\text{leg})^2}$	Find the missing leg of the following triangle. leg $= \sqrt{(25)^2 - (16)^2}$ leg $= \sqrt{625 - 256}$ $= \sqrt{369} = 19.21$ 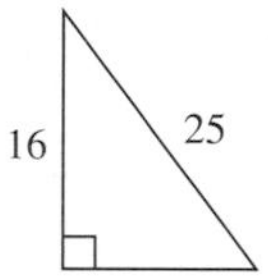
8.9 Finding the Missing Lengths of Similar Triangles	Use the fact that in similar triangles, the ratios of the lengths of corresponding sides are equal.	Find the lengths BC and AC in the smaller triangle. $\frac{10}{8} = \frac{5}{BC}$ $\frac{10}{12} = \frac{5}{AC}$ $10BC = 40$ $10AC = 60$ $BC = 4$ $AC = 6$ 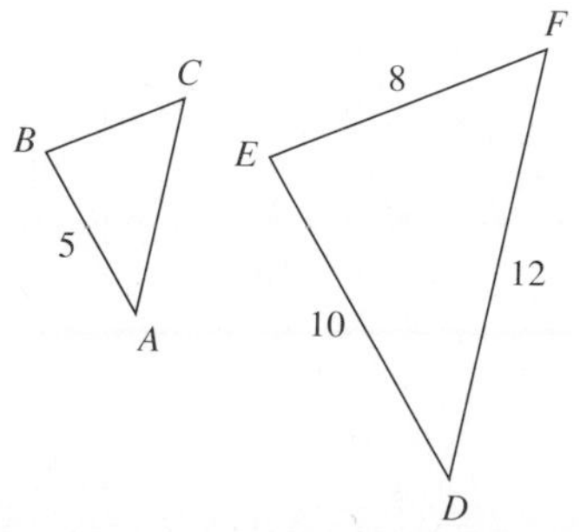

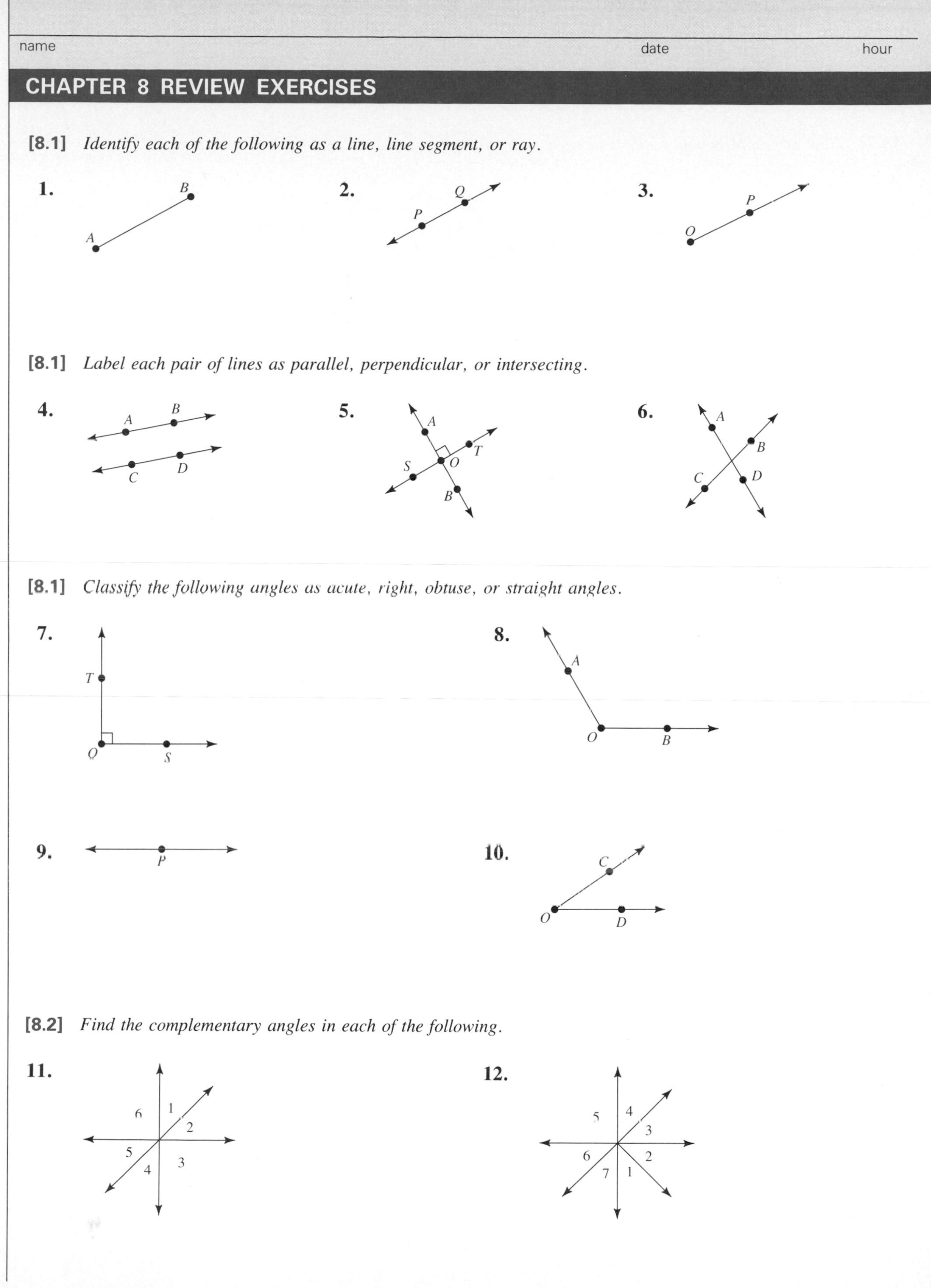

name date hour

CHAPTER 8 REVIEW EXERCISES

[8.1] *Identify each of the following as a line, line segment, or ray.*

1.

2.

3.

[8.1] *Label each pair of lines as parallel, perpendicular, or intersecting.*

4.

5.

6.

[8.1] *Classify the following angles as acute, right, obtuse, or straight angles.*

7.

8.

9.

10.

[8.2] *Find the complementary angles in each of the following.*

11.

12.

[8.2] *Find the supplementary angles in each of the following.*

13. **14.**

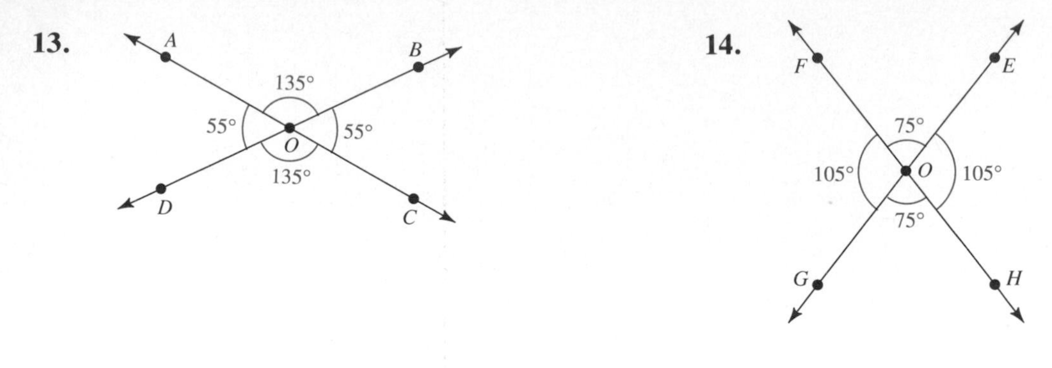

[8.2] *Identify the vertical angles in each of the following.*

15.

16.

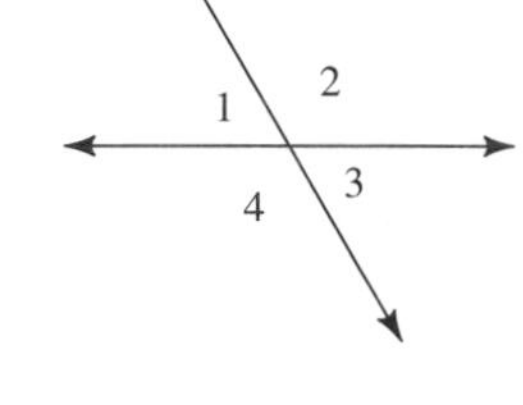

[8.3] *Find the perimeter of each of the following figures.*

17. a rectangle 7.2 m by 11.8 m

18. a square 42.7 cm on a side

19.

20.

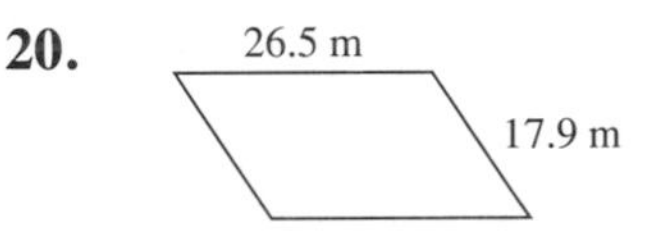

Find the area of each rectangle or square. Round all answers to the nearest tenth.

21. **22.** **23.**

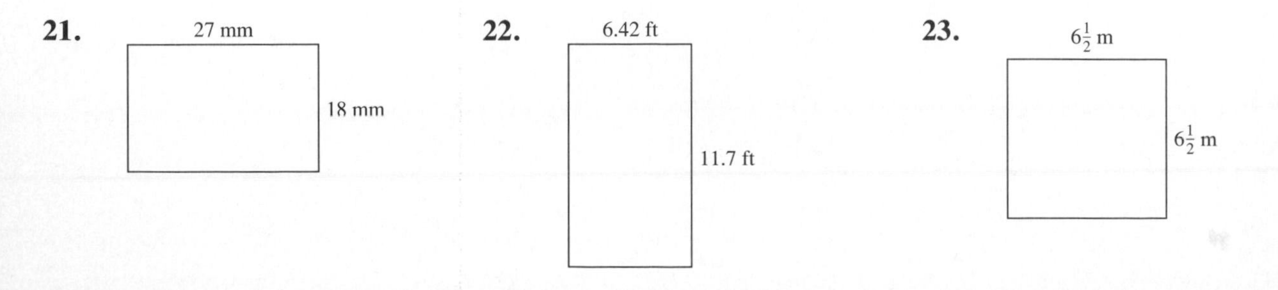

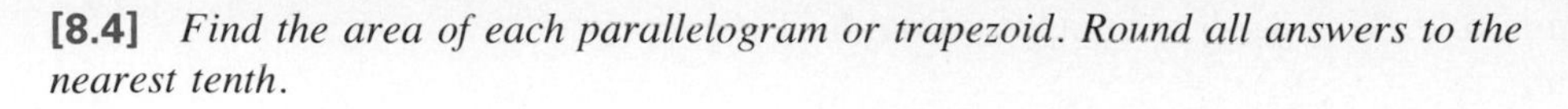

[8.4] *Find the area of each parallelogram or trapezoid. Round all answers to the nearest tenth.*

24. **25.** **26.**

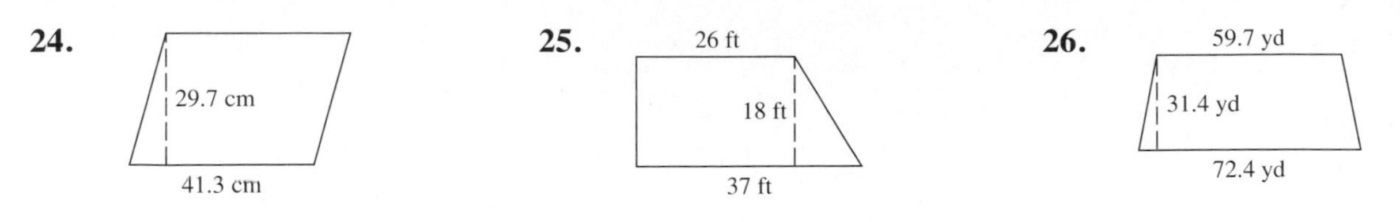

[8.5] *Find the perimeter of each triangle.*

27. a triangle with sides 7 m, 12 m, and 11 m

28. a triangle with sides 9.4 cm, 17.2 cm, and 16.8 cm

Find the area of each triangle.

29. **30.** **31.**

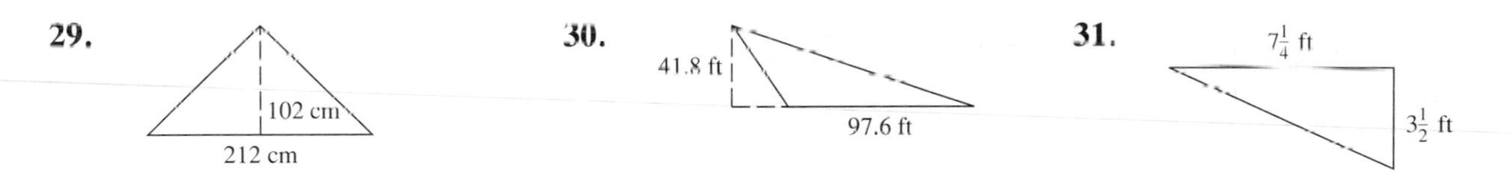

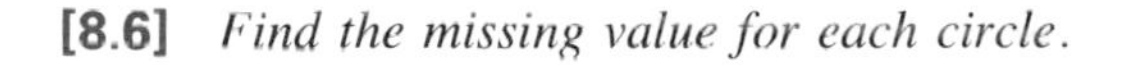

[8.6] *Find the missing value for each circle.*

32. radius is 72.8 m, find the diameter

33. diameter is 34 m, find the radius

Find the circumference and area for each of the following circles. Use 3.14 as an approximation for π. Round all answers to the nearest tenth.

34. **35.** **36.**

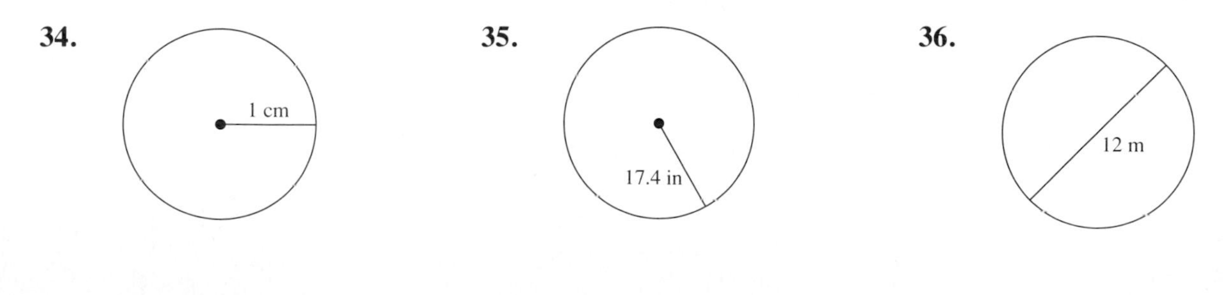

37. Find the area of this semicircle. Use 3.14 as an approximation for π. Round to the nearest tenth.

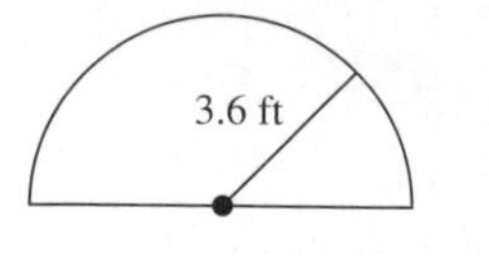

[8.7] *Find each volume. Use* 3.14 *as an approximation for* π. *Round each answer to the nearest tenth.*

38. a box 7 m by 2 m by 5 m

39. a box 9.42 cm by 3.87 cm by 2.04 cm

40. a box 57 mm by 174 mm by 86 mm

41. a sphere with a radius of 5.2 cm

42. a sphere with a diameter of 4.8 cm

43. a cylinder with a radius of 3.7 cm and a height of 5.4 cm

44. a cylinder with a radius of 12 m and a height of 3 m

45. a cylinder with a radius of 50.6 in and a height of 11.8 in

name date hour

Find each volume. Use 3.14 *as an approximation for* π. *Round each answer to the nearest tenth.*

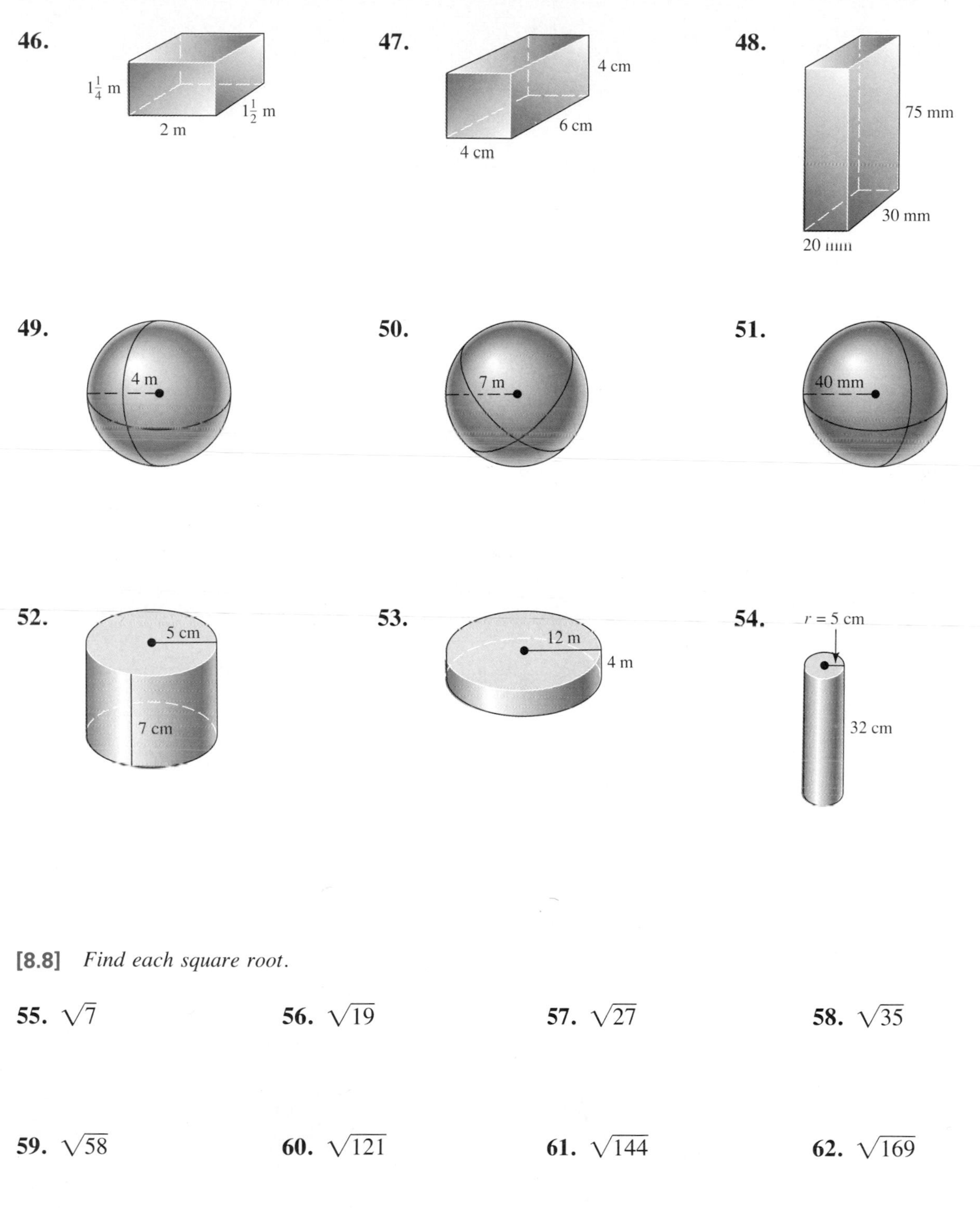

[8.8] *Find each square root.*

55. $\sqrt{7}$ **56.** $\sqrt{19}$ **57.** $\sqrt{27}$ **58.** $\sqrt{35}$

59. $\sqrt{58}$ **60.** $\sqrt{121}$ **61.** $\sqrt{144}$ **62.** $\sqrt{169}$

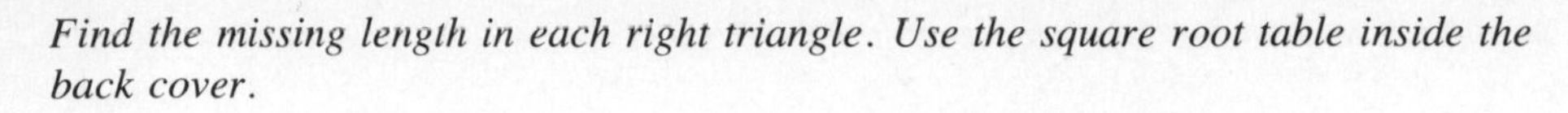

Find the missing length in each right triangle. Use the square root table inside the back cover.

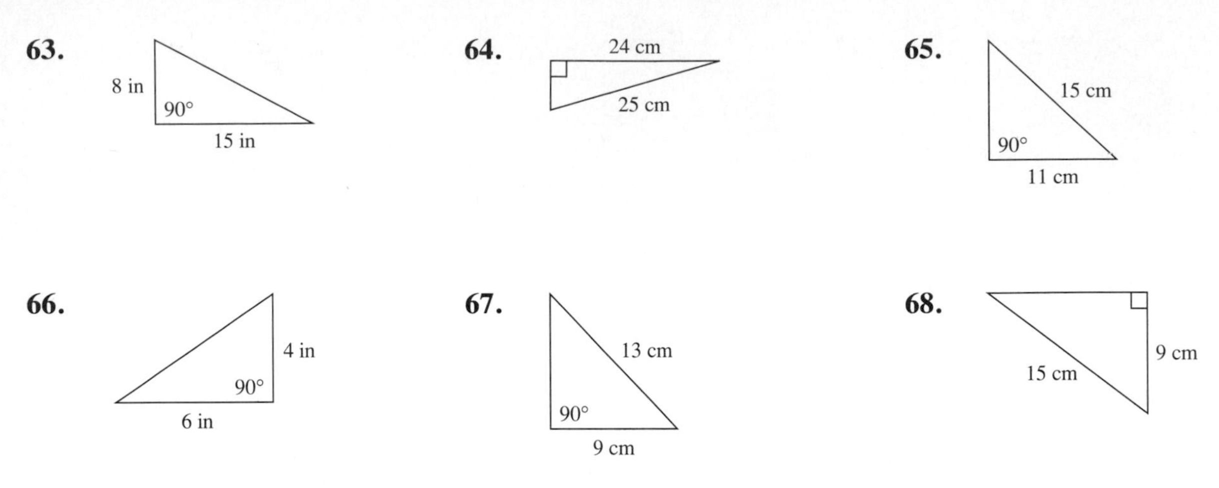

[8.9] *Find the lengths of the missing sides in each pair of similar triangles.*

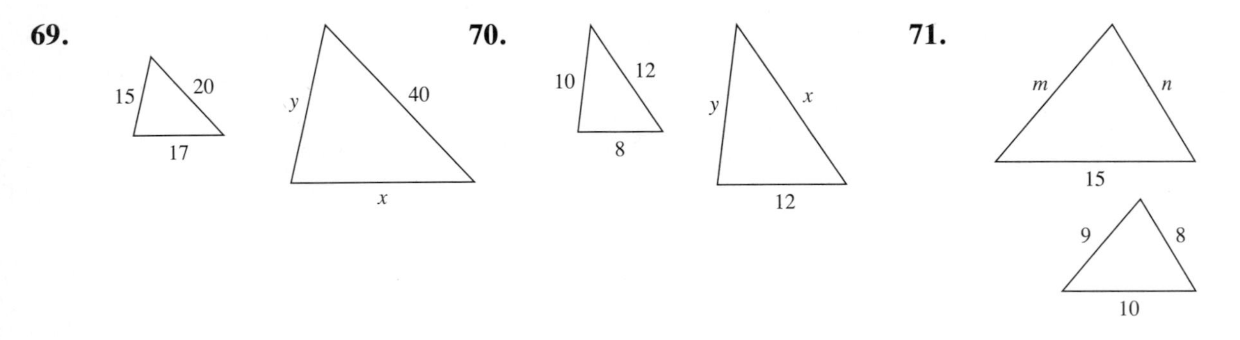

72. Find the height of the tower.

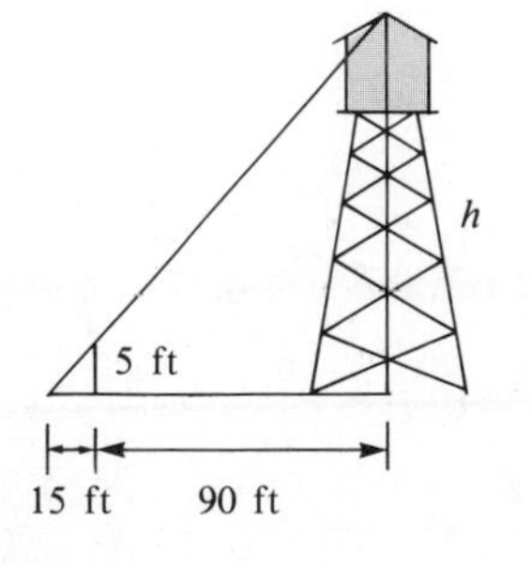

73. Find the length of the lake.

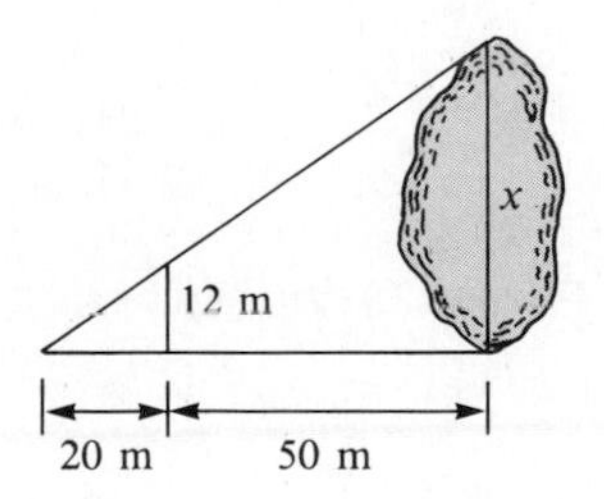

MIXED REVIEW EXERCISES

Find the perimeter and area of the following figures.

74. **75.**

Find the volume of each of the following figures. Use 3.14 as an approximation for π. Round answers to the nearest tenth.

76. **77.** **78.** **79.** **80.**

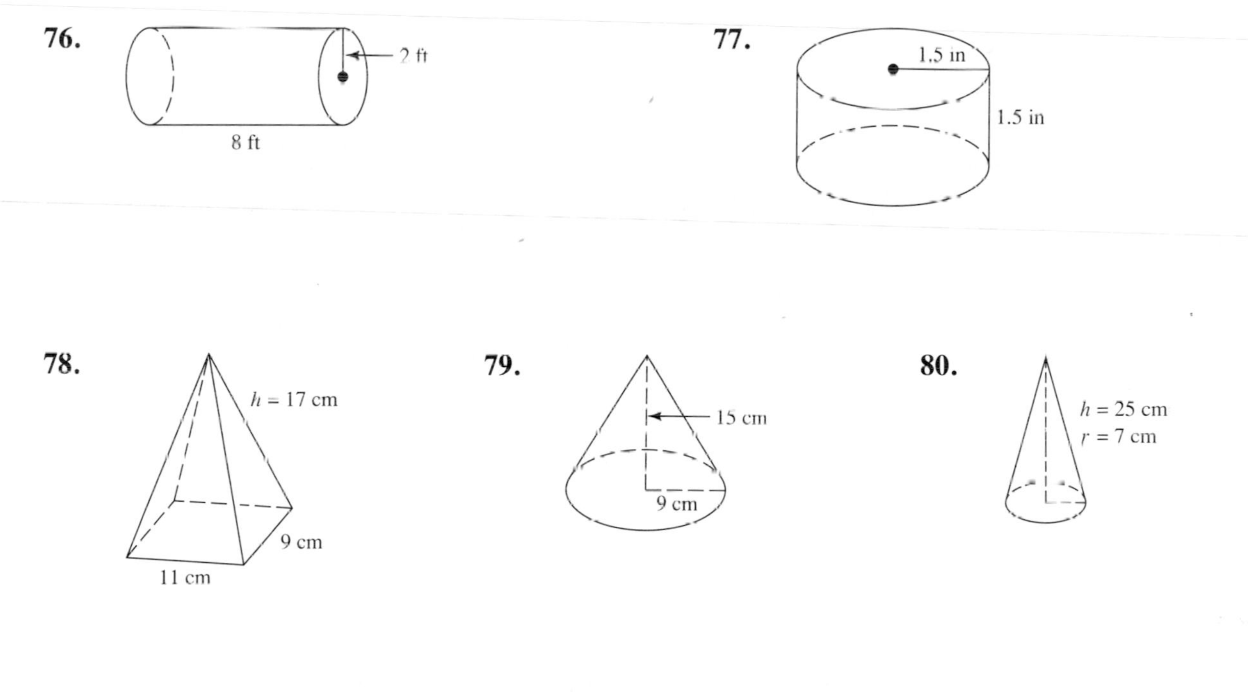

81. The cylinder shown has a base area of 25 cm^2 and a height of 12 cm. What is the volume in cubic centimeters?

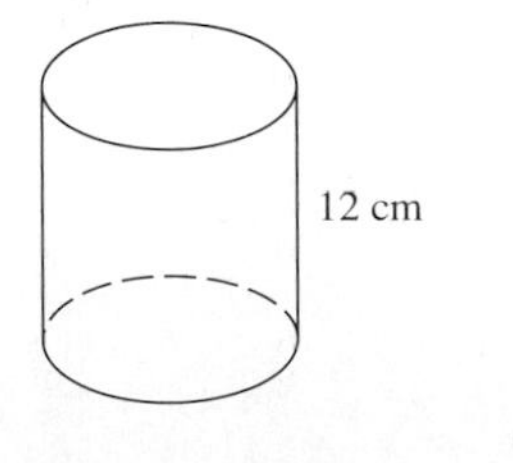

82. In the figure shown, find the length of AB.

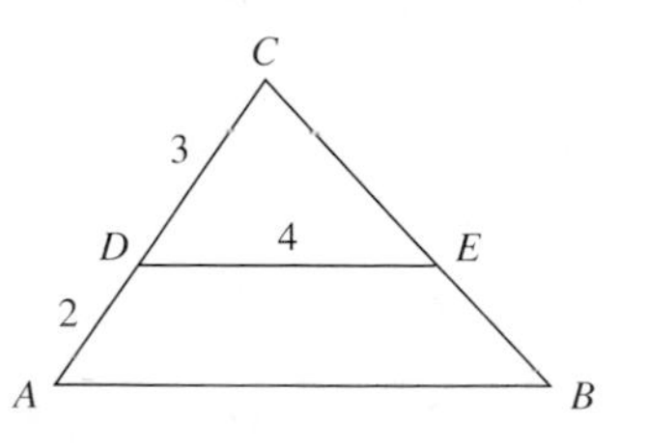

83. Find the volume of the solid in the figure shown.

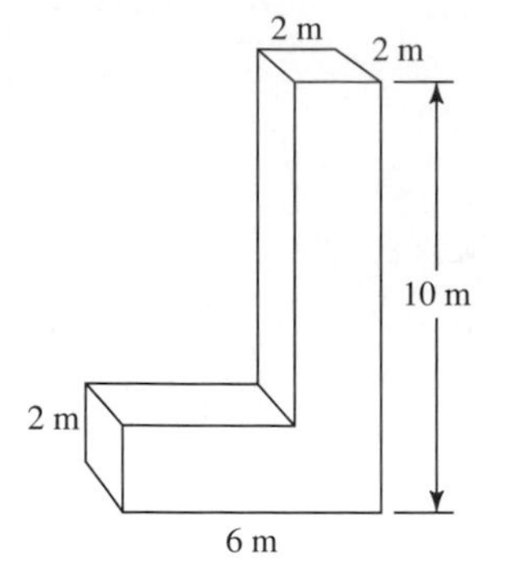

84. If the edge of a cube is doubled in length, then the volume of the cube is multiplied by what factor?

Find each shaded area. For the circles, use 3.14 *as an approximation for* π. *Round all answers to the nearest tenth.*

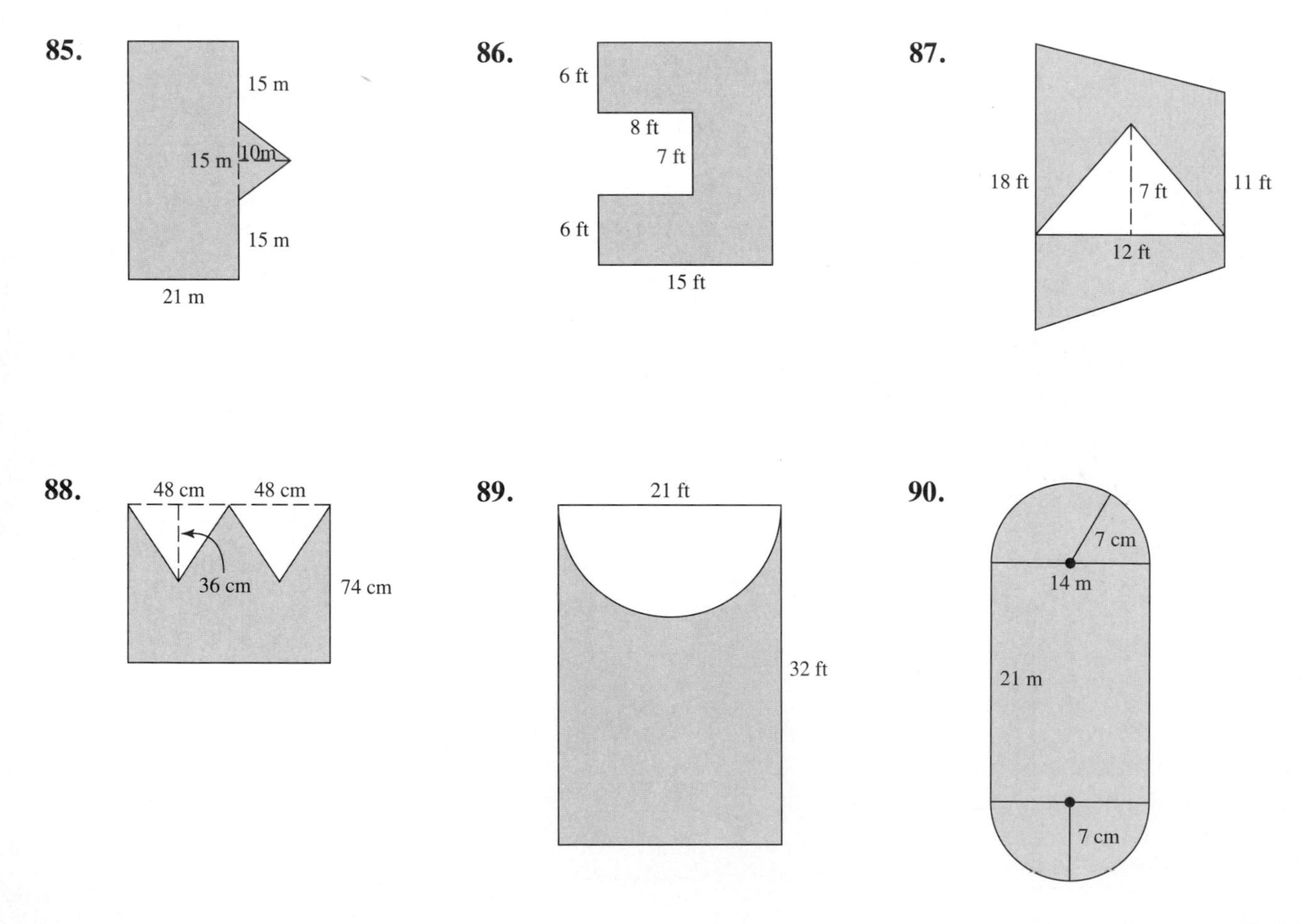

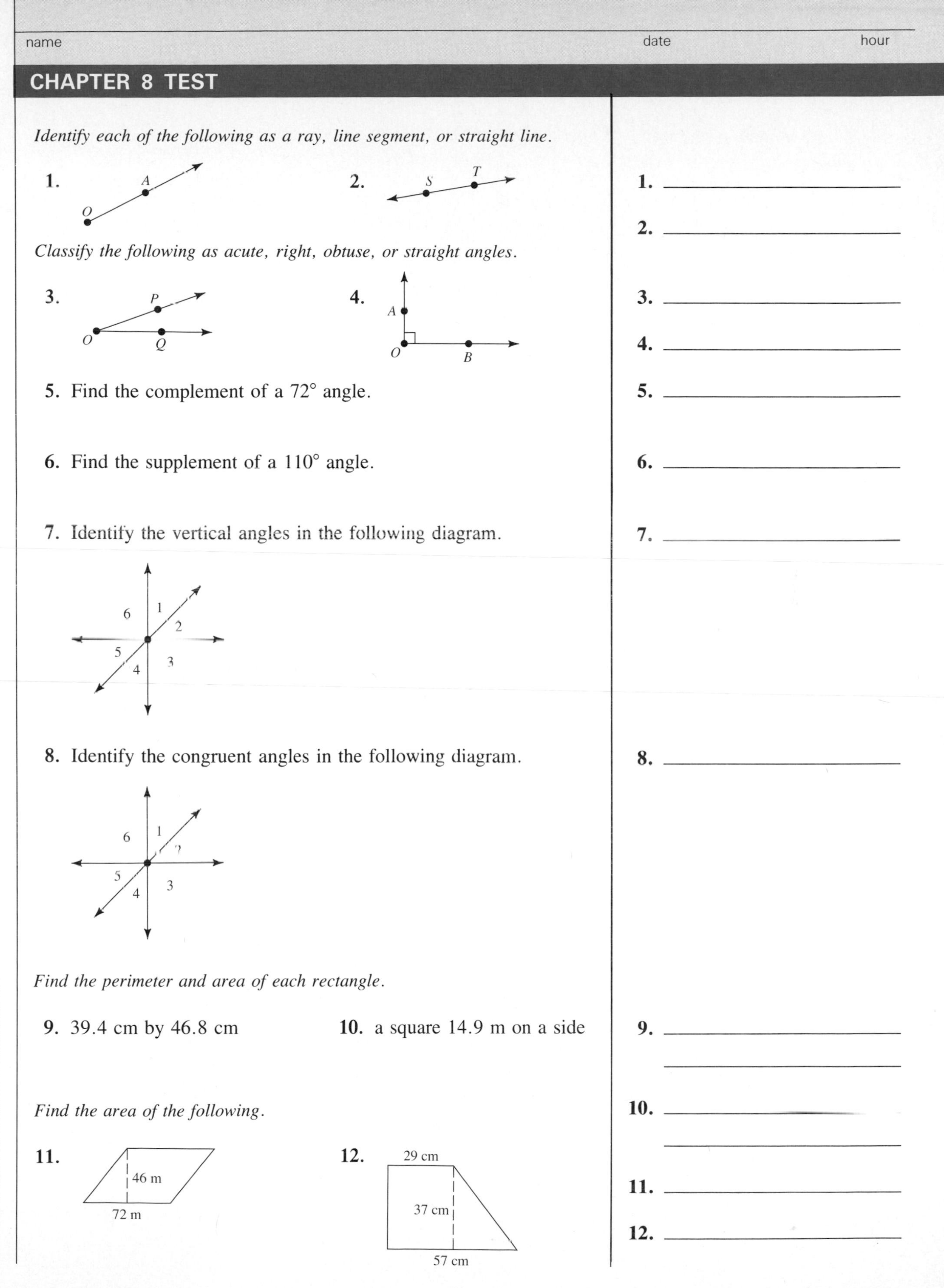

name date hour

CHAPTER 8 TEST

Identify each of the following as a ray, line segment, or straight line.

1.

2.

Classify the following as acute, right, obtuse, or straight angles.

3.

4.

5. Find the complement of a 72° angle.

6. Find the supplement of a 110° angle.

7. Identify the vertical angles in the following diagram.

8. Identify the congruent angles in the following diagram.

Find the perimeter and area of each rectangle.

9. 39.4 cm by 46.8 cm

10. a square 14.9 m on a side

Find the area of the following.

11.

12.

1. ______

2. ______

3. ______

4. ______

5. ______

6. ______

7. ______

8. ______

9. ______

______

10. ______

______

11. ______

12. ______

13. ______

14. ______

15. ______

16. ______

17. ______

18. ______

19. ______

20. ______

21. ______

22. ______

23. ______

24. ______

25. ______

26. ______

Find the area of each triangle.

13. 8 m, 12 m

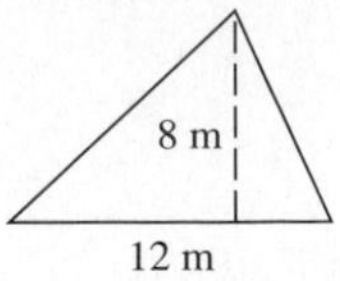

14. height 15.4 cm, base 11.8 cm

15. Find the perimeter of a triangle with sides 51.4 m, 29.7 m, and 38.2 m.

Find the area of each figure. Use 3.14 as an approximation for π. Round each answer to the nearest tenth.

16. a circle with radius 12 m

17. a circle with diameter 39.4 cm

18. a semicircle with radius 6 m

19. Find the circumference of a circle with diameter 25 in.

Find the volume of each figure.

20. 12 m, 18 m, 30 m

21. 9 m

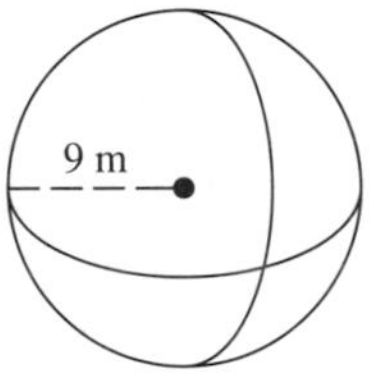

22. 8 m, 6 m

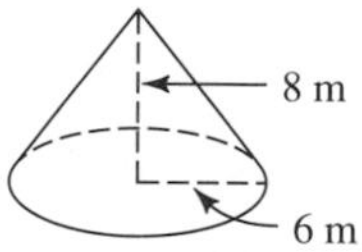

Find the missing length in each right triangle.

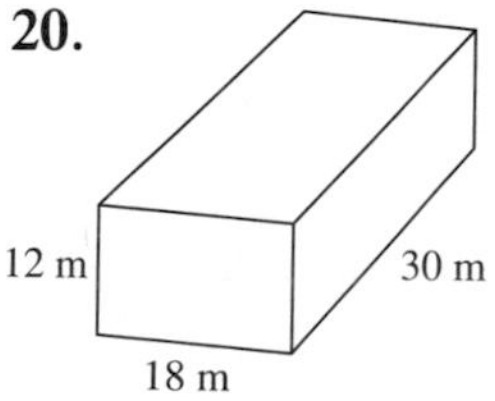

23. 9 cm, 15 cm, 90°

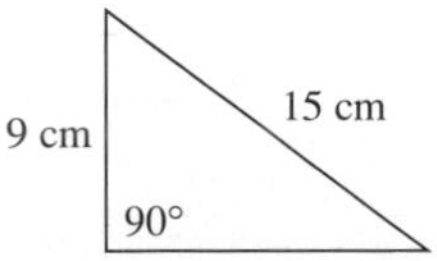

24. 6 cm, 7 cm

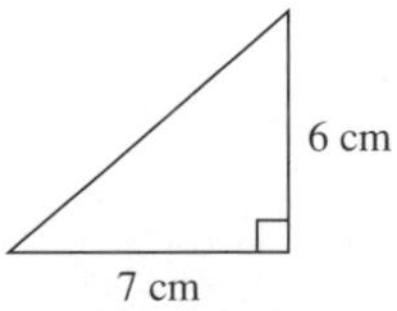

Find the missing lengths in each pair of similar triangles.

25. 5, 4, 6; q, 8, p

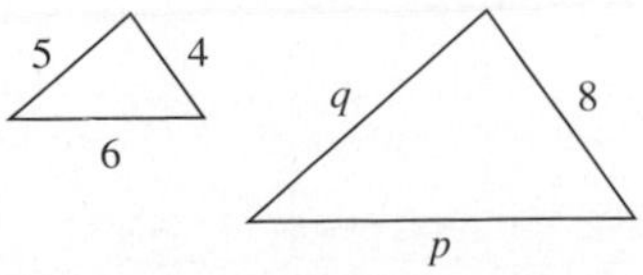

26. 18, 15, 9; y, 10, z

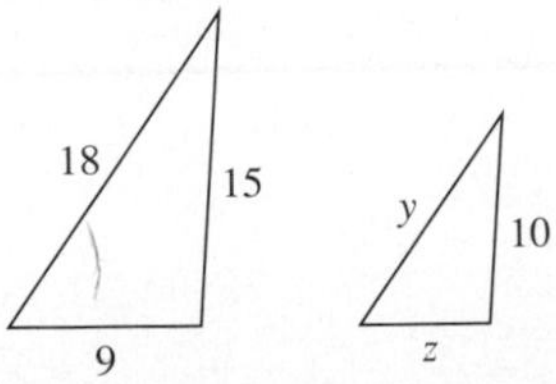

9

BASIC ALGEBRA

9.1 SIGNED NUMBERS

All the numbers studied so far in this book have been either 0 or greater than 0. Numbers greater than 0 are called *positive numbers*. For example, recall the *whole* numbers from Chapter 1:

whole numbers 0, 1, 2, 3, 4, 5, 6, ...

All the whole numbers, *except 0,* are positive.

1 Not all numbers are positive or 0. For example, "15 degrees below 0" or "a loss of $500" are expressed with numbers smaller than 0. Numbers smaller than 0 are called *negative numbers*.

Write negative numbers with a *negative sign,* −.

For example, "15 degrees below 0" is written with a negative sign, as −15. Also, "a loss of $500" is written −500.

WORK PROBLEM 1 AT THE SIDE.

2 In an earlier chapter, the positive numbers and zero were graphed on a number line. Negative numbers can also be shown on a number line. The positive numbers, the negative numbers, and zero make up the integers.

integers . . ., −4, −3, −2, −1, 0, 1, 2, 3, 4, . . .

The negative number −5 is read "negative five."

−6 −5 −4 −3 −2 −1 0 1 2 3 4 5 6 7 8

As mentioned earlier, positive numbers are greater than 0, and negative numbers are smaller than 0.

NOTE Every positive number corresponds to a negative number on the other side of 0 on a number line.

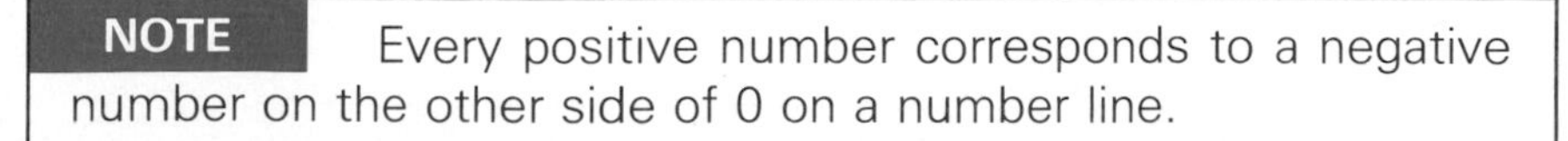

OBJECTIVES

1. Write negative numbers.
2. Use number lines.
3. Graph numbers.
4. Tell which of two numbers is smaller.
5. Find absolute value.
6. Find the opposite of a number.

1. Write each number.

(a) a temperature 34 degrees below 0

(b) a loss of 850 dollars

(c) the altitude of a place 284 feet below sea level

ANSWERS
1. (a) −34° (b) −$850 (c) −284 ft

2. Write *positive, negative,* or *neither* for each number.

(a) -5

(b) -26

(c) 1

(d) 0

3. Graph each list of numbers.

(a) $-1, 2, 3, -4, -3$

–4 –3 –2 –1 0 1 2 3 4

(b) $-2, 4, 0, -1, -4$

–4 –3 –2 –1 0 1 2 3 4

ANSWERS

2. (a) negative (b) negative (c) positive (d) neither

3. (a) –4 –3 –2 –1 0 1 2 3

(b) –4 –3 –2 –1 0 1 2 3 4

Zero is neither a positive number nor a negative number. Positive numbers, negative numbers, and zero are called **signed numbers.**

NOTE Numbers that have no sign indicated are assumed to be positive.

WORK PROBLEM 2 AT THE SIDE.

3 The next example shows how to graph signed numbers.

EXAMPLE 1 Graphing Signed Numbers

Graph -4, 3, -1, 0, and $1\frac{1}{4}$.

SOLUTION

Place a dot at the correct location for each number. ■

–4 –3 –2 –1 0 1 2 3 4

WORK PROBLEM 3 AT THE SIDE.

4 As shown on the following number line, 3 is to the left of 5.

0 3 5

Also, 3 is *smaller than* 5.

Recall the following symbols for comparing two numbers.

$<$ means "is smaller than"

$>$ means "is greater than"

Use these symbols to write "3 is smaller than 5" as follows.

$$3 < 5$$

3 is smaller than 5

As this example suggests,

The smaller of two numbers is the one farther to the left on a number line.

EXAMPLE 2 Using the Symbols $<$ and $>$

Use this number line and $>$ or $<$ to make true statements.

–9 –8 –7 –6 –5 –4 –3 –2 –1 0 1 2 3 4 5 6

(a) On the number line, 2 is to the left of 6, so 2 is smaller than 6, or $2 < 6$.

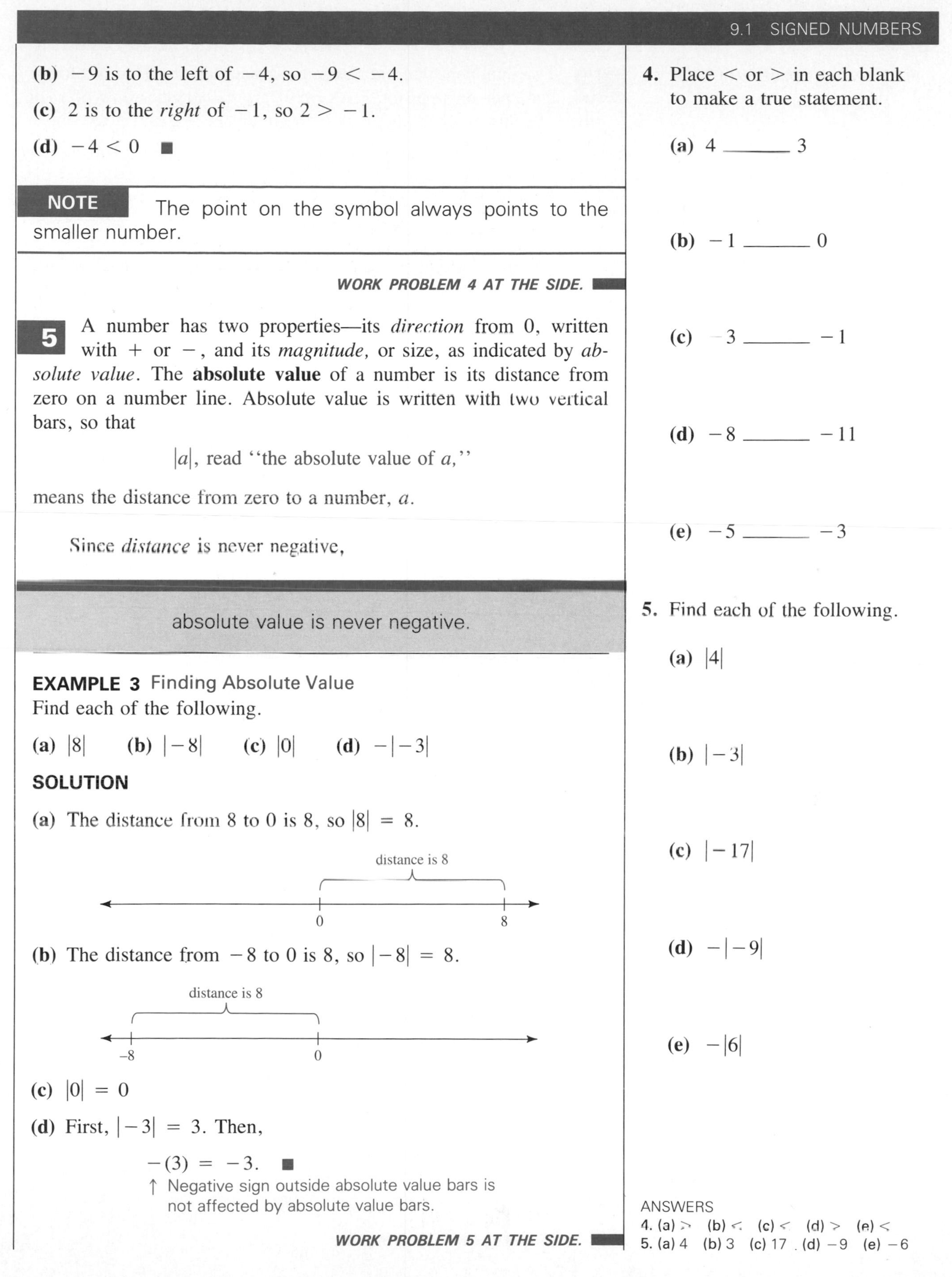

(b) -9 is to the left of -4, so $-9 < -4$.

(c) 2 is to the *right* of -1, so $2 > -1$.

(d) $-4 < 0$ ■

NOTE The point on the symbol always points to the smaller number.

WORK PROBLEM 4 AT THE SIDE.

5 A number has two properties—its *direction* from 0, written with + or −, and its *magnitude,* or size, as indicated by *absolute value*. The **absolute value** of a number is its distance from zero on a number line. Absolute value is written with two vertical bars, so that

$$|a|, \text{ read ``the absolute value of } a\text{,''}$$

means the distance from zero to a number, a.

Since *distance* is never negative,

absolute value is never negative.

EXAMPLE 3 Finding Absolute Value

Find each of the following.

(a) $|8|$ **(b)** $|-8|$ **(c)** $|0|$ **(d)** $-|-3|$

SOLUTION

(a) The distance from 8 to 0 is 8, so $|8| = 8$.

distance is 8

0 8

(b) The distance from -8 to 0 is 8, so $|-8| = 8$.

distance is 8

−8 0

(c) $|0| = 0$

(d) First, $|-3| = 3$. Then,

$$-(3) = -3. \quad ■$$

↑ Negative sign outside absolute value bars is not affected by absolute value bars.

WORK PROBLEM 5 AT THE SIDE.

4. Place $<$ or $>$ in each blank to make a true statement.

(a) 4 _____ 3

(b) -1 _____ 0

(c) -3 _____ -1

(d) -8 _____ -11

(e) -5 _____ -3

5. Find each of the following.

(a) $|4|$

(b) $|-3|$

(c) $|-17|$

(d) $-|-9|$

(e) $-|6|$

ANSWERS

4. (a) $>$ (b) $<$ (c) $<$ (d) $>$ (e) $<$

5. (a) 4 (b) 3 (c) 17 (d) -9 (e) -6

6. Give the opposite of each number.

(a) 4

(b) 12

(c) 49

(d) $\frac{3}{8}$

7. Find the opposite of each number.

(a) -5

(b) -11

(c) -25

(d) -1.9

(e) -2.8

(f) $-\frac{3}{4}$

ANSWERS

6. (a) -4 (b) -12 (c) -49 (d) $-\frac{3}{8}$

7. (a) 5 (b) 11 (c) 25 (d) 1.9 (e) 2.8 (f) $\frac{3}{4}$

6 Two numbers that are the same distance from 0 on a number line, but on opposite sides of 0, are called *opposites* of each other. As this number line shows, the numbers -3 and 3 are opposites of each other.

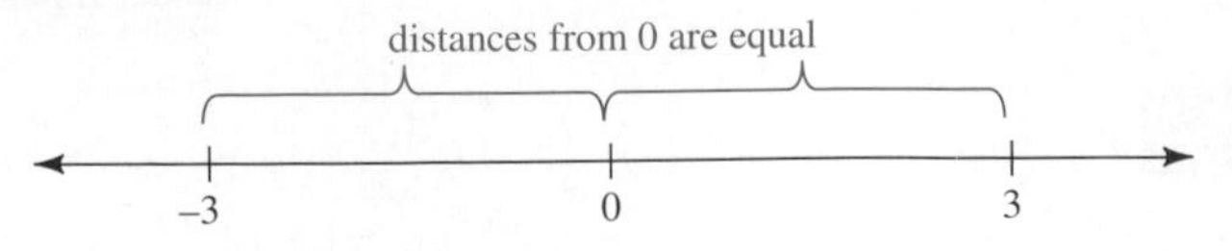

The opposite of a number can be found by attaching a negative sign in front of the number.

EXAMPLE 4 Finding Opposites

Give the opposite of each number.

number	*opposite*
5	-5 ← Attach minus sign.
9	-9
$\frac{4}{5}$	$-\frac{4}{5}$
0	$-0 = 0$

The opposite of 0 is -0, but

$$-0 = 0. \blacksquare$$

WORK PROBLEM 6 AT THE SIDE.

Some numbers have two negative signs, such as

$$-(-3).$$

The negative sign in front of -3 means the *opposite* of -3. The opposite of -3 is 3, so

$$-(-3) = 3.$$

Use the following rule to find the opposite of a negative number.

DOUBLE NEGATIVE RULE

$$-(-x) = x$$

EXAMPLE 5 Finding Opposites

Find the opposite of each number.

number	*opposite*	
-2	$-(-2) = 2$	by double negative rule
-9	$-(-9) = 9$	
$-\frac{1}{2}$	$-\left(-\frac{1}{2}\right) = \frac{1}{2}$ ■	

WORK PROBLEM 7 AT THE SIDE.

name date hour

9.1 EXERCISES

Write a signed number for each of the following.

1. a temperature of 12 degrees above zero

2. a profit of $920

3. The price fell $12.

4. a lake 140 feet below sea level

5. The plane is 18,000 feet above sea level.

6. The submarine is 192 feet below sea level.

Write positive, negative, *or* neither *for each of the following numbers.*

7. 24

8. -8

9. -3

10. -15

11. 0

12. $+6$

13. $\frac{7}{8}$

14. $-\frac{5}{4}$

Graph each of the following lists of numbers.

Example: $-3, -\frac{2}{3}, -5, 1, 2, \frac{5}{8}$

Solution: Place a dot on a number line for each number.

$-\frac{2}{3}$ $\frac{5}{8}$

−5 −4 −3 −2 −1 0 1 2 3 4 5

15. $4, -1, 2, 3, 0, -2$

−5 −4 −3 −2 −1 0 1 2 3 4 5

16. $-5, -3, 1, 4, 0$

−5 −4 −3 −2 −1 0 1 2 3 4 5

17. $-\frac{1}{2}, -3, -5, \frac{1}{4}, 1\frac{7}{8}, 3$

−5 −4 −3 −2 −1 0 1 2 3 4 5

18. $-4, -\frac{3}{4}, -2, 4, 1, 3$

−5 −4 −3 −2 −1 0 1 2 3 4 5

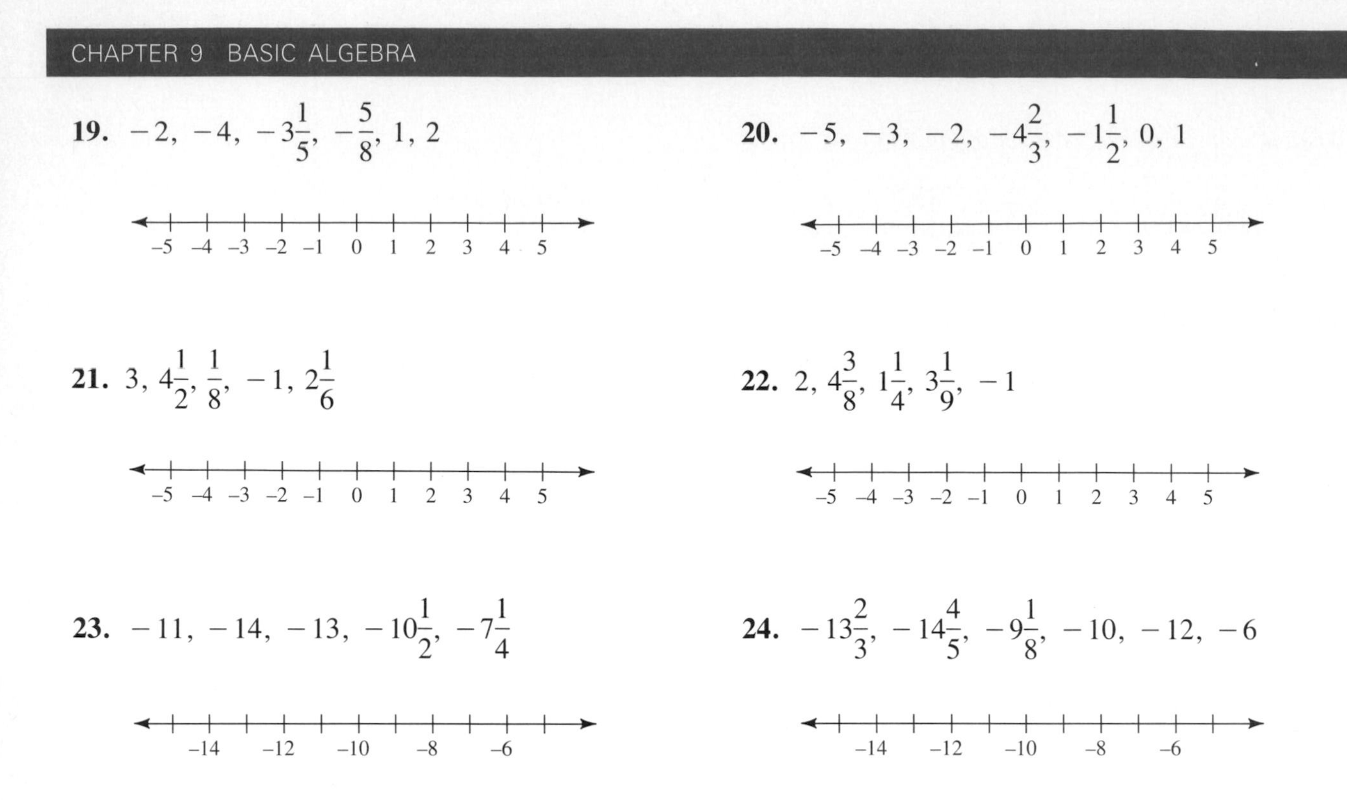

19. $-2, -4, -3\frac{1}{5}, -\frac{5}{8}, 1, 2$

20. $-5, -3, -2, -4\frac{2}{3}, -1\frac{1}{2}, 0, 1$

21. $3, 4\frac{1}{2}, \frac{1}{8}, -1, 2\frac{1}{6}$

22. $2, 4\frac{3}{8}, 1\frac{1}{4}, 3\frac{1}{9}, -1$

23. $-11, -14, -13, -10\frac{1}{2}, -7\frac{1}{4}$

24. $-13\frac{2}{3}, -14\frac{4}{5}, -9\frac{1}{8}, -10, -12, -6$

Place < or > in each of the following to make a true statement.

Examples:

$-4 \quad 2$ $\qquad -5 \quad -9$

Solutions:

Since -4 is to the left of 2 on a number line, -4 is smaller than 2.

$-4 < 2$

Since -5 is to the right of -9 on a number line, -5 is greater than -9.

$-5 > -9$

25. 9 14 **26.** 6 11 **27.** 5 4 **28.** 8 3

29. -6 3 **30.** -4 7 **31.** 8 -6 **32.** 5 -3

33. -11 -2 **34.** -5 -1 **35.** -8 -3 **36.** -12 -14

37. 2 -1 **38.** 4 -9 **39.** 0 -2 **40.** -5 0

name date hour

Find each of the following.

Examples: $|8|$, $|-7|$, $-|-2|$
Solutions: $|8| = \mathbf{8}$ $\quad |-7| = \mathbf{7}$ $\quad -|-2| = \mathbf{-2}$

41. $|5|$ **42.** $|12|$ **43.** $|-5|$ **44.** $|-28|$ **45.** $|-32|$

46. $|-14|$ **47.** $|251|$ **48.** $|397|$ **49.** $|0|$ **50.** $|23|$

51. $\left|-\frac{1}{2}\right|$ **52.** $\left|-\frac{9}{5}\right|$ **53.** $|-9.5|$ **54.** $|-11.2|$ **55.** $|8.3|$

56. $|4.7|$ **57.** $\left|\frac{3}{4}\right|$ **58.** $\left|\frac{7}{3}\right|$ **59.** $\left|-\frac{9}{7}\right|$ **60.** $\left|-\frac{8}{5}\right|$

61. $-|-9|$ **62.** $-|-12|$ **63.** $-|4|$ **64.** $-|15|$

Give the opposite of each number.

Examples: **Solutions:**

number	*opposite*
8	**−8**
−5	$-(-5) = \mathbf{5}$
0	**0**

65. 2 **66.** 7 **67.** −5 **68.** −9

69. −11 **70.** −24 **71.** 16 **72.** 30

73. $\frac{4}{3}$ **74.** $\frac{5}{8}$ **75.** $-\frac{1}{2}$ **76.** $-\frac{7}{9}$

77. 5.2 **78.** 3.7 **79.** -1.4 **80.** -9.8

81. What is the opposite of $-(-2)$?

82. Find the opposite of $-|-2|$.

CHALLENGE EXERCISES

Write true or false for each statement.

83. $-8 < -4$ **84.** $-|-12| > |-15|$ **85.** $0 < -(-6)$

86. $-9 < -(-6)$ **87.** $-|-7| < -(3)$

REVIEW EXERCISES

Add the following numbers. (For help, see ***Sections 3.3 and 3.4****.)*

88. $16 + 21 + 11$ **89.** $\frac{3}{4} + \frac{1}{2} + \frac{3}{5}$ **90.** $2\frac{1}{2} + 3\frac{2}{3}$

91. $2 + 1 + \frac{2}{5} + 1\frac{1}{10}$ **92.** $6 + 2\frac{4}{5} + 1\frac{3}{10}$

9.2 ADDITION AND SUBTRACTION OF SIGNED NUMBERS

A positive number can be represented on a number line by an arrow pointing to the right. In the following examples both arrows represent 4.

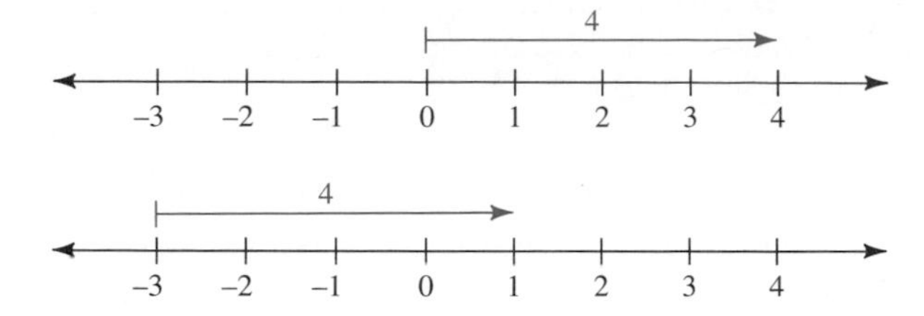

Use arrows pointing to the left for negative numbers. Both of the following arrows represent -3.

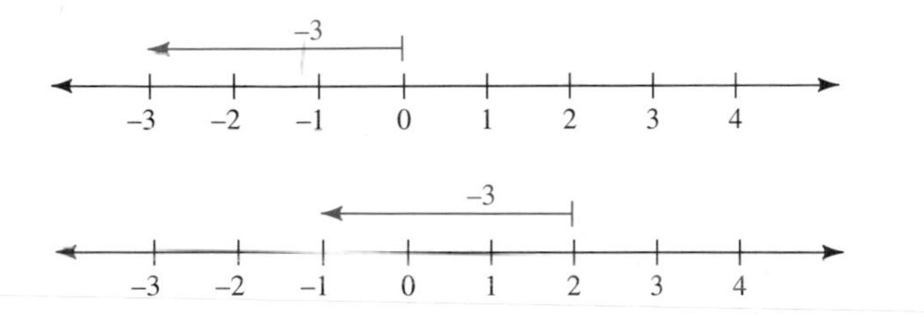

WORK PROBLEM 1 AT THE SIDE.

OBJECTIVES

1. Add signed numbers by using a number line.
2. Add signed numbers without using a number line.
3. Find the additive inverse of a number.
4. Subtract signed numbers.
5. Add or subtract a series of numbers.

1. Complete each arrow so it represents the indicated number.

(a) 3

−1 0 1 2 3 4

(b) 5

−2 −1 0 1 2 3 4

(c) −4

−2 −1 0 1 2 3 4

(d)

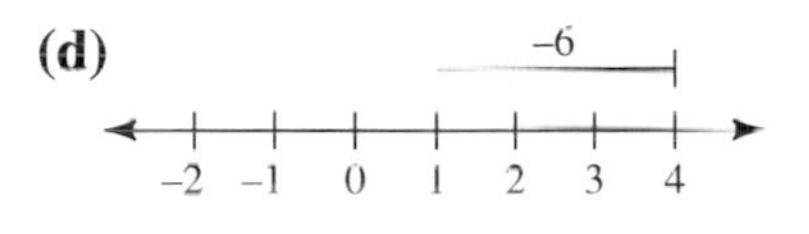

1 A number line can be used to add signed numbers. For example, the next number line shows how to add 2 and 3.

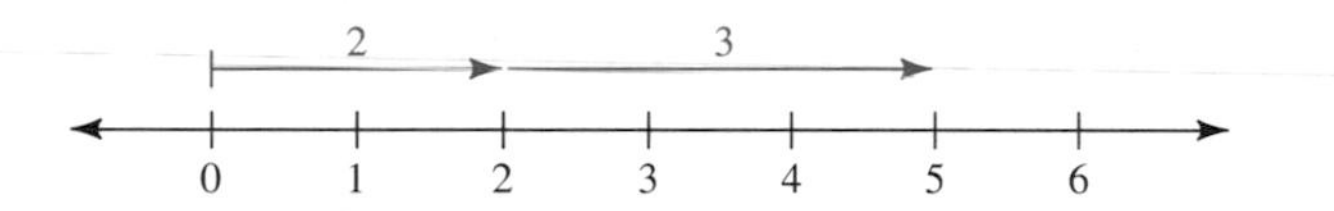

Add 2 and 3 by starting at zero and drawing an arrow 2 units to the right. From the end of this arrow, draw another arrow 3 units to the right. This second arrow ends at 5, showing that

$$2 + 3 = 5.$$

EXAMPLE 1 Adding Signed Numbers by Using a Number Line
Add by using a number line.

(a) $4 + (-1)$ **(b)** $-6 + 2$ **(c)** $-3 + (-5)$

SOLUTION

(a) $4 + (-1)$

−1
4
0 1 2 3 4 5 6

(*Note:* Always start at 0.)

Start at zero and draw an arrow 4 units to the right. From the end of this arrow, draw an arrow 1 unit to the *left*. (Remember to go to the left for a negative number.) This second arrow ends at 3, so

$$4 + (-1) = 3.$$

ANSWERS

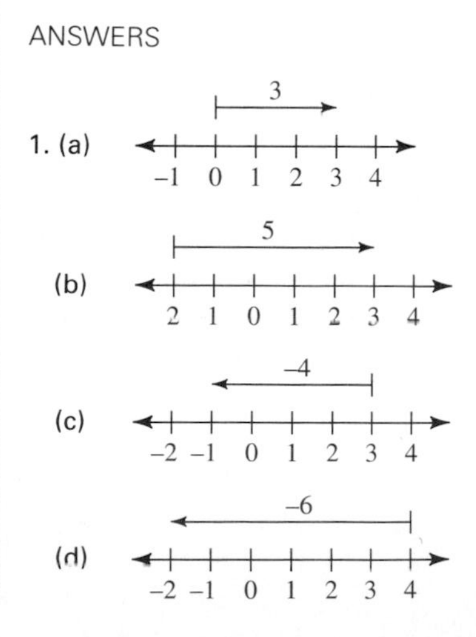

2. Draw arrows to find each of the following.

(a) $5 + (-2)$

0 1 2 3 4 5

(b) $-4 + 1$

−4 −3 −2 −1 0

(c) $-3 + 7$

−3 −2 −1 0 1 2 3 4

(d) $-4 + (-2)$

−6 −4 −2 0

ANSWERS

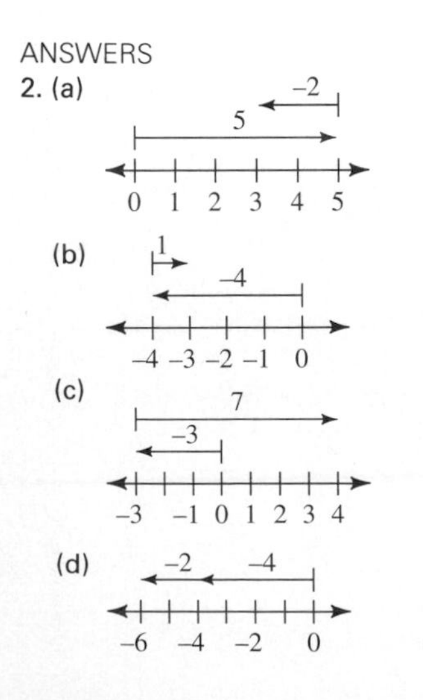

(b) $-6 + 2$

2
−6
−7 −6 −5 −4 −3 −2 −1 0

Draw an arrow from zero going 6 units to the left. From the end of this arrow, draw an arrow 2 units to the right. This second arrow ends at -4, so

$$-6 + 2 = -4.$$

(c) $-3 + (-5)$

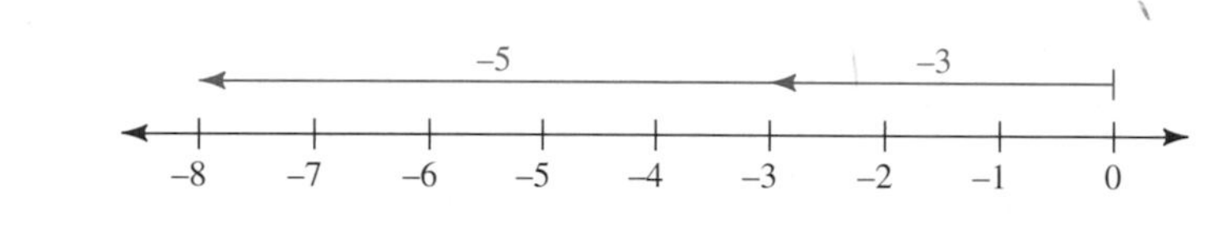

As the arrows along the number line show,

$$-3 + (-5) = -8. \quad \blacksquare$$

WORK PROBLEM 2 AT THE SIDE.

2 This method of addition with a number line suggests rules that permit addition of signed numbers without a number line. The first rule tells how to add two negative numbers.

> *Add two negative numbers* by adding the absolute values of the numbers. Then place a negative sign before the sum.

EXAMPLE 2 Adding Signed Numbers with the Same Sign

Add (without number lines).

(a) $-4 + (-12)$ **(b)** $-5 + (-25)$ **(c)** $-11 + (-46)$

(d) $-\frac{3}{4} + \left(-\frac{1}{2}\right)$

SOLUTION

(a) $-4 + (-12)$

The absolute value of -4 is $|-4| = 4$.

The absolute value of -12 is $|-12| = 12$.

Add the absolute values.

$$4 + 12 = 16$$

Place a negative sign before the sum.

$$-4 + (-12) = -16 \leftarrow$$ Place a negative sign before 16.

(b) $-5 + (-25) = -30$ ← Sum of absolute values, with a negative sign placed before 30.

(c) $-11 + (-46) = -57$ **(d)** $-\frac{3}{4} + \left(-\frac{1}{2}\right)$

The absolute value of $-\frac{3}{4}$ is $\left|-\frac{3}{4}\right| = \frac{3}{4}$, and the absolute value of $-\frac{1}{2}$ is $\left|-\frac{1}{2}\right| = \frac{1}{2}$. Add the absolute values.

$$\frac{3}{4} + \frac{1}{2} = \frac{3}{4} + \frac{2}{4} = \frac{5}{4}$$

Place a negative sign before the sum.

$$-\frac{3}{4} + \left(-\frac{1}{2}\right) = -\frac{5}{4}$$ ← Attach a negative sign. ■

WORK PROBLEM 3 AT THE SIDE.

Use the following rule to add two numbers with *different* signs.

Add two numbers with different signs (one positive and one negative) by first *subtracting* the absolute values of the numbers; the smaller number should be subtracted from the larger. Next, place the sign of the number with the *larger* absolute value before the difference.

EXAMPLE 3 Adding Signed Numbers with Different Signs

Add: **(a)** $8 + (-3)$ **(b)** $-12 + 4$ **(c)** $15 + (-21)$

(d) $-13 + 9$ **(e)** $-\frac{1}{2} + \frac{2}{3}$.

SOLUTION

(a) $8 + (-3)$

Find this sum with a number line as follows.

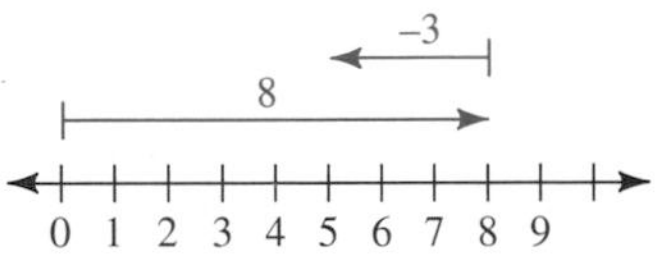

Since the top arrow ends at 5.

$$8 + (-3) = 5.$$

Find the sum by using the rule as follows. First, find the absolute value of each number.

$$|8| = 8 \qquad |-3| = 3$$

3. Add.

(a) $-9 + (-4)$

(b) $-7 + (-2)$

(c) $-31 + (-5)$

(d) $-10 + (-8)$

(e) $-\frac{9}{10} + \left(-\frac{3}{5}\right)$

ANSWERS
3. (a) -13 (b) -9 (c) -36 (d) -18 (e) $-\frac{3}{2}$

4. Add.

(a) $-12 + 5$

(b) $-7 + 8$

(c) $-11 + 11$

(d) $23 + (-32)$

(e) $\frac{3}{4} + \left(-\frac{5}{8}\right)$

ANSWERS

4. (a) -7 (b) 1 (c) 0 (d) -9 (e) $\frac{1}{8}$

Subtract the absolute value of the smaller number from the larger number.

$$8 - 3 = 5$$

Here the positive number 8 has the larger absolute value, so the answer is positive.

$$8 + (-3) = 5$$

(b) $-12 + 4$

First, find absolute values.

$$|-12| = 12 \qquad |4| = 4$$

Subtract:

$$12 - 4 = 8.$$

The negative number -12 has the larger absolute value, so the answer is negative.

$$-12 + 4 = -8$$

↑ Place negative sign before 8, since the negative number has the larger absolute value.

(c) $15 + (-21) = -6$

(d) $-13 + 9 = -4$

(e) $-\frac{1}{2} + \frac{2}{3}$

The absolute value of $-\frac{1}{2}$ is $\frac{1}{2}$, and the absolute value of $\frac{2}{3}$ is $\frac{2}{3}$. Subtract the absolute values.

$$\frac{2}{3} - \frac{1}{2} = \frac{4}{6} - \frac{3}{6} = \frac{1}{6}$$

Since the positive number $\frac{2}{3}$ has the larger absolute value, the answer is positive.

$$-\frac{1}{2} + \frac{2}{3} = \frac{1}{6} \leftarrow \text{positive answer}$$ ■

WORK PROBLEM 4 AT THE SIDE.

Recall that the opposite of 9 is -9, and the opposite of -4 is $-(-4)$, or 4. Add these opposites as follows.

$$9 + (-9) = 0 \quad \textit{and} \quad -4 + 4 = 0$$

The sum of a number and its opposite is always 0.

3 For this reason, opposites are also called *additive inverses* of each other.

Two numbers are **additive inverses** of each other if their sum is zero.

EXAMPLE 4 Finding the Additive Inverse

The following chart shows several numbers and the additive inverse of each.

number	*additive inverse*
6	-6
-8	$-(-8)$ or 8 ← $-(-x) = x$
4	-4
-3	$-(-3)$ or 3
$\frac{5}{8}$	$-\frac{5}{8}$
0	0 ■

WORK PROBLEM 5 AT THE SIDE.

4 Subtraction, like addition, can be shown on a number line. The number line that follows shows the subtraction of 4 from 6.

Start at zero on the number line, and draw an arrow to 6. From the end of this arrow, draw another arrow 4 units to the left. This second arrow ends at 2, showing that

$$6 - 4 = 2.$$

The same answer would be found by *adding* 6 and -4, the additive inverse of 4. This suggests the following definition of subtraction.

The difference of two numbers, a and b, is

$$\boldsymbol{a - b = a + (-b).}$$

Subtract two numbers by adding the first number and the additive inverse of the second.

5. Give the additive inverse of each number.

(a) 12

(b) -7

(c) -11

(d) $-\frac{7}{10}$

(e) 0

ANSWERS

5. (a) -12 (b) 7 (c) 11 (d) $\frac{7}{10}$ (e) 0

6. Subtract.

(a) $-6 - 5$

(b) $2 - 9$

(c) $5 - (-8)$

(d) $2 - (-4)$

(e) $-7 - (-15)$

(f) $-\frac{2}{3} - \left(-\frac{5}{12}\right)$

ANSWERS
6. (a) -11 (b) -7 (c) 13 (d) 6 (e) 8
(f) $-\frac{1}{4}$

A simple procedure for subtracting signed numbers is described below.

Step 1 Change the subtraction sign to addition sign.
Step 2 Change the sign of the number following the subtraction sign.
Step 3 Proceed as in addition.

EXAMPLE 5 Subtracting Signed Numbers
Subtract: **(a)** $8 - 11$ **(b)** $4 - 17$ **(c)** $-9 - 15$
(d) $5 - (-7)$ **(e)** $7.6 - (-8.3)$ **(f)** $\frac{5}{8} - \left(-\frac{1}{2}\right)$.

SOLUTION

(a) $8 - 11$

Write the first number, unchanged. Change subtraction sign to addition sign and write the additive inverse of the second number.

$$8 - 11$$

unchanged ↓ additive inverse of 11 ↓

$$8 + (-11)$$

↑ changed

Now add.

$$8 + (-11) = -3$$
$$8 - 11 = -3$$

(b) $4 - 17$

$$4 + (-17) = -13$$

(c) $-9 - 15$

$$-9 + (-15) = -24$$

(d) $5 - (-7)$

$$5 + 7 = 12$$

(e) $7.6 - (-8.3)$

$$7.6 + 8.3 = 15.9$$

(f) $\frac{5}{8} - \left(-\frac{1}{2}\right) = \frac{5}{8} + \left(\frac{1}{2}\right) = \frac{9}{8} = 1\frac{1}{8}$ ■

NOTE The pattern in each example is: first number − second number = first number + additive inverse of second number.

WORK PROBLEM 6 AT THE SIDE.

5 If a problem involves both addition and subtraction, use the order of operations and proceed from left to right.

EXAMPLE 6 Combining Addition and Subtraction of Signed Numbers

Perform the operations from left to right.

(a) $-6 + (-11) - (-5)$ **(b)** $4 - (-3) + (-9)$

SOLUTION

(a) $\underbrace{-6 + (-11)} - (-5)$

$-17 \quad - (-5)$

$\underbrace{-17 \quad +5}$

-12

(b) $\underbrace{4 - (-3)} + (-9)$

$\underbrace{7 \quad + (-9)}$

-2 ■

WORK PROBLEM 7 AT THE SIDE.

EXAMPLE 7 Using the Order of Operations to Combine More Than Two Numbers

Find each sum.

(a) $\underbrace{-7 + 12} + (-3)$

$\underbrace{5 \quad + (-3)}$

2

(b) $\underbrace{14 + (-9)} - (-8) + 10$

$= \underbrace{5 \quad - (-8)} + 10$

$= \underbrace{13 \quad + 10}$

$= \quad 23$

(c)

$$\begin{array}{r} -6.3 \\ -14.9 \\ 8.5 \\ -7.4 \\ \underline{5.2} \end{array}$$

Start at the top.

$$\left.\begin{array}{r} -6.3 \\ -14.9 \end{array}\right\} \rightarrow \left.\begin{array}{r} -21.2 \\ 8.5 \end{array}\right\} \rightarrow \left.\begin{array}{r} -12.7 \\ -7.4 \end{array}\right\} \rightarrow -20.1$$

$$\begin{array}{rrrr} 8.5 & & & \\ -7.4 & -7.4 & & \\ \underline{5.2} & \underline{5.2} & \underline{5.2} & \underline{5.2} \\ & & & -14.9 \end{array}$$ ■

WORK PROBLEM 8 AT THE SIDE.

7. Perform the operations from left to right.

(a) $9 - (-8) + 6$

(b) $-4 + (-11) - (-6)$

(c) $-3 - (-9) - (-5)$

(d) $8 - (-2) + (-6)$

8. Add.

(a) $-11 + 4 + (-6)$

(b) $24 + (-15) - (-19)$

(c) $-6 + (-15) - (-19) + (-25)$

(d)

$$\begin{array}{r} -19.2 \\ -6.7 \\ 15.8 \\ 17.1 \\ \underline{-5.4} \end{array}$$

ANSWERS

7. (a) 23 (b) −9 (c) 11 (d) 4

8. (a) −13 (b) 28 (c) −27 (d) 1.6

name date hour

9.2 EXERCISES

Add by using the number line.

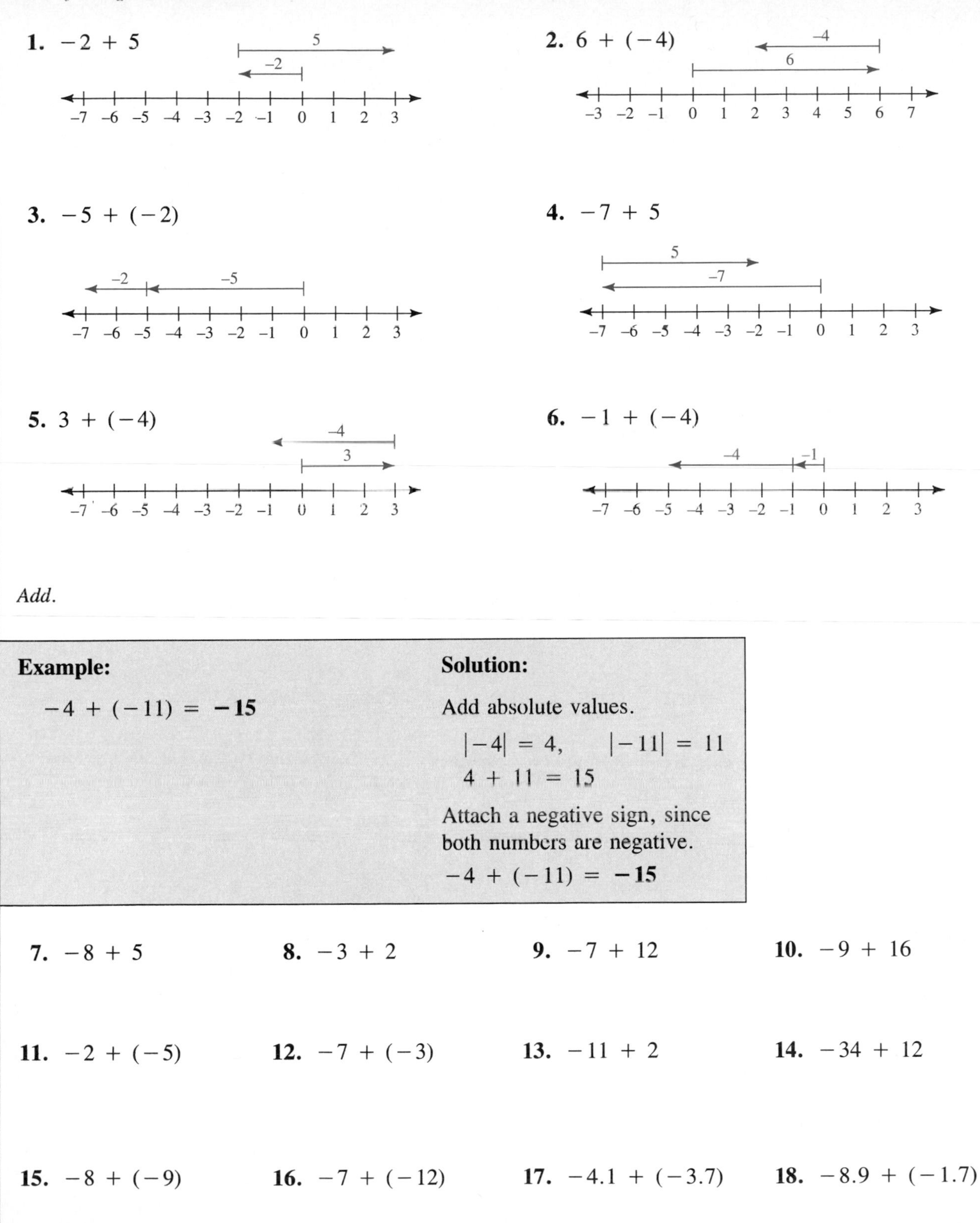

1. $-2 + 5$

2. $6 + (-4)$

3. $-5 + (-2)$

4. $-7 + 5$

5. $3 + (-4)$

6. $-1 + (-4)$

Add.

Example:

$-4 + (-11) = \mathbf{-15}$

Solution:

Add absolute values.

$|-4| = 4, \quad |-11| = 11$

$4 + 11 = 15$

Attach a negative sign, since both numbers are negative.

$-4 + (-11) = \mathbf{-15}$

7. $-8 + 5$

8. $-3 + 2$

9. $-7 + 12$

10. $-9 + 16$

11. $-2 + (-5)$

12. $-7 + (-3)$

13. $-11 + 2$

14. $-34 + 12$

15. $-8 + (-9)$

16. $-7 + (-12)$

17. $-4.1 + (-3.7)$

18. $-8.9 + (-1.7)$

19. $7.8 + (-14.6)$ **20.** $4.9 + (-8.1)$ **21.** $-\frac{1}{2} + \frac{3}{4}$ **22.** $-\frac{2}{3} + \frac{5}{6}$

23. $-\frac{5}{8} + \frac{1}{4}$ **24.** $-\frac{11}{15} + \frac{2}{5}$ **25.** $-\frac{7}{3} + \left(-\frac{5}{9}\right)$ **26.** $-\frac{8}{5} + \left(-\frac{3}{10}\right)$

27. $\begin{array}{r} -9 \\ \underline{-2} \end{array}$ **28.** $\begin{array}{r} -11 \\ \underline{-7} \end{array}$ **29.** $\begin{array}{r} -15 \\ \underline{9} \end{array}$ **30.** $\begin{array}{r} -8 \\ \underline{2} \end{array}$

Give the additive inverse of each number.

number	***additive inverse***	***number***	***additive inverse***	***number***	***additive inverse***	***number***	***additive inverse***
31. 3		**32.** 4		**33.** -9		**34.** -14	
35. -8		**36.** -12		**37.** 154		**38.** 393	
39. $\frac{1}{4}$		**40.** $\frac{7}{15}$					

Subtract.

Examples:	$-8 - (-2)$	$7 - (-11)$
Solutions:	$-8 - (-2) = -8 + 2 = \mathbf{-6}$	$7 - (-11) = 7 + 11 = \mathbf{18}$

41. $19 - 5$ **42.** $24 - 11$ **43.** $6 - 9$ **44.** $1 - 4$

45. $7 - 19$ **46.** $2 - 17$ **47.** $5 - 15$ **48.** $7 - 13$

49. $9 - 14$ **50.** $3 - 11$ **51.** $-2 - 15$ **52.** $-15 - 17$

53. $6 - (-14)$ **54.** $8 - (-1)$ **55.** $7 - (-9)$ **56.** $10 - (-1)$

57. $-\frac{7}{10} - \frac{4}{5}$ **58.** $-\frac{8}{15} - \frac{3}{10}$ **59.** $-4 - (-9)$

60. $-8 - (-4)$ **61.** $-6.4 - (-2.8)$ **62.** $-7.9 - (-11.2)$

Add.

63. $\begin{array}{r} -6 \\ -2 \\ \underline{5} \end{array}$ **64.** $\begin{array}{r} 3 \\ -4 \\ \underline{-7} \end{array}$ **65.** $\begin{array}{r} -8 \\ 9 \\ \underline{7} \end{array}$

66. $\begin{array}{r} -11 \\ 5 \\ \underline{-2} \end{array}$ **67.** $\begin{array}{r} -9 \\ -8 \\ 2 \\ 5 \\ \underline{6} \end{array}$ **68.** $\begin{array}{r} -3 \\ -11 \\ -4 \\ -6 \\ \underline{12} \end{array}$

Perform the operations from left to right in each problem

Example:	**Solution:**
$-6 + (-9) - (-10)$	$= -15 - (-10)$
	$= -15 + 10$
	$= \mathbf{-5}$

69. $-2 + (-11) - (-3)$ **70.** $-5 - (-2) + (-6)$ **71.** $4 - (-13) + (-5)$

72. $7 - (-9) + (-12)$ **73.** $-2 - (-8) - (-3)$ **74.** $-4 - (-15) - (-3)$

75. $\frac{1}{2} - \frac{2}{3} + \left(-\frac{5}{6}\right)$

76. $\frac{2}{5} - \frac{7}{10} + \left(-\frac{3}{2}\right)$

77. $-5.7 - (-9.4) - 8.1$

78. $-6.5 - (-11.2) - 1.4$

CHALLENGE EXERCISES

Add or subtract the following.

79. $-2 + (-11) + |-2|$

80. $|-7 + 2| + (-2) + 4$

81. $-3 + (-2 + 4) + (-5)$

82. $-3 - 2 - [-(-4)]$

83. $2\frac{1}{2} + 3\frac{1}{4} - \left(-1\frac{3}{8}\right) - 2\frac{3}{8}$

84. $\frac{5}{8} - \left(-\frac{1}{2} - \frac{3}{4}\right)$

85. $-9 + [(3 - 2) - (-4 + 2)]$

REVIEW EXERCISES

Multiply or divide the following. (For help, see ***Sections 1.4, 2.5, 2.8, 4.5, or 4.6.****)*

86. $23 \cdot 46$

87. $\frac{8}{11} \cdot \frac{3}{5}$

88. $71.20 \cdot 21.25$

89. $3\frac{1}{2} \cdot 2\frac{3}{4}$

90. $1235 \div 5$

91. $1\frac{2}{3} \div 2\frac{7}{9}$

92. $\frac{7}{9} \div \frac{14}{27}$

9.3 MULTIPLICATION AND DIVISION OF SIGNED NUMBERS

How are two numbers with different signs multiplied? Look first at the following list of products.

$$\begin{aligned} \mathbf{4} \cdot 2 &= 8 \\ \mathbf{3} \cdot 2 &= 6 \\ \mathbf{2} \cdot 2 &= 4 \\ \mathbf{1} \cdot 2 &= 2 \\ \mathbf{0} \cdot 2 &= 0 \\ \mathbf{-1} \cdot 2 &= ? \end{aligned}$$

As the numbers in bold type decrease by 1, the numbers in color decrease by 2. This pattern can be maintained by replacing ? with a number 2 less than 0, which is -2. Therefore,

$$-1 \cdot 2 = -2.$$

1 As suggested by this example,

the product of two numbers with different signs is negative.

EXAMPLE 1 Multiplying Numbers with Different Signs

Multiply.

(a) $-8 \cdot 4$ **(b)** $6 \cdot (-3)$ **(c)** $-5 \cdot (11)$ **(d)** $12 \cdot (-7)$

SOLUTION

(a) $-8 \cdot 4 = -32$ ← The product is negative.

(b) $6 \cdot (-3) = -18$

(c) $-5 \cdot (11) = -55$

(d) $12 \cdot (-7) = -84$ ■

WORK PROBLEM 1 AT THE SIDE.

For two numbers with the same signs, look at this pattern.

$$\begin{aligned} \mathbf{4} \cdot (-2) &= -8 \\ \mathbf{3} \cdot (-2) &= -6 \\ \mathbf{2} \cdot (-2) &= -4 \\ \mathbf{1} \cdot (-2) &= -2 \\ \mathbf{0} \cdot (-2) &= 0 \\ \mathbf{-1} \cdot (-2) &= ? \end{aligned}$$

This time, as the numbers in bold type decrease by 1, the products increase by 2. Maintain the pattern by replacing ? with a number 2 *greater than 0,* or 2.

$$-1 \cdot (-2) = 2$$

2 This example suggests that

the product of two numbers with the same sign is positive.

OBJECTIVES

1. Multiply or divide two numbers with opposite signs.
2. Multiply or divide two numbers with the same sign.

1. Multiply.

(a) $7 \cdot (-4)$

(b) $-9 \cdot (15)$

(c) $5 \cdot (-7)$

(d) $-6 \cdot (10)$

(e) $\left(-\frac{7}{8}\right)\left(\frac{4}{3}\right)$

ANSWERS

1. (a) -28 (b) -135 (c) -35 (d) -60 (e) $-\frac{7}{6}$

EXAMPLE 2 Multiplying Two Numbers with the Same Sign
Multiply.

(a) $(-9)(-2)$

$$(-9)(-2) = 18 \leftarrow \text{The product is positive.}$$

(b) $-7 \cdot (-4) = 28$ **(c)** $(-6)(-2) = 12$

(d) $(-10)(-5) = 50$ **(e)** $7 \cdot 5 = 35$ ■

WORK PROBLEM 2 AT THE SIDE.

Division of signed numbers follows exactly the same rules as does multiplication of signed numbers.

When two nonzero numbers with different signs are divided, the result is negative.

When two nonzero numbers with the same sign are divided, the result is positive.

EXAMPLE 3 Dividing Signed Numbers
Divide.

(a) $\frac{-15}{5}$

$$\frac{-15}{5} = -3 \leftarrow \text{Numbers have different signs, so answer is negative.}$$

(b) $\frac{-8}{-4}$

$$\frac{-8}{-4} = 2 \leftarrow \text{Numbers have same sign, so answer is positive.}$$

(c) $\frac{-75}{-25} = 3$ **(d)** $\frac{-6}{3} = -2$

(e) $\frac{15}{5} = 3$ **(f)** $\frac{-90}{9} = -10$

(g) $\frac{-\frac{2}{3}}{-\frac{5}{9}} = -\frac{2}{3} \cdot \left(-\frac{9}{5}\right)$

$$= -\frac{2}{\not{3}_1} \cdot \left(-\frac{\not{9}^3}{5}\right)$$

$$= \frac{6}{5} = 1\frac{1}{5} \quad ■$$

WORK PROBLEM 3 AT THE SIDE.

2. Multiply.

(a) $(-9) \cdot (-7)$

(b) $(-2) \cdot (-3)$

(c) $-8 \cdot (-4)$

(d) $3 \cdot 12$

(e) $\left(-\frac{2}{3}\right)\left(-\frac{6}{5}\right)$

3. Divide.

(a) $\frac{-20}{4}$

(b) $\frac{-50}{-5}$

(c) $\frac{22}{2}$

(d) $\frac{-42}{6}$

(e) $\frac{-\frac{1}{2}}{-\frac{3}{4}}$

(f) $\frac{-\frac{3}{5}}{\frac{9}{10}}$

ANSWERS
2. (a) 63 (b) 6 (c) 32 (d) 36 (e) $\frac{4}{5}$
3. (a) -5 (b) 10 (c) 11 (d) -7 (e) $\frac{2}{3}$ (f) $-\frac{2}{3}$

name date hour

9.3 EXERCISES

Multiply.

Examples:	$-6 \cdot 5$	$8 \cdot (-2)$	$(-4)(-3)$	$-\frac{3}{4} \cdot \left(-\frac{2}{3}\right)$
Solutions:	$-6 \cdot 5 = \mathbf{-30}$	$8 \cdot (-2) = \mathbf{-16}$	$(-4)(-3) = \mathbf{12}$	$-\frac{3}{4} \cdot \left(-\frac{2}{3}\right) = \mathbf{\frac{1}{2}}$

1. $-7 \cdot 2$

2. $-5 \cdot 8$

3. $-9 \cdot 3$

4. $-4 \cdot 9$

5. $8 \cdot (-5)$

6. $3 \cdot (-2)$

7. $10 \cdot (-5)$

8. $5 \cdot (-11)$

9. $12 \cdot (-4)$

10. $7 \cdot (-15)$

11. $-8 \cdot (-4)$

12. $-3 \cdot (-9)$

13. $-5 \cdot (-15)$

14. $-10 \cdot (-5)$

15. $-19 \cdot (-7)$

16. $-21 \cdot (-3)$

17. $-18 \cdot (-4)$

18. $15 \cdot 9$

19. $-\frac{3}{4} \cdot 12$

20. $-\frac{5}{8} \cdot 32$

21. $-\frac{1}{4} \cdot (-6)$

22. $-\frac{3}{5} \cdot (-6)$

23. $-10 \cdot \left(\frac{2}{5}\right)$

24. $-25 \cdot \left(-\frac{7}{10}\right)$

25. $\frac{1}{2} \cdot \left(-\frac{2}{3}\right)$

26. $\frac{3}{8} \cdot \left(-\frac{4}{9}\right)$

27. $-\frac{7}{5} \cdot \frac{10}{3}$

28. $-\frac{9}{10} \cdot \frac{5}{4}$

29. $-\frac{7}{15} \cdot \frac{25}{14}$ **30.** $-\frac{5}{9} \cdot \frac{18}{25}$ **31.** $-\frac{7}{4} \cdot \left(-\frac{8}{3}\right)$ **32.** $-\frac{10}{3} \cdot \left(-\frac{7}{5}\right)$

33. $9 \cdot (-4.7)$ **34.** $15 \cdot (-6.3)$ **35.** $-11.4 \cdot (-18)$ **36.** $-42.3 \cdot (-12)$

37. $-6.2 \cdot (5.1)$ **38.** $-4.3 \cdot (9.7)$ **39.** $-11.2 \cdot (-4.2)$ **40.** $5.8 \cdot 9.7$

Divide.

Examples: $\frac{-10}{2}$ $\frac{-32}{8}$ $\frac{-21}{-3}$

Solutions: $\frac{-10}{2} = -5$ $\frac{-32}{-8} = 4$ $\frac{-21}{-3} = 7$

41. $\frac{-14}{7}$ **42.** $\frac{-8}{2}$ **43.** $\frac{-25}{5}$ **44.** $\frac{-42}{7}$

45. $\frac{-28}{4}$ **46.** $\frac{-40}{4}$ **47.** $\frac{-86}{2}$ **48.** $\frac{-93}{3}$

49. $\frac{-20}{-2}$ **50.** $\frac{-80}{-4}$ **51.** $\frac{-72}{-36}$ **52.** $\frac{-35}{-7}$

name date hour

53. $-\frac{2}{3} \div \frac{5}{6}$

54. $-\frac{5}{8} \div \frac{10}{7}$

55. $-\frac{1}{9} \div \left(-\frac{5}{18}\right)$

56. $\dfrac{-\frac{3}{11}}{-\frac{7}{22}}$

57. $\dfrac{-\frac{5}{7}}{-\frac{15}{14}}$

58. $\dfrac{-\frac{3}{4}}{-\frac{9}{16}}$

59. $-\frac{2}{3} \div (-2)$

60. $-\frac{3}{4} \div (-9)$

61. $5 \div \left(-\frac{5}{8}\right)$

62. $7 \div \left(-\frac{14}{15}\right)$

63. $\dfrac{-10.72}{2}$

64. $\dfrac{-25.11}{3}$

65. $\dfrac{-18.92}{-4}$

66. $\dfrac{-22.75}{-7}$

67. $\dfrac{-9.75}{2.5}$

68. $\dfrac{-17.02}{7.4}$

69. $\dfrac{-45.58}{-8.6}$

70. $\dfrac{-59.28}{-11.4}$

Work left to right in each of the following.

71. $(-4) \cdot (-6) \cdot \frac{1}{2}$

72. $(-9) \cdot (-3) \cdot \frac{2}{3}$

73. $(.6) \cdot (-.2) \cdot (-.3)$

74. $(.5) \cdot (-.8) \cdot \left(-\frac{2}{3}\right)$

75. $\left(-\frac{1}{2}\right) \cdot \left(-\frac{2}{5}\right) \cdot \left(\frac{7}{8}\right)$

76. $\left(\frac{3}{4}\right) \cdot \left(-\frac{5}{6}\right) \cdot \left(-\frac{2}{3}\right)$

CHALLENGE EXERCISES

Evaluate the following.

77. $(-2) \cdot (-4) \cdot [(-3) \cdot (-2) \cdot (4)]$

78. $-(-3 \cdot 4) \cdot (-5) \cdot (4)$

79. $(-12) \div (-2) \div (4 \div 2)$

80. $-8 \div (-4) \div [6 \div (-3)]$

REVIEW EXERCISES

Evaluate the following. (For help, ***see Section 3.5.****)*

81. $8 + 4 \cdot 2 \div 8$

82. $16 \div 2 \cdot 4 + (15 - 2 \cdot 3)$

83. $7 + 6 \div 2 \cdot 4 - 9$

84. $9 \div 3 + 8 \div 4 \cdot 2 - 7$

9.4 ORDER OF OPERATIONS

The last few sections showed how to simplify a series of operations by starting at the left. For example,

$$\underbrace{\underbrace{-6 + (-4)}_{-10} + (-15)}_{-25}$$

and

$$\underbrace{\underbrace{-8 - (-6)}_{-2} + (-11)}_{-13.}$$

Also,

$$\underbrace{\underbrace{(-8) \cdot (-3)}_{24} \cdot 6}_{144.}$$

WORK PROBLEM 1 AT THE SIDE.

OBJECTIVES

1. Use the order of operations.
2. Use the order of operations with exponents.

1. Perform each operation.

(a) $-9 + (-15) + (-3)$

(b) $-8 + (-2) + (-6)$

(c) $-2 - (-7) - (-4)$

(d) $-3 - (-6) - 5$

(e) $\frac{1}{2} - \frac{3}{4} + \left(-\frac{1}{6}\right)$

1 These examples use the order of operations as discussed in Chapter 1.

ORDER OF OPERATIONS

1. Work inside **parentheses.**
2. Simplify expressions with exponents, and find any **square roots.**
3. Multiply or divide from **left to right.**
4. Add or subtract from **left to right.**

NOTE a *slash, /,* is commonly used in algebra to indicate division.

EXAMPLE 1 Using the Order of Operations

Use the order of operations to simplify each of the following.

(a) 6 / 3 + 4

Steps 1 and 2 in the order of operations do not apply, so start with Step 3. Divide first; next, add.

$$\underbrace{6 / 3}_{2} + 4$$

$$6$$

ANSWERS

1. (a) -27 (b) -16 (c) 9 (d) -2 (e) $-\frac{5}{12}$

2. Use the order of operations to simplify each of the following.

(a) $8/4 + 6$

(b) $9 \div (-3) + 4$

(c) $-3 + (-5) \cdot 2$

(d) $-2 \cdot (4 - 7)$

(e) $(9 - 17) \cdot (-3 - 1)$

ANSWERS
2. (a) 8 (b) 1 (c) −13 (d) 6 (e) 32

(b) $-12 + (-3) \cdot (-6)$
Start by multiplying.

$$\underbrace{-12 + \underbrace{(-3) \cdot (-6)}_{18}}_{6}$$

(c) $8 \cdot (5 - 7) - 9$
Work inside parentheses first.

$$\underbrace{8 \cdot \underbrace{(5 - 7)}_{(-2)}}_{} - 9$$

Now multiply. Next, add.

$$\underbrace{(-16) + (-9)}_{-25}$$ ■

WORK PROBLEM 2 AT THE SIDE.

2 Remember that 2^3 means 2 is used as a factor 3 times:

$$2^3 = 2 \cdot 2 \cdot 2 = 8.$$

The 3 is called an *exponent*. Exponents are also used with signed numbers. For example,

$$(-3)^2 = (-3) \cdot (-3) = 9$$

$$(-4)^3 = (-4) \cdot (-4) \cdot (-4) = -64$$

$$\left(-\frac{1}{2}\right)^4 = \left(-\frac{1}{2}\right) \cdot \left(-\frac{1}{2}\right) \cdot \left(-\frac{1}{2}\right) \cdot \left(-\frac{1}{2}\right) = \frac{1}{16}$$

and

$$(.6)^2 = (.6) \cdot (.6) = .36.$$

Be very careful with exponents and signed numbers. Although

$$(-3)^2 = (-3) \cdot (-3) = 9,$$

the expression -3^2, with no parentheses, means

$$-3^2 = -(3 \cdot 3) = -9.$$

So

$$(-3)^2 \text{ is not the same as } -3^2.$$

EXAMPLE 2 Using the Order of Operations
Simplify each of the following.

(a) $4^2 - (-3)^2$

Since $4^2 = 4 \cdot 4 = 16$ and $(-3)^2 = (-3) \cdot (-3) = 9$,

$$4^2 - (-3)^2 = 16 - 9 = 7.$$

(b) $-5^2 - (4 - 6)^2 \cdot (-3)$

Work inside parentheses first.

$-5^2 - (4 - 6)^2 \cdot (-3)$

$-5^2 - (-2)^2 \cdot (-3)$

$-25 - 4 \cdot (-3)$ Use the exponents.

$-25 - (-12)$ Multiply.

$-25 + 12$ Change to addition.

-13 Add.

(c) $\left(\frac{2}{3} - \frac{1}{6}\right)^2 \div \left(-\frac{3}{8}\right)$

Work inside parentheses first.

$$\left(\frac{2}{3} - \frac{1}{6}\right)^2 \div \left(-\frac{3}{8}\right) = \left(\frac{1}{2}\right)^2 \div \left(-\frac{3}{8}\right)$$

$$= \frac{1}{4} \div \left(-\frac{3}{8}\right) \qquad \left(\frac{1}{2}\right)^2 = \frac{1}{2} \cdot \frac{1}{2} = \frac{1}{4}$$

$$= \frac{1}{4} \cdot \left(-\frac{8}{3}\right) \qquad \text{Invert.}$$

$$= -\frac{2}{3} \qquad \text{Multiply.}$$ ■

WORK PROBLEM 3 AT THE SIDE.

3. Simplify each of the following.

(a) $2^3 - 3^2$

(b) $4^2 - 3^2 \cdot (5 - 2)$

(c) $-18 \div (-3) \cdot 2^3$

(d) $3^2 - (-2)^3$

(e) $\left(\frac{1}{4} - \frac{1}{2}\right) \div \frac{3}{32}$

ANSWERS
3. (a) -1 (b) -11 (c) 48 (d) 17 (e) $\frac{8}{3}$

name date hour

9.4 EXERCISES

Simplify each of the following.

Examples: $-6 - 3^2$ $\qquad$ $4 \cdot (6 - 11)^2 - (-8)$

Solutions: $-6 - 3^2 = -6 - 9 = \mathbf{-15}$

$4 \cdot (6 - 11)^2 - (-8) = 4 \cdot (-5)^2 - (-8)$
$= 4 \cdot 25 - (-8)$
$= 100 - (-8)$
$= 100 + 8$
$= \mathbf{108}$

1. $15 / 5 + 7$

2. $20 / 4 + 3$

3. $-8 + 7 + (-9) \cdot 2$

4. $3 + (-6) + 3 \cdot (-4)$

5. $4^2 + 3^2$

6. $7^2 + 5^2$

7. $9 - 5^2$

8. $16 - 3^3$

9. $(-3)^3 + 9$

10. $(-2)^4 - 7$

11. $4^2 + 3^2 + (-8)$

12. $5^2 + 2^2 + (-12)$

13. $2 - (-5) + 3^2$

14. $6 - (-9) + 2^3$

15. $(-4)^2 + (-3)^2 + 5$

16. $(-5)^2 + (-6)^2 + 12$

17. $3 + 5 \cdot (6 - 2)$

18. $4 + 3 \cdot (8 - 3)$

19. $-7 + 6 \cdot (8 - 14)$

20. $-3 + 5 \cdot (9 - 12)$

21. $-6 + (-5) \cdot (9 - 14)$

22. $-2 + (-7) \cdot (10 - 18)$

23. $(-5) \cdot (7 - 13) \div (-10)$

24. $(-4) \cdot (9 - 17) \div (-8)$

25. $36 / (-2)^2 + (-4)$

26. $-48 / (-4)^2 + 3$

27. $2 - (-5) \cdot (-3)^2$

28. $7 - (-3) \cdot (-2)^3$

29. $(-2) \cdot (-7) + 3 \cdot 9$

30. $4 \cdot (-2) + (-3) \cdot (-5)$

31. $36 \div (-4) - 25 \div (-5)$

32. $8 \div (-4) - 42 \div (-7)$

33. $2 \cdot 5 - 3 \cdot 4 + 5 \cdot 3$

34. $4 \cdot 7 - 8 \cdot 2 + 6 \cdot 5$

35. $4 \cdot 3^2 + 7 \cdot (3 + 9) - (-6)$

36. $5 \cdot 4^2 - 6 \cdot (1 + 4) - (-3)$

37. $-5 \cdot (-2) + 4^2 - (-9)$

38. $-6 \cdot (-5) + 3^2 - (-7)$

39. $2 \cdot 3^2 - 4 \cdot (7 - 3) - 5^2$

40. $3 \cdot 4^2 - 5 \cdot (9 - 2) - 6^2$

41. $5^2 \cdot (7 - 13) \div (9 + 1) - (-9)$

42. $8^2 \cdot (9 - 14) \div (3 - 13) - (-2)$

43. $(-4)^2 \cdot (7 - 9)^2 \div 2^3$

44. $(-5)^2 \cdot (9 - 17)^2 \div (-10)^2$

45. $(-.8)^2 + (-.4)^3 + 1.7$

46. $(.3)^3 - (-.5)^2 + 5.4$

47. $(-2.1) \cdot (1.8 - 2.7)^2$

48. $(-.9) \cdot (7.6 - 8.7)^2$

49. $(.3)^2 \cdot (-7) - (9.2)$

50. $(.2)^3 \cdot (-4) - (-3.1)$

51. $\frac{2}{3} \div \left(-\frac{5}{6}\right) - \frac{1}{2}$

52. $\frac{5}{8} \div \left(-\frac{10}{3}\right) - \frac{3}{4}$

53. $\left(-\frac{1}{2}\right)^2 - \left(\frac{3}{4} - \frac{7}{4}\right)$

54. $\left(-\frac{2}{3}\right)^2 - \left(\frac{1}{6} - \frac{11}{6}\right)$

55. $\frac{3}{5} \cdot \left(-\frac{7}{6}\right) - \left(\frac{1}{6} - \frac{5}{3}\right)$

56. $\frac{2}{7} \cdot \left(-\frac{14}{5}\right) - \left(\frac{4}{3} - \frac{13}{9}\right)$

57. $5^2 \cdot (9 - 11) \cdot (-3) \cdot (-2)^3$

58. $4^2 \cdot (13 - 17) \cdot (-2) \cdot (-3)^2$

59. $4.2 \cdot (-1.6) \div (-.56) \div 2^2$

60. $8.5 \cdot (-3.2) \div (-.4) \div (-2)^2$

61. $(6 - 19) \cdot (5 - 7)^2 \cdot (.4)$

62. $(9 - 15)^2 \cdot (8 - 9) \cdot (-.3)$

63. $\frac{1}{3} \cdot \left(-\frac{1}{2}\right) + \left(-\frac{5}{8}\right) \cdot \left(-\frac{7}{10}\right)$

64. $\left(-\frac{2}{5}\right) \cdot \left(\frac{1}{4}\right) + \left(-\frac{3}{4}\right) \cdot \left(-\frac{6}{5}\right)$

CHALLENGE EXERCISES

Simplify each of the following.

65. $5 - 4 \cdot 12 \div 3 \cdot 2$

66. $4 + 27 \div 3 \cdot 2 - 6$

67. $-7 \cdot \left(6 - \frac{5}{8} \cdot 24 + 3 \cdot \frac{8}{3}\right)$

68. $(-.3)^2 \cdot (-5 \cdot 3) + (6 \div 2 \cdot .4)$

69. $|-12| \div 4 + 2 \cdot 3^2 \div 6$

70. $6 - (2 - 3 \cdot 4) + 5^2 \div \left(-2 \cdot \frac{5}{2}\right) + (2)^2$

REVIEW EXERCISES

Evaluate the following. (For help, see ***Section 3.5.****)*

71. $\frac{6}{7} \cdot \frac{14}{9}$

72. $\frac{17}{15} \div \frac{34}{5}$

73. $\frac{7}{4} + \frac{3}{4} \cdot \frac{12}{15}$

74. $\frac{6}{7} \div \frac{12}{14} \cdot \frac{3}{5} \div \frac{1}{10}$

9.5 EVALUATING EXPRESSIONS AND FORMULAS

In formulas, numbers are often represented by letters. For example, the formula for interest was shown earlier.

$$I = p \cdot r \cdot t$$

In this formula, p represents the amount of money borrowed, r is the rate of interest, and t is the time in years. In algebra, it is common to write products without the multiplication dots; so leaving out the dots in the interest formula gives

$$I = prt.$$

1 Letters (such as I, p, r, or t here) that represent numbers are called **variables.** A combination of letters and numbers is an **expression.** Some examples of expressions are

$$9 + p, \qquad 8r, \qquad \text{and} \qquad 7k - 2m.$$

2 **An expression takes on different numerical values for different values of the variables.** To find the value of an expression, replace the variables with their values.

EXAMPLE 1 Finding the Value of an Expression
What is the value of $5x - 3y$, if $x = 2$ and $y = 7$?

SOLUTION
Replace x with 2 and y with 7.

$$\begin{array}{c} \quad\quad\quad\;\; x = 2 \quad y = 7 \\ \quad\quad\quad\;\; \downarrow \qquad \downarrow \\ 5x - 3y = 5(2) - 3(7) \end{array}$$

(*Note:* Here, $5x$ means 5 times x and $3y$ means 3 times y.) Next, multiply, and then add, using the order of operations.

$$= 10 - 21$$
$$= -11 \quad ■$$

WORK PROBLEM 1 AT THE SIDE.

EXAMPLE 2 Finding the Value of an Expression
What is the value of $7m - 8n + p$, if $m = -2$, $n = 4$, and $p = 3$?

SOLUTION
Replace m with -2, n with 4, and p with 3.

$$7m - 8n + p = 7(-2) - 8(4) + 3$$

Multiply.

$$= -14 - 32 + 3$$
$$= -46 + 3$$
$$= -43 \quad ■$$

WORK PROBLEMS 2 AND 3 AT THE SIDE.

OBJECTIVES

1 Define variable and expression.

2 Find the value of an expression when values of the variable are given.

1. Find the value of $5x - 3y$ if

(a) $x = 1$, $y = 2$.

(b) $x = 5$, $y = -3$.

(c) $x = 0$, $y = 6$.

2. Find the value of $7m - 8n + p$ if

(a) $m = 1$, $n = 2$, $p = 5$.

(b) $m = -4$, $n = -3$, $p = -7$.

(c) $m = -5$, $n = 0$, $p = -1$.

3. What is the value of $4x - 2y$, if

(a) $x = 1$, $y = -1$?

(b) $x = -2$, $y = 1$?

(c) $x = -1$, $y = -2$?

ANSWERS
1. (a) -1 (b) 34 (c) -18
2. (a) -4 (b) -11 (c) -36
3. (a) 6 (b) -10 (c) 0

EXAMPLE 3 Finding the Value of an Expression

What is the value of $5x - y$, if $x = 2$ and $y = -3$?

SOLUTION

Replace x with 2 and y with -3.

$$\begin{aligned} 5x - y &= 5(2) - (-3) \\ &= 10 + 3 \\ &= 13 \quad ■ \end{aligned}$$

EXAMPLE 4 Finding the Value of an Expression

What is the value of $\frac{6k + 2r}{5s}$, if $k = -2$, $r = 5$, and $s = -1$?

SOLUTION

Replace k with -2, r with 5, and s with -1.

$$\frac{6k + 2r}{5s} = \frac{6(-2) + 2(5)}{5(-1)}$$

$$= \frac{-12 + 10}{-5} \qquad \text{Multiply.}$$

$$= \frac{-2}{-5} \qquad \text{Add in numerator.}$$

$$= \frac{2}{5} \quad ■ \qquad \text{Quotient of two numbers with the same sign is positive.}$$

WORK PROBLEM 4 AT THE SIDE.

EXAMPLE 5 Finding the Value of an Expression

Recall the formula for the area of a triangle.

$$A = \frac{1}{2}bh$$

In this formula, b is the length of the base and h is the height. What is A, if $b = 9$ cm and $h = 24$ cm?

SOLUTION

Replace b with 9 and h with 24.

$$A = \frac{1}{2}bh$$

$$A = \frac{1}{2}(9)(24)$$

$$A = \frac{1}{\cancel{2}_1}(9)(\cancel{24}^{12})$$

$$A = 108$$

The area of the triangle is 108 cm^2. ■

NOTE The units are "square cm," since area is being measured.

WORK PROBLEM 5 AT THE SIDE.

4. What is the value of $\frac{6k + 2r}{5s}$, if

(a) $k = 1$, $r = 1$, $s = 2$?

(b) $k = 8$, $r = -2$, $s = -4$?

(c) $k = -3$, $r = 1$, $s = -2$?

5. Find the value of the missing variable.

(a) $A = \frac{1}{2}bh$; $b = 6$, $h = 12$

(b) $P = 2l + 2w$; $l = 10$, $w = 8$

(c) $d = rt$; $r = 4$, $t = 80$

(d) $C = 2\pi r$; $\pi = 3.14$, $r = 6$

ANSWERS

4. (a) $\frac{8}{10} = \frac{4}{5}$ (b) $\frac{44}{-20} = \frac{-11}{5}$ (c) $\frac{-16}{-10} = \frac{8}{5}$

5. (a) $A = 36$ (b) $P = 36$ (c) $d = 320$ (d) $C = 37.68$

name date hour

9.5 EXERCISES

Find the value of the expression $2r + 4s$ *for each of the following values of r and s.*

Example: $r = 3,\ s = -5$

Solution: Replace r with 3 and s with -5.

$$\begin{aligned} 2r + 4s &= 2(3) + 4(-5) \\ &= 6 + (-20) \\ &= \mathbf{-14} \end{aligned}$$

1. $r = 4,\ s = 2$

2. $r = 6,\ s = 1$

3. $r = 5,\ s = -3$

4. $r = 7,\ s = -2$

5. $r = -8,\ s = 6$

6. $r = -3,\ s = 5$

7. $r = -2,\ s = -1$

8. $r = -3,\ s = -5$

9. $r = 0,\ s = -2$

10. $r = -7,\ s = 0$

Use the given values of the variables to find the value of each expression.

11. $8x - y;\quad x = 1,\quad y = 2$

12. $7a - 3b;\quad a = 3,\quad b = 4$

13. $6k + 2s;\quad k = 1,\quad s = -2$

14. $9p + 5q;\quad p = -3,\quad q - 1$

15. $\dfrac{-m + 7n}{s + 1}$; $m = 2$, $n = -1$, $s = 2$

16. $\dfrac{2y - z}{x - 2}$; $y = 0$, $z = 5$, $x = 1$

17. $8m - 2n$; $m = -\dfrac{1}{2}$, $n = \dfrac{3}{4}$

18. $7k - 3r$; $k = \dfrac{2}{3}$, $r = -\dfrac{1}{3}$

In each of the following, use the given formula and values of the variables to find the value of the remaining variable.

Example: $P = 2l + 2w$; $l = 21$, $w = 16$

Solution: Replace l with 21 and w with 16.

$$P = 2 \cdot (21) + 2 \cdot (16)$$
$$P = 42 + 32$$
$$P = \mathbf{74}$$

19. $P = 4s$; $s = 7$

20. $P = a + b + c$; $a = 8$, $b = 9$, $c = 6$

21. $P = 2l + 2w$; $l = 9$, $w = 5$

22. $P = 2l + 2w$; $l = 14$, $w = 10$

23. $A = \frac{1}{2}bh; \quad b = 15, \quad h = 20$

24. $A = \frac{1}{2}bh; \quad b = 23, \quad h = 12$

25. $A = \frac{1}{2}bh; \quad b = 7, \quad h = 9$

26. $A = \frac{1}{2}bh; \quad b = 5, \quad h = 11$

27. $V = \frac{1}{3}bh; \quad b = 40, \quad h = 6$

28. $V = \frac{1}{3}bh; \quad b = 105, \quad h = 5$

29. $d = rt; \quad r = 90, \quad t = 8$

30. $d = rt; \quad r = 180, \quad t = 5$

31. $C = 2\pi r; \quad \pi = 3.14, \quad r = 7$

32. $C = 2\pi r; \quad \pi = 3.14, \quad r = 18$

CHALLENGE EXERCISES

Find the value of each expression by using the given values of the variable.

33. $A = \frac{1}{2}(b + B)h$; $b = 10$, $B = 12$, $h = 3$

34. $C = \frac{5}{9}(F - 32)$; $F = 104$

35. $F = \frac{9}{5}C + 32$; $C = -40$

36. $c^2 = a^2 + b^2$; $a = 3$, $b = 4$

REVIEW EXERCISES

Simplify the following. (For help, see ***Chapter 1****.)*

37. $-\frac{1}{9} \cdot 9$

38. $4 \cdot \left(\frac{1}{4}\right)$

39. $14 - 21 + 7$

40. $17 + 2 - 2$

9.6 SOLVING EQUATIONS

An **equation** is a statement that says two expressions are equal. Examples of equations include

$$x + 1 = 9, \qquad 5k = 20, \qquad \text{and} \qquad 6r - 1 = 17.$$

The equal sign in an equation divides the equation into two parts, the *left side* and the *right side*. In $6r - 1 = 17$, the **left side** is $6r - 1$, and the **right side** is 17.

$$\underbrace{6r - 1}_{\text{left side}} = \underbrace{17}_{\text{right side}}$$

Solve an equation by finding all numbers that can be substituted for the variable to make the equation true. These numbers are called **solutions** of the equation.

1 To tell whether a number is a solution of the equation, substitute the number in the equation to see whether the result is true.

EXAMPLE 1 Determining If a Number Is a Solution of an Equation

Is the number 7 a solution of the following equations?

(a) $x + 5 = 12$ **(b)** $2y + 1 = 16$

SOLUTION

(a) Replace x with 7.

$$x + 5 = 12$$
$$7 + 5 = 12 \quad \text{Let } x = 7.$$
$$12 = 12 \quad \text{true}$$

Because the statement is true, 7 is a solution of $x + 5 = 12$.

(b) Replace y with 7.

$$2y + 1 = 16$$
$$2(7) + 1 = 16$$
$$14 + 1 = 16 \quad \text{false}$$

The false statement shows that 7 is *not* a solution of $2y + 1 = 16$. ■

WORK PROBLEM 1 AT THE SIDE.

2 If the equation $a = b$ is true, and if a number c is added to both a and b, the new equation is also true. The rule, or property, used to add the same number to both sides of an equation is called the *addition property of equality*.

ADDITION PROPERTY OF EQUALITY

If $a = b$, then $a + c = b + c$. (The same number can be added or subtracted on both sides of an equation.)

OBJECTIVES

1. Tell whether a number is a solution of an equation.
2. Solve equations with the addition property of equality.
3. Solve equations using the multiplication property of equality.

1. Decide if the given number is a solution of the given equation.

(a) $p + 1 = 8$; 7

(b) $6r = 12$; 2

(c) $3k - 2 = 4$; 3

(d) $8y + 1 = 17$; 2

ANSWERS

1. (a) solution (b) solution (c) not a solution (d) solution

2. Solve each equation. Check each solution.

(a) $m + 2 = 6$

(b) $y - 5 = 2$

(c) $3 = z + 1$

(d) $9 = k + 17$

(e) $2 = y + 9$

(f) $r + \frac{1}{2} = \frac{3}{4}$

ANSWERS
2. (a) 4 (b) 7 (c) 2 (d) -8 (e) -7
(f) $\frac{1}{4}$

Use the addition property of equality to solve an equation by rewriting the equation with the variable on one side and any numbers on the other side.

EXAMPLE 2 Solving an Equation by Using the Addition Property
Solve each equation.

(a) $k - 4 = 6$

This equation can be solved by rewriting it with k alone on the left side. Do this by adding the *additive inverse* of -4, which is 4, to each side.

$$k - 4 = 6 \leftarrow \text{original equation}$$

Add 4 to each side.

$$\begin{aligned} k - 4 + 4 &= 6 + 4 \\ k + 0 &= 10 \\ k &= 10 \end{aligned}$$

The solution is 10. Check by replacing k with 10 in the original equation.

$$\begin{aligned} k - 4 &= 6 \\ 10 - 4 &= 6 \quad \text{Let } k = 10. \\ 6 &= 6 \quad \text{true} \end{aligned}$$

This result is true, so 10 checks.

(b) $2 = z + 8$

Solve this equation by rewriting it as *number* $= z$. Do this by subtracting 8 from each side.

$$2 = z + 8 \leftarrow \text{original equation}$$

Subtract.

$$\begin{aligned} 2 - 8 &= z + 8 - 8 \\ -6 &= z \end{aligned}$$

Check the solution. ■

WORK PROBLEM 2 AT THE SIDE.

In summary, if a and b are specific numbers and x is unknown, use the following rules to solve equations by the addition property of equality.

Solve $x - a = b$ or $b = x - a$ by adding a to each side.

Solve $x + a = b$ or $b = x + a$ by subtracting a from each side.

3 Not all equations can be solved by using the addition property; some must be solved by using the *multiplication property of equality*.

MULTIPLICATION PROPERTY OF EQUALITY

If $a = b$ and c does not equal 0, then

$$a \cdot c = b \cdot c \text{ and } \frac{a}{c} = \frac{b}{c}.$$

(The same nonzero number can be multiplied or divided on both sides of an equation.)

EXAMPLE 3 Solving an Equation by Using the Multiplication Property

Solve each equation.

(a) $9p = 63$

This equation can be solved by rewriting it as $p =$ *number*. The expression $9p$ on the left means $9 \cdot p$. To get p alone on the left, *divide* each side by 9.

$$9p = 63$$

$$\frac{9p}{9} = \frac{63}{9} \quad \text{Divide each side by 9.}$$

$$p = 7$$

Check.

$$9p = 63$$

$$9 \cdot 7 = 63 \quad \text{Let } p = 7.$$

$$63 = 63 \quad \text{true}$$

Since this result is true, 7 is the solution.

(b) $-4r = 24$

Divide each side by -4 to get r alone on the left.

$$\frac{-4r}{-4} = \frac{24}{-4} \quad \text{Divide each side by } -4.$$

$$r = -6$$

Check this solution.

(c) $-11m = -55$

Divide by -11.

$$\frac{-11m}{-11} = \frac{-55}{-11}$$

$$m = 5$$

Check this solution. ■

WORK PROBLEM 3 AT THE SIDE.

3. Solve each equation. Check each solution.

(a) $2y = 14$

(b) $6p = 54$

(c) $-8a = 32$

(d) $3r = 15$

(e) $-9k = -72$

ANSWERS

3. (a) 7 (b) 9 (c) −4 (d) 5 (e) 8

4. Solve each equation. Check each solution.

(a) $\frac{a}{4} = 2$

(b) $\frac{y}{7} = -3$

(c) $\frac{k}{5} = -9$

(d) $-\frac{3}{4}z = 6$

(e) $-\frac{5}{8}p = -10$

EXAMPLE 4 Solving an Equation by Using the Multiplication Property

Solve each equation.

(a) $\frac{x}{2} = 9$

Dividing by 2 is the same as multiplying by $\frac{1}{2}$.

$$\frac{x}{2} = 9$$

$$\frac{1}{2}x = 9$$

Get x alone, not $\frac{1}{2}x$, on the left by multiplying each side by 2.

$$\frac{1}{2}x = 9$$

$$2 \cdot \frac{1}{2}x = 2 \cdot 9 \quad \text{Multiply each side by 2.}$$

$$1x = 18$$

$$x = 18$$

Check.
$$\frac{x}{2} = 9$$

$$\frac{1}{2}x = 9$$

$$\frac{1}{2} \cdot 18 = 9 \quad \text{Let } x = 18.$$

$$9 = 9 \quad \text{true}$$

The solution is 18.

(b) $-\frac{2}{3}r = 4$

Multiply each side by $-\frac{3}{2}$ (since the product of $-\frac{3}{2}$ and $-\frac{2}{3}$ is 1).

$$-\frac{2}{3}r = 4$$

$$-\frac{3}{2} \cdot \left(-\frac{2}{3}r\right) = -\frac{3}{2} \cdot 4$$

$$r = -6$$

Check this solution. ■

WORK PROBLEM 4 AT THE SIDE.

In summary, if a and b are specific nonzero numbers and x is unknown, use the following rules to solve equations by multiplication or division.

Solve the equation $ax = b$ by dividing each side by a.

Solve the equation $\frac{a}{b}x = s$ by multiplying each side by $\frac{b}{a}$.

ANSWERS
4. (a) 8 (b) −21 (c) −45 (d) −8 (e) 16

name date hour

9.6 EXERCISES

Decide whether the given number is a solution of the given equation.

1. $x + 7 = 11$; 4

2. $k - 2 = 7$; 9

3. $3y = 27$; 9

4. $5p = 30$; 6

5. $2z - 1 - -15$; -8

6. $3r - 5 = -15$; -4

Solve each equation by addition. Check each solution.

Example: $m - 2 = 7$

Solution:

Add 2 on each side.

$$m - 2 + 2 = 7 + 2$$
$$m = \mathbf{9}$$

Check:

$$m - 2 = 7$$
$$9 - 2 = 7$$
$$\mathbf{7 = 7} \quad \text{true}$$

7. $p + 5 = 9$

8. $a + 3 = 12$

9. $k + 15 = 26$

10. $y + 11 = 15$

11. $z - 5 = 3$

12. $x - 2 = 5$

13. $8 = r - 2$

14. $3 - b - 5$

15. $-4 = n + 2$

16. $2 = a + 11$

17. $7 = r + 13$

18. $z + 11 = 5$

19. $k + 15 = 2$

20. $-9 + y = 7$

21. $-3 + x = 5$

22. $-2 + m = -1$

23. $-7 + r = -8$

24. $-6 = -2 + y$

25. $-11 = -8 + d$

26. $x + \frac{1}{2} = 4$

27. $z + \frac{5}{8} = 2$

28. $m - \frac{3}{4} = 21$

29. $k - \frac{2}{3} = 5$

30. $t - 1 = \frac{1}{4}$

31. $m - 1\frac{4}{5} = 2\frac{1}{10}$

32. $z - 5\frac{1}{3} = 3\frac{5}{9}$

33. $x - 1.72 = 4.86$

34. $a - 3.82 = 7.99$

35. $4.76 + r = 3.25$

36. $11.38 + b = 9.46$

Solve each equation. Check each solution.

> **Example:** $5k = 60$
>
> **Solution:**
>
> Divide each side by 5.
>
> $\frac{5k}{5} = \frac{60}{5}$
>
> $k = \mathbf{12}$
>
> Check:
>
> $5 \cdot 12 = 60$
>
> $\mathbf{60 = 60}$ true

37. $6z = 12$

38. $8k = 24$

39. $12r = 48$

40. $11m = 77$

41. $3y = -24$

42. $5a = -25$

43. $-6k = 36$

44. $-10y = 50$

45. $-2p = -16$

46. $-7r = -35$

47. $-1.2m = 8.4$

48. $-5.4z = 27$

49. $-8.4p = -9.24$

50. $-3.2y = -16.64$

Solve each equation. Check each solution.

Example: $\frac{p}{4} = -3$

Solution:

Multiply each side by 4.

$4 \cdot \frac{p}{4} = 4 \cdot (-3)$

$p = \mathbf{-12}$

Check:

$\frac{-12}{4} = -3$

$\mathbf{-3 = -3}$ true

51. $\frac{k}{2} = 17$

52. $\frac{y}{3} = 5$

53. $\frac{a}{5} = 10$

54. $\frac{m}{7} = 5$

55. $\frac{r}{3} = -12$

56. $\frac{z}{9} = -3$

57. $\frac{1}{2}p = 5$

58. $\frac{1}{3}k = 7$

59. $-\frac{1}{4}m = 2$

60. $\frac{1}{10}b = -3$

61. $\frac{3}{8}x = 6$

62. $\frac{2}{3}a = 4$

63. $-\frac{4}{7}p = 16$

64. $-\frac{5}{6}r = 20$

65. $-\frac{9}{8}k = -18$

66. $\frac{3}{5}a = \frac{1}{8}$

67. $\frac{y}{1.7} = .8$

68. $\frac{k}{2.4} = .5$

69. $\frac{z}{-3.8} = 1.3$

70. $\frac{m}{-5.2} = 2.1$

CHALLENGE EXERCISES

Solve the following equations.

71. $x - 17 = 5 - 3$

72. $x + 5\frac{1}{2} = 2 - 2\frac{1}{4}$

73. $3 = x + 9 - 15$

74. $6x - 3x = 8 - 6$

75. $\frac{7}{2}x = \frac{4}{3}$

76. $\frac{3}{4}x - \frac{5}{3} = 0$

REVIEW EXERCISES

Use the order of operations to simplify the following. (For help, see ***Sections 3.5 and 9.4.****)*

77. $2\frac{1}{5} \div \left(3\frac{1}{3} - 4\frac{1}{5}\right)$

78. $6 \div 2 \cdot 3$

79. $\frac{1}{2} + \frac{3}{4} \cdot \frac{8}{9} - \frac{1}{6}$

80. $2 - \left(3 \cdot \frac{1}{2} \div \frac{2}{3}\right) + \frac{7}{4}$

9.7 SOLVING EQUATIONS WITH SEVERAL STEPS

The equation $5m + 1 = 16$ cannot be solved by just adding a number to each side, or by just dividing each side by the same number.

1 Instead, solve this equation with the following steps.

Step 1 Add or subtract the same expression on each side, so that just the term with the variable is on one side.

Step 2 Multiply or divide each side by the same number to find the solution.

Step 3 Check the solution.

EXAMPLE 1 Solving Equations with Several Steps

Solve $5m + 1 = 16$.

First subtract 1 from each side to get $5m$ alone.

$$5m + 1 - 1 = 16 - 1$$
$$5m = 15$$

Next, divide each side by 5.

$$\frac{5m}{5} = \frac{15}{5}$$
$$m = 3$$

Check.

$$5m + 1 = 16$$
$$5 \cdot 3 + 1 = 16 \quad \text{Let } m = 3.$$
$$15 + 1 = 16$$
$$16 = 16 \quad \text{true}$$

The solution is 3. ■

WORK PROBLEM 1 AT THE SIDE.

2 Use the order of operations to simplify

$$2(6 + 8) \quad \text{and} \quad 2 \cdot 6 + 2 \cdot 8.$$

On the left, $2(6 + 8) = 2(14) = 28$.

On the right, $2 \cdot 6 + 2 \cdot 8 = 12 + 16 = 28$.

Since both answers are the same,

$$2(6 + 8) = 2 \cdot 6 + 2 \cdot 8$$

is an example of the distributive property:

$$a(b + c) = ab + ac.$$

OBJECTIVES

1. Solve equations with several steps.
2. Use the distributive property.
3. Combine like terms.
4. Solve more difficult equations.

1. Solve each equation. Check each solution.

(a) $2r + 7 = 13$

(b) $8y - 1 = 23$

(c) $7m + 9 = 44$

(d) $-3p + 8 = 20$

(e) $-10z - 9 = 11$

ANSWERS

1. (a) 3 (b) 3 (c) 5 (d) −4 (e) −2

2. Use the distributive property.

(a) $3(2 + 6)$

(b) $8(k - 3)$

(c) $-4(r - 2)$

(d) $-9(s - 8)$

3. Combine like terms.

(a) $5y + 11y$

(b) $7a + 25a$

(c) $31r - 17r$

(d) $9k - 15k$

ANSWERS
2. (a) $3 \cdot 2 + 3 \cdot 6 = 24$
(b) $8k - 24$
(c) $-4r + 8$
(d) $-9s + 72$
3. (a) $16y$ (b) $32a$ (c) $14r$ (d) $-6k$

EXAMPLE 2 Using the Distributive Property
Simplify each expression by using the distributive property.

(a) $9(4 + 2) = 9 \cdot 4 + 9 \cdot 2 = 36 + 18 = 54$

The 9 on the outside of the parentheses is *distributed* over the 4 and the 2 on the inside of the parentheses.

(b) $-3(k + 9) = -3k + (-3)(9) = -3k - 27$

(c) $4x + 8x = (4 + 8)x = 12x$ ■

WORK PROBLEM 2 AT THE SIDE.

3 A single letter or number, or the product of a variable and a number, make up a term. For example,

$$3y, \quad 5, \quad -9, \quad 8r, \quad 10r^2, \quad \text{and} \quad a$$

are all examples of terms. **Terms with exactly the same variable and the same exponent are called *like terms*.** For example, $5x$ and $3x$ are like terms, but $5x$ and $3m$ are not like terms because the variables are different; also, $5x^2$ and $5x^3$ have different exponents and are not like terms.

The distributive property can be used to simplify a sum of like terms such as $6r + 3r$, as shown in Example 2c.

$$6r + 3r = (6 + 3)r = 9r$$

This process is called **combining like terms.**

EXAMPLE 3 Combining Like Terms
Combine like terms.

(a) $5k + 11k$ **(b)** $10m - 14m$

SOLUTION

(a) $5k + 11k = (5 + 11)k = 16k$

(b) $10m - 14m = (10 - 14)m = -4m$ ■

WORK PROBLEM 3 AT THE SIDE.

4 The next example shows how to solve more difficult equations.

EXAMPLE 4 Solving Equations
Solve each equation.

(a) $6r + 3r = 36$
Since $6r + 3r = 9r$, the equation becomes

$$9r = 36.$$

Next, divide both sides by 9.

$$\frac{9r}{9} = \frac{36}{9}$$

$$r = 4$$

Check.

$$6r + 3r = 36$$
$$6 \cdot 4 + 3 \cdot 4 = 36 \quad \text{Let } r = 4.$$
$$24 + 12 = 36$$
$$36 = 36 \quad \text{true}$$

The solution is 4.

(b) $2k - 5 = 5k - 11$

First, to get the terms with variables on one side, subtract $5k$ from each side.

$$2k - 5 - 5k = 5k - 11 - 5k$$
$$2k - 5k - 5 = 5k - 5k - 11$$
$$-3k - 5 = -11$$

Next, add 5 to each side.

$$-3k - 5 + 5 = -11 + 5$$
$$-3k = -6$$

Finally, divide each side by -3.

$$\frac{-3k}{-3} = \frac{-6}{-3}$$

$$k = 2$$

Check this solution. ■

WORK PROBLEM 4 AT THE SIDE

4. Solve each equation. Check each solution.

(a) $3y - 1 = 2y + 7$

(b) $9a + 8 = 8a - 3$

(c) $4p - 7 = 9p - 2$

ANSWERS
4. **(a)** $y = 8$ **(b)** $a = -11$ **(c)** $p = -1$

name date hour

9.7 EXERCISES

Solve each equation. Check each solution.

Examples:

$9p - 7 = 11$

$-3m + 2 = 8$

Solutions:

Add 7 to each side.

$$9p - 7 + 7 = 11 + 7$$
$$9p = 18$$

Divide each side by 9.

$$\frac{9p}{9} = \frac{18}{9}$$
$$p = \mathbf{2}$$

Check:

$$9 \cdot 2 - 7 = 11$$
$$18 - 7 = 11$$
$$\mathbf{11 = 11} \quad \text{true}$$

Subtract 2 from each side.

$$-3m + 2 - 2 = 8 - 2$$
$$-3m = 6$$

Divide each side by -3.

$$\frac{-3m}{-3} = \frac{6}{-3}$$
$$m = \mathbf{-2}$$

Check:

$$-3(-2) + 2 = 8$$
$$6 + 2 = 8$$
$$\mathbf{8 = 8} \quad \text{true}$$

1. $5p - 3 = 2$

2. $6k + 3 = 15$

3. $2y - 8 = 14$

4. $3r - 4 = 2$

5. $-3m + 1 = -5$

6. $-4k + 5 = -7$

7. $-8a + 7 = 23$

8. $-2p + 5 = 11$

9. $-5x - 4 = 16$

10. $-12a - 3 = 21$

11. $-\frac{1}{2}z + 2 = -1$

12. $-\frac{5}{8}r + 4 = -6$

Use the distributive property to simplify.

13. $8(2 + 9)$

14. $6(3 + 7)$

15. $5(r + 3)$

16. $9(k + 7)$

17. $-3(m + 6)$

18. $-5(a + 2)$

19. $-2(y - 3)$

20. $-4(r - 7)$

21. $-5(z - 9)$

Combine like terms.

22. $11r + 6r$

23. $2m + 5m$

24. $8z + 7z$

25. $12y + 6y$

26. $28m + 5m$

27. $9z - 4z$

28. $4k - 2k$

29. $15a - 25a$

30. $7y - 12y$

Solve each equation. Check each solution.

31. $4k + 6k = 50$

32. $3a + 2a = 15$

33. $17m - 12m = 35$

34. $8z - 5z = 9$

35. $2b - 6b = 24$

36. $3r - 9r = 18$

37. $5y - 12y = -14$

38. $11z - 17z = -24$

39. $6p - 2 = 4p + 6$

name date hour

40. $5y - 5 = 2y + 10$

41. $7z + 9 = 9z + 13$

42. $4a + 8 = 2a + 2$

43. $2.5y + 6 = 4.5y + 10$

44. $5.2x + 4 = 3.2x - 6$

45. $-3.6m + 1 = 2.4m + 7$

46. $-8.2p + 7 = 1.8p - 3$

47. $\frac{1}{2}y - 2 = \frac{1}{4}y + 3$

48. $\frac{2}{3}z + 1 = \frac{1}{2}z - 3$

CHALLENGE EXERCISES

Solve each equation.

49. $8 - 6x + 8 = 2x$

50. $4x - 7 = 3(2x + 5) - 2$

51. $-4(p - 8) + 3(2p + 1) = 7$

52. $7(2p + 6) = 9(p + 3) + 5$

REVIEW EXERCISES

Solve the following word problems. Round to the nearest cent. (For help, see ***Sections 6.7 and 6.8.****)*

53. If the sales tax rate is 6% and the sales are $420, what is the amount of sales tax?

54. A VCR sells for $450 plus 4% sales tax. Find the price of the VCR including sales tax.

55. John Harrison borrows $2500 at 11% for 6 months. Find the interest John must pay.

56. Joan Krebs lends $1500 at 9% for 18 months. How much money should Joan receive when the loan is due?

9.8 APPLICATIONS

It is rare for an application to be presented as an equation. Usually, a problem is given in words. These words must then be *translated* into an equation that can be solved.

OBJECTIVES

1. Translate word phrases by using variables.
2. Solve word problems.

1 The following examples show how to translate word phrases into algebra.

EXAMPLE 1 Translating Word Phrases by Using Variables
Write in symbols by using x as the variable.

words	*algebra*
a number plus 2	$x + 2$
the sum of a number and 8	$x + 8$
5 more than a number	$x + 5$
-35 added to a number	$x + (-35)$
9 less than a number	$x - 9$
3 subtracted from a number	$x - 3$
a number decreased by 4	$x - 4$ ■

WORK PROBLEM 1 AT THE SIDE.

1. Write in symbols by using x as the variable.

(a) the sum of 7 and a number

(b) 12 more than a number

(c) a number increased by 13

(d) a number minus 8

(e) 15 subtracted from a number

EXAMPLE 2 Translating Word Phrases by Using Variables
Write in symbols by using x as the variable.

words	*algebra*
8 times a number	$8x$
the product of 12 and a number	$12x$
double a number	$2x$
the quotient of 6 and a number	$\frac{6}{x}$
a number divided by 10	$\frac{x}{10}$
one third of a number	$\frac{1}{3}x$
the result is	$=$ ■

WORK PROBLEM 2 AT THE SIDE.

2. Write in symbols by using x as the variable.

(a) -6 times a number

(b) the product of 23 and a number

(c) the quotient of 15 and a number

2 The next examples show how to solve word problems. Notice that each solution begins by selecting a variable to represent the unknown.

ANSWERS
1. (a) $x + 7$ (b) $x + 12$ (c) $x + 13$ (d) $x - 8$ (e) $x - 15$
2. (a) $-6x$ (b) $23x$ (c) $\frac{15}{x}$

3. (a) If 4 times a number is added to 19, the result is 27. Find the number.

(b) If -6 times a number is added to 5, the result is -13. Find the number.

EXAMPLE 3 Solving Word Problems
If 5 times a number is added to 11, the result is 26. Find the number.

SOLUTION
Let x represent the unknown number.
Use the information in the problem to write an equation.

$$\underbrace{\text{5 times a number}}_{5x}\ \underbrace{\text{added to}}_{+}\ \text{11 is 26.}$$

$$5x + 11 = 26$$

NOTE The phrase "the result is" translates to "=."

Next, solve the equation. First subtract 11 from each side.

$$5x + 11 - 11 = 26 - 11$$
$$5x = 15$$

Divide each side by 5.

$$x = 3$$

Check by using the words of the original problem. First take five times the answer $(5 \cdot 3)$, getting 15. This result, added to 11, gives $15 + 11 = 26$ as needed. The number, therefore is 3. ■

EXAMPLE 4 Solving Word Problems
Michael has 5 less than three times as many lab experiments completed as David. If Michael has completed 13 experiments, how many lab experiments has David completed?

SOLUTION
Let x represent the unknown number.

$$\underbrace{\text{3 times a number}}_{3x}\ \text{minus}\ \underbrace{5}\ \text{is}\ \underbrace{13}$$

$$3x - 5 = 13$$

Next, solve the equation. First add 5 to both sides.

$$3x - 5 + 5 = 13 + 5$$
$$3x = 18$$

Divide each side by 3.

$$x = 6$$

David has completed 6 lab experiments. Check by using the words of the original problem. First take three times the answer $(3 \cdot 6)$, getting 18. This result, decreased by 5, is 13 as needed. ■

WORK PROBLEM 3 AT THE SIDE.

ANSWERS
3. (a) $4x + 19 = 27$
$x = 2$
(b) $-6x + 5 = -13$
$x = 3$

The steps in solving a word problem are summarized below.

Step 1 Choose a variable to represent the unknown.

Step 2 Use the information in the problem to write an equation that relates known information and the unknown.

Step 3 Solve the equation.

Step 4 Answer the question raised in the problem.

Step 5 Check the solution with the original words of the problem.

EXAMPLE 5 Solving Word Problems

In a given amount of time, Sheila drove 72 km more than Russell. The total distance traveled by them both was 232 km. Find the distance traveled by Russell.

SOLUTION

Let x be the distance traveled by Russell.

Since Sheila went 72 km more, the distance she traveled is $x + 72$ km. Next, write an equation.

distance for Russell	plus	distance for Sheila	is	total distance
x	$+$	$x + 72$	$=$	232

Since $x = 1x$, the sum $x + x$ is $1x + 1x = 2x$. The equation becomes

$$2x + 72 = 232.$$

Subtract 72 from each side.

$$2x + 72 - 72 = 232 - 72$$
$$2x = 160$$
$$x = 80$$

Check by using the words of the original problem. If Russell drove 80 km, Sheila drove 72 km more than Russell, or $80 + 72 = 152$ km. The total distance traveled by both was $152 + 80 = 232$ km. This checks with the statement of the problem. ■

WORK PROBLEM 4 AT THE SIDE.

4. (a) In a day of work, a mother makes \$12 more than her daughter. Together they make \$132. Find the amount made by the daughter.

(b) A board is 33 cm long. It is cut into two pieces, so that one piece is 3 cm longer than the other. Find the length of the shorter piece.

ANSWERS
4. (a) \$60 (b) 15 cm

EXAMPLE 6 Solving a Geometry Problem
The length of a rectangle is 2 cm more than the width. The perimeter is 68 cm. Find the length.

SOLUTION
Let x be the width of the rectangle.

Since the length is 2 cm more than the width, the length is $x + 2$ Thus,

$$\text{width} = x$$
$$\text{length} = x + 2$$
$$\text{perimeter} = 68.$$

Use the formula $2l + 2w = p$.

$2(x + 2) + 2x = 68$	
$2x + 4 + 2x = 68$	distributive property
$4x + 4 = 68$	Combine like terms.
$4x + 4 - 4 = 68 - 4$	Subtract 4 from each side.
$4x = 64$	
$\frac{4x}{4} = \frac{64}{4}$	Divide both sides by 4.
$x = 16$	

This does not answer the question in the problem, since x represents the width of the rectangle and the problem asks for the length, which is $x + 2$. The length is $16 + 2$ or 18 cm. Check by using the words of the problem. The width is 16 cm, and the length is 18 cm. Thus, the perimeter is $2(16) + 2(18)$ or $32 + 36 = 68$ cm. This checks with the statement of the problem. ■

name date hour

9.8 EXERCISES

Write in symbols by using x as the variable.

Examples:	**Solutions:**
the sum of 7 and a number	$7 + x$
12 added to a number	$x + 12$
a number subtracted from 57	$57 - x$
three times a number	$3x$
the sum of 15 and twice a number	$15 + 2x$

1. 14 plus a number

2. the sum of 9 and a number

3. 4 added to a number

4. a number increased by -10

5. the sum of a number and 6

6. the total of a number and 2

7. 9 less than a number

8. a number subtracted from 2

9. subtract 4 from a number

10. 3 fewer than a number

11. the product of a number and 2

12. double a number

13. triple a number

14. half a number

15. a number divided by 2

16. 4 divided by a number

17. twice a number added to 3

18. the sum of three times a number and 4

19. five times a number added to four times the number

20. eight times a number subtracted from ten times the number

Solve each word problem. Use the five steps given on page 611.

Example: If a number is multiplied by 5 and the product is added to 2, the result is -13. Find the number.

Solution:

Let x represent the unknown number.

$$\underbrace{\text{a number multiplied by 5}}_{5x}\ \underbrace{\text{added to}}_{+}\ \underset{\downarrow}{2}\ \underbrace{\text{result is}}_{=}\ \underset{\downarrow}{-13}$$

$$5x + 2 = -13$$

Solve the equation.

$$5x + 2 + (-2) = -13 + (-2)$$
$$5x = -15$$
$$x = -3$$

The number is **-3.**
Check with the words of the original problem.

21. If four times a number is decreased by 2, the result is 26. Find the number.

22. When 8 is added to twice a number, the result is 14. Find the number.

23. If twice a number is added to four times the number, the result is 48. Find the number.

24. When three times a number is subtracted from seven times the number, the result is 12. Find the number.

25. If half a number is added to twice the number, the answer is 50. Find the number.

26. If one third of a number is added to three times the number, the result is 30. Find the number.

27. A board is 78 cm long. It is to be cut into two pieces, with one piece 10 cm longer than the other. Find the length of the shorter piece.

28. Ed and Marge were opposing candidates for city council. Marge won, with 93 more votes than Ed. The total number of votes received by both candidates was 587. Find the number of votes received by Ed.

29. A chain saw can be rented for a one-time $7 sharpening fee plus $6 a day rental. The bill for a rental was $49. For how many days was the saw rented?

30. A rental car costs $28 per day plus 25¢ per mile. The bill for a one-day rental was $58. How many miles was the car driven?

In the next two exercises, use the formula for the perimeter of a rectangle, $P = 2l + 2w$.

31. The perimeter of a rectangle is 48 m. The width is 5 m. Find the length.

32. The length of a rectangle is 27 cm, and the perimeter is 74 cm. Find the width of the rectangle.

CHALLENGE EXERCISES

33. A fence is 706 m long. It is to be cut into three parts. Two parts are the same length, and the third part is 25 m longer than the other two. Find the length of each part.

34. A wooden railing is 82 m long. It is to be divided into four pieces. Three pieces will be the same length, and the fourth piece will be 2 m longer than each of the other three. Find the length that each of the three pieces of equal length will be.

In the following exercises, use the formula for interest, $I = prt$.

35. For how long must \$800 be deposited at 12% per year to earn \$480 interest?

36. How much money must be deposited at 12% per year for 7 years to earn \$1008 interest?

REVIEW EXERCISES

*Solve the following problems. (For help, see **Section 6.4.**)*

37. 80% of what number is 256?

38. What is 15% of 360?

39. 225 is 75% of what number?

40. 90 is what percent of 720?

CHAPTER 9 SUMMARY

KEY TERMS

9.1	**negative numbers**	Negative numbers are numbers that are less than zero.
	integers	Integers are the "counting numbers" or positive numbers, their opposites or negative numbers, and zero.
	signed numbers	Signed numbers are positive numbers, negative numbers, and zero.
	absolute value	Absolute value is the magnitude of a number or the distance of a number from zero on a number line.
	opposite of a number	The opposite of a number is a number the same distance from zero as the original number but on the opposite side of zero on a number line.
9.2	**additive inverse**	The additive inverse is the opposite of a number.
9.5	**variables**	Variables are letters that represent numbers.
	expression	An expression is a combination of letters and numbers.
9.6	**equation**	An equation is a statement that says two expressions are equal.
	solution	The solution is a number that can be substituted for the variable in an equation, so that the equation is true.
	addition property of equality	The addition property of equality states that the same number can be added to both sides of an equation.
	multiplication property of equality	The multiplication property of equality states that the same nonzero number can be multiplied or divided on both sides of an equation.
9.7	**like terms**	Like terms are terms with exactly the same variable and the same exponent.

QUICK REVIEW

Section Number and Topic	Approach	Example
9.1 Graphing Signed Numbers	Place a dot at the correct location on the number line.	Graph -2, 1, 0, and $2\frac{1}{2}$. 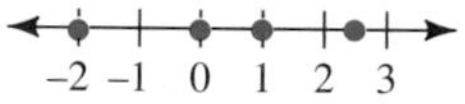
9.1 Identifying the Smaller of Two Numbers	Place the symbols < (smaller than) or > (greater than) between two numbers to make the statement true.	Use the symbol < or > to make the following statements true. $2 \underline{\ >\ } 1$ $-3 \underline{\ >\ } -5$ $-6 \underline{\ <\ } 2$
9.1 Finding the Absolute Value of a Number	Determine the distance from 0 to the given number on the number line.	Find each of the following. **(a)** $\lvert 8\rvert$; $\lvert 8\rvert = 8$ **(b)** $\lvert -7\rvert$; $\lvert -7\rvert = 7$

Section Number and Topic	Approach	Example
9.1 Finding the Opposite of a Number	Determine the number that is the same distance from the given number on the opposite side of a number line.	Find the opposite of each of the following. **(a)** -6 **(b)** $+9$ $-(-6) = 6$ $-(+9) = -9$
9.2 Adding Two Signed Numbers by Using a Number Line	Draw an arrow to represent the first signed number (to the right for positive numbers and to the left for negative numbers). From the end of the first arrow draw a second arrow representing the second number. The location of the end of the last arrow is the sum.	Add $2 + (-3)$. −3 2 −1 0 1 2 The second arrow ends at -1, so $2 + (-3) = -1$.
9.2 Adding Two Numbers Without Using a Number Line	Case 1: *Two positive numbers* Add the numerical values. Case 2: *Two negative numbers* Add the absolute values and attach a negative sign to the sum. Case 3: *Two numbers with different signs* Subtract the absolute values and attach the sign of the number with the larger absolute value.	Add the following. **(a)** $8 + 6 = 14$ **(b)** $(-8) + (-6)$ **SOLUTION** Find absolute values. $\|-8\| = 8;$ $\|-6\| = 6$ Add absolute values. $8 + 6 = 14$ Attach negative sign: -14. So, $(-8) + (-6) = -14$ **(c)** $(-7) + 5$ **SOLUTION** Find absolute values $\|-7\| = 7;$ $\|5\| = 5$ Subtract absolute values. $7 - 5 = 2$ Attach sign of number with larger absolute value: -7. So, $(-7) + (5) = -2$.
9.2 Subtracting Signed Numbers	Change the subtraction sign to addition. Change the sign of the number following the subtraction sign. Proceed as in addition.	Subtract. **(a)** $-6 - 5 = -6 + (-5) = -11$ **(b)** $5 - (-8) = 5 + (+8) = 13$ **(c)** $7 - 9 = 7 + (-9) = -2$.

Section Number and Topic	Approach	Example
9.3 Multiplying Signed Numbers	Use the rules: The product of two numbers with the same sign is positive; or, the product of two numbers with different signs is negative.	Multiply the following. **(a)** $7 \cdot 3 = 21$ **(b)** $(-3) \cdot 4 = -12$ **(c)** $(-9) \cdot (-6) = 54$
9.4 Using the Order of Operations to Evaluate Numerical Expressions	Use the following procedure to evaluate numerical expressions: Work inside parentheses first. Simplify expressions with exponents and find any square roots. Multiply or divide from left to right. Add or subtract from left to right.	Simplify the following. **(a)** $-4 + 6 \div (-2) = -4 + (-3)$ $= -7$ **(b)** $3^2 \cdot 4 + (8 \div 2 \cdot 3)$ $= 9 \cdot 4 + (4 \cdot 3)$ $= 9 \cdot 4 + 12$ $= 36 + 12$ $= 48$ **(c)** $9 \cdot 2^3 - (-4 \cdot 3)$ $= 9 \cdot 8 - (-12)$ $= 72 + 12$ $= 84$
9.5 Evaluating Expressions	Replace the variables in the expression with the numerical values. Use the order of operations to evaluate.	What is the value of $6p - 5s$, if $p = -3$ and $s = -4$? $6p - 5s = 6(-3) - 5(-4)$ $= -18 + 20$ $= 2$
9.6 Determining if a Number is a Solution of an Equation	Substitute the number for the variable in an equation. If the equation is true, the number is a solution of the equation.	Is the number 4 a solution of the following equation? $3x - 5 = 7$ Replace x with 4. $3(4) - 5 = 7$ $12 - 5 = 7$ $7 = 7$ true 4 is the solution.
9.6 Using the Addition Property of Equality to Solve an Equation	Add or subtract the same number to both sides of the original equation, so that the equation becomes variable = number.	Solve each equation. **(a)** $x - 6 = 9$ $x - 6 + 6 = 9 + 6$ $x + 0 = 15$ $x = 15$ **(b)** $-7 = x + 9$ $-7 - 9 = x + 9 - 9$ $-16 = x + 0$ $-16 = x$

Section Number and Topic	Approach	Example
9.6 Using the Multiplication Property of Equality to Solve an Equation	Multiply or divide both sides of the original equation by the same nonzero number, so that the equation becomes variable = number.	Solve each equation. **(a)** $6x = 54$ $\frac{6x}{6} = \frac{54}{6}$ $x = 9$ **(b)** $\frac{1}{3}x = 8$ $3 \cdot \frac{1}{3}x = 3 \cdot 8$ $1x = 24$ $x = 24$
9.7 Solving Equations with Several Steps	Use the following steps: *Step 1* Add or subtract the same expression on each side, so that only the variable term remains on one side and only a number remains on the other side. *Step 2* Multiply or divide each side by the same number to find the solution. *Step 3* Check the solution.	Solve: $2p - 3 = 9$. $2p - 3 + 3 = 9 + 3$ $2p = 12$ $\frac{2p}{2} = \frac{12}{2}$ $p = 6$ Check: $2p - 3 = 9$ $2(6) - 3 = 9$ Let $p = 6$. $12 - 3 = 9$ $9 = 9$ true
9.7 Using the Distributive Property	To simplify expressions, use the following property: $a(b + c) = ab + ac$.	Simplify: $-2(3x + 4)$ $= (-2)(3x) + (-2)(+4)$ $= -6x - 8$
9.8 Combining Like Terms	If terms are like, combine the numbers that multiply each variable.	Combine like terms in the following. **(a)** $6p + 7p$ $6p + 7p = (6 + 7)p = 13p$ **(b)** $8m - 11m$ $8m - 11m = (8 - 11)m = -3m$
9.8 Translating Word Phrases to Algebra	Use x as a variable and symbolize the operations described by the words of the problem.	Translate the following word phrases into algebra. **(a)** Two more than a number $x + 2$ **(b)** A number decreased by 8 $x - 8$ **(c)** The product of a number and 15 $15x$ **(d)** A number divided by 9 $\frac{x}{9}$

name date hour

CHAPTER 9 REVIEW EXERCISES

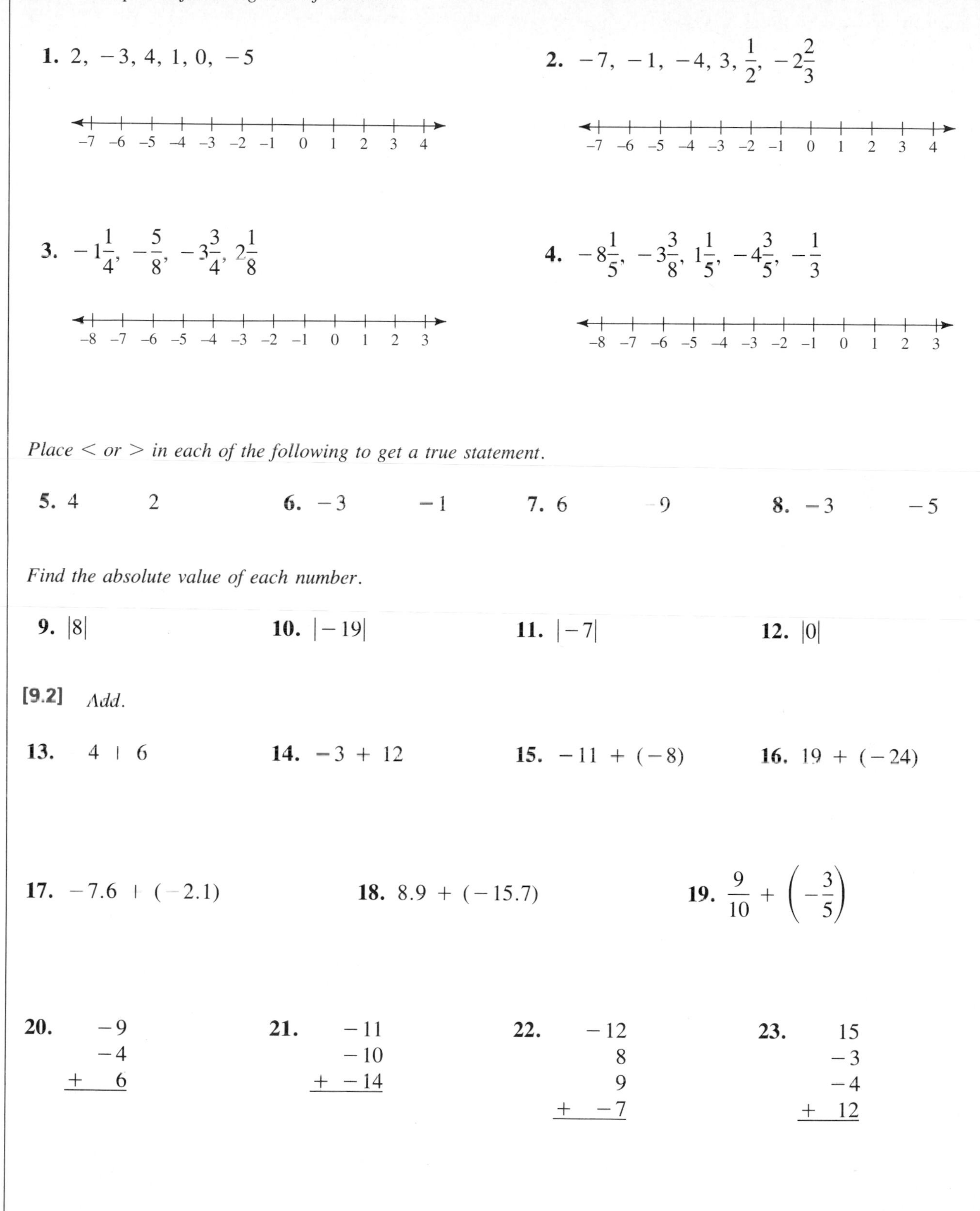

[9.1] *Graph the following lists of numbers.*

1. $2, -3, 4, 1, 0, -5$

2. $-7, -1, -4, 3, \frac{1}{2}, -2\frac{2}{3}$

3. $-1\frac{1}{4}, -\frac{5}{8}, -3\frac{3}{4}, 2\frac{1}{8}$

4. $-8\frac{1}{5}, -3\frac{3}{8}, 1\frac{1}{5}, -4\frac{3}{5}, -\frac{1}{3}$

Place < *or* > *in each of the following to get a true statement.*

5. $4 \quad 2$ **6.** $-3 \quad -1$ **7.** $6 \quad -9$ **8.** $-3 \quad -5$

Find the absolute value of each number.

9. $|8|$ **10.** $|-19|$ **11.** $|-7|$ **12.** $|0|$

[9.2] *Add.*

13. $4 + 6$ **14.** $-3 + 12$ **15.** $-11 + (-8)$ **16.** $19 + (-24)$

17. $-7.6 + (-2.1)$ **18.** $8.9 + (-15.7)$ **19.** $\frac{9}{10} + \left(-\frac{3}{5}\right)$

20.
$$\begin{array}{r} -9 \\ -4 \\ +\ \ 6 \\ \hline \end{array}$$

21.
$$\begin{array}{r} -11 \\ -10 \\ +\ -14 \\ \hline \end{array}$$

22.
$$\begin{array}{r} -12 \\ 8 \\ 9 \\ +\ \ -7 \\ \hline \end{array}$$

23.
$$\begin{array}{r} 15 \\ -3 \\ -4 \\ +\ \ 12 \\ \hline \end{array}$$

[9.3] *Give the additive inverse of each number.*

24. 6

25. -9

26. $\frac{2}{3}$

Simplify.

27. $-(-4)$

28. $-(-15)$

29. $-\left(-\frac{5}{8}\right)$

Subtract.

30. $6 - 9$

31. $-9 - 2$

32. $3 - (-3)$

33. $-3 - (-8)$

34. $-\frac{3}{4} - \frac{1}{8}$

[9.3] *Multiply or divide.*

35. $-4 \cdot 6$

36. $5 \cdot (-4)$

37. $-3 \cdot (-5)$

38. $-\frac{2}{3} \cdot \left(-\frac{5}{8}\right)$

39. $8.9 \cdot (-4.2)$

40. $\frac{-9}{3}$

41. $\frac{-25}{5}$

42. $\frac{-120}{-6}$

[9.4] *Use the order of operations to simplify each of the following.*

43. $2 \cdot (-5) - 11$

44. $(-4) \cdot (-8) - 9$

45. $48 \div (-2)^3 - (-5)$

46. $-36 \div (-3)^2 - (-2)$

47. $5 \cdot 4 - 7 \cdot 6 + 3 \cdot (-4)$

48. $2 \cdot 8 - 4 \cdot 9 + 2 \cdot (-6)$

49. $-4 \cdot 3^3 - 2 \cdot (5 - 9)$

50. $6 \cdot (-4)^2 - 3 \cdot (7 - 14)$

51. $4^2 \cdot (9 - 16) \div (-5 - 3)$

52. $(.8)^2 \cdot (.2) - (-1.2)$

53. $(\ \ .3)^2 \cdot (-.1) - 2.4$

54. $\frac{2}{9} \cdot \left(-\frac{6}{7}\right) - \left(-\frac{1}{3}\right)$

[9.5] *Find the value of the expression* $2k + 4m$ *for each of the following values of* k *and* m.

55. $k = 4, \quad m = 3$

56. $k = -6, \quad m = 2$

57. $k = -8, \quad m = -5$

Use the given values of the variables to find the value of each expression.

58. $2p + q; \quad p = -1, \quad q = 4$

59. $\frac{5a - 7y}{2 + m}; \quad a = -1, \quad y = 4, \quad m = -3$

In each of the following, use the formula and the values of the variables to find the value of the remaining variable.

60. $P = a + b + c; \quad a = 9, \quad b = 12, \quad c = 14$

61. $A = \frac{1}{2}bh; \quad b = 6, \quad h = 9$

[9.6–9.7] *Solve each equation. Check each solution.*

62. $y + 2 = 11$

63. $6 = a - 4$

64. $z + 5 = 2$

65. $-8 = -9 + r$

66. $x + \frac{1}{6} = \frac{5}{3}$

67. $12.92 + k = 4.87$

68. $8r = 56$

69. $-3p = 24$

70. $\frac{z}{4} = 5$

71. $\frac{a}{5} = -11$

72. $-\frac{11}{10}y = 22$

73. $-\frac{5}{8}m = \frac{5}{12}$

74. $2y + 6 = 12$

75. $-3p + 5 = -1$

Use the distributive property to simplify.

76. $6(r - 5)$

77. $11(p + 7)$

78. $-9(z - 3)$

Combine like terms.

79. $3r + 8r$

80. $10z - 15z$

81. $3p - 12p$

Solve each equation. Check each solution.

82. $4z + 2z = 42$

83. $9k - 2k = -35$

84. $8r - 16r = 24$

name date hour

[9.8] *Write in symbols by using x to represent the variable.*

85. 18 plus a number

86. half a number

87. the sum of four times a number and 6

88. three times a number added to 7

Solve each word problem.

89. If eight times a number is subtracted from eleven times the number, the result is -9. Find the number.

90. A snowmobile can be rented for $45 for the first day and $35 for each additional day. The bill for a rental was $255. For how many days was it rented?

91. The perimeter of a rectangle is 124 cm. The width is 25 cm. Find the length. (Use the formula for the perimeter of a rectangle, $P = 2l + 2w$.)

92. For how long must $1500 be deposited at 10% per year to earn $600 interest. (Use the formula for interest, $I = prt$.)

MIXED REVIEW EXERCISES

Simplify the following.

93. $-\frac{5}{8} \div \left(-\frac{3}{16}\right)$

94. $-7.2 - (-8.6)$

95. $3^2 \cdot (7 - 9) : (5 - 1)$

96. $-\frac{7}{12} + \frac{7}{18}$

97. $|-6| + 2 - (-3) \cdot (-8) + 5$

98. $-\frac{3}{4} \cdot \left(-\frac{8}{5}\right) + \left(-\frac{2}{3}\right)$

Solve each equation.

99. $-\frac{1}{2}y - 3 = -5$

100. $-\frac{2}{3}m + 2 = -2$

101. $6z - 3 = 3z + 9$

102. $4 + 2p + 6p = 9 + 4p + 1$

Write in symbols by using x to represent the variable.

103. twice a number decreased by 8

104. one third of a number increased by 7

Solve each word problem.

105. The length of a rectangle is 3 more than twice the width. The perimeter is 36 inches. Find the length (use $P = 2l + 2w$).

106. A customer purchases a television for \$404.25 including 5% sales tax. Find the price of the television.

name date hour

CHAPTER 9 TEST

Work each of the following problems.

1. Graph the numbers -4, -1, $1\frac{1}{4}$, $3\frac{3}{5}$, and 0.

2. Place < or > in the blank to make a true statement: -9 ______ -6.

3. Find $|7|$.

Add or subtract.

4. $-8 + 7$

5. $-11 + (-2)$

6. $-\frac{3}{8} + \left(-1\frac{1}{4}\right)$

7. $8 - 15$

8. $4 - (-12)$

9. $-\frac{1}{2} - \left(\frac{3}{4}\right)$

Multiply or divide.

10. $8(-4)$

11. $-7(-12)$

12. $\frac{-100}{4}$

Use the order of operations to simplify each of the following.

13. $(-5) \cdot (-3)^2 - (-2)$

14. $(-5) \cdot (3 - 9) - (-4)$

Find the value of $8k - 3m$, *given the following.*

15. $k = -4, \quad m = 2$

16. $k = 7, \quad m = 9$

1. 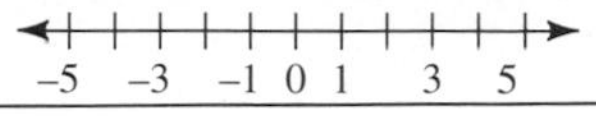

______

2. ______

3. ______

4. ______

5. ______

6. ______

7. ______

8. ______

9. ______

10. ______

11. ______

12. ______

13. ______

14. ______

15. ______

16. ______

17. ____________

18. ____________

19. ____________

20. ____________

21. ____________

22. ____________

23. ____________

24. ____________

25. ____________

26. ____________

17. The formula for the area of a triangle is $A = \frac{1}{2}bh$. Find A, if $b = 20$ and $h = 11$.

Solve each equation.

18. $x - 9 = -4$

19. $-2 + r = 5$

20. $11k = -99$

21. $-\frac{1}{4}p = 2$

Solve each equation.

22. $8r - 3r = -25$

23. $3m - 5 = 7m - 13$

Solve each word problem.

24. If seven times a number is added to 3, the result is 17. Find the number.

25. A board is 118 cm long. It is cut into two pieces, with one piece 4 cm longer than the other. Find the length of the shorter piece.

26. If $1200 is deposited for 3 years and produces $324 in interest, what is the rate of interest?

name date hour

CUMULATIVE REVIEW: CHAPTERS 7–9

[7.1] *Convert each of the following.*

1. 1 foot = inches **2.** feet = 1 yard **3.** $2\frac{1}{2}$ pounds = ounces

[7.2] *Solve the following and simplify the answers.*

4.
```
  5 yards 2 feet 11 inches
+ 2 yards 1 foot   7 inches
```

5.
```
  3 weeks 5 days 7 hours
− 2 weeks 8 days 9 hours
```

6.
```
  4 gallons 3 quarts 1 pint
×                         9
```

7. $4\overline{)3 \text{ weeks } 5 \text{ days } 28 \text{ hours}}$

[7.3–7.4] *Convert each of the following as indicated.*

8. 5978 mm to m **9.** 4317 g to kg **10.** 2.83 kl to ℓ

[7.5] *Use the table on page 453 to make the following conversions. Round answers to the nearest tenth of a unit.*

11. 15 quarts to ℓ **12.** 96 feet to m **13.** 120 km to miles

Convert Celsius temperatures to Fahrenheit and convert Fahrenheit temperatures to Celsius. Round to the nearest degree.

14. 70°F **15.** 420°F **16.** 150°C

[8.1] *Find the perimeter and area of each rectangle or square. Round answers to the nearest tenth.*

17. a rectangle 5.7 m by 11.8 m **18.** a square 14.8 cm on a side

[8.2] *Find the area of each parallelogram or trapezoid. Round answers to the nearest tenth.*

19. a parallelogram with height 15.4 m and base 7.2 m

20. a trapezoid with bases 28.4 m and 19.2 m, and height 11.6 m

[8.3] *Find the perimeter of each triangle.*

21. sides 74.8 m, 63.7 m, and 81.7 m

22. sides 3.47 ft, 2.99 ft, and 5.03 ft

Find the area of each triangle. Round answers to the nearest tenth.

23. base 23.4 m and height 9.7 m

24. base 892 cm and height 541 cm

[8.4] *Find the missing value for each circle.*

25. radius is 9.48 m, find the diameter

26. diameter is 13.4 ft, find the radius

Find the circumference and area of each circle. Use 3.14 *as an approximation for* π. *Round answers to the nearest tenth.*

27. radius 7.6 m

28. diameter 143 m

Find the area of each semicircle. Use 3.14 *as an approximation for* π. *Round each answer to the nearest tenth.*

29. radius 32.8 cm

30. diameter 7.6 cm

name date hour

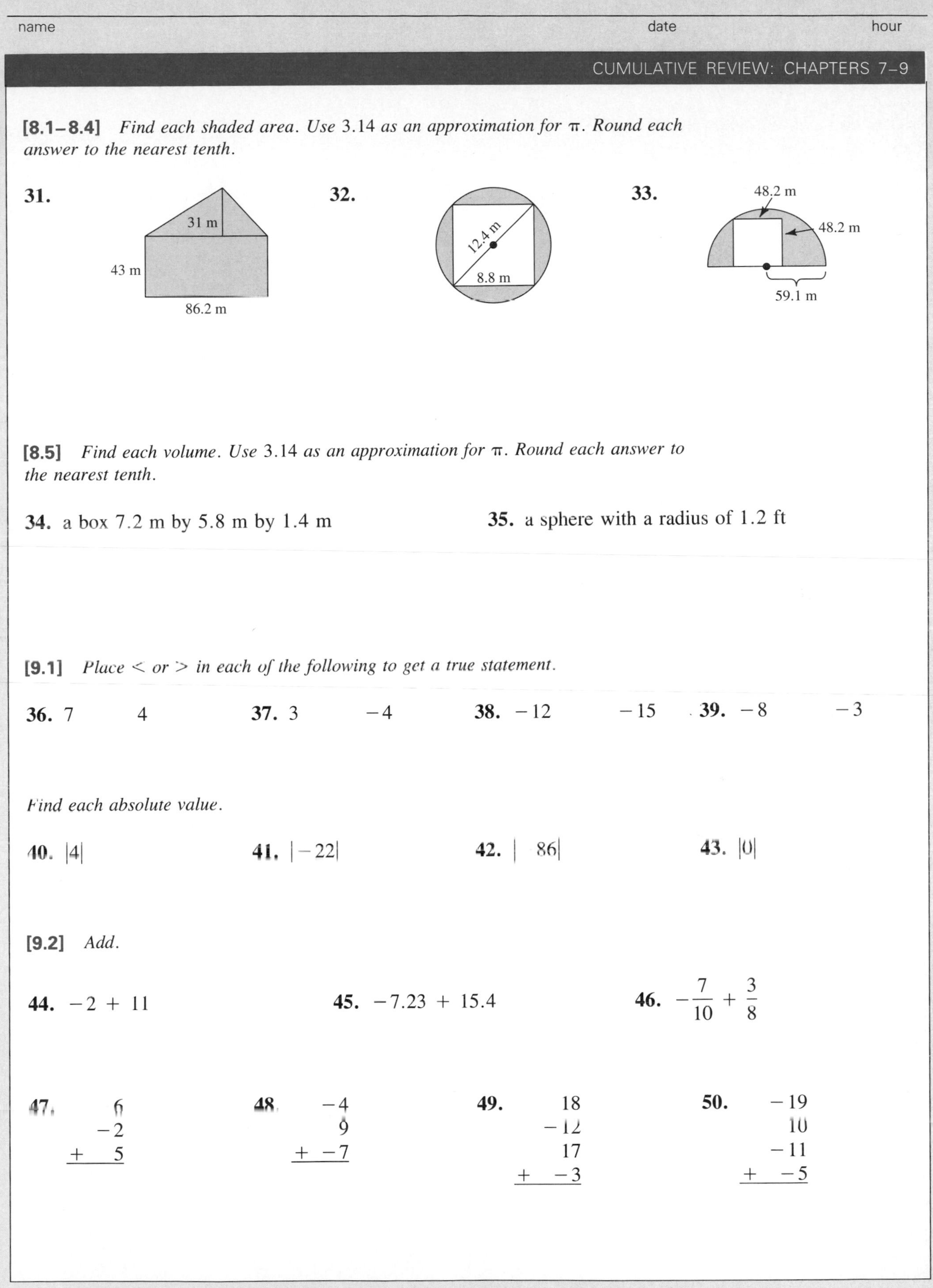

CUMULATIVE REVIEW: CHAPTERS 7–9

[8.1–8.4] *Find each shaded area. Use 3.14 as an approximation for π. Round each answer to the nearest tenth.*

31.

32.

33.

[8.5] *Find each volume. Use 3.14 as an approximation for π. Round each answer to the nearest tenth.*

34. a box 7.2 m by 5.8 m by 1.4 m

35. a sphere with a radius of 1.2 ft

[9.1] *Place < or > in each of the following to get a true statement.*

36. 7 4

37. 3 −4

38. −12 −15

39. −8 −3

Find each absolute value.

40. $|4|$

41. $|-22|$

42. $|\ 86|$

43. $|0|$

[9.2] *Add.*

44. $-2 + 11$

45. $-7.23 + 15.4$

46. $-\frac{7}{10} + \frac{3}{8}$

47.
$$\begin{array}{r} 6 \\ -2 \\ +\ \ 5 \\ \hline \end{array}$$

48.
$$\begin{array}{r} -4 \\ 9 \\ +\ -7 \\ \hline \end{array}$$

49.
$$\begin{array}{r} 18 \\ -12 \\ 17 \\ +\ \ -3 \\ \hline \end{array}$$

50.
$$\begin{array}{r} -19 \\ 10 \\ -11 \\ +\ \ -5 \\ \hline \end{array}$$

Give the additive inverse of each number.

51. 11

52. -19

53. $\frac{5}{8}$

54. 0

Subtract.

55. $8 - 15$

56. $-11 - 6$

57. $9 - (-6)$

58. $15 - (-19)$

59. $-\frac{5}{9} - \left(-\frac{1}{3}\right)$

60. $-15.7 - 9.8$

[9.3] *Multiply or divide as indicated.*

61. $-9 \cdot 2$

62. $-15 \cdot 8$

63. $7 \cdot (-11)$

64. $-5(-9)$

65. $-8(-7)$

66. $(-6.2)(-3)$

67. $-\frac{8}{11} \cdot \left(\frac{5}{4}\right)$

68. $\left(-\frac{3}{10}\right)\left(-\frac{5}{9}\right)$

69. $\frac{-28}{7}$

70. $\frac{-15}{-3}$

71. $\frac{-540}{27}$

72. $\frac{-\frac{7}{15}}{-\frac{14}{3}}$

[9.4] *Simplify by using the order of operations.*

73. $-2 \cdot (-8) - (-3)$

74. $(-7)^2 \cdot (-3)$

75. $(-.2)^2 \cdot (8 - 15)$

[9.5] *Find the value of the expression* $7x - 5y$ *for each of the following values of* x *and* y.

76. $x = 2, \quad y = 1$

77. $x = -4, \quad y = -3$

Use the given values of the variables to find the value of each expression.

78. $8a + b; \quad a = 4, \quad b = -7$

79. $5r - 3s + 4t; \quad r = 1, \quad s = -2, \quad t = -1$

Use the given formula and the given values of the variables to find the values of the remaining variables.

80. $P = 2l + 2w; \ l = 17.4, \ w = 13.9$

81. $V = \frac{1}{3}bh; \ b = 540, \ h = 32$

[9.6–9.7] *Solve each equation. Check each solution.*

82. $m + 7 = 15$

83. $12 = r - 3$

84. $a + 6 = 2$

85. $a + \frac{2}{3} = \frac{5}{12}$

86. $12m = 72$

87. $\frac{k}{2} = 13$

88. $\frac{3}{5}p = -12$

89. $-\frac{5}{9}z = 10$

90. $-\frac{2}{3}r = -6$

91. $-\frac{7}{8}k = \frac{3}{5}$

92. $3r + 2 = 8$

93. $-\frac{1}{2}a + 3 = 5$

94. $2p + 5p = -49$

95. $8a + 7 = 3a + 2$

Simplify by using the distributive property.

96. $4(a + 6)$

97. $-4(m - 3)$

98. $-6(r - 7)$

[9.8] *Solve each word problem.*

99. If seven times a number is added to -12, the result is 23. Find the number.

100. A board is 167 cm long. It is cut into two pieces, with one piece 23 cm longer than the other. Find the length of the shorter piece.

10

STATISTICS

The word *statistics* comes from words that mean *state numbers*. State numbers refer to numerical information, or **data,** gathered by the government such as the number of births, deaths, or marriages in a population. Today the word *statistics* has a much broader application; data from economics, social science, science, and business can all be organized and studied under the branch of mathematics called **statistics.**

10.1 CIRCLE GRAPHS

OBJECTIVES

1. Read a circle graph.
2. Use a circle graph.
3. Draw a circle graph.

1 It can be hard to understand a large collection of data. The various types of graphs described in this section can be used to help make sense of such data. For example, a **circle graph** is used to show how a total amount is divided into parts. The circle graph below shows how all the activities of a college student are divided during a 24 hour period.

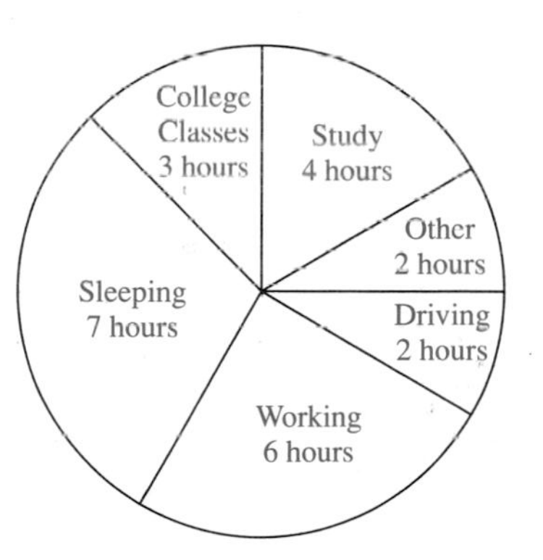

2 The circle graph uses pie-shaped pieces called **sectors** to show the amount of time spent on each activity; a circle graph can therefore be used to compare the time spent on one activity to the total number of hours in the day.

1. Use the circle graph to find the following ratios.

(a) hours spent studying to whole day

(b) hours spent working to whole day

(c) hours spent driving to whole day

(d) hours spent sleeping to whole day

2. Use the circle graph to find the following ratios.

(a) hours spent in class to hours spent working

(b) hours spent working to hours spent sleeping

(c) hours spent driving to hours spent studying

(d) hours spent for *other* to hours spent in class

ANSWERS

1. (a) $\frac{1}{6}$ (b) $\frac{1}{4}$ (c) $\frac{1}{12}$ (d) $\frac{7}{24}$

2. (a) $\frac{1}{2}$ (b) $\frac{6}{7}$ (c) $\frac{1}{2}$ (d) $\frac{2}{3}$

EXAMPLE 1 Using a Circle Graph

Find the ratio of time spent in college classes to the total number of hours in a day.

SOLUTION

The circle graph shows that 3 of the 24 hours in a day are spent in class. The ratio of class time to the hours in a day is

$$\frac{3 \text{ hours (college classes)}}{24 \text{ hours (whole day)}} = \frac{3 \text{ hours}}{24 \text{ hours}} = \frac{1}{8}. \quad ■$$

WORK PROBLEM 1 AT THE SIDE.

This circle graph can also be used to find the ratio of the time spent on one activity to the time spent on any other activity.

EXAMPLE 2 Finding a Ratio from a Circle Graph

Find the ratio of driving time to working time.

SOLUTION

The circle graph shows 2 hours spent driving and 6 hours spent working. The ratio of driving time to working time is

$$\frac{2 \text{ hours (driving)}}{6 \text{ hours (working)}} = \frac{2 \text{ hours}}{6 \text{ hours}} = \frac{1}{3}. \quad ■$$

WORK PROBLEM 2 AT THE SIDE.

A circle graph often shows data as percents. For example, a family with an annual income of $36,000 kept track of its expenses for a year. The next circle graph shows how their income was spent. The circle represents all the income, and each sector represents an expense as a percent of the annual income.

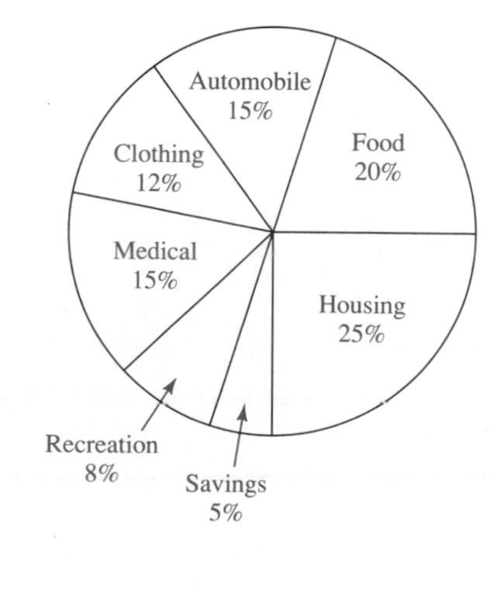

EXAMPLE 3 Calculating by Using a Circle Graph
Use the circle graph on income to find the amount spent on housing for the year.

SOLUTION
Recall the percent equation:

$$\text{amount} = \text{percent} \cdot \text{base}$$

or

$$a = p \cdot b.$$

The total income is \$36,000, so $b = 36{,}000$. The percent is 25, so $p = .25$. Find a.

$$\underset{a}{\overset{\text{amount}}{\downarrow}} = \underset{.25}{\overset{\text{percent}}{\downarrow}} \cdot \underset{36{,}000}{\overset{\text{base}}{\downarrow}} = 9000$$

The amount spent on housing is \$9000. ■

WORK PROBLEM 3 AT THE SIDE.

3 A swimming instructor tests all the students who will take swimming lessons. She places the students at various skill levels, as follows.

Skill Level	*Percent of Total*
Guppies	20%
Super Guppies	15%
Beginners	25%
Intermediates	25%
Advanced	15%

The swimming instructor can show these percents by using a circle graph. A circle has 360 degrees (written 360°). The 360° represents the entire swimming class.

EXAMPLE 4 Drawing a Circle Graph
Using the data on *skill levels,* find the number of degrees in the sector that would represent the Guppies, and begin constructing a graph.

SOLUTION
Since Guppies make up 20% of the total number of students, the number of degrees needed for the Guppy sector of the circle graph is

$$360° \times 20\% = 360° \times .2 = 72°.$$

Use a tool called a **protractor** to make a graph. First, draw a line from the center of a circle to the left edge. Place the hole in the protractor over the center of the circle, making sure that zero on the

3. Use the circle graph on income to find the following:

(a) the amount spent on clothing

(b) the amount spent on medical costs

(c) the amount spent on food

(d) the amount saved

ANSWERS
3. (a) \$1320 (b) \$5400 (c) \$7200 (d) \$1800

4. Using the information on the swimming skill levels in the table, find the number of degrees needed for each of the following and complete the circle graph.

(a) Beginners

(b) Intermediates

(c) Advanced

ANSWERS
4. (a) 90° (b) 90° (c) 54°

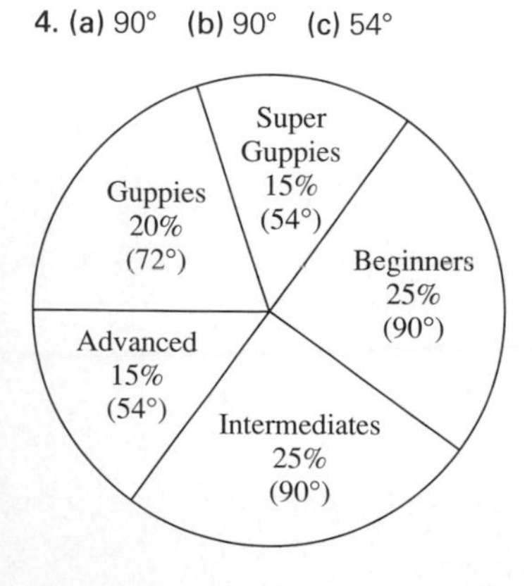

protractor lines up with the line that was drawn. Find 72° as shown in the illustration, and draw a line from the center of the circle to the 72° mark at the edge of the circle.

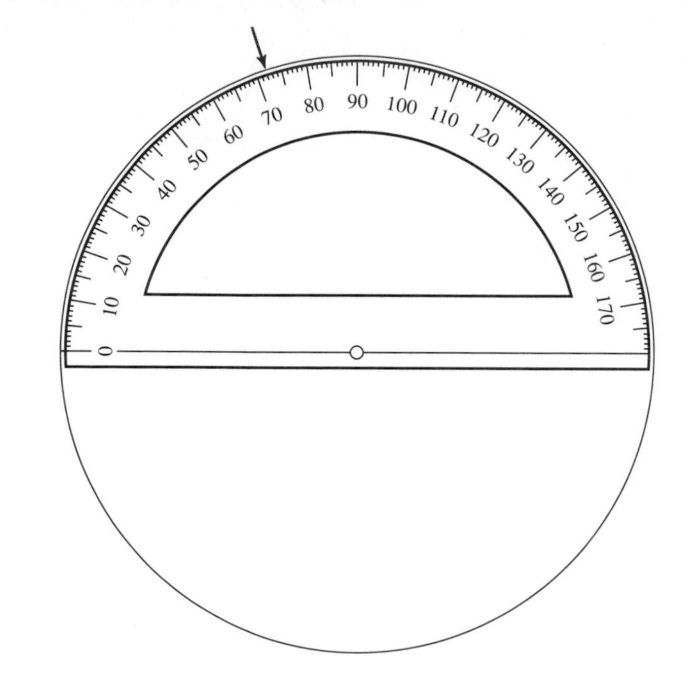

To draw the Super Guppies sector, begin by finding the number of degrees in the sector.

$$360° \times 15\% = 360° \times .15 = 54°$$

Again, place the hole of the protractor at the center of the circle, but this time align zero on the second line that was drawn. Make a mark at 54° and draw a line as before. This sector is 54° and represents the Super Guppies. ■

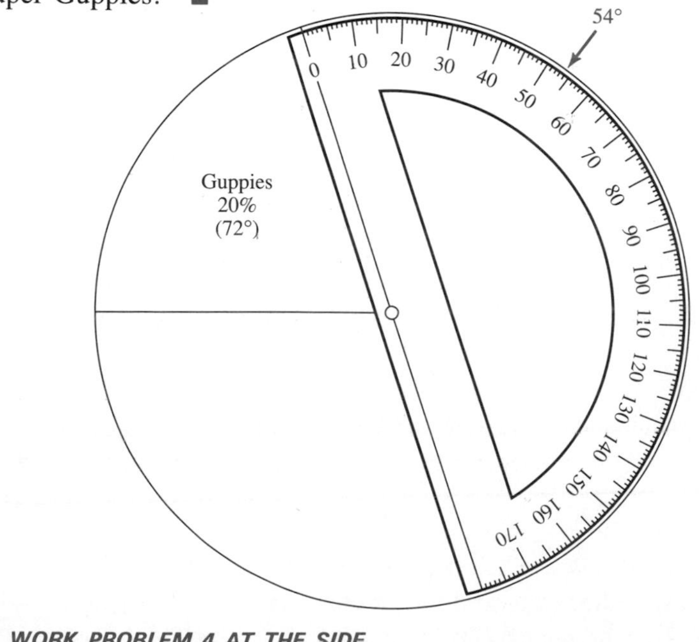

WORK PROBLEM 4 AT THE SIDE.

name date hour

10.1 EXERCISES

Use the circle graph below, which shows the cost of remodeling a kitchen, to find each of the following.

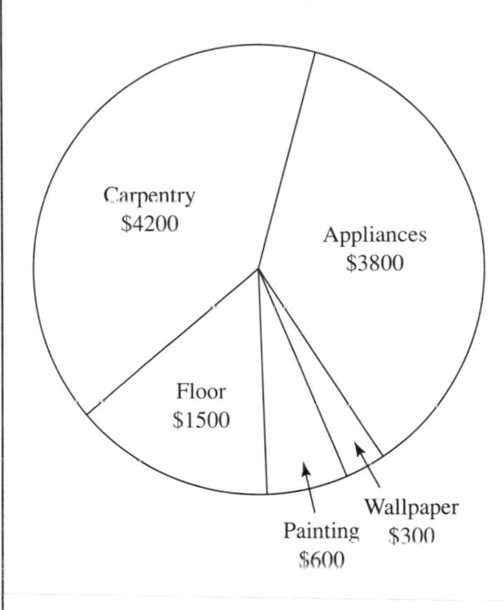

1. Find the total cost of remodeling the kitchen.

2. What is the largest single expense in remodeling the kitchen?

3. Find the ratio of the cost of appliances to the total remodeling cost.

4. Find the ratio of the cost of painting to the total remodeling cost.

5. Find the ratio of the cost of wallpaper to the cost of carpentry.

6. Find the ratio of the cost of painting to the cost of carpentry.

7. Use the circle graph at the right, which shows the number of students at a college who are enrolled in certain majors, to find which major has the least number of students enrolled.

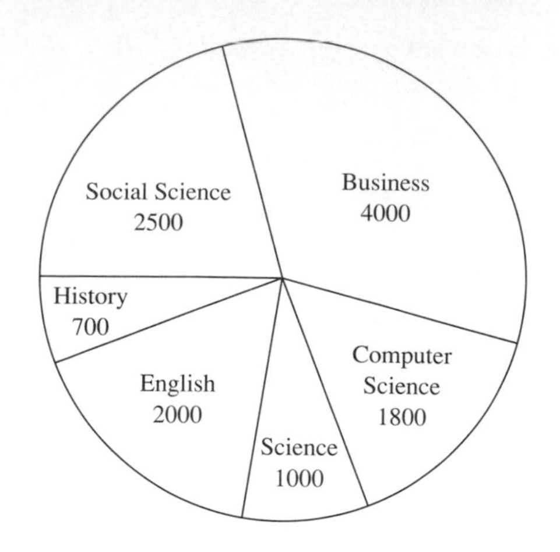

Use the circle graph in Exercise 7 to find the ratio of the first number to the second number for each of the following.

8. business majors to the total number of students

9. English majors to the total number of students

10. computer science majors to the number of English majors

11. history majors to the number of social science majors

12. science majors to the number of business majors

The following circle graph shows the expenses involved in keeping a sales force on the road. Each expense item is expressed as a percent of the total sales force cost of \$580,000. *Use the graph to find the dollar amount spent for each of the following.*

13. car and plane

14. lodging

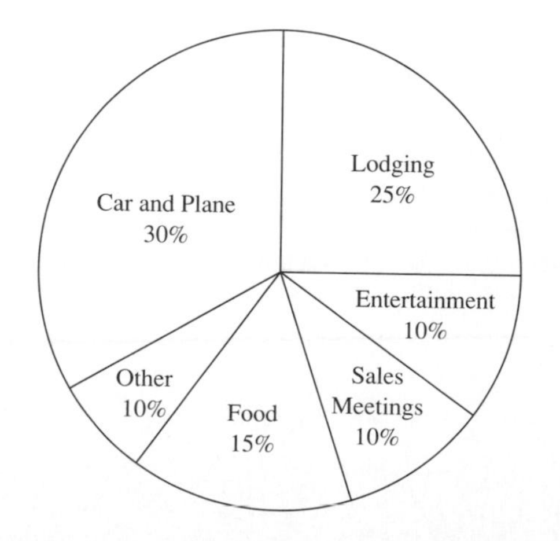

15. entertainment

16. sales meetings

17. food

18. other

The circle graph below shows how the Recycling Club's income of \$19,600 *is budgeted. Use the graph to find the dollar amount budgeted for each of the following.*

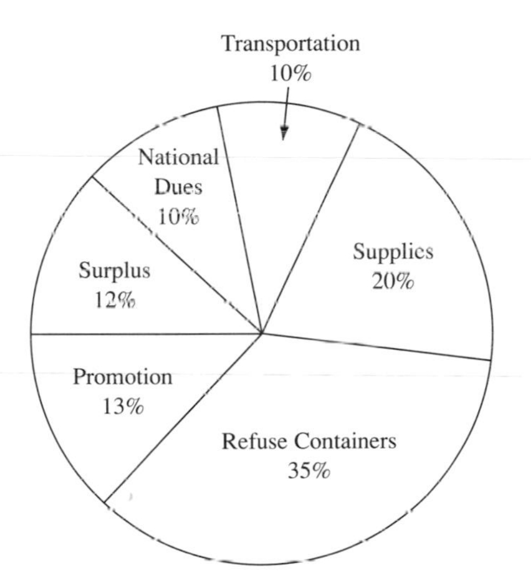

19. supplies

20. refuse containers

21. promotion

22. surplus

23. national dues

24. transportation

During one quarter Marge Prullage, a student, spent $1400 *for expenses as shown in the following chart. Find all numbers missing from the chart.*

	item	***dollar amount***	***percent of total***	***degrees of a circle***
	food	$350	25%	90°
25.	rent	$280	________	72°
26.	clothing	$210	________	________
27.	books	$140	10%	________
28.	entertainment	$210	________	54°
29.	savings	$70	________	________
30.	other	________	________	36°

31. Draw a circle graph by using the above information.

CHALLENGE EXERCISES

32. Brasher Construction Company has its annual sales divided into five categories, as follows.

Item	*Annual Sales*
Parts	$25,000
Hand tools	$80,000
Bench tools	$120,000
Brass fittings	$100,000
Cabinet hardware	$75,000

(a) Find the total sales for a year.

(b) Find the number of degrees in a circle graph for each item.

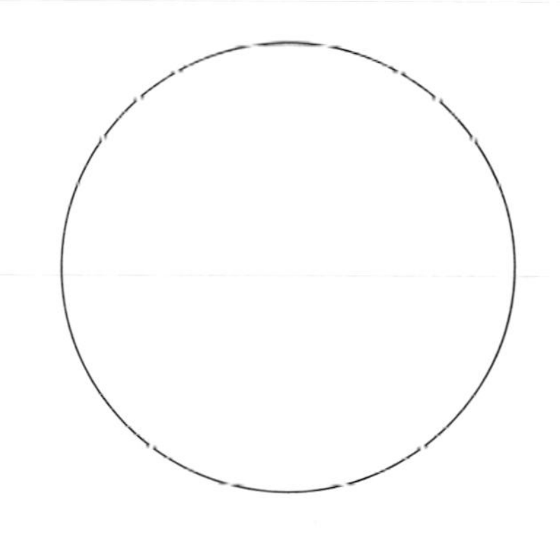

(c) Make a circle graph showing this information.

33. A book publisher had 25% of total sales in mysteries, 10% in biographies, 15% in cookbooks, 15% in romantic novels, 20% in science, and the rest in business books.

(a) Find the number of degrees in a circle graph for each type of book.

(b) Draw a circle graph.

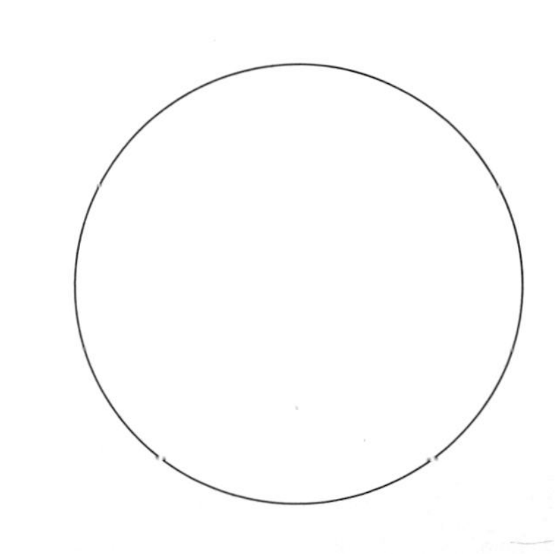

34. A family kept track of its expenses for a year and recorded the following results.

Item	*Percent of Total*
Housing	30%
Food	21%
Automobile	14%
Clothing	10%
Medical	5%
Savings	8%
Other	12%

Find the number of degrees in a circle graph for each item, and then draw a circle graph.

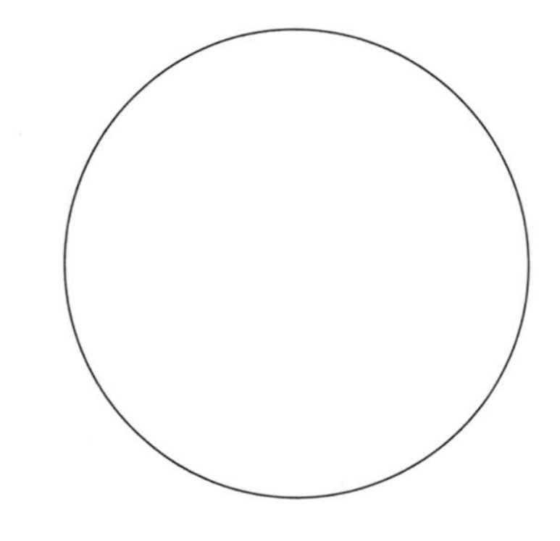

REVIEW EXERCISES

Write < or > to make a true statement. (For help, see ***Section 3.5****.)*

35. $\frac{1}{4}$ $\frac{3}{8}$

36. 28.6% 25.3%

37. .4219 .4230

38. $\frac{2}{5}$ $\frac{3}{7}$

39. 38.25% 38.29%

40. .863 .836

41. 70,485 70,510

42. 450,500 452,000

OBJECTIVES

Read and understand

1 a bar graph;

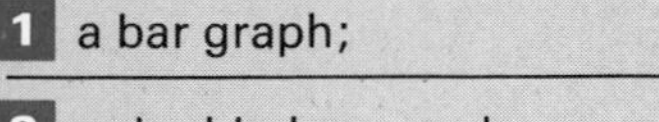

2 a double-bar graph;

3 a line graph;

4 a comparison line graph.

1 **Bar graphs** are useful when showing comparisons. For example, the bar graph below shows the comparison of the number of licensed real estate agents in a state each year during a five-year period.

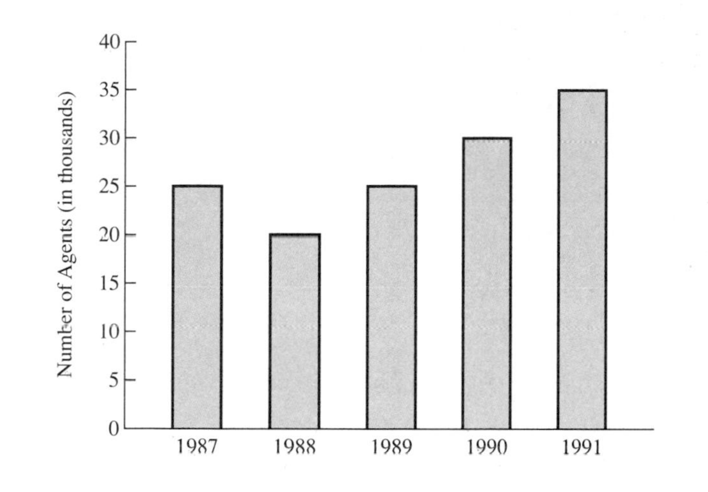

EXAMPLE 1 Using a Bar Graph

Use the bar graph to find the number of real estate agents in the state in 1989.

SOLUTION

The bar for 1989 rises to 25, showing there were 25,000 licensed real estate agents in 1989. ■

WORK PROBLEM 1 AT THE SIDE.

1. Use the bar graph in the text to find the number of licensed real estate agents in the state in each of the following years.

(a) 1987

(b) 1988

(c) 1990

(d) 1991

2 A **double-bar graph** can be used to compare two sets of data. The following double bar graph shows the number of new utility hookups each quarter for two different years.

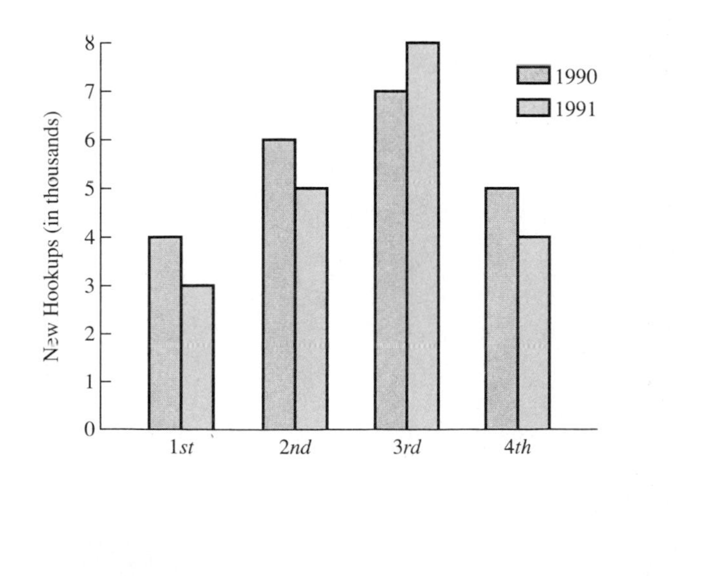

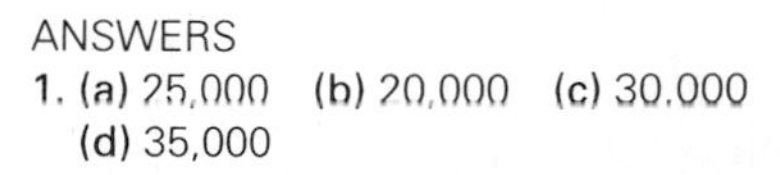

ANSWERS

1. (a) 25,000 (b) 20,000 (c) 30,000 (d) 35,000

2. Use the double-bar graph in the text to find the number of new utility hookups in 1990 and 1991 for each of the following quarters.

(a) 1st quarter

(b) 3rd quarter

(c) 4th quarter

(d) Find the greatest number of hookups. Identify the quarter and the year in which they occurred.

3. Use the line graph in the text to find the number of trout stocked in each of the following months.

(a) April

(b) June

(c) July

(d) August

ANSWERS

2. (a) 4000; 3000 (b) 7000; 8000 (c) 5000; 4000 (d) 8000; 3rd quarter of 1991

3. (a) 40,000 (b) 50,000 (c) 60,000 (d) 20,000

EXAMPLE 2 Reading a Double-Bar Graph

Use the double-bar graph to find each of the following.

(a) the number of new utility hookups in the second quarter of 1990
The bar on the left for the second quarter (the bar for 1990) rises to 6000, so there were 6000 new hookups for the second quarter in 1990.

(b) the number of new utility hookups in the second quarter of 1991
In the second quarter of 1991, there were 5000 new hookups. ■

WORK PROBLEM 2 AT THE SIDE.

3 A **line graph** is often useful for showing a trend. The line graph that follows shows the number of trout stocked over a five-month period. Each dot over a month indicates the number of trout stocked during that month.

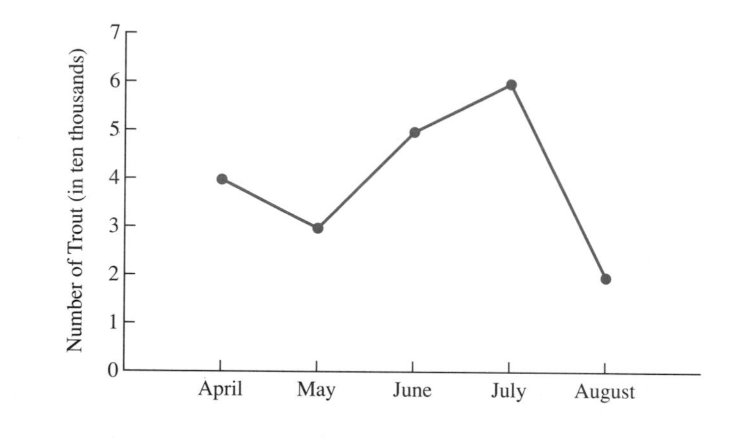

EXAMPLE 3 Understanding a Line Graph

Use the line graph to find the following.

(a) In which month were the least number of trout stocked?
The least number of trout were stocked in August.

(b) How many trout were stocked in May?
There were 30,000 ($3 \times 10{,}000$) trout stocked in May. ■

WORK PROBLEM 3 AT THE SIDE.

4 Two sets of data can also be compared by picturing two line graphs together as a **comparison line graph.** For example, the following comparison line graph shows the number of telephones leased by customers and the number of telephones owned by customers during each year of a five-year period.

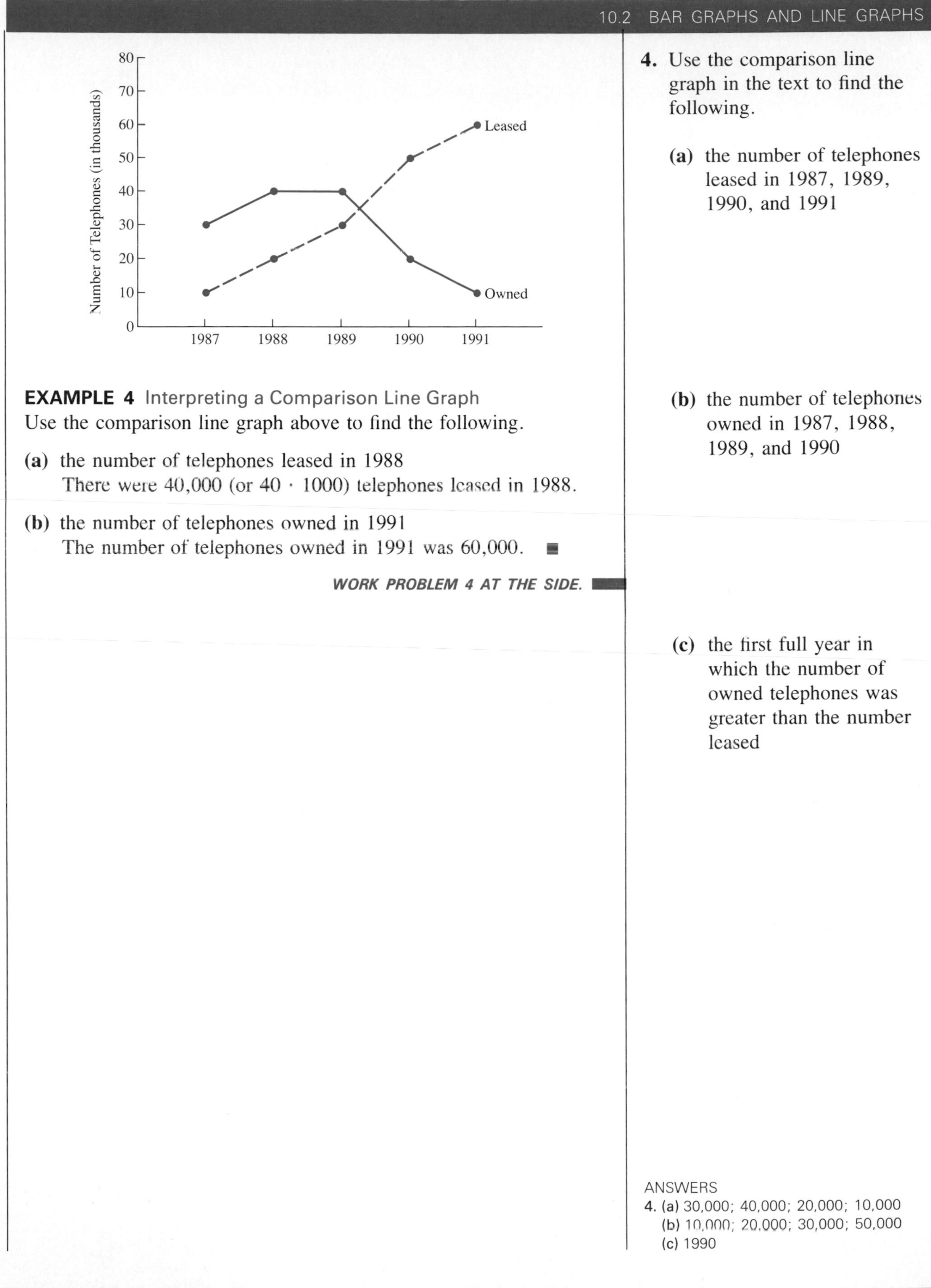

EXAMPLE 4 Interpreting a Comparison Line Graph
Use the comparison line graph above to find the following.

(a) the number of telephones leased in 1988
There were 40,000 (or 40 · 1000) telephones leased in 1988.

(b) the number of telephones owned in 1991
The number of telephones owned in 1991 was 60,000. ■

WORK PROBLEM 4 AT THE SIDE.

4. Use the comparison line graph in the text to find the following.

(a) the number of telephones leased in 1987, 1989, 1990, and 1991

(b) the number of telephones owned in 1987, 1988, 1989, and 1990

(c) the first full year in which the number of owned telephones was greater than the number leased

ANSWERS
4. (a) 30,000; 40,000; 20,000; 10,000
(b) 10,000; 20,000; 30,000; 50,000
(c) 1990

name date hour

10.2 EXERCISES

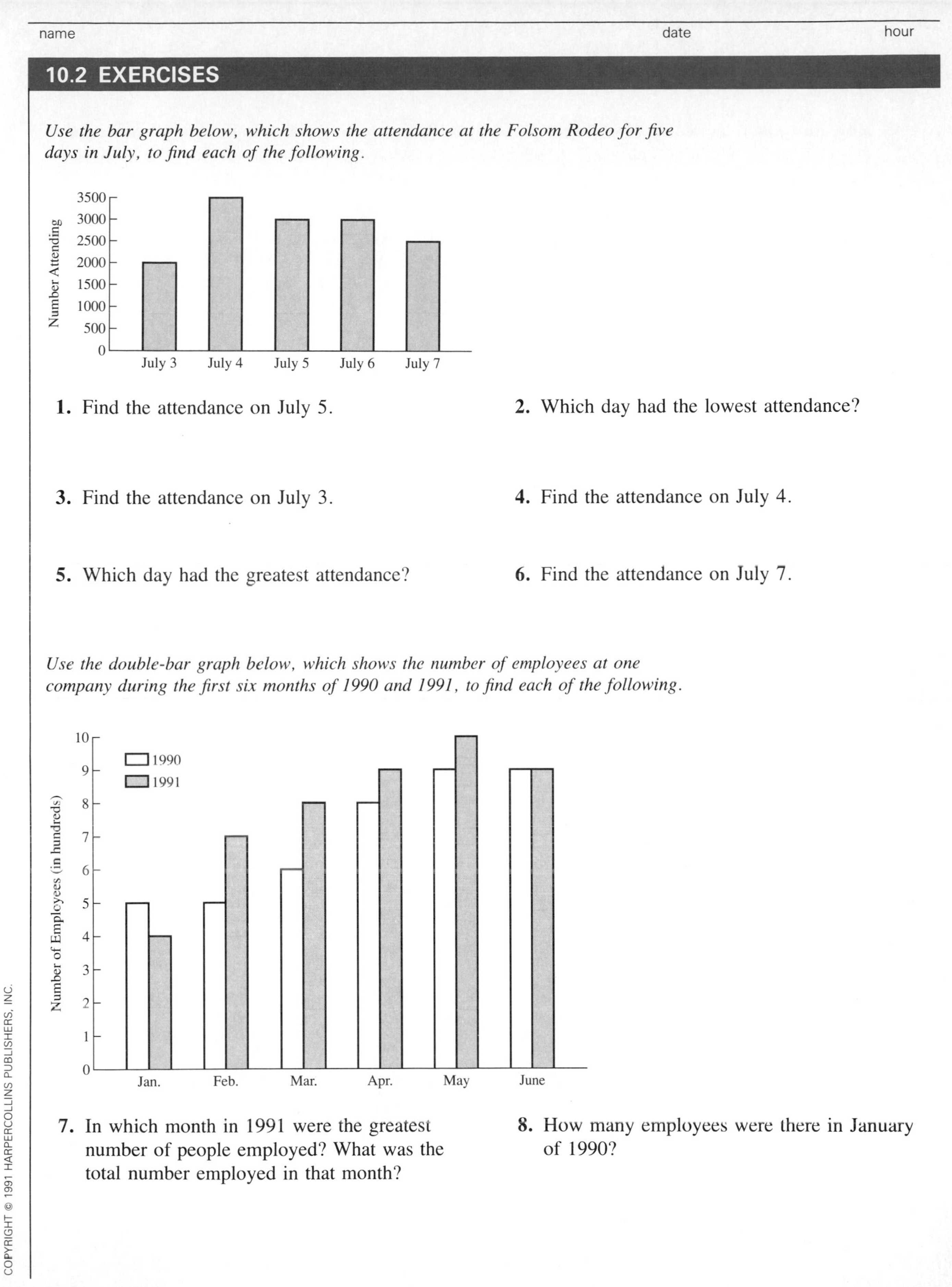

Use the bar graph below, which shows the attendance at the Folsom Rodeo for five days in July, to find each of the following.

1. Find the attendance on July 5.

2. Which day had the lowest attendance?

3. Find the attendance on July 3.

4. Find the attendance on July 4.

5. Which day had the greatest attendance?

6. Find the attendance on July 7.

Use the double-bar graph below, which shows the number of employees at one company during the first six months of 1990 and 1991, to find each of the following.

7. In which month in 1991 were the greatest number of people employed? What was the total number employed in that month?

8. How many employees were there in January of 1990?

9. How many more employees were there in February of 1991 than in February of 1990?

10. How many more employees were there in March of 1991 than in March of 1990?

11. Find the increase in the number of employees from February 1990 to April 1990.

12. Find the increase in the number of employees from January 1991 to June 1991.

Use the double-bar graph below, which shows sales of unleaded and supreme unleaded gasoline at a service station for a five-year period, to find each of the following.

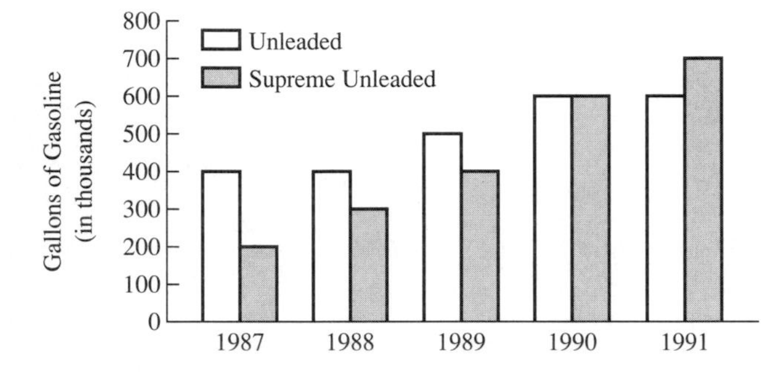

13. how many gallons of supreme unleaded gasoline were sold in 1987?

14. How many gallons of unleaded gasoline were sold in 1990?

15. The greatest difference in sales between unleaded and supreme unleaded gasoline was in which year?

16. In which year did the sales of supreme unleaded gasoline surpass the sales of unleaded gasoline?

17. Find the increase in supreme unleaded gasoline sales from 1987 to 1991.

18. Find the increase in unleaded gasoline sales from 1987 to 1991.

name date hour

Use the line graph below, which shows the number of burglaries in a community during the first six months of a year, to find each of the following.

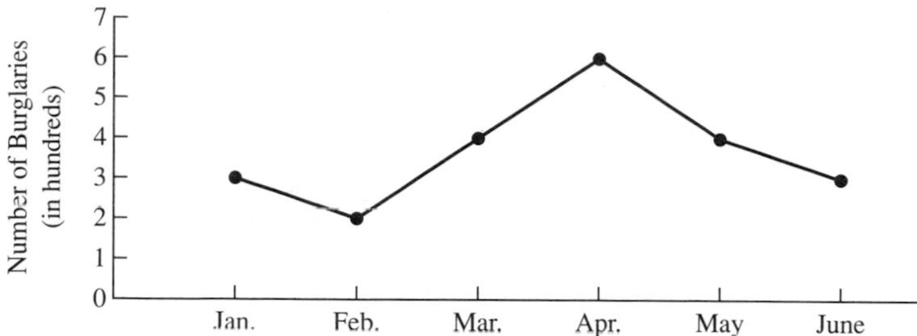

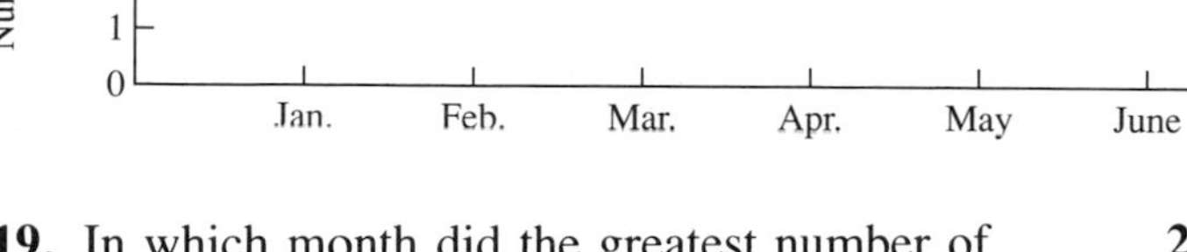

19. In which month did the greatest number of burglaries occur?

20. In which month did the least number of burglaries occur?

21. Find the number of burglaries in June.

22. Find the number of burglaries in March.

23. Find the increase in the number of burglaries from March to April.

24. Find the decrease in the number of burglaries from April to May.

Use the comparison line graph below, which shows annual sales for two different stores during a five-year period, to find the annual sales in each of the following years.

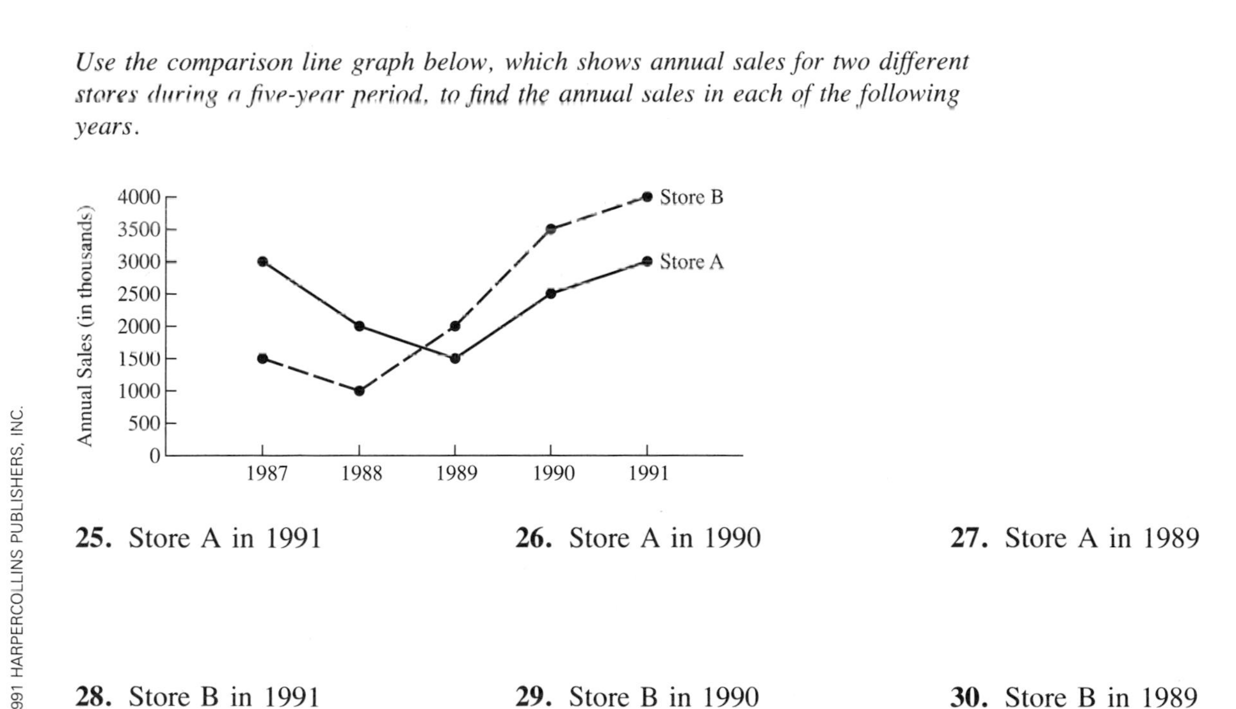

25. Store A in 1991

26. Store A in 1990

27. Store A in 1989

28. Store B in 1991

29. Store B in 1990

30. Store B in 1989

CHALLENGE EXERCISES

Use the comparison line graph below, which shows sales and profits of a fast food operation for a five-year period, to find each of the following.

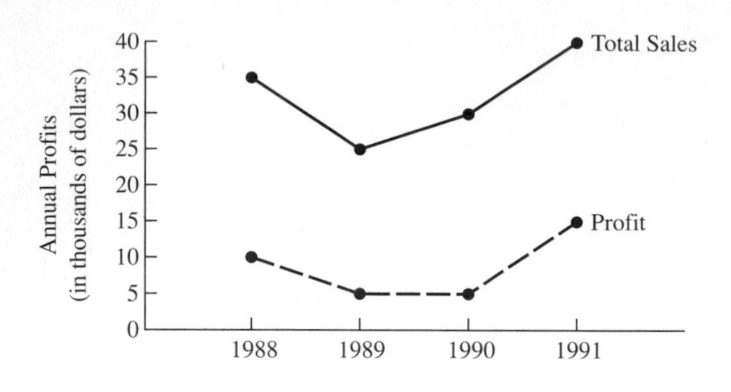

31. total sales in 1991

32. total sales in 1990

33. total sales in 1989

34. profit in 1991

35. profit in 1990

36. profit in 1989

REVIEW EXERCISES

*The circle graph below shows where the money collected for parking tickets on campus actually goes. The amount collected is $8400. Use the graph to find the amount going to each place. (For help, see **Section 10.1.**)*

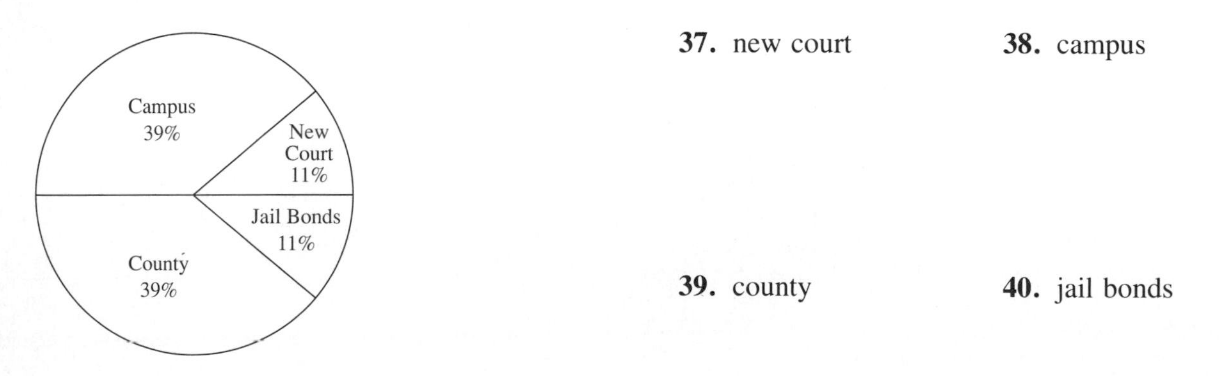

37. new court

38. campus

39. county

40. jail bonds

10.3 HISTOGRAMS AND FREQUENCY POLYGONS

OBJECTIVES

1. Arrange data in class intervals.
2. Understand class frequency.
3. Read and understand a histogram.
4. Read, understand, and construct a frequency polygon.

An employer asked her 30 employees how many college units each had completed. The list of responses looked like this:

74	133	4	127	20	30
103	27	139	118	138	121
149	132	64	141	130	76
42	50	95	56	65	104
4	140	12	88	119	64

1 A long list of numbers can be confusing. Make the data easier to read by dividing the numbers into smaller groups, or **class intervals,** such as *0 to 24 units* or *25–49 units*.

2 Next, use a **tally** column, as shown below, to find the **class frequency,** or number of employees in each class interval.

Class Interval (Number of Units)	*Tally*	*Class Frequency (Number of Employees)*
0–24	IIII	4
25–49	III	3
50–74	卌 I	6
75–99	III	3
100–124	卌	5
125–149	卌 IIII	9

3 The results in the table have been used to draw the special bar graph, called a **histogram,** that follows. In a histogram, the width of each bar represents a range of numbers (*class interval*). The height of each bar in a histogram gives the *class frequency,* the number of occurrences in each class interval.

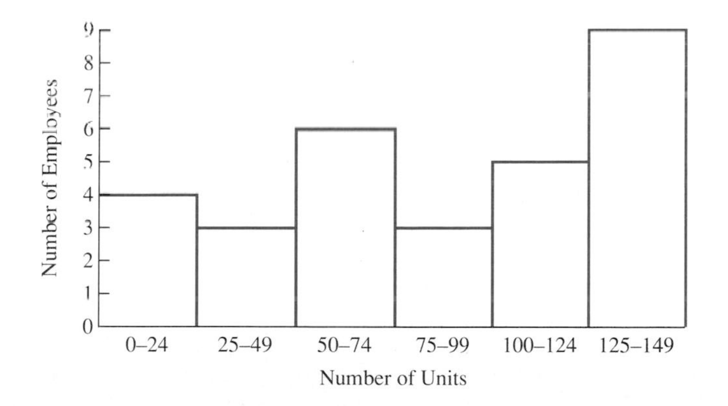

EXAMPLE 1 Using a Histogram

Use the histogram to find the number of employees who have completed fewer than 50 college units.

SOLUTION

Since 4 employees have completed 0–24 units and 3 employees have completed 25–49 units, the number of employees who have completed fewer than 50 units is $4 + 3 = 7$. ■

WORK PROBLEM 1 AT THE SIDE.

1. Use the histogram in the text to find the following.

(a) the number of employees who have completed 75 units or more

(b) the number of employees who have completed 50 to 124 units

(c) the number of employees who have completed fewer than 125 units

ANSWERS

1. (a) 17 (b) 14 (c) 21

2. Use the frequency polgyon in the text to find the number of customers who used their charge accounts.

(a) more than 60 times in the year.

(b) 80 or fewer times in the year.

(c) from 21 to 60 times in the year.

ANSWERS
2. (a) $11 + 7 = 18$
(b) $18 + 35 + 29 + 11 = 93$
(c) $35 + 29 = 64$

4 Sometimes the results of dividing data into class intervals are shown on a graph different from a histogram. For example, suppose the amount of charge account use of 100 customers is studied, and the results are listed in the following table. The middle of each class interval, called the **class midpoint,** is included in the table.

Class Interval (Charge Account Use)	*Class Midpoint*	*Class Frequency*
0–20	10	18
21–40	30	35
41–60	50	29
61–80	70	11
81–100	90	7

A histogram for this data is shown below. The class midpoints are located, and dots are placed at the corresponding points at the tops of the bars of the histogram. These dots are then connected, forming a **frequency polygon.**

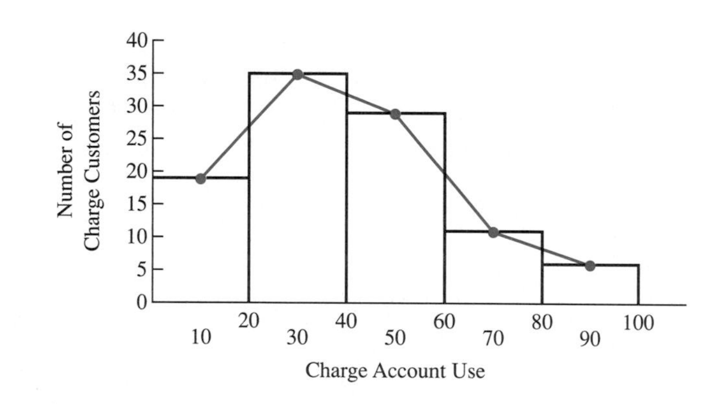

EXAMPLE 2 Using a Frequency Polygon

Use the frequency polygon above to find the number of customers who used their charge accounts 60 or fewer times in the year.

SOLUTION

The number of customers who used their charge accounts 60 or fewer times was $29 + 35 + 18 = 82$. ■

WORK PROBLEM 2 AT THE SIDE.

name date hour

10.3 EXERCISES

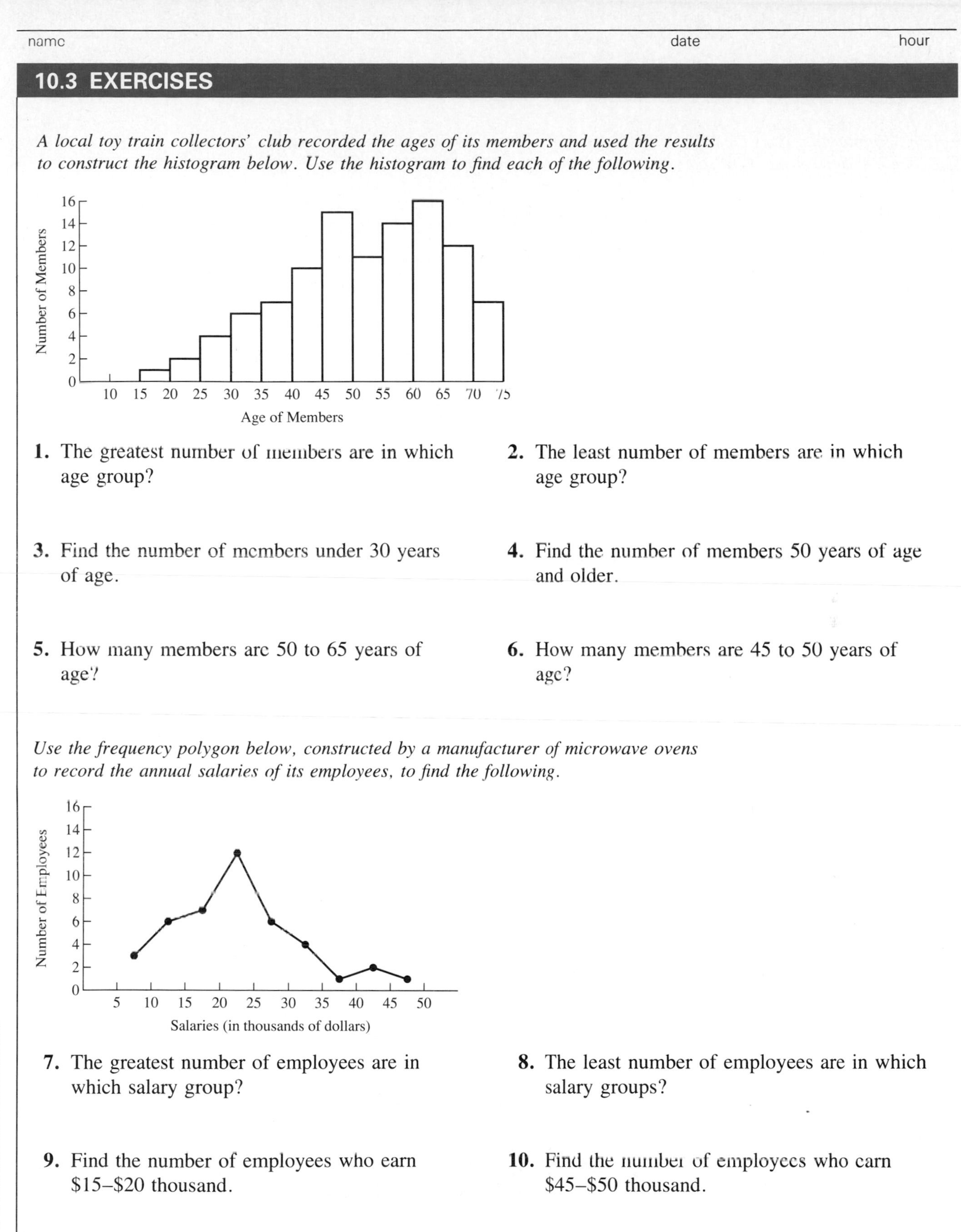

A local toy train collectors' club recorded the ages of its members and used the results to construct the histogram below. Use the histogram to find each of the following.

1. The greatest number of members are in which age group?

2. The least number of members are in which age group?

3. Find the number of members under 30 years of age.

4. Find the number of members 50 years of age and older.

5. How many members are 50 to 65 years of age?

6. How many members are 45 to 50 years of age?

Use the frequency polygon below, constructed by a manufacturer of microwave ovens to record the annual salaries of its employees, to find the following.

7. The greatest number of employees are in which salary group?

8. The least number of employees are in which salary groups?

9. Find the number of employees who earn \$15–\$20 thousand.

10. Find the number of employees who earn \$45–\$50 thousand.

11. How many employees earn \$25,000 or less?

12. How many employees earn \$30,000 or more?

Use the following list, which shows the number of sets of encyclopedias sold annually by the members of the local sales staff, to complete the following table.

120	130	144	132	147	158	174
135	142	155	174	162	151	178
145	151	139	128	147	134	146

	number of sets	*tally*	*frequency*
13.	120–129	________	________
14.	130–139	________	________
15.	140–149	________	________
16.	150–159	________	________
17.	160–169	________	________
18.	170–179	________	________

The daily high temperatures for Phoenix, Arizona, during June of one year follow. (The numbers are in chronological order from left to right. For example, the temperature on June 5 was 102°, and on June 11, it was 104°.) Use these numbers to complete the following table.

79°	84°	88°	96°	102°	104°	110°	108°	106°	106°
104°	99°	97°	92°	94°	90°	82°	74°	72°	83°
85°	92°	100°	99°	101°	107°	111°	102°	97°	94°

	temperature	*tally*	*frequency*
19.	70°–74°	________	________
20.	75°–79°	________	________
21.	80°–84°	________	________
22.	85°–89°	________	________
23.	90°–94°	________	________
24.	95°–99°	________	________
25.	100°–104°	________	________
26.	105°–109°	________	________
27.	110°–114°	________	________

28. Construct a histogram by using the data in Exercises 19–27.

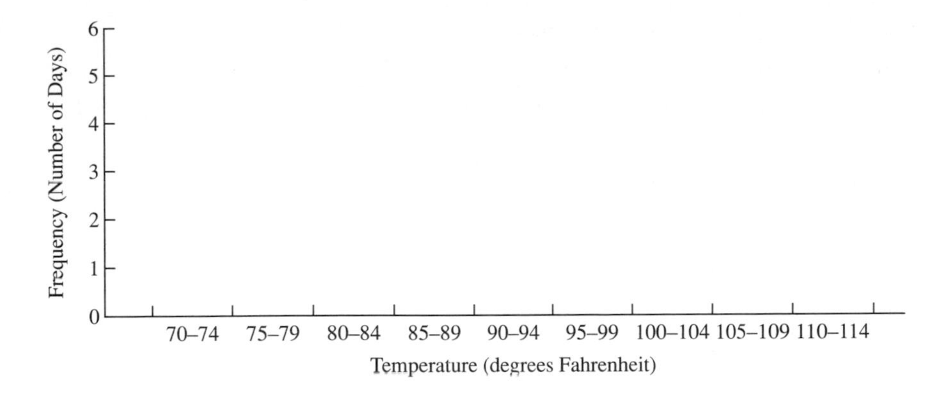

CHALLENGE EXERCISES

Use the following numbers, which show the scores of 80 *students on a mathematics test, to complete the following.*

79	60	74	59	55	98	61	67	83	71
71	46	63	66	69	42	75	62	71	77
78	65	87	57	78	91	82	73	94	48
87	65	62	81	63	66	65	49	45	51
69	56	84	93	63	60	68	51	73	54
50	88	76	93	48	70	39	76	95	57
63	94	82	54	89	64	77	94	72	69
51	56	67	88	81	70	81	54	66	87

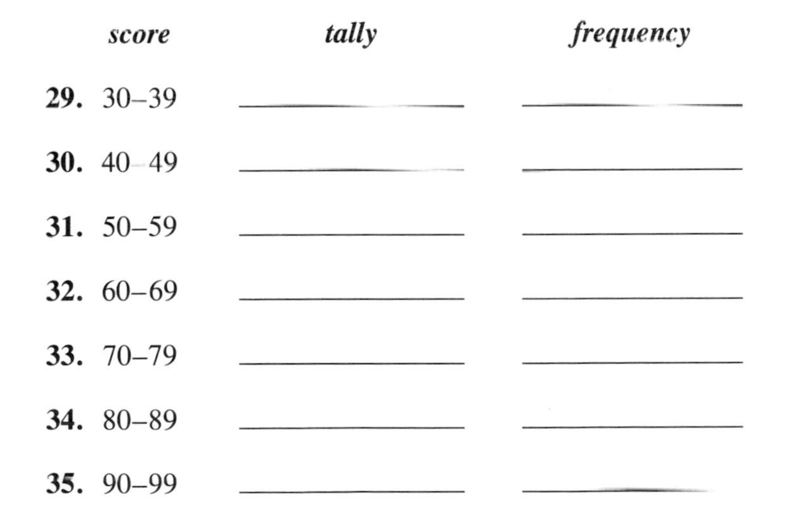

	score	*tally*	*frequency*
29.	30–39	________	________
30.	40–49	________	________
31.	50–59	________	________
32.	60–69	________	________
33.	70–79	________	________
34.	80–89	________	________
35.	90–99	________	________

36. Complete a histogram by using the data in Exercises 29–35.

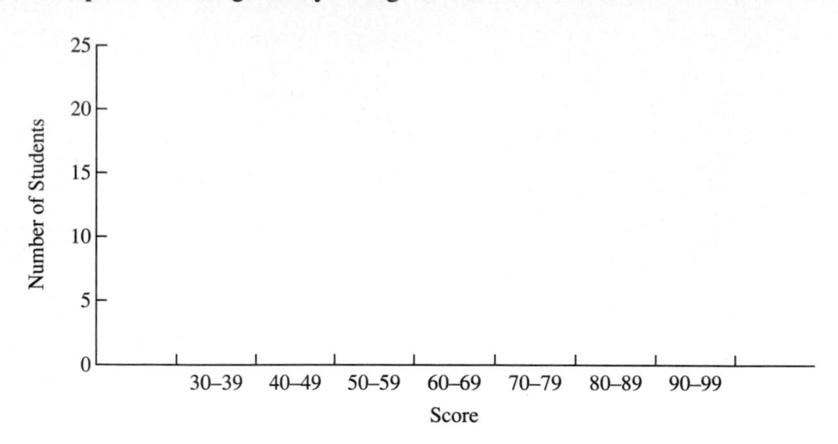

REVIEW EXERCISES

Solve each problem by using the order of operations. (For help, see ***Section 1.8.****)*

37. $(22 + 36 + 58 + 27) \div 4$

38. $(16 + 24 + 32 + 38 + 31) \div 5$

39. $(5 \cdot 2) + (7 \cdot 3) \div 5$

40. $(9 \cdot 2) + (12 \cdot 4) \div 6$

41. $(4 \cdot 3) + (3 \cdot 9) + (2 \cdot 3) \div 15$

42. $(4 \cdot 3) + (3 \cdot 6) + (2 \cdot 6) + (3 \cdot 3) \div 18$

10.4 MEAN, MEDIAN, AND MODE

OBJECTIVES

1. Find the mean of a list of numbers.
2. Find a weighted mean.
3. Find the median.
4. Find the mode.

1 When analyzing data, one of the first things to look for is a *measure of central tendency*—a single number that represents the entire list of numbers. One such measure is the *average* or **mean.** Find the mean with the following formula.

$$\text{mean} = \frac{\text{sum of all values}}{\text{number of values}}$$

EXAMPLE 1 Finding the Mean

A student had test scores of 84, 90, 95, 98, and 88. Find the average or mean of these scores.

SOLUTION

Use the formula for finding mean. Divide the sum of all test scores by the number of tests.

$$\text{mean} = \frac{84 + 90 + 95 + 98 + 88 \leftarrow \text{sum of test scores}}{5 \leftarrow \text{number of tests}}$$

$$= \frac{455}{5}$$

$$= 91 \quad \text{divide}$$

The mean score is 91. ■

WORK PROBLEM 1 AT THE SIDE.

EXAMPLE 2 Applying the Average or Mean

The sales of carnations at Tom's Flower Shop for each day last week were

$86, $103, $118, $117, $126, $158, and $149.

For Tom's Flower Shop, the mean (rounded to the nearest cent) is

$$\text{mean} = \frac{86 + 103 + 118 + 117 + 126 + 158 + 149}{7}$$

$$= \frac{857}{7}$$

$$= \$122.43. \quad ■$$

WORK PROBLEM 2 AT THE SIDE.

2 Some items in a list might appear more than once. In this case, find a **weighted mean,** in which each value is "weighted" by multiplying it by the number of times it occurs.

1. A student had test scores of 96, 98, 84, 88, 82, and 92. Find the average or mean score.

2. Find the mean for each list of numbers.

(a) $25.12, $42.58, $76.19, $32, $81.11, $26.41, $19.76, $59.32, $71.18, $21.03

(b) A list of the sales for one year at eight different restaurants follows: $374,910; $382,740; $321,872; $412,111; $242,943; $334,089; $351,147; $262,900

ANSWER

1. 90

2. (a) $\frac{\$454.70}{10} = \$45.47$

(b) $\frac{\$2,682,712}{8} = \$335,339$

3. Find the weighted mean for the numbers given in the following table.

Value	*Frequency*
6	2
7	3
8	3
9	4
10	6

ANSWER
3. 8.5

EXAMPLE 3 Understanding the Weighted Mean
Find the weighted mean for the numbers in the following table.

Value	*Frequency*
3	4
5	2
7	1
8	5
9	3
10	2
12	1
13	2

SOLUTION
Several values occur more than once: for example, the value 5 occurs twice and 8 occurs five times. Other values, such as 12, occur once. Find the mean by multiplying each value by its frequency and adding the products. Next, add the numbers in the *frequency* column to find the total number of values.

Value	*Frequency*	*Product*
3	4	12
5	2	10
7	1	7
8	5	40
9	3	27
10	2	20
12	1	12
13	2	26
Totals	**20**	**154**

Finally, divide the totals.

$$\text{mean} = \frac{154}{20} = 7.7. \blacksquare$$

WORK PROBLEM 3 AT THE SIDE.

Weighted averages are used to find a student's *grade point average,* as shown by the next example.

EXAMPLE 4 Applying the Weighted Mean
Find the grade point average for a student earning the following grades. Assume A = 4, B = 3, C = 2, D = 1, and F = 0.

Course	*Units*	*Grade*	*Grade × Units*
Mathematics	3	A (= 4)	4 × 3 = 12
Retailing	4	C (= 2)	2 × 4 = 8
English	3	B (= 3)	3 × 3 = 9
Data Processing	2	A (= 4)	4 × 2 = 8
Lab for Data Processing	2	D (= 1)	1 × 2 = 2
Totals	**14**		**39**

SOLUTION
The grade point average for this student is

$$\frac{39}{14} = 2.8. \quad ■$$

NOTE It is common to round grade point averages to the nearest tenth.

WORK PROBLEM 4 AT THE SIDE.

3 Since it can be affected by extremely high or low numbers, the mean is often a poor indicator of the central tendency of a list of numbers. In cases like this, another measure of central tendency, called the **median,** can be used. The *median* divides a group of numbers in half; half the numbers lie at or above the median, and half lie at or below the median.

Find the median by listing the numbers in order from smallest to largest. If the list contains an *odd* number of items, the median is the *middle number*.

EXAMPLE 5 Using the Median
Find the median for the following list of numbers.

7, 23, 15, 6, 18, 12, 24

SOLUTION
First arrange the numbers in numerical order.

smallest → 6, 7, 12, 15, 18, 23, 24 ← greatest

Next, find the middle number in the list.

6, 7, 12, 15, 18, 23, 24
↓
middle number

The median is 15. ■

WORK PROBLEM 5 AT THE SIDE.

4. Find the grade point average for a student earning the following grades. Assume A = 4, B = 3, C = 2, D = 1, and F = 0.

Course	*Units*	*Grade*
Mathematics	3	A (= 4)
P.E.	1	C (= 2)
English	3	D (= 1)
Data Entry	2	B (= 3)
Electronics	4	B (= 3)

5. Find the median for the following list of numbers.
25, 23, 17, 21, 29, 40, 49, 15, 20

ANSWERS
4. 2.7 rounded to the nearest tenth
5. 23 (the fifth number when arranged in numerical order)

6. Find the median for the following list of numbers.
147, 159, 132, 174, 181, 253

7. Find the mode for each list of numbers.

(a) 17, 28, 15, 36, 28

(b) 228, 179, 675, 228, 389, 179

(c) 1706, 1289, 1653, 1892, 1301, 1782

ANSWERS

6. $166\frac{1}{2}$

7. (a) 28 (b) bimodal, 179 and 228 (this list has two modes) (c) no mode (no number occurs more than once)

If a list contains an *even* number of items, there is no single middle number. In this case, the median is defined as the mean (average) of the *middle two* numbers.

EXAMPLE 6 Finding the Median
Find the median for the following list of numbers.

7, 13, 15, 25, 28, 32, 47, 59, 68, 74

SOLUTION
The numbers are already in numerical order, so find the middle two numbers.

smallest → 7, 13, 15, 25, 28, 32, 47, 59, 68, 74 ← greatest

(28, 32: middle two numbers)

The median is the mean of these two numbers.

$$\text{median} = \frac{28 + 32}{2} = \frac{60}{2} = 30$$ ■

WORK PROBLEM 6 AT THE SIDE.

4 The last important statistical measure is the **mode,** the number that occurs most often in a list of numbers. For example, if 10 students earned scores of

74, 81, 39, 74, 82, 80, 100, 92, 74, and 85

on a business law examination, then the mode is 74, since more students obtained this score than any other score.

A list can have two modes; such a list is sometimes called **bimodal.** If no number occurs more than once in a list, the list has *no mode*.

EXAMPLE 7 Finding the Mode
Find the mode for each list of numbers.

(a) 51, 32, 49, 73, 49, 90
The number 49 occurs more often than any other number; therefore, 49 is the mode. (It is not necessary to place the numbers in numerical order when looking for the mode.)

(b) 482, 485, 483, 485, 487, 487, 489
Since both 485 and 487 occur twice, each is a mode.

(c) 10,708, 11,519, 10,972, 12,546, 13,905, 12,182
No number occurs more than once. This list has no mode. ■

WORK PROBLEM 7 AT THE SIDE.

name date hour

10.4 EXERCISES

Find the mean for each list of numbers. Round to the nearest tenth.

1. 6, 8, 14, 19, 23

2. 51, 48, 32, 43, 74, 58

3. 40, 51, 59, 62, 68, 73, 49, 80

4. 31, 37, 43, 51, 58, 64, 79, 83

5. 21,900, 22,850, 24,930, 29,710, 28,340, 40,000

6. 38,500, 39,720, 42,183, 21,982, 43,250

Solve the following word problems.

7. A driver had receipts for the following purchases at gasoline stations: $9.40, $11.30, $10.50, $7.40, $9.10, $8.40, $9.70, $5.20, $1.10, $4.70. Find the mean for these purchases.

8. In one evening, a waitress collected the following checks from her dinner customers: $30.10, $42.80, $91.60, $51.20, $88.30, $21.90, $43.70, $51.20. Find the mean of these checks.

Find the weighted mean for the following. Round to the nearest tenth.

9.

Value	*Frequency*
3	4
5	2
9	1
12	3

10.

Value	*Frequency*
9	3
12	5
15	1
18	1

11.

Value	*Frequency*
12	4
13	2
15	5
19	3
22	1
23	5

12.

Value	*Frequency*
25	1
26	2
29	5
30	4
32	3
33	5

Find the median for the following lists of numbers.

13. 12, 18, 32, 51, 58, 92, 106

14. 596, 604, 612, 683, 719

15. 100, 114, 125, 135, 150, 172

16. 298, 346, 412, 501, 515, 521, 528, 621

name date hour

Find the mode or modes for each list of numbers.

17. 4, 9, 8, 6, 9, 2, 1, 3

18. 21, 32, 46, 32, 49, 32, 49

19. 74, 68, 68, 68, 75, 75, 74, 74, 70

20. 12, 15, 17, 18, 21, 29, 32, 74, 80

CHALLENGE EXERCISES

Find the grade point average for students earning the following grades. Assume $A = 4$, $B = 3$, $C = 2$, $D = 1$, *and* $F = 0$. *Round to the nearest tenth.*

21.

Units	*Grade*
4	B
2	A
5	C
1	F
3	B

22.

Units	*Grade*
3	A
3	B
4	B
2	C
4	D

Find the median for the following lists of numbers.

23. 32, 58, 97, 21, 49, 38, 72, 46, 53

24. 1072, 1068, 1093, 1042, 1056, 1005, 1009

Find the mode or modes for each list of numbers.

25. 5, 9, 17, 3, 2, 8, 19, 1, 4, 20

26. 158, 161, 165, 162, 165, 157, 163

REVIEW EXERCISES

Solve the following word problems. Round to the nearest cent. (For help, see ***Section 6.8.****)*

27. Michelle Dyas lends $8500 for 1 year at 12%. How much interest will she earn?

28. A loan of $4050 will be paid back with 12% interest at the end of 9 months. Find the total amount due.

29. An investor deposits $6000 at 8% for 2 years. How much interest will be earned?

30. Kara Usta deposits $790 at 7% for 5 months. Find the interest she will earn.

CHAPTER 10 SUMMARY

KEY TERMS

	Term	Definition
10.1	**circle graph**	A circle graph is a circle broken up into various parts or sectors, based on percents of 360°.
	protractor	A protractor is a device (usually in the shape of a half-circle) used to measure degrees or parts of a circle.
10.2	**bar graph**	A bar graph is a graph that uses bars to show quantity or frequency.
	double-bar graph	A double-bar graph is a graph used to show two sets of data; this graph has two sets of bars.
	line graph	A line graph is a graph that uses dots connected by lines to show a trend.
	comparison line graph	A comparison line graph is one graph that shows how several different items relate to each other.
10.3	**histogram**	A histogram is a bar graph in which the width of each bar represents a range of numbers (class interval) and the height represents the quantity or frequency.
	frequency polygon	A frequency polygon is the result when dots are placed at the top of the bars in a histogram and are connected with lines.
10.4	**mean**	The mean is the sum of all the values divided by the number of values.
	weighted mean	The weighted mean is a mean calculated so that each value is multiplied by its frequency.
	median	The median is the middle number in a group of values. It divides a group of values in half. In an evenly numbered group of values, median is the mean of the two middle values.
	mode	The mode is the most common value in a group of values.

QUICK REVIEW

Section Number and Topics	Approach	Example
10.1 Constructing a Circle Graph	1. Determine the percent of the total for each item. 2. Find the number of degrees of a circle each percent represents. 3. Draw the circle.	Construct a circle graph for the following table, which lists expenses for a business trip.

Item	*Amount*
Car	\$200
Lodging	\$300
Food	\$250
Entertainment	\$150
Other	\$100
Total	\$1000

Item	*Amount*	*Percent of Total*
Car	\$200	$\frac{\$200}{\$1000} = \frac{1}{5} = 20\% = 360 \cdot 20\% = 360 \cdot .20 = 72°$
Lodging	\$300	$\frac{\$300}{\$1000} = \frac{3}{10} = 30\% = 360 \cdot 30\% = 360 \cdot .30 = 108°$
Food	\$250	$\frac{\$250}{\$1000} = \frac{1}{4} = 25\% = 360 \cdot 25\% = 360 \cdot .25 = 90°$
Entertainment	\$150	$\frac{\$150}{\$1000} = \frac{3}{20} = 15\% = 360 \cdot 15\% = 360 \cdot .15 = 54°$
Other	\$100	$\frac{\$100}{\$1000} = \frac{1}{10} = 10\% = 360 \cdot 10\% = 360 \cdot .10 = 36°$

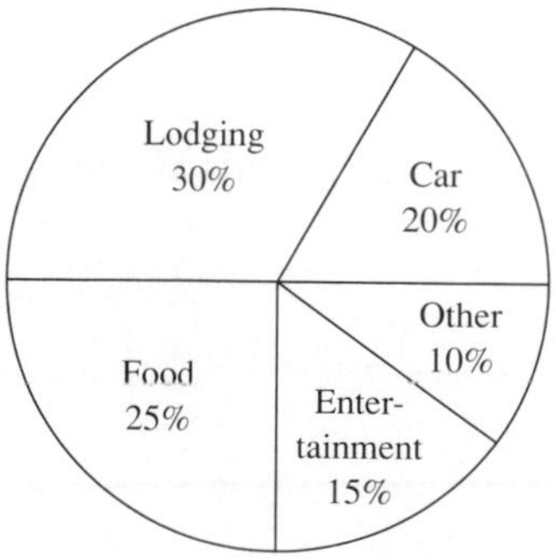

Section Number and Topics	Approach	Example
10.2 Reading a Bar Graph	The height of the bar is used to show the quantity or frequency (number) in a specific category.	Use the bar graph below to determine the number in each category. 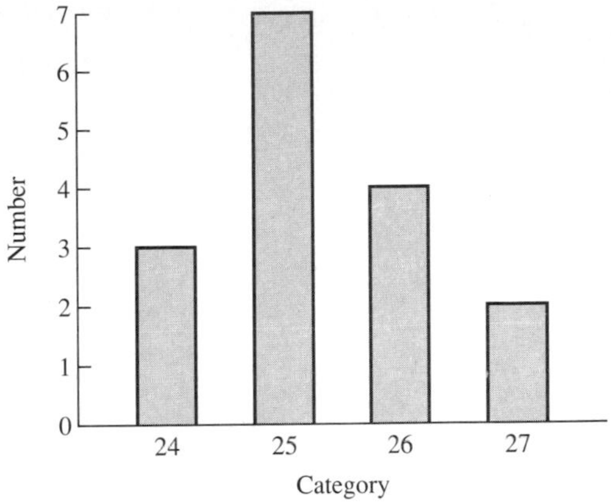

Category	*Number*
24	3
25	7
26	4
27	2

Section Number and Topics	Approach	Example
10.2 Reading a Line Graph	A point is used to show the number or quantity in a specific class. The points are connected with lines. This kind of graph is used to show a trend.	The line graph below shows the sales volume on the left for different years across the bottom. Find the sales in each year.

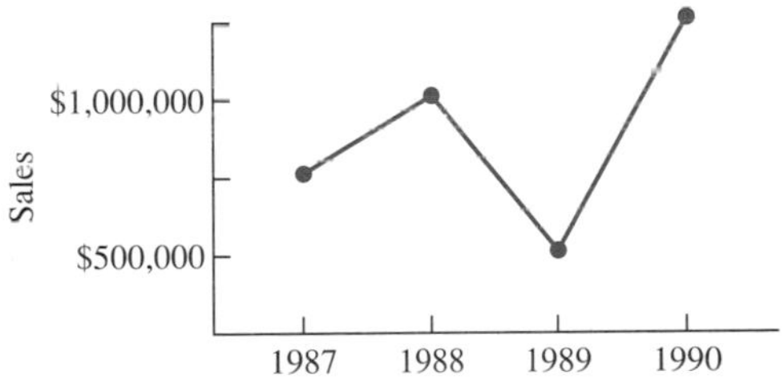

Year	*Total Sales*
1987	$750,000
1988	$1,000,000
1989	$500,000
1990	$1,500,000

Section Number and Topics	Approach	Example
10.3 Constructing a Histogram and a Frequency Polygon from Raw Data	1. Construct a table listing each value, and the number of times this value occurs. 2. Divide the data into groups, categories, or classes. 3. Graphed bars representing these groups produce a histogram. 4. When the midpoints of the bars are connected with lines, the result is a frequency polygon.	For the following data, construct a frequency polygon by using the four classes given below. 12 15 15 14 13 20 10 12 11 9 10 12 17 20 16 17 14 18 19 13

Data	*Tally*	*Frequency*
9	|	1
10	||	2
11	|	1
12	|||	3
13	||	2
14	||	2
15	||	2
16	|	1
17	||	2
18	|	1
19	|	1
20	||	2

Classes	*Frequency*
9–11	4
12–14	7
15–17	5
18–20	4

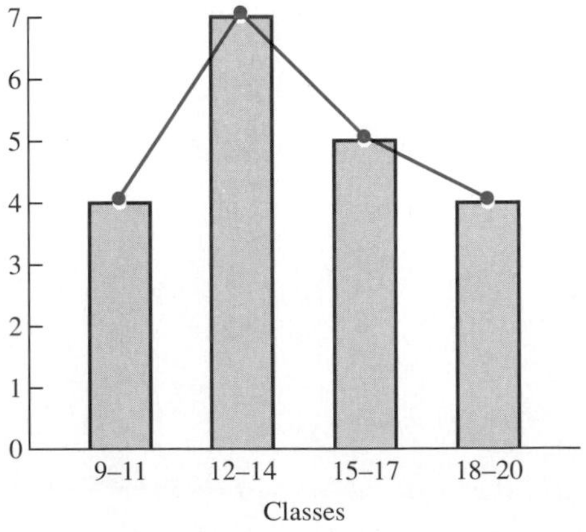

Section Number and Topics	Approach	Example
10.4 Finding the Mean of a Set of Numbers	1. Add all values to obtain a total. 2. Divide the total of values by the number of values.	The quiz scores for Pat Phelan in her business math course were as follows: 85 76 93 91 78 82 87 85 Find Pat's quiz average.

$$\text{Mean} = \frac{85 + 76 + 93 + 91 + 78 + 82 + 87 + 85}{8}$$

$$= \frac{677}{8} = 84.63$$

Section Number and Topics	Approach	Example
10.4 Finding the Weighted Mean	1. Multiply frequency by value. 2. Add all products obtained in Step 1. 3. Divide the sum in Step 2 by the total number of pieces of data.	The following shows the distribution of the number of school-age children in a survey of 30 families.

Number of School-Age Children	*Frequency*
0	12
1	6
2	7
3	3
4	2
	30

Find the mean number of school-age children per family.

Number	*Frequency*	*Number ·Frequency*
0	12	0
1	6	6
2	7	14
3	3	9
4	2	8
		37

$$\text{Average} = \frac{37}{30} = 1.23$$

Section Number and Topics	Approach	Example
10.4 Finding the Median of a Set of Numbers	1. Arrange the data from smallest to greatest. 2. Select the middle value or the average of the two middle values, if there is an even number of values.	Find the median for Pat Phelan's grades from the previous example. The data arranged from smallest to greatest is as follows: 76 78 82 85 85 87 91 93 The middle two values are 85 and 85. The average of these two values is $\frac{85 + 85}{2} = 85$
10.4 Determining the Mode of a Set of Values	Determine the most frequently occurring value.	Find the mode for Pat's grades in the previous example. The most frequently occurring score is 85 (it occurs twice). Therefore, the mode is 85.

name date hour

CHAPTER 10 REVIEW EXERCISES

[10.1]

1. Use this circle graph, which shows the cost of a family vacation, to find the largest single expense of the vacation.

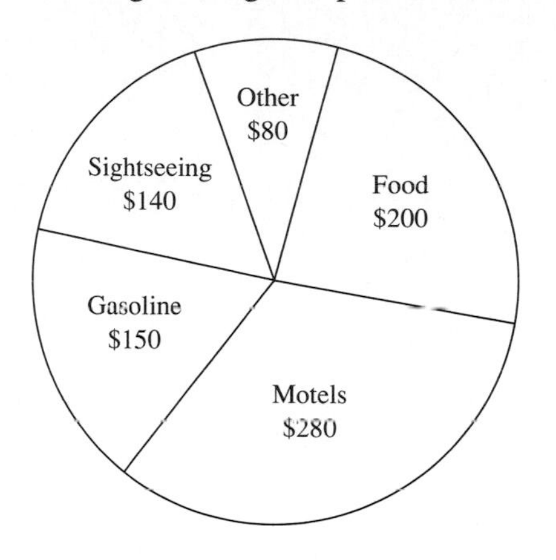

Using the circle graph in Exercise 1, find each of the following ratios.

2. cost of the food to the total cost of the vacation

3. cost of the gasoline to the total cost of the vacation

4. cost of sightseeing to the total cost of the vacation

5. cost of *other* to the cost of the gasoline

6. cost of the food to the cost of the motels

Use the following circle graph, which shows the inventory of equipment and supplies for a park district, to find the dollar value of each item listed. The total value of the inventory is $88,500.

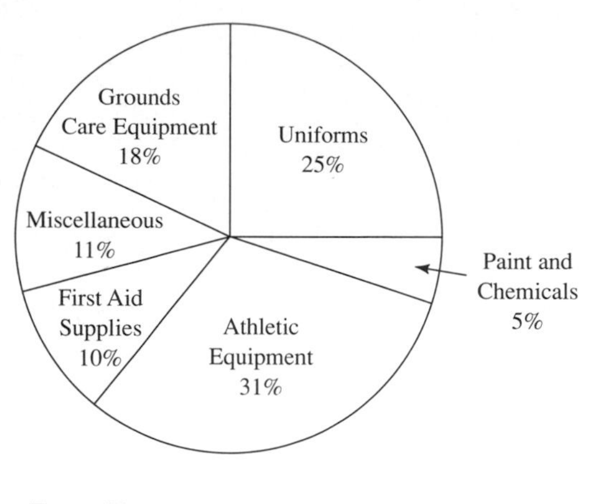

7. uniforms

8. paint and chemicals

9. athletic equipment

10. first aid supplies

11. miscellaneous

12. grounds care equipment

[10.2] *Use the double-bar graph below, which shows the number of acre-feet of water in Lake Natoma for the first six months of 1990 and 1991, to find each of the following.*

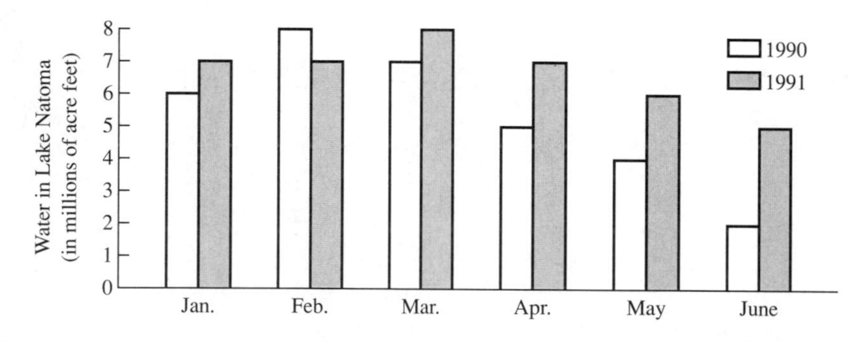

13. During which month in 1991 was the greatest amount of water in the lake?

14. During which month in 1990 was the least amount of water in the lake?

15. How many acre-feet of water were in the lake in March of 1991?

16. How many acre-feet of water were in the lake in May of 1990?

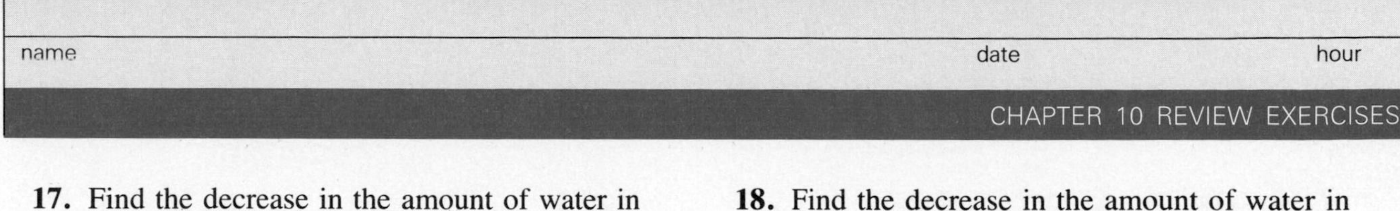

17. Find the decrease in the amount of water in the lake from March, 1990 to June, 1990.

18. Find the decrease in the amount of water in the lake from April, 1991 to June, 1991.

Use the comparison line graph below, which shows the annual book purchases of two different library districts for a five-year period, to find the amount of annual book purchases in each of the following.

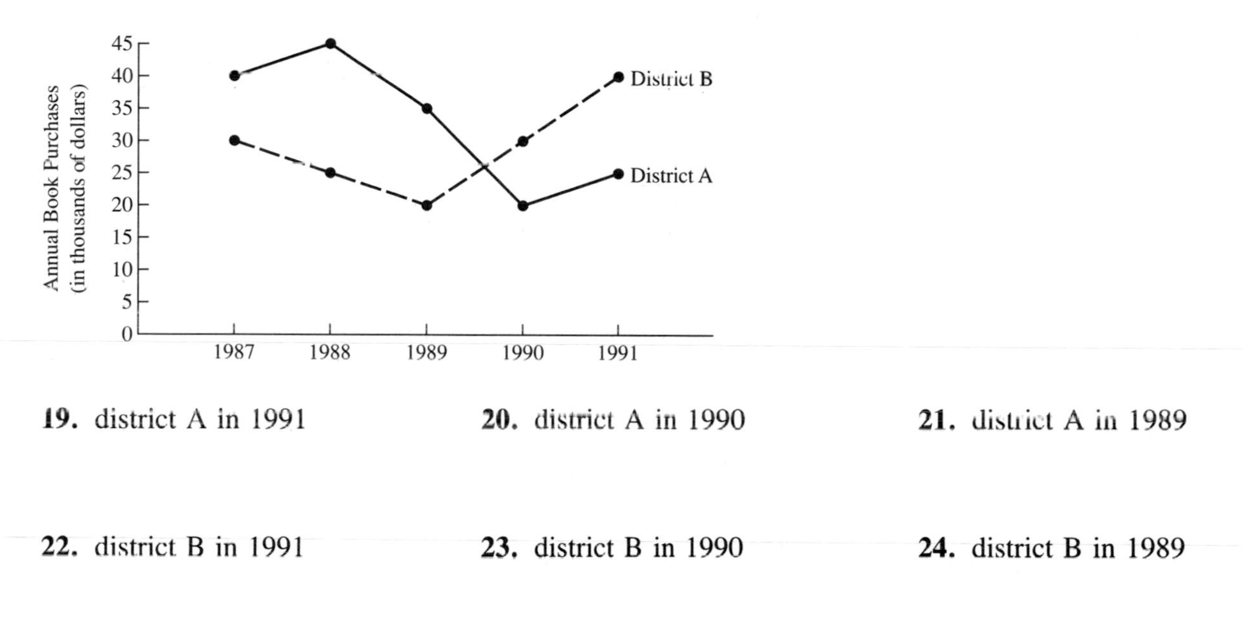

19. district A in 1991

20. district A in 1990

21. district A in 1989

22. district B in 1991

23. district B in 1990

24. district B in 1989

[10.4] *Find the mean for each list of numbers. Round to the nearest tenth.*

25. 6, 4, 5, 8, 3, 14, 18, 29, 7, 1

26. 31, 9, 8, 22, 46, 51, 48, 42, 53, 42

Find the weighted mean for each of the following. Round to the nearest tenth.

27.

Value	*Frequency*
42	3
47	7
53	2
55	3
59	5

28.

Value	*Frequency*
243	1
247	3
251	5
255	7
263	4
271	2
279	2

Find the median for each list of numbers.

29. 32, 42, 29, 27, 34, 51, 48, 74, 29

30. 576, 578, 542, 551, 559, 565, 525, 590

Find the mode or modes for each list of numbers.

31. 80, 72, 64, 64, 72, 53, 64

32. 29, 42, 38, 29, 71, 72, 71, 29

MIXED REVIEW EXERCISES

A home economics department spent $5600 *for the new semester as shown on the following chart. Find all the missing numbers in Exercises 33–37.*

	item	*dollar amount*	*percent of total*	*degrees of circle*
33.	spices	$560	10%	______
34.	equipment	$1960	______	126°
35.	books	$1120	______	72°
36.	supplies	$560	10%	______
37.	small appliances	$1400	______	90°

38. Draw a circle graph by using the information in Exercises 33–37.

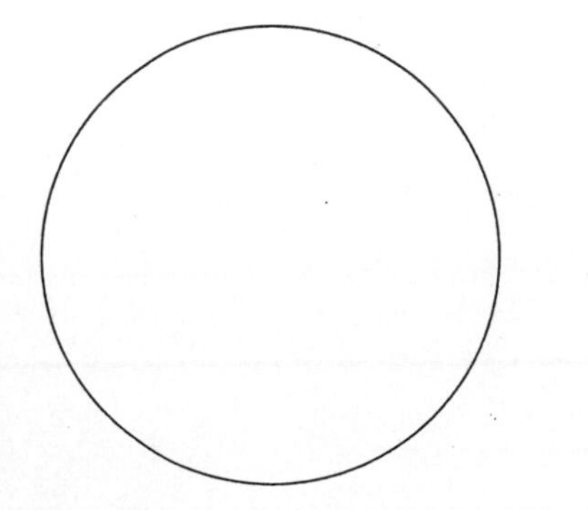

name date hour

Find the mean for each list of numbers. Round to the nearest tenth.

39. 12, 18, 13, 37, 45

40. 122, 135, 146, 159, 128, 147, 168, 139, 158

Find the mode or modes for each list of numbers.

41. 97, 95, 94, 95, 94, 97, 97

42. 58, 62, 62, 74, 83, 91, 62, 83, 83

Find the median for each list of numbers.

43. 4.7, 3.2, 2.9, 5.3, 7.1, 8.2, 9.4, 1.0

44. 7, 15, 28, 3, 14, 18, 46, 59, 1, 2, 9, 21

Use these numbers, which show the scores of 40 students on a data processing exam, to complete the following table.

78	89	36	59	78	99	92	86
73	78	85	57	99	95	82	76
63	93	53	76	92	79	72	62
74	81	77	76	59	84	76	94
58	37	76	54	80	30	45	38

score	***tally***	***frequency***
45. 30–39	________	________
46. 40–49	________	________
47. 50–59	________	________
48. 60–69	________	________
49. 70–79	________	________
50. 80–89	________	________
51. 90–99	________	________

52. Construct a frequency polygon by using the data in Exercises 45–51.

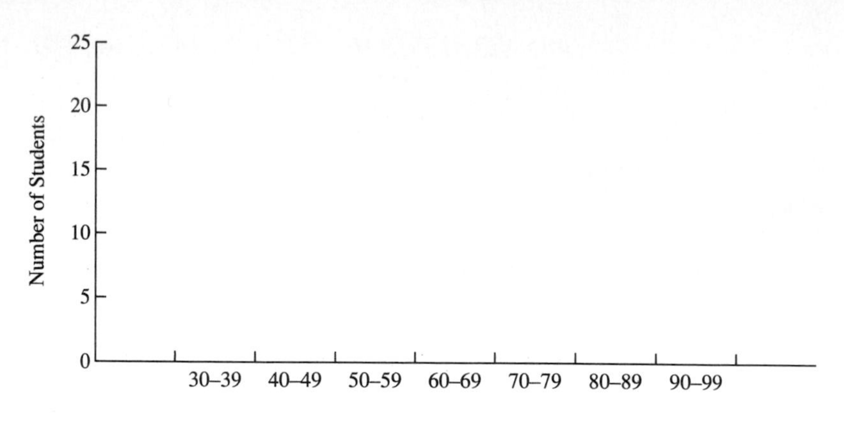

Find the weighted mean for each of the following. Round to the nearest tenth.

53.

Value	*Frequency*
12	1
14	6
17	2
23	7
25	4

54.

Value	*Frequency*
104	6
112	14
115	21
119	13
123	22
127	6
132	9

name date hour

CHAPTER 10 TEST

Use the following circle graph, which shows the operating budget for the Rocklin Water District, to find the dollar amount budgeted for each category. The total budget of the district is $2,800,000.

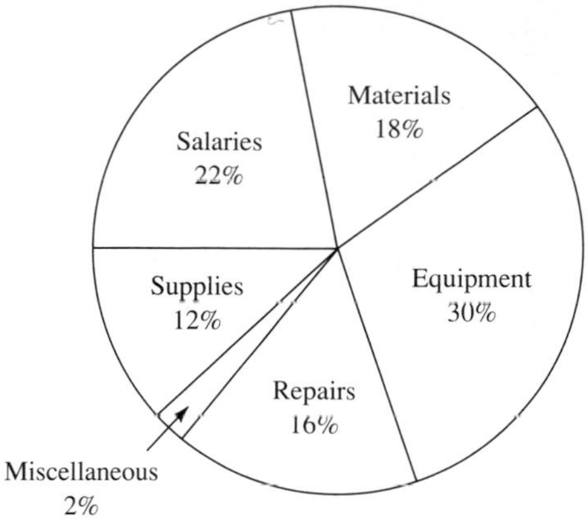

Salaries 22%
Materials 18%
Equipment 30%
Repairs 16%
Supplies 12%
Miscellaneous 2%

1. salaries

2. materials

3. equipment

4. repairs

5. miscellaneous

6. supplies

1. ______
2. ______
3. ______
4. ______
5. ______
6. ______

During a one-year period, the Daily Blat newspaper had the following expenses. Find all numbers missing from the chart.

item	*dollar amount*	*percent of total*	*degrees of a circle*
7. newsprint	$12,000	20%	______
8. ink	$6000	______	36°
9. wire service	$18,000	30%	______
10. salaries	$18,000	30%	______
11. other	$6000	10%	______

7. ______
8. ______
9. ______
10. ______
11. ______

12. Draw a circle graph by using the information in Exercises 7–11.

Use the following 20 numbers, which give the profits for each of the past 20 weeks at Jo's Hamburger Haven, to complete the following table.

$142 $137 $125 $132 $147 $129 $151 $172 $175 $129
$159 $148 $173 $160 $152 $174 $169 $163 $149 $173

	profit	***number of weeks***
13.	$120–129	______
14.	$130–139	______
15.	$140–149	______
16.	$150–$159	______
17.	$160–$169	______
18.	$170–$179	______

13. ______

14. ______

15. ______

16. ______

17. ______

18. ______

name date hour

19. Use the numbers in the table for Exercises 13–18 to draw a histogram.

Frequency in Weeks: 0, 1, 2, 3, 4, 5

120–129, 130–139, 140–149, 150–159, 160–169, 170–179

Profit in Dollars

Find the mean for each of the following. Round to the nearest tenth if necessary.

20. 42, 51, 58, 59, 63, 65, 69, 74, 81, 88

20. ____________________

21. 12, 18, 14, 17, 19, 22, 23, 25

21. ____________________

22. 458, 432, 496, 491, 500, 508, 512, 396, 492, 504

22. ____________________

Find the weighted mean for each of the following. Round to the nearest tenth.

23.

Value	*Frequency*
6	7
10	3
11	4
14	2
19	3
24	1

23. ____________________

24. ______________

24.

Value	*Frequency*
150	15
160	17
170	21
180	28
190	19
200	7

Find the median for each list of numbers.

25. ______________

25. 15, 18, 19, 27, 29, 31, 42

26. ______________

26. 41, 39, 45, 47, 38, 42, 51, 38

27. ______________

27. 7.6, 9.3, 21.8, 10.4, 4.2, 5.3, 7.1, 9.0, 8.3

Find the mode or modes for each list of numbers.

28. ______________

28. 51, 47, 48, 32, 47, 71, 82, 47

29. ______________

29. 32, 51, 74, 19, 25, 43, 75, 82, 98, 100

30. ______________

30. 96, 104, 103, 104, 103, 104, 91, 74, 103

11

CONSUMER MATHEMATICS

Consumers must make many decisions about savings, taxes, and consumer credit—and about major purchases like automobiles, houses, and insurance. A basic understanding of topics in this chapter will help with such decisions.

11.1 COMPOUND INTEREST

The interest studied so far in this book is *simple interest* (interest only on the principal; see **Section 6.8**). A more common type of interest is **compound interest,** or interest paid on past interest as well as on principal.

Suppose Barbara Greenwald makes a single deposit of $1000 in a savings account that earns 6% per year. What will happen to her savings over three years? At the end of the first year, one year's interest on the original deposit is found.

OBJECTIVES

1. Use a compound interest table.
2. Find the compound amount and the amount of compound interest.
3. Find interest compounded more often than annually.

Interest = principal · rate · time

Year 1 $1000 · .06 · 1 = $60

Add the interest to the $1000. $1000 + **$60** = $1060

The interest for the second year is found on $1060; that is, the interest is **compounded.**

Year 2 $1060 · .06 · 1 = $63.60

Add this interest to the $1060. $1060 + **$63.60** = $1123.60

The interest for the third year is found on $1123.60.

Year 3 $1123.60 · .06 · 1 = $67.42

Add this interest to the $1123.60. $1123.60 + **$67.42** = $1191.02

At the end of three years, Barbara Greenwald will have $1191.02 in her savings account.

1. Find the compound amount.

(a) \$1 at 4% for 12 years

(b) \$1 at 8% for 24 years

(c) \$1 at 6% for 9 years

The interest earned during the second year is greater than that earned during the first year, and the interest earned during the third year is greater than that earned during the second year.

This happens because the interest earned each year is added to the principal, and that total is used to find the amount of interest in the next year.

Interest paid on principal plus interest is compound interest.

1 Compound interest tables are used to calculate compound interest. The table on page 685 gives the **compound amount** (principal and interest) on a \$1 deposit for given lengths of time and interest rates.

EXAMPLE 1 Using a Compound Interest Table

Find the compound amount.

(a) \$1 is deposited at a 5% interest rate for 10 years.

Look down the column headed 5%, and across to row 10 (10 years = 10 time periods). At the intersection of the column and row, read the compound amount, 1.6289, which can be rounded to \$1.63.

(b) \$1 is deposited at 6% for 21 years.

The intersection of the 6% column and row 21 shows 3.3996 as the compound amount. Round this to \$3.40. ■

WORK PROBLEM 1 AT THE SIDE.

2 Find the compound amount and interest as follows.

Find the compound amount for any amount of principal by multiplying the principal by the compound amount for \$1.

Find the amount of interest earned on a deposit by subtracting the amount originally deposited from the compound amount.

EXAMPLE 2 Finding Compound Interest

Find the compound amount and the interest.

(a) \$1000 at $5\frac{1}{2}\%$ interest for 12 years

Look in the table for $5\frac{1}{2}\%$ and 12 periods, finding the number 1.9012. Multiply this number and the principal of \$1000.

$$\$1000 \cdot 1.9012 = \$1901.20$$

The account will contain \$1901.20 after 12 years.

ANSWERS

1. (a) \$1.60 (b) \$6.34 (c) \$1.69

Compound Interest

time periods	2.00%	2.50%	3.00%	3.50%	4.00%	4.50%	5.00%	5.50%	6.00%	8.00%	10.00%	12.00%
1	1.0200	1.0250	1.0300	1.0350	1.0400	1.0450	1.0500	1.0550	1.0600	1.0800	1.1000	1.1200
2	1.0404	1.0506	1.0609	1.0712	1.0816	1.0920	1.1025	1.1130	1.1236	1.1664	1.2100	1.2544
3	1.0612	1.0769	1.0927	1.1087	1.1249	1.1412	1.1576	1.1742	1.1910	1.2597	1.3310	1.4049
4	1.0824	1.1038	1.1255	1.1475	1.1699	1.1925	1.2155	1.2388	1.2625	1.3605	1.4641	1.5735
5	1.1041	1.1314	1.1593	1.1877	1.2167	1.2462	1.2763	1.3070	1.3382	1.4693	1.6105	1.7623
6	1.1262	1.1597	1.1941	1.2293	1.2653	1.3023	1.3401	1.3788	1.4185	1.5869	1.7716	1.9738
7	1.1487	1.1887	1.2299	1.2723	1.3159	1.3609	1.4071	1.4547	1.5036	1.7138	1.9487	2.2107
8	1.1717	1.2184	1.2668	1.3168	1.3686	1.4221	1.4775	1.5347	1.5938	1.8509	2.1436	2.4760
9	1.1951	1.2489	1.3048	1.3629	1.4233	1.4861	1.5513	1.6191	1.6895	1.9990	2.3579	2.7731
10	1.2190	1.2801	1.3439	1.4106	1.4802	1.5530	1.6289	1.7081	1.7908	2.1589	2.5937	3.1058
11	1.2434	1.3121	1.3842	1.4600	1.5395	1.6229	1.7103	1.8021	1.8983	2.3316	2.8531	3.4785
12	1.2682	1.3449	1.4258	1.5111	1.6010	1.6959	1.7959	1.9012	2.0122	2.5182	3.1384	3.8960
13	1.2936	1.3785	1.4685	1.5640	1.6651	1.7722	1.8856	2.0058	2.1329	2.7196	3.4523	4.3635
14	1.3195	1.4130	1.5126	1.6187	1.7317	1.8519	1.9799	2.1161	2.2609	2.9372	3.7975	4.8871
15	1.3459	1.4483	1.5580	1.6753	1.8009	1.9353	2.0789	2.2325	2.3966	3.1722	4.1772	5.4736
16	1.3728	1.4845	1.6047	1.7340	1.8730	2.0224	2.1829	2.3553	2.5404	3.4259	4.5950	6.1304
17	1.4002	1.5216	1.6528	1.7947	1.9479	2.1134	2.2920	2.4848	2.6928	3.7000	5.0545	6.8660
18	1.4282	1.5597	1.7024	1.8575	2.0258	2.2085	2.4066	2.6215	2.8543	3.9960	5.5599	7.6900
19	1.4568	1.5987	1.7535	1.9225	2.1068	2.3079	2.5270	2.7656	3.0256	4.3157	6.1159	8.6128
20	1.4859	1.6386	1.8061	1.9898	2.1911	2.4117	2.6533	2.9178	3.2071	4.6610	6.7275	9.6463
21	1.5157	1.6796	1.8603	2.0594	2.2788	2.5202	2.7860	3.0782	3.3996	5.0338	7.4002	10.8038
22	1.5460	1.7216	1.9161	2.1315	2.3699	2.6337	2.9253	3.2475	3.6035	5.4365	8.1403	12.1003
23	1.5769	1.7646	1.9736	2.2061	2.4647	2.7522	3.0715	3.4262	3.8197	5.8715	8.9543	13.5523
24	1.6084	1.8087	2.0328	2.2833	2.5633	2.8760	3.2251	3.6146	4.0489	6.3412	9.8497	15.1786
25	1.6406	1.8539	2.0938	2.3632	2.6658	3.0054	3.3864	3.8134	4.2919	6.8485	10.8347	17.0001

Find the amount of interest by subtracting the original deposit from the compound amount.

compound amount → $1901.20 − original amount → $1000 = amount of interest → $901.20

$$\$1901.20 - \$1000 = \$901.20$$

(b) $6400 at 8% for 7 years.

Look in the table for 8% and 7 periods, finding 1.7138. Multiply.

$$\$6400 \cdot 1.7138 = \$10{,}968.32$$

Subtract the deposit from the compound amount.

$$\$10{,}968.32 - \$6400 = \$4568.32$$

A total of $4568.32 in interest was earned. ■

WORK PROBLEM 2 AT THE SIDE.

3 In the previous examples, interest was calculated at the end of each year, or **compounded annually.** It is common for banks and other financial institutions to compound interest more often, such as every six months, or every quarter, or even every day.

2. Find the compound amount and the interest.

(a) $5000 at 6% for 14 years

(b) $11,300 at 8% for 14 years

(c) $22,700 at 12% for 8 years

ANSWERS
2. (a) $11,304.50; $6304.50
(b) $33,190.36; $21,890.36
(c) $56,205.20; $33,505.20

EXAMPLE 3 Compounding Semiannually

Find the compound amount and the amount of interest earned on a deposit of $850 at 6% compounded semiannually for 8 years.

APPROACH

"Semiannually" means twice a year. At 6% per year, the interest earned each six months is 6% ÷ 2 = 3%. In 8 years, there are 8 · 2 = 16 semiannual periods. Thus, we look in the table for 3% and 16 periods, finding 1.6047.

SOLUTION

$$\$850 \cdot 1.6047 = \$1364 \qquad \text{(rounded)}$$

The amount of interest earned is

$$\$1364 - \$850 = \$514. \qquad \blacksquare$$

WORK PROBLEM 3 AT THE SIDE.

3. Find the compound amount and the amount of interest.

(a) $4000 at 8% compounded semiannually for 7 years

(b) $29,500 at 10% compounded semiannually for 9 years

EXAMPLE 4 Compounding Quarterly

Find the compound amount and the amount of interest earned on a deposit of $1500 at 12% compounded quarterly for 5 years.

APPROACH

Since there are four quarters in one year, the interest earned per quarter is 12% ÷ 4 = 3% per quarter. In five years, there are 5 · 4 = 20 quarterly periods.

SOLUTION

Look for 3% and 20 periods in the table, finding 1.8061. The compound amount is

$$\$1500 \cdot 1.8061 = \$2709.15$$

and the amount of interest earned is

$$\$2709.15 - \$1500 = \$1209.15. \qquad \blacksquare$$

WORK PROBLEM 4 AT THE SIDE.

4. Find the compound amount and the amount of interest.

(a) $1600 at 8% compounded quarterly for 6 years

(b) $25,000 at 10% for 5 years compounded quarterly

(c) $42,750 at 12% compounded quarterly for 6 years

ANSWERS

3. **(a)** $6926.80; $2926.80
(b) $70,994.70; $41,494.70

4. **(a)** $2573.44; $973.44
(b) $40,965; $15,965
(c) $86,902.20; $44,152.20

name date hour

11.1 EXERCISES

Use the table on page 685 to find the compound amount for the following. Interest is compounded annually.

Example: \$500 at 4% for 8 years

Solution: 4% column, row 8 of the table gives 1.3686.

Multiply. $\$500 \cdot 1.3686 =$ **\$684.30**

1. \$1000 at 6% for 4 years

2. \$1000 at 8% for 12 years

3. \$4000 at 5% for 9 years

4. \$7500 at 6% for 7 years

5. \$8428.17 at $4\frac{1}{2}\%$ for 6 years

6. \$10,472.88 at $5\frac{1}{2}\%$ for 20 years

Find the compound amount for the following.

Example: \$600 at 6% compounded semiannually for 7 years

Solution: Look up 3% ($6\% \div 2$) and $2 \cdot 7 = 14$ time periods. Find 1.5126.

Multiply. $\$600 \cdot 1.5126 =$ **\$907.56**

7. \$1000 at 8% compounded semiannually for 9 years

8. \$1000 at 10% compounded semiannually for 5 years

9. \$800 at 12% compounded semiannually for 7 years

10. \$9000 at 12% compounded semiannually for 12 years

11. \$2800 at 8% compounded quarterly for 5 years

12. \$10,000 at 10% compounded quarterly for 6 years

13. \$25,800 at 12% compounded quarterly for 4 years

14. \$35,670 at 10% compounded quarterly for 6 years

Find the missing numbers.

Example: \$1250 at 8% compounded quarterly for 3 years

Solution: 8% ÷ 4 is 2% and 3 · 4 = 12 quarters. Look up 2%, 12 periods to find 1.2682.

Multiply. \$1250 · 1.2682 = **\$1585.25** compound amount
\$1585.25 − \$1250 = **\$335.25** compound interest

	principal	*rate*	*compounded*	*time in years*	*compound amount*	*compound interest*
15.	\$1000	8%	quarterly	4	______	______
16.	\$1000	10%	semiannually	12	______	______
17.	\$975	12%	semiannually	8	______	______
18.	\$1150	12%	quarterly	5	______	______
19.	\$1480	8%	quarterly	4	______	______
20.	\$2370	$5\frac{1}{2}$%	annually	16	______	______
21.	\$7500	12%	quarterly	6	______	______
22.	\$10,900	10%	semiannually	12	______	______

name date hour

Find the compound amount when interest is compounded **(a)** *annually,* **(b)** *semiannually, and* **(c)** *quarterly.*

Example: $2000 8% 2 years		
Solution:	**(a)** annually	8%, 2 periods from the table—1.1664
		$2000 · 1.1664 = **$2332.80**
	(b) semiannually	4%, 4 periods from the table—1.1699
		$2000 · 1.1699 = **$2339.80**
	(c) quarterly	2%, 8 periods from the table—1.1717
		$2000 · 1.1717 = **$2343.40**

	principal	*rate*	*time*	**(a)** *annually*	**(b)** *semiannually*	**(c)** *quarterly*
23.	$1000	12%	3 years	______	______	______
24.	$4000	12%	6 years	______	______	______
25.	$8000	10%	5 years	______	______	______
26.	$15,000	10%	6 years	______	______	______

27. Which period of compounding gives the highest return?

28. Which period of compounding gives the lowest return?

Solve each word problem. Round answers to the nearest cent.

29. Al Grenard lends $7500 to the owner of a new restaurant. He will be repaid at the end of 6 years at 8% interest compounded annually. Find the total amount that he should be repaid.

30. Glenda Wong deposits $8270 in a bank that pays 10% interest compounded semiannually. Find the total amount she will have at the end of 5 years.

CHALLENGE EXERCISES

Solve each word problem.

31. There are two banks in Citrus Heights. One pays 8% interest compounded annually, and the other pays 8% compounded quarterly.

(a) If Stan deposits $10,000 in each bank, how much will he have in each bank at the end of 6 years?

(b) How much more will he have in the bank that pays more interest?

32. Which yields more interest for Barker Aluminum: $5000 deposited for 7 years at 10% simple interest or $4000 deposited for 7 years at 10% interest compounded semiannually? What is the difference in interest?

33. Jennifer Del Campo deposits $10,000 at 8% compounded quarterly. Two years after she makes the first deposit, she adds another $20,000, also at 8% compounded quarterly. What total amount will she have five years after her first deposit?

34. Scott Striver invests $9000 at 10% compounded quarterly. Three years after he makes the first deposit, he adds another $15,000, also at 10% compounded quarterly. What total amount will he have five years after his first deposit?

REVIEW EXERCISES

Simplify by using the order of operations. (For help, see ***Section 4.6****.)*

35. $6.3 \div 4.2 \cdot 3.1$

36. $18.304 \div 8.32 \cdot 3$

37. $19.3 + (6.7 - 5.2) \cdot 58$

38. $2.12 + (9.7 - 7.9) \cdot 4.5$

39. $4.34 - 2.6 \cdot 5.2 \div 2.6$

40. $51.5 - 22.8 \cdot 15 \div 5.7$

11.2 CONSUMER CREDIT—TRUE ANNUAL INTEREST

OBJECTIVES

1. Find the true annual interest rate of a loan.
2. Find typical interest rates.

Since different lenders compute interest by using different methods, it is important for the consumer to learn these methods to compare the cost of borrowing money.

1 To help consumers compare interest rates, all lenders are required to reveal the **true annual interest rate** of borrowing from them. This rate is found by using tables similar to the one here.

True Annual Interest Rate Table

number of payments	14%	$14\frac{1}{2}$%	15%	$15\frac{1}{2}$%	16%	$16\frac{1}{2}$%	17%
	(finance charge per $100 of amount financed)						
6	$ 4.12	4.27	4.42	4.57	4.72	4.87	5.02
12	7.74	8.03	8.31	8.59	8.88	9.16	9.45
18	11.45	11.87	12.29	12.72	13.14	13.57	13.99
24	15.23	15.80	16.37	16.94	17.51	18.09	18.66
30	19.10	19.81	20.54	21.26	21.99	22.72	23.45
36	23.04	23.92	24.80	25.68	26.57	27.46	28.35
42	27.06	28.10	29.15	30.19	31.25	32.31	33.37
48	31.17	32.37	33.59	34.81	36.03	37.27	38.50

EXAMPLE 1 Finding the True Annual Interest Rate

Suppose Herb Moyer used his charge account to buy new carpeting and drapes costing $2800. The store's credit department tells him that his monthly payment will be $80.79 per month for 48 months.

The total he will repay is

$$\underset{\text{monthly payment}}{\$80.79} \cdot \underset{\text{months}}{48} = \underset{\text{total repaid}}{\$3877.92.}$$

Find the interest charge by subtracting the cost of the carpet and drapes (the amount financed or borrowed) from this total.

$$\underset{\text{total repaid}}{\$3877.92} - \underset{\text{cost}}{\$2800} = \underset{\text{interest charge}}{\$1077.92}$$

Find the true annual interest rate by dividing the interest charge by the amount financed and then multiplying by $100.

$$\frac{\text{interest charge}}{\text{amount financed}} \cdot \$100 = \frac{\$1077.92}{\$2800} \cdot \$100$$

$$= .38497 \cdot 100$$

$$= \$38.50 \quad \text{(rounded)}$$

1. What is the true annual interest rate when $1100 is financed with payments of $54 each for 24 months?

This amount, $38.50, represents the interest charge per $100 of the amount financed. Since Moyer is paying off the loan in 48 months, look across the "48 payments row" in the table. Find the number closest to $38.50. In this case, $38.50 is in the table. Read up the column to find the true annual interest rate. It is 17%. ■

EXAMPLE 2 Using the True Annual Interest Rate Table
Pam Seaver has decided to buy a 120-mile-per-gallon moped. After a down payment, she owes a balance of $700, which she will repay in 18 monthly payments of $44 each. Find the true annual interest rate of the loan.

APPROACH
The interest charge per $100 of the amount financed must be found first. This amount is then found in the table by using the "18 payments row" of the table.

SOLUTION
To find the interest charge, she must find her total payments. She will pay $44 for 18 months, or a total of

$$\$44 \cdot 18 = \$792. \quad \text{total paid}$$

She financed $700, so the interest charge is

$$\$792 - \$700 = \$92. \quad \text{interest charge}$$

Next, divide the interest charge by the amount financed and multiply by $100.

$$\frac{\$92}{\$700} \cdot \$100 = \$13.14 \quad \text{(rounded)}$$

Find the "18 payments row" of the table. Read across that row, finding the number closest to $13.14. Next, read up and find the true annual interest rate, 16%. ■

WORK PROBLEM 1 AT THE SIDE.

2 The Truth-in-Lending Act requires that the borrower be told the true annual interest rate on all loans. This act, however, does not limit the interest rates that may be charged on different types of loans. The individual states set the maximum allowable interest rates, and these vary from state to state. Some typical interest rates are shown here.

ANSWER
1. $16\frac{1}{2}\%$

Typical Interest Rates

type of loan	*interest rate per month*	*true annual interest*
retail credit	$1\frac{1}{2}\%$ on first $1000; 1% on amount over $1000	18%
personal credit union	1.1%	13.2%
new car credit union	1%	12%
life insurance policy		5%
finance company	$2\frac{1}{2}\%$ on first $200	30%
	2% on $201 to $500	24%
	$1\frac{1}{2}\%$ on $501 to $700	18%
	1% over $700	12%
car dealer's	1% add-on	$21\frac{1}{2}\%$
new home		12%

EXAMPLE 3 Finding the Annual Interest Rate

Using these typical interest rates, find the true annual interest rate for each of the following.

(a) car dealer's loan $21\frac{1}{2}\%$ Look in the previous table.

(b) retail credit loan on $825 18% ■

WORK PROBLEM 2 AT THE SIDE.

2. Using the typical interest rates, find the true annual interest rate for each of the following.

(a) new home loan

(b) life insurance policy loan

ANSWERS
2. (a) 12% **(b)** 5%

name date hour

11.2 EXERCISES

Use the table on page 691 to find the true annual interest rate for each of the following.

Example: \$2500 financed \$121.88 monthly payment 24 payments

Solution: \$121.88 (monthly payment) · 24 payments = \$2925.12 total paid
\$2925.12 − \$2500 = \$425.12 interest charge

$\frac{\$425.12}{\$2500} \cdot \$100 = \$17$ (rounded)

24 months, \$17 from the table gives **$15\frac{1}{2}\%$** true annual interest rate

	amount financed	*monthly payment*	*number of payments*	*true annual interest rate*
1.	\$1200	\$108	12	________
2.	\$4800	\$298.67	18	________
3.	\$5000	\$243.75	24	________
4.	\$4900	\$195.83	30	________
5.	\$1500	\$ 73.11	24	________

	amount financed	*monthly payment*	*number of payments*	*true annual interest rate*
6.	$4000	$140.72	36	__________
7.	$3950	$162.61	30	__________
8.	$6000	$170	48	__________

Solve each word problem.

9. Ed Harper bought a small television set for $222. He made no down payment and paid for the set in 12 monthly installments of $20 each. Find the interest charge that he paid. What was his true annual interest rate?

10. Wei-Jen Luan bought a used car for $5750. She paid $3000 down and made payments of $96 per month for 36 months. Find the total interest charge she paid. Find the true annual interest rate she paid.

11. On a twelve-month loan of $1000 payable in equal monthly installments, the Mouse House advertises only $92 interest. What is the true annual interest rate on this loan?

12. Jack Armstrong borrows $2500 to construct a hot tub and pays $808.75 in interest over 42 months. Find his true annual rate of interest.

name date hour

Using the typical rates given in the text, find the true annual rate for each loan.

13. $200 from a finance company

14. new car loan from a credit union

15. television set costing $600 bought on credit at an appliance store

16. television set bought with money borrowed as a personal loan from a credit union

17. loan against life insurance

18. new home loan

CHALLENGE EXERCISES

Solve each word problem.

19. House Television and Appliance wants to advertise a table model color television for $400, with 25% down and monthly payments of only $27.35 for 12 months. They must also include the true annual interest rate in the ad. Find the true annual interest rate.

20. A department store has a sofa on sale for $750. A buyer paying 20% down may finance the balance by paying $29.50 per month for 2 years. Find the true annual interest rate that the store must include in its advertising.

REVIEW EXERCISES

Multiply by using a shortcut with multiples of ten. (For help, see ***Section 1.4.****)*

21. 36 · 10

22. 14 · 10

23. 728 · 100

24. 631 · 100

25. $36.80 · 1000

26. $23.50 · 1000

27. $27.90 · 1000

28. $158.60 · 1000

11.3 BUYING A HOME

There are three things most people need to know about the cost of a home or condominium—the sale price, the down payment, and the monthly payment. The sale price depends on many factors. Location, size, age, and condition are just some of the things that affect the price of a home.

The down payment is also affected by many factors. These include the type and length of the loan, the credit worthiness of the borrower, the age of the home, and general conditions in the money market. Common down payments range from 5% to 25% of the sale price. With a lower down payment, the interest rate and the monthly payment are often higher.

1 The monthly payment is determined by the amount of the loan, the length of the loan, and the interest rate. The amount that you will pay each month toward principal and interest can be found from a table. A portion of such a table is shown here.

Monthly Payment to Repay a $1000 Loan

number of years	10%	$10\frac{1}{2}$%	11%	$11\frac{1}{2}$%	12%	$12\frac{1}{2}$%	13%
20	$9.66	9.98	10.32	10.67	11.01	11.36	11.72
25	$9.09	9.44	9.80	10.17	10.53	10.90	11.28
30	$8.78	9.15	9.52	9.91	10.29	10.67	11.06

EXAMPLE 1 Using a Table to Find the Monthly Payment

Find the monthly payment necessary to pay off the following home loans.

(a) $88,000 at 10% for 25 years

Find the 10% column in the table and read down to the row for 25 years, finding the number 9.09. This is the amount necessary to pay off a loan of $1000. For a loan of $88,000, or 88 · 1000, the monthly payment is

$$88 \cdot 9.09 = \$799.92.$$

(b) $64,280 at 11% for 30 years

The amount owed, $64,280, is 64.28 · 1000. Look in the table at the 11% column, 30-year row, to find 9.52. The monthly payment is

$$64.28 \cdot 9.52 = \$611.95. \quad \text{(rounded)}$$ ■

WORK PROBLEM 1 AT THE SIDE.

With many home loans, the borrower pays one-twelfth of the annual taxes and insurance along with the monthly payment of principal and interest. The lender then pays the bills for taxes and insurance as they become due.

OBJECTIVES

1. Find house payments.
2. Find the maximum possible house payment.

1. Find the monthly payment necessary to pay off the following home loans.

(a) $72,000 at $11\frac{1}{2}$% for 25 years

(b) $49,300 at $10\frac{1}{2}$% for 30 years

(c) $86,500 at 13% for 25 years

ANSWERS
1. (a) $732.24 (b) $451.10 (c) $975.72

2. Find the total monthly payment, including taxes and insurance.

amount of loan	\$72,000
interest rate	12%
term of loan	25 years
annual taxes	\$904
annual insurance	\$380

EXAMPLE 2 Finding the Total Monthly House Payment

Find the monthly payment, including taxes and insurance, for a home costing \$108,000. Assume a 10%, 30-year loan with a 20% down payment. The annual taxes are \$780, and fire insurance is \$453.

APPROACH

First, find the monthly payment for principal and interest on the loan. Next, divide the annual taxes (\$780) and the annual fire insurance premium (\$453) by 12 to find the monthly amounts. Finally, add the principal and interest loan payment, the monthly taxes, and the monthly insurance amounts to find the total monthly payment.

SOLUTION

The down payment is 20% of the cost, or

$$20\% \cdot \$108{,}000 = 20 \cdot \$108{,}000 = \$21{,}600.$$

The amount of the loan is

$$\underset{\text{cost}}{\$108{,}000} - \underset{\text{down payment}}{\$21{,}600} = \underset{\text{loan}}{\$86{,}400}.$$

Since the loan is \$86,400 (or 86.4 · 1000), the monthly payment is found by finding 10% and 30 years in the table, or 8.78, and multiplying.

$$86.4 \cdot 8.78 = \$758.59 \quad \text{(rounded) monthly payment}$$

Next, divide annual taxes and annual insurance by 12.

$$\$780 \text{ taxes} \div 12 = \$65 \quad \text{monthly taxes}$$
$$\$453 \text{ insurance} \div 12 = \$37.75 \quad \text{monthly insurance}$$

Adding principal, interest, taxes, and insurance gives

$$\$758.59 + \$65 + \$37.75 = \$861.34 \quad \text{total monthly payment}$$ ■

WORK PROBLEM 2 AT THE SIDE.

2 How much can a person afford to spend on a home? Lenders in some areas use the following rule to help decide.

Step 1 Find the total monthly income before tax deductions.

Step 2 Subtract all monthly bills that will not be paid off within six months.

Step 3 Divide the result by 4. (A monthly home payment should not exceed this amount.)

ANSWER
2. \$865.16

EXAMPLE 3 Determining the Maximum Monthly Payment
The Williams family has a total monthly income of $2700. They have a car payment of $230 per month and a school loan payment of $135 per month. Neither bill will be paid off in 6 months. Find the largest monthly home payment they should make.

APPROACH
The total of all monthly bills must be subtracted from the total monthly income. The result is then divided by 4 to find the maximum monthly home payment.

SOLUTION
Add the monthly payments.

$$\$230 + \$135 = \$365$$

Subtract $365 from $2700, the total income.

$$\$2700 - \$365 = \$2335$$

Divide the result by 4.

$$\$2335 \div 4 = \$583.75$$

Thus, $583.75 is the largest monthly home payment the Williams family should make. ■

WORK PROBLEM 3 AT THE SIDE.

3. A couple has earnings of $2800 per month and monthly bills of $142 and $278. (The bills will not be paid off in six months.) Find the maximum monthly payment they should make.

ANSWER
3. $595

name date hour

11.3 EXERCISES

Use the table on page 699 to find the monthly payment for principal and interest on the following loans.

	loan amount	*interest rate*	*term of loan in years*	*monthly payment*
1.	$80,000	10%	30	________
2.	$55,000	$11\frac{1}{2}\%$	20	________
3.	$30,000	12%	25	________
4.	$143,000	13%	20	________
5.	$96,500	$10\frac{1}{2}\%$	30	________

Find the total monthly payment, including taxes and insurance, on the following loans.

Example:		**Solution:**
loan	$53,000	
interest rate	12%	From the table, 12%, 30 years is 10.29.
term of loan	30 years	$53 \cdot 10.29 = \$545.37$ principal and interest
annual taxes	$575	$\frac{\$575}{12} = \$47.92$ monthly taxes
annual insurance	$280	$\frac{\$280}{12} = \$23.33$ monthly insurance
total monthly payment	**$616.62**	$\$545.37 + \$47.92 + \$23.33 = \$616.62$ total monthly payment

	loan	*interest rate*	*term of loan*	*annual taxes*	*annual insurance*	*total monthly payment*
6.	$98,700	13%	30 years	$ 509	$176	________
7.	$134,500	12%	25 years	$ 660	$312	________

	loan	*interest rate*	*term of loan*	*annual taxes*	*annual insurance*	*total monthly payment*
8.	$59,200	$12\frac{1}{2}\%$	25 years	$ 775	$287	________
9.	$32,200	11%	30 years	$ 172	$165	________
10.	$89,890	$10\frac{1}{2}\%$	25 years	$1080.19	$423.74	________

Find the maximum monthly home payment for each income.

Example:	**Solution:**	
$2000 gross monthly income	$2000 − $248 = $1752	gross earnings after payments
$248 total monthly payments extending beyond 6 months	$\frac{1752}{4}$ = $438	maximum monthly payment

	gross monthly income	*total current monthly payments extending beyond 6 months*	*maximum monthly house payment*
11.	$1720	$120	________
12.	$2080	$160	________
13.	$2500	$300	________

name date hour

	gross monthly income	*total current monthly payments extending beyond 6 months*	*maximum monthly house payment*
14.	$1800	$125	________
15.	$3200	$400	________
16.	$2115	$180	________

CHALLENGE EXERCISES

Suppose $90,000 *is owed on a house after a down payment of* $10,000 *is made. The monthly payment for principal and interest at* 11% *for* 30 *years is* 90 · 9.52 = $856.80.

17. How many monthly payments will be made?

18. What is the total amount that will be paid for principal and interest?

19. The total interest charged is the total amount paid minus the amount financed. What is the total interest?

20. Which is more—the house price or the total interest paid?

REVIEW EXERCISES

Multiply each of the following. (For help, see ***Section 4.5****.)*

21. $\begin{array}{r} \$7.92 \\ \times \quad 12 \\ \hline \end{array}$

22. $\begin{array}{r} \$6.72 \\ \times \quad 10 \\ \hline \end{array}$

23. $\begin{array}{r} \$11.63 \\ \times \quad 15 \\ \hline \end{array}$

24. $\begin{array}{r} \$9.62 \\ \times \quad 20 \\ \hline \end{array}$

25. $15.75 · 16.3

26. $23.18 · 25

27. $42.10 · 30.5

28. $17.20 · 51.6

11.4 LIFE INSURANCE

OBJECTIVES

1. Understand the different types of life insurance.
2. Find premiums for life insurance.

A **life insurance policy** is an agreement with a life insurance company. You agree to pay a certain amount of money, called the **premium,** and the company agrees to pay your survivors a lump sum of money, called the **face value** of the policy, when you die. In addition, some kinds of policies pay retirement benefits.

There are several kinds of life insurance. The following summary gives a brief description of the most common kinds.

Whole Life Most insurance salespeople will try to sell you whole life insurance, also called **straight life** or **ordinary life.** One reason is that the commission to the salesperson is substantial—about 55% of the total premium paid during the first year.

A whole life policy combines life insurance protection with a savings plan. You pay a constant premium until death or retirement, whichever is first. On retirement, the insured receives monthly payments from the insurance company.

Whole life insurance builds up **cash value,** the money later used to pay the retirement income. These cash vales can also be borrowed against, at favorable interest rates.

Term Insurance Term insurance provides protection for a fixed period of time, such as one year, five years, or ten years, at a much lower premium than whole life. The premium is constant for the stated period but increases, if you renew the policy for an additional period. Term insurance provides no retirement benefits or cash value, but it does provide the greatest amount of life insurance coverage for the premium dollar.

Endowment Policies Endowment policies guarantee the payment of a fixed amount of money to a given individual at a specified time, whether or not the insured lives. Policies of this type are often taken out by parents to guarantee their children's education.

2 Insurance premiums are determined by life expectancy. Typical annual premium rates are shown in this table.

Annual Premium Rates per $1000 of Life Insurance*

age	*10-year term*	*whole life*	*20-year endowment*
20	5.11	12.05	33.69
21	5.22	12.33	33.80
22	5.35	12.59	33.86
23	5.45	12.80	33.89
24	5.62	13.05	33.99
25	5.77	13.36	34.11
30	6.59	14.97	34.45
35	7.54	17.44	35.22
40	9.13	20.34	36.42
45	10.82	23.95	38.44
50		29.27	41.04
55		35.45	44.95
60		44.95	52.38

*The ages given are for males. For females, subtract 3 years. For example, rates for a 28-year-old female would be the 25-year-old rate from the table.

With the preceding table, use this formula.

premium = number of thousands · rate per $1000

EXAMPLE 1 Finding the Cost of Life Insurance

Silvio DiLoreto is 45 years old and wants to buy a life insurance policy with a face value of $30,000. Find his annual premium for **(a)** a 10-year term policy, **(b)** a whole life policy, and **(c)** a 20-year endowment.

APPROACH

The number of thousands of dollars of life insurance coverage is multiplied by the premium cost per $1000 of life insurance coverage. This rate is based on the specific type of policy purchased.

SOLUTION

Since the table gives the rate per $1000 of insurance, first find

$$\$30{,}000 \div \$1000 = 30 \text{ thousands.}$$

(a) For a 10-year term policy, the rate per $1000 for a 45-year-old male is $10.82. Use the formula for premiums previously given.

$$\text{annual premium} = 30 \cdot \$10.82 = \$324.60$$

(b) For a whole life policy, the rate is $23.95.

$$\text{annual premium} = 30 \cdot \$23.95 = \$718.50$$

(c) For a 20-year endowment, the rate is $38.44.

$$\text{annual premium} = 30 \cdot \$38.44 = \$1153.20$$

WORK PROBLEM 1 AT THE SIDE.

NOTE Remember to subtract 3 years from the age of females before using the rate table.

EXAMPLE 2 Finding the Life Insurance Premium

Rosemary Doty is 38 years old and wants to buy a $25,000 life insurance policy. Find her annual premium for **(a)** a 10-year term policy, **(b)** a whole life policy, and **(c)** a 20-year endowment.

APPROACH

Find the number of thousands of dollars of life insurance coverage and multiply by the proper rate from the table. Be certain to subtract 3 years from the age of a female.

SOLUTION

There are 25 thousands.

(a) The rate for females is found by subtracting 3 years from the actual age. For a 10-year term policy, the $1000 rate for a 35-year old (38 − 3) is $7.54.

$$\text{annual premium} = 25 \cdot \$7.54 = \$188.50$$

(b) For a whole life policy the rate is $17.44.

$$\text{annual premium} = 25 \cdot \$17.44 = \$436$$

(c) For a 20-year endowment the rate is $35.22.

$$\text{annual premium} = 25 \cdot \$35.22 = \$880.50$$

WORK PROBLEM 2 AT THE SIDE.

1. Milton Kimura is 25 years old and wants to buy a $50,000 life insurance policy. Find his annual premium for

(a) a 10-year term policy;

(b) a whole life policy;

(c) a 20-year endowment policy.

2. Barbara Allen is 27 years old and wants to buy a $40,000 life insurance policy. Find her annual premium for

(a) a 10-year term policy;

(b) a whole life policy;

(c) a 20-year endowment policy.

ANSWERS
1. (a) $288.50 (b) $668 (c) $1705.50
2. (a) $224.80 (b) $522 (c) $1359.60

name	date	hour

11.4 EXERCISES

Use the table on page 707 to find the annual premium for each of the following.

Example:

face value of policy	*age of insured*	*sex*	*type of policy*	*annual premium*
$35,000	22	M	endowment	$1185.10

Solution:

annual premium = 35 · $33.86 = $1185.10

	face value of policy	*age of insured*	*sex*	*type of policy*	*annual premium*
1.	$40,000	26	F	whole life	________
2.	$50,000	38	F	endowment	________
3.	$40,000	25	M	10-year term	________
4.	$70,000	40	M	whole life	________
5.	$27,500	53	F	endowment	________
6.	$75,000	21	M	10-year term	________
7.	$26,500	24	F	whole life	________
8.	$83,900	27	F	10-year term	________
9.	$65,750	50	M	endowment	________
10.	$100,000	45	M	whole life	________

Solve each word problem.

11. Harry Lane buys an endowment policy at age 30. The policy has a face value of $40,000. Find his annual premium.

12. Joanna Greere buys a $32,000 whole life policy at age 28. Find her annual premium.

13. Valley Trucking feels that it would suffer greatly if the firm's head dispatcher died. Therefore, the company takes out a $50,000 policy on the dispatcher's life. The dispatcher is a 35-year-old male, and the company buys a 10-year term policy. Find the annual premium.

14. Find the annual premium for an endowment policy with a face value of $20,000. The insured is a 24-year-old female.

15. Louis Martinez takes out a whole life policy with a face value of $32,000. He is 45 years old. Find the annual premium.

16. Susan Broom feels that a 10-year term policy of $125,000 would be needed by her family. She is 38 years old. Find the annual premium of this policy.

CHALLENGE EXERCISES

Solve each word problem.

17. Find the total premium paid over 20 years for a 20-year endowment policy with a face value of $10,000. Assume the policy is taken out by a 40-year-old male.

18. Mary Luan is 28 years old and takes out a 10-year term life policy for $50,000. Find the total premium paid over the 10-year period.

CHAPTER 11 SUMMARY

KEY TERMS

	Term	Definition
11.1	**compound interest**	Compound interest is interest paid on past interest as well as on principal.
	compounding	Interest that is compounded once each year is compounded annually, interest that is compounded twice each year is compounded semiannually, and interest that is compounded four times each year is compounded quarterly.
11.2	**true annual interest rate**	The true annual interest rate is the actual rate of interest. It is used to compare the rates of various loans.
11.3	**monthly payment table**	A monthly payment table is a table used to find the monthly payment on a loan.
11.4	**whole life policy**	A whole life policy combines life insurance and a savings plan. It is also called straight life or ordinary life.
	term insurance	Term insurance is a policy of life insurance that provides protection for a certain period of time.
	endowment policy	An endowment policy guarantees the payment of a fixed amount of money at a specified time while also providing life insurance during this time period.

QUICK REVIEW

Section Number and Topic	Approach	Example
11.1 Finding Compound Amount and Compound Interest	Find the number of compounding periods and the interest rate per period. Use the table to find the interest on \$1. Multiply table value by the principal to obtain compound amount. Subtract principal from compound amount to obtain interest.	What is the compound amount and interest, if \$1500 is deposited at 12% interest compounded quarterly for 6 years. interest of 12% per year $= \frac{12\%}{4}$ $= 3\%$ per period Interest compounded quarterly means there are $6 \cdot 4 = 24$ total periods. Locate 3% across the top of the table and 24 periods at left. Table value is 2.0328. compound amount $= \$1500 \cdot 2.0328$ $= \$3049.20$ interest $= \$3049.20 - \$1500$ $= \$1549.20$

Section Number and Topic	Approach	Example
11.2 Finding the True Annual Interest Rate by Using a Table	First determine the interest charge per \$100 of amount financed by using the formula $\frac{\text{interest charge}}{\text{amount financed}} \cdot \$100.$ Next, read down the left column of the table to find the proper number of payments. Go across the columns to the number closest to the number found above. Read up the column to find the true annual interest rate.	Use the table to find the true annual interest rate, if \$5650 is financed for 36 months and the interest charged is \$1500. interest charge = \$1500 amount financed = \$5650 interest charge per \$100 $= \frac{\$1500}{\$5650} \cdot \$100$ $= \$26.55$ number of payments = 36 table value = \$26.55 true annual interest rate = 16%
11.3 Finding the Amount of Monthly Home Loan Payments Including Taxes and Insurance	The monthly loan payment is determined by finding the interest rate and length of loan in the table. This amount is multiplied by the number of thousands of dollars in the loan amount. The total monthly payment is found by adding this amount to $\frac{1}{12}$ of the annual taxes and insurance costs.	What is the total monthly payment on an \$85,000, 10%, 25-year loan, if annual taxes are \$780 and insurance is \$336? The value from the table for a 10%, 25-year loan is \$9.09. There are $\frac{85{,}000}{1000} = 85$ thousands in \$85,000. monthly payment = $85 \cdot 9.09$ = \$772.65 monthly taxes = \$780 ÷ 12 = \$65 monthly insurance = \$336 ÷ 12 = \$28 Therefore, total payment = \$772.65 + \$65 + \$28 = \$865.65
11.4 Finding the Annual Life Insurance Premium	There are several types of life policies. Use the table and multiply by the number of \$1000s of coverage. Subtract 3 years from the age of females. premium = $\left(\begin{matrix}\text{number of}\\ \text{thousands}\end{matrix}\right) \cdot \left(\begin{matrix}\text{rate per}\\ \$1000\end{matrix}\right)$	Find the following premiums on a \$50,000 policy for a 30-year-old male. **(a)** 10-year term 50 · \$6.59 = \$329.50 **(b)** ordinary life 50 · \$14.97 = \$748.50 **(c)** 20-year endowment 50 · \$34.45 = \$1722.50

name date hour

CHAPTER 11 REVIEW EXERCISES

Use the appropriate tables from this chapter to solve the review exercises.

[11.1] *Find the amount on deposit for the following, if interest is compounded annually.*

1. \$500 at 6% for 2 years

2. \$1500 at 8% for 10 years

3. \$2150 at $5\frac{1}{2}$% for 20 years

4. \$6380.50 at 6% for 15 years

Find the compound amount and the compound interest for each of the following.

	principal	*rate*	*compounded*	*time*	*final amount*	*compound interest*
5.	\$400	8%	semiannually	2 years	________	________
6.	\$850	12%	quarterly	3 years	________	________

Solve the following word problem. Round answer to the nearest cent.

7. Helen Rice deposits her savings of \$15,000 for 6 years compounded quarterly at 8%. Find the total amount in her account at the end of 6 years.

[11.2] *Find the true annual interest rate for each of the following.*

	amount financed	*monthly payment*	*number of payments*	*true annual interest rate*
8.	\$600	\$54	12	________
9.	\$3000	\$85	48	________

Solve the following word problem.

10. Bethany Hart bought a pet raccoon and necessary equipment for $550. She made a down payment of $75 and made payments of $43.10 per month for 12 months. Find

(a) the total interest charge she paid:

(b) the true annual interest rate she paid.

[11.3] *Find the monthly payment for principal and interest on each home loan.*

	loan amount	*interest rate*	*term of loan in years*	*monthly payment*
11.	$135,000	13%	20	__________
12.	$82,750	$11\frac{1}{2}\%$	25	__________

Find the total monthly payment, including taxes and insurance, on each home loan.

	loan	*interest rate*	*term of loan*	*annual taxes*	*annual insurance*	*total monthly payment*
13.	$90,000	10%	20	$108	$204	__________
14.	$47,500	12%	20	$252	$180	__________

name date hour

Suppose \$75,000 is owed on a condominium after the down payment is made. This amount is financed at $11\frac{1}{2}$% for 25 years.

15. What is the monthly payment?

16. How many monthly payments will be made?

17. What is the total amount that will be paid for the principal and interest?

18. The total interest charged is the total amount paid minus the amount financed. What is the total interest?

19. What is more—the amount financed or the total interest paid?

[11.4] *Find the annual premium for each life insurance policy.*

	face value of policy	*age of insured*	*sex*	*type of policy*	*annual premium*
20.	\$30,000	25	F	whole life	__________
21.	\$40,000	53	F	endowment	__________

Solve each word problem.

22. Orville Duncan buys a 10-year term policy at age 40. The policy has a face value of $50,000. Find his annual premium.

23. Find the annual premium for a whole life policy with a face value of $25,000. The insured is a 28-year-old female.

24. Manuel Enriquez takes out an endowment policy with a face value of $40,000. He is 35 years old. Find his annual premium.

CHALLENGE EXERCISES

Solve each word problem.

25. Linda Wacaser is 38 years old and buys a $40,000 whole life policy. Find the annual premium.

26. If $8225 is invested at 8% compounded quarterly for 6 years, find **(a)** the compound amount; **(b)** the compound interest.

name date hour

27. Fair Oaks Building and Thrift pays 8% compounded quarterly. Carmichael Federal Loan pays 8% compounded semiannually.

(a) If a customer deposits $5000 in each institution, how much will she have in each at the end of 5 years?

(b) How much more will she have in the institution that pays more interest?

28. If $7900 is financed for 30 months with payments of $325.22, find the true annual interest rate.

Find the total monthly payment, including taxes and insurance, on this home loan.

	loan	*interest rate*	*term of loan*	*annual taxes*	*annual insurance*	*total monthly payments*
29.	$84,500	$10\frac{1}{2}\%$	30	$1020	$324	________

Find the monthly payment for principal and interest on this home loan.

	loan amount	*interest rate*	*term of loan in years*	*monthly payment*
30.	$95,500	10%	30	________

Solve each word problem.

31. Chris Prentiss bought a used jeep for $2800 with no down payment. He paid for the jeep in 18 monthly installments of $177.32. Find each of the following.

(a) the interest charge he paid

(b) the true annual interest rate he paid

32. John Eason buys a 10-year term policy at age 45. The policy has a face value of $70,000. Find his annual premium.

name date hour

CHAPTER 11 TEST

Find the compound amount for each of the following. Use the table on page 685.

1. $1000 deposited at 6% compounded annually for 11 years

2. $8500 deposited at 8% compounded semiannually for 7 years

3. $15,000 deposited at 12% compounded quarterly for 4 years

Use the table on page 691 to find the true annual interest rate for each loan.

4. amount financed $600, 18 monthly payments of $38

5. amount financed $3000, 48 monthly payments of $83.25

6. Newark Hardware wants to include financing terms in their advertising. If the price of a floor waxer is $150 paid off in 6 monthly payments of $26.25 each, find the true annual interest rate.

Find the monthly payment for the following home loans. Use the table on page 699.

7. $80,000 borrowed at 11% for 50 years

8. $71,500 borrowed at $10\frac{1}{2}$% for 25 years

1. ______
2. ______
3. ______
4. ______
5. ______
6. ______
7. ______
8. ______

9. ______________________

9. Mr. and Mrs. Hardy plan to buy a new house. After a down payment, their balance owed is $122,000, which will be paid off at 13% over 30 years. The taxes are $960 per year, and fire insurance is $252 per year. Find the total monthly payment, including taxes and insurance.

Find the maximum allowable monthly house payment for the following people.

10. ______________________

10. a couple with monthly gross income of $2000 and total current monthly payments of $80

11. ______________________

11. a family with monthly gross income of $2800 and total current monthly payments of $150

Find the annual premium for each life insurance policy. Use the table on page 707.

12. ______________________

12. Chuck Brasher, whole life, $40,000 face value, age 25

13. ______________________

13. Judy Lewis, 10-year term, $55,000 face value, age 48

14. ______________________

14. Bud Payne, 20-year endowment, $30,000 face value, age 35

name date hour

FINAL EXAMINATION

The numbers in brackets tell the chapter in which each type of problem is explained.

[1]

1. Add.
$$\begin{array}{r} 574 \\ 891 \\ 3725 \\ +\ 7806 \\ \hline \end{array}$$

2. Subtract.
$$\begin{array}{r} 3059 \\ -2874 \\ \hline \end{array}$$

3. Round 28,746 to the nearest **(a)** ten; **(b)** hundred.

4. Divide. $74\overline{)9859}$

5. Simplify. $10\sqrt{64} - 32 \div 4$

[2]

6. Write $\frac{125}{225}$ in lowest terms.

7. Multiply. $\frac{5}{9} \cdot \frac{12}{7}$

8. Find the area of a rectangle $\frac{2}{3}$ inch by $\frac{6}{5}$ inch.

Divide.

9. $\dfrac{\frac{5}{12}}{\frac{10}{3}}$

10. $\dfrac{3\frac{2}{5}}{1\frac{1}{10}}$

[3]

11. Find the least common denominator for 3, 6, 8, and 15.

Add or subtract. Write all answers in lowest terms.

12. $\frac{3}{5} + \frac{1}{10}$

13. $\frac{5}{8} - \frac{1}{3}$

14.
$$\begin{array}{r} 5\frac{3}{4} \\ +\ 7\frac{2}{3} \\ \hline \end{array}$$

15.
$$\begin{array}{r} 8\frac{3}{8} \\ -2\frac{9}{10} \\ \hline \end{array}$$

1. ____________
2. ____________
3. (a) ____________
(b) ____________
4. ____________
5. ____________
6. ____________
7. ____________
8. ____________
9. ____________
10. ____________
11. ____________
12. ____________
13. ____________
14. ____________
15. ____________

16. (a) ______
(b) ______
17. ______
18. ______
19. ______
20. ______
21. ______
22. ______
23. ______
24. ______
25. ______
26. ______
27. ______

[4]

16. Round 42.0845 to the nearest **(a)** thousandth; **(b)** hundredth.

17. Add. 5.4 + 19.769 + 385.214

18. Subtract. 48.2 − 36.0941

19. Divide. Round the answer to the nearest thousandth.

$4.21\overline{)25.87}$

20. Write in order, from smallest to largest.

$.58, \frac{3}{5}, \frac{9}{16}, .579, .5803$

[5] *Write each ration in lowest terms.*

21. 18 men to 12 women

22. 5 days to 3 weeks

23. Is the proportion $\frac{55}{15} = \frac{88}{24}$ true?

24. Find the missing number in the following proportion.

$$\frac{3\frac{1}{2}}{9} = \frac{14}{k}$$

25. A car can go 308 miles on 11 gallons of gas. How far could it go on 27 gallons of gas?

[6]

26. Write $\frac{5}{8}$ as a percent.

27. Find 28% of 950.

name date hour

28. 48 is 15% of what number? **28.** ________

29. 520 is what percent of 650? **29.** ________

30. A loan of $2500 is due in 8 months with interest of 15%. Find the total amount that will be due. **30.** ________

[7]

31. Convert 5 pints to quarts. **31.** ________

32. Simplify. 5 weeks 15 days 28 hours **32.** ________

33. Add. **33.** ________

  7 yards 2 feet 10 inches
+ 5 yards 2 feet 7 inches

34. Convert 85°C to Fahrenheit. **34.** ________

[8] *Find the area of each figure. Use 3.14 as an approximation for π. Round answers to the nearest tenth.*

35. a rectangle 6.2 m by 9.4 m **35.** ________

36. a trapezoid with bases 5.4 cm and 9.8 cm and height 4.1 cm **36.** ________

37. a triangle with base 4.6 ft and height 5 ft **37.** ________

38. a circle with a radius of 9.7 cm **38.** ________

Solve the following.

39. Find the volume of a cylinder with radius 5.2 cm and height 2.9 cm. **39.** ________

40. Find the length of the missing side in this right triangle. **40.** ________

90°
8
17

41. ____________________

42. ____________________

43. ____________________

44. ____________________

45. ____________________

46. ____________________

47. ____________________

48. ____________________

49. ____________________

50. ____________________

[9]

41. Add. $-2 + (-8)$

42. Subtract. $-3 - (-11)$

43. Find the value of $8k - 9z$ if $k = -\frac{1}{2}$ and $z = -3$.

Solve each equation.

44. $3p + 1 = 19$

45. $8r - 9 = 12r - 1$

[10] *Find the mean, the median, and the mode for each of the following. Round to the nearest tenth if necessary.*

46. 18, 37, 26, 19, 22, 64, 13, 26, 28

47. 7.8, 24.2, 12.5, 5.9, 4.2, 8.0, 6.3, 10.4, 12.5

[11] *Find the compound amount for each of the following. Use the table on page 685.*

48. $10,500 deposited at 8% compounded semiannually for 5 years

49. $15,000 deposited at 12% compounded quarterly for 3 years

50. Ralph and Alice Cramden plan to buy a new house. After a down payment, their balance owed is $128,000, which will be paid off at 12% over 30 years. The taxes are $1152 per year and fire insurance is $384 per year. Find the total monthly payment, including taxes and insurance. Use the table on page 699.

APPENDICES

APPENDIX A: CALCULATORS

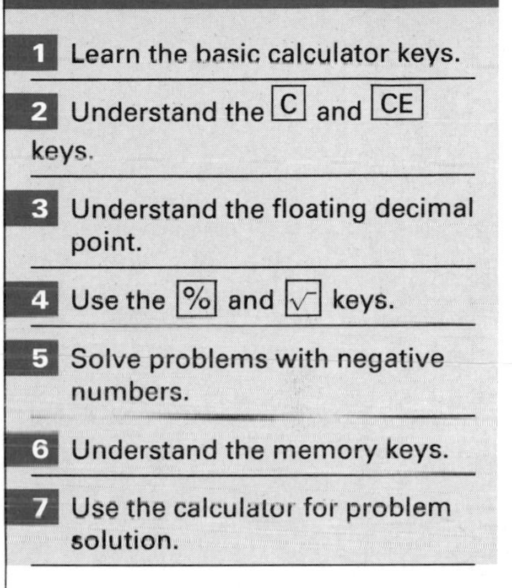

OBJECTIVES

1 Learn the basic calculator keys.

2 Understand the [C] and [CE] keys.

3 Understand the floating decimal point.

4 Use the [%] and [√] keys.

5 Solve problems with negative numbers.

6 Understand the memory keys.

7 Use the calculator for problem solution.

Calculators are among the more popular inventions of the last two decades. Each year better calculators are developed and costs drop. A machine that cost $200 a few years ago could add, subtract, multiply, and divide decimals (but could not locate the decimal point in a division problem). Today, these same calculations are performed quite well on a calculator costing less than this textbook. And today's $200 pocket calculators have more ability to solve problems than some of the early computers.

Many colleges allow students to use calculators in basic mathematics courses. There are many types of calculators available, from the inexpensive basic calculator to the more complex financial and programmable models. The discussion here is confined to the common 8-digit four-function (add, subtract, multiply, and divide), percent key, square root key, and memory function models. Any explanation needed for specific calculator models or special function keys is best gained by referring to the booklet supplied with the calculator.

1 Most calculators use algebraic logic. These calculators can be recognized by the

[+] and [−]

keys. On these calculators, the problem 9 + 8 would be entered as

9 [+] 8 [−]

and 17 would appear as the answer. Enter 17 − 8 as

17 [−] 8 [=]

and 9 appears as the answer. If your calculator does not work problems in this way, check its instruction book to see how to proceed.

2 All calculators have a

[C]

key. Pressing this key erases everything in the calculator and prepares the calculator to begin a new problem. Some calculators also have a

[CE]

key. Pressing this key erases *only* the number displayed and allows the person using the calculator to correct a mistake without having to start the problem over.

Many calculators combine the [C] key and [CE] key and use an [ON/C] key. This key turns the calculator on and is also used to erase the calculator display. If the [ON/C] is pressed after the [=] or one of the operation keys ([+], [−], [×], [÷]), everything in the calculator is erased. If the wrong operation key is pressed, pressing the correct key fixes the error. For example, 7 [+] [−] 3 [=] 4. Pressing the [−] key cancels out the previous [+] key entry.

3 Most calculators have a floating decimal that locates the decimal point in the final result. For example, to buy 55.75 square yards of Armstrong Solarian at $18.99 per square yard, proceed as follows.

55.75 [×] 18.99 [=] 1058.6925

The decimal point is automatically placed in the answer. You should **round** money answers to the nearest cent.

digit to the right (less than 5)
↓
1058.6925
↑
cent position (hundredths)

Since the digit to the right of the position being rounded is smaller than 5 (1, 2, 3, or 4), the position being rounded remains the same and everything to the right is dropped. If the digit to the right had been a 5 or greater (5, 6, 7, 8, 9), then 1 would have been added to the cent position. The answer is rounded to $1058.69.

When using a machine with a floating decimal, enter the decimal point as needed. For example, enter $47 as

47

with no decimal point, but enter 95¢ as

[·] 95

One problem using a floating decimal is shown by the following example (adding $21.38 and $1.22).

21.38 [+] 1.22 [=] 22.6

The final 0 is left off. Remember that the problem dealt with dollars and cents, and write the answer as $22.60.

4 The percent key moves the decimal point two places to the left when used following multiplication or division. The problem, 8% of $4205 is solved as follows.

4205 [×] 8 [%]   336.4 = $336.40

The square root key calculates the square root of the number that appears in the display. For example, the $\sqrt{144}$ is found by entering 144 and the square root key.

144 [√] 12

The square root of 144 is 12.

The square root of 20 is

20 [√] 4.4721359

which may be rounded to the desired position.

5 Negative numbers may be entered by using the [−] before entering the number. For example, solve $\frac{-10 + 22}{3}$ as follows.

[−] 10 [+] 22 [÷] 3 [=] 4

6 Many calculators feature memory keys, which are a sort of electronic scratch paper. These memory keys are used to store intermediate steps in a calculation. On some calculators, a key labeled [M] is used to store the numbers in the display, with [MR] used to recall the numbers from memory.

Other calculators have [M+] and [M−] keys. The [M+] key adds the number displayed to the number already in memory. For example, if the memory contains the number 0 at the beginning of a problem, and the calculator display contains the number 29.4, then pushing [M+] will cause 29.4 to be stored in the memory (the result of adding 0 and 29.4). If 57.8 is then entered into the display, pushing [M+] will cause

$$29.4 + 57.8 = 87.2$$

to be stored. If 11.9 is then entered into the display, with [M−] pushed, the memory will contain

$$87.2 - 11.9 = 75.3.$$

The [MR] key is used to recall the number in memory as needed, with [MC] used to clear the memory.

NOTE Always clear the memory before starting a problem—forgetting to do so is a very common error.

Memory keys are very useful when working long calculations, called *chain calculations*. For example, to find

$$\frac{8 \times 19.4}{15.7 + 11.8 \times 4.6}$$

calculate the denominator first and follow the order of operations.

Start by calculating 11.8×4.6.

$$11.8 \boxed{\times} 4.6 \boxed{=} 54.28$$

Next, add 15.7.

$$\boxed{+} 15.7 \boxed{=} 69.98$$

Store this result in memory, using $\boxed{\text{M}}$ or $\boxed{\text{M+}}$, depending on the machine. Then find the numerator.

$$8 \boxed{\times} 19.4 \boxed{=} 155.2$$

To divide, push $\boxed{\div}$ and then $\boxed{\text{MR}}$ $\boxed{=}$, giving the final quotient

$$\frac{8 \times 19.4}{15.7 + 11.8 \times 4.6} = 2.218. \qquad \text{(rounded)}$$

7 A calculator is especially helpful when multiplying and dividing large numbers. Example 4 in Section 4.6 is $27.69 \div .3$. To solve this problem by using a calculator, proceed as follows.

$$27.69 \boxed{\div} .3 \boxed{=} 92.3$$

This is much faster than using long division.

The calculator can be used to solve Example 5 in Section 6.7:

A furniture store has a sofa with an original price of $470 on sale at 15% off. Find the sale price of the sofa.

$$470 \boxed{\times} 15 \boxed{\%} \boxed{=} 70.50$$

Store this result in memory by using $\boxed{\text{M}}$ or $\boxed{\text{M+}}$, depending on the machine.
Next, find the actual price.

$$470 \boxed{-} \boxed{\text{MR}} \boxed{=} 399.50 = \$399.50$$

Example 3 in Section 11.1 involves compound interest:

Find the compounded amount and the compound interest earned on a deposit of $850 at 6% compounded semiannually for 8 years. Multiply $850 by the figure from the table.

$$850 \boxed{\times} 1.6047 \boxed{=} 1363.995 = \$1364 \qquad \text{compound amount (rounded)}$$

Finally, subtract the deposit from the compound amount.

$$1364 \boxed{-} 850 \boxed{=} 514 = \$514 \qquad \text{interest}$$

Several steps are involved in solving Example 2 in Section 11.3:

Find the monthly payment, including taxes and insurance, for a home costing $108,000. Assume a 10% 30-year loan with a 20% down payment. The annual taxes are $780, and fire insurance is $453.

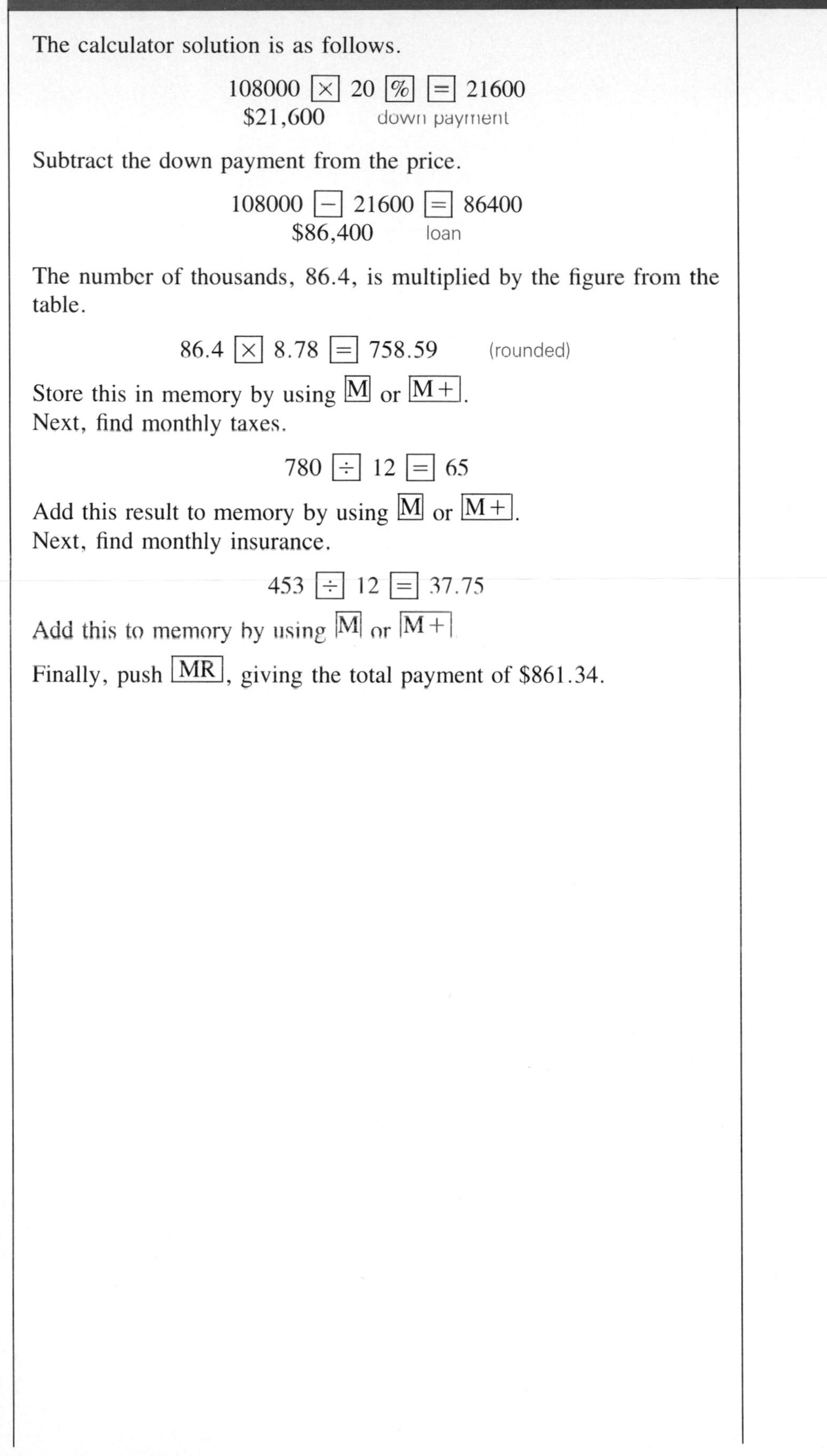

The calculator solution is as follows.

108000 [×] 20 [%] [=] 21600
$21,600 down payment

Subtract the down payment from the price.

108000 [−] 21600 [=] 86400
$86,400 loan

The number of thousands, 86.4, is multiplied by the figure from the table.

86.4 [×] 8.78 [=] 758.59 (rounded)

Store this in memory by using [M] or [M+].
Next, find monthly taxes.

780 [÷] 12 [=] 65

Add this result to memory by using [M] or [M+].
Next, find monthly insurance.

453 [÷] 12 [=] 37.75

Add this to memory by using [M] or [M+].

Finally, push [MR], giving the total payment of $861.34.

name date hour

APPENDIX A EXERCISES

Solve the following problems on a calculator. Round answers to the nearest hundredth.

1. $\begin{array}{r} 28.96 \\ 34.25 \\ 19.78 \\ +\ 21.59 \\ \hline \end{array}$

2. $\begin{array}{r} 758.42 \\ 76.98 \\ +\ 217.91 \\ \hline \end{array}$

3. $\begin{array}{r} 28{,}974.2 \\ 15{,}892 \\ +\ \quad 38.55 \\ \hline \end{array}$

4. $\begin{array}{r} 21.5 \\ -\ 13.82 \\ \hline \end{array}$

5. $\begin{array}{r} 769.2 \\ -\ \ 35.75 \\ \hline \end{array}$

6. $12 - 10.798$

7. $\begin{array}{r} 89.7 \\ \times\ \ .63 \\ \hline \end{array}$

8. $\begin{array}{r} 47.82 \\ \times\ \ .13 \\ \hline \end{array}$

9. $\begin{array}{r} 3409 \\ \times\ .006 \\ \hline \end{array}$

10. $\dfrac{9225}{25}$

11. $\dfrac{4594.2}{78}$

12. $\dfrac{3676.5}{142.9}$

13. $1.705\overline{).8359}$

14. $.0028\overline{).00719}$

15. $\dfrac{222.34}{1.67}$

16. $\dfrac{79{,}693.8}{365}$

Solve each of the following "chain" calculations. Round each answer to the nearest hundredth.

17. $\dfrac{9 \times 9}{2 \times 5}$

18. $\dfrac{15 \times 8 \times 3}{11 \times 7 \times 4}$

19. $\dfrac{87 \times 24 \times 47.2}{13.6 \times 12.8}$

20. $\dfrac{731.25 \times 360}{11{,}700 \times 150}$

21. $\dfrac{1155 \times 360}{16{,}500 \times 120}$

22. $\dfrac{4200 \times .12 \times 90}{365}$

23. $\dfrac{74{,}500 \times .14 \times 200}{360}$

24. $\dfrac{47 \times 1256}{14.93 + 85.77 \times .663}$

25. $\dfrac{633 \times .0299}{8.911 + 525 \times .399}$

Solve each of the following word problems by using a calculator. Round each answer to the nearest cent.

26. Bucks County Community College Bookstore bought 397 copies of this book at a net cost of $23.86 each, 125 copies of an accounting book at $28.74 each, and 740 copies of a real estate text at $11.76 each. Find the total amount paid by the bookstore.

27. To find the monthly interest due on a certain home mortgage, multiply the mortgage balance by .007292. Find the monthly interest on a mortgage having a balance of $42,798.46.

28. Find the monthly interest on a mortgage having a balance of $37,908.42. (See Exercise 27.)

29. Judy Martinez needs to file her expense account claim. She spent 5 nights at the Macon Holiday Inn at $47.46 per night, 4 nights at the Charlotte Sheraton at $51.62 per night, and rented a car for 7.6 days at $29.95 per day. She drove the car 916 miles with a charge of 24¢ per mile. Find her total expenses.

30. In Virginia City, the sales tax is 6.5%. Find the tax on **(a)** a new car costing $17,908.43; **(b)** an office typewriter costing $1463.58.

APPENDIX B: DEDUCTIVE AND INDUCTIVE REASONING

OBJECTIVES

1 Use inductive reasoning to analyze patterns.

2 Use deductive reasoning to analyze arguments.

3 Use deductive reasoning to solve problems.

1 In many scientific experiments, conclusions are drawn from specific outcomes. After many repetitions and similar outcomes, the findings are generalized into statements that appear to be true. When general conclusions are drawn from specific observations, we are using a type of reasoning called **inductive reasoning.** In the next several examples, this type of reasoning will be illustrated.

EXAMPLE 1 Using Inductive Reasoning
Find the next number in the sequence 3, 7, 11, 15, ...

SOLUTION
To discover a pattern, calculate the difference between each pair of successive numbers.

$$7 - 3 = 4$$
$$11 - 7 = 4$$
$$15 - 11 = 4$$

As shown, the difference is 4. Each number is 4 greater than the previous one. Thus, the next number in the pattern is 15 + 4, or 19. ■

WORK PROBLEM 1 AT THE SIDE.

1. Find the next number in the sequence 2, 6, 10, 14, ...

EXAMPLE 2 Using Inductive Reasoning
Find the number that comes next in the sequence.

7, 11, 8, 12, 9, 13, ...

SOLUTION
The pattern in this example can be determined as follows.

$$7 \mathbf{+ 4} = 11$$
$$11 \mathbf{- 3} = 8$$
$$8 \mathbf{+ 4} = 12$$
$$12 \mathbf{- 3} = 9$$
$$9 \mathbf{+ 4} = 13$$

To get the second number, we add 4 to the first number. To get the third number, we subtract 3 from the second number. To obtain subsequent numbers, this pattern is continued. The next number is 13 − 3, or 10. ■

WORK PROBLEM 2 AT THE SIDE.

2. Find the next number in the sequence

6, 11, 7, 12, 8, 13, ...

EXAMPLE 3 Using Inductive Reasoning
Find the next number in the sequence 1, 2, 4, 8, 16, ...

SOLUTION
Each number after the first is obtained by multiplying the previous number by 2. So the next number would be $16 \cdot 2 = 32$. ■

ANSWERS
1. 18 2. 9

3. Find the next number in the sequence

2, 6, 18, 54, ...

4. Find the next shape in the following sequence.

NOTE The sequence in Example 3 is called a geometric sequence.

WORK PROBLEM 3 AT THE SIDE.

EXAMPLE 4 Using Inductive Reasoning
Find the next geometric shape in the following sequence.

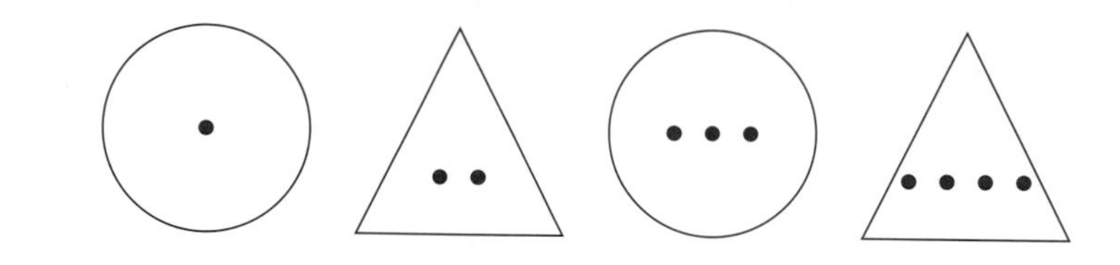

SOLUTION
In this sequence, the figures alternate between a circle and a triangle. In addition, the number of dots increases by 1 in each subsequent figure. Thus, the next figure should be a circle with five dots contained in it, or

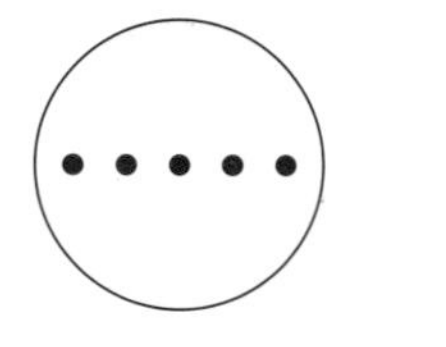

■

EXAMPLE 5 Using Inductive Reasoning
Find the next geometric shape in the following sequence.

SOLUTION
The first two shapes consist of vertical lines with horizontal lines at the bottom facing left and right. The third shape is a vertical line with a horizontal line at the top facing to the left. The fourth shape should be a vertical line with a horizontal line at the top facing to the right, or

■

WORK PROBLEM 4 AT THE SIDE.

2 In the previous discussion, specific cases were used to find patterns and predict the next event. There is another type of rea-

ANSWERS
3. 162
4.

soning called **deductive reasoning**, which moves from general cases to specific conclusions.

EXAMPLE 6 Using Deductive Reasoning
Does the conclusion follow from the premises in this argument?

All Buicks are automobiles.
All automobiles have horns.
$\therefore$ All Buicks have horns.

SOLUTION
In this example, the first two statements are called *premises* and the third statement (below the line) is called a conclusion. The symbol $\therefore$ is a mathematical symbol meaning "therefore." The entire set of statements is called an *argument*. The focus of deductive reasoning is to determine if the conclusion follows (is valid) from the premises. A series of circles called **Euler circles** is used to analyze the argument. In Example 6, the statement "All Buicks are automobiles" can be represented by two circles, one for Buicks and one for automobiles.

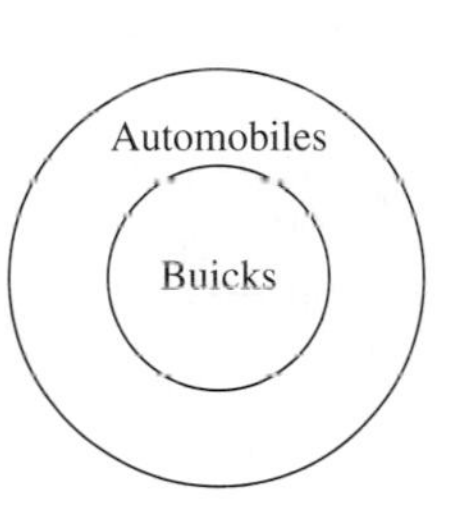

Note that the circle representing Buicks is totally inside the circle representing automobiles.

If a circle representing the second statement is added, a circle representing vehicles with horns must surround the circle representing automobiles.

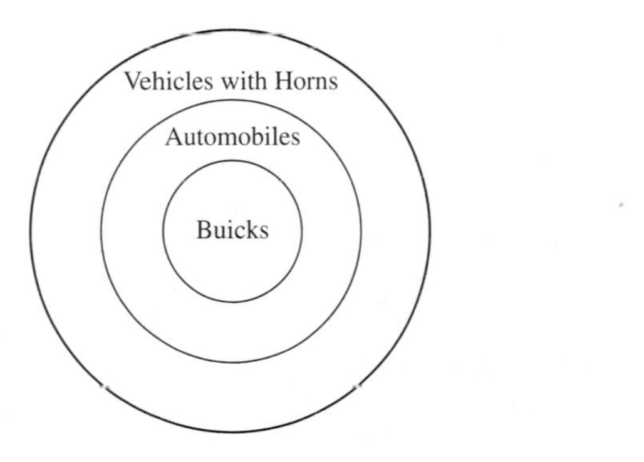

Notice that the circle representing Buicks is completely inside the circle representing vehicles with horns. It must follow that

all Buicks have horns. 

WORK PROBLEM 5 AT THE SIDE.

5. Does the conclusion follow from the premises in the following argument?

All cars have four wheels.
All Fords are cars.
$\therefore$ All Fords have four wheels.

ANSWER
5. The conclusion follows from the premises.

6. Does the conclusion follow from the premises?

All animals are wild.
All cats are animals.
∴ All cats are wild.

ANSWER
6. The conclusion follows from premises.

EXAMPLE 7 Using Deductive Reasoning

Does the conclusion follow from the premises in this argument?

All tables are round.
All glasses are round.
∴ All glasses are tables.

SOLUTION

Using Euler circles, a circle representing tables is drawn inside a circle representing round objects.

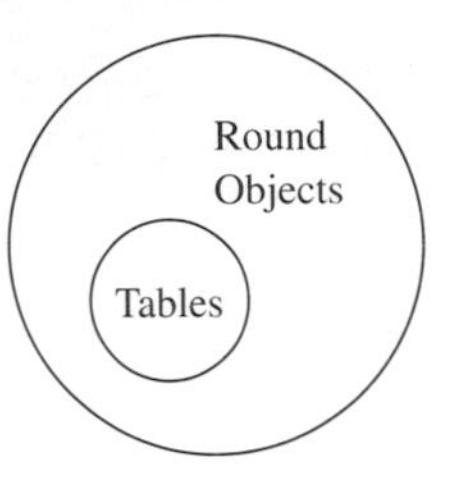

The second statement requires that a circle representing glasses must now be drawn inside the circle representing round objects but not necessarily inside the circle representing tables.

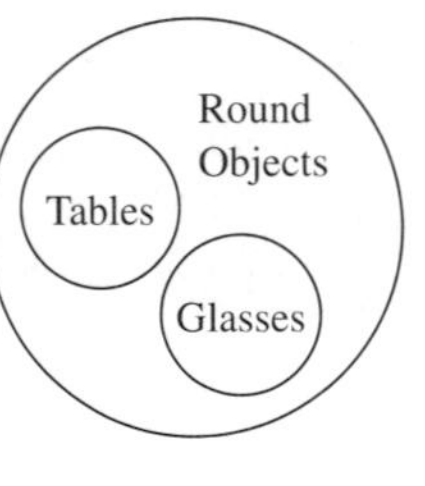

The conclusion does not follow from the premises. This means that the conclusion is invalid or untrue. ■

WORK PROBLEM 6 AT THE SIDE.

3 Another type of deductive reasoning problem occurs when a set of facts is given in a problem and a conclusion must be drawn by using these facts.

EXAMPLE 8 Using Deductive Reasoning

There were 25 students enrolled in a ceramics class. During the class, 10 of the students made an ashtray and 8 students made a birdbath. Three students made both an ashtray and a birdbath. How many students did not make either an ashtray or a birdbath?

SOLUTION

This type of problem is best solved by organizing the data by using a device called a *Venn diagram*. Two overlapping circles are drawn, with each circle representing one item made by students.

Venn diagram: Ashtray circle containing 7, overlap region containing 3, Birdbath circle containing 6.

In the region where the circles overlap, place the number that represents the number of students who made both items, namely 3. In the remaining portion of the birdbath circle, write the number 5, which when added to 3 will give the total number of students who made a birdbath, namely 8. In a similar manner, write 7 in the remaining portion of the ashtray circle, since $7 + 3 = 10$, the total number of students who made an ashtray. The three numbers that have been written in the regions total 15. Since there are 25 students in the class, this means $25 - 15$ or 10 students did not make either a birdbath or an ashtray. ■

WORK PROBLEM 7 AT THE SIDE.

EXAMPLE 9 Using Deductive Reasoning

Four cars in a race finish first, second, third, and fourth. The following facts are known

(a) Car A beat Car C.

(b) Car D finished between Cars C and B.

(c) Car C beat Car B.

In which order did the cars finish?

SOLUTION

To solve this type of problem, it is helpful to use a line diagram.

1. *Write A before C,* since Car A beat Car C (fact a).

A C

2. *Write B after C,* since Car C beat Car B (fact c).

A C B

3. *Write D between C and B,* since Car D finished between Cars C and B (fact b).

So

A C D B

is the correct order of finish. ■

WORK PROBLEM 8 AT THE SIDE.

7. In a college class of 100 students, 35 take both math and history, 50 take history, and 40 take math. How many take neither math nor history?

8. A Chevy, BMW, Cadillac, and Oldsmobile are parked side by side.

(a) The Oldsmobile is on the right end.

(b) The BMW is next to the Cadillac.

(c) The Chevy is between the Oldsmobile and the Cadillac.

Which car is parked on the left end?

ANSWERS

7. 45 **8.** BMW

name date hour

APPENDIX B: EXERCISES

Find the next number in each of the following sequences.

1. 2, 9, 16, 23, 30, ...

2. 5, 8, 11, 14, 17, ...

3. 1, 6, 11, 16, 21, ...

4. 3, 6, 12, 24, 48, ...

5. 1, 2, 4, 8, ...

6. 1, 8, 27, 64, ...

7. 1, 3, 9, 27, 81, ...

8. 3, 5, 7, 9, 11, ...

9. 1, 4, 9, 16, 25, ...

10. 6, 7, 9, 12, 16, ...

Find the next shape in each of the following sequences.

11.

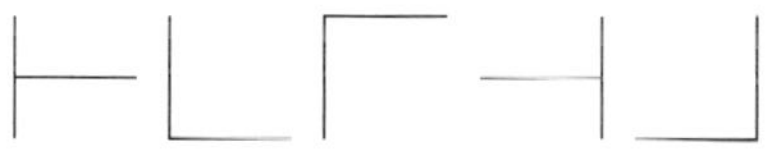

12.

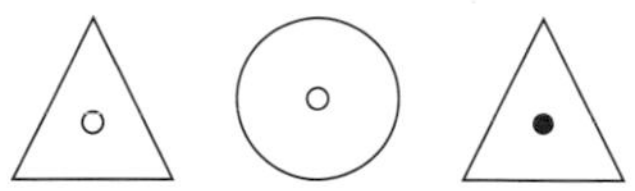

13.

14.

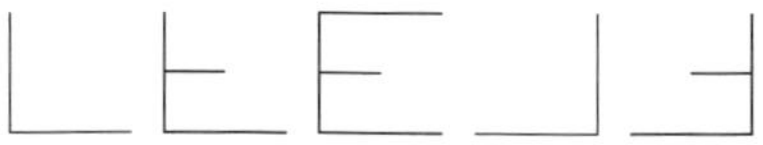

In each of the following, state whether or not the conclusion follows from the premises.

15. All animals are wild.
All lions are animals.
∴ All lions are wild.

16. All students are hard workers.
All business majors are students.
∴ All business majors are hard workers.

17. All teachers are serious.
All mathematicians are serious.
∴ All mathematicians are teachers.

18. All boys ride bikes.
All Americans ride bikes.
∴ All Americans are boys.

Solve the following problems.

19. In a given 30-day period, a man watched television 20 days and his wife watched television 25 days. If they watched television together 18 days, how many days did neither watch television?

20. In a class of 40 students, 21 students take both calculus and physics. If 30 students take calculus and 25 students take physics, how many do not take either calculus or physics?

21. Tom, Dick, Mary, and Joan all work for the same company. One is a secretary, one is a computer operator, one is a receptionist, and one is a mail clerk.
(a) Tom and Joan eat dinner with the computer operator.
(b) Dick and Mary carpool with the secretary.
(c) Mary works on the same floor as the computer operator and the mail clerk.
Who is the computer operator?

22. Four cars—a Ford, a Buick, a Mercedes, and an Audi—are parked in a garage in four spaces.
(a) The Ford is in the last space.
(b) The Buick and Mercedes are next to each other.
(c) The Audi is next to the Ford but not next to the Buick.
Which car is in the first space?

ANSWERS TO SELECTED EXERCISES

The solutions to selected odd-numbered exercises are given in the section beginning on page A-17.

Diagnostic Pretest (page xvii)

1. thousands; tens **2.** eight hundred seventy-four **3.** 8295 **4.** 2001 **5.** 3620 **6.** 47 R4 **7.** 40,000 **8.** 16 **9.** 10,780 **10.** $\frac{59}{10}$ **11.** $\frac{8}{35}$ **12.** 28 students **13.** 9 **14.** $\frac{35}{42}$ **15.** $\frac{23}{24}$ **16.** $\frac{7}{36}$ **17.** $23\frac{1}{2}$ **18.** $64\frac{7}{12}$ **19.** .69 **20.** 18.296 **21.** 20.786 **22.** .00126 **23.** 92.3 **24.** 10.52 **25.** $\frac{15}{8}$ **26.** 25 miles per gallon **27.** no **28.** $5\frac{1}{3}$ **29.** .76 **30.** 60% **31.** 200 **32.** 25% **33.** $399.50 **34.** $847 **35.** 6 yards 2 feet 5 inches **36.** 37,000 meters **37.** 66 feet; 156 square feet **38.** 517 m^2 **39.** 221.6 m^2 **40.** 67.0 cm^3 **41.** $a = 60$; $b = 120$ **42.** -33 **43.** 5 **44.** 50 **45.** 20 **46.** -43 **47.** $k = 2$ **48.** 12.2 **49.** 102 **50.** 18

CHAPTER 1

Section 1.1 (page 5)

1. 1; 8 **3.** 1; 0 **5.** 4; 2 **7.** 71; 105 **9.** 60; 0; 502; 109 **11.** nine thousand, six hundred thirteen **13.** seven hundred twenty-five thousand, six hundred fifty-nine **15.** seventy-five million, seven hundred fifty-six thousand, six hundred sixty-five **17.** 3135 **19.** 785,223 **21.** 2070 **23.** 13,112 **25.** 800,621,020,215

Section 1.2 (page 13)

1. 39 **3.** 79 **5.** 69 **7.** 498 **9.** 7785 **11.** 789 **13.** 7676 **15.** 78,446 **17.** 15,776 **19.** 129,224 **21.** 127 **23.** 142 **25.** 121 **27.** 161 **29.** 102 **31.** 1781 **33.** 970 **35.** 413 **37.** 1771 **39.** 1125 **41.** 9253 **43.** 6624 **45.** 9611 **47.** 7954 **49.** 10,648 **51.** 16,858 **53.** 3557 **55.** 12,078 **57.** 4250 **59.** 12,268 **61.** correct **63.** 759—incorrect; should be 769 **65.** correct **67.** 11,377—incorrect; should be 11,577 **69.** correct **71.** 32 miles **73.** 33 miles **75.** $36 **77.** 620 people **79.** 5051 books **81.** 260 inches **83.** 708 feet

Section 1.3 (page 23)

1. 25 **3.** 41 **5.** 17 **7.** 62 **9.** 12 **11.** 122 **13.** 301 **15.** 665 **17.** 1111 **19.** 4110 **21.** 513 **23.** 2213 **25.** 8314 **27.** 13,160 **29.** 41,110 **31.** correct **33.** 63—incorrect; should be 62 **35.** 131—incorrect; should be 121 **37.** correct **39.** 7212—incorrect; should be 7222 **41.** 7 **43.** 29 **45.** 16 **47.** 362 **49.** 168 **51.** 459 **53.** 8279 **55.** 7589 **57.** 15,899 **59.** 8859 **61.** 7 **63.** 23 **65.** 62 **67.** 311 **69.** 2833 **71.** 214 **73.** 7775 **75.** 503 **77.** 156 **79.** 1942 **81.** 5687 **83.** 19,038 **85.** 15,778 **87.** 2129 **89.** 5584 **91.** correct **93.** correct **95.** correct **97.** correct **99.** 2553—incorrect; should be 2453 **101.** 29,231—incorrect; should be 29,221 **103.** 22 dogs **105.** 121 passengers **107.** 312 miles **109.** 1243 people **111.** $144 refund **113.** $702 **115.** 758 people

Section 1.4 (page 33)

1. 24 **3.** 56 **5.** 0 **7.** 27 **9.** 36 **11.** 0 **13.** 440 **15.** 252 **17.** 2048 **19.** 1872 **21.** 8612 **23.** 10,084 **25.** 36,060 **27.** 131,010 **29.** 325,036 **31.** 80 **33.** 350 **35.** 2220 **37.** 2000 **39.** 3750 **41.** 25,100 **43.** 320,000 **45.** 86,000,000 **47.** 48,500 **49.** 450,000 **51.** 1,940,000 **53.** 783 **55.** 2376 **57.** 3735 **59.** 2378 **61.** 6120 **63.** 18,192 **65.** 22,085 **67.** 14,564 **69.** 82,320 **71.** 354,090 **73.** 1,172,583 **75.** 2,678,325 **77.** 778,662 **79.** 54,135 **81.** 65,163 **83.** 592,596 **85.** 88,476 **87.** 17,668,533 **89.** 3,198,328 **91.** 8,354,745 **93.** 24,000 pages **95.** 108 cans **97.** 216 miles **99.** $350 **101.** $3074 **103.** $3996 **105.** 50,568 **107.** 140 miles **109.** 1058 calories **111.** 55 employees

Section 1.5 (page 45)

1. $\frac{21}{7} = 3$; $7\overline{)21}$ with quotient 3 **3.** $45 \div 9 = 5$; $9\overline{)45}$ with quotient 5 **5.** $16 \div 2 = 8$; $\frac{16}{2} = 8$ **7.** 1 **9.** 5 **11.** meaningless **13.** 0 **15.** 0 **17.** meaningless **19.** 0 **21.** 8 **23.** 26 **25.** 46 **27.** 447 **29.** 401 **31.** 627 R1

33. 224 R8 **35.** 1522 R5 **37.** 309 **39.** 1006 **41.** 5006 **43.** 811 R1 **45.** 2589 R2 **47.** 9785 R2 **49.** 3689 R1 **51.** 2745 **53.** 5841 **55.** 12,458 R3 **57.** 10,258 R5 **59.** 10,253 R5 **61.** 11,255 **63.** correct **65.** 1908 R2—incorrect; should be 1908 R1 **67.** 650 R2—incorrect; should be 670 R2 **69.** 3568 R2—incorrect; should be 3568 R1 **71.** correct **73.** correct **75.** 9628 R7—incorrect; should be 9628 R3 **77.** correct **79.** 52,136 R5—incorrect; should be 52,136 R4 **81.** 12¢ **83.** $8252 **85.** $3277 **87.** √ √ √ √ **89.** √ X X X **91.** X X √ X **93.** X √ X X **95.** X √ X X **97.** X X X X **99.** $4050

Section 1.6 (page 53)

1. 35 **3.** 47 R10 **5.** 37 R7 **7.** 307 R4 **9.** 38 **11.** 274 R33 **13.** 2407 R1 **15.** 1239 R15 **17.** 640 R8 **19.** 9746 R1 **21.** 7746 R20 **23.** 3331 R82 **25.** 21 **27.** 1114 R196 **29.** 170 **31.** 106 R17—incorrect; should be 106 R7 **33.** 658 R9—incorrect; should be 658 **35.** 62 R3—incorrect; should be 62 **37.** 25 hours **39.** $3906 **41.** $108 **43.** 1680 circuits **45.** $10,512

Section 1.7 (page 61)

1. 7900 **3.** 1280 **5.** 710 **7.** 86,810 **9.** 42,500 **11.** 8000 **13.** 15,800 **15.** 31,100 **17.** 6000 **19.** 53,000 **21.** 500,000 **23.** 9,000,000 **25.** 7460; 7500; 7000 **27.** 4280; 4300; 4000 **29.** 5050; 5000; 5000 **31.** 3130; 3100; 3000 **33.** 3650; 3600; 4000 **35.** 28,170; 28,200; 28,000 **37.** 23,500; 23,500; 24,000 **39.** 180 **41.** 40 **43.** 2100 **45.** 1900 **47.** 300 **49.** 360,000 **51.** 622,000,000; 622,000,000; 622,000,000

Section 1.8 (page 67)

1. 2 **3.** 4 **5.** 12 **7.** 11 **9.** 8 **11.** 36 **13.** 144 **15.** 225 **17.** 324; 324 **19.** 784; 784 **21.** 1225; 1225 **23.** 1600; 1600 **25.** 2916; 2916 **27.** 84 **29.** 21 **31.** 5 **33.** 20 **35.** 24 **37.** 63 **39.** 23 **41.** 77 **43.** 24 **45.** 34 **47.** 20 **49.** 10 **51.** 63 **53.** 33 **55.** 6 **57.** 7 **59.** 26 **61.** 28 **63.** 104 **65.** 19 **67.** 9 **69.** 44 **71.** 30 **73.** 5 **75.** 3 **77.** 7 **79.** 20 **81.** 11 **83.** 25 **85.** 16 **87.** 20 **89.** 209

Section 1.9 (page 75)

1. $8605 **3.** 2625 tickets **5.** $4083 **7.** 155 yards **9.** 1332 miles **11.** $252 **13.** 12,125 pounds **15.** 174,240 square feet **17.** 76,529,550 gallons **19.** 63 miles **21.** $61,080 **23.** 768 deer **25.** $1932 **27.** $102 **29.** $213 **31.** 352 machines **33.** 20 seats

Chapter 1 Review Exercises (page 83)

1. 7; 816 **2.** 78; 915 **3.** 206; 792 **4.** 1; 768; 710; 618 **5.** seven hundred twenty-five **6.** seventeen thousand, six hundred fifteen **7.** sixty-two million, five hundred thousand, five **8.** 8120 **9.** 600,015,759 **10.** 103 **11.** 120 **12.** 1237 **13.** 4656 **14.** 15,657 **15.** 9179 **16.** 6979 **17.** 40,602 **18.** 11 **19.** 22 **20.** 39 **21.** 3803 **22.** 4327 **23.** 1017 **24.** 224 **25.** 25,866 **26.** 64 **27.** 0 **28.** 24 **29.** 54 **30.** 36 **31.** 36 **32.** 56 **33.** 81 **34.** 48 **35.** 45 **36.** 72 **37.** 8 **38.** 0 **39.** 81 **40.** 35 **41.** 0 **42.** 130 **43.** 644 **44.** 72 **45.** 89 **46.** 234 **47.** 5467 **48.** 1396 **49.** 45,815 **50.** 14,518 **51.** 22,400 **52.** 465,525 **53.** 174,984 **54.** 330 **55.** 1872 **56.** 1176 **57.** 46,020 **58.** 13,755 **59.** 30,184 **60.** 241,536 **61.** 357,294 **62.** $576 **63.** $312 **64.** $13,344 **65.** $1512 **66.** 350 **67.** $30,400 **68.** 300,800 **69.** 128,000,000 **70.** 90,300,000 **71.** 604,000,000 **72.** 2 **73.** 5 **74.** 7 **75.** 9 **76.** 9 **77.** 3 **78.** 9 **79.** 0 **80.** meaningless **81.** 0 **82.** 81 **83.** 36 **84.** 6251 **85.** 352 **86.** 96 **87.** 150 R4 **88.** 124 R25 **89.** 120 **90.** 16,700 **91.** 20,000 **92.** 70,000 **93.** 1500; 1500; 1000 **94.** 10,060; 10,100; 10,000 **95.** 98,200; 98,200; 98,000 **96.** 352,120; 352,100; 352,000 **97.** 2 **98.** 12 **99.** 9 **100.** 14 **101.** 2; 3; 9 **102.** 3; 5; 125 **103.** 5; 3; 243 **104.** 5; 4; 1024 **105.** 72 **106.** 20 **107.** 9 **108.** 4 **109.** 9 **110.** 6 **111.** $20,340 **112.** 84,000 revolutions **113.** 1440 words **114.** 320,000 nails **115.** 24,000 hours **116.** 400 miles **117.** 576 cans **118.** $405 **119.** $199 **120.** 33 hours **121.** 50 pounds **122.** $247 **123.** 23 acres **124.** 1629 feet **125.** 320 **126.** 9 **127.** 115 **128.** 40 **129.** 1041 **130.** 2088 **131.** 32,062 **132.** 3 **133.** 30 **134.** 7 **135.** 93,625 **136.** 5607 **137.** meaningless **138.** 138 **139.** 5501 **140.** 1,079,040 **141.** 5 **142.** 108 **143.** 115,713 **144.** two hundred eight-six thousand, seven hundred fifty-three **145.** 7200 **146.** 1,000,000 **147.** 6 **148.** 10 **149.** $2128 **150.** $31,320 **151.** $2220 **152.** $15,782 **153.** 468 cards **154.** 10,000 boxes **155.** 3650 hours **156.** $49,363 **157.** $789 **158.** 52 miles **159.** $3850 **160.** $12

Chapter 1 Test (page 91)

[1.1] **1.** three thousand twenty-two **2.** fifty-two thousand eight **3.** 138,008 [1.2] **4.** 16,706 **5.** 112,630 [1.3] **6.** 2116 **7.** 2542 [1.4] **8.** 120 **9.** 147,000 **10.** 1350 **11.** 4,450,743 [1.5–1.6] **12.** 7747 **13.** meaningless **14.** 458 **15.** 170 [1.7] **16.** 7850 **17.** 76,700 **18.** 46,000 [1.8] **19.** 51 **20.** 28 [1.9] **21.** $1140 **22.** 123 days **23.** $165 **24.** 969 employees **25.** 1028 ovens

CHAPTER 2

Section 2.1 (page 97)

1. $\frac{5}{8}$ **3.** $\frac{2}{3}$ **5.** $\frac{7}{5}$ **7.** $\frac{3}{11}$ **9.** $\frac{14}{25}$ **11.** $\frac{13}{71}$
13. 5; 7 **15.** 9; 8 **17.** proper: $\frac{1}{4}, \frac{3}{8}, \frac{7}{12}$; improper: $\frac{9}{7}, \frac{11}{4}$ **19.** proper: $\frac{3}{4}, \frac{9}{11}, \frac{7}{15}$; improper: $\frac{3}{2}, \frac{19}{18}$
21. 9; 17 **23.** 18 **25.** 105 **27.** 7 **29.** 8

Section 2.2 (page 101)

1. $\frac{15}{8}$ **3.** $\frac{19}{5}$ **5.** $\frac{11}{4}$ **7.** $\frac{31}{9}$ **9.** $\frac{18}{11}$
11. $\frac{19}{3}$ **13.** $\frac{23}{2}$ **15.** $\frac{43}{4}$ **17.** $\frac{27}{8}$ **19.** $\frac{44}{5}$
21. $\frac{54}{11}$ **23.** $\frac{206}{9}$ **25.** $\frac{233}{13}$ **27.** $\frac{269}{15}$
29. $\frac{146}{19}$ **31.** $4\frac{1}{2}$ **33.** $1\frac{4}{5}$ **35.** $1\frac{3}{11}$ **37.** $3\frac{1}{9}$
39. $4\frac{2}{5}$ **41.** $5\frac{2}{3}$ **43.** $11\frac{3}{5}$ **45.** $5\frac{2}{9}$ **47.** $7\frac{1}{7}$
49. $16\frac{4}{5}$ **51.** $21\frac{1}{5}$ **53.** $26\frac{1}{7}$ **55.** $\frac{2041}{8}$
57. $\frac{1000}{3}$ **59.** $171\frac{8}{15}$ **61.** 13 **63.** 66
65. 16

Section 2.3 (page 109)

1. 1, 2, 3, 6 **3.** 1, 2, 4, 8 **5.** 1, 3, 7, 21
7. 1, 2, 3, 6, 9, 18 **9.** 1, 2, 4, 5, 8, 10, 20, 40
11. 1, 2, 4, 8, 16, 32, 64 **13.** composite
15. prime **17.** composite **19.** prime
21. prime **23.** composite **25.** composite
27. composite **29.** $2 \cdot 3$ **31.** 2^4 **33.** $3 \cdot 7$
35. 2^5 **37.** $3 \cdot 13$ **39.** $2^3 \cdot 11$ **41.** $3 \cdot 5^2$
43. $2^2 \cdot 5^2$ **45.** $2 \cdot 3 \cdot 5^2$ **47.** $3^2 \cdot 5^2$
49. $2^6 \cdot 5$ **51.** $2^3 \cdot 3^2 \cdot 5$ **53.** 125 **55.** 512
57. 1296 **59.** 432 **61.** 1125 **63.** 576
65. $2 \cdot 3^2 \cdot 5^2$ **67.** $2^6 \cdot 3 \cdot 5$ **69.** $2^6 \cdot 5^2$
71. 5 **73.** 70 **75.** 12 **77.** 9

Section 2.4 (page 117)

1. $\frac{3}{4}$ **3.** $\frac{1}{2}$ **5.** $\frac{5}{8}$ **7.** $\frac{6}{7}$ **9.** $\frac{9}{10}$ **11.** $\frac{6}{7}$
13. $\frac{4}{7}$ **15.** $\frac{1}{50}$ **17.** $\frac{11}{12}$ **19.** $\frac{2 \cdot 5}{2 \cdot 2 \cdot 2 \cdot 2} = \frac{5}{8}$
21. $\frac{5 \cdot 5}{3 \cdot 3 \cdot 5} = \frac{5}{9}$ **23.** $\frac{2 \cdot 2 \cdot 3 \cdot 5}{2 \cdot 3 \cdot 5 \cdot 5} = \frac{2}{5}$
25. $\frac{2 \cdot 2 \cdot 3 \cdot 3}{2 \cdot 2 \cdot 3} = 3$ **27.** $\frac{7 \cdot 11}{2 \cdot 2 \cdot 2 \cdot 3 \cdot 11} = \frac{7}{24}$
29. equivalent **31.** not equivalent **33.** not equivalent **35.** equivalent **37.** not equivalent
39. equivalent **41.** $\frac{5}{12}$ **43.** $\frac{8}{25}$ **45.** 1, 2, 4, 8
47. 1, 2, 4, 8, 16, 32, 64

Section 2.5 (page 125)

1. $\frac{3}{20}$ **3.** $\frac{1}{54}$ **5.** $\frac{9}{10}$ **7.** $\frac{3}{10}$ **9.** $\frac{5}{12}$
11. $\frac{99}{160}$ **13.** $\frac{1}{2}$ **15.** $\frac{24}{35}$ **17.** $\frac{21}{128}$ **19.** 6
21. 40 **23.** 5 **25.** $10\frac{1}{2}$ **27.** 15 **29.** 375
31. 75 **33.** 540 **35.** 810 **37.** $\frac{21}{40}$ square inch
39. 4 square yards **41.** 4 square inches **43.** $1\frac{1}{3}$ square yards **45.** 2070 rafts

Section 2.6 (page 133)

1. $\frac{5}{6}$ square yard **3.** $\frac{1}{4}$ square inch **5.** 380 items are taxable **7.** \$2700 **9.** 360 employees have given **11.** 75 men **13.** \$8000 **15.** \$2000
17. 75 **19.** 1500 votes **21.** $\frac{1}{32}$ of the estate
23. 74 cartons

Section 2.7 (page 141)

1. $\frac{8}{9}$ **3.** $\frac{15}{32}$ **5.** $\frac{4}{9}$ **7.** $2\frac{1}{10}$ **9.** $\frac{35}{108}$
11. $\frac{24}{25}$ **13.** 18 **15.** 20 **17.** $\frac{1}{14}$ **19.** $\frac{2}{9}$ acre
21. 35 shakers **23.** 56 vials **25.** 60 trips
27. 512 pages **29.** 36 miles **31.** \$120,000
33. $\frac{17}{6}$ **35.** $\frac{38}{3}$ **37.** $\frac{54}{11}$

Section 2.8 (page 149)

1. 25 **3.** $4\frac{1}{2}$ **5.** 4 **7.** $72\frac{1}{2}$ **9.** 138
11. 78 **13.** $1\frac{5}{21}$ **15.** $\frac{2}{3}$ **17.** $\frac{8}{33}$ **19.** $2\frac{2}{5}$
21. $\frac{3}{10}$ **23.** 34 **25.** 52 yards **27.** 9 homes
29. $21\frac{7}{8}$ ounces **31.** 5000 dictionaries **33.** 157 rolls **35.** $\frac{3}{4}$ **37.** $\frac{5}{8}$ **39.** $\frac{9}{10}$

Chapter 2 Review Exercises (page 155)

1. $\frac{1}{4}$ **2.** $\frac{5}{8}$ **3.** $\frac{3}{4}$ **4.** proper: $\frac{3}{4}, \frac{2}{3}, \frac{1}{8}$; improper: $\frac{2}{1}, \frac{6}{5}$ **5.** proper: $\frac{15}{16}, \frac{1}{8}$; improper: $\frac{6}{5}, \frac{16}{13}, \frac{5}{3}$ **6.** $\frac{3}{2}$
7. $\frac{181}{16}$ **8.** $2\frac{2}{5}$ **9.** $13\frac{6}{13}$ **10.** 1, 2, 3, 6

11. 1, 2, 3, 4, 6, 12 **12.** 1, 5, 11, 55 **13.** 1, 2, 3, 5, 6, 9, 10, 15, 18, 30, 45, 90 **14.** 5^2 **15.** $2^3 \cdot 3 \cdot 5$ **16.** $3^2 \cdot 5^2$ **17.** 25 **18.** 36 **19.** 2700 **20.** 3888 **21.** $\frac{2}{3}$ **22.** $\frac{5}{7}$ **23.** $\frac{35}{38}$ **24.** $\frac{5 \cdot 5}{2 \cdot 2 \cdot 3 \cdot 5}$; $\frac{5}{12}$ **25.** $\frac{2 \cdot 2 \cdot 89}{2 \cdot 2 \cdot 2 \cdot 2 \cdot 2 \cdot 3 \cdot 5}$; $\frac{89}{120}$ **26.** equivalent **27.** equivalent **28.** $\frac{4}{25}$ **29.** $\frac{9}{40}$ **30.** $\frac{1}{7}$ **31.** 6 **32.** 15 **33.** 625 **34.** $\frac{8}{9}$ **35.** $\frac{5}{3} = 1\frac{2}{3}$ **36.** $\frac{5}{2} = 2\frac{1}{2}$ **37.** 2 **38.** 16 **39.** $9\frac{1}{3}$ **40.** $1\frac{1}{7}$ **41.** $\frac{2}{15}$ **42.** $\frac{4}{13}$ **43.** $\frac{63}{128}$ square foot **44.** $\frac{7}{12}$ square inch **45.** 4 square yards **46.** 25 square feet **47.** $3\frac{15}{16}$ **48.** $21\frac{3}{8}$ **49.** $6\frac{1}{6}$ **50.** $\frac{35}{64}$ **51.** 16 cans **52.** $\frac{2}{15}$ of the estate **53.** 18 blankets **54.** $\$97\frac{1}{2}$ or $97.50 **55.** 30 pounds **56.** $\frac{3}{10}$ mile **57.** $\frac{5}{32}$ of the profits **58.** $\frac{1}{3}$ **59.** $1\frac{1}{6}$ **60.** $20\frac{5}{6}$ **61.** $\frac{7}{48}$ **62.** $\frac{71}{112}$ **63.** $1\frac{1}{2}$ **64.** 30 **65.** $2\frac{1}{6}$ **66.** $2\frac{1}{2}$ **67.** $\frac{13}{7}$ **68.** $49\frac{1}{2}$ **69.** $\frac{271}{21}$ **70.** $\frac{2 \cdot 2 \cdot 2}{2 \cdot 2 \cdot 3} = \frac{2}{3}$ **71.** $\frac{2 \cdot 2 \cdot 3 \cdot 3 \cdot 3}{2 \cdot 3 \cdot 5 \cdot 7} = \frac{18}{35}$ **72.** $\frac{1}{3}$ **73.** $\frac{3}{8}$ **74.** $\frac{1}{3}$ **75.** $152\frac{4}{9}$ ounces **76.** 56 gallons **77.** 96 offices **78.** $\frac{1}{2}$ square inch

Chapter 2 Test (page 161)

[2.1] **1.** $\frac{3}{8}$ **2.** $\frac{1}{6}$ **3.** $\frac{5}{8}, \frac{7}{16}, \frac{2}{3}, \frac{3}{14}$ **[2.2]** **4.** $\frac{29}{8}$ **5.** $17\frac{7}{8}$ **[2.3]** **6.** 1, 3, 5, 15 **7.** $2^3 \cdot 5$ **8.** $3 \cdot 5^2$ **9.** $2^2 \cdot 5^3$ **[2.4]** **10.** $\frac{4}{5}$ **11.** $\frac{2}{3}$ **[2.5]** **12.** $\frac{5}{18}$ **13.** $\frac{4}{21}$ **14.** $4\frac{4}{5}$ **15.** 18 **[2.5–2.6]** **16.** $\frac{1}{4}$ square inch **17.** 161 students **[2.7]** **18.** $\frac{3}{4}$ **19.** $15\frac{3}{4}$ **20.** 125 trucks **[2.8]** **21.** $17\frac{23}{32}$ **22.** $7\frac{17}{18}$ **23.** $4\frac{4}{15}$ **24.** $2\frac{17}{20}$ **25.** $10\frac{5}{8}$

CHAPTER 3

Section 3.1 (page 167)

1. $\frac{2}{3}$ **3.** $\frac{7}{10}$ **5.** $\frac{1}{2}$ **7.** $\frac{1}{3}$ **9.** $\frac{5}{6}$ **11.** $\frac{13}{20}$ **13.** $\frac{10}{17}$ **15.** $\frac{3}{4}$ **17.** $\frac{11}{27}$ **19.** $\frac{5}{11}$ **21.** $\frac{8}{21}$ **23.** $\frac{3}{5}$ **25.** $\frac{2}{3}$ **27.** $\frac{1}{5}$ **29.** $\frac{1}{4}$ **31.** $\frac{1}{9}$ **33.** $\frac{3}{5}$ **35.** $\frac{1}{2}$ mile **37.** $\frac{1}{2}$ mile **39.** $\frac{3}{4}$ acre **41.** $2 \cdot 5$ **43.** $2 \cdot 2 \cdot 5 \cdot 5$ **45.** $3 \cdot 5 \cdot 5$

Section 3.2 (page 175)

1. 12 **3.** 60 **5.** 72 **7.** 200 **9.** 180 **11.** 180 **13.** 120 **15.** 360 **17.** 120 **19.** 300 **21.** 72 **23.** 300 **25.** $\frac{18}{36}$ **27.** $\frac{27}{36}$ **29.** $\frac{12}{36}$ **31.** $\frac{9}{24}$ **33.** $\frac{36}{40}$ **35.** $\frac{49}{56}$ **37.** $\frac{21}{70}$ **39.** $\frac{60}{76}$ **41.** $\frac{72}{56}$ **43.** $\frac{136}{51}$ **45.** $\frac{96}{132}$ **47.** 7200 **49.** 10,584 **51.** $1\frac{2}{5}$ **53.** $1\frac{4}{5}$ **55.** $3\frac{6}{7}$

Section 3.3 (page 183)

1. $\frac{4}{5}$ **3.** $\frac{3}{4}$ **5.** $\frac{13}{20}$ **7.** $\frac{19}{22}$ **9.** $\frac{23}{36}$ **11.** $\frac{13}{18}$ **13.** $\frac{29}{36}$ **15.** $\frac{17}{20}$ **17.** $\frac{14}{15}$ **19.** $\frac{11}{12}$ **21.** $\frac{5}{6}$ **23.** $\frac{2}{3}$ **25.** $\frac{1}{2}$ **27.** $\frac{1}{6}$ **29.** $\frac{1}{3}$ **31.** $\frac{19}{45}$ **33.** $\frac{5}{36}$ **35.** $\frac{1}{15}$ **37.** $\frac{23}{24}$ ton **39.** $\frac{7}{12}$ acre **41.** $\frac{9}{16}$ of the units **43.** 25 **45.** $4\frac{1}{2}$ **47.** $\frac{2}{5}$

Section 3.4 (page 191)

1. $70\frac{4}{7}$ **3.** $80\frac{3}{4}$ **5.** $61\frac{5}{8}$ **7.** $98\frac{2}{5}$ **9.** $152\frac{7}{10}$ **11.** $304\frac{11}{40}$ **13.** $42\frac{1}{8}$ **15.** $71\frac{3}{4}$ **17.** $3\frac{1}{2}$ **19.** $6\frac{17}{20}$ **21.** $6\frac{13}{48}$ **23.** $12\frac{7}{8}$ **25.** $20\frac{19}{24}$ **27.** $13\frac{59}{72}$ **29.** $162\frac{5}{12}$ **31.** $322\frac{11}{24}$ **33.** $15\frac{1}{8}$ **35.** $52\frac{1}{16}$ **37.** $8\frac{5}{8}$ hours **39.** $20\frac{1}{6}$ feet **41.** 38 hours **43.** $2\frac{3}{8}$ cubic yards **45.** $8\frac{7}{8}$ hours **47.** $14\frac{23}{24}$ tons **49.** $4\frac{11}{16}$ **51.** 84 **53.** 63 **55.** 27

Section 3.5 (page 201)

1.–7. $\frac{1}{4}$ $\frac{7}{9}$ $\frac{13}{6}$ $\frac{7}{3}$

0 1 2 3 4

1. 3. 5. 7.

9. < **11.** < **13.** < **15.** < **17.** < **19.** > **21.** $\frac{16}{25}$ **23.** $\frac{49}{225}$ **25.** $\frac{8}{27}$ **27.** $\frac{8}{729}$ **29.** $\frac{81}{16} = 5\frac{1}{16}$ **31.** $\frac{1}{32}$ **33.** 5 **35.** 67 **37.** 1 **39.** $\frac{3}{16}$ **41.** $\frac{1}{4}$ **43.** $\frac{1}{3}$ **45.** $1\frac{3}{5}$ **47.** $\frac{3}{8}$ **49.** $\frac{1}{4}$ **51.** $1\frac{1}{2}$ **53.** $\frac{1}{2}$ **55.** 3 **57.** $\frac{23}{112}$ **59.** $\frac{1}{10}$ **61.** eight thousand, four hundred thirty-six **63.** four million, seventy-one thousand, two hundred eighty

Chapter 3 Review Exercises (page 207)

1. 1 **2.** $\frac{7}{8}$ **3.** $\frac{1}{3}$ **4.** $\frac{5}{14}$ **5.** $\frac{1}{5}$ **6.** $\frac{1}{8}$ **7.** $\frac{13}{31}$ **8.** $\frac{19}{180}$ **9.** $\frac{5}{8}$ of journey **10.** $\frac{1}{7}$ **11.** 40 **12.** 60 **13.** 60 **14.** 180 **15.** 120 **16.** 3600 **17.** $\frac{10}{25}$ **18.** $\frac{28}{48}$ **19.** $\frac{85}{102}$ **20.** $\frac{45}{81}$ **21.** $\frac{63}{144}$ **22.** $\frac{12}{88}$ **23.** $\frac{5}{7}$ **24.** $\frac{7}{15}$ **25.** 1 **26.** $\frac{15}{16}$ **27.** $\frac{11}{12}$ **28.** $\frac{23}{36}$ **29.** $\frac{31}{48}$ **30.** $\frac{2}{9}$ **31.** $\frac{7}{16}$ **32.** $\frac{3}{16}$ **33.** $\frac{7}{24}$ **34.** $\frac{7}{30}$ **35.** $\frac{23}{24}$ cubic yard **36.** $\frac{109}{120}$ of the bricks **37.** 18 **38.** $42\frac{1}{8}$ **39.** $96\frac{2}{7}$ **40.** $31\frac{43}{80}$ **41.** $5\frac{1}{6}$ **42.** $6\frac{1}{6}$ **43.** $17\frac{1}{3}$ **44.** $85\frac{3}{4}$ **45.** $8\frac{5}{6}$ gallons **46.** $16\frac{7}{15}$ tons **47.** $15\frac{23}{24}$ pounds **48.** $4\frac{1}{16}$ acres

49.–52. $\frac{3}{8}$ $\frac{7}{4}$ $\frac{11}{5}$ $\frac{8}{3}$

0 1 2 3

49. 50. 52. 51.

53. < **54.** < **55.** > **56.** > **57.** < **58.** < **59.** < **60.** > **61.** $\frac{1}{9}$ **62.** $\frac{9}{64}$ **63.** $\frac{27}{125}$ **64.** $\frac{81}{4096}$ **65.** $\frac{2}{3}$ **66.** $2\frac{1}{4}$ **67.** $\frac{4}{49}$ **68.** 2 **69.** $\frac{3}{16}$ **70.** $2\frac{3}{8}$ **71.** $\frac{13}{15}$ **72.** $\frac{5}{8}$ **73.** $\frac{67}{86}$ **74.** $\frac{11}{16}$ **75.** $3\frac{17}{24}$ **76.** $24\frac{1}{4}$ **77.** $\frac{13}{40}$ **78.** $31\frac{43}{80}$ **79.** $317\frac{3}{4}$ **80.** $\frac{8}{11}$ **81.** $\frac{1}{250}$ **82.** $\frac{1}{2}$ **83.** $\frac{2}{9}$ **84.** < **85.** < **86.** < **87.** > **88.** 132 **89.** 1008 **90.** $\frac{240}{560}$ **91.** $\frac{108}{144}$ **92.** $28\frac{11}{16}$ inches **93.** $3\frac{13}{24}$ positions

Chapter 3 Test (page 211)

[3.1] 1. $\frac{3}{4}$ **2.** $\frac{4}{5}$ **3.** $\frac{1}{5}$ **4.** $\frac{1}{5}$ **[3.2] 5.** 16 **6.** 105 **7.** 315 **[3.3] 8.** $\frac{53}{48}$ or $1\frac{5}{48}$ **9.** $\frac{43}{35}$ or $1\frac{8}{35}$ **10.** $\frac{7}{18}$ **11.** $\frac{1}{56}$ **[3.4] 12.** $4\frac{3}{4}$ **13.** $8\frac{5}{8}$ **14.** $5\frac{9}{20}$ **15.** $2\frac{3}{4}$ **16.** $40\frac{29}{60}$ **17.** $148\frac{5}{8}$ **18.** $18\frac{1}{4}$ hours **19.** $5\frac{7}{12}$ tons **[3.5] 20.** < **21.** < **22.** $\frac{1}{2}$ **23.** $\frac{25}{144}$ **24.** $1\frac{1}{4}$ **25.** $1\frac{1}{2}$

Cumulative Review: Chapters 1–3 (page 213)

1. 9; 6 **2.** 8; 4 **3.** 26 **4.** 123 **5.** 77,321 **6.** 1,255,609 **7.** 176 **8.** 263 **9.** 3744 **10.** 2,801,695 **11.** 2850; 2800; 3000 **12.** 59,800; 59,800; 60,000 **13.** 189 **14.** 80 **15.** 216 **16.** 632 **17.** 2146 **18.** 274,625 **19.** 528,360 **20.** 4240 **21.** 14,500 **22.** 233,400 **23.** 800 boxes **24.** 90,000 revolutions **25.** 8 **26.** 158 **27.** 8975 R2 **28.** 582 **29.** 4710 **30.** 56 R42 **31.** 32 R166 **32.** \$2345 **33.** 580 cans **34.** $2 \cdot 3 \cdot 5$ **35.** $2^2 \cdot 5^2$ **36.** $2 \cdot 5^3$ **37.** 144 **38.** 400 **39.** 675 **40.** 3 **41.** 8 **42.** 15 **43.** 48 **44.** 44 **45.** proper **46.** proper **47.** improper **48.** $\frac{5}{8}$ **49.** $\frac{19}{25}$ **50.** $\frac{7}{20}$ **51.** $\frac{5}{6}$ **52.** $\frac{5}{22}$ **53.** 15 **54.** $\frac{3}{10}$ square inch **55.** \$4000 **56.** $2\frac{1}{2}$ **57.** $2\frac{3}{16}$ **58.** 8 **59.** $\frac{1}{2}$ **60.** $\frac{1}{15}$ **61.** $\frac{3}{5}$ **62.** 150 **63.** 300 **64.** 144 **65.** $\frac{40}{72}$ **66.** $\frac{77}{132}$ **67.** $\frac{27}{168}$ **68.** $\frac{60}{84}$ **69.** $\frac{7}{9}$ **70.** $\frac{15}{16}$ **71.** $\frac{43}{75}$ **72.** $5\frac{7}{8}$ **73.** $26\frac{7}{24}$ **74.** $2\frac{3}{8}$

75.–78. $\frac{1}{9}$ $\frac{3}{4}$ $\frac{5}{3}$ $\frac{10}{3}$

0 1 2 3 4

76. 75. 77. 78.

79. $<$ **80.** $<$ **81.** $<$ **82.** $\frac{1}{48}$ **83.** $\frac{9}{10}$

84. $\frac{32}{192}$

CHAPTER 4

Section 4.1 (page 223)

1. 4; 0 **3.** 9; 1 **5.** 4; 7 **7.** 2; 6 **9.** 1; 8

11. 6; 2 **13.** 6: tenths; 2: hundredths
15. 9: tenths; 6: hundredths; 5: thousandths **17.** 4: tens; 7: ones; 6: tenths; 9: hundredths; 1: thousandths
19. $\frac{7}{10}$ **21.** $\frac{2}{5}$ **23.** $\frac{9}{20}$ **25.** $\frac{22}{25}$ **27.** $\frac{41}{100}$
29. $\frac{1}{50}$ **31.** $\frac{81}{200}$ **33.** $\frac{919}{1000}$ **35.** $\frac{343}{500}$
37. seven tenths **39.** sixty-four hundredths
41. one hundred sixty-five thousandths **43.** twelve and four tenths **45.** one and seventy-two hundredths
47. 3.7 **49.** 27.32 **51.** 420.308 **53.** 760.921
55. four thousand three hundred twenty-two and five hundred thirty-one ten thousandths **57.** $625\frac{1071}{2500}$
59. 8240; 8200; 8000 **61.** 46,810; 46,800; 47,000

Section 4.2 (page 231)

1. 16.9 **3.** 965.498 **5.** 42.40 **7.** 27.906
9. 899.5 **11.** .0986 **13.** \$69 **15.** \$140
17. \$11,563 **19.** 78.414; 78.41; 78.4 **21.** .784; .78; .8 **23.** .088; .09; .1 **25.** \$18,765
27. \$109 **29.** \$380 **31.** \$705 **33.** 35.615048; 35.61505; 35.6150 **35.** 614.789915; 614.78992; 614.7899 **37.** 1000.005003; 1000.00500; 1000.0050
39. 22,223 **41.** 34,760 **43.** 110,053

Section 4.3 (page 235)

1. 186.09 **3.** 1259.14 **5.** 82.11 **7.** 59.132
9. 42.83 **11.** 240.694 **13.** 199.861
15. 59.323 **17.** problem: 63.65; estimate: 64
19. problem: 543.49; estimate; 544 **21.** problem: 1586.308; estimate: 1587 **23.** problem: 608.339; estimate: 609 **25.** \$41.64 **27.** 14.49 days
29. \$34.70 **31.** 7914.9 miles **33.** 41.1 hours
35. 40.5144 yd **37.** 321.165 yd **39.** 228.2152 ft
41. 211 **43.** 6569

Section 4.4 (page 241)

1. 54.3 **3.** 15.7 **5.** 264.77 **7.** 32.566
9. 38.554 **11.** 48.041 **13.** 12.848
15. 45.236 **17.** 393.507 **19.** 17.104
21. 12.38 **23.** 508.69 **25.** problem: 13.16; estimate: 13 **27.** problem: 4.849; estimate: 5
29. problem: 0.019; estimate: 0 **31.** problem: 384.615; estimate: 385 **33.** 48.0749 **35.** 62.0918
37. 26.15 hours **39.** \$10.88 **41.** \$32.42
43. 2475.8 miles **45.** 206.33869 **47.** 2.511317
49. \$75.53 **51.** 19.925 feet **53.** .932 inch
55. 2.812 inches **57.** 592,596 **59.** 8,354,745

Section 4.5 (page 247)

1. answer: 190.08; estimate: 200 **3.** answer: 124.2882; estimate: 144 **5.** answer: 300.38; estimate: 325 **7.** answer: 1924.165; estimate: 1967
9. .1344 **11.** 132.37 **13.** .7345 **15.** 1.08801
17. 28.534 **19.** .0032112 **21.** 10.0936
23. .0022288 **25.** .0445718 **27.** .109764
29. 36.54 **31.** 5.094 **33.** .0000252
35. .000006 **37.** \$181.20 **39.** \$187.17
41. \$209.28 **43.** \$444.68 **45.** \$10.20
47. \$25.80 **49.** \$86.70 **51.** \$140.80
53. \$37.70 **55.** \$695.04 **57.** \$1237
59. 42.96 gallons **61.** \$388.34 **63.** 21 R33
65. 2407 R1 **67.** 640 R8

Section 4.6 (page 257)

1. 4.14 **3.** 4.002 **5.** 3.685 **7.** .002
9. 9.000 **11.** 11.69 **13.** 3.038 **15.** .004
17. 1.246 **19.** .847 **21.** .009 **23.** 3000
25. 2.16 **27.** 1.918 **29.** 4000 **31.** .004
33. 3.957 **35.** 3.036 **37.** .4 **39.** 56
41. 36.859 **43.** 31.286 **45.** .541 **47.** .03
49. \$1.00 **51.** \$.42 **53.** \$.39 **55.** \$.05
57. 21.2 miles per gallon **59.** \$.30 **61.** \$4.71 per hour **63.** \$408.66 **65.** \$.085 **67.** \$.095
69. 10.25 **71.** 3.3 **73.** 73.4 **75.** 1.205
77. \$112.50 **79.** \$237.25 **81.** $<$ **83.** $>$
85. $<$

Section 4.7 (page 267)

1. .5 **3.** .333 **5.** .167 **7.** 5.125 **9.** .1
11. .667 **13.** .8 **15.** 3.88 **17.** .688
19. .857 **21.** .722 **23.** 15.472 **25.** .25
27. $\frac{2}{5}$ **29.** .875 **31.** $\frac{7}{8}$ **33.** $\frac{7}{20}$ **35.** .35
37. $\frac{13}{20}$ **39.** .833 **41.** $\frac{1}{10}$ **43.** $\frac{3}{20}$ **45.** .6
47. $<$ **49.** $<$ **51.** $>$ **53.** $<$ **55.** $>$
57. $<$ **59.** .062 **61.** $\frac{1}{2}$ **63.** .0909 **65.** $\frac{1}{12}$
67. $\frac{2}{13}$ **69.** $\frac{7}{8}$ **71.** .5399, .54, .5455
73. 5.4443, 5.79, 5.8, 5.804 **75.** .609, $\frac{5}{8}$, .628, .62812 **77.** $\frac{7}{8}$, .8751, .876, .8902 **79.** .043, $\frac{1}{20}$, .051, .506 **81.** $\frac{3}{4}$, .762, .7781, $\frac{12}{15}$ **83.** $\frac{6}{11}$, $\frac{5}{9}$, .571, $\frac{4}{7}$ **85.** .25, $\frac{4}{15}$, $\frac{3}{11}$, $\frac{1}{3}$ **87.** $\frac{3}{16}$, .188, $\frac{1}{5}$, $\frac{1}{4}$
89. $\frac{3}{4}$ **91.** $\frac{3}{4}$ **93.** $\frac{8}{11}$

Chapter 4 Review Exercises (page 277)

1. 5; 8 **2.** 7; 3 **3.** 6; 1 **4.** 5; 0 **5.** 7; 8 **6.** $\frac{9}{10}$ **7.** $\frac{3}{4}$ **8.** $\frac{3}{100}$ **9.** $\frac{7}{8}$ **10.** $\frac{3079}{5000}$ **11.** $\frac{1779}{2000}$ **12.** eight tenths **13.** forty-five hundredths **14.** twelve and eighty-seven hundredths **15.** three hundred thirty-five and seven hundred eight thousandths **16.** forty-two and one hundred five thousandths **17.** 275.6 **18.** 72.79 **19.** 896.25 **20.** .024 **21.** 87 **22.** $16 **23.** $82 **24.** $17,626 **25.** $79 **26.** $37 **27.** $19 **28.** 140.39 **29.** 44.85 **30.** 46.23 **31.** 134.855 **32.** 136.41 **33.** 160.056 **34.** 11.2 **35.** 10.7 **36.** 11.008 **37.** .931 **38.** 10.4 hours **39.** $171.64 **40.** $22.95 **41.** 16.97 miles **42.** 1.7472 **43.** .227106 **44.** .96807 **45.** .112 **46.** $100.86 **47.** $4.22 **48.** $128.52 **49.** $476.64 **50.** 14.5 **51.** 4.715 **52.** 15,500 **53.** .4 **54.** 88 shares **55.** $6.40 **56.** 19 **57.** .47 **58.** .8 **59.** .64 **60.** .667 **61.** .111 **62.** .67295, .68, .6821 **63.** .209, .2102, .215, .22 **64.** $\frac{3}{20}$, .159, $\frac{1}{6}$, .17 **65.** 122.623 **66.** 254.8 **67.** 3583.261 **68.** 34.532 **69.** 2.742 **70.** .121 **71.** 175.675 **72.** 9.04 **73.** 17.259 **74.** .027 **75.** .813 **76.** 80.872 **77.** 6.674 **78.** 3.7 **79.** .4 **80.** .55 **81.** .3 **82.** .143 **83.** .3749, $\frac{3}{8}$, .381, $\frac{2}{5}$ **84.** $\frac{8}{23}$, .348, $\frac{7}{20}$, .375 **85.** 15.5 miles per gallon **86.** $449.84 **87.** 18 months **88.** $146.74 **89.** $17.56 **90.** $208.80

Chapter 4 Test (page 283)

[4.1] **1.** $\frac{937}{1000}$ **2.** $\frac{3053}{10,000}$ **3.** $\frac{7}{8}$ **4.** $\frac{3}{400}$ **[4.2]** **5.** 725.609 **6.** .7051 **7.** $611 **8.** $7860 **[4.3]** **9.** 122.2087 **10.** 839.896 **[4.4]** **11.** 55.498 **12.** 666.819 **[4.5]** **13.** 1.3737 **14.** .0000483 **[4.6]** **15.** 6400 **16.** 605.03 **17.** .508, .5108, .516 **18.** .44, $\frac{9}{20}$, .451, .4606 **19.** .5983, $\frac{3}{5}$, .602, $\frac{2}{3}$ **[4.2–4.6]** **20.** 35.49 **21.** $64.76 **22.** 405.45 gallons **23.** 27.7 miles per gallon **24.** $2.78 per pound **25.** 147.2 therms

CHAPTER 5

Section 5.1 (page 291)

1. $\frac{8}{9}$ **3.** $\frac{3}{4}$ **5.** $\frac{3}{4}$ **7.** $\frac{8}{5}$ **9.** $\frac{3}{7}$ **11.** $\frac{13}{8}$ **13.** $\frac{6}{5}$ **15.** $\frac{5}{6}$ **17.** $\frac{8}{5}$ **19.** $\frac{1}{12}$ **21.** $\frac{5}{16}$ **23.** $\frac{3}{14}$ **25.** $\frac{7}{2}$ **27.** $\frac{4}{9}$ **29.** $\frac{98}{33}$ **31.** $\frac{7}{5}$ **33.** $\frac{2}{1}$ **35.** $\frac{38}{17}$ **37.** $\frac{34}{35}$ **39.** $\frac{1}{2}$ **41.** 43.01 **43.** 552.192 **45.** 9.465

Section 5.2 (page 299)

1. $\frac{2 \text{ miles}}{1 \text{ minute}}$ **3.** $\frac{5 \text{ yards}}{1 \text{ second}}$ **5.** $\frac{1 \text{ person}}{2 \text{ dresses}}$ **7.** $\frac{7 \text{ gallons}}{1 \text{ hour}}$ **9.** $\frac{1 \text{ chapter}}{12 \text{ pages}}$ **11.** $\frac{18 \text{ miles}}{1 \text{ gallon}}$ **13.** 12 dollars per hour **15.** 90 dollars per day **17.** 2.25 dollars per pound **19.** 20.2 miles per gallon **21.** 1.5 pounds per person **23.** 103.3 dollars per day **25.** 4 ounces for $.89 **27.** 20 ounces for $2.11 **29.** 18 ounces for $1.43 **31.** $1\frac{1}{3}$ pounds per person **33.** $12.26 per hour **35.** $11.50 **37.** $\frac{1}{3}$ crate per minute; 3 min/crate **39.** $11.25 **41.** $72\frac{1}{2}$ **43.** 552 **45.** $3\frac{1}{2}$

Section 5.3 (page 305)

1. $\frac{13}{15} = \frac{26}{30}$ **3.** $\frac{20}{45} = \frac{4}{9}$ **5.** $\frac{120}{150} = \frac{8}{10}$ **7.** $\frac{32}{48} = \frac{22}{33}$ **9.** $\frac{1\frac{1}{2}}{8} = \frac{6}{32}$ **11.** true **13.** true **15.** false **17.** true **19.** true **21.** false **23.** true **25.** false **27.** false **29.** true **31.** false **33.** true **35.** true **37.** false **39.** true **41.** $\frac{2}{3}$ **43.** $1\frac{1}{4}$ **45.** $\frac{4}{7}$ **47.** $6\frac{7}{8}$

Section 5.4 (page 316)

1. 4 **3.** 21 **5.** 20 **7.** 40 **9.** 5 **11.** 10 **13.** 5 **15.** 31 **17.** 200 **19.** $3\frac{1}{2}$ **21.** $2\frac{2}{3}$ **23.** $2\frac{5}{8}$ **25.** $3\frac{3}{8}$ **27.** $5\frac{1}{3}$ **29.** 6 **31.** $3\frac{1}{2}$ **33.** 21 **35.** 0 **37.** $\frac{6}{5} = \frac{18}{x}$; 15 **39.** $\frac{42}{a} = \frac{30}{60}$; 84 **41.** $\frac{k}{8} = \frac{\frac{4}{3}}{2}$; $5\frac{1}{3}$ **43.** $\frac{10}{4} = \frac{40}{16}$ **45.** $\frac{6}{9} = \frac{10}{15}$

Section 5.5 (page 321)

1. $75 **3.** $35 **5.** 7 lbs **7.** $153.45 **9.** 1080 miles **11.** $220 **13.** $270.90 **15.** $30 **17.** 1240 pounds **19.** $481 **21.** 12 **23.** 287 **25.** .0193

Chapter 5 Review Exercises (page 327)

1. $\frac{6}{11}$ **2.** $\frac{3}{2}$ **3.** $\frac{25}{2}$ **4.** $\frac{5}{9}$ **5.** $\frac{1}{2}$ **6.** $\frac{2}{3}$
7. 3 **8.** 50 **9.** 5 **10.** $\frac{3}{20}$ **11.** $\frac{1}{2}$ **12.** $\frac{1}{2}$
13. $\frac{1}{4}$ **14.** $\frac{24}{5}$ **15.** 6 **16.** $\frac{3}{4}$ **17.** $\frac{5}{18}$
18. $\frac{12}{5}$ **19.** $\frac{5}{14}$ **20.** $\frac{21}{8}$ **21.** $\frac{7}{10}$ **22.** $\frac{7}{6}$
23. 16 ounces for \$2.80 **24.** 30 ounces for \$1.50
25. $\frac{5}{10} = \frac{20}{40}$ **26.** $\frac{7}{2} = \frac{35}{10}$ **27.** $\frac{1\frac{1}{4}}{5} = \frac{3}{12}$
28. true **29.** false **30.** false **31.** true
32. true **33.** false **34.** true **35.** true
36. true **37.** 2 **38.** 20 **39.** 400 **40.** $12\frac{1}{2}$
41. $14\frac{2}{3}$ **42.** $3\frac{1}{8}$ **43.** $\frac{5}{8} = \frac{x}{16}$; 10 **44.** $\frac{7}{10} = \frac{14}{y}$; 20 **45.** 64 files **46.** $68\frac{3}{4}$ miles **47.** 18 gallons **48.** \$570 **49.** \$132.50 **50.** \$157.50
51. $\frac{110}{140} = \frac{22}{28}$ **52.** $\frac{8}{3\frac{1}{2}} = \frac{32}{14}$ **53.** $\frac{15}{5} = \frac{21}{7}$
54. $\frac{p}{6} = \frac{8}{15}$; $3\frac{1}{5}$ **55.** $\frac{4}{w} = \frac{15}{2}$; $\frac{8}{15}$ **56.** 105
57. 0 **58.** 128 **59.** false **60.** false
61. true **62.** $24\frac{1}{2}$ **63.** 28 **64.** $2\frac{7}{10}$ **65.** $\frac{8}{5}$
66. $\frac{33}{80}$ **67.** $\frac{15}{4}$ **68.** $\frac{3}{8}$ **69.** $\frac{1}{4}$ **70.** $\frac{13}{45}$
71. 20 ounces **72.** 9 gallons **73.** $2\frac{1}{2}$ bars
74. 50 minutes

Chapter 5 Test (page 333)

[5.1–5.2] **1.** $\frac{9}{11}$ **2.** $\frac{128}{225}$ **3.** $\frac{3}{2}$ **4.** $\frac{17}{56}$
5. 28 ounces for \$2.15 **6.** $\frac{1}{240}$ **7.** $\frac{1}{20}$
[5.3] **8.** $\frac{7}{15} = \frac{21}{45}$ **9.** $\frac{12}{7} = \frac{36}{21}$ **10.** true
11. false **12.** true **13.** true **[5.4]** **14.** 25
15. $2\frac{2}{3}$ **16.** $32\frac{1}{2}$ **[5.5]** **17.** 576 words
18. $184\frac{1}{2}$ miles **19.** $22\frac{1}{2}$ yards **20.** $13\frac{1}{2}$ miles

CHAPTER 6

Section 6.1 (page 339)

1. .50 or .5 **3.** .25 **5.** .65 **7.** 1.40 or 1.4
9. .07 **11.** .149 **13.** .008 **15.** .0025
17. 60% **19.** 75% **21.** 2% **23.** 12.5%
25. 45.3% **27.** 382% **29.** 150% **31.** 3.12%
33. 372.5% **35.** .75% **37.** .06 **39.** .65
41. 6.5% **43.** 260% **45.** .5% **47.** .478
49. 95%; 5% **51.** 30%; 70% **53.** 75%; 25%
55. .5 **57.** .75 **59.** .7 **61.** .8

Section 6.2 (page 349)

1. $\frac{1}{10}$ **3.** $\frac{1}{4}$ **5.** $\frac{17}{20}$ **7.** $\frac{3}{8}$ **9.** $\frac{1}{16}$ **11.** $\frac{1}{6}$
13. $\frac{1}{15}$ **15.** $\frac{1}{250}$ **17.** $1\frac{1}{5}$ **19.** $2\frac{1}{4}$ **21.** 75%
23. 70% **25.** 57% **27.** 20% **29.** 62.5%
31. 87.5% **33.** 48% **35.** 74% **37.** 55%
39. 83.3% **41.** 55.6% **43.** 14.3% **45.** $\frac{1}{10}$; 10% **47.** $\frac{1}{8}$; .125 **49.** $\frac{1}{5}$; 20% **51.** .3; 30%
53. $\frac{2}{5}$; 40% **55.** $\frac{3}{5}$; .6 **57.** .667; 66.7%
59. .75; 75% **61.** .833; 83.3% **63.** $\frac{9}{10}$; .9
65. .005; .5% **67.** $\frac{3}{1000}$; .003 **69.** $2\frac{1}{2}$; 250%
71. 3.25; 325% **73.** $\frac{15}{20} = \frac{3}{4}$; .75; 75%
75. $\frac{100}{500} = \frac{1}{5}$; .2; 20% **77.** $\frac{3}{5}$; .6; 60%
79. $\frac{18}{40} = \frac{9}{20}$; .45; 45% **81.** $\frac{1}{10}$; .1; 10%
83. 40 **85.** 5 **87.** 10

Section 6.3 (page 355)

1. 160 **3.** 900 **5.** 20 **7.** 416.7 **9.** 50
11. 300 **13.** $33\frac{1}{3}$ **15.** 13 **17.** 10.8
19. 13.2 **21.** 356 **23.** 25 **25.** 8.5
27. 12.25 **29.** 75% **31.** 57% **33.** 87.5%
35. 71.4%

Section 6.4 (page 359)

1. 65; 1000; 650 **3.** 81; what number; 748
5. 12; 48; what number **7.** 72; what number; 18
9. what percent; 104; 52 **11.** what percent; 50; 30
13. what percent; 508; 29.81 **15.** .68; 487; what number **17.** what percent; 24; 18 **19.** what percent; \$200; \$8 **21.** 7; 1500; what number
23. 15; 30; what number **25.** 16.8; what number; 504 **27.** $\frac{21}{x} = \frac{15}{30}$; 42 **29.** $\frac{y}{16} = \frac{\frac{5}{4}}{4}$; 5

Section 6.5 (page 369)

1. 121 **3.** 109.2 **5.** 4.8 **7.** 25 **9.** 247.5
11. 3.84 **13.** 742 **15.** 42.625 **17.** 21.6
19. 500 **21.** 240 **23.** 1200 **25.** 680

27. 2800 **29.** 50% **31.** 52% **33.** 2% **35.** 1.5% **37.** 12.9% **39.** 9.2% **41.** 54 children **43.** 836 drivers **45.** $185,500 **47.** 500 applicants **49.** 3% **51.** 1.5% **53.** 77.5% **55.** 150 students **57.** $850,000 **59.** $113.75 **61.** 40.8425 **63.** 10.0936 **65.** 3600 **67.** .25

Section 6.6 (page 379)

1. 248 **3.** 765 **5.** 83.2 **7.** 11.2 **9.** 90 **11.** 1029.2 **13.** 3 **15.** 1.44 **17.** 80 **19.** 500 **21.** 1250 **23.** 400 **25.** 200 **27.** 40% **29.** 60% **31.** 35% **33.** .25% **35.** 144% **37.** 1155 miles **39.** 27% **41.** 120 units **43.** 29.4 miles per gallon **45.** $p = 15$, $b = 375$; $a = 56.25$ **47.** $p = 72$; $b =$ what number; $a = 36$ **49.** $p =$ what percent; $b = 650$; $a = 146.15$

Section 6.7 (page 387)

1. $2; $102 **3.** $2.50; $52.50 **5.** $12.90; $227.90 **7.** $.40 $10.40; **9.** $10 **11.** $220 **13.** $1152 **15.** $622.50 **17.** $25; $75 **19.** $273; $507 **21.** $4.38; $13.12 **23.** $1.25; $11.25 **25.** $17.90 **27.** $31.50 **29.** 4% **31.** 32.5% **33.** $13 **35.** 6% **37.** $681.72 **39.** .5% **41.** $1560 **43.** $336; $624 **45.** $4276.97 **47.** $10,416 **49.** 62 **51.** 2.88 **53.** 800 **55.** 32%

Section 6.8 (page 395)

1. $10 **3.** $240 **5.** $8 **7.** $1170 **9.** $190 **11.** $116.25 **13.** $16 **15.** $55 **17.** $131.20 **19.** $104 **21.** $19.50 **23.** $676.67 **25.** $333 **27.** $803.40 **29.** $1725 **31.** $2287.35 **33.** $1869 **35.** $20,141.33 **37.** $108.80 **39.** $1560 **41.** $1280 **43.** $337.50 **45.** $38.06 **47.** $210.94 **49.** $\frac{6}{8}$ **51.** $\frac{20}{48}$ **53.** $\frac{45}{60}$ **55.** $\frac{60}{76}$

Chapter 6 Review Exercises (page 403)

1. .5 **2.** 2.5 **3.** .176 **4.** .00036 **5.** 280% **6.** 9% **7.** 37.5% **8.** .2% **9.** $\frac{19}{50}$ **10.** $\frac{1}{16}$ **11.** $\frac{2}{3}$ **12.** $\frac{1}{4000}$ **13.** 75% **14.** 87.5% **15.** 70% **16.** .5% **17.** .125 **18.** 12.5% **19.** $\frac{1}{20}$ **20.** 5% **21.** $\frac{7}{8}$ **22.** .875 **23.** 1000 **24.** 96 **25.** 40; 150; 60 **26.** what percent; 90; 73 **27.** 46; 1040; what number **28.** 30; what number; 418 **29.** what percent; 8; 3 **30.** 78; 640; unknown **31.** 60 **32.** 2833.6 **33.** 43.2 **34.** 2.8 **35.** 500 **36.** 2560 **37.** 242 **38.** 17,000 **39.** 50% **40.** 1.3% **41.** 9.5% **42.** 20.8% **43.** $8640 **44.** $77.49 **45.** 38.64 **46.** 36.48 **47.** .4% **48.** 125% **49.** 120 **50.** 143.75 **51.** $3; $103 **52.** $1.14; $58.14 **53.** $90 **54.** $1137 **55.** $10; $90 **56.** $87.83; $497.67 **57.** $14 **58.** $2268 **59.** $7.50 **60.** $172.50 **61.** $268 **62.** $1462.60 **63.** 35 **64.** 1400 **65.** 23.28 **66.** 150% **67.** .51 **68.** 312.5 **69.** 40% **70.** 249.4 **71.** .25 **72.** 1 **73.** 50% **74.** 680% **75.** .085 **76.** 71.9% **77.** .0025 **78.** .06% **79.** $\frac{9}{20}$ **80.** 50% **81.** $\frac{3}{8}$ **82.** 37.5% **83.** $\frac{29}{200}$ **84.** 20% **85.** $\frac{1}{200}$ **86.** .3% **87.** $975 **88.** $1037.50 **89.** 90% **90.** 5.1% **91.** $3750 **92.** 7.6% **93.** .6% **94.** $7990 **95.** $4403 **96.** 7.3%

Chapter 6 Test (page 409)

[6.1] **1.** .75 **2.** .0005 **3.** 25% **4.** 37.5% **[6.2]** **5.** $\frac{3}{8}$ **6.** $\frac{1}{200}$ **7.** 50% **8.** 62.5% **[6.3]** **9.** 270 **10.** 137.5 **[6.4]** **11.** $a = 138$; $b = 920$; $p = 15\%$ **[6.5]** **12.** 16 **13.** 750 **14.** 25% **[6.6]** **15.** $16,800 **[6.7]** **16.** $1.12; $29.12 **17.** $3.65; $76.65 **18.** 6% **19.** $393.60 **20.** 35% **21.** $3.84; $44.16 **22.** $65.63; $109.37 **[6.8]** **23.** $180 **24.** $210 **25.** $196 **26.** $19,092

Cumulative Review: Chapter 4–6 (page 411)

1. 6; 4 **2.** 4; 8 **3.** $\frac{3}{4}$ **4.** $\frac{1}{8}$ **5.** $\frac{1}{25}$ **6.** $\frac{7}{8}$ **7.** 61.6 **8.** .660 **9.** $25 **10.** $183 **11.** $4730 **12.** 90.94 **13.** 960.329 **14.** 58.26 **15.** 11.14 **16.** 804.824 **17.** 62.64 **18.** 2.6752 **19.** 22.17655 **20.** .000474 **21.** 18.36 **22.** 12.35 **23.** 110.5 **24.** .4 **25.** .875 **26.** .85 **27.** .857 **28.** $\frac{1}{2}$ **29.** $\frac{1}{2}$ **30.** $\frac{1}{8}$ **31.** $\frac{2}{15}$ **32.** $\frac{1}{4}$ **33.** true **34.** false **35.** false **36.** true **37.** true **38.** true **39.** 3 **40.** 3 **41.** 18 **42.** $9\frac{3}{8}$ **43.** 28 watches **44.** $59\frac{1}{2}$ ounces **45.** .25 **46.** 1.397 **47.** .00025 **48.** .0262 **49.** 68% **50.** 271% **51.** 2.3% **52.** $\frac{1}{8}$ **53.** $\frac{3}{8}$ **54.** 87.5% **55.** .5% **56.** 1000 **57.** 24 **58.** $p = 25$; $b = 240$; $a = 60$ **59.** p is unknown; $b = 300$; $a = 18$ **60.** p is unknown; 272; 204 **61.** 125 **62.** 64.8 **63.** 921 **64.** 195 **65.** 1600

66. 50% **67.** 40% **68.** 25% **69.** $.50; $25.50 **70.** $13.72; $209.72 **71.** $731.10 **72.** $2151.36 **73.** $26.60; $49.40 **74.** $53.66; $184.84 **75.** 35% **76.** 35% **77.** $485.52 **78.** $19,863.88

CHAPTER 7

Section 7.1 (page 423)

1. 36 inches **3.** 2 pints **5.** 5280 feet **7.** 2000 pounds **9.** 2 yards **11.** 2 gallons **13.** 1 ton **15.** 1.5 or $1\frac{1}{2}$ feet **17.** 14 quarts; $3\frac{1}{2}$ gallons **19.** 8.5 or $8\frac{1}{2}$ pounds **21.** 6000 pounds **23.** 88 ounces **25.** 186 inches **27.** 10,560 feet **29.** 5.25 or $5\frac{1}{4}$ miles **31.** 8.5 or $8\frac{1}{2}$ quarts = 2.125 or $2\frac{1}{8}$ gallons **33.** 18 pints = 9 quarts **35.** 4 miles **37.** 10,000 pounds **39.** 14,250 pounds **41.** 72 hours **43.** 29,700 seconds **45.** 95,040 inches **47.** $<$ **49.** $>$ **51.** $>$

Section 7.2 (page 433)

1. 2 feet 4 inches **3.** 2 pounds 4 ounces **5.** 4 quarts 1 pint or 1 gallon 1 pint **7.** 2 miles 1970 feet **9.** 10 days 12 hours or 1 week 3 days 12 hours **11.** 8 gallons 2 quarts **13.** 3 yards 1 foot 6 inches **15.** 3 weeks 4 days 12 hours **17.** 6 gallons 3 quarts 1 pint **19.** 11 feet 1 inch **21.** 10 quarts 1 pint or 2 gallons 2 quarts 1 pint **23.** 11 pounds 6 ounces **25.** 11 weeks 1 day 13 hours **27.** 1 gallon 1 quart **29.** 2 tons 300 pounds **31.** 2 feet **33.** 6 days 16 hours 57 minutes **35.** 12 feet 8 inches **37.** 32 tons 800 pounds **39.** 756 yards 2 feet **41.** 15 yards **43.** 2 gallons 1 quart **45.** 162 pounds 10 ounces **47.** 3 yards 1 foot **49.** 6 days 17 hours **51.** 6 gallons 2 quarts 1 pint **53.** 10 weeks 6 days 7 hours 54 minutes **55.** 0; 7 **57.** 7; 1

Section 7.3 (page 445)

1. 100 centimeters **3.** 1000 millimeters **5.** .01 meters **7.** answer varies—about 8 centimeters **9.** answer varies—about 7 millimeters **11.** 8000 millimeters **13.** 1500 millimeters **15.** .525 meters **17.** 85 meters **19.** 1186 millimeters **21.** 5300 meters **23.** 27.5 kilometers **25.** 240,000 centimeters **27.** 820 millimeters **29.** 480 meters **31.** smaller **33.** 981 centimeters **35.** .0370316 kilometers **37.** $\frac{7}{8}$ **39.** $\frac{2}{25}$ **41.** $\frac{3}{8}$

Section 7.4 (page 451)

1. 600 centiliters **3.** 8700 milliliters **5.** 9.25 liters **7.** 8.974 liters **9.** .0084 kiloliters **11.** 8640 liters **13.** 8 kilograms **15.** 5200 grams **17.** 4200 milligrams **19.** 76.34 grams **21.** unreasonable **23.** unreasonable **25.** reasonable **27.** unreasonable **29.** 200 nickels **31.** .2 grams **33.** 8.172 **35.** 38.6232 **37.** 661.0184

Section 7.5 (page 457)

1. 21.9 yards **3.** 262.4 feet **5.** 4.9 meters **7.** 897.9 meters **9.** 21,428.8 grams **11.** 30.3 quarts **13.** 106.7 liters **15.** answer varies **17.** answer varies **19.** $4.17 **21.** $9.72 **23.** 20°C **25.** 40°C **27.** 37°C **29.** 95°F **31.** 77°F **33.** 59°F **35.** 58°C **37.** 932°F **39.** 32 **41.** 81 **43.** 41

Chapter 7 Review Exercises (page 463)

1. 16 ounces **2.** 3 feet **3.** 2000 pounds **4.** 4 quarts **5.** 5280 feet **6.** 36 inches **7.** 4 yards **8.** 168 hours **9.** 21,120 feet **10.** 2 feet 7 inches **11.** 3 miles 720 feet **12.** 4 yards 2 feet 3 inches **13.** 3 quarts 1 pint 1 cup **14.** 8 feet 3 inches **15.** 4 pounds 14 ounces **16.** 28 pounds 2 ounces **17.** 5 feet 4 inches **18.** 100 centimeters **19.** 1000 meters **20.** 800 centimeters **21.** 3.781 meters **22.** 56 millimeters **23.** 1270 meters **24.** 300 centiliters **25.** .68 kilograms **26.** 4850 milligrams **27.** 17,600 liters **28.** 530 centigrams **29.** 5 kilograms **30.** 10.9 yards **31.** 55.1 inches **32.** 67.1 miles **33.** 1287.4 kilometers **34.** 21.8 liters **35.** 44.0 quarts **36.** 88.5 kilometers per hour **37.** $7.60 **38.** 25°C **39.** 80°C **40.** 33.3°C **41.** 70.6°C **42.** 122°F **43.** 536°F **44.** 212°F **45.** 82.2°C **46.** 7 quarts 1 pint **47.** 4 weeks 5 days 18 hours **48.** 21 weeks 2 days 12 hours **49.** 1 gallon 1 quart 1 pint **50.** 27.5 millimeters **51.** 42.885 meters **52.** 130,000 centimeters **53.** 783.5 centigrams **54.** .0675 kilograms **55.** 89.4 liters **56.** 80 ounces **57.** $7\frac{1}{2}$ feet **58.** 52 weeks **59.** 3 days 3 hours 30 minutes **60.** 4 miles 799 feet 2 inches **61.** 7.3 meters **62.** 5.3 meters **63.** 198.0 liters **64.** 1589 gallons **65.** 52.8¢ **66.** 264 gallons **67.** 91.4 meters **68.** 2760°C

Chapter 7 Test (page 467)

[7.1] **1.** 4 **2.** 12 **3.** 9.5 or $9\frac{1}{2}$ **4.** 5.25 or $5\frac{1}{4}$ **5.** 8 **6.** 7.2 or $7\frac{1}{5}$ **[7.2]** **7.** 17 feet 1 inch **8.** 3 gallons 1 pint **9.** 34 pounds 14 ounces **10.** 1 yard 2 feet 2 inches **[7.3–7.4]** **11.** 2.5 meters **12.** 460 centimeters **13.** 40,000 centimeters **14.** 8.412 kilograms **15.** 900 centiliters **16.** 156 milliliters **17.** .198 kilometers **18.** 2610 grams **[7.5]** **19.** 5.6 meters **20.** 32.7 kilograms **21.** 2.1 pounds **22.** 10.3 kilometers **23.** $1.25 **24.** 3.5 kilograms **25.** 24.4°C **26.** 341.6°F

CHAPTER 8

Section 8.1 (page 473)

1. line **3.** ray **5.** ray **7.** perpendicular **9.** parallel **11.** parallel **13.** $\angle AOS$ or $\angle SOA$ **15.** $\angle POS$ or $\angle SOP$ **17.** $\angle AOC$ or $\angle COA$ **19.** right **21.** acute **23.** straight **25.** obtuse **27.** false **29.** false **31.** 81 **33.** 90 **35.** 69

Section 8.2 (page 479)

1. 50° **3.** 25° **5.** 50° **7.** 150° **9.** $\angle EOD$ and $\angle COD$; $\angle AOB$ and $\angle BOC$
11. $\angle COA$ and $\angle AOD$; $\angle DOF$ and $\angle COF$
$\angle COA$ and $\angle COF$; $\angle AOD$ and $\angle DOF$
13. $\angle SON \cong \angle TOM$; $\angle TOS \cong \angle MON$ **15.** 63° **17.** 37° **19.** 54° **21.** 126°
23. $\angle ABC \cong \angle BCD$; $\angle DCG \cong \angle EBA$ **25.** 32 **27.** 36

Section 8.3 (page 487)

1. $P = 28$ yards; $A = 48$ square yards
3. $P = 38$ feet; $A = 70$ square feet
5. $P = 22$ feet; $A = 10$ square feet
7. $P = 68.2$ meters; $A = 273.48$ square meters
9. $P = 196.2$ feet; $A = 1674.2$ square feet
11. $P = 33.6$ yards; $A = 70.56$ square yards
13. $P = 38$ meters; $A = 39$ square meters
15. $P = 98$ meters; $A = 492$ square meters
17. $P = 52$ centimeters; $A = 150$ square centimeters
19. 720 square inches **21.** \$624 **23.** \$1413.12 **25.** \$69.12 **27.** .3 meters **29.** 600 inches **31.** 75 yards

Section 8.4 (page 493)

1. 208 meters **3.** 207.2 meters **5.** 752.8 centimeters **7.** 984 square yards **9.** 19.25 square feet **11.** 3099.6 square centimeters **13.** \$2968 **15.** \$940.50 **17.** 4670 square centimeters **19.** 25,344 square feet **21.** 52 **23.** $\frac{200}{3}$

Section 8.5 (page 499)

1. 27 yards **3.** 37 yards **5.** 302.46 square centimeters **7.** 198 square meters **9.** 1196 square meters **11.** 868 square centimeters
13. **(a)** 32 square meters **(b)** 13.5 square meters
15. 50.24 **17.** .2414

Section 8.6 (page 507)

1. $d = 94$ meters **3.** $r = 29.45$ meters **5.** $C = 69.1$ feet; $A = 379.9$ square feet **7.** $C = 157$ meters; $A = 1962.5$ square meters **9.** $C = 47.1$ centimeters; $A = 176.6$ square centimeters **11.** $C = 23.6$ meters; $A = 44.2$ square meters **13.** $C = 57.3$ centimeters; $A = 261.2$ square centimeters **15.** 57 square centimeters **17.** 4221.3 square meters **19.** 12.6 meters **21.** 49,062.5 square feet **23.** 73.0 square feet (rounded) **25.** \$1170.33 (rounded) **27.** $\frac{3}{2}$ **29.** $\frac{2}{7}$

Section 8.7 (page 517)

1. 528 cubic centimeters **3.** 1728 cubic inches
5. 44,579.6 cubic meters **7.** 3617.3 cubic inches
9. 471 cubic feet **11.** 5086.8 cubic feet
13. 513 cubic centimeters **15.** 175 cubic meters
17. 418.7 cubic meters **19.** 1969.1 cubic centimeters
21. 427.3 cubic centimeters **23.** 301.4 cubic meters
25. 9 cubic yards **27.** 83.468 **29.** 8.93

Section 8.8 (page 525)

1. 5 **3.** 8 **5.** 3.317 **7.** 6.708 **9.** 8.544 **11.** 10.296 **13.** 13.784 **15.** 13.153
17. $a^2 = 36$; $b^2 = 64$; $c^2 = 100$; side $c = 10$; true
19. 13 centimeters **21.** 17 inches **23.** 8 inches
25. $\sqrt{40} = 6.325$ inches **27.** $\sqrt{65} = 8.062$ inches
29. $\sqrt{195} = 13.964$ centimeters
31. $\sqrt{14.8} = 3.847$ inches **33.** $\sqrt{65.01} = 8.063$ centimeters **35.** $\sqrt{358.8} = 18.942$ inches
37. $\sqrt{65} = 8.062$ feet **39.** $\sqrt{346} = 18.601$ miles
41. $\sqrt{58} = 7.616$ centimeters **43.** $BC = 8$ feet; $BD = 6.245$ feet **45.** 8 **47.** $\frac{15}{16}$

Section 8.9 (page 533)

1. similar **3.** not similar **5.** similar
7. B and Q; C and R; A and P; AB and PQ; BC and QR; AC and PR **9.** P and S; N and R; M and Q; MP and QS; MN and QR; NP and RS **11.** $\frac{9}{6}$; $\frac{12}{8}$; $\frac{15}{10}$; yes
13. $a = 5$; $b = 3$ **15.** $a = 6$; $b = 7.5$
653 **17.** $x = 6$ **19.** $n = 24$ **21.** $n = 110$
23. 87.5 meters; 75 meters **25.** $y = 12$
27. $m = 85.3$ **29.** 8 **31.** 21

Chapter 8 Review Exercises (page 543)

1. line segment **2.** line **3.** ray **4.** parallel
5. perpendicular **6.** intersecting **7.** right
8. obtuse **9.** straight **10.** acute
11. $\angle 1$ and $\angle 2$; $\angle 4$ and $\angle 5$
$\angle 1$ and $\angle 5$; $\angle 2$ and $\angle 4$
12. $\angle 1$ and $\angle 2$; $\angle 3$ and $\angle 4$
$\angle 6$ and $\angle 7$; $\angle 4$ and $\angle 6$
$\angle 3$ and $\angle 7$
13. $\angle AOB$ and $\angle BOC$; $\angle BOC$ and $\angle COD$
$\angle COD$ and $\angle DOA$; $\angle DOA$ and $\angle AOB$
14. $\angle EOH$ and $\angle HOG$; $\angle HOG$ and $\angle GOF$
$\angle GOF$ and $\angle FOE$; $\angle FOE$ and $\angle EOH$
15. $\angle 1$ and $\angle 3$; $\angle 2$ and $\angle 4$ **16.** $\angle 1$ and $\angle 3$; $\angle 2$ and $\angle 4$ **17.** 38 meters **18.** 170.8 centimeters

19. 120 meters **20.** 88.8 meters **21.** 486 square millimeters **22.** 75.1 square feet **23.** $42\frac{1}{4}$ square meters or 42.3 square meters to the nearest tenth **24.** 1226.6 square centimeters **25.** 567 square feet **26.** 2074.0 square yards **27.** 30 meters **28.** 43.4 centimeters **29.** 10,812 square centimeters **30.** 2039.84 square feet **31.** $12\frac{11}{16}$ square feet **32.** 145.6 meters **33.** 17 meters **34.** $C = 6.3$ centimeters; $A = 3.1$ square centimeters **35.** $C = 109.3$ inches; $A = 950.7$ square inches **36.** $C = 37.7$ meters; $A = 113.0$ square meters **37.** 20.3 square feet **38.** 70 cubic meters **39.** 74.4 cubic centimeters **40.** 852,948 cubic millimeters **41.** 588.7 cubic centimeters **42.** 57.9 cubic centimeters **43.** 232.1 cubic centimeters **44.** 1356.5 cubic meters **45.** 94,866.5 cubic meters **46.** $3\frac{3}{4}$ cubic meters **47.** 96 cubic centimeters **48.** 45,000 cubic millimeters **49.** 267.9 cubic meters **50.** 1436.0 cubic meters **51.** 267,946.7 cubic millimeters **52.** 549.5 cubic centimeters **53.** 1808.6 cubic meters **54.** 2512 cubic centimeters **55.** 2.646 **56.** 4.359 **57.** 5.196 **58.** 5.916 **59.** 7.616 **60.** 11 **61.** 12 **62.** 13 **63.** 17 inches **64.** 7 centimeters **65.** 10.198 centimeters **66.** 7.211 inches **67.** 9.381 centimeters **68.** 12 centimeters **69.** $y = 30$; $x = 34$ **70.** $y = 15$; $x = 18$ **71.** $m = 13.5$; $n = 12$ **72.** 35 ft **73.** 42 meters **74.** $P = 90$ meters; $A = 92$ square meters **75.** $P = 282$ centimeters; $A = 4190$ square centimeters **76.** 100.5 cubic feet **77.** 10.6 cubic inches **78.** 561 cubic centimeters **79.** 1271.7 cubic centimeters **80.** 1282.2 cubic centimeters **81.** 300 cubic centimeters **82.** $\frac{20}{3}$ **83.** 56 cubic meters **84.** 8 **85.** 1020 square meters **86.** 229 square feet **87.** 132 square feet **88.** 5376 square centimeters **89.** 498.9 square feet **90.** 447.9 square centimeters

Chapter 8 Test (page 551)

[8.1] **1.** ray **2.** line **3.** acute **4.** right
[8.2] **5.** 18° **6.** 70°
7. ∠1 and ∠4; ∠2 and ∠5; ∠3 and ∠6
8. ∠1 and ∠4; ∠2 and ∠5; ∠3 and ∠6
[8.3] **9.** $P = 172.4$ centimeters; $A = 1843.92$ square centimeters **10.** $P = 59.6$ meters; $A = 222.01$ square meters **[8.4]** **11.** 3312 square meters **12.** 1591 square centimeters **[8.5]** **13.** 48 square meters **14.** 90.86 square centimeters **15.** 119.3 meters **[8.6]** **16.** 452.2 square meters **17.** 1218.6 square centimeters **18.** 56.5 square meters **19.** 78.5 inches **[8.7]** **20.** 6480 cubic meters **21.** 3052.08 cubic meters **22.** 301.44 cubic meters **[8.8]** **23.** 12 centimeters **24.** 9.220 centimeters **[8.9]** **25.** $q = 10$; $p = 12$ **26.** $y = 12$; $z = 6$

CHAPTER 9

Section 9.1 (page 557)

1. +12 **3.** −12 **5.** +18,000 **7.** positive **9.** negative **11.** neither **13.** positive

15.
−5 −4 −3 −2 −1 0 1 2 3 4 5

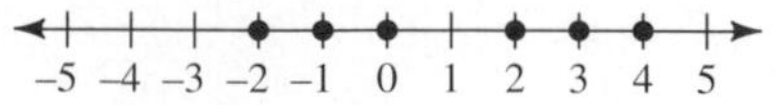

17.
−5 −4 −3 −2 −1 0 1 2 3 4 5

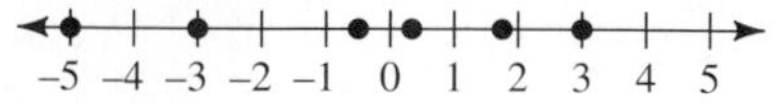

19.
−5 −4 −3 −2 −1 0 1 2 3 4 5

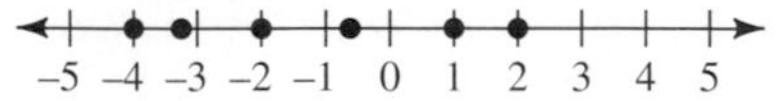

21.
−5 −4 −3 −2 −1 0 1 2 3 4 5

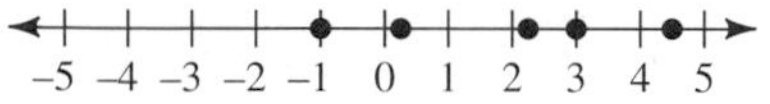

23.
−15 −13 −11 −9 −7 −5

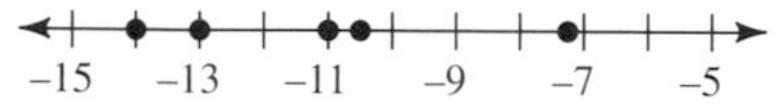

25. < **27.** > **29.** < **31.** > **33.** < **35.** < **37.** > **39.** > **41.** 5 **43.** 5 **45.** 32 **47.** 251 **49.** 0 **51.** $\frac{1}{2}$ **53.** 9.5 **55.** 8.3 **57.** $\frac{3}{4}$ **59.** $\frac{9}{7}$ **61.** −9 **63.** −4 **65.** −2 **67.** 5 **69.** 11 **71.** −16 **73.** $-\frac{4}{3}$ **75.** $\frac{1}{2}$ **77.** −5.2 **79.** 1.4 **81.** −2 **83.** true **85.** true **87.** true **89.** $\frac{37}{20}$ or $1\frac{17}{20}$ **91.** $4\frac{1}{2}$

Section 9.2 (page 569)

1. $-2 + 5 = 3$

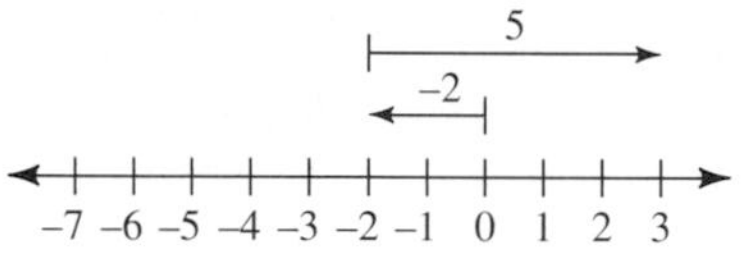

5
−2
−7 −6 −5 −4 −3 −2 −1 0 1 2 3

3. $-5 + (-2) = -7$

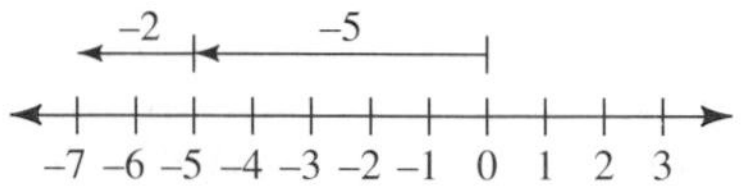

−2 −5
−7 −6 −5 −4 −3 −2 −1 0 1 2 3

5. $3 + (-4) = -1$

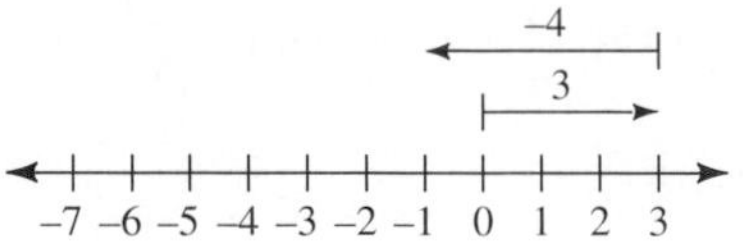

−4
3
−7 −6 −5 −4 −3 −2 −1 0 1 2 3

7. −3 **9.** 5 **11.** −7 **13.** −9 **15.** −17 **17.** −7.8 **19.** −6.8 **21.** $\frac{1}{4}$ **23.** $-\frac{3}{8}$

25. $-\frac{26}{9}$ or $-2\frac{8}{9}$ 27. -11 29. -6 31. -3
33. 9 35. 8 37. -154 39. $-\frac{1}{4}$ 41. 14
43. -3 45. -12 47. -10 49. -5
51. -17 53. 20 55. 16 57. $-\frac{3}{2} = -1\frac{1}{2}$
59. 5 61. -3.6 63. -3 65. -6
67. -4 69. -10 71. 12 73. 9 75. -1
77. -4.4 79. -11 81. -6 83. $4\frac{3}{4}$
85. -6 87. $\frac{24}{25}$ 89. $\frac{77}{8}$ 91. $\frac{3}{5}$

Section 9.3 (page 575)

1. -14 3. -27 5. -40 7. -50
9. -48 11. 32 13. 75 15. 133 17. 72
19. -9 21. $\frac{3}{2}$ or $1\frac{1}{2}$ 23. -4 25. $-\frac{1}{3}$
27. $-\frac{14}{3}$ or $-4\frac{2}{3}$ 29. $-\frac{5}{6}$ 31. $\frac{14}{3} = 4\frac{2}{3}$
33. -42.3 35. 205.2 37. -31.62
39. 47.04 41. -2 43. -5 45. -7
47. -43 49. 10 51. 2 53. $-\frac{4}{5}$ 55. $\frac{2}{5}$
57. $\frac{2}{3}$ 59. $\frac{1}{3}$ 61. -8 63. -5.36
65. 4.73 67. -3.9 69. 5.3 71. 12
73. .036 75. $\frac{7}{40}$ 77. 192 79. 3 81. 9
83. 10

Section 9.4 (page 583)

1. 10 3. -19 5. 25 7. -16 9. -18
11. 17 13. 16 15. 30 17. 23 19. -43
21. 19 23. -3 25. 5 27. 47 29. 41
31. -4 33. 13 35. 126 37. 35
39. -23 41. -6 43. 8 45. 2.276
47. -1.701 49. -9.83 51. $-1\frac{3}{10}$ 53. $1\frac{1}{4}$
55. $\frac{4}{5}$ 57. -1200 59. 3 61. -20.8
63. $\frac{13}{48}$ 65. -27 67. 7 69. 6 71. $\frac{4}{3}$
73. $\frac{47}{20}$ or $2\frac{7}{20}$

Section 9.5 (page 589)

1. 16 3. -2 5. 8 7. -8 9. -8
11. 6 13. 2 15. -3 17. $-5\frac{1}{2}$ 19. 28
21. 28 23. 150 25. $31\frac{1}{2}$ 27. 80 29. 720
31. 43.96 33. 33 35. -40 37. -1
39. 0

Section 9.6 (page 597)

1. yes 3. yes 5. no 7. 4 9. 11
11. 8 13. 10 15. -6 17. -6 19. -13
21. 8 23. -1 25. -3 27. $\frac{11}{8}$ or $1\frac{3}{8}$
29. $5\frac{2}{3}$ 31. $3\frac{9}{10}$ 33. 6.58 35. -1.51
37. 2 39. 4 41. -8 43. -6 45. 8
47. -7 49. 1.1 51. 34 53. 50 55. -36
57. 10 59. -8 61. 16 63. -28 65. 16
67. 1.36 69. -4.94 71. 19 73. 9
75. $\frac{8}{21}$ 77. $-\frac{33}{13}$ or $-2\frac{7}{13}$ 79. 1

Section 9.7 (page 605)

1. 1 3. 11 5. 2 7. -2 9. -4 11. 6
13. 88 15. $5r + 15$ 17. $-3m - 18$
19. $-2y + 6$ 21. $-5z + 45$ 23. $7m$
25. $18y$ 27. $5z$ 29. $-10a$ 31. 5 33. 7
35. -6 37. 2 39. 4 41. -2 43. -2
45. -1 47. 20 49. 2 51. -14
53. \$25.20 55. \$137.50

Section 9.8 (page 613)

1. $14 + x$ 3. $4 + x$ 5. $x + 6$ 7. $x - 9$
9. $x - 4$ 11. $2x$ 13. $3x$ 15. $\frac{x}{2}$
17. $2x + 3$ 19. $5x + 4x$ 21. 7 23. 8
25. 20 27. 34 centimeters 29. 7 days 31. 19 meters 33. 227 meters, 227 meters, 252 meters
35. 5 years 37. 320 39. 300

Chapter 9 Review Exercises (page 621)

1. −7 −6 −5 −4 −3 −2 −1 0 1 2 3 4

2. −7 −6 −5 −4 −3 −2 −1 0 1 2 3 4

3. −7 −6 −5 −4 −3 −2 −1 0 1 2 3 4

4. −8 −7 −6 −5 −4 −3 −2 −1 0 1 2 3

5. $>$ 6. $<$ 7. $>$ 8. $>$ 9. 8 10. 19
11. 7 12. 0 13. 2 14. 9 15. -19
16. -5 17. -9.7 18. -6.8 19. $\frac{3}{10}$
20. -7 21. -35 22. -2 23. 20
24. -6 25. 9 26. $-\frac{2}{3}$ 27. 4 28. 15
29. $\frac{5}{8}$ 30. -3 31. -11 32. 6 33. 5
34. $-\frac{7}{8}$ 35. -24 36. -20 37. 15
38. $\frac{5}{12}$ 39. -37.38 40. -3 41. -5

42. 20 **43.** −21 **44.** 23 **45.** −1
46. −2 **47.** −34 **48.** −32 **49.** −100
50. 117 **51.** 14 **52.** 1.328 **53.** −2.409
54. $\frac{1}{7}$ **55.** 20 **56.** −4 **57.** −36 **58.** 2
59. 33 **60.** 35 **61.** 27 **62.** 9 **63.** 10
64. −3 **65.** 1 **66.** $1\frac{1}{2}$ **67.** −8.05 **68.** 7
69. −8 **70.** 20 **71.** −55 **72.** −20
73. $-\frac{2}{3}$ **74.** 3 **75.** 2 **76.** $6r - 30$
77. $11p + 77$ **78.** $-9z + 27$ **79.** $11r$
80. $-5z$ **81.** $-9p$ **82.** 7 **83.** −5
84. −3 **85.** $18 + x$ **86.** $\frac{1}{2}x$ **87.** $4x + 6$
88. $3x + 7$ **89.** −3 **90.** 7 days **91.** 37 cm
92. 4 years **93.** $\frac{10}{3}$ **94.** 1.4 **95.** 3
96. $-\frac{7}{36}$ **97.** −11 **98.** $\frac{8}{15}$ **99.** 4 **100.** 6
101. 4 **102.** $\frac{3}{2}$ or $1\frac{1}{2}$ **103.** $2x - 8$
104. $\frac{1}{3}x + 7$ **105.** 13 inches **106.** $385

Chapter 9 Test (page 627)

[9.1] **1.** 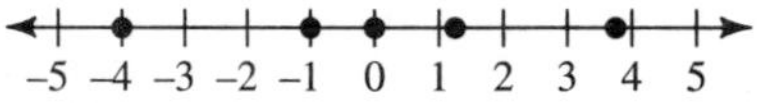

2. < **3.** 7 [9.2] **4.** −1 **5.** −13
6. $-1\frac{5}{8}$ **7.** −7 **8.** 16 **9.** $\frac{1}{4}$ [9.3]
10. −32 **11.** 84 **12.** −25 [9.4] **13.** −43
14. 34 [9.5] **15.** −38 **16.** 29 **17.** 110
[9.6] **18.** 5 **19.** 7 **20.** −9 **21.** −8
[9.7] **22.** −5 **23.** 2 [9.8] **24.** 2 **25.** 57 centimeters **26.** 9%

Cumulative Review: Chapters 7–9 (page 629)

1. 12 inches **2.** 3 feet **3.** 40 ounces **4.** 8 yards 1 foot 6 inches **5.** 3 days 22 hours **6.** 43 gallons 3 quarts 1 pint **7.** 6 days 19 hours **8.** 5.978 meters **9.** 4.317 kilograms **10.** 2830 liters **11.** 14.2 liters **12.** 29.3 meters **13.** 74.4 miles **14.** 21°C **15.** 216°C **16.** 302°F **17.** P = 35 meters; A = 67.3 square meters **18.** P = 59.2 centimeters; A = 219 square centimeters **19.** 110.9 square meters **20.** 276.1 square meters **21.** 220.2 meters **22.** 11.49 feet **23.** 113.5 square meters **24.** 241,286 square centimeters **25.** 18.96 meters **26.** 6.7 feet **27.** C = 47.7 meters; A = 181.4 square meters **28.** C = 449.0 meters; A = 16,052.5 square meters **29.** 1689.1 square meters **30.** 22.7 square centimeters **31.** 5042.7 square meters **32.** 43.3 square meters **33.** 3160.5 square meters **34.** 58.5 cubic meters **35.** 7.2 cubic feet **36.** > **37.** >
38. > **39.** < **40.** 4 **41.** 22 **42.** 86
43. 0 **44.** 9 **45.** 8.17 **46.** $-\frac{13}{40}$ **47.** −3
48. −2 **49.** 20 **50.** −25 **51.** −11
52. 19 **53.** $-\frac{5}{8}$ **54.** 0 **55.** −7 **56.** −17
57. 15 **58.** 34 **59.** $-\frac{2}{9}$ **60.** −25.5
61. −18 **62.** −120 **63.** −77 **64.** 45
65. 56 **66.** 18.6 **67.** $-\frac{10}{11}$ **68.** $\frac{1}{6}$ **69.** −4
70. 5 **71.** −20 **72.** $\frac{1}{10}$ **73.** 19
74. −147 **75.** −.28 **76.** 9 **77.** −13
78. 25 **79.** 7 **80.** 62.6 **81.** 5760 **82.** 8
83. 15 **84.** −4 **85.** $-\frac{1}{4}$ **86.** 6 **87.** 26
88. −20 **89.** −18 **90.** 9 **91.** $-\frac{24}{35}$
92. 2 **93.** −4 **94.** −7 **95.** −1
96. $4a + 24$ **97.** $-4m + 12$ **98.** $-6r + 42$
99. 5 **100.** 72 centimeters

CHAPTER 10

Section 10.1 (page 639)

1. $10,400 **3.** $\frac{3800}{10400} = \frac{19}{52}$ **5.** $\frac{300}{4200} = \frac{1}{14}$
7. history **9.** $\frac{2000}{12000} = \frac{1}{6}$ **11.** $\frac{700}{2500} = \frac{7}{25}$
13. $174,000 **15.** $58,000 **17.** $87,000
19. $3920 **21.** $2548 **23.** $1960 **25.** 20%
27. 36° **29.** 5%; 18°

31.

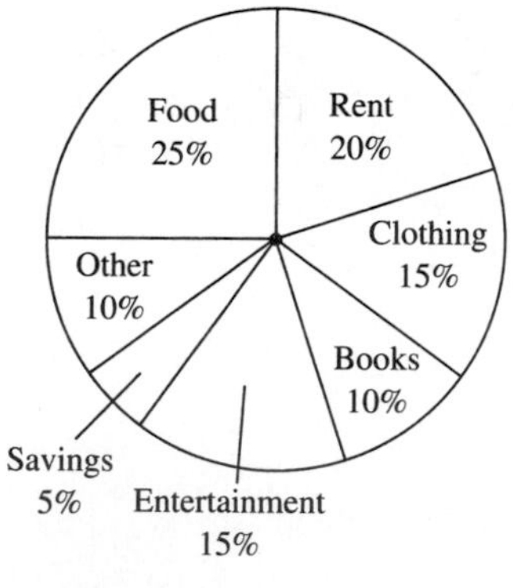

33. **(a)** 90°; 36°; 54°; 54°; 72°; 54° **(b)**

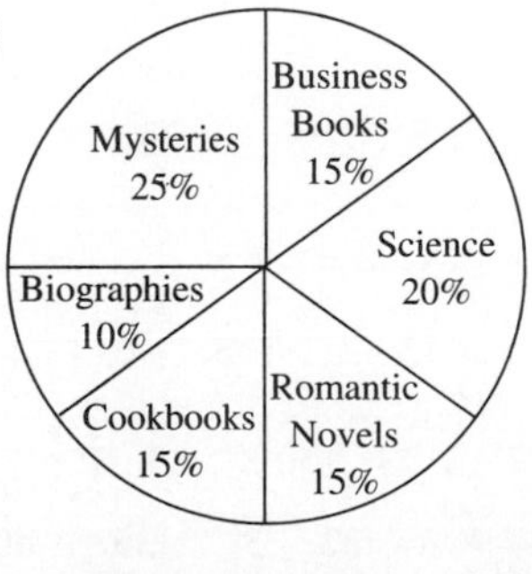

35. < **37.** < **39.** < **41.** <

Section 10.2 (page 649)

1. 3000 **3.** 2000 **5.** July 4 **7.** May; 1000 **9.** 200 **11.** 300 **13.** 200,000 **15.** 1987 **17.** 500,000 gallons **19.** April **21.** 300 **23.** 200 **25.** $3,000,000 **27.** $1,500,000 **29.** $3,500,000 **31.** $40,000 **33.** $25,000 **35.** $5000 **37.** $924 **39.** $3276

Section 10.3 (page 655)

1. 60–65 years **3.** 7 members **5.** 41 members **7.** $20–$25 thousand **9.** 7 employees **11.** 28 employees **13.** ||; 2 **15.** ~~||||~~ |; 6 **17.** |; 1 **19.** ||; 2 **21.** |||; 3 **23.** ~~||||~~; 5 **25.** ~~||||~~ |; 6 **27.** ||; 2 **29.** |; 1 **31.** ~~||||~~ ~~||||~~ |||; 13 **33.** ~~||||~~ ~~||||~~ ~~||||~~ ||; 17 **35.** |||| |||; 8 **37.** 35.75 **39.** 14.2 **41.** 39.4

Section 10.4 (page 663)

1. 14 **3.** 60.3 **5.** 27,955 **7.** $7.68 **9.** 6.7 **11.** 17.2 **13.** 51 **15.** 130 **17.** 9 **19.** 68 and 74 **21.** 2.6 **23.** 49 **25.** no mode **27.** $1020 **29.** $960

Section 10 Review Exercises (page 673)

1. motels **2.** $\frac{200}{850} = \frac{4}{17}$ **3.** $\frac{150}{850} = \frac{3}{17}$ **4.** $\frac{140}{850} = \frac{14}{85}$ **5.** $\frac{80}{150} = \frac{8}{15}$ **6.** $\frac{200}{280} = \frac{5}{7}$ **7.** $22,125 **8.** $4425 **9.** $27,435 **10.** $8850 **11.** $9735 **12.** $15,930 **13.** March **14.** June **15.** 8 million **16.** 4 million **17.** 5 million **18.** 2 million **19.** $25,000 **20.** $20,000 **21.** $35,000 **22.** $40,000 **23.** $30,000 **24.** $20,000 **25.** 9.5 **26.** 35.2 **27.** 51.1 **28.** 257.3 **29.** 34 **30.** 562 **31.** 64 **32.** 29 **33.** 36° **34.** 35% **35.** 20% **36.** 36°

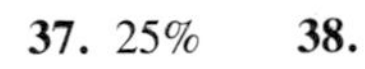

37. 25% **38.**

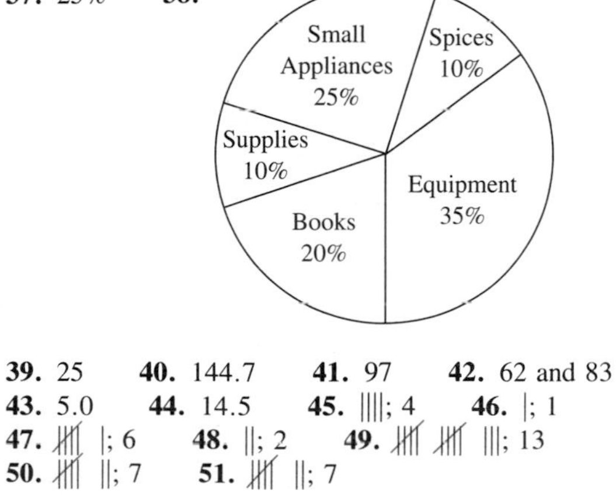

39. 25 **40.** 144.7 **41.** 97 **42.** 62 and 83 **43.** 5.0 **44.** 14.5 **45.** ||||; 4 **46.** |; 1 **47.** ~~||||~~ |; 6 **48.** ||; 2 **49.** ~~||||~~ ~~||||~~ |||; 13 **50.** ~~||||~~ ||; 7 **51.** ~~||||~~ ||; 7

52.

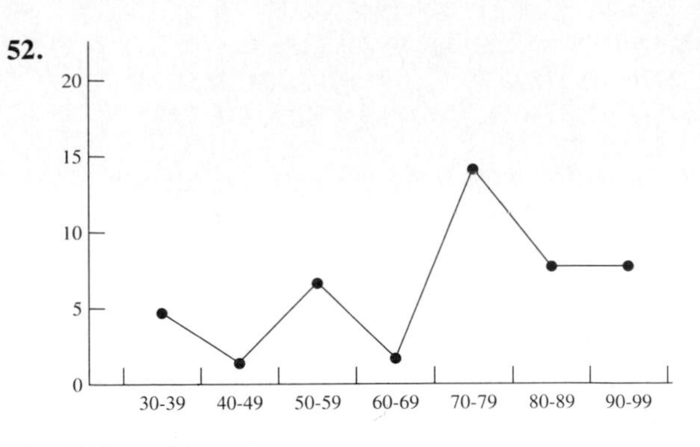

53. 19.6 **54.** 118.8

Chapter 10 Test (page 679)

[10.1] **1.** $616,000 **2.** $504,000 **3.** $840,000 **4.** $448,000 **5.** $56,000 **6.** $336,000 **7.** 72° **8.** 10% **9.** 108° **10.** 108° **11.** 36°

12.

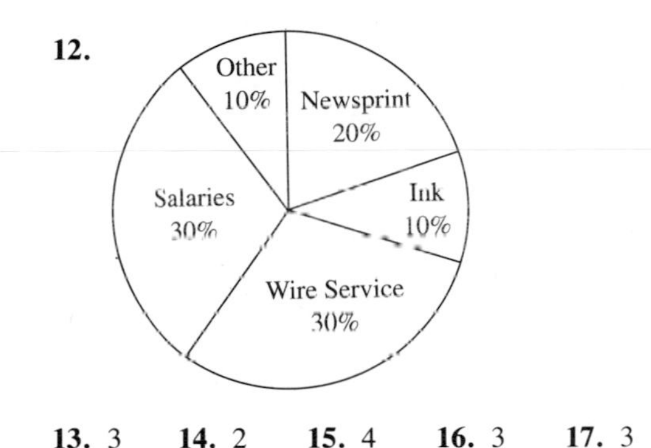

13. 3 **14.** 2 **15.** 4 **16.** 3 **17.** 3 **18.** 5

[10.3] **19.**

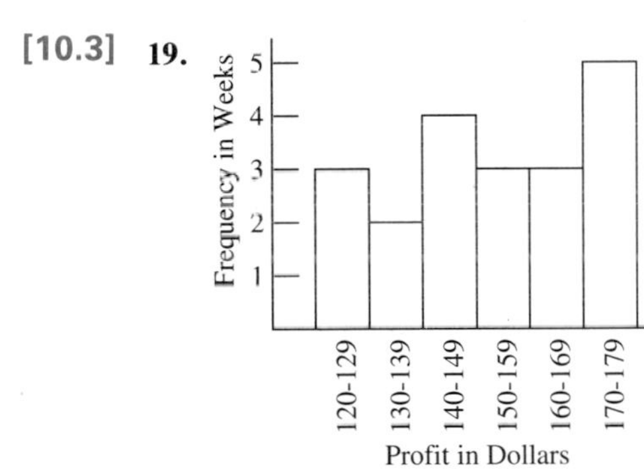

[10.4] **20.** 65 **21.** 18.8 **22.** 478.9 **23.** 11.3 **24.** 173.7 **25.** 27 **26.** 41.5 **27.** 8.3 **28.** 47 **29.** no mode **30.** 103 and 104

CHAPTER 11

Section 11.1 (page 687)

1. $1262.50 **3.** $6205.20 **5.** $10,976.01 **7.** $2025.80 **9.** $1808.72 **11.** $4160.52

13. \$41,401.26 **15.** \$1372.80; \$372.80 **17.** \$2476.89; \$1501.89 **19.** \$2031.74; \$551.74 **21.** \$15,246; \$7746 **23.** \$1404.90; \$1418.50; \$1425.80 **25.** \$12,884; \$13,031.20; \$13,108.80 **27.** quarterly **29.** \$11,901.75 **31.** **(a)** \$15,869; \$16,084 **(b)** \$215 **33.** \$40,223.50 **35.** 4.65 **37.** 106.3 **39.** −.86

Section 11.2 (page 695)

1. $14\frac{1}{2}\%$ **3.** $15\frac{1}{2}\%$ **5.** $15\frac{1}{2}\%$ **7.** 17% **9.** \$18; $14\frac{1}{2}\%$ **11.** $16\frac{1}{2}\%$ **13.** 30% **15.** 18% **17.** 5% **19.** 17% **21.** 360 **23.** 72,800 **25.** \$36,800 **27.** \$27,900

Section 11.3 (page 703)

1. \$702.40 **3.** \$315.90 **5.** \$882.98 **7.** \$1497.29 **9.** \$334.62 **11.** \$400 **13.** \$550 **15.** \$700 **17.** 360 **19.** \$218,448 **21.** \$95.40 **23.** \$174.45 **25.** \$256.73 **27.** \$1284.05

Section 11.4 (page 709)

1. \$512 **3.** \$230.80 **5.** \$1128.60 **7.** \$326.75 **9.** \$2698.38 **11.** \$1378 **13.** \$377 **15.** \$766.40 **17.** \$7284

Chapter 11 Review Exercises (page 713)

1. \$561.80 **2.** \$3238.35 **3.** \$6273.27 **4.** \$15,291.51 **5.** \$467.96; \$67.96 **6.** \$1211.93; \$361.93 **7.** \$24,126 **8.** $14\frac{1}{2}\%$ **9.** 16% **10.** (a) \$42.20 **(b)** 16% **11.** \$1582.20 **12.** \$841.57 **13.** \$895.40 **14.** \$558.98 **15.** \$762.75 **16.** 300 payments **17.** \$228,825 **18.** \$153,825 **19.** interest **20.** \$377.70 **21.** \$1641.60 **22.** \$456.50 **23.** \$334 **24.** \$1408.80 **25.** \$697.60 **26.** **(a)** \$13,229.09 **(b)** \$5004.09 **27.** **(a)** \$7429.50 in Fair Oaks; \$7401 in Carmichael **(b)** \$28.50 **28.** 17% **29.** \$885.18 **30.** \$838.49 **31.** **(a)** \$391.76 **(b)** 17% **32.** \$757.40

Chapter 11 Test (page 719)

[11.1] **1.** \$1898.30 **2.** \$14,719.45 **3.** \$24,070.50 [11.2] **4.** 17% **5.** 15% **6.** 17% [11.3] **7.** \$761.60 **8.** \$674.96 **9.** \$1450.32 **10.** \$480 **11.** \$662.50 [11.4] **12.** \$534.40 **13.** \$595.10 **14.** \$1056.60

Final Examination (page 721)

[1] **1.** 12,996 **2.** 185 **3.** **(a)** 28,750 **(b)** 28,700 **4.** 133 R17 **5.** 72 [2] **6.** $\frac{5}{9}$ **7.** $\frac{20}{21}$ **8.** $\frac{4}{5}$ square inch **9.** $\frac{1}{8}$ **10.** $3\frac{1}{11}$ [3] **11.** 120 **12.** $\frac{7}{10}$ **13.** $\frac{7}{24}$ **14.** $13\frac{5}{12}$ **15.** $5\frac{19}{40}$ [4] **16.** **(a)** 42.085 **(b)** 42.08 **17.** 410.383 **18.** 12.1059 **19.** 6.145 **20.** $\frac{9}{16}$; .579; .58; .5803; $\frac{3}{5}$ [5] **21.** $\frac{3}{2}$ **22.** $\frac{5}{21}$ **23.** yes **24.** 36 **25.** 756 miles [6] **26.** 62.5% **27.** 266 **28.** 320 **29.** 80% **30.** \$2750 [7] **31.** $2\frac{1}{2}$ quarts **32.** 7 weeks 2 days 4 hours **33.** 13 yards 2 feet 5 inches **34.** 185°F [8] **35.** 58.3 square meters **36.** 31.2 square centimeters **37.** 11.5 square feet **38.** 295.4 square centimeters **39.** 246.2 cubic centimeters **40.** 15 [9] **41.** −10 **42.** 8 **43.** 23 **44.** 6 **45.** −2 [10] **46.** 28.1; 26; 26 **47.** 10.2; 8.0; 12.5 [11] **48.** \$15,542.10 **49.** \$21,387 **50.** \$1445.12

Appendix A: Calculators (page 731)

1. 104.58 **3.** 44,904.75 **5.** 733.45 **7.** 56.51 **9.** 20.45 **11.** 58.90 **13.** .49 **15.** 133.14 **17.** 8.1 **19.** 566.14 **21.** .21 **23.** 5794.4 **25.** .09 **27.** \$312.09 **29.** \$891.24

Appendix B: Inductive and Deductive Reasoning (page 739)

1. 37 **3.** 26 **5.** 16 **7.** 243 **9.** 36

11. ┐ **13.** ⟍＿

15. Conclusion follows. **17.** Conclusion does not follow. **19.** 3 **21.** Dick

SOLUTIONS

SOLUTIONS TO SELECTED EXERCISES

For the answers to all odd-numbered section exercises, all chapter review exercises, all cumulative review exercises, and all chapter tests, see the section beginning on page A–1.

If you would like to see more solutions, you may order the *Student's Solutions Manual* for *Basic College Mathematics* from your college bookstore. It contains solutions to the odd-numbered exercises that do not appear in this section, plus solutions to all chapter review exercises, chapter tests, cumulative review exercises, and the final exams.

CHAPTER 1

Section 1.1 (page 5)

1. 1: thousands; 8: tens
5. 4: millions; 2: thousands
9. 60: billions; 0: millions; 502: thousands; 109: ones
13. seven hundred twenty-five thousand, six hundred fifty-nine
17. 3153
19. 2070
25. 800,621,020,215

Section 1.2 (page 13)

1. $15 + 24 = 39$ (ones added; tens added)

5. $17 + 52 = 69$ (ones added; tens added)

9. $6310 + 252 + 1223 = 7785$ (ones added; tens added; hundreds added; thousands added)

13. $1251 + 4311 + 2114 = 7676$ (ones added; tens added; hundreds added; thousands added)

17. $8642 + 7134 = 15{,}776$

21. $89 + 38 = 127$ — 9 + 8 = 17; write 7, carry 1.

25. (carry 1) $47 + 74 = 121$ — 7 + 4 = 11; write 1, carry 1.

29. (carry 1) $73 + 29 = 102$

33. (carry 1) $201 + 769 = 970$ — 1 + 9 = 10; write 0, carry 1.

37. (carry 1) $928 + 843 = 1771$ — 8 + 3 = 11; write 1, carry 1.

41. (carries 111) $7968 + 1285 = 9253$
- 8 + 5 = 13; write 3, carry 1.
- 1 + 6 + 8 = 15; write 5, carry 1.
- 1 + 9 + 2 = 12; write 2, carry 1.

45. (carries 111) $1625 + 7986 = 9611$

49. (carries 112) $18 + 708 + 9286 + 636 = 10{,}648$

53. (carries 1 1) $1321 + 603 + 8 + 21 + 1604 = 3557$

57.
```
  2 3 3
   553
    97
  2772
   437
    63
+  328
  4250 <- 3 + 7 + 2 + 7 + 3 + 8 = 30; write 0,
             carry 3.
       <- 3 + 5 + 9 + 7 + 3 + 6 + 2 = 35; write 5,
             carry 3.
```

61.
```
   1260   correct
    729
    372
+   159
   1260
```

65.
```
   1420   correct
    713
     28
    615
+    64
   1420
```

69.
```
   14,332   correct
    4 714
       27
       77
    8 878
+     636
   14,332
```

73. Southtown and Austin

Southtown to Murphy	22
Murphy to Austin	+ 11
Southtown to Austin	33

Any other route is longer.

77. Add men and women to get total number of people.
```
   275 women
 + 345 men
   620 people
```

81.
```
   2
   72
   58
   72
 + 58
  260 inches
```
(It does not matter in what order you add the lengths—remember the commutative property.)

Section 1.3 (page 23)

1.
```
   37   check    25
 - 12          + 12
   25            37   match
```

5.
```
   77   check    60
 - 60          + 17
   17            77   match
```

9.
```
   73   check    61
 - 61          + 12
   12            73   match
```

13.
```
   602   check    301
 - 301          + 301
   301            602   match
```

17.
```
   2318   check    1207
 - 1207          + 1111
   1111            2318   match
```

21.
```
   1875   check    1362
 - 1362          +  513
    513            1875   match
```

25.
```
   8625   check     311
 -  311          + 8314
   8314            8625   match
```

29.
```
   46,253   check     5,143
 -  5,143          + 41,100
   41,110            46,253   match
```

33.
```
   89      27
 - 27    + 63
   63      90   incorrect; answer should be 62
```

37.
```
   2984      1321
 - 1321    + 1663
   1663      2984   check
```

41.
```
   2 12
   3̸ 2̸
 - 2  5
      7
```

45.
```
   3 15
   4̸ 5̸
 - 2  9
   1  6
```

49.
```
     11
   2  1̸ 16
   3̸  2̸  6̸
 - 1  5  8
   1  6  8
```

53.
```
           14
      7    4̸  12
   8̸  8̸  5̸  2̸
 -     5  7  3
   8  2  7  9
```

57.
```
      11   15  10
   1  1̸    5̸   0̸  18
   2̸  2̸,  6̸   1̸  8̸
 -     6,  7   1  9
   1  5,  8   9  9
```

61.
```
   2 10
   3̸ 0̸
 - 2  3
      7
```

65.
```
      9
   0  1̸0̸ 10
   1̸  0̸  0̸
 -     3  8
       6  2
```

69.
```
   3 10  3 11
   4̸  0̸  4̸  1̸
 - 1  2  0  8
   2  8  3  3
```

73.
```
     12
   8  2̸ 10
   9̸  3̸  0̸  5
 - 1  5  3  0
   7  7  7  5
```

77.
```
      9
   1  1̸0̸ 10
   2̸  0̸  0̸  6
 - 1  8  5  0
      1  5  6
```

81.
```
   7 14 9 13
   8  5 0  3   (borrowing: 10)
 − 2  8 1  6
   5  6 8  7
```

85.
```
 2 12  9  9 10
 3  3, 0  0  0
−1  7, 2  2  2
 1  5, 7  7  8
```

89.
```
    8  9 17 10
 1  9, 0  8  0
−1  3, 4  9  6
    5  5  8  4
```

93.
```
   1
  1638
+  191
  1829   match; correct
```

97.
```
  2230
+  538
  2768   match; correct
```

101.
```
   1 11
  18,679
+ 29,231
  47,910   incorrect; answer should be 29,221

   3 17  8  9 10
   4  7, 9  0  0
 − 1  8, 6  7  9
   2  9, 2  2  1
```

105.
```
  254   passengers
− 133   get off
  121   passengers remain
```

109.
```
  3754   last election
− 2511   this election
  1243   more in last
```

113.
```
  $439   payment now
+ $263   increase
  $702   new payment
```

Section 1.4 (page 33)

1. $2 \times 3 \times 4$

$2 \times 3 = 6 \qquad 6 \times 4 = 24$

5. $8 \cdot 9 \cdot 0$

$8 \cdot 9 = 72 \qquad 72 \cdot 0 = 0$

9. $(2)(3)(6)$

$(2)(3) = 6 \qquad 6(6) = 36$

13.
```
  4
  55
×  8
 440
```

17.
```
  512
×   4
 2048
```

21.
```
  21
 2153
×   4
 8612
```

25.
```
  1 1
 7 212
×    5
36,060
```

29.
```
   1 23
 81 259
×     4
325,036
```

33.
```
  50      5     50
×  7    × 7   ×  7
         35    350 ← Attach 1 zero.
```

37.
```
  500     5     500
×   4   × 4   ×   4
         20    2000 ← Attach 2 zeros.
```

41.
```
            11
  1255     1255     1255
×   20   ×    2   ×   20
           2510   25,100 ← Attach 1 zero.
```

45.
```
  43,000     43       43000
×   2000   ×  2     ×  2000
             86   86,000,000 ← Attach 6 zeros.
```

49.
```
  500      5      500
× 900    × 9    × 900
          45   450,000 ← Attach 4 zeros.
```

53.
```
  1
  6
  29
× 27
 203
 58
 783
```

57.
```
  1
  1
  83
× 45
 415
332
3735
```

61.
```
  1
  1
  72
× 85
 360
576
6120
```

65.
```
   631
×   35
  3 155
 18 93
 22,085
```

69.
```
   735
×  112
  1 470
  7 35
 73 5
 82,320
```

73.
```
     1233
×     951
    1 233
   61 65
 1 109 7
 1,172,583
```

77.
```
     1629
   × 478
   13 032
  114 03
  651 6
  778,662
```

81.
```
   321
 × 203
   963
 64 20
 65,163
```

85.
```
   219
 × 404
   876
 87 60
 88,476
```

89.
```
     1592
   × 2009
   14 328
 3 184 00
 3,198,328
```

93.
```
   800   pages each volume       8
 ×  30   volumes               × 3
                                24
   800
    30
 24,000 ← Attach 3 zeros; 24,000 total pages.
```

97.
```
  27   miles per gallon
 × 8   gallons
 216   miles
```

101.
```
    58   lockers
 × $53   price per locker
   174
   290
 $3074   total cost
```

105. $21 \cdot 43 \cdot 56$

$21 \cdot 43 = 903 \qquad 903 \cdot 56 = 50{,}568$

109.
```
  1406   large meal
 − 348   small meal
  1058   more calories
```

Section 1.5 (page 45)

1. $\frac{21}{7} = 3$, $7\overline{)21}$ with quotient 3

5. $16 \div 2 = 8$; $\frac{16}{2} = 8$

9. $\frac{10}{2} = 5$

13. $\frac{0}{4} = 0$

17. $0\overline{)15}$ meaningless

21. $\frac{8}{1} = 8$

25.
```
    4 6
 7)32⁴2
```
check
```
  ⁴
  46
 × 7
 322   match
```

29.
```
   401
 5)2005
```
check
```
  401
 ×  5
 2005   match
```

33.
```
   2 2 4 R8
 9)20²2⁴4
```
check
```
  224
 ×  9
 2016
 +  8
 2024   match
```

37.
```
   30 9
 6)185⁵4
```
check
```
  309
 ×  6
 1854   match
```

41.
```
   5 00 6
 6)30,03³6
```
check
```
   5006
 ×    6
 30,036   match
```

45.
```
   2 5 8 9 R2
 5)12,²9⁴4⁴7
```
check
```
   2589
 ×    5
  12945
      2
 12,947   match
```

49.
```
   3 6 8 9 R1
 4)14,²7³5³7
```
check
```
   2 3 3
   3689
 ×    4
  14756
 +    1
 14,757   match
```

53.
```
   5 8 41
 9)52⁷5³69
```
check
```
    7 3
   5841
 ×    9
 52,569   match
```

57.
```
   10, 2 5 8 R5
 9)92,²3⁵2⁷7
```
check
```
    2 5 7
   10 258
 ×      9
   92 322
 +      5
   92,327   match
```

61.
```
   11 2 5 5
 7)78,¹7³8³5
```
check
```
   1 3 3
   11,255
 ×      7
   78,785   match
```

65. 1908
× 3
5724
+ 2
5726 incorrect

$$\begin{array}{r} 1908\text{ R1} \\ 3\overline{)5^{2}72^{2}5} \end{array}$$

check 1908
× 3
5724
+ 1
5725 match

69. 3 568
× 6
21 408
+ 2
21,410 incorrect

$$\begin{array}{r} 3\,5\,6\,8\text{ R1} \\ 6\overline{)21^{3}4^{4}0^{4}9} \end{array}$$

check 3 568
× 6
21 408
+ 1
21,409 match

73. 11,523
× 6
69,138
+ 2
69,140 correct

77. 27,822
× 8
222,576 correct

81.
$$\begin{array}{r} 12¢ \\ 8\overline{)96¢} \end{array}$$

check 12
× 8
96¢ match

12¢ per ounce

85.
$$\begin{array}{r} 3\,2\,7\,7 \\ 9\overline{)29^{2}4^{6}9^{6}3} \end{array}$$

check 3 277
× 9
29,493 match

$3277 per month

89. 184 is

divisible (√) by 2, since it ends in 4;

not divisible (X) by 3, since 1 + 8 + 4 = 13 which is not divisible by 3;

not divisible (X) by 5, since it does not end in 0 or 5;

not divisible (X) by 10, since it does not end in 0.

93. 903 is

not divisible (X) by 2, since it does not end in 0, 2, 4, 6, or 8;

divisible (√) by 3, since 9 + 0 + 3 = 12 which is divisible by 3;

not divisible (X) by 5, since it does not end in 0 or 5;

not divisible (X) by 10, since it does not end in 0

97. 21,763 is

not divisible (X) by 2, since it does not end in 0, 2, 4, 6, or 8;

not divisible (X) by 3, since 2 + 1 + 7 + 6 + 3 = 19 which is not divisible by 3;

not divisible (X) by 5, since it does not end in 0 or 5;

not divisible (X) by 10, since it does not end in 0.

Section 1.6 (page 53)

1.
35
21)735
63
105
105
0

check 35
21
35
70
735 match

5.
37 R7
59)2190
177
420
413
7

check 37
× 59
333
185
2183

2183 + 7 = 2190 match

9.
38
58)2204
174
464
464
0

check 38
× 58
304
190
2204 match

13.
2 407 R1
26)62,583
52
10 5
10 4
183
182
1

check 2 407
× 26
14 442
48 14
62,582

62,582 + 1 = 62,583 match

17.
640 R8
38)24,328
22 8
1 52
1 52
8

check 640
× 38
5 120
19 20
24,320

24,320 + 8 = 24,328 match

21.
7 746 R20
32)247,892
224
23 8
22 4
1 49
1 28
212
192
20

check 7746
× 32
15 492
232 38
247,872

247,872 + 20 = 247,892 match

SOLUTIONS

25. $821\overline{)17{,}241}$ = 21
```
        21
821)17,241     check      21
    16 42               × 821
       821                 21
       821                42
         0              16 8
                        17,241   match
```

29.
```
        170
900)153,000    check     170
     90 0              × 900
     63 00               000
     63 00              0 00
         0             153 0
                       153,000   match
```

33.
```
                            658
     658               28)18,424   check    658
   ×  28                  16 8            ×  28
   5 264                   1 62           5 264
  13 16                    1 40          13 16
  18,424                    224          18,424
 +     9                    224           match
  18,433   incorrect          0
```

37.
```
       25
54)1350     check     25
   108              × 54
    270              100
    270             125
      0             1350   match; answer: 25 hours
```

41.
```
      $108
36)$3888    check     108
   36               ×  36
    288               648
    288              324
      0              3888   match; answer: $108
```

45.
```
                             292    match; answer: $10,512
   $292    check   36)10,512
 ×   36                7 2
  1 752                3 31
  8 76                 3 24
$10,512                  72
                         72
                          0
```

Section 1.7 (page 61)

1. 7900 (7 8 [6] 2 — 6 is 5 or larger) ↑ hundreds

5. 710 (7 1 [4] — 4 is 4 or less) ↑ tens

9. 42,500 (4 2, 4 [9] 5 — 9 is 5 or larger) ↑ hundreds

13. 15,800 (1 5, 7 [5] 8 — 5 is 5 or larger) ↑ hundreds

17. 6000 (5 [8] 4 7 — 8 is 5 or larger) ↑ thousands

21. 500,000 (4 9 [6] , 1 9 2 — 6 is 5 or larger) ↑ ten thousands (9 + 1 = 10, write 0, carry 1 to hundred thousands)

25. ten: 7460 (7 4 5 [9] — 9 is 5 or larger)
hundred: 7500 (7 4 [5] 9 — 5 is 5 or larger)
thousand: 7000 (7 [4] 5 9 — 4 is 4 or less)

29. ten: 5050 (5 0 4 [9] — 9 is 5 or larger)
hundred: 5000 (5 0 [4] 9 — 4 is 4 or less)
thousand: 5000 (5 [0] 4 9 — 0 is 4 or less)

33. ten: 3650 (3 6 4 [5] — 5 is 5 or larger) ↑ tens — Round to 3650.
hundred: 3600 (3 6 [4] 5 — 4 is 4 or less) ↑ hundreds — Round to 3600.
thousand: 4000 (3 [6] 4 5 — 6 is 5 or larger) ↑ thousands — Round to 4000.

37. ten: 23,500 (2 3, 5 0 [2] — 2 is 4 or less)
hundred: 23,500 (2 3, 5 [0] 2 — 0 is 4 or less)
thousand: 24,000 (2 3, [5] 0 2 — 5 is 5 or more)

41.
```
         rounded
  57        60
− 24      − 20
            40   estimate
```

45.
```
         rounded
  623       600
  362       400
  189       200
+ 736     + 700
           1900   estimate
```

49.
```
         rounded
  649       600
× 594     × 600
         360,000   estimate
```

Section 1.8 (page 67)

1. 2, since $2^2 = 2 \cdot 2 = 4$
5. 12, since $12^2 = 12 \cdot 12 = 144$
9. 8, since $2^3 = 2 \cdot 2 \cdot 2 = 8$
13. 144, since $12^2 = 12 \cdot 12 = 144$
17. $18^2 = 324$ so $\sqrt{324} = 18$
21. $35^2 = 1225$ so $\sqrt{1225} = 35$
25. $54^2 = 2916$ so $\sqrt{2916} = 54$

29. $6 \cdot 4 - \frac{9}{3}$

$24 - \frac{9}{3}$ Multiply first.

21 Subtract last.

33. $25 \div 5(8 - 4)$

$25 \div 5(4)$ Work inside parentheses first.

$5 \cdot 4$ Divide next.

20 Multiply last.

37. $4 \cdot 1 + 8 \cdot (9 - 2) + 3$

$4 \cdot 1 + 8(7) + 3$ Work inside parentheses first.

$4 + 56 + 3$ Multiply before adding, *left to right.*

$60 + 3$ Add *left to right.*

63 Add again.

41. $2^3 \cdot 3^2 + (15 - 10)$

$2^3 \cdot 3^2 + 5$ Work inside parentheses first.

$8 \cdot 9 + 5$ Do exponents next, *left to right.* Do $\times$ before $+$.

$72 + 5$ Add last.

77

45. $6 \cdot 2 + 9 \cdot 3 - 5$

$12 + 27 - 5$ Multiply first.

34 Add and subtract, left to right.

49. $8 + 10 \div 5 + \frac{0}{3}$

$8 + 2 + 0$ Divide first.

10 Add last.

53. $7 \cdot \sqrt{81} - 5 \cdot 6$

$7 \cdot 9 - 5 \cdot 6$ Simplify the square root first.

$63 - 30$ Multiply, left to right.

33 Subtract last.

57. $4 \cdot \sqrt{49} - 7(5 - 2)$

$4 \cdot \sqrt{49} - 7(3)$ Work inside parentheses first.

$4 \cdot 7 - 7(3)$ Simplify the square root next.

$28 - 21$ Multiply, left to right.

7 Subtract last.

61. $6^2 + 2^2 - 6 \cdot 2$

$36 + 4 - 6 \cdot 2$ Do exponents first.

$36 + 4 - 12$ Multiply next.

28 Add and subtract, left to right.

65. $7 + 6 \div 3 + 5 \cdot 2$

$7 + 6 \div 3 + 10$ Multiply first (multiply and divide before adding or subtracting), left to right.

$7 + 2 + 10$ Divide next.

19 Add last.

69. $6^2 - 2^2 + 3 \cdot 4$

$36 - 4 + 3 \cdot 4$ Do exponents first.

$36 - 4 + 12$ Multiply next.

$32 + 12$ Subtract (+ and − left to right).

44 Add last.

73. $3 \cdot \sqrt{16} - 7 \cdot 1$

$3 \cdot 4 - 7 \cdot 1$ Simplify square root first.

$12 - 7 \cdot 1$ Multiply, left to right.

$12 - 7$ Multiply before subtracting.

5 Subtract last.

77. $7 \div 1 \cdot 8 \cdot 2 \div (21 - 5)$

$7 \div 1 \cdot 8 \cdot 2 \div 16$ Work inside parentheses first.

$7 \cdot 8 \cdot 2 \div 16$ Multiply or divide, left to right.

$56 \cdot 2 \div 16$ Multiply, left to right.

$112 \div 16$ Multiply, left to right.

7 Divide last.

81. $5 \cdot \sqrt{9} - 2 \cdot \sqrt{4}$

$5 \cdot 3 - 2 \cdot \sqrt{4}$ Simplify square roots first, *left to right.*

$5 \cdot 3 - 2 \cdot 2$ Simplify square roots first, *left to right.*

$15 - 2 \cdot 2$ Multiply *left to right* before subtracting.

$15 - 4$ Multiply *left to right* before subtracting.

11 Subtract last.

85. $8 \cdot 9 \div \sqrt{36} - 4 \div 2 + (14 - 8)$

$8 \cdot 9 \div \sqrt{36} - 4 \div 2 + 6$ Work inside parentheses first.

$8 \cdot 9 \div 6 - 4 \div 2 + 6$ Do square roots.

$72 \div 6 - 4 \div 2 + 6$ Multiply next (multiply and divide before adding or subtracting, *left to right*).

$12 - 4 \div 2 + 6$ Divide next (do $\times$ and $\div$, left to right).

$12 - 2 + 6$ Divide before subtracting or adding.

$10 + 6$ Subtract next (do + and − *left to right*)

16 Add last.

89. $6 \cdot \sqrt{25} \cdot \sqrt{100} \div 3 \cdot \sqrt{4} + 9$

$6 \cdot 5 \cdot \sqrt{100} \div 3 \cdot \sqrt{4} + 9$ Simplify square roots first, *left to right.*

$6 \cdot 5 \cdot 10 \div 3 \cdot \sqrt{4} + 9$ Simplify square roots first, *left to right.*

$6 \cdot 5 \cdot 10 \div 3 \cdot 2 + 9$ Simplify square roots first.

$30 \cdot 10 \div 3 \cdot 2 + 9$ Do $\times$ and $\div$ *left to right* before +.

$300 \div 3 \cdot 2 + 9$ Do $\times$ and $\div$ *left to right* before +.

$100 \cdot 2 + 9$ Do $\times$ and $\div$ *left to right* before +.

$200 + 9$ Do $\times$ before +

209 Add last.

SOLUTIONS

Section 1.9 (page 75)

1. *Step 1* Find yesterday's sales.
Step 2 Since today's sales and the decrease from yesterday's sales are known, add.

today's sales
+ decrease
yesterday's sales

or today's sales + decrease = yesterday's sales

Step 3 Yesterday's sales should be about $2000 more than today's sales, or $7000 + $2000 = $9000

Step 4 Add

$6975 today's sales
+ 1630 decrease
$8605 yesterday's sales

$8605 is reasonable (it is close to the estimate of $9000)

check
$8605 yesterday's sales
− 1630 decrease
$6975 today's sales—match

Answer: $8605

5. *Step 1* Find the total Carl owes on the loan.
Step 2 He owes an amount plus the interest. To find the total owed, add:

amount
+ interest
total owed.

Step 3 The total should be about 3800 plus 300, or $4100.
Step 4 Add.

$3815
+ 268
$4083

$4083 is reasonably close to the estimate of $4100.

check 4083
3815
268
4083 match

Answers: $4083

9. *Step 1* Find the number of miles Lois can go on 37 gallons of gas.
Step 2 Lois gets 36 miles for each gallon and will use 37 gallons. To find the number of miles she can go, multiply:

gallons
× miles per gallon
miles total.

Step 3 The answer should be about 40 × 40, or 1600 miles.
Step 4 Multiply.

2
4
37
× 36
222
111
1332

1332 miles is reasonably close to 1600 (the estimate). Note that estimates are less accurate when they involve multiplying.

check 36)1332 = 37 match
108
252
252
0

Answer: 1332 miles

13. *Step 1* Find the weight of the firewood.
Step 2 Since the weight of the empty truck and how much it weighs with the firewood are known, subtract:

weight with firewood
− weight empty
weight of firewood.

Step 3 The answer should be about 21,000 − 9000, which is 12,000.
Step 4 Subtract.

21,375
− 9,250
12,125 pounds

12,125 is close to the estimate of 12,000.

check
12,125
+ 9,250
21,375 match

Answer: 12,125 pounds

17. *Step 1* Find the number of gallons polluted in a year.
Step 2 The total in a year will be 365 times the amount polluted each day.

polluted each day
× 365
polluted in a year

Step 3 The answer should be about 200,000 × 400, or 80,000,000.
Step 4 Multiply.

209,670
× 365
1 048 350
12 580 20
62 901 0
76,529,550 gallons

76,529,550 is close to the estimate of 80,000,000.

check 365)76,529,550 = 209,670 match
73 0
3 529
3 285
244 5
219 0
25 55
25 55
0

Answer: 76,529,550 gallons

21. *Step 1* The value of 60 feet of beachfront property must be found.
Step 2 The value of one foot of beachfront property must be multiplied by the number of beachfront feet.
Step 3 The value of the beachfront lot is about $1000 × 60 = $60,000.
Step 4 Compute.

$1 018
× 60 feet
$61,080 value of lot

The answer is close to the estimate of $60,000.
Answer: $61,080

25. *Step 1* Find the amount remaining in the checking account.

Step 2 Since there are three checks written, the amount of these checks must be subtracted from the beginning balance in the checking account.

beginning balance − checks written = balance remaining

Step 3 The remaining balance is about

$3000 − $300 − $600 − $200 = $1900

Step 4 Subtract.

$3010 − $280 − $620 − $178 = $1932 balance remaining

This answer is close to the estimate of $1900

check $1932 + $178 + $620 + $280 = $3010 match

Answer: $1932 balance remaining

29. *Step 1* Find total amount of tax deductions (federal and state) each month.

Step 2 Since both federal and state annual tax deductions are given, they must be added to find the total. This total must be divided by 12 to find the monthly deductions.

(federal taxes + state taxes) ÷ 12 = total monthly tax deduction

Step 3 The answer should be about

($2200 + $400) ÷ 12 = $217 (rounded).

Step 4 Compute.

($2184 + $372) ÷ 12 = $213 monthly tax deduction

The answer is close to the estimate of $217.

check $213 × 12 = $2556

$2556 − $372 = $2184 match

Answer: $213 monthly tax deductions

33. *Step 1* Find the number of seats in each row of the balcony.

Step 2 The number of seats on the main floor must be subtracted from the total number of seats to find the number of seats in the balcony. The number of seats in the balcony is divided by the number of rows of seats in the balcony.

Step 3 The answer should be about

1300 − 800 = 500 ÷ 25 = 25 seats.

Step 4 Compute

1250 − (30 × 25) = 500 seats in balcony

500 ÷ 25 rows = 20 seats

20 seats is close to our estimate of 25 seats.

Answer: 20 seats in each balcony row

CHAPTER 2

Section 2.1 (page 97)

1. $\frac{5}{8}$ (There are 8 parts, and 5 are shaded.)

5. $\frac{7}{5}$ (The object is divided into 5 parts, and 7 of these parts are shaded.)

9. $\frac{14}{25}$ (There are 25 students and 14 are female.)

13. numerator: 5; denominator: 7

17. Proper fraction has numerator (top) smaller than denominator (bottom).

proper fractions: $\frac{1}{4}, \frac{3}{8}, \frac{7}{12}$; improper fractions: $\frac{9}{7}, \frac{11}{4}$

21. 9 (numerator); 17 (denominator)

25. $5 \cdot 3 \cdot 7 = 15 \cdot 7 = 105$

29. $72 \div 9 = 8$

Section 2.2 (page 101)

1. $1\frac{7}{8}$ $\quad 1 \cdot 8 = 8 \quad 8 + 7 = 15 \quad \frac{15}{8}$

5. $2\frac{3}{4}$ $\quad 2 \cdot 4 = 8 \quad 8 + 3 = 11 \quad \frac{11}{4}$

9. $1\frac{7}{11}$ $\quad 1 \cdot 11 = 11 \quad 11 + 7 = 18 \quad \frac{18}{11}$

13. $11\frac{1}{2}$ $\quad 11 \cdot 2 = 22 \quad 22 + 1 = 23 \quad \frac{23}{2}$

17. $3\frac{3}{8}$ $\quad 3 \cdot 8 = 24 \quad 24 + 3 = 27 \quad \frac{27}{8}$

21. $4\frac{10}{11}$ $\quad 4 \cdot 11 = 44 \quad 44 + 10 = 54 \quad \frac{54}{11}$

25. $17\frac{12}{13}$ $\quad 17 \cdot 13 = 221 \quad 221 + 12 = 233 \quad \frac{233}{13}$

29. $7\frac{13}{19}$ $\quad 7 \cdot 19 = 133 \quad 133 + 13 = 146 \quad \frac{146}{19}$

33. $\frac{9}{5}$ $\quad 5\overline{)9}$ = 1 R4; 9 − 5 = 4 $\quad 1\frac{4}{5}$

37. $\frac{28}{9}$ $\quad 9\overline{)28}$ = 3 R1; 28 − 27 = 1 $\quad 3\frac{1}{9}$

41. $\frac{17}{3}$ $\quad 3\overline{)17}$ = 5 R2; 17 − 15 = 2 $\quad 5\frac{2}{3}$

45. $\frac{47}{9}$ $\quad 9\overline{)47}$ = 5 R2; 47 − 45 = 2 $\quad 5\frac{2}{9}$

49. $\frac{84}{5}$ $\quad 5\overline{)84}$ = 16 R4; 8 − 5 = 3, 34 − 30 = 4 $\quad 16\frac{4}{5}$

53. $\frac{183}{7}$ $\quad 7\overline{)183}$ = 26 R1; 18 − 14 = 4, 43 − 42 = 1 $\quad 26\frac{1}{7}$

57. $333\frac{1}{3}$ $333 \cdot 3 = 999$ $999 + 1 = 1000$ $\frac{1000}{3}$

61. $2^2 + 3^2 = 4 + 9 = 13$

65. $5 \cdot 2^2 - 4 = 5 \cdot 4 - 4 = 20 - 4 = 16$

Section 2.3 (page 109)

1. Factorizations of 6:

$6 \cdot 1 \quad 2 \cdot 3$

The factors of 6 are 1, 2, 3, 6.

5. Factorizations of 21:

$21 \cdot 1 \quad 3 \cdot 7$

The factors of 21 are 1, 3, 7, 21.

9. The factorizations of 40 are

$40 \cdot 1 \quad 2 \cdot 20 \quad 4 \cdot 10 \quad 8 \cdot 5.$

Use the divisibility rules.

The factors of 40 are 1, 2, 4, 5, 8, 10, 20, 40.

13. 8 is composite, since it is divisible by 2.

17. 9 is composite, since it is divisible by 3.

21. 19 is prime. Try dividing it by prime numbers. $19 \div 2 = 9$ R1 $19 \div 3 = 6$ R1 $19 \div 5 = 3$ R4 $19 \div 7 = 2$ R5 $19 \div 11 = 1$ R8

25. 34 is composite, since it is divisible by 2.

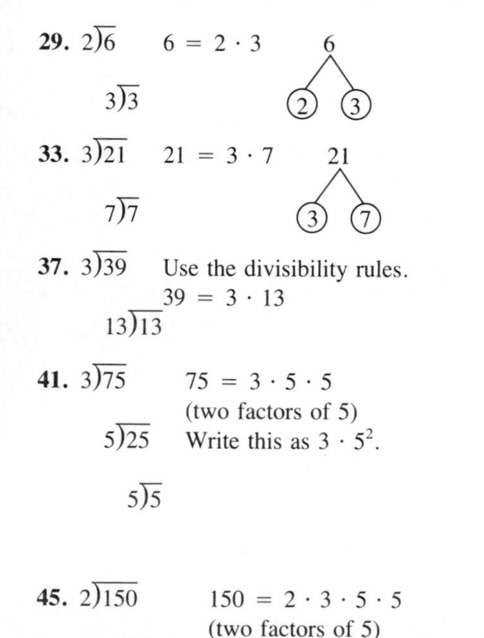

29. $2\overline{)6}$ $3\overline{)3}$ $6 = 2 \cdot 3$

33. $3\overline{)21}$ $7\overline{)7}$ $21 = 3 \cdot 7$

37. $3\overline{)39}$ $13\overline{)13}$ Use the divisibility rules. $39 = 3 \cdot 13$

41. $3\overline{)75}$ $5\overline{)25}$ $5\overline{)5}$ $75 = 3 \cdot 5 \cdot 5$ (two factors of 5) Write this as $3 \cdot 5^2$.

45. $2\overline{)150}$ $3\overline{)75}$ $5\overline{)25}$ $5\overline{)5}$ $150 = 2 \cdot 3 \cdot 5 \cdot 5$ (two factors of 5) Write this as $2 \cdot 3 \cdot 5^2$.

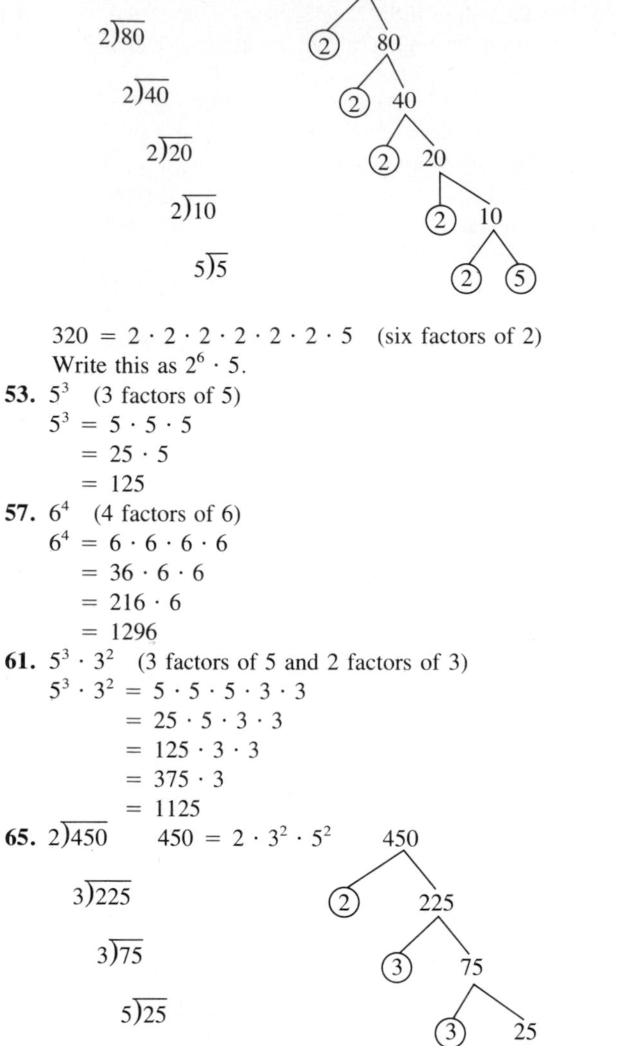

49. $2\overline{)320}$ $2\overline{)160}$ $2\overline{)80}$ $2\overline{)40}$ $2\overline{)20}$ $2\overline{)10}$ $5\overline{)5}$

$320 = 2 \cdot 2 \cdot 2 \cdot 2 \cdot 2 \cdot 2 \cdot 5$ (six factors of 2)
Write this as $2^6 \cdot 5$.

53. 5^3 (3 factors of 5)

$5^3 = 5 \cdot 5 \cdot 5$
$= 25 \cdot 5$
$= 125$

57. 6^4 (4 factors of 6)

$6^4 = 6 \cdot 6 \cdot 6 \cdot 6$
$= 36 \cdot 6 \cdot 6$
$= 216 \cdot 6$
$= 1296$

61. $5^3 \cdot 3^2$ (3 factors of 5 and 2 factors of 3)

$5^3 \cdot 3^2 = 5 \cdot 5 \cdot 5 \cdot 3 \cdot 3$
$= 25 \cdot 5 \cdot 3 \cdot 3$
$= 125 \cdot 3 \cdot 3$
$= 375 \cdot 3$
$= 1125$

65. $2\overline{)450}$ $3\overline{)225}$ $3\overline{)75}$ $5\overline{)25}$ $5\overline{)5}$ $450 = 2 \cdot 3^2 \cdot 5^2$

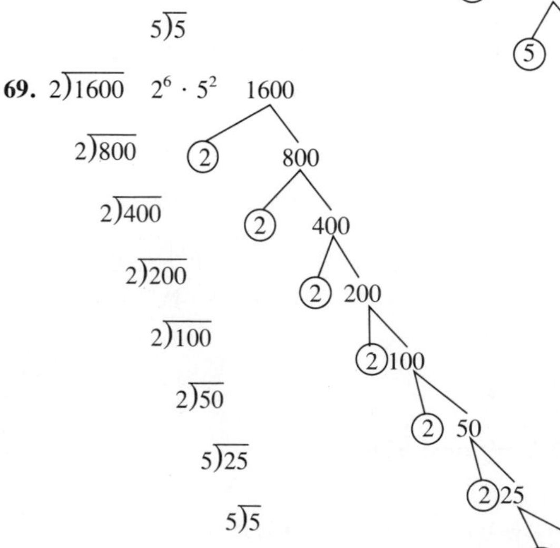

69. $2\overline{)1600}$ $2\overline{)800}$ $2\overline{)400}$ $2\overline{)200}$ $2\overline{)100}$ $2\overline{)50}$ $5\overline{)25}$ $5\overline{)5}$ $2^6 \cdot 5^2$

73. $7 \cdot 5 \cdot 2 = 35 \cdot 2 = 70$

77. $45 \div 5 = 9$

Section 2.4 (page 117)

1. $\frac{6}{8} = \frac{6 \div 2}{8 \div 2} = \frac{3}{4}$

5. $\frac{25}{40} = \frac{25 \div 5}{40 \div 5} = \frac{5}{8}$

9. $\frac{63}{70} = \frac{63 \div 7}{70 \div 7} = \frac{9}{10}$

13. $\frac{36}{63} = \frac{36 \div 9}{63 \div 9} = \frac{4}{7}$

17. $\frac{165}{180} = \frac{165 \div 5}{180 \div 5} = \frac{33}{36} = \frac{33 \div 3}{36 \div 3} = \frac{11}{12}$

or

$\frac{165 \div 15}{180 \div 15} = \frac{11}{12}$

21. $\frac{25}{45} = \frac{5 \cdot \overset{1}{\cancel{5}}}{3 \cdot 3 \cdot \underset{1}{\cancel{5}}} = \frac{5}{9}$

25. $\frac{36}{12} = \frac{\overset{1}{\cancel{2}} \cdot \overset{1}{\cancel{2}} \cdot \overset{1}{\cancel{3}} \cdot 3}{\underset{1}{\cancel{2}} \cdot \underset{1}{\cancel{2}} \cdot \underset{1}{\cancel{3}}} = 3$

29. $\frac{3}{4} \times \frac{18}{24}$ cross products $3 \cdot 24 = 72$, $4 \cdot 18 = 72$

The cross products are equal, so the fractions are equivalent.

33. $\frac{5}{8} \times \frac{3}{4}$ cross products $5 \cdot 4 = 20$, $8 \cdot 3 = 24$

The cross products are not equal, so the fractions are not equivalent.

37. $\frac{7}{52} \times \frac{9}{40}$ cross products $7 \cdot 40 = 280$, $9 \cdot 52 = 468$

The cross products are not equal, so the fractions are not equivalent.

41. $\frac{105}{252} = \frac{\overset{1}{\cancel{3}} \cdot \overset{1}{\cancel{7}} \cdot 5}{2 \cdot 2 \cdot \underset{1}{\cancel{3}} \cdot 3 \cdot \underset{1}{\cancel{7}}} = \frac{5}{12}$

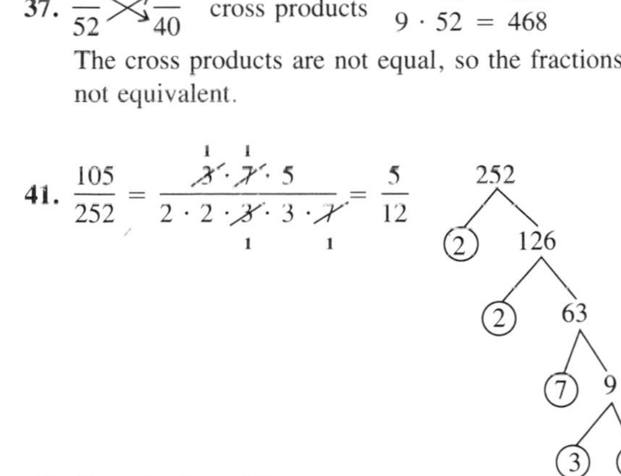

45. Factorization of 8:

$8 \cdot 1 \quad 2 \cdot 4$

The factors of 8 are 1, 2, 4, 8.

Section 2.5 (page 125)

1. $\frac{1}{4} \cdot \frac{3}{5} = \frac{1 \cdot 3}{4 \cdot 5} = \frac{3}{20}$

5. $\frac{3}{\underset{2}{\cancel{8}}} \cdot \frac{\overset{3}{\cancel{12}}}{5} = \frac{3 \cdot 3}{2 \cdot 5} = \frac{9}{10}$

9. $\frac{\overset{1}{\cancel{3}}}{\underset{2}{\cancel{4}}} \cdot \frac{5}{\underset{2}{\cancel{6}}} \cdot \frac{\overset{1}{\cancel{2}}}{3} = \frac{5}{12}$

13. $\frac{\overset{1}{\overset{\cancel{3}}{\cancel{21}}}}{\underset{2}{\underset{\cancel{6}}{\cancel{30}}}} \cdot \frac{\overset{1}{\cancel{5}}}{\underset{1}{\cancel{7}}} = \frac{1 \cdot 1}{2 \cdot 1} = \frac{1}{2}$ or $\frac{\overset{1}{\overset{\cancel{7}}{\cancel{21}}}}{\underset{2}{\underset{\cancel{10}}{\cancel{30}}}} \cdot \frac{\overset{1}{\cancel{5}}}{\underset{1}{\cancel{7}}} = \frac{1 \cdot 1}{2 \cdot 1} = \frac{1}{2}$

17. $\frac{\overset{1}{\cancel{16}}}{\underset{1}{\underset{\cancel{5}}{\cancel{25}}}} \cdot \frac{\overset{7}{\cancel{35}}}{\underset{2}{\cancel{32}}} \cdot \frac{\overset{3}{\cancel{15}}}{64} = \frac{21}{128}$

21. $72 \cdot \frac{5}{9} = \frac{\overset{8}{\cancel{72}}}{1} \cdot \frac{5}{\underset{1}{\cancel{9}}} = \frac{8 \cdot 5}{1 \cdot 1} = \frac{40}{1} = 40$

25. $21 \cdot \frac{5}{7} \cdot \frac{7}{10} = \frac{\overset{3}{\cancel{21}}}{1} \cdot \frac{\overset{1}{\cancel{5}}}{\underset{1}{\cancel{7}}} \cdot \frac{7}{\underset{2}{\cancel{10}}} = \frac{3 \cdot 1 \cdot 7}{1 \cdot 1 \cdot 2} = \frac{21}{2} = 10\frac{1}{2}$

29. $\frac{3}{4} \cdot 500 = \frac{3}{\underset{1}{\cancel{4}}} \cdot \frac{\overset{125}{\cancel{500}}}{1} = \frac{3 \cdot 125}{1 \cdot 1} = \frac{375}{1} = 375$

33. $\frac{27}{32} \cdot 640 = \frac{27}{\underset{1}{\cancel{32}}} \cdot \frac{\overset{20}{\cancel{640}}}{1} = \frac{27 \cdot 20}{1 \cdot 1} = \frac{540}{1} = 540$

37. Multiply the length and the width.

$\frac{\overset{7}{\cancel{14}}}{\underset{5}{\cancel{15}}} \cdot \frac{\overset{3}{\cancel{9}}}{\underset{8}{\cancel{16}}} = \frac{7 \cdot 3}{5 \cdot 8} = \frac{21}{40}$ square inches

41. Multiply the length and the width.

$16 \cdot \frac{1}{4} = \frac{\overset{4}{\cancel{16}}}{1} \cdot \frac{1}{\underset{1}{\cancel{4}}} = \frac{4 \cdot 1}{1 \cdot 1} = 4$ square inches

45. *Step 1* Find the number of rafts rented in 9 days.

Step 2 Since the number of rafts rented each day is given and the number of rafts rented in 9 days must be found, multiply.

number of rafts rented in one day × number of days = total number of rafts rented

Step 3 The number of rafts rented should be about 200 × 10, or 2000.

Step 4 Multiply.

230 rented in one day × 9 days = 2070 rafts

2070 rafts is reasonable. (It is close to the estimate of 2000 rafts)

Answer: 2070 rafts

Section 2.6 (page 133)

1. Multiply the length and the width.

$\frac{5}{\underset{2}{\cancel{4}}} \cdot \frac{\overset{1}{\cancel{2}}}{3} = \frac{5 \cdot 1}{2 \cdot 3} = \frac{5}{6}$ square yard

5. $\frac{1}{5}$ of 1900 items are taxable.

↓ ↓ ↓

$$\frac{1}{5} \cdot 1900 = \frac{1}{\cancel{5}_1} \cdot \frac{\cancel{1900}^{380}}{1} = \frac{1 \cdot 380}{1 \cdot 1}$$

$$= \frac{380}{1} = 380 \text{ items are taxable}$$

9. $\frac{9}{14}$ of 560 employees have given.

↓ ↓ ↓

$$\frac{9}{14} \cdot 560 = \frac{9}{\cancel{14}_1} \cdot \frac{\cancel{560}^{40}}{1} = \frac{9 \cdot 40}{1 \cdot 1}$$

$$= \frac{360}{1} = 360 \text{ employees have given}$$

13. $\frac{1}{4}$ of income went to taxes.

↓ ↓ ↓

$$\frac{1}{4} \cdot 32{,}000 = \frac{1}{\cancel{4}_1} \cdot \frac{\cancel{32000}^{8000}}{1} = \frac{1 \cdot 8000}{1 \cdot 1}$$

$$= \frac{8000}{1} = \$8000 \text{ taxes}$$

17. $\frac{5}{8}$ of \$120 is earned.

↓ ↓ ↓

$$\frac{5}{8} \cdot 120 = \frac{5}{\cancel{8}_1} \cdot \frac{\cancel{120}^{15}}{1} = \frac{5 \cdot 15}{1 \cdot 1}$$

$$= \frac{75}{1} = \$75 \text{ earned}$$

21. $\frac{1}{4}$ of $\frac{1}{8}$ of an estate goes to the American Cancer Society.

↓ ↓ ↓

$$\frac{1}{4} \cdot \frac{1}{8} = \frac{1 \cdot 1}{4 \cdot 8} = \frac{1}{32} \text{ of the estate}$$

Section 2.7 (page 141)

1. $\frac{2}{3} \div \frac{3}{4} = \frac{2}{3} \cdot \frac{4}{3} = \frac{2 \cdot 4}{3 \cdot 3} = \frac{8}{9}$

5. $\frac{5}{9} \div \frac{5}{4} = \frac{5}{9} \cdot \frac{4}{5} = \frac{\cancel{5}^1 \cdot 4}{9 \cdot \cancel{5}_1} = \frac{1 \cdot 4}{9 \cdot 1} = \frac{4}{9}$

9. $\frac{7}{9} \div \frac{12}{5} = \frac{7}{9} \cdot \frac{5}{12} = \frac{7 \cdot 5}{9 \cdot 12} = \frac{35}{108}$

13. $9 \div \frac{1}{2} = \frac{9}{1} \cdot \frac{2}{1} = \frac{9 \cdot 2}{1 \cdot 1} = \frac{18}{1} = 18$

17. $\frac{\frac{4}{7}}{8} = \frac{4}{7} \div \frac{8}{1} = \frac{4}{7} \cdot \frac{1}{8} = \frac{\cancel{4}^1 \cdot 1}{7 \cdot \cancel{8}_2} = \frac{1 \cdot 1}{7 \cdot 2} = \frac{1}{14}$

21. "How many $\frac{4}{5}$'s out of 28" is solved by dividing

$28 \div \frac{4}{5}$. (If this doesn't make sense, think of whole numbers: if each shaker took 2 pounds, we could fill 14: $28 \div 2 = 14$.)

$$28 \div \frac{4}{5} = \frac{28}{1} \div \frac{4}{5} = \frac{\cancel{28}^7}{1} \cdot \frac{5}{\cancel{4}_1} = \frac{7 \cdot 5}{1 \cdot 1} = \frac{35}{1}$$

$$= 35 \text{ shakers}$$

25. The pick-up truck carries $\frac{2}{3}$-cord of firewood, so 40 cords must be divided by $\frac{2}{3}$ to find the number of trips needed to move the firewood.

$$40 \div \frac{2}{3} = \frac{40}{1} \cdot \frac{3}{2} = \frac{\cancel{40}^{20} \cdot 3}{1 \cdot \cancel{2}_1} = \frac{60}{1} = 60 \text{ trips}$$

29. $\frac{6}{7}$ of the trip is 216 miles. 216 is 6 parts of the total of 7. 216 must be divided by $\frac{6}{7}$ to find the total trip. Then subtract the miles already driven to find the miles remaining.

$$216 \div \frac{6}{7} = \frac{216}{1} \cdot \frac{7}{6} = \frac{\cancel{216}^{36} \cdot 7}{1 \cdot \cancel{6}_1} = \frac{36 \cdot 7}{1 \cdot 1}$$

$$= \frac{252}{1} = 252 \text{ miles}$$

252 miles (total trip) − 216 miles (already traveled) = 36 miles

33. $2\frac{5}{6}$ $\quad 2 \cdot 6 = 12 \quad 12 + 5 = 17 \quad \frac{17}{6}$

37. $4\frac{10}{11}$ $\quad 4 \cdot 11 = 44 \quad 44 + 10 = 54 \quad \frac{54}{11}$

Section 2.8 (page 149)

1. $7\frac{1}{2} \cdot 3\frac{1}{3} = \frac{15}{2} \cdot \frac{10}{3} = \frac{\cancel{15}^5}{\cancel{2}_1} \cdot \frac{\cancel{10}^5}{\cancel{3}_1} = \frac{5 \cdot 5}{1 \cdot 1} = \frac{25}{1} = 25$

5. $3\frac{1}{9} \cdot 1\frac{2}{7} = \frac{28}{9} \cdot \frac{9}{7} = \frac{\cancel{28}^4}{\cancel{9}_1} \cdot \frac{\cancel{9}^1}{\cancel{7}_1} = \frac{4 \cdot 1}{1 \cdot 1} = \frac{4}{1} = 4$

9. $9\frac{1}{5} \cdot 15 = \frac{46}{5} \cdot \frac{15}{1} = \frac{46}{\cancel{5}_1} \cdot \frac{\cancel{15}^3}{1} = \frac{46 \cdot 3}{1 \cdot 1} = \frac{138}{1} = 138$

13. $3\frac{1}{4} \div 2\frac{5}{8} = \frac{13}{4} \div \frac{21}{8} = \frac{13}{4} \cdot \frac{8}{21} = \frac{13 \cdot \cancel{8}^2}{\cancel{4}_1 \cdot 21}$

$$= \frac{13 \cdot 2}{1 \cdot 21} = \frac{26}{21} = 1\frac{5}{21}$$

17. $1\frac{1}{3} \div 5\frac{1}{2} = \frac{4}{3} \div \frac{11}{2} = \frac{4}{3} \cdot \frac{2}{11} = \frac{4 \cdot 2}{3 \cdot 11} = \frac{8}{33}$

21. $\frac{3}{8} \div 1\frac{1}{4} = \frac{3}{8} \div \frac{5}{4} = \frac{3}{8} \cdot \frac{4}{5} = \frac{3 \cdot \overset{1}{\cancel{4}}}{\underset{2}{\cancel{8}} \cdot 5}$

$= \frac{3 \cdot 1}{2 \cdot 5} = \frac{3}{10}$

25. $16 \cdot 3\frac{1}{4} = \frac{16}{1} \cdot \frac{13}{4} = \frac{\overset{4}{\cancel{16}} \cdot 13}{1 \cdot \underset{1}{\cancel{4}}} = \frac{4 \cdot 13}{1 \cdot 1}$

↑ dresses ↑ material per dress

$= \frac{52}{1} = 52$ yards of material needed

29. $1\frac{3}{4} \cdot 12\frac{1}{2} = \frac{7}{4} \cdot \frac{25}{2} = \frac{7 \cdot 25}{4 \cdot 2}$

ounces of chemical per gallon; gallons of water

$= \frac{175}{8} = 21\frac{7}{8}$ ounces of chemical

33. $4\frac{1}{4} \cdot 28 = \frac{17}{4} \cdot \frac{28}{1} = \frac{17 \cdot \overset{7}{\cancel{28}}}{\underset{1}{\cancel{4}} \cdot 1}$

film per wedding; weddings

$= \frac{17 \cdot 7}{1 \cdot 1} = \frac{119}{1} = 119$ rolls of film at wedding

$2\frac{3}{8} \cdot 16 = \frac{19}{8} \cdot \frac{16}{1} = \frac{19 \cdot \overset{2}{\cancel{16}}}{\underset{1}{\cancel{8}} \cdot 1}$

film per retirement party; parties

$= \frac{19 \cdot 2}{1 \cdot 1} = 38$ rolls of film at retirement parties

$119 + 38 = 157$ total rolls of film needed

CHAPTER 3

Section 3.1 (page 167)

1. $\frac{1}{3} + \frac{1}{3} = \frac{1 + 1}{3} = \frac{2}{3}$

5. $\frac{1}{4} + \frac{1}{4} = \frac{1 + 1}{4} = \frac{\overset{1}{\cancel{2}}}{\underset{2}{\cancel{4}}} = \frac{1}{2}$

9. $\frac{7}{12} + \frac{3}{12} = \frac{7 + 3}{12} = \frac{\overset{5}{\cancel{10}}}{\underset{6}{\cancel{12}}} = \frac{5}{6}$

13. $\frac{3}{17} + \frac{2}{17} + \frac{5}{17} = \frac{3 + 2 + 5}{17} = \frac{10}{17}$

17. $\frac{2}{54} + \frac{8}{54} + \frac{12}{54} = \frac{2 + 8 + 12}{54} = \frac{\overset{11}{\cancel{22}}}{\underset{27}{\cancel{54}}} = \frac{11}{27}$

21. $\frac{16}{21} - \frac{8}{21} = \frac{16 - 8}{21} = \frac{8}{21}$

25. $\frac{14}{15} - \frac{4}{15} = \frac{14 - 4}{15} = \frac{\overset{2}{\cancel{10}}}{\underset{3}{\cancel{15}}} = \frac{2}{3}$

29. $\frac{43}{72} - \frac{25}{72} = \frac{43 - 25}{72} = \frac{\overset{1}{\cancel{18}}}{\underset{4}{\cancel{72}}} = \frac{1}{4}$

33. $\frac{746}{400} - \frac{506}{400} = \frac{746 - 506}{400} = \frac{\overset{3}{\cancel{240}}}{\underset{5}{\cancel{400}}} = \frac{3}{5}$

37. $\frac{11}{12} - \frac{5}{12} = \frac{11 - 5}{12} = \frac{\overset{1}{\cancel{6}}}{\underset{2}{\cancel{12}}} = \frac{1}{2}$ mile

(total) − (walked already) = (left)

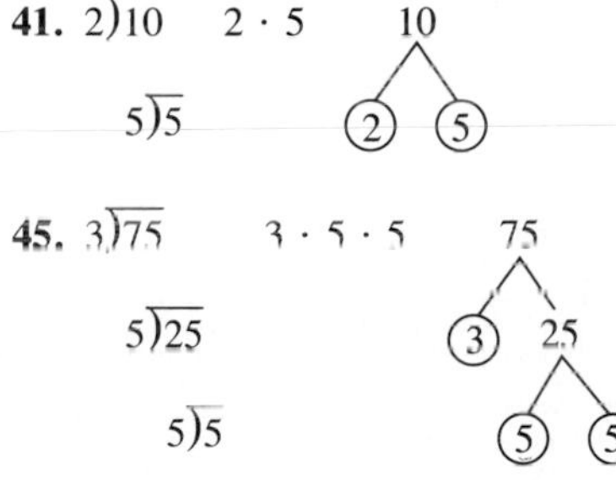

41. $2\overline{)10}$ $5\overline{)5}$ $2 \cdot 5$

45. $3\overline{)75}$ $5\overline{)25}$ $5\overline{)5}$ $3 \cdot 5 \cdot 5$

Section 3.2 (page 175)

1. Prime factorizations:

$6 = 2 \cdot 3$
$12 = 2 \cdot 2 \cdot 3$

Complete the table and circle the largest product in each column.

prime	2	3
6 =	2 ·	(3)
12 =	(2 · 2) ·	3
LCM	(2 · 2)	(3)

Least common multiple: $2 \cdot 2 \cdot 3 = 12$

5. Prime factorizations:

$18 = 2 \cdot 3 \cdot 3$
$24 = 2 \cdot 2 \cdot 2 \cdot 3$

prime	2	3
18 =	2 ·	(3 · 3)
24 =	(2 · 2 · 2) ·	3
LCM	(2 · 2 · 2)	(3 · 3)

Least common multiple:

$2 \cdot 2 \cdot 2 \cdot 3 \cdot 3 = 72$

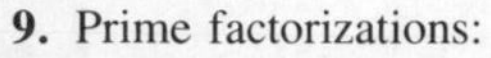

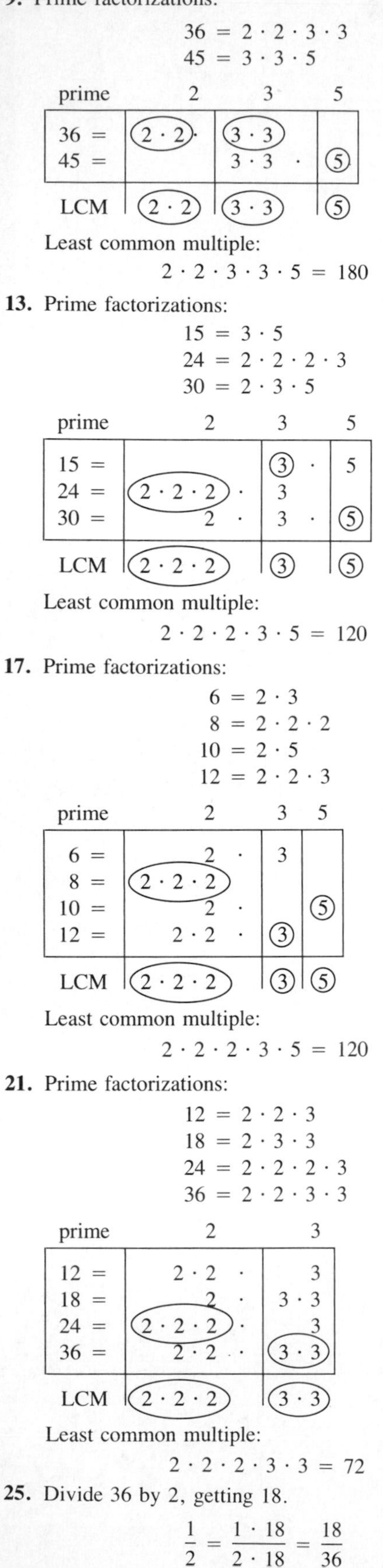

9. Prime factorizations:

$$36 = 2 \cdot 2 \cdot 3 \cdot 3$$
$$45 = 3 \cdot 3 \cdot 5$$

prime	2	3	5
36 =	(2 · 2) ·	(3 · 3)	
45 =		3 · 3 ·	(5)
LCM	(2 · 2)	(3 · 3)	(5)

Least common multiple:

$$2 \cdot 2 \cdot 3 \cdot 3 \cdot 5 = 180$$

13. Prime factorizations:

$$15 = 3 \cdot 5$$
$$24 = 2 \cdot 2 \cdot 2 \cdot 3$$
$$30 = 2 \cdot 3 \cdot 5$$

prime	2	3	5
15 =		(3) ·	5
24 =	(2 · 2 · 2) ·	3	
30 =	2 ·	3 ·	(5)
LCM	(2 · 2 · 2)	(3)	(5)

Least common multiple:

$$2 \cdot 2 \cdot 2 \cdot 3 \cdot 5 = 120$$

17. Prime factorizations:

$$6 = 2 \cdot 3$$
$$8 = 2 \cdot 2 \cdot 2$$
$$10 = 2 \cdot 5$$
$$12 = 2 \cdot 2 \cdot 3$$

prime	2	3	5
6 =	2 ·	3	
8 =	(2 · 2 · 2)		
10 =	2 ·		(5)
12 =	2 · 2 ·	(3)	
LCM	(2 · 2 · 2)	(3)	(5)

Least common multiple:

$$2 \cdot 2 \cdot 2 \cdot 3 \cdot 5 = 120$$

21. Prime factorizations:

$$12 = 2 \cdot 2 \cdot 3$$
$$18 = 2 \cdot 3 \cdot 3$$
$$24 = 2 \cdot 2 \cdot 2 \cdot 3$$
$$36 = 2 \cdot 2 \cdot 3 \cdot 3$$

prime	2	3
12 =	2 · 2 ·	3
18 =	2 ·	3 · 3
24 =	(2 · 2 · 2) ·	3
36 =	2 · 2 ·	(3 · 3)
LCM	(2 · 2 · 2)	(3 · 3)

Least common multiple:

$$2 \cdot 2 \cdot 2 \cdot 3 \cdot 3 = 72$$

25. Divide 36 by 2, getting 18.

$$\frac{1}{2} = \frac{1 \cdot 18}{2 \cdot 18} = \frac{18}{36}$$

29. Divide 36 by 9, getting 4.

$$\frac{3}{9} = \frac{3 \cdot 4}{9 \cdot 4} = \frac{12}{36}$$

33. Divide 40 by 10, getting 4.

$$\frac{9}{10} = \frac{9 \cdot 4}{10 \cdot 4} = \frac{36}{40}$$

37. Divide 70 by 10, getting 7.

$$\frac{3}{10} = \frac{3 \cdot 7}{10 \cdot 7} = \frac{21}{70}$$

41. Divide 56 by 7, getting 8.

$$\frac{9}{7} = \frac{9 \cdot 8}{7 \cdot 8} = \frac{72}{56}$$

45. Divide 132 by 11, getting 12.

$$\frac{8}{11} = \frac{8 \cdot 12}{11 \cdot 12} = \frac{96}{132}$$

49. Prime factorizations:

$$1512 = 2 \cdot 2 \cdot 2 \cdot 3 \cdot 3 \cdot 3 \cdot 7$$
$$392 = 2 \cdot 2 \cdot 2 \cdot 7 \cdot 7$$

prime	2	3	7
1512 =	2 · 2 · 2 ·	(3 · 3 · 3) ·	7
392 =	(2 · 2 · 2) ·		(7 · 7)
LCM	(2 · 2 · 2)	(3 · 3 · 3)	(7 · 7)

Least common multiple:

$$2 \cdot 2 \cdot 2 \cdot 3 \cdot 3 \cdot 3 \cdot 7 \cdot 7 = 10{,}584$$

53. $\frac{9}{5}$ $\quad 5\overline{)9}$ with quotient 1 R4: $9 - 5 = 4$ $\quad 1\frac{4}{5}$

Section 3.3 (page 183)

1. $\frac{3}{5} + \frac{1}{5} = \frac{3+1}{5} = \frac{4}{5}$

5. $\frac{7}{20} + \frac{3}{10} = \frac{7}{20} + \frac{6}{20} = \frac{13}{20}$

Least common denominator is 20.

9. $\frac{1}{12} + \frac{5}{9}$

2	12	9
2	6	9
3	3	9
3	1	3
	1	1

Least common denominator is $2 \cdot 2 \cdot 3 \cdot 3 = 36$.

$\frac{1}{12} = \frac{3}{36}$ and $\frac{5}{9} = \frac{20}{36}$,

so $\frac{1}{12} + \frac{5}{9} = \frac{3}{36} + \frac{20}{36} = \frac{3+20}{36} = \frac{23}{36}$

13. $\frac{1}{4}+\frac{2}{9}+\frac{1}{3}$

$$\begin{array}{r|rrr} 2 & 4 & 9 & 3 \\ 2 & 2 & 9 & 3 \\ 3 & 1 & 9 & 3 \\ 3 & 1 & 3 & 1 \\ \hline & 1 & 1 & 1 \end{array}$$

Least common denominator is $2 \cdot 2 \cdot 3 \cdot 3 = 36$.

$\frac{1}{4}=\frac{9}{36}, \frac{2}{9}=\frac{8}{36}, \frac{1}{3}=\frac{12}{36},$

so $\frac{1}{4}+\frac{2}{9}+\frac{1}{3}=\frac{9}{36}+\frac{8}{36}+\frac{12}{36}=\frac{29}{36}$

17. $\frac{7}{10}+\frac{1}{15}+\frac{1}{6}$

$$\begin{array}{r|rrr} 2 & 10 & 15 & 6 \\ 3 & 5 & 15 & 3 \\ 5 & 5 & 5 & 1 \\ \hline & 1 & 1 & 1 \end{array}$$

Least common denominator is $2 \cdot 3 \cdot 5 = 30$.

$\frac{7}{10}=\frac{21}{30}, \frac{1}{15}=\frac{2}{30}, \frac{1}{6}=\frac{5}{30},$

so $\frac{7}{10}+\frac{1}{15}+\frac{1}{6}=\frac{21}{30}+\frac{2}{30}+\frac{5}{30}=\frac{28}{30}=\frac{14}{15}$

21. Least common denominator is 30.

$$\begin{array}{rcl} \frac{8}{15} & = & \frac{16}{30} \\ +\ \frac{3}{10} & = & \frac{9}{30} \\ \hline & & \frac{25}{30} = \frac{5}{6} \end{array}$$

25. $\frac{2}{3}-\frac{1}{6}$ Least common denominator is 6, and $\frac{2}{3}=\frac{4}{6}$.

$\frac{2}{3}-\frac{1}{6}=\frac{4}{6}-\frac{1}{6}=\frac{4-1}{6}=\frac{3}{6}=\frac{1}{2}$

29. $\frac{3}{4}-\frac{5}{12}=\frac{9}{12}-\frac{5}{12}=\frac{9-5}{12}=\frac{4}{12}=\frac{1}{3}$

33. Least common denominator is 36.

$$\begin{array}{rcl} \frac{5}{9} & = & \frac{20}{36} \\ -\ \frac{5}{12} & = & \frac{15}{36} \\ \hline & & \frac{5}{36} \end{array}$$

$\leftarrow$ Subtract numerators.

37. Add the fractions of tons of grain bought.

$\frac{3}{8}+\frac{1}{4}+\frac{1}{3}=\frac{9}{24}+\frac{6}{24}+\frac{8}{24}=\frac{23}{24}$ ton bought

41. Add the units already completed. $\frac{5}{16}+\frac{1}{8}=\frac{5}{16}+\frac{2}{16}$ $=\frac{7}{16}$ unit already completed. Subtract $\frac{7}{16}$ from 1 to find the fraction of units still needed.

$1-\frac{7}{16}=\frac{16}{16}-\frac{7}{16}=\frac{9}{16}$ of the units still needed

45. $1\frac{2}{3}\cdot 2\frac{7}{10}=\frac{5}{3}\cdot\frac{27}{10}=\frac{\overset{1}{\cancel{5}}}{\underset{1}{\cancel{3}}}\cdot\frac{\overset{9}{\cancel{27}}}{\underset{2}{\cancel{10}}}=\frac{1\cdot 9}{1\cdot 2}=\frac{9}{2}=4\frac{1}{2}$

Section 3.4 (page 191)

1.
$$\begin{array}{r} 21\frac{1}{7} \\ +\ 49\frac{3}{7} \\ \hline 70\frac{4}{7} \end{array}$$

5.
$$\begin{array}{rcl} 46\frac{3}{8} & = & 46\frac{3}{8} \\ +\ 15\frac{1}{4} & = & 15\frac{2}{8} \\ \hline & & 61\frac{5}{8} \end{array}$$

9.
$$\begin{array}{rcl} 126\frac{4}{5} & = & 126\frac{8}{10} \\ +\ 25\frac{9}{10} & = & 25\frac{9}{10} \\ \hline & & 151\frac{17}{10} = 151 + 1\frac{7}{10} = 152\frac{7}{10} \end{array}$$

13.
$$\begin{array}{rcl} 7\frac{1}{4} & = & 7\frac{2}{8} \\ 25\frac{3}{8} & = & 25\frac{3}{8} \\ +\ 9\frac{1}{2} & = & 9\frac{4}{8} \\ \hline & & 41\frac{9}{8} = 41 + 1\frac{1}{8} = 42\frac{1}{8} \end{array}$$

17.
$$\begin{array}{r} 9\frac{3}{4} \\ -\ 6\frac{1}{4} \\ \hline 3\frac{2}{4} = 3\frac{1}{2} \end{array}$$

21.
$$\begin{array}{rcl} 14\frac{11}{16} & = & 14\frac{33}{48} \\ -\ 8\frac{5}{12} & = & 8\frac{20}{48} \\ \hline & & 6\frac{13}{48} \end{array}$$

25.
$$\begin{array}{rcl} 47\frac{3}{8} & = & 47\frac{9}{24} \\ -\ 26\frac{7}{12} & = & 26\frac{14}{24} \end{array}$$

Borrow.

$47\frac{9}{24} = 46 + 1 + \frac{9}{24}$

$= 46 + \frac{24}{24} + \frac{9}{24} = 46\frac{33}{24}$

Subtract.
$$\begin{array}{r} 46\frac{33}{24} \\ -\ 26\frac{14}{24} \\ \hline 20\frac{19}{24} \end{array}$$

29.
$$\begin{array}{r} 374\frac{2}{8} = 374\frac{6}{24} \\ -\ 211\frac{5}{6} = 211\frac{20}{24} \\ \hline \end{array}$$

Borrow.

$$374\frac{6}{24} = 373 + 1 + \frac{6}{24}$$
$$= 373 + \frac{24}{24} + \frac{6}{24} = 373\frac{30}{24}$$

Subtract.
$$\begin{array}{r} 373\frac{30}{24} \\ -\ 211\frac{20}{24} \\ \hline 162\frac{10}{24} = 162\frac{5}{12} \end{array}$$

33.
$$\begin{array}{r} 21 \\ -\ 5\frac{7}{8} \\ \hline \end{array}$$

Borrow.

$$21 = 20 + 1 = 20 + \frac{8}{8}$$

Subtract.
$$\begin{array}{r} 20\frac{8}{8} \\ -\ 5\frac{7}{8} \\ \hline 15\frac{1}{8} \end{array}$$

37. Russell worked $15\frac{1}{8}$ hours on the weekend. If he worked $6\frac{1}{2}$ hours on Saturday, he worked $15\frac{1}{8} - 6\frac{1}{2}$ hours on Sunday.

$$\begin{array}{r} 15\frac{1}{8} = 15\frac{1}{8} \\ -\ 6\frac{1}{2} = 6\frac{4}{8} \\ \hline \end{array}$$

Borrow.

$$15\frac{1}{8} = 14 + 1 + \frac{1}{8}$$
$$= 14 + \frac{8}{8} + \frac{1}{8} = 14\frac{9}{8}$$

Subtract.
$$\begin{array}{r} 14\frac{9}{8} \\ -\ 6\frac{4}{8} \\ \hline 8\frac{5}{8} \text{ hours} \end{array}$$

41. Marty worked $6\frac{3}{8} + 7\frac{1}{2} + 8\frac{3}{4} + 7\frac{3}{8} + 8 = 6\frac{9}{24} + 7\frac{12}{24} + 8\frac{18}{24} + 7\frac{9}{24} + 8 = 36\frac{48}{24} = 38$. He worked 38 hours.

45. On the first four days of the week, she worked

$$\begin{array}{r} 8\frac{1}{4} = 8\frac{2}{8} \\ 6\frac{3}{8} = 6\frac{3}{8} \\ 7\frac{3}{4} = 7\frac{6}{8} \\ +\ 8\frac{3}{4} = 8\frac{6}{8} \\ \hline 29\frac{17}{8} = 29 + 2\frac{1}{8} = 31\frac{1}{8} \text{ hours.} \end{array}$$

She worked a total of 40 hours for the week, so on Friday she worked

$$\begin{array}{r} 40 = 39\frac{8}{8} \\ -\ 31\frac{1}{8} = 31\frac{1}{8} \\ \hline 8\frac{7}{8} \text{ hours.} \end{array}$$

49. The total width is $9\frac{7}{16}$. The known portions of the width are $2\frac{3}{8} + 2\frac{3}{8} = 2\frac{6}{16} + 2\frac{6}{16} = 4\frac{12}{16}$. Subtract to find x (the unknown portion of the width).

$$\begin{array}{r} 9\frac{7}{16} = 8\frac{23}{16} \\ -\ 4\frac{12}{16} = 4\frac{12}{16} \\ \hline 4\frac{11}{16} \end{array} \qquad x = 4\frac{11}{16}$$

53. $4 \cdot 1 + 8 \cdot 7 + 3$
$4 + 8 \cdot 7 + 3$ Multiply first, left to right.
$4 + 56 + 3$ Multiply, left to right.
$60 + 3$ Add, left to right.
63 Add last.

Section 3.5 (page 201)

1. and **5.**

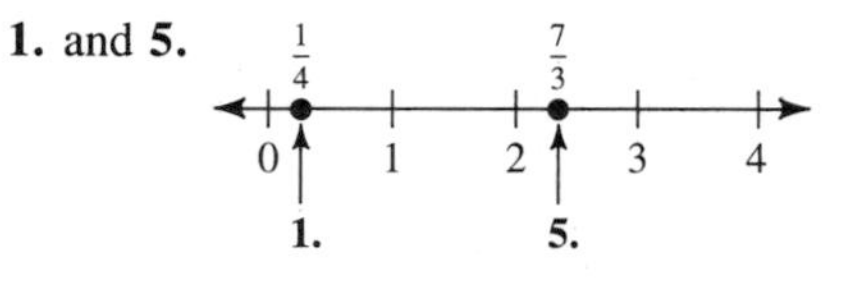

9. The least common multiple for 2 and 4 is 4.

$$\frac{1}{2} = \frac{2}{4} \text{ and } \frac{3}{4} = \frac{3}{4}\text{; therefore, } \frac{1}{2} < \frac{3}{4}$$

13. The least common multiple for 8 and 12 is 24.

$$\frac{3}{8} = \frac{9}{24} \text{ and } \frac{5}{12} = \frac{10}{24}\text{; therefore, } \frac{3}{8} < \frac{5}{12}$$

17. The least common multiple for 24 and 36 is 72.

$$\frac{11}{24} = \frac{33}{72} \text{ and } \frac{19}{36} = \frac{38}{72}\text{; therefore, } \frac{11}{24} < \frac{19}{36}$$

21. $\left(\frac{4}{5}\right)^2 = \frac{4}{5} \cdot \frac{4}{5} = \frac{16}{25}$

25. $\left(\frac{2}{3}\right)^3 = \frac{2}{3} \cdot \frac{2}{3} \cdot \frac{2}{3} = \frac{8}{27}$

29. $\left(\frac{3}{2}\right)^4 = \frac{3}{2} \cdot \frac{3}{2} \cdot \frac{3}{2} \cdot \frac{3}{2} = \frac{81}{16} = 5\frac{1}{16}$

33. $2^2 + 3 - 2$
$4 + 3 - 2$ Do exponent first.
$7 - 2$ Add, left to right.
5 Subtract last.

37. $\left(\frac{1}{2}\right)^2 \cdot 4$
$\frac{1}{4} \cdot \frac{4}{1}$ Do exponent first.
1 Multiply.

41. $\left(\frac{3}{4}\right)^2 \cdot \left(\frac{2}{3}\right)^2$
$\frac{9}{16} \cdot \frac{4}{9}$ Do exponents first.
$\frac{1}{4}$ Multiply.

45. $\frac{1}{2} \cdot \frac{4}{5} + \frac{2}{3} \cdot \frac{9}{5}$
$\frac{2}{5} + 1\frac{1}{5}$ Multiply first.
$1\frac{3}{5}$ Add last.

49. $\left(\frac{1}{3} + \frac{1}{6}\right) \cdot \frac{1}{2}$
$\frac{3}{6} \cdot \frac{1}{2}$ Work inside parentheses first.
$\frac{1}{4}$ Multiply.

53. $\left(\frac{5}{6} - \frac{1}{12}\right) \div \frac{3}{2}$
$\frac{3}{4} \div \frac{3}{2}$ Work inside parentheses first.
$\frac{1}{2}$ Divide.

57. $\left(\frac{3}{4}\right)^2 - \left(\frac{3}{4} - \frac{1}{8}\right) \div \frac{7}{4}$
$\left(\frac{3}{4}\right)^2 - \frac{5}{8} \div \frac{7}{4}$ Work inside parentheses first.
$\frac{9}{16} - \frac{5}{8} \div \frac{7}{4}$ Do exponents.
$\frac{9}{16} - \frac{5}{14}$ Divide.
$\frac{23}{112}$ Subtract.

61. 8436 eight thousand, four hundred thirty-six

CHAPTER 4

Section 4.1 (page 223)

1. 62.407
tenths: 4 (first digit to right of decimal point)
hundredths: 0 (second digit to right of decimal point)

5. .51472
thousandths: 4 (third digit to right of decimal point)
ten-thousandths: 7 (fourth digit to right of decimal point)

9. 149.0832
hundreds: 1 (third digit to left of decimal point)
hundredths: 8 (second digit to right of decimal point)

13. .62
6: tenths (first digit to right of decimal point)
2: hundredths (second digit to right of decimal point)

17. 47.691
4: tens (second digit to left of decimal point)
7: units or ones (first digit to left of decimal point)
6: tenths (first digit to right of decimal point)
9: hundredths (second digit to right of decimal point)
1: thousandths (third digit to right of decimal point)

21. $.4 = \frac{4}{10} = \frac{2}{5}$

25. $.88 = \frac{88}{100} = \frac{22}{25}$

29. $.02 = \frac{2}{100} = \frac{1}{50}$

33. $.919 = \frac{919}{1000}$

37. seven tenths

41. one hundred sixty-five thousandths

45. one and seventy-two hundredths

49. 27.32

53. 760.921

57. $625.4284 = 625\frac{4284}{10000} = 625\frac{1071}{2500}$

61. ten: 46,810 (4 6 8 0 [5] 5 is 5 or larger)
hundred: 46,800 (4 6 8 [0] 5 0 is 4 or less)
thousand: 47,000 (4 6 [8] 0 5 8 is 5 or larger)

Section 4.2 (page 231)

1. 16.8974 to the nearest tenth
16.8974 5 or more, so add 1 to 8, getting 9. Drop off digits to right of 8 (974).
tenths
Answer: 16.9

5. 42.399 to the nearest hundredth
42.399 5 or more, so add 1 to 9, getting 10. Write 0 and carry 1 to the tenths position. Drop off digits to right of 9 (9).
hundredths
Answer: 42.40

SOLUTIONS

9. 899.498 to the nearest tenth

899.498 — 5 or more, so add 1 to 4, getting 5. Drop off digits to right of 4 (98).
↑ tenths

Answer: 899.5

13. $69.13 to the nearest dollar

69.13 — 4 or less, so leave 9 alone. Drop digits to the right of 9 (13).
↑ one dollar

Answer: $69

17. $11,562.59 to the nearest dollar

11,562.59 — 5 or more, so add 1 to 2, getting 3. Drop off digits to right of 2 (59).
↑ one dollar

Answer: $11,563

21. .7837

nearest thousandth: .784 (7 is 5 or more so add 1 to thousandths)

nearest hundredth: .78 (3 is 4 or less, so hundredth stays 8)

nearest tenth: .8 (8 is 5 or more, so add 1 to tenths)

25. $18,765.48 to the nearest dollar

18,765.48 — less than 5
↑ one dollar

Answer: $18,765

29. $379.82 to the nearest dollar

379.82 — 5 or more. Write 0 and carry 1 to the tens position.
↑ one dollar

Answer: $380

33. 35.6150479

nearest millionth: 35.615048

(9 is 5 or more, so add 1 to millionths)

nearest hundred-thousandths: 35.61505

(7 is 5 or more, so add 1 to hundred-thousandths)

nearest ten-thousandths: 35.6150

(4 is 4 or less, so ten-thousandths stays as 0)

37. 1000.0050028

nearest millionth: 1000.005003

(8 is 5 or more, so add 1 to millionths)

nearest hundred-thousandth: 1000.00500

(2 is 4 or less, so hundred-thousandths stays as 0)

nearest ten-thousandth: 1000.0050

(0 is 4 or less, so ten-thousandths stays as 0)

41.
```
   11 12
       2
     816
      43
   7,591
+ 26,308
  34,760
```

Section 4.3 (page 235)

1.
```
  22
  48.96
  37.42
+ 99.71
 186.09
```

5.
```
  21 1
  28.76
  14.10 ← 0 attached
+ 39.25
  82.11
```

9.
```
  32
   9.71
   4.80 ← 0 attached
   3.60 ← 0 attached
   5.20 ← 0 attached
+ 19.52
  42.83
```

13. 14.23 + 28 + 74.63 + 18.715 + 64.286

Write numbers in a column, with decimal points lined up. Attach 0's as needed.
```
  21 11
  14.230
  28.000
  74.630
  18.715
+ 64.286
 199.861
```

17.
```
  37.25   estimate    2   problem    21
  18.9               37            37.25
+  7.5               19            18.90
                   +  8 ← close → + 7.50
                     64            63.65
```

21.
```
  382.504
  591.089
+ 612.715

estimate   1     problem    1 1 11
         383               382.504
         591               591.089
       + 613             + 612.715
        1587 ← close →    1586.308
```

25.
```
          2  1
         $17.14  books
          19.36  blouse
        +  5.14  record
Answer:  $41.64  total spent
```

29.
```
          1  2
         $ 7.42  apples
          10.09  peaches
        + 17.19  pears
Answer:  $34.70  total spent
```

33.
```
         23
          8.6  hours
          3.7  hours
         11.3  hours
          2.9  hours
       + 14.6  hours
Answer:  41.1  total hours
```

37.

$$\begin{array}{r} \scriptstyle 222 \\ 107.345 \\ 74.000 \\ 82.900 \\ +\ \ 56.920 \\ \hline 321.165 \end{array}$$

Answer: 321.165 yards

41.

$$\begin{array}{r} 315 \\ -\ 104 \\ \hline 211 \end{array}$$

Section 4.4 (page 241)

1.

$$\begin{array}{r} \scriptstyle 6\ 13 \\ \not{7}\ \not{3}.\ 5 \\ -\ 1\ \ 9.\ 2 \\ \hline 5\ \ 4.\ 3 \end{array}$$

5.

$$\begin{array}{r} \scriptstyle 12\ \ 14 \\ \scriptstyle 7\ \ \not{2}\ \ \not{4}\ \ 14 \\ 2\ \not{8}\ \not{3}.\ \not{5}\ \not{4} \\ -\ \ 1\ 8.\ 7\ 7 \\ \hline 2\ 6\ 4.\ 7\ 7 \end{array}$$

9.

$$\begin{array}{r} \scriptstyle 17 \\ \scriptstyle 4\ \ \not{7}\ \ 12 \\ \not{5}\ \not{8}.\ \not{2}\ 5\ 4 \\ -\ 1\ 9.\ 7\ 0\ 0 \\ \hline 3\ 8.\ 5\ 5\ 4 \end{array}$$

← Attach 00.

13.

$$\begin{array}{r} \scriptstyle 16\ \ 9 \\ \scriptstyle 4\ \ \not{6}\ \ \not{10}\ \ 10 \\ 1\ \not{5}.\ \not{7}\ \not{0}\ \not{0} \\ 2.\ 8\ 5\ 2 \\ \hline 1\ 2.\ 8\ 4\ 8 \end{array}$$

← Attach 00.

17.

$$\begin{array}{r} \scriptstyle 12\ \ 9 \\ \scriptstyle 3\ \ 17\ \ 8\ \ \not{2}\ \ \not{10}\ \ 10 \\ \not{4}\ \not{7}\ \not{9}.\ \not{3}\ \not{0}\ \not{0} \\ -\ \ 8\ 5.\ 7\ 9\ 3 \\ \hline 3\ 9\ 3.\ 5\ 0\ 7 \end{array}$$

← Attach 00.

21. 39.8 − 27.42

$$\begin{array}{r} \scriptstyle 7\,10 \\ 39.\not{8}\not{0} \\ +\ 27.42 \\ \hline 12.38 \end{array}$$

25.

$$\begin{array}{r} 19.74 \\ -\ \ 6.58 \end{array}$$

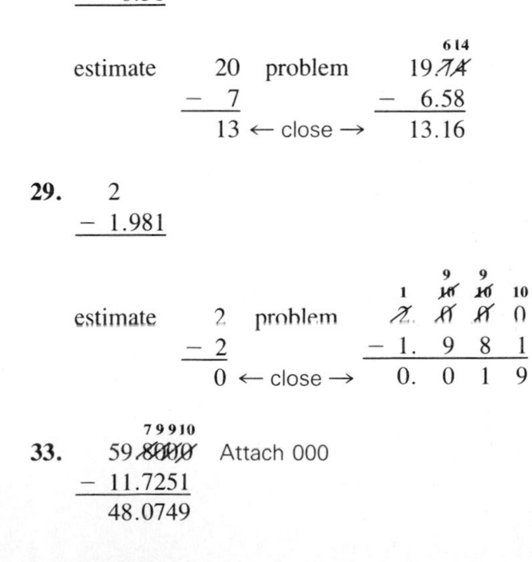

37.

$$\begin{array}{r} \scriptstyle 4\,10 \\ 42.\not{5}\not{0} \\ -\ 16.35 \\ \hline 26.15 \end{array}$$

agreed to work / already worked / left to work

41.

	\$72.18	total bill
−	39.76	tune-up
	\$32.42	gas expense

45. 386.02100 − 179.68231 = 206.33869 (attach 00)

$$\begin{array}{r} \scriptstyle 15\ \ 9\,11\,10\,9 \\ \scriptstyle 7\,\not{8}\ \ \not{10}\,\not{1}\ \not{9}\,\not{10}\,10 \\ 3\not{8}\not{6}.\not{0}\not{2}\not{1}\not{0}\not{0} \\ -\ 179.68231 \\ \hline 206.33869 \end{array}$$

49.

	\$ 129.86	balance (Sept. 1)
+	1749.82	deposit
	\$1879.68	after deposit
−	1802.15	checks written
	\$ 77.53	
−	2.00	service charge
	\$ 75.53	amount in account

53.

$$\begin{array}{r} \scriptstyle 2\ 11 \\ 1.009 \\ 1.662 \\ 1.897 \\ +\ 1.500 \\ \hline 6.068 \end{array}$$

total lengths given

	7.000	distance around
−	6.068	sides given
	.932 in	length of *m*

57.

$$\begin{array}{r} 837 \\ \times\ 708 \\ \hline 6\,696 \\ 585\,90\ \ \\ \hline 592{,}596 \end{array}$$

Section 4.5 (page 247)

1.

$$\begin{array}{r} 39.6 \\ \times\ \ 4.8 \end{array}$$

estimate

$$\begin{array}{r} 40 \\ \times\ 5 \\ \hline 200 \end{array}$$

problem

$$\begin{array}{r} \scriptstyle 3\,2 \\ \scriptstyle 7\,4 \\ 39.6 \\ \times\ \ 4.8 \\ \hline 316\ 8 \\ 1584\ \ \\ \hline 190.08 \end{array}$$

200 ← close → 190.08

1 decimal place
+ 1 decimal place
2 in answer

(Remember: in multiplying, our estimates will usually be a little further off than for adding or subtracting.)

SOLUTIONS

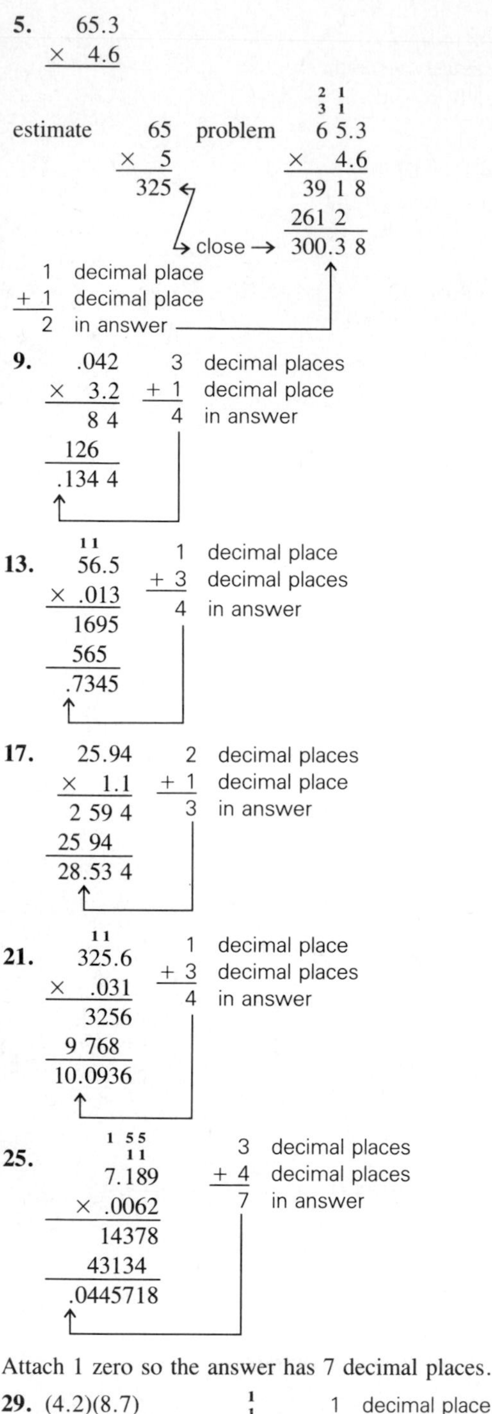

Attach 1 zero so the answer has 7 decimal places.

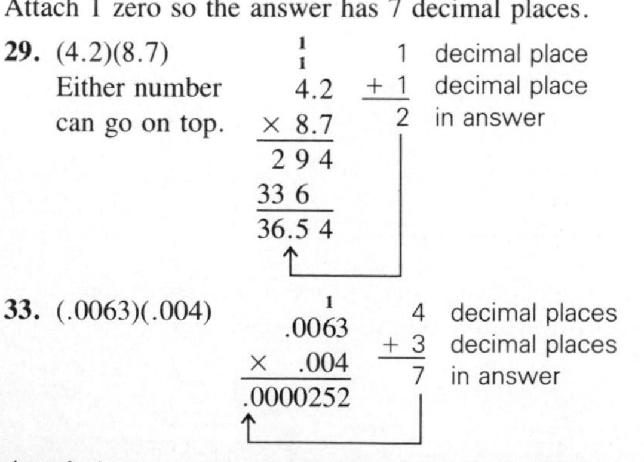

Attach 4 zeros so the answer has 7 decimal places.

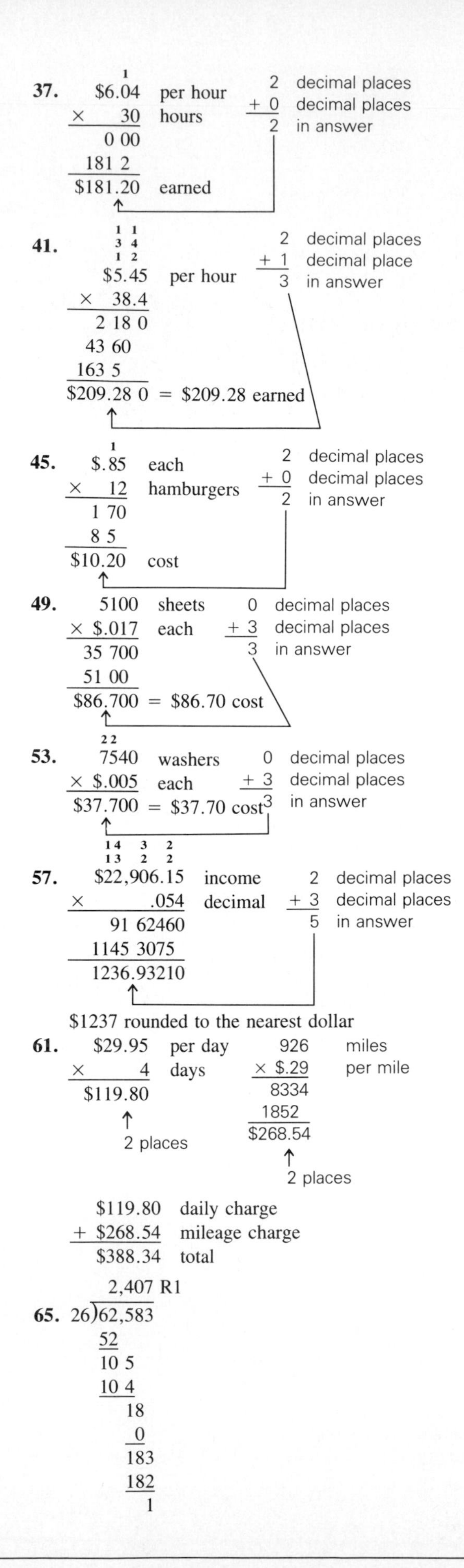

Section 4.6 (page 257)

Decimal point comes straight up.

```
       4.14
1. 6)24.84
     24
      0 8
        6
        24
        24
         0
```

```
       3.6847
5. 4)14.7389
     12
      2 7
      2 4
        33
        32
         18
         16
          29
          28
           1
```

No need to add zeros and divide further, since 4 is in thousandths place already and there is one more digit.

Round: 3.6847 — 5 or more, so add 1 to 4, getting 5; drop digit to right of 4(7)

(4 is in the thousandths place)

Answer: 3.685

Decimal point comes straight up.

```
       9.0001
9. 9)81.0009
     81
      0 0
        0
        00
         0
         00
          0
          09
           9
           0
```

Round: 9.0001 — less than 5, so drop digits to right

(thousandth)

Answer: 9.000

```
         3.038
13. 22)66.836
       66
        0 8
        0 0
          83
          66
          176
          176
            0
```

```
          1.2458
17. 35)43.6050
       35
        8 6
        7 0
        1 60
        1 40
          205
          175
          300  <- zero added to dividend, then brought down
          280
           20  <- stop
```

Round: 1.2458 — 5 or more, so add 1 to thousandths

(thousandth)

Answer: 1.246

Decimal point comes straight up.

```
         .009
21. 3).027
       0
       02
        0
        27
        27
         0
```

```
           2.16
25. 7.5)16.2 00
        15 0
         1 2 0  <- zero added to dividend, then brought down
           7 5
           4 50  <- zero added to dividend
           4 50
              0
```

zeros added to dividend to fill places to decimal point

```
              4000.
29. .004)16.000
         16
          0
```

```
             3.9569
33. 2.43)9.61 5300
         7 29
         2 32 5
         2 18 7
           13 83
           12 15
            1 680   <- zero added to dividend, then brought down
            1 458
              2220  <- zero added to dividend
              2187
                33  <- stop
```

Round: 3.9569 — 5 or more, so add 1 to thousandth

(thousandth)

Answer: 3.957

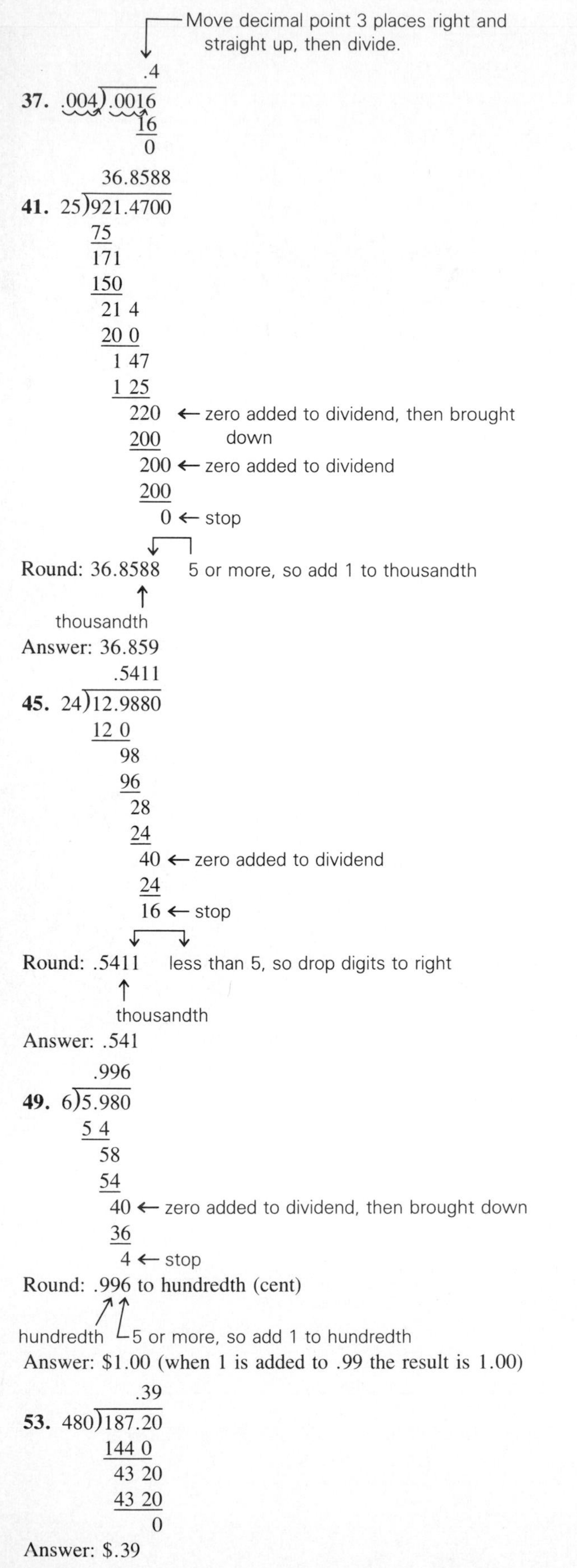

Move decimal point 3 places right and straight up, then divide.

37. $.004\overline{).0016}$ = .4
16
0

41. $25\overline{)921.4700}$ = 36.8588
75
171
150
21 4
20 0
1 47
1 25
220 ← zero added to dividend, then brought down
200
200 ← zero added to dividend
200
0 ← stop

Round: 36.8588 5 or more, so add 1 to thousandth
thousandth
Answer: 36.859

45. $24\overline{)12.9880}$ = .5411
12 0
98
96
28
24
40 ← zero added to dividend
24
16 ← stop

Round: .5411 less than 5, so drop digits to right
thousandth
Answer: .541

49. $6\overline{)5.980}$ = .996
5 4
58
54
40 ← zero added to dividend, then brought down
36
4 ← stop

Round: .996 to hundredth (cent)
hundredth 5 or more, so add 1 to hundredth
Answer: $1.00 (when 1 is added to .99 the result is 1.00)

53. $480\overline{)187.20}$ = .39
144 0
43 20
43 20
0

Answer: $.39

57. Divide: miles ÷ gallons = miles per gallon

$16.3\overline{)346.200}$ = 21.23
326
20 2
16 3
3 9 0 ← zero added to dividend
3 2 6
6 40 ← zero added to dividend
4 89
1 51 ← Stop here. There are enough digits to round.

Round: 21.23 less than 5, so tenth stays 2. Drop digit to right of 2 (3).
tenth
Answer: 21.2 miles per gallon

61. Divide: earnings ÷ hours = earnings per hour

$50\overline{)235.50}$ = 4.71
200
35 5
35 0
50
50
0

Answer: $4.71 per hour

65. $144\overline{)12.240}$ = .085
11 52
720
720
0

Answer: $.085

69. $3.5^2 + 5.2 - 7.2$
$12.25 + 5.2 - 7.2$ Do exponents first.
10.25 Add and subtract, left to right.

73. $38.6 + (13.4 - 10.4) \cdot 11.6$
$38.6 + 3 \cdot 11.6$ Work inside parentheses first.
$38.6 + 34.8$ Multiply.
73.4 Add.

77. Subtract then divide:
(cost − down payment) ÷ number of monthly payments
= amount of monthly payment
($1250 − $350) ÷ 8 = 900 ÷ 8 = $112.50
Answer: $112.50 per month

81. The least common multiple for 4 and 3 is 12.
$\frac{1}{4} = \frac{3}{12}$ and $\frac{2}{3} = \frac{8}{12}$; therefore, $\frac{1}{4} < \frac{2}{3}$

85. The least common multiple for 24 and 36 is 72.
$\frac{12}{24} = \frac{36}{72}$ and $\frac{23}{36} = \frac{46}{72}$; therefore, $\frac{12}{24} < \frac{23}{36}$

SOLUTIONS

Section 4.7 (page 267)

1. $2\overline{)1.0}$ = .5

1 0 ← zero added to dividend
0

5. $6\overline{)1.0000}$ = .1666

6
40 ← zero added to dividend
36
40 ← zero added to dividend
36
40 ← zero added to dividend
36
4 ← stop

Round: .1666 — 5 or more, add 1 to thousandth (↑ thousandth)

Answer: .167

9. $10\overline{)1.0}$ = .1

1 0 ← zero added to dividend
0

13. $5\overline{)4.0}$ = .8

4 0 ← zero added to dividend
0

17. $16\overline{)11.0000}$ = .6875 ← zero added to dividend

9 6
1 40 ← zero added to dividend
1 28
120 ← zero added to dividend
112
80 ← zero added to dividend
80
0

.6875 = .688 (rounded)

21. $18\overline{)13.0000}$ = .7222 ← zero added to dividend

12 6
40 ← zero added to dividend
36
40 ← zero added to dividend
36
40 ← zero added to dividend
36

.7222 = .722 rounded

25. $4\overline{)1.00}$ = .25

8
20 ← zero added to dividend
20
0

29. $8\overline{)7.000}$ = .875

6 4
60 ← zero added to dividend
56
40
40 ← zero added to dividend
0

33. $.35 = \frac{\cancel{35}^{7}}{\cancel{100}_{20}} = \frac{7}{20}$

37. $.65 = \frac{\cancel{65}^{13}}{\cancel{100}_{20}} = \frac{13}{20}$

41. $.1 = \frac{1}{10}$

45. $5\overline{)3.0}$ = .6 ← zero added to dividend

3 0
0

49. $\frac{1}{4} = .25$

$\frac{1}{4} < .28$

53. $\frac{7}{8} = .875;\ .90 = .9000$

$\frac{7}{8} < .90$

57. $\frac{1}{6} = .167$ (rounded); $.18 = .180$

$\frac{1}{6} < .18$

61. $\frac{1}{2} = .500;\ \frac{5}{8} = .625$

$\frac{1}{2}$ is smaller than .506 or $\frac{5}{8}$

65. $\frac{1}{12} = .08333$ (rounded); . 084 = .08400

$\frac{1}{12}$ is smaller than .08395 or .084

69. $\frac{7}{8} = .8750;\ .893 = .8930$

$\frac{7}{8}$ is smaller than .8925 or .893

73. 5.8, 5.79, 5.4443, 5.804
5.8000, 5.7900, 5.4443, 5.8040
Answer: 5.4443, 5.79, 5.8, 5.804

77. .8751, .876, .875, .8902
.8751, .8760, .8750, .8902

Answer: $\frac{7}{8}$, .8751, .876, .8902

81. .75, .8, .762, .7781
.7500, .8000, .7620, .7781

Answer: $\frac{3}{4}$, .762, .7781, $\frac{12}{15}$

85. $\frac{3}{11}$, $\frac{4}{15}$, .25, $\frac{1}{3}$

.273, .267, .250, .333

Answer: .25, $\frac{4}{15}$, $\frac{3}{11}$, $\frac{1}{3}$

89. $\frac{9}{12} = \frac{9 \div 3}{12 \div 3} = \frac{3}{4}$

93. $\frac{96}{132} = \frac{96 \div 12}{132 \div 12} = \frac{8}{11}$

CHAPTER 5

Section 5.1 (page 291)

1. $\frac{8}{9}$

5. $\frac{90 \text{ cents}}{120 \text{ cents}} = \frac{3}{4}$

9. $\frac{\$30}{\$70} = \frac{3}{7}$

13. $\frac{3}{2\frac{1}{2}} = \frac{3}{\frac{5}{2}} = 3 \div \frac{5}{2} = \frac{3}{1} \cdot \frac{2}{5} = \frac{6}{5}$

17. $\frac{4 \text{ feet}}{30 \text{ inches}} = \frac{4 \cdot 12 \text{ inches}}{30 \text{ inches}} = \frac{48}{30} = \frac{8}{5}$

21. $\frac{15 \text{ hours}}{2 \text{ days}} = \frac{15 \text{ hours}}{2 \cdot 24 \text{ hours}} = \frac{15}{48} = \frac{5}{16}$

25. $\frac{35}{10} = \frac{35 \div 5}{10 \div 5} = \frac{7}{2}$

29. $\frac{24\frac{1}{2} \text{ ft}}{8\frac{1}{4} \text{ ft}} = \frac{\frac{49}{2}}{\frac{33}{4}} = \frac{49}{2} \div \frac{33}{4} = \frac{49}{\cancel{2}_1} \cdot \frac{\cancel{4}^2}{33} = \frac{98}{33}$

33. $\frac{4\frac{1}{2}}{2\frac{1}{4}} = \frac{\frac{9}{2}}{\frac{9}{4}} = \frac{9}{2} \div \frac{9}{4} = \frac{\cancel{9}^1}{\cancel{2}_1} \times \frac{\cancel{4}^2}{\cancel{9}_1} = \frac{2}{1}$

37. $\frac{59\frac{1}{2} \text{ days}}{8\frac{3}{4} \text{ weeks}} = \frac{59\frac{1}{2} \div 7 \text{ days}}{8\frac{3}{4} \text{ weeks}} = \frac{8\frac{1}{2} \text{ weeks}}{8\frac{3}{4} \text{ weeks}}$

$\frac{\frac{17}{2}}{\frac{35}{4}} = \frac{17}{2} \div \frac{35}{4} = \frac{17}{\cancel{2}_1} \times \frac{\cancel{4}^2}{35} = \frac{34}{35}$

41.
```
   18.7
 ×  2.3
   5 61
  37 4
  43.01
```

45.
```
        9.4647
.71)6.72 0000
    6 39
      33 0
      28 4
       4 60
       4 26
         340
         284
          560
          497
```
9.4647 = 9.465 (rounded)

Section 5.2 (page 299)

1. $\frac{54 \text{ miles}}{27 \text{ minutes}} = \frac{2 \text{ miles}}{1 \text{ minute}}$

5. $\frac{14 \text{ people}}{28 \text{ dresses}} = \frac{1 \text{ person}}{2 \text{ dresses}}$

9. $\frac{11 \text{ chapters}}{132 \text{ pages}} = \frac{1 \text{ chapter}}{12 \text{ pages}}$

13. $\frac{60 \text{ dollars}}{5 \text{ hours}} = \frac{12 \text{ dollars}}{1 \text{ hour}} = \$12/\text{h}$

17. $\frac{\$101.25}{45 \text{ pounds}} = \frac{\$2.25}{1 \text{ pound}} = \$2.25/\text{lb}$

21. $\frac{7\frac{1}{2} \text{ pounds}}{5 \text{ people}}$ Divide $7\frac{1}{2}$ by 5.

$\frac{15}{2} \div \frac{5}{1} = \frac{\cancel{15}^3}{2} \cdot \frac{1}{\cancel{5}_1} = \frac{3}{2} = 1\frac{1}{2}$

$1\frac{1}{2}$ pounds per person

25.

size	*unit cost*
4 ounces	$\frac{.89}{.4} = .2225$
8 ounces	$\frac{2.13}{8} = .26625$

Answer: 4 ounces for $.89

29.

size	*unit cost*
12 ounces	$\frac{1.09}{12} = .0908$
18 ounces	$\frac{1.43}{18} = .0794$
28 ounces	$\frac{2.29}{28} = .0818$
40 ounces	$\frac{3.19}{40} = .0798$

Answer: 18 ounces for $1.43

33. $85.82 in 7 hours. Find earnings (dollars) per hour.

$$\frac{85.82 \text{ dollars} \div 7}{7 \text{ hours} \div 7} = \frac{12.26 \text{ dollars}}{1 \text{ hour}} = \$12.26 \text{ per hour}$$

37. 6 crates packed in 18 minutes. Find crates per minute and minutes per crate.

$$\frac{6 \text{ crates} \div 18}{18 \text{ minutes} \div 18} = \frac{\frac{1}{3} \text{ crate}}{1 \text{ minute}} = \frac{1}{3} \text{ crate per minute}$$

$$\frac{18 \text{ minutes} \div 6}{6 \text{ crates} \div 6} = \frac{3 \text{ minutes}}{\text{crate}} = 3 \text{ minutes per crate}$$

41. $10 \cdot 7\frac{1}{4} = \frac{\overset{5}{\cancel{10}}}{1} \cdot \frac{29}{\underset{2}{\cancel{4}}} = \frac{145}{2} = 72\frac{1}{2}$

45. $1\frac{1}{6} \cdot 3 = \frac{7}{\underset{2}{\cancel{6}}} \cdot \frac{\overset{1}{\cancel{3}}}{1} = \frac{7}{2} = 3\frac{1}{2}$

Section 5.3 (page 305)

1. 13 is to 15 as 26 is to 30

$$\frac{13}{15} = \frac{26}{30} \qquad \frac{13}{15} = \frac{26}{30}$$

5. 120 is to 150 as 8 is to 10

$$\frac{120}{150} = \frac{8}{10} \qquad \frac{120}{150} = \frac{8}{10}$$

9. $1\frac{1}{2}$ is to 8 as 6 is to 32

$$\frac{1\frac{1}{2}}{8} = \frac{6}{32} \qquad \frac{1\frac{1}{2}}{8} = \frac{6}{32}$$

13. $\frac{5}{8} = \frac{25}{40}$ $\quad \frac{5}{8}$ is reduced

true $\quad \frac{25}{40} = \frac{5}{8}$ $\quad$ equal

17. $\frac{42}{15} = \frac{28}{10}$ $\quad \frac{42}{15} = \frac{14}{5}$

true $\quad \frac{28}{10} = \frac{14}{5}$ $\quad$ equal

21. $\frac{7}{6} = \frac{54}{48}$ $\quad \frac{7}{6}$ is reduced

false $\quad \frac{54}{48} = \frac{9}{8}$ $\quad$ not equal

25. $\frac{7}{10} = \frac{82}{120}$ cross products:

$7 \cdot 120 = 840$

false $\quad 10 \cdot 82 = 820$ $\quad$ not equal

29. $\frac{3\frac{1}{2}}{4} = \frac{7}{8}$ cross products:

$3\frac{1}{2} \cdot 8 = \frac{7}{\underset{1}{\cancel{2}}} \cdot \frac{\overset{4}{\cancel{8}}}{1} = 28$

true $\quad 4 \cdot 7 = 28$ $\quad$ equal

33. $\frac{6}{3\frac{2}{3}} = \frac{18}{11}$ cross products:

$6 \cdot 11 = 66$

true $\quad 3\frac{2}{3} \cdot 18 = \frac{11}{\underset{1}{\cancel{3}}} \cdot \frac{\overset{6}{\cancel{18}}}{1} = 66$ $\quad$ equal

37. $\frac{8.15}{2.03} = \frac{6.09}{24.45}$ cross products:

$8.15 \cdot 24.45 = 199.2675$

false $\quad 2.03 \cdot 6.09 = 12.3627$ $\quad$ not equal

41. $\frac{16}{24} = \frac{16 \div 8}{24 \div 8} = \frac{2}{3}$

45. $\frac{36}{63} = \frac{36 \div 9}{63 \div 9} = \frac{4}{7}$

Section 5.4 (page 315)

1. $\frac{1}{3} = \frac{r}{12}$

$1 \cdot 12 = 3 \cdot r$ $\quad$ Write cross products.

$12 = 3 \cdot r$ $\quad$ Multiply 2 numbers on one side.

$\frac{12}{3} = \frac{3 \cdot r}{3}$ $\quad$ Divide by number next to *r*.

$4 = r$ $\quad$ Answer.

5. $\frac{16}{b} = \frac{4}{5}$

$16 \cdot 5 = b \cdot 4$ $\quad$ Write cross products.

$80 = b \cdot 4$ $\quad$ Multiply 2 numbers on one side.

$\frac{80}{4} = \frac{b \cdot 4}{4}$ $\quad$ Divide by number next to *b*.

$20 = b$ $\quad$ Answer

9. $\frac{n}{25} \div \frac{4}{20}$

$n \cdot 20 = 25 \cdot 4$

$n \cdot 20 = 100$

$\frac{n \cdot 20}{20} = \frac{100}{20}$

$n = 5$

13. $\frac{15}{42} = \frac{h}{14}$

$15 \cdot 14 = 42 \cdot h$

$210 = 42 \cdot h$

$\frac{210}{42} = \frac{42 \cdot h}{42}$

$5 = h$

17.
$$\frac{50}{7} = \frac{r}{28}$$
$$50 \cdot 28 = 7 \cdot r$$
$$1400 = 7 \cdot r$$
$$\frac{1400}{7} = \frac{7 \cdot r}{7}$$
$$200 = r$$

21.
$$\frac{2}{9} = \frac{p}{12}$$
$$2 \cdot 12 = 9 \cdot p$$
$$24 = 9 \cdot p$$
$$\frac{24}{9} = \frac{9 \cdot p}{9}$$
$$\frac{8}{3} = p \quad \text{(reduced)}$$
$$2\frac{2}{3} = p$$

25.
$$\frac{3}{q} = \frac{8}{9}$$
$$3 \cdot 9 = 8 \cdot q$$
$$27 = 8 \cdot q$$
$$\frac{27}{8} = \frac{8 \cdot q}{8}$$
$$3\frac{3}{8} = q$$

29.
$$\frac{2\frac{1}{4}}{3} = \frac{b}{8}$$
$$8 \cdot 2\frac{1}{4} = 3 \cdot b$$
$$8 \cdot \frac{9}{4} = 3 \cdot b$$
$$18 = 3 \cdot b$$
$$\frac{18}{3} = \frac{3 \cdot b}{3}$$
$$6 = b$$

33.
$$\frac{3}{p} = \frac{.6}{4.2}$$
$$4.2 \cdot 3 = .6 \cdot p$$
$$12.6 = .6 \cdot p$$
$$\frac{12.6}{.6} = \frac{.6 \cdot p}{.6}$$
$$21 = p$$

37. 6 is to 5 as 18 is to x

$$\frac{6}{5} = \frac{18}{x}$$
$$6 \cdot x = 5 \cdot 18$$
$$6 \cdot x = 90$$
$$\frac{6 \cdot x}{6} = \frac{90}{6}$$
$$x = 15$$

41. k is to 8 as $\frac{4}{3}$ is to 2

$$\frac{k}{8} = \frac{\frac{4}{3}}{2}$$
$$k \cdot 2 = 8 \cdot \frac{4}{3}$$
$$k \cdot 2 = \frac{8}{1} \cdot \frac{4}{3}$$
$$k \cdot 2 = \frac{32}{3}$$
$$\frac{k \cdot 2}{2} = \frac{\frac{32}{3}}{2}$$
$$k = \frac{32}{3} \div 2$$
$$k = \frac{32}{3} \cdot \frac{1}{2}$$
$$k = \frac{16}{3} = 5\frac{1}{3}$$

45. 6 is to 9 as 10 is to 15

$$\frac{6}{9} = \frac{10}{15} \qquad \frac{6}{9} = \frac{10}{15}$$

Section 5.5 (page 321)

1. $\dfrac{\$30}{50 \text{ sq ft}} = \dfrac{\$d}{125 \text{ sq ft}}$

↑ known ratio of cost to square feet; ↑ desired cost for 125 square feet

$$30 \cdot 125 = 50 \cdot d$$
$$3750 = 50 \cdot d$$
$$\frac{3750}{50} = \frac{50 \cdot d}{50}$$
$$75 = d$$

Answer: \$75

5. $\dfrac{5 \text{ lb}}{3500 \text{ sq ft}} = \dfrac{n \text{ lb}}{4900 \text{ sq ft}}$

↑ known ratio of pounds to square feet; ↑ desired pounds of seed for 4900 square feet

$$5 \cdot 4900 = 3500 \cdot n$$
$$24{,}500 = 3500 \cdot n$$
$$\frac{24{,}500}{3500} = \frac{3500 \cdot n}{3500}$$
$$7 = n$$

Answer: 7 lbs

9. $\dfrac{5 \text{ inches}}{600 \text{ miles}} = \dfrac{9 \text{ inches}}{m \text{ miles}}$

↑ known ratio of inches to miles on map (scale); ↑ desired distance between 2 cities 9 inches apart on map

$$5 \cdot m = 600 \cdot 9$$
$$5 \cdot m = 5400$$
$$\frac{5 \cdot m}{5} = \frac{5400}{5}$$
$$m = 1080$$

Answer: 1080 miles

13. $\dfrac{\$150.50}{10 \text{ ft wide}} = \dfrac{\$d}{18 \text{ ft wide}}$

↑ known ratio of cost to width of deck
↑ desired cost of deck 18 feet wide

$$150.5 \cdot 18 = 10 \cdot d$$
$$2709 = 10 \cdot d$$
$$\frac{2709}{10} = \frac{10 \cdot d}{10}$$
$$270.9 = d$$

Answer: $270.90

17. $\dfrac{248 \text{ pounds ascorbic acid}}{575 \text{ pounds binder and filler}} = \dfrac{p \text{ pounds ascorbic acid}}{2875 \text{ pounds binder and filler}}$

↑ known ratio of ascorbic acid to binder and filler
↑ desired pounds of ascorbic acid

$$248 \cdot 2875 = 575 \cdot p$$
$$713{,}000 = 575 \cdot p$$
$$\frac{713{,}000}{575} = \frac{575 \cdot p}{575}$$
$$1240 = p$$

Answer: 1240 pounds

21. $.12 \times 100 = .12 = 12$ move decimal point 2 places right

25. $1.93 \div 100 = 01.93 = .0193$ ↑ move decimal point 2 places left

CHAPTER 6

Section 6.1 (page 339)

1. $50\% = 50. = .5$ (2 1) Drop % sign and move decimal 2 places to left.

5. $65\% = 65. = .65$ (2 1)

9. $7\% = 07. = .07$ (2 1) Add 1 zero to left so decimal point can be moved 2 places to left.

13. $.8\% = 00.8 = .008$ (2 1) Add 2 zeros to left.

17. $.60 = .60\ \% = 60\%$ (1 2) Move decimal 2 places to right and attach % sign.

21. $.02 = .02\ \% = 2\%$ (1 2)

25. $.453 = .45\ 3\% = 45.3\%$ (1 2)

29. $1.5 = 1.50\ \% = 150\%$ (1 2) Add 1 zero to right so decimal point can be moved 2 places to right.

33. $3.725 = 3.72\ 5\% = 372.5\%$ (1 2)

37. Sales tax is $6\% = 06.\% = .06.$ (2 1)

41. Sales tax is $.065 = .065\% = 6.5\%.$ (1 2)

45. .005 of total has the defect $= .005\% = .5\%.$ (1 2)

49. 95 of 100 shaded $= \dfrac{95}{100} = .95 = .95\ \%$ (1 2) $= 95\%$ shaded

5 of 100 unshaded $= \dfrac{5}{100} = .05 = .05\ \%$ (1 2) $= 5\%$ unshaded

53. 3 of 4 shaded $= \dfrac{3}{4} = .75 = .75\ \%$ (1 2) $= 75\%$ shaded

1 of 4 unshaded $= \dfrac{1}{4} = .25 = .25$ (1 2) $= 25\%$ unshaded

57. $4\overline{)3.00}$ = .75
28
20 ← zero added to dividend
20
0

61. $5\overline{)4.0}$ = .8
40 ← zero added to dividend
0

Section 6.2 (page 349)

1. $10\% = \dfrac{10}{100} = \dfrac{1}{10}$

5. $85\% = \dfrac{85}{100} = \dfrac{17}{20}$

9. $6.25\% = \dfrac{6.25}{100} = \dfrac{6.25 \cdot 100}{100 \cdot 100} = \dfrac{625}{10{,}000} = \dfrac{1}{16}$

13. $6\frac{2}{3}\% = \dfrac{6\frac{2}{3}}{100}$
$$= 6\frac{2}{3} \div 100$$
$$= \frac{20}{3} \cdot \frac{1}{100}$$
$$= \frac{\cancel{20}^{1}}{3} \cdot \frac{1}{\cancel{100}_{5}} = \frac{1}{15}$$

17. $120\% = \dfrac{120}{100} = \dfrac{6}{5} = 1\frac{1}{5}$

21.
$$\begin{aligned} \frac{3}{4} &= \frac{p}{100} \\ 100 \cdot 3 &= 4 \cdot p \\ 300 &= 4p \\ \frac{300}{4} &= p \\ 75 &= p \\ \frac{3}{4} &= 75\% \end{aligned}$$

25. $\frac{57}{100} = 57\%$

29.
$$\begin{aligned} \frac{5}{8} &= \frac{p}{100} \\ 100 \cdot 5 &= 8 \cdot p \\ 500 &= 8p \\ \frac{500}{8} &= p \\ 62.5 &= p \\ \frac{5}{8} &= 62.5\% \text{ or } 62\tfrac{1}{2}\% \end{aligned}$$

33.
$$\begin{aligned} \frac{12}{25} &= \frac{p}{100} \\ 100 \cdot 12 &= 25 \cdot p \\ 1200 &= 25p \\ \frac{1200}{25} &= p \\ 48 &= p \\ \frac{12}{25} &= 48\% \end{aligned}$$

37.
$$\begin{aligned} \frac{11}{20} &= \frac{p}{100} \\ 100 \cdot 11 &= 20 \cdot p \\ 1100 &= 20p \\ \frac{1100}{20} &= p \\ 55 &= p \\ \frac{11}{20} &= 55\% \end{aligned}$$

41.
$$\begin{aligned} \frac{5}{9} &= \frac{p}{100} \\ 100 \cdot 5 &= 9 \cdot p \\ 500 &= 9p \\ \frac{500}{9} &= p \\ 55.6 &= p \quad \text{(rounded)} \\ \frac{5}{9} &= 55.6\% \quad \text{(rounded)} \end{aligned}$$

45. Decimal is .1 (given).

$.1 = .\underbrace{1\,0}_{1\ 2}\% = 10\%$ (percent)

$.1 = \frac{1}{10}$ (fraction)

49. Decimal is .2 (given).

$.2 = .\underbrace{2\,0}_{1\ 2}\% = 20\%$ (percent)

$.2 = \frac{2}{10} = \frac{1}{5}$ (fraction)

53. Decimal is .4 (given).

$.4 = .\underbrace{4\,0}_{1\ 2}\% = 40\%$ (percent)

$.4 = \frac{4}{10} = \frac{2}{5}$ (fraction)

57. Fraction is $\frac{2}{3}$ (given).

$\frac{2}{3} = .667$ (decimal rounded)

$\frac{2}{3} = .667 = .\underbrace{6\,6}_{1\ 2}7\% = 66.7\%$ (percent)

61. Fraction is $\frac{5}{6}$ (given).

$\frac{5}{6} = .833$ (decimal rounded)

$\frac{5}{6} = .833 = .\underbrace{8\,3}_{1\ 2}3\% = 83.3\%$ (percent)

65. Fraction is $\frac{1}{200}$ (given).

$\frac{1}{200} = .005$ (decimal)

$\frac{1}{200} = .005 = .\underbrace{0\,0}_{1\ 2}5\% = 0.5\%$ (percent)

69. Decimal is 2.5 (given).

$2.5 = 2.\underbrace{5\,0}_{1\ 2}\% = 250\%$ (percent)

$2.5 = 2\frac{5}{10} = 2\frac{1}{2}$ (fraction)

73. 15 out of 20

fraction: $\frac{15}{20} = \frac{3}{4}$

decimal: $\frac{3}{4} = .75$

percent: $.75 = .\underbrace{7\,5}_{1\ 2}\% = 75\%$

77. 3 out of 5

fraction: $\frac{3}{5}$

decimal: $\frac{3}{5} = .6$

percent: $.6 = .\underbrace{6\,0}_{1\ 2}\% = 60\%$

81. 342 out of 380 do not have reaction.
38 out of 380 ($380 - 342 = 38$) do have reaction.

fraction: $\frac{38}{380} = \frac{1}{10}$

decimal: $\frac{1}{10} = .1$

percent: $.1 = .\underbrace{1\,0}_{1\ 2}\% = 10\%$

85.
$$\begin{aligned} \frac{4}{y} &= \frac{12}{15} \\ 4 \cdot 15 &= 12y \\ 60 &= 12y \\ \frac{60}{12} &= \frac{12y}{12} \\ 5 &= y \end{aligned}$$

Section 6.3 (page 355)

1. $a = 40, p = 25$
$$\begin{aligned} \frac{40}{b} &= \frac{25}{100} \\ 40 \cdot 100 &= b \cdot 25 \\ 4000 &= b \cdot 25 \\ \frac{4000}{25} &= \frac{b \cdot 25}{25} \\ 160 &= b \end{aligned}$$

5. $a = 8, p = 40$
$$\begin{aligned} \frac{8}{b} &= \frac{40}{100} \\ 8 \cdot 100 &= b \cdot 40 \\ 800 &= b \cdot 40 \\ \frac{800}{40} &= \frac{b \cdot 40}{40} \\ 20 &= b \end{aligned}$$

9. $a = 70, b = 140$
$$\begin{aligned} \frac{70}{140} &= \frac{p}{100} \\ 70 \cdot 100 &= 140 \cdot p \\ 7000 &= 140 \cdot p \\ \frac{7000}{140} &= \frac{140 \cdot p}{140} \\ 50 &= p \end{aligned}$$

13. $a = 1\frac{1}{2}, b = 4\frac{1}{2}$
$$\begin{aligned} \frac{1\frac{1}{2}}{4\frac{1}{2}} &= \frac{p}{100} \\ 1\frac{1}{2} \cdot 100 &= 4\frac{1}{2} \cdot p \\ \frac{3}{2} \cdot 100 &= 4\frac{1}{2} \cdot p \\ 150 &= 4\frac{1}{2} \cdot p \\ \frac{150}{4\frac{1}{2}} &= \frac{4\frac{1}{2} \cdot p}{4\frac{1}{2}} \\ \frac{150}{1} \div \frac{9}{2} &= p \\ \frac{\overset{50}{\cancel{150}}}{1} \cdot \frac{2}{\underset{3}{\cancel{9}}} &= p \\ \frac{100}{3} &= p \\ p &= 33\frac{1}{3} \end{aligned}$$

17. $b = 72, p = 15$
$$\begin{aligned} \frac{a}{72} &= \frac{15}{100} \\ a \cdot 100 &= 72 \cdot 15 \\ a \cdot 100 &= 1080 \\ \frac{a \cdot 100}{100} &= \frac{1080}{100} \\ a &= 10.8 \end{aligned}$$

21. $a = 89, p = 25$
$$\begin{aligned} \frac{89}{b} &= \frac{25}{100} \\ 89 \cdot 100 &= b \cdot 25 \\ 8900 &= b \cdot 25 \\ \frac{8900}{25} &= \frac{b \cdot 25}{25} \\ 356 &= b \end{aligned}$$

25. $b = 1850, a = 157.25$
$$\begin{aligned} \frac{157.25}{1850} &= \frac{p}{100} \\ 157.25 \cdot 100 &= p \cdot 1850 \\ 15{,}725 &= p \cdot 1850 \\ \frac{15{,}725}{1850} &= \frac{p \cdot 1850}{1850} \\ 8.5 &= p \end{aligned}$$

29. Fraction is $\frac{3}{4}$ (given).

$\frac{3}{4} = .75 = .\underbrace{75}_{1\ 2}\% = 75\%$

33. Fraction is $\frac{7}{8}$ (given).

$\frac{7}{8} = .875 = .\underbrace{87}_{1\ 2}5\% = 87.5\%$

Section 6.4 (page 359)

1. $\underbrace{65\%}_{p}$ of $\underbrace{1000}_{b}$ is $\underbrace{650}_{a}$. $p = 65$; $b = 1000$; $a = 650$

5. $\underbrace{\text{What}}_{a}$ is $\underbrace{12\%}_{p}$ of $\underbrace{48}_{b}$? $p = 12$; $b = 48$; $a =$ unknown

9. $\underbrace{52}_{a}$ is $\underbrace{\text{what percent}}_{p}$ of $\underbrace{104}_{b}$? $p =$ unknown; $b = 104$; $a = 52$

13. $\underbrace{29.81}_{a}$ is $\underbrace{\text{what percent}}_{p}$ of $\underbrace{508}_{b}$? $p =$ unknown; $b = 508$; $a = 29.81$

17. $p =$ an unknown (percent of games won);

$b = 24$ (total games); $a = 18$ (games won)

21. $p = 7$ (percent not healthy); $b = 1500$ (number shipped)

$a =$ an unknown (number not healthy)

25. $p = 16.8$ (percent late);

$b =$ an unknown (total number patients)

$a = 504$ (number late patients)

29. y is to 16 as $\frac{5}{4}$ is to 4

$$\frac{y}{16} = \frac{\frac{5}{4}}{4}$$

$$y \cdot 4 = 16 \cdot \frac{5}{4}$$
$$y \cdot 4 = \frac{16}{1} \cdot \frac{5}{4}$$
$$y \cdot 4 = \frac{80}{4}$$
$$y \cdot 4 = 20$$
$$\frac{y \cdot 4}{4} = \frac{20}{4}$$
$$y = 5$$

Section 6.5 (page 369)

1. 25% of 484 = .25 · .484 = 121

5. 5% of 96 = .05 · 96 = 4.8

9. 22.5% of 1100 = .225 · 1100 = 247.5

13. 200% of 371 = 2.00 · 371 = 742

17. .9% of 2400 = .009 · 2400 = 21.6

21.
$$\frac{84}{b} = \frac{35}{100}$$
$$84 \cdot 100 = b \cdot 35$$
$$8400 = b \cdot 35$$
$$\frac{8400}{35} = \frac{b \cdot 35}{35}$$
$$240 = b$$

25.
$$\frac{748}{b} = \frac{110}{100}$$
$$748 \cdot 100 = b \cdot 110$$
$$74{,}800 = b \cdot 110$$
$$\frac{74{,}800}{110} = \frac{b \cdot 110}{110}$$
$$680 = b$$

29.
$$\frac{35}{70} = \frac{p}{100}$$
$$35 \cdot 100 = 70 \cdot p$$
$$3500 = 70 \cdot p$$
$$\frac{3500}{70} = \frac{70 \cdot p}{70}$$
$$50 = p$$
50%

33.
$$\frac{8}{400} = \frac{p}{100}$$
$$8 \cdot 100 = 400 \cdot p$$
$$800 = 400 \cdot p$$
$$\frac{800}{400} = \frac{400 \cdot p}{400}$$
$$2 = p$$
2%

37.
$$\frac{16}{124} = \frac{p}{100}$$
$$16 \cdot 100 = 124 \cdot p$$
$$1600 = 124 \cdot p$$
$$\frac{1600}{124} = \frac{124 \cdot p}{124}$$
$$12.9 = p$$
12.9% (rounded)

41. $p = 20$; $b = 270$

20% of 270 = .20 · 270 = 54 children

45. $p = 2.6\%$ (percent that is not paid back 100% − 97.4%);

b = an unknown (total amount played)

a = \$4823 (amount retained)

$$\frac{4823}{b} = \frac{2.6}{100}$$
$$4823 \cdot 100 = b \cdot 2.6$$
$$482{,}300 = b \cdot 2.6$$
$$\frac{482{,}300}{2.6} = \frac{b \cdot 2.6}{2.6}$$
$$185{,}500 = b$$
\$185,000

49. p = an unknown (% non-paid circulation);

b = 180,000 (total daily circulation)

a = 5400 (non-paid circulation)

$$\frac{5400}{180{,}000} = \frac{p}{100}$$
$$5400 \cdot 100 = p \cdot 180{,}000$$
$$540{,}000 = p \cdot 180{,}000$$
$$\frac{540{,}000}{180{,}000} = \frac{p \cdot 180{,}000}{180{,}000}$$
$$3 = p$$
3%

53. p = an unknown (percent of goal achieved);

b = 2,380,000 (goal); a = 1,844,500 (sales)

$$\frac{1{,}844{,}500}{2{,}380{,}000} = \frac{p}{100}$$
$$1{,}844{,}500 \cdot 100 = 2{,}380{,}000 \cdot p$$
$$184{,}450{,}000 = 2{,}380{,}000 \cdot p$$
$$\frac{184{,}450{,}000}{2{,}380{,}000} = \frac{2{,}380{,}000 \cdot p}{2{,}380{,}000}$$
$$77.5 = p$$
77.5%

57. $p = 52.5 \begin{pmatrix}\text{percent agency} \\ \text{owned by one} \\ \text{person}\end{pmatrix};$

$b = \text{an unknown} \begin{pmatrix}\text{total value} \\ \text{of agency}\end{pmatrix}$

$a = \$446{,}250 \begin{pmatrix}\text{amount owned} \\ \text{by one person}\end{pmatrix}$

$$\frac{446{,}250}{b} = \frac{52.5}{100}$$
$$446{,}250 \cdot 100 = 52.5 \cdot b$$
$$44{,}625{,}000 = 52.5b$$
$$\frac{44{,}625{,}000}{52.5} = \frac{52.5b}{52.5}$$
$$850{,}000 = b$$
$$\$850{,}000$$

61.

$$\begin{array}{r} 48.05 \\ \times\ \ .85 \\ \hline 2\ 4025 \\ 38\ 440\ \\ \hline 40.8425 \end{array}$$

2 decimal places
+ 2 decimal places
4 in answer

65.

$$\begin{array}{r} 3\ 600. \\ .085\overline{)306.000} \\ 255 \\ \hline 510 \\ 510 \\ \hline 0 \end{array}$$

zero added to dividend, then brought down

Section 6.6 (page 379)

1. amount = percent · base
$a = .4 \cdot 620$ Write 40% as the decimal .4.
$a = 248$

5. amount = percent · base
$a = .16 \cdot 520$
$a = 83.2$

9. $a = 1.25 \cdot 72$
$a = 90$

13. $a = .004 \cdot 750$
$a = 3$

17. amount = percent · base
$32 = .4 \cdot b$
$\frac{32}{.4} = \frac{.4 \cdot b}{.4}$
$80 = b$

21. $500 = .4 \cdot b$
$\frac{500}{.4} = \frac{.4 \cdot b}{.4}$
$1250 = b$

25. $5 = .025 \cdot b$
$\frac{5}{.025} = \frac{.025 \cdot b}{.025}$
$200 = b$

29. amount = percent · base
$75 = p \cdot 125$
$\frac{75}{125} = \frac{p \cdot 125}{125}$
$.6 = p \text{ percent} = 60\%$

33. $3 = p \cdot 1200$
$\frac{3}{1200} = \frac{p \cdot 1200}{1200}$
$.0025 = p \text{ percent} = .25\%$

37. amount = percent · base
$a = .35 \cdot 3300$
$a = 1155 = 1155$ miles

41. amount = percent · base
$90 = .75 \cdot b$
$\frac{90}{.75} = \frac{.75 \cdot b}{.75}$
$120 = b = 120$ units

45. $\underbrace{15\%}_{p}$ of $\underbrace{375}_{b}$ is $\underbrace{56.25}_{a}$.
$p = 15;\ b = 375;\ a = 56.25$

49. $\underbrace{\text{What percent}}_{p}$ of $\underbrace{650}_{b}$ is $\underbrace{146.15}_{a}$?
$p = \text{unknown};\ b = 650;\ a = 146.15$

Section 6.7 (page 387)

1. $100 \cdot 2\% = 100 \cdot .02 = \$2$ tax
total cost $= 100 + 2 = \$102$

5. $215 \cdot 6\% = 215 \cdot .06 = \$12.90$ tax
total cost $= 215 + 12.90 = \$227.90$

9. $\$100 \cdot 10\% = 100 \cdot .10 = \$10$ commission

13. $\$3200 \cdot 36\% = 3200 \cdot .36 = \$1152$ commission

17. $100 \cdot 25\% = 100 \cdot .25$
$= \$25$ discount
$100 - 25 = \$75$ price after discount

21. $17.50 \cdot 25\% = 17.50 \cdot .25$
$= \$4.38$ discount (rounded)
$17.50 - 4.38 = \$13.12$ price after discount

25. $358 \cdot 5\% = 358 \cdot .05 = \$17.90$ tax

29. $58 \div 1450 = .04 = 4\%$

33. $20 \cdot 35\% = 20 \cdot .35$
$= \$7$ discount
$20 - 7 = \$13$ price after discount

37. $10{,}907.58 \cdot 6.25\% = 10{,}907.58 \cdot .0625$
$= 681.72375$
Round to \$681.72 tax.

41. $78{,}000 \cdot 2\% = 78{,}000 \cdot .02 = \$1560$ tax

45. $129{,}605 \cdot 6\% = 129{,}605 \cdot .06$
$= \$7776.30$ commission charged
$7776.30 \cdot 55\% = 7776.3 \cdot .55$
$= \$4276.97$ paid to agent (rounded)

49. amount = percent · base
$a = .2 \cdot 310$
$a = 62$

53. amount = percent · base
$50 = 6\frac{1}{4}\% \cdot b$
$50 = .0625 \cdot b$
$\frac{50}{.0625} = \frac{.0625 \cdot b}{.0625}$
$800 = b$

SOLUTIONS

Section 6.8 (page 395)

1. $I = p \cdot r \cdot t$
$I = 100 \cdot 10\% \cdot 1$
$= 100 \cdot .10 \cdot 1$
$= 10 \cdot 1$
$= \$10$

5. $I = 80 \cdot 10\% \cdot 1$
$= 80 \cdot .10 \cdot 1$
$= 8 \cdot 1$
$= \$8$

9. $I = 760 \cdot 10\% \cdot 2\frac{1}{2}$
$= 760 \cdot .10 \cdot \frac{5}{2}$ (or 2.5)
$= 76 \cdot \frac{5}{2}$ (or 2.5)
$= \$190$

13. $I = 200 \cdot 16\% \cdot \frac{6}{12}$
$= 200 \cdot .16 \cdot \frac{1}{2}$ (or .5)
$= 32 \cdot \frac{1}{2}$ (or .5)
$= \$16$

17. $I = 820 \cdot 8\% \cdot \frac{24}{12}$
$= 820 \cdot .08 \cdot 2$
$= 65.6 \cdot 2$
$= \$131.20$

21. $I = 650 \cdot 12\% \cdot \frac{3}{12}$
$= 650 \cdot .12 \cdot \frac{1}{4}$ (or .25)
$= 78 \cdot \frac{1}{4}$ (or .25)
$= \$19.50$

25. First: interest $= 300 \cdot 11\% \cdot 1 = 300 \cdot .11 \cdot 1$
$= 33 \cdot 1 = \$33$
Second: Add interest to principal:
$300 + 33 = \$333$

29. First: interest $= 1500 \cdot 10\% \cdot \frac{18}{12}$
$= 1500 \cdot .10 \cdot 1.5 = \$225$
Second: Add interest to principal:
$1500 + 225 = \$1725.$

33. First: interest $= 1780 \cdot 10\% \cdot \frac{6}{12}$
$= 1780 \cdot .10 \cdot \frac{1}{2}$
$= 178 \cdot \frac{1}{2} = \$89$
or $1780 \cdot .10 \cdot .5 = \$89$
Second: Add interest to principal:
$1780 + 89 = \$1869.$

37. $I = 680 \cdot .16 \cdot 1 = \$108.80$

41. $I = 4000 \cdot .16 \cdot 2 = \$1280$

45. $I = 2100 \cdot 7\frac{1}{4}\% \cdot \frac{1}{4}$
$= 2100 \cdot .0725 \cdot .25$
$= \$38.06$ (rounded)

49. Divide 8 by 4, getting 2.
$$\frac{3}{4} = \frac{3 \cdot 2}{4 \cdot 2} = \frac{6}{8}$$

53. Divide 60 by 4, getting 15.
$$\frac{3}{4} = \frac{3 \cdot 15}{4 \cdot 15} = \frac{45}{60}$$

CHAPTER 7

Section 7.1 (page 423)

1. 1 yard = 3 feet = $3 \cdot 12$ inches = 36 inches

5. 1 mile = 5280 feet

9. 6 feet = $6 \cancel{\text{feet}} \cdot \frac{1 \text{ yard}}{3 \cancel{\text{feet}}} = \frac{6}{3}$ yards = 2 yards

13. $2000 \cancel{\text{pounds}} \cdot \frac{1 \text{ ton}}{2000 \cancel{\text{pounds}}} = 1$ ton

17. 28 pints = $28 \cancel{\text{pints}} \cdot \frac{1 \text{ quart}}{2 \cancel{\text{pints}}} = \frac{28}{2}$ quarts
= 14 quarts
= 3.5 or $3\frac{1}{2}$ gallons

21. $3 \cancel{\text{tons}} \cdot \frac{2000 \text{ pounds}}{1 \cancel{\text{ton}}} = 6000$ pounds

25. $15\frac{1}{2}$ feet $= \frac{31}{\cancel{2}_1} \cancel{\text{feet}} \cdot \frac{\cancel{12}^6 \text{ inches}}{1 \cancel{\text{foot}}} = \frac{31 \cdot 6}{1}$ inches
= 186 inches

29. $27{,}720 \cancel{\text{feet}} \cdot \frac{1 \text{ mile}}{5280 \cancel{\text{feet}}} = \frac{27{,}720}{5280}$ miles
$= 5\frac{1320}{5280}$ miles = 5.25 miles or $5\frac{1}{4}$ miles

33. $\cancel{36}^{18} \cancel{\text{cups}} \cdot \frac{1 \text{ pint}}{\cancel{2}_1 \cancel{\text{cups}}} = 18$ pints = 9 quarts

37. $5 \cancel{\text{tons}} \cdot \frac{2000 \text{ pounds}}{1 \cancel{\text{ton}}} = 10{,}000$ pounds

41. $3 \cancel{\text{days}} \cdot \frac{24 \text{ hours}}{1 \cancel{\text{day}}} = 72$ hours

45. $1\frac{1}{2} \cancel{\text{miles}} \cdot \frac{5280 \text{ feet}}{1 \cancel{\text{miles}}} = 7920$ feet
$= 7920 \cancel{\text{feet}} \cdot \frac{12 \text{ inches}}{1 \cancel{\text{foot}}} = 95{,}040$ inches

49. $4 \cancel{\text{hours}} \cdot \frac{60 \text{ minutes}}{1 \cancel{\text{hour}}} = 240$ minutes
4 hours $\geq$ 185 minutes

Section 7.2 (page 433)

1. 1 foot 16 inches = 1 foot + 1 foot 4 inches
= 2 feet 4 inches

5. 1 quart 7 pints = 1 quart + 3 quarts 1 pint
= 4 quarts 1 pint or 1 gallon 1 pint

9. 9 days 36 hours = 9 days + 1 day 12 hours
= 10 days 12 hours or 1 week 3 days 12 hours

13. 2 yards 3 feet 18 inches
= 2 yards + 3 feet + 1 foot 6 inches
= 2 yards + 4 feet + 6 inches
= 2 yards + 1 yard 1 foot + 6 inches
= 3 yards 1 foot 6 inches

17. 5 gallons 6 quarts 3 pints
= 5 gallons + 6 quarts + 1 quart 1 pint
= 5 gallons + 7 quarts + 1 pint
= 5 gallons + 1 gallon 3 quarts + 1 pint
= 6 gallons 3 quarts 1 pint

21.
 6 quarts 2 pints
+ 3 quarts 1 pint
 9 quarts 3 pints
= 9 quarts + 1 quart 1 pint
= 10 quarts 1 pint or 2 gallons 2 quarts 1 pint

25.
 7 weeks 2 days 19 hours
+ 3 weeks 5 days 18 hours
 10 weeks 7 days 37 hours
= 10 weeks + 7 days + 1 day 13 hours
= 10 weeks + 8 days + 13 hours
= 10 weeks + 1 week 1 day + 13 hours
= 11 weeks 1 day 13 hours

29.
 6 tons 800 pounds
− 4 tons 500 pounds
 2 tons 300 pounds

33.
 (14) ~~15~~ days (38) ~~14~~ ~~15~~ hours (99) ~~39~~ minutes
− 8 days 22 hours 42 minutes
 6 days 16 hours 57 minutes

37.
 5 tons 800 pounds
× 6
 30 tons 4800 pounds
= 30 tons + 2 tons 800 pounds
= 32 tons 800 pounds

41.
 2 yards 1 foot 6 inches
× 6
 12 yards 6 feet 36 inches
= 12 yards + 6 feet + 3 feet
= 12 yards + 9 feet
= 12 yards + 3 yards
= 15 yards

45.
```
    162 pounds 10 ounces
4)650 pounds  8 ounces
  4
  25
  24
   10
    8
    2 pounds = + 32 ounces
                 40
                 40
                 00
```

49.
```
           6 days      17 hours
5)4 weeks  5 days      13 hours
  └──> = + 28 days
           33
           30 days
            3 days = + 72 hours
                       85 hours
                       85
                        0
```

53.
 3 weeks 4 days 10 hours 38 minutes
× 3
 9 weeks 12 days 30 hours 114 minutes
30 hours + 1 hour 54 minutes
12 days + 1 day 7 hours 54 minutes
9 weeks + 1 week 6 days 7 hours 54 minutes
10 weeks 6 days 7 hours 54 minutes

57. 7250.618
thousands: 7 (fourth digit to left of decimal point)
hundredths: 1 (second digit to right of decimal point)

Section 7.3 (page 445)

1. 1 m = 100 cm

5. 1 cm = 1 · .01 m = .01 m or $\frac{1}{100}$ m

9. Place one side of the width of a pencil at the zero mark and count the number of mm to the other side. The width of a pencil is about 7 mm.

13. 1.5 m = 1.5 · 1000 mm = 1500 mm or
$1.5\,\cancel{m} \cdot \frac{1000\text{ mm}}{1\,\cancel{m}} = 1500\text{ mm}$

SOLUTIONS

17. 8500 cm = 8500 · .01 m = 85 m or

$8500\,\cancel{cm} \cdot \dfrac{1\text{ m}}{100\,\cancel{cm}} = 85\text{ m}$

21. 5.3 km = 5.3 · 1000 m = 5300 m or

$5.3\,\cancel{km} \cdot \dfrac{1000\text{ m}}{1\,\cancel{km}} = 5300\text{ m}$

25. cm is 5 places to right of km; move decimal point 5 places to right. 2.4 km = 240,000 cm

29. .48 km = .48 · 1000 m = 480 m or

$.48\,\cancel{km} \cdot \dfrac{1000\text{ m}}{1\,\cancel{km}} = 480\text{ m}$

33. .00981 km = .00981 · 1000 m = 9.81 m
9.81 m = 9.81 · 100 cm = 981 cm
or

$.00981\,\cancel{km} \cdot \dfrac{1000\text{ m}}{1\,\cancel{km}} = 9.81\text{ m}$

$9.81\,\cancel{m} \cdot \dfrac{100\text{ cm}}{1\,\cancel{m}} = 981\text{ cm}$

37. $.875 = \dfrac{875}{1000} = \dfrac{7}{8}$

41. $.375 = \dfrac{375}{1000} = \dfrac{3}{8}$

Section 7.4 (page 451)

1. 6 ℓ = 6 ℓ · 100 cl = 600 cl
5. 925 cl = 925 · .01 ℓ = 9.25 ℓ
9. 8.4 ℓ = 8.4 · .001 k; = .0084 kl
13. 8000 g = 8000 · .001 kg = 8 kg
17. 4.2 g = 4.2 · 1000 mg = 4200 mg
21. 2.1 kiloliters is 2.1 · 1000 liters = 2100 liters. This is *unreasonable,* since most of us drink about .2 to .5 liters of fluid with each meal.
25. 15 ml of cough syrup is *reasonable,* since it is 15 · .001 ℓ = .015 ℓ. Note that 1 tablespoon = .021 quarts, and a quart and a liter are close in size.
29. 5 grams each, total

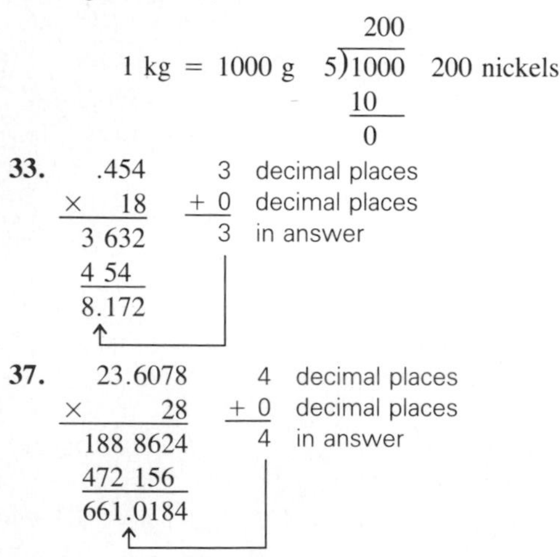

Section 7.5 (page 457)

1. 20 m · 1.094 = 21.88 yards = 21.9 yards
5. 16 ft · .305 = 4.880 m = 4.88 m round: 4.9 m
9. 47.2 lbs · 454 = 21,428.8 g
13. 28.2 gal · 3.785 = 106.737 ℓ round: 106.7 ℓ
17. Multiply your weight by 454. For example,
150 lb = 150 × 454 = 68,100 g.
21. First: 4 ℓ = 4 · .264 = 1.056 gal
Second: 1.056 · \$9.20 = \$9.7152 = \$9.72
25. 104°F First: 104 − 32 = 72
Second: 72 · 5 = 360
Third: 360 ÷ 9 = 40
40°C
29. 35°C First: 35 · 9 = 315
Second: 315 ÷ 5 = 63
Third: 63 + 32 = 95
95°F
33. 15°C First: 15 · 9 = 135
Second: 135 ÷ 5 = 27
Third: 27 + 32 = 59
59°F
37. 500°C First: 500 · 9 = 4500
Second: 4500 ÷ 5 = 900
Third: 900 + 32 = 932
932°F
41. $9^2 = 9 \cdot 9 = 81$

CHAPTER 8

Section 8.1 (page 473)

1. A line is a straight collection of points that goes on forever in opposite directions. This is a line.
5. A ray is a part of a line that has only one endpoint and continues forever in one direction. This is a ray.
9. Lines that never intersect are called parallel lines. These are parallel lines.
13. The vertex is named in the middle. The highlighted angle is named $\angle AOS$ or $\angle SOA$.
17. The highlighted angle is named $\angle AOC$ or $\angle COA$.
21. An acute angle is an angle whose measure is between 0° and 90°. This is an acute angle.
25. An obtuse angle is an angle whose measure is between 90° and 180°. This is an obtuse angle.
29. $\angle SQP$ and $\angle QST$ are both right angles. They are the same measure (90°).
Answer: false
33. 180 − 75 − 15 = 105 − 15 = 90

Section 8.2 (page 479)

1. Two angles are complementary if the sum of their measure is 90°. The complement of a 40° angle is a 50° angle.
Answer: 50°
5. Two angles are supplementary if the sum of their measure is 180°. The supplement of a 130° angle is a 50° angle.
Answer: 50°
9. The 15° angle plus the 75° angle equals 90°. The 20° angle plus the 70° angle equals 90°.
Answer: $\angle EOD$ and $\angle DOC$;
$\angle COB$ and $\angle BOA$
13. If two angles are vertical angles they are congruent.
Answer: $\angle SON \cong \angle TOM$;
$\angle TOS \cong \angle MON$

17. $\angle EOF$ and $\angle AOH$ are vertical angles. Vertical angles are congruent.
$\angle AOH \cong \angle EOF$
$\angle AOH = 37°$
Answer: $\angle EOF = 37°$

21. $\angle COB + \angle BOD = 180°$ (straight line)
Since $\angle BOD = 54°$,
$\angle COB = 180° - 54°$
$\angle COB = 126°$.

25. $16 \cdot 4 \div 2$
$64 \div 2$ Multiply left to right.
32 Divide last.

Section 8.3 (page 487)

1. $P = 2l + 2w$
$= 2 \cdot 8 + 2 \cdot 6$
$= 16 + 12 = 28$ yd
$A = lw$
$= 8 \cdot 6 = 48$ yd^2

5. $P = 2l + 2w$
$= 2 \cdot 10 + 2 \cdot 1$
$= 20 + 2 = 22$ ft
$A = lw$
$= 10 \cdot 1 = 10$ ft^2

9. $P = 2l + 2w$
$= 2 \cdot 76.1 + 2 \cdot 22$
$= 152.2 + 44 = 196.2$ ft
$A = lw$
$= 76.1 \cdot 22 = 1674.2$ ft^2

13.
7 m
3 m
5 m
(7 – 2)
12 m
9 m
2 m

$P = 7 + 3 + 5 + 9 + 2 + 12 = 38$ m
A (top) $= 7 \cdot 3 = 21$
A (bottom) $= 2 \cdot 9 = 18$
A (total) $= 21 + 18 = 39$ m^2

17.

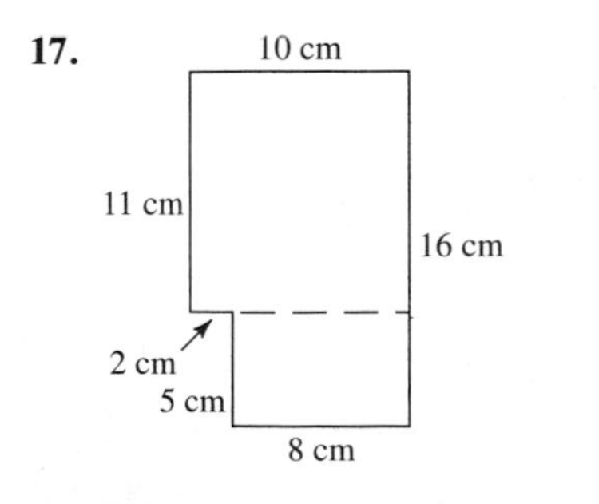

$P = 11 + 10 + 16 + 8 + 5 + 2 = 52$ cm
A (top) $= 11 \cdot 10 = 110$
A (bottom) $= 5 \cdot 8 = 40$
A (total) $= 110 + 40 = 150$ cm^2

21. $A = lw$
First: We need area in yd^2, so change feet to yards.
24 ft = 8 yd
18 ft = 6 yd
$A = 8 \cdot 6 = 48$ yd^2
Carpet: $\underbrace{48}_{\text{yd}^2} \cdot \underbrace{\$13}_{\text{per yd}^2} = \$624$

25.

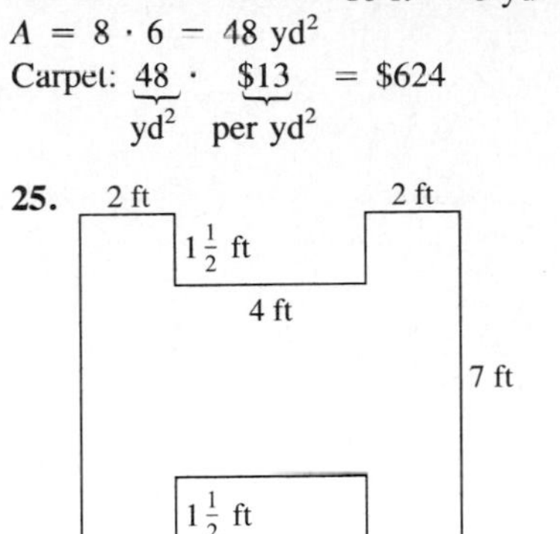

First: We need perimeter.

$$P = 2 + 1\frac{1}{2} + 4 + 1\frac{1}{2} + 2 + 7 + 2$$
$$+ 1\frac{1}{2} + 4 + 1\frac{1}{2} + 2 + 7 = 36 \text{ ft}$$

Cost of weather stripping is
$\$1.92 \times 36$ ft $= \$69.12$.

29. $50 \text{ ft} = \frac{50 \cancel{\text{ft}}}{1} \cdot \frac{12 \text{ inches}}{1 \cancel{\text{ft}}} = 600 \text{ inches}$

Section 8.4 (page 493)

1. $P = 46 + 58 + 46 + 58 = 208$ m

5. $P = 287.1 + 72 + 304 + 89.7 = 752.8$ cm

9. $A = b \cdot h$
$= 3\frac{1}{2} \cdot 5\frac{1}{2}$
$= \frac{7}{2} \cdot \frac{11}{2} = \frac{77}{4} = 19\frac{1}{4}$ ft^2
or $3.5 \cdot 5.5 = 19.25$ ft^2

13. First: $A = \frac{1}{2} \cdot 32 \cdot (47 + 59)$
$A = 1696$ ft^2
Carpet: $\underbrace{1696}_{\text{ft}^2} \cdot \underbrace{\$1.75}_{\text{per ft}^2} = \$2968$

17.
A = 55 · 62 = 3410
21 cm
55 cm
60 cm
60 cm
62 cm
A = 21 · 60 = 1260
A (total) = 3410 + 1260 = 4670 cm²

A (total) $= 3410 + 1260 = 4670$ cm^2

21. $6\frac{1}{2} \cdot 8 = \frac{13}{\cancel{2}_1} \cdot \frac{\cancel{8}^4}{1} = 52$

Section 8.5 (page 499)

1. $P = 7 + 11 + 9 = 27$ yd

5. $A = \frac{1}{2} \cdot b \cdot h = \frac{1}{2} \cdot 21.3 \cdot 28.4 = 302.46 \text{ cm}^2$

9. $A \text{ (total)} = 52 \cdot 37 = 1924$

$A \text{ (triangle)} = \frac{1}{2} \cdot 28 \cdot 52 = 728$

$A \text{ (shaded)} = A \text{ (total)} - A \text{ (triangle)}$
$= 1924 - 728 = 1196 \text{ m}^2$

13. **(a)** $A \text{ (triangle)} = \frac{1}{2}bh$
$= \frac{1}{2} \cdot 8 \cdot 8$
$= 32 \text{ m}^2$

(b) $A \text{ (triangle)} = \frac{1}{2} \cdot 3 \cdot 9$
$= 13.5 \text{ m}^2$

17.

```
     .71   2  decimal places
   × .34   2  decimal places
     284   4  in answer
    213
   .2414
```

Section 8.6 (page 507)

1. $r = 47$ m; $d = 2 \cdot r = 2 \cdot 47 = 94$ m

5. $r = 11$ ft;
$C = 2\pi r = 2 \cdot 3.14 \cdot 11$
$= 69.08$ Round to 69.1 ft.
$A = \pi r^2 = 3.14 \cdot 11 \cdot 11$
$= 379.94$ Round to 379.9 ft^2.

9. $d = 15$ cm;
$C = \pi d = 3.14 \cdot 15 = 47.1$ cm
$A = \pi r^2$;
$r = \frac{1}{2} \cdot d = \frac{1}{2} \cdot 15 = 7.5$
$A = 3.14 \cdot 7.5 \cdot 7.5 = 176.625$ Round to 176.6 cm^2.

13. $d = 18.24$ cm;
$C = \pi d = 3.14 \cdot 18.24 = 57.3$ cm (rounded)
$A = \pi r^2$;
$r = \frac{1}{2} \cdot d = \frac{1}{2} \cdot 18.24 = 9.12$
$A = 3.14 \cdot 9.12 \cdot 9.12 = 261.1676$ Round to 261.2 cm^2.

17. $r = 37$ m
$A \text{ (semicircle)} = \frac{1}{2}\pi r^2 = \frac{1}{2} \cdot 3.14 \cdot 37 \cdot 37$
$= 2149.3 \text{ m}^2$

$A \text{ (parallelogram)} = bh$

Step 1 Solve for diameter which is the base of the parallelogram.
$d = 2 \cdot r$
$d = 2 \cdot 37$
$d = 74$

Step 2 Solve for the area of the parallelogram.
$A = bh$
$A = 74 \cdot 28$
$A = 2072 \text{ m}^2$

total area $= 2149.3 \text{ m}^2 + 2072 \text{ m}^2 = 4221.3 \text{ m}^2$

21. $A \text{ (circle)} = \pi r^2$
radius $= 250 \text{ ft} \div 2 = 125$ radius
$A = 3.14 \cdot 125 \cdot 125 = 49{,}062.5 \text{ ft}^2$

25. $A \text{ (rectangle)} = b \cdot h$
$A = 29 \cdot 16 = 464 \text{ ft}^2$ (rectangle)
$A \text{ (circle)} = \pi r^2$ both semicircles are same so they equal 1 whole circle
$A = 3.14 \cdot 8 \cdot 8 = 200.96 \text{ ft}^2$
$A \text{ (total)} = 464 + 200.96 = 664.96 \text{ ft}^2$
$\$1.76 \cdot 664.96 = \$1170.33$ cost of sod (rounded)

29. $\dfrac{\frac{8}{7}}{4} = \frac{8}{7} \div \frac{4}{1} = \frac{\overset{2}{\cancel{8}}}{7} \times \frac{1}{\underset{1}{\cancel{4}}} = \frac{2}{7}$

Section 8.7 (page 517)

1. $l = 12$, $w = 11$, $h = 4$
$V = l \cdot w \cdot h = 12 \cdot 11 \cdot 4 = 528 \text{ cm}^3$

5. $r = 22$
$V = \frac{4}{3}\pi r^3 = \frac{4}{3} \cdot 3.14 \cdot 22 \cdot 22 \cdot 22$
$= 44{,}579.627$ Round to $44{,}579.6 \text{ m}^3$.

9. $r = 5$, $h = 6$;
$V = \pi r^2 h$
$= 3.14 \cdot 5 \cdot 5 \cdot 6 = 471 \text{ ft}^3$

13. $V \text{ (top box)} = 2 \cdot 12 \cdot 9 = 216$
$V \text{ (bottom box)} = 3 \cdot 11 \cdot 9 = 297$
$V \text{ (total)} = 216 + 297 = 513 \text{ cm}^3$

17. $V = \frac{1}{3}\pi r^2 h$
$V = \frac{1}{3} \cdot 3.14 \cdot 5 \cdot 5 \cdot 16$
$= 418.66$ Round to 418.7 m^3.

21. First: $r = d \div 2 = 7.2 \div 2 = 3.6$
$V = \pi r^2 h$
$V = 3.14 \cdot 3.6 \cdot 3.6 \cdot 10.5$
$= 427.29$ Round to 427.3 cm^3.

25. First: 3 inches $= .25$ ft
$A = 54 \cdot 18 \cdot .25 = 243 \text{ ft}^3$
Second: 1 cubic yard $= 3 \cdot 3 \cdot 3 = 27 \text{ ft}^3$
$243 \text{ ft}^3 \div 27 \text{ ft}^3 = 9 \text{ yd}^3$
Answer: 9 cubic yards

29.

```
     12
   1  2   11
   2̸  3̸. 1̸  65
 − 1  4.  2  35
      8.  9  30
```

Answer: 8.93

Section 8.8 (page 525)

1. Since $5 \cdot 5 = 25$, $\sqrt{25} = 5$.

5. Find 11 in the n column. Then look across to the square root column, finding 3.317.

9. Find 73 in the n column. Then look across to the square root column, finding 8.544.

13. Find 190 in the n column. Then look across to the square root column, finding 13.784.

17. Counting the squares on each side,

$a = 36$ $\quad$ $b = 64$ $\quad$ $c = 100$

$$\begin{aligned} \text{hypotenuse} &= \sqrt{(\text{leg})^2 + (\text{leg})^2} \\ &= \sqrt{(6)^2 + (8)^2} \\ &= \sqrt{36 + 64} \\ &= \sqrt{100} \\ &= 10 \quad \text{The Pythagorean formula holds true.} \end{aligned}$$

21.
$$\begin{aligned} \text{hypotenuse} &= \sqrt{(\text{leg})^2 + (\text{leg})^2} \\ &= \sqrt{(15)^2 + (8)^2} \\ &= \sqrt{225 + 64} \\ &= \sqrt{289} \\ &= 17 \text{ in} \end{aligned}$$

25.
$$\begin{aligned} \text{leg} &= \sqrt{(\text{hypotenuse})^2 - (\text{leg})^2} \\ &= \sqrt{(7)^2 - (3)^2} \\ &= \sqrt{49 - 9} \\ &= \sqrt{40} \\ &= 6.325 \text{ in} \quad \text{(rounded)} \end{aligned}$$

29.
$$\begin{aligned} \text{leg} &= \sqrt{(\text{hypotenuse})^2 - (\text{leg})^2} \\ &= \sqrt{(22)^2 - (17)^2} \\ &= \sqrt{484 - 289} \\ &= \sqrt{195} \\ &= 13.964 \text{ cm} \quad \text{(rounded)} \end{aligned}$$

33.
$$\begin{aligned} \text{leg} &= \sqrt{(\text{hypotenuse})^2 - (\text{leg})^2} \\ &= \sqrt{(11.5)^2 - (8.2)^2} \\ &= \sqrt{132.25 - 67.24} \\ &= \sqrt{65.01} \\ &= 8.063 \text{ cm} \quad \text{(rounded)} \end{aligned}$$

37.
$$\begin{aligned} \text{hypotenuse} &= \sqrt{(\text{leg})^2 + (\text{leg})^2} \\ &= \sqrt{(4)^2 + (7)^2} \\ &= \sqrt{16 + 49} \\ &= \sqrt{65} \\ &= 8.062 \text{ ft} \quad \text{(rounded)} \end{aligned}$$

41.
$$\begin{aligned} \text{hypotenuse} &= \sqrt{(\text{leg})^2 + (\text{leg})^2} \\ &= \sqrt{(7)^2 + (3)^2} \\ &= \sqrt{49 + 9} \\ &= \sqrt{58} \\ &= 7.616 \text{ cm} \quad \text{(rounded)} \end{aligned}$$

45.
$$\begin{aligned} \frac{2}{9} &= \frac{x}{36} \\ 2 \cdot 36 &= 9 \cdot x \\ 72 &= 9x \\ \frac{72}{9} &= \frac{9x}{9} \\ 8 &= x \end{aligned}$$

Section 8.9 (page 533)

1. Similar

5. Similar

9.

P	and	S
N	and	R
M	and	Q
MP	and	QS
MN	and	QR
NP	and	RS

13. Solve for side a.

$$\begin{aligned} \frac{\overset{1}{\cancel{6}}}{\underset{2}{\cancel{12}}} &= \frac{a}{10} \\ 2a &= 10 \\ \frac{2a}{2} &= \frac{10}{2} \\ a &= 5 \end{aligned}$$

Solve for side b.

$$\begin{aligned} \frac{\overset{1}{\cancel{6}}}{\underset{2}{\cancel{12}}} &= \frac{b}{6} \\ 2b &= 6 \\ \frac{2b}{2} &= \frac{6}{2} \\ b &= 3 \end{aligned}$$

17.
$$\begin{aligned} \frac{\overset{2}{\cancel{6}}}{\underset{3}{\cancel{9}}} &= \frac{4}{x} \\ 2x &= 12 \\ \frac{2x}{2} &= \frac{12}{2} \\ x &= 6 \end{aligned}$$

21.
$$\begin{aligned} \frac{n}{50} &= \frac{\overset{11}{\cancel{220}}}{\underset{5}{\cancel{100}}} \\ 5n &= 550 \\ \frac{5n}{5} &= \frac{550}{5} \\ n &= 110 \end{aligned}$$

25.
$$\begin{aligned} \frac{y}{60} &= \frac{\overset{1}{\cancel{40}}}{\underset{5}{\cancel{200}}} \\ 5y &= 60 \\ \frac{5y}{5} &= \frac{60}{5} \\ y &= 12 \end{aligned}$$

29. $16 \div 4 \cdot 2$

$4 \cdot 2$ $\quad$ Divide, left to right.

8 $\quad$ Multiply.

CHAPTER 9

Section 9.1 (page 557)

1. 12 degrees above zero $= +12$.

5. 18,000 feet above sea level $= +18{,}000$.

9. -3 is negative.

13. $\frac{7}{8}$ is positive.

17.

21.

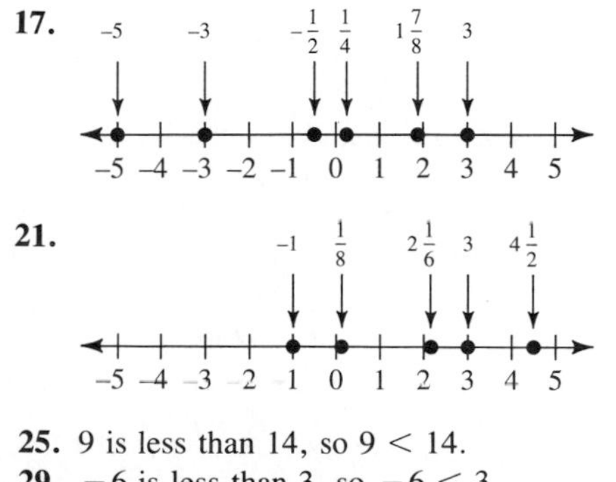

25. 9 is less than 14, so $9 < 14$.

29. -6 is less than 3, so $-6 < 3$.

33. -11 is less than -2, so $-11 < -2$.

37. 2 is greater than -1, so $2 > -1$.

41. $|5| = 5$; the absolute value is 5.

45. $|-32| = 32$; the absolute value is 32.

49. $|0| = 0$; the absolute value is 0.

53. $|-9.5| = 9.5$; the absolute value is 9.5.

57. $\left|\frac{3}{4}\right| = \frac{3}{4}$; the absolute value is $\frac{3}{4}$.

61. $|-9| = 9$, so $-|-9| = -9$.

65. The opposite of 2 is -2.

69. The opposite of -11 is 11.

73. The opposite of $\frac{4}{3}$ is $-\frac{4}{3}$.

77. The opposite of 5.2 is -5.2.

81. $-(-2) = 2$, so the opposite of $-(-2) =$ the opposite of $2 = -2$.

85. 0 is less than $-(-6)$, so $0 < -(-6)$ is true.

89. Least common denominator is 20.

$$\begin{array}{r} \frac{3}{4} = \frac{15}{20} \\ \frac{1}{2} = \frac{10}{20} \\ +\ \frac{3}{5} = \frac{12}{20} \\ \hline \frac{37}{20} = 1\frac{17}{20} \end{array}$$

Section 9.2 (page 569)

1. Answer: $+3$ or 3

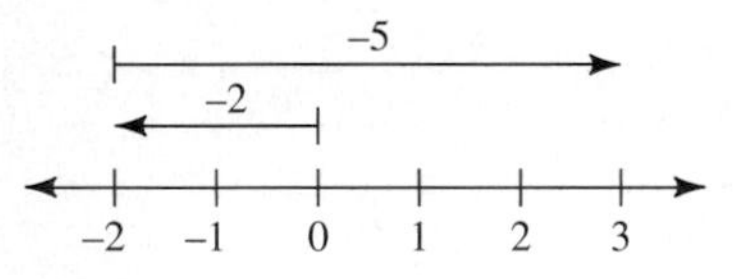

5. Answer: -1

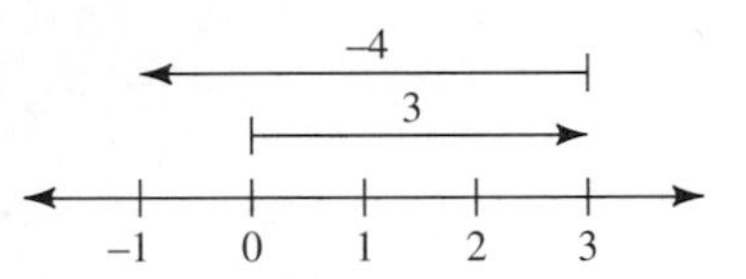

9. $-7 + 12$ Different signs, so subtract absolute value:
$|12| - |-7| = 12 - 7 = 5$.
Use sign of larger absolute value $+12$.
Answer: $+5$ or 5

13. $-11 + 2$ Different signs, so subtract absolute values:
$|-11| - |2| = 11 - 2 = 9$.
Use sign of larger absolute value -11.
Answer: -9

17. $-4.1 + (-3.7)$ Same signs, so add absolute values:
$|-4.1| + |-3.7| = 4.1 + 3.7 = 7.8$. Keep minus sign.
Answer: -7.8

21. $-\frac{1}{2} + \frac{3}{4}$ Different signs, so subtract absolute values:
$\left|\frac{3}{4}\right| - \left|-\frac{1}{2}\right| = \frac{3}{4} - \frac{1}{2} = \frac{3}{4} - \frac{2}{4} = \frac{1}{4}$.
Use sign of larger absolute value $\frac{3}{4}$.
Answer: $\frac{1}{4}$.

25. $-\frac{7}{3} + \left(-\frac{5}{9}\right)$ Same sign, so add absolute values:
$\left|-\frac{7}{3}\right| + \left|-\frac{5}{9}\right| = \frac{7}{3} + \frac{5}{9} = \frac{21}{9} + \frac{5}{9} = \frac{26}{9}$.
Keep minus sign.
Answer: $-\frac{26}{9}$ or $-2\frac{8}{9}$

29. $\begin{array}{r} -15 \\ 9 \\ \hline -6 \end{array}$ (different signs, so subtract absolute value; use sign of larger)

33. The additive inverse of -9 is 9 (change sign).

37. The additive inverse of 154 is -154 (change sign).

41. $19 - 5 = 19 + (-5)$ Add the additive inverse of second number.
$= 14$

45. $7 - 19 = 7 + (-19)$ Add the additive inverse of second number.
$= -12$

49. $9 - 14 = 9 + (-14)$ Add the additive inverse of second number.
$= -5$

53. $6 - (-14) = 6 + (+14)$ Add the additive inverse of second number.
$= 20$

57. $-\frac{7}{10} - \frac{4}{5} = -\frac{7}{10} + \left(-\frac{4}{5}\right)$
$= -\frac{7}{10} + \left(-\frac{8}{10}\right)$
$= -\frac{15}{10} = -\frac{3}{2}$ or $-1\frac{1}{2}$

61. $-6.4 - (-2.8) = -6.4 + (+2.8) = -3.6$

65. Add from the top. $\left.\begin{array}{r} -8 \\ 9 \end{array}\right\} \rightarrow \begin{array}{r} 1 \\ -7 \\ \hline -6 \end{array}$
-7

69. $-2 + (-11) - (-3) = -13 - (-3)$
$= -13 + 3$
$= -10$

73. $-2 - (-8) - (-3) = -2 + (+8) + (+3) = 9$

77. $-5.7 - (-9.4) - 8.1 = -5.7 + (+9.4) + (-8.1)$
$= 3.7 + (-8.1)$
$= -4.4$

81. $[-3 + (-2 + 4) + (-5)] = -3 + (+2) + (-5)$
$= -3 + 2 + (-5)$
$= -1 + (-5)$
$= -6$

85. $-9 + [(3 - 2) - (-4 + 2)] = -9 + [1 - (-2)]$
$= -9 + (+3)$
$= -6$

89. $3\frac{1}{2} \cdot 2\frac{3}{4} = \frac{7}{2} \cdot \frac{11}{4} = \frac{77}{8}$

Section 9.3 (page 575)

1. $-7 \cdot 2 = -14$ (opposite signs, product is negative)

5. $8 \cdot (-5) = -40$ (opposite signs, product is negative)

9. $12 \cdot (-4) = -48$

13. $-5 \cdot (-15) = 75$ (same signs, product is positive)

17. $-18 \cdot (-4) = 72$ (same signs, product is positive)

21. $-\frac{1}{4} \cdot (-6) = -\frac{1}{\cancel{4}_{2}} \cdot \left(-\frac{\cancel{6}^{3}}{1}\right) = \frac{3}{2} = 1\frac{1}{2}$

25. $\frac{1}{2} \cdot \left(-\frac{2}{3}\right) = -\frac{1 \cdot \cancel{2}^{1}}{\cancel{2}_{1} \cdot 3} = -\frac{1}{3}$

29. $-\frac{7}{15} \cdot \frac{25}{14} = -\frac{\cancel{7}^{1} \cdot \cancel{25}^{5}}{\cancel{15}_{3} \cdot \cancel{14}_{2}} = -\frac{5}{6}$

33. $9 \cdot (-4.7) = -42.3$

37. $6.2 \cdot (5.1) = -31.62$

41. $\frac{-14}{7} = -2$ (opposite signs, quotient is negative)

45. $\frac{-28}{4} = -7$ (opposite signs, quotient is negative)

49. $\frac{-20}{-2} = 10$ (same signs, quotient is positive)

53. $-\frac{2}{3} \div \frac{5}{6} = -\frac{2}{3} \cdot \frac{6}{5} = -\frac{2 \cdot \cancel{6}^{2}}{\cancel{3}_{1} \cdot 5} = -\frac{4}{5}$

57. $-\frac{5}{7} \div \left(-\frac{15}{14}\right) = -\frac{5}{7} \cdot \left(-\frac{14}{15}\right) = \frac{\cancel{5}^{1} \cdot \cancel{14}^{2}}{\cancel{7}_{1} \cdot \cancel{15}_{3}} = \frac{2}{3}$

61. $5 \div \left(-\frac{5}{8}\right) = \frac{5}{1} \cdot \left(-\frac{8}{5}\right) = -\frac{\cancel{5}^{1} \cdot 8}{1 \cdot \cancel{5}_{1}} = -8$

65. $\frac{-18.92}{-4} = 4.73$

69. $\frac{-45.58}{-8.6} = 5.3$

73. $(.6) \cdot (-.2) \cdot (-.3) = (-.12) \cdot (-.3)$
$= .036$

77. $(-2) \cdot (-4) \cdot [(-3) \cdot (-2) \cdot (4)]$
$= (-2) \cdot (-4) \cdot [(6) \cdot (4)]$ Work inside brackets first.
$= (-2) \cdot (-4) \cdot [24]$ Multiply next.
$= 8 \cdot 24 = 192$ Multiply, left to right.

81. $8 + 4 \cdot 2 \div 8 = 8 + 8 \div 8$ Multiply first.
$= 8 + 1 = 9$ Divide next, then add.

Section 9.4 (page 583)

1. $15 / 5 + 7$
$3 + 7$ Divide first.
10 Add.

5. $4^2 + 3^2$
$16 + 9$ Do exponents first.
25 Add.

9. $(-3)^3 + 9$
$-27 + 9$ Do exponents first.
-18 Add.

13. $2 - (-5) + 3^2$
$2 - (-5) + 9$ Do exponents first.
$7 + 9$ Add and subtract, *left to right.*
16

17. $3 + 5 \cdot (6 - 2)$
$3 + 5 \cdot 4$ Work inside parentheses first.
$3 + 20$ Multiply next.
23 Add.

21. $-6 + (-5) \cdot (9 - 14)$
$-6 + (-5) \cdot (-5)$ Work inside parentheses.
$-6 + 25$ Multiply next.
19 Add.

25. $36 / (-2)^2 + (-4)$
$36 / 4 + (-4)$ Do exponents first.
$9 + (-4)$ Divide next.
5 Add.

29. $(-2) \cdot (-7) + 3 \cdot 9$
$14 + 27$ Multiply first.
41 Add.

33. $2 \cdot 5 - 3 \cdot 4 + 5 \cdot 3$
$10 \quad 12 + 15$ Multiply first.
$-2 + 15$ Add and subtract,
13 *left to right.*

37. $-5 \cdot (-2) + 4^2 - (-9)$
$-5 \cdot (-2) + 16 - (-9)$ Do exponents first.
$10 + 16 - (-9)$ Multiply next,
35 then add and subtract.

41. $5^2 \cdot (7 - 13) \div (9 + 1) - (-9)$
$5^2 \cdot (-6) \div (10) - (-9)$ Work inside parentheses first.
$25 \cdot (-6) \div 10 - (-9)$ Do exponents next.
$-150 \div 10 - (-9)$ Multiply and divide, *left to right.*
$-15 - (-9)$
-6 Finally, subtract.

45. $(-.8)^2 + (-.4)^3 + 1.7$
$.64 + (-.064) + 1.7$ Do exponents first.
$.576 + 1.7$ Add and subtract, *left to right.*
2.276

49. $(.3)^2 \cdot (-7) - (9.2)$
$.09 \cdot (-7) - (9.2)$ Do exponents first.
$-.63 - 9.2$ Multiply next.
-9.83 Subtract.

53. $\left(-\frac{1}{2}\right)^2 - \left(\frac{3}{4} - \frac{7}{4}\right)$

$\left(-\frac{1}{2}\right)^2 - \left(-\frac{4}{4}\right)$ Work inside parentheses first.

$\frac{1}{4} - \left(-\frac{4}{4}\right)$ Do exponents next.

$1\frac{1}{4}$ Subtract.

57. $5^2 \cdot (9 - 11) \cdot (-3) \cdot (-2)^3$

$5^2 \cdot (-2) \cdot (-3) \cdot (-2)^3$ Work inside parentheses first.

$25 \cdot (-2) \cdot (-3) \cdot (-8)$ Do exponents.

-1200 Multiply, *left to right.*

61. $(6 - 19) \cdot (5 - 7)^2 \cdot (.4)$

$(-13) \cdot (-2)^2 \cdot (.4)$ Work inside parentheses first.

$(-13) \cdot (4) \cdot (.4)$ Do exponents next.

-20.8 Multiply, *left to right.*

65. $5 - 4 \cdot 12 \div 3 \cdot 2$

$5 - 48 \div 3 \cdot 2$ Multiply and divide, left to right.

$5 - 16 \cdot 2$

$5 - 32$

-27 Subtract.

69. $|-12| \div 4 + 2 \cdot 3^2 \div 6$

$|-12| \div 4 + 2 \cdot 9 \div 6$ Do exponents first.

$12 \div 4 + 2 \cdot 9 \div 6$ Change to absolute value.

$3 + 2 \cdot 9 \div 6$ Multiply and divide, left to right.

$3 + 18 \div 6$

$3 + 3$

6 Add.

73. $\frac{7}{4} + \frac{3}{4} \cdot \frac{12}{15}$

$\frac{7}{4} + \frac{36}{60}$ Multiply first.

$\frac{47}{20}$ Add.

Answer: $\frac{47}{20}$ or $2\frac{7}{20}$

Section 9.5 (page 589)

1. $2r + 4s \quad r = 4, s = 2$

$2 \cdot 4 + 4 \cdot 2 = 8 + 8 = 16$

5. $2r + 4s \quad r = -8, s = 6$

$2 \cdot (-8) + 4 \cdot 6 = -16 + 24 = 8$

9. $2r + 4s \quad r = 0, s = -2$

$2 \cdot 0 + 4(-2) = 0 + (-8) = -8$

13. $6k + 2s \quad k = 1, s = -2$

$6 \cdot 1 + 2(-2) = 6 + (-4) = 2$

17. $8m - 2n \quad m = -\frac{1}{2}, n = \frac{3}{4}$

$8 \cdot \left(-\frac{1}{2}\right) - 2\left(\frac{3}{4}\right) = -4 - \frac{3}{2}$

$= -\frac{8}{2} - \frac{3}{2} = -\frac{11}{2}$ or $-5\frac{1}{2}$

21. $P = 2l + 2w; l = 9, w = 5$

$P = 2 \cdot 9 + 2 \cdot 5$

$= 18 + 10 = 28$

25. $A = \frac{1}{2}bh; b = 7, h = 9$

$A = \frac{1}{2} \cdot 7 \cdot 9 = \frac{63}{2} = 31\frac{1}{2}$

29. $d = rt; r = 90, t = 8$

$d = 90 \cdot 8 = 720$

33. $A = \frac{1}{2}(b + B)h \quad b = 10, B = 12, h = 3$

$= \frac{1}{2}(10 + 12)3$

$= \frac{1}{2}(22)3$

$= 11 \cdot 3$

$= 33$

37. $-\frac{1}{9} \cdot 9 = -\frac{1}{\cancel{9}_1} \cdot \frac{\cancel{9}^1}{1} = -\frac{1}{1} = -1$

Section 9.6 (page 597)

1. $4 + 7 = 11$ yes

↑

Substitute for *x*.

4 is a solution

5. $2(-8) - 1 = -15$

↑

Substitute for *z*.

$-16 - 1 = -15$ No

-8 is not a solution

9. $k + 15 = 26$

$k + 15 - 15 = 26 - 15$ Subtract 15 from each side.

$k = 11$

check $k + 15 = 26$

$11 + 15 = 26$

$26 = 26$ correct

13. $8 = r - 2$

$8 + 2 = r - 2 + 2$ Add the inverse of -2 to each side.

$10 = r$

check $8 = r - 2$

$8 = 10 - 2$

$8 = 8$ correct

17. $7 = r + 13$

$7 - 13 = r + 13 - 13$ Subtract 13 from each side.

$-6 = r$

check $7 = r + 13$

$7 = -6 + 13$

$7 = 7$ correct

21. $-3 + x = 5$

$-3 + x + 3 = 5 + 3$ Add the inverse of -3 to each side.

$x = 8$

check $-3 + x = 5$

$-3 + 8 = 5$

$5 = 5$ correct

25. $-11 = -8 + d$

$-11 + 8 = -8 + d + 8$ Add the inverse of -8 to each side.

$-3 = d$

check $-11 = -8 + d$

$-11 = -8 + (-3)$

$-11 = -11$ correct

29. $k - \frac{2}{3} = 5$

$k - \frac{2}{3} + \frac{2}{3} = 5 + \frac{2}{3}$ Add the inverse of $-\frac{2}{3}$ to each side.

$k = 5\frac{2}{3}$

check $k - \frac{2}{3} = 5$

$5\frac{2}{3} - \frac{2}{3} = 5$

$5 = 5$ correct

33. $x - 1.72 = 4.86$

$x - 1.72 + 1.72 = 4.86 + 1.72$

$x = 6.58$

check $x - 1.72 = 4.86$

$6.58 - 1.72 = 4.86$

$4.86 = 4.86$ correct

37. $6z = 12$

$\frac{6z}{6} = \frac{12}{6}$ Divide each side by 6.

$z = 2$

check $6 \cdot 2 = 12$

$12 = 12$ correct

41. $3y = -24$

$\frac{3y}{3} = \frac{-24}{3}$ Divide each side by 3.

$y = -8$

check $3y = -24$

$3(-8) = -24$

$-24 = -24$ correct

45. $-2p = -16$

$\frac{-2p}{-2} = \frac{-16}{-2}$

$p = 8$

check $-2p = -16$

$-2(8) = -16$

$-16 = -16$ correct

49. $-8.4p = -9.24$

$\frac{-8.4p}{-8.4} = \frac{-9.24}{-8.4}$

$p = 1.1$

check $-8.4p = -9.24$

$-8.4 \cdot (1.1) = -9.24$

$-9.24 = -9.24$ correct

53. $\frac{a}{5} = 10$

$5 \cdot \frac{a}{5} = 10 \cdot 5$ Multiply each side by 5.

$a = 50$

check $\frac{a}{5} = 10$

$\frac{50}{5} = 10$

$10 = 10$ correct

57. $\frac{1}{2}p = 5$

$2 \cdot \frac{1}{2}p = 5 \cdot 2$ Multiply each side by 2.

$p = 10$

check $\frac{1}{2}p = 5$

$\frac{1}{2} \cdot 10 = 5$

$5 = 5$ correct

61. $\frac{3}{8}x = 6$

$8 \cdot \frac{3}{8}x = 6 \cdot 8$ Multiply each side by 8.

$3x = 48$

$\frac{3x}{3} = \frac{48}{3}$ Divide each side by 3.

$x = 16$

check $\frac{3}{8}x = 6$

$\frac{3}{8} \cdot 16 = 6$

$6 = 6$ correct

65. $-\frac{9}{8}k = -18$

$8 \cdot \left(-\frac{9}{8}k\right) = -18 \cdot 8$

$-9k = -144$

$k = 16$

check $-\frac{9}{8}k = -18$

$-\frac{9}{8} \cdot 16 = -18$

$-18 = -18$

69. $\frac{z}{-3.8} = 1.3$

$-3.8 \cdot \frac{z}{-3.8} = 1.3 \cdot (-3.8)$

$z = -4.94$

check $\frac{z}{-3.8} = 1.3$

$\frac{-4.94}{-3.8} = 1.3$

$1.3 = 1.3$

73. $3 = x + 9 - 15$

$3 + 15 = x + 9 - 15 + 15$

$18 = x + 9$

$18 - 9 = x + 9 - 9$

$9 = x$

check $3 = x + 9 - 15$

$3 = 9 + 9 - 15$

$3 = 3$ correct

77. $2\frac{1}{5} \div \left(3\frac{1}{3} - 4\frac{1}{5}\right)$

$2\frac{1}{5} \div \left(3\frac{5}{15} - 4\frac{3}{15}\right)$

$2\frac{1}{5} \div \left(-\frac{13}{15}\right) = \frac{11}{5} \div \left(-\frac{13}{15}\right) = \frac{11}{\cancel{5}_1} \cdot \left(-\frac{\cancel{15}^3}{13}\right)$

$= -\frac{33}{13}$ or $-2\frac{7}{13}$

Section 9.7 (page 605)

1. $5p - 3 = 2$

$5p - 3 + 3 = 2 + 3$ Add 3 to both sides.

$5p = 5$

$\frac{5p}{5} = \frac{5}{5}$ Divide both sides by 5.

$p = 1$

check $5p - 3 = 2$

$5 \cdot 1 - 3 = 2$

$5 - 3 = 2$

$2 = 2$ correct

5. $-3m + 1 = -5$

$-3m + 1 - 1 = -5 - 1$ Subtract 1 from both sides.

$-3m = -6$

$\frac{-3m}{-3} = \frac{-6}{-3}$ Divide both sides by -3.

$m = 2$

check $-3m + 1 = -5$

$-3(2) + 1 = -5$

$-6 + 1 = -5$

$-5 = -5$ correct

9. $-5x - 4 = 16$

$-5x - 4 + 4 = 16 + 4$

$-5x = 20$

$\frac{-5x}{-5} = \frac{20}{-5}$

$x = -4$

check $-5x - 4 = 16$

$-5(-4) - 4 = 16$

$20 - 4 = 16$

$16 = 16$ correct

13. $8(2 + 9) = 8 \cdot 2 + 8 \cdot 9$

$= 16 + 72$

$= 88$

17. $-3(m + 6) = -3 \cdot m + (-3) \cdot 6$

$= -3m - 18$

21. $-5(z - 9) = -5 \cdot z - (-5) \cdot 9$

$= -5z + 45$

25. $12y + 6y = (12 + 6)y = 18y$

29. $15a - 25a = (15 - 25)a = -10a$

33. $17m - 12m = 35$

$5m = 35$

$\frac{5m}{5} = \frac{35}{5}$

$m = 7$

check $17m - 12m = 35$

$17 \cdot 7 - 12 \cdot 7 = 35$

$119 - 84 = 35$

$35 = 35$ correct

37. $5y - 12y = -14$

$-7y = -14$

$\frac{-7y}{-7} = \frac{-14}{-7}$

$y = 2$

check $5y - 12y = -14$

$5 \cdot 2 - 12 \cdot 2 = -14$

$10 - 24 = -14$

$-14 = -14$ correct

41. $7z + 9 = 9z + 13$

$7z + 9 - 9 = 9z + 13 - 9$ Subtract 9 from both sides to get rid of number on left side.

$7z = 9z + 4$

$7z - 9z = 9z - 9z + 4$ Subtract $9z$ from both sides to get z on same side.

$-2z = 4$

$\frac{-2z}{-2} = \frac{4}{-2}$

$z = -2$

check $7z + 9 = 9z + 13$

$7 \cdot (-2) + 9 = 9(-2) + 13$

$-14 + 9 = -18 + 13$

$-5 = -5$ correct

45. $-3.6m + 1 = 2.4m + 7$

$-3.6m + 1 - 1 = 2.4m + 7 - 1$

$-3.6m = 2.4m + 6$

$-3.6m - 2.4m = 2.4m - 2.4m + 6$

$-6m = 6$

$\frac{-6m}{-6} = \frac{6}{-6}$

$m = -1$

check $-3.6m + 1 = 2.4m + 7$

$-3.6(-1) + 1 = 2.4(-1) + 7$

$3.6 + 1 = -2.4 + 7$

$4.6 = 4.6$ correct

49. $8 - 6x + 8 = 2x$

$8 - 6x + 8 + 6x = 2x + 6x$

$8 + 8 = 8x$

$16 = 8x$

$\frac{16}{8} = \frac{8x}{8}$

$2 = x$

check $8 - 6x + 8 = 2x$

$8 - (6 \cdot 2) + 8 = 2 \cdot 2$

$8 - 12 + 8 = 4$

$4 = 4$ correct

53. sales tax rate = 6% of sales

sales = \$420

6% · \$420 = .06 · 420

= \$25.20 sales tax

Section 9.8 (page 613)

1. $14 + x$

5. $x + 6$

9. $x - 4$

13. $3x$

17. $2x + 3$

21. First: x is the number.

Second: Translate:

$\underbrace{\text{four times a number}}_{4x}$ is $\underbrace{\text{decreased by 2,}}_{-2}$

$\underbrace{\text{the result is 26.}}_{= 26}$

Third: Solve:
$$\begin{aligned} 4x - 2 &= 26 \\ 4x - 2 + 2 &= 26 + 2 \\ 4x &= 28 \\ \frac{4x}{4} &= \frac{28}{4} \\ x &= 7. \end{aligned}$$

Fourth: check 4 times 7 is 28 and 28 decreased by 2 is 26. correct

25. First: x is the number.
Second: half a number is added to twice the number, the answer is 50

$$\underbrace{\tfrac{1}{2}x}_{\text{half a number}} \quad \underbrace{+2x}_{\text{added to twice the number}} \quad \underbrace{= 50}_{\text{the answer is 50}}$$

Third:
$$\begin{aligned} \frac{1}{2}x + 2x &= 50 \\ \left(\frac{1}{2} + 2\right)x &= 50 \\ \frac{5}{2}x &= 50 \\ \frac{2}{5} \cdot \frac{5}{2}x &= 50 \cdot \frac{2}{5} \\ x &= 20 \end{aligned}$$

Fourth: Half of 20 is 10.
Twice 20 is 40.
Added together they give 50. correct

29. First: x = number of days
Second: one-time \$7 fee plus \$6 a day was \$49

$$7 \quad + \quad 6 \cdot x \quad = \quad 49$$

Third:
$$\begin{aligned} 7 + 6x &= 49 \\ 7 - 7 + 6x &= 49 - 7 \\ 6x &= 42 \\ \frac{6x}{6} &= \frac{42}{6} \\ x &= 7 \text{ days} \end{aligned}$$

Fourth: check: \$7 + \$6 for 7 days = \$7 + \$42 = \$49 correct

33. First. x = length of each equal part; the third part is $x + 25$
Second: $x + x + x + 25 = 706$
Third:
$$\begin{aligned} 3x + 25 &= 706 \\ 3x + 25 - 25 &= 706 - 25 \\ 3x &= 681 \\ x &= 227 \text{ m} \end{aligned}$$

Answer: two parts are 227 m and the third is 252 m (227 + 25).

Fourth: Two parts are 227 m and the third is 252 m.
227 + 227 + 252 = 706 correct

37.
$$\begin{aligned} \frac{256}{b} &= \frac{80}{100} \\ 256 \cdot 100 &= b \cdot 80 \\ 25{,}600 &= b \cdot 80 \\ \frac{25{,}600}{80} &= \frac{b \cdot 80}{80} \\ 320 &= b \end{aligned}$$

CHAPTER 10

Section 10.1 (page 639)

1. The total cost of remodeling the kitchen is \$3800 + \$300 + \$600 + \$1500 + \$4200 = \$10,400.

5. $\dfrac{\text{wallpaper}}{\text{carpentry}} = \dfrac{300}{4200} = \dfrac{1}{14}$

9. total students = 4000 + 1800 + 1000 + 2000 + 700 + 2500 = 12,000;

$\dfrac{\text{English majors}}{\text{total students}} = \dfrac{2000}{12{,}000} = \dfrac{1}{6}$

13. total sales force cost = \$580,000;
car and plane expense = 30% of total cost
$30\% \cdot 580{,}000 = .3 \cdot 580{,}000$
= \$174,000 car and plane expense

17. total sales force cost = \$580,000;
food expense = 15% of total cost
$15\% \cdot \$580{,}000 = .15 \cdot 580{,}000$
= \$87,000 food expense

21. total income = \$19,600.
promotion = 13% of total income
$13\% \cdot \$19{,}600 = .13 \cdot 19{,}600$
= \$2548 promotion budget

25. total student expenses = \$1400;
rent = \$280

percent of total $= \dfrac{280}{1400} = 20\%$

29. total student expenses = \$1400;
savings = \$70

percent of total $= \dfrac{70}{1400} = 5\%$

number of degrees is $= 5\% \cdot 360° = .05 \cdot 360° = 18°$

33. **(a)** 25% are mysteries.
$25\% \cdot 360° = .25 \cdot 360° = 90°$

10% are autobiographies.
$10\% \cdot 360° = .1 \cdot 360° = 36°$

15% are cookbooks.
$15\% \cdot 360° = .15 \cdot 360° = 54°$

15% are romantic novels.
$15\% \cdot 360° = .15 \cdot 360° = 54°$

20% are science.
$20\% \cdot 360° = .2 \cdot 360° = 72°$

Balance of books are
$100\% - 25\% - 10\% - 15\% - 15\% - 20\% = 15\%$
$15\% \cdot 360° = .15 \cdot 360° = 54°.$

(b) See graph in the answer section.

37. .4219 < .4230

41. 70,485 < 70,510

Section 10.2 (page 649)

1. Attendance on July 5 was 3000.
5. The greatest attendance was on July 4.
9. Feb. 1990—500 employees
Feb. 1991—700 employees
$700 - 500 = 200$
There were 200 more employees in February 1991 than in February 1990.
13. 200,000 gallons of supreme unleaded gasoline were sold in 1987.
17.
$$\begin{array}{rl} 700{,}000 & \text{sales in 1991} \\ -\ 200{,}000 & \text{sales in 1987} \\ \hline 500{,}000 & \text{gallon increase} \end{array}$$
21. There were 300 burglaries in June.
25. Store A had \$3,000,000 annual sales in 1991.
29. Store B had \$3,500,000 annual sales in 1990.
33. Total sales in 1989 were \$25,000.
37. new court = 11% money collected = \$8400
$11\% \cdot \$8400 = \$924$ to new court

Section 10.3 (page 655)

1. The greatest number of members are 60–65 years of age.
5. $11 + 14 + 16 = 41$ members are 50–65 years of age.
9. 7 employees earn \$15–\$20 thousand.
13. 2 is the frequency for 120–129 sets.
17. 1 is the frequency for 160–169 sets.
21. 3 is the frequency for 80°–84°.
25. 6 is the frequency for 100°–104°.
29. 1 is the frequency for a score of 30–39.
33. 17 is the frequency for a score of 70–79.
37. $(22 + 36 + 58 + 27) \div 4$
$143 \div 4$ Work inside parentheses first.
35.75 Divide last.
41. $(4 \cdot 3) + (3 \cdot 9) + (2 \cdot 3) \div 15$
$12 + 27 + 6 \div 15$ Work inside parentheses first.
$12 + 27 + .4$ Divide next.
39.4 Add last.

Section 10.4 (page 663)

1. $\dfrac{6 + 8 + 14 + 19 + 23}{5} = \dfrac{70}{5} = 14$

5. $\dfrac{21{,}900 + 22{,}850 + 24{,}930 + 29{,}710 + 28{,}340 + 40{,}000}{6}$
$= \dfrac{167{,}730}{6} = 27{,}955$

9.

value	*frequency*	
3	4	$3 \cdot 4 = 12$
5	2	$5 \cdot 2 = 10$
9	1	$9 \cdot 1 = 9$
12	3	$12 \cdot 3 = 36$
	10	$67 \div 10 = 6.7$

13. 12, 18, 32, 51, 58, 92, 106
↑ The median is the middle number.
The median is 51.
17. 4, 9, 8, 6, 9, 2, 1, 3
→ 9 is the mode (no other number is duplicated)
21.

Units	*Grade*	
4	B	$4 \cdot 3 = 12$
2	A	$2 \cdot 4 = 8$
5	C	$5 \cdot 2 = 10$
1	F	$1 \cdot 0 = 0$
3	B	$3 \cdot 3 = 9$
15		$39 \div 15 = 2.6$

25. There is no mode, since no number is duplicated.
29. $I = 6000 \cdot 8\% \cdot 2$
$= 480 \cdot 2$
$= \$960$

CHAPTER 11

Section 11.1 (page 687)

1. \$1000 at 6% for 4 years
First: from table: 1.2625
Second: multiply: $1000 \cdot 1.2625 = 1262.5$
Answer: \$1262.50
5. 8428.17 at $4\frac{1}{2}\%$ for 6 years
First: from table: 1.3023
Second: multiply: $8428.17 \cdot 1.3023$
$= 10{,}976.005791$
round: \$10,976.01
Answer: \$10,976.01
9. \$800 at 12% compounded semiannually for 7 years
First: $7 \cdot 2 = 14$ periods
$12\% \div 2 = 6\%$ per period
Second: from table: 2.2609
Third: multiply: $800 \cdot 2.2609 = 1808.72$
Answer: \$1808.72
13. \$25,800 at 12% compounded quarterly for 4 years
First: $4 \cdot 4 = 16$ periods
$12\% \div 4 = 3\%$ per period
Second: from table: 1.6047
Third: multiply: $25{,}800 \cdot 1.6047 = 41{,}401.26$
Answer: \$41,401.26
17. \$975 at 12% compounded semiannually for 8 years
First: $2 \cdot 8 = 16$ periods
$12\% \div 2 = 6\%$ per period
Second: from table: 2.5404
Third: multiply: $\$975 \cdot 2.5404 = 2476.89$
Answer: \$2476.89 compound amount
$\$2476.89 - \$975 = \$1501.89$ compound interest
21. \$7500 at 12% compounded quarterly for 6 years
First: $6 \cdot 4 = 24$ periods
$12\% \div 4 = 3\%$ per period
Second: from table: 2.0328
Third: multiply: $7500 \cdot 2.0328 = 15{,}246$
Answer: \$15,246 compound amount
$\$15{,}246 - \$7500 = \$7746$ compound interest

25. (a) $8000 at 10% compounded annually for 5 years
First: from table: 1.6105
Second: multiply: $8000 \cdot 1.6105 = 12{,}884$
Answer: $12,884
(b) $8000 at 10% compounded semiannually for 5 years
First: $5 \cdot 2 = 10$ periods
$10\% \div 2 = 5\%$ per period
Second: from table: 1.6289
Third: multiply: $8000 \cdot 1.6289 = 13{,}031.20$
Answer: $13,031.20
(c) $8000 at 10% compounded quarterly for 5 years
First: $5 \cdot 4 = 20$ periods
$10 \div 4 = 2\frac{1}{2}\%$ per period
Second: from table: 1.6386
Third: multiply: $8000 \cdot 1.6386 = 13{,}108.80$
Answer: $13,108.80

29. $7500 at 8% compounded annually for 6 years
First: from table: 1.5869
Second: multiply: $\$7500 \cdot 1.5869 = \$11{,}901.75$
Answer: $11,901.75

33. $10,000 at 8% compounded quarterly for 2 years
First: $2 \cdot 4 = 8$ periods
$8\% \div 4 = 2\%$ per period
Second: from table: 1.1717
Third: multiply: $\$10{,}000 \cdot 1.1717 = \$11{,}717$
Fourth: $\$11{,}717 + \$20{,}000 = \$31{,}717$ principal at 8% compounded quarterly for 3 years
Fifth: $3 \cdot 4 = 12$ periods
$8\% \div 4 = 2\%$ per period
Sixth: from table: 1.2682
Seventh: multiply: $\$31{,}717 \cdot 1.2682 = \$40{,}223.50$ (rounded)

37. $19.3 + (6.7 - 5.2) \cdot 58$
$19.3 + 1.5 \cdot 58$ Work inside parentheses first.
$19.3 + 87$ Multiply next.
106.3 Add last.

Section 11.2 (page 695)

1. First: amount paid $= 108 \cdot 12 = \$1296$
Second: interest charge $= 1296 - 1200 = \$96$
Third: $\frac{96}{1200} \cdot 100 = \$8$
Fourth: from table: $14\frac{1}{2}\%$ (Closest figure in 12 payment row is 8.03, under $14\frac{1}{2}\%$.)

5. First: amount paid $= 73.11 \cdot 24 = 1754.64$
Second: interest charge $= 1754.64 - 1500$
$= 254.64$
Third: $\frac{254.64}{1500} \cdot 100 = 16.976$
Fourth: from table: $15\frac{1}{2}\%$ (Closest figure in 24 payment row is 16.94, under $15\frac{1}{2}\%$.)

9. First: amount paid $= 20 \cdot 12 = 240$
Second: interest charge $= 240 - 222 = \$18$
Third: $\frac{18}{222} \cdot 100 = 8.11$ (rounded)
Fourth: from table: $14\frac{1}{2}\%$ (Closest figure in 12 payment row is 8.03, under $14\frac{1}{2}\%$.)

13. finance company: 30% on $200

17. life insurance: 5%

21. $36 \cdot 10 = 360$ ← Attach zero (move decimal point one place to right).

25. $\$36.80 \cdot 1000 = \$36{,}800$ Move decimal point 3 places to right.

Section 11.3 (page 703)

1. $80,000 at 10% for 30 years
First: from table: 8.78
Second: multiply: $80 \cdot 8.78 = 702.4$
$702.40 monthly payment

5. $96,500 at $10\frac{1}{2}\%$ for 30 years
First: from table: 9.15
Second: multiply: $96.5 \cdot 9.15 = 882.98$ (rounded)
$882.98 monthly payment (rounded)

9. First: from table, 11%, 30, find 9.52
Second: multiply: $32.2 \cdot 9.52 = 306.54$ (rounded)
Third: monthly taxes are $\frac{\$172}{12} = \$14.33$ (rounded)
Fourth: monthly insurance: $\frac{\$165}{12} = \$13.75$
Fifth: total $= \$306.54 + \$14.33 + \$13.75$
$= \$334.62.$

13. monthly income = $2500; monthly payments = $300
$2500 - 300 = 2200$
$\frac{2200}{4} = 550$
The maximum house payment is $550.

17. number of monthly payments
= 30 years · 12 payments per year
= 360 payments

21.
```
   $7.95      2  decimal places
 ×    12    + 0  decimal places
   15 90      2  in answer
   79 5
  $95.40
```

25.
```
   $15.75     2  decimal places
 ×   16.3   + 1  decimal place
    4725      3  in answer
   9450
  1575
 $256.725
```
Answer: $256.73 (rounded)

Section 11.4 (page 709)

1. \$40,000 whole life policy, female, age 26
Use 26 − 3 = 23 year row of table.
$40 \cdot 12.80 = \$512$ annual premium

5. \$27,500 endowment policy, female, age 53
Use 53 − 3 = 50 year row of table.
$27.5 \cdot 41.04 = \$1128.60$ annual premium

9. \$65,750 endowment policy, male, age 50
$65.75 \cdot 41.04 = \$2698.38$ annual premium

13. \$50,000 10-year policy, male, age 35
$50 \cdot 7.54 = 377$ annual premium

17. \$10,000 endowment policy, male, age 40
$10 \cdot 36.42 = \$364.20$ annual premium
total premium paid over 20 years
$= \$364.20 \cdot 20$
$= \$7284$

Appendix A: Calculators (page 731)

1. 28.96 [+] 34.25 [+] 19.78 [+] 21.59 [=] 104.58

5. 769.2 [−] 35.75 [=] 733.45

9. 3409 [×] .006 [=] 20.45

13. .8359 [÷] 1.705 [=] .49

17. 9 [×] 9 [÷] 2 [÷] 5 [=] 8.1

21. 1155 [×] 360 [÷] 16,500 [÷] 120 [=] .21

25. 525 [×] .399 [+] 8.911 [M+] 633 [×] .0299
[÷] [MR] [=] .0866662
Answer: .09 (rounded)

29. 5 [×] 47.46 [M+] 4 [×] 51.62 [M+]
7.6 [×] 29.95 [M+] 916 [×] .24 [M+] [MR]
891.24

Answer: \$891.24

Appendix B: Inductive and Deductive Reasoning (page 739)

1. 2, 9, 16, 23, 30, ...
The number in the sequence are the result of adding 7 to the preceding number.
When 7 is added to 30 ($30 + 7 = 37$), the result is 37.

5. 1, 2, 4, 8, ...
Each number in the sequence is twice (2 times) the size of the preceding number. When 8 is multiplied by 2 ($8 \times 2 = 16$), the result is 16.

9. 1, 4, 9, 16, 25, ...
Each number in the sequence is the result of adding the odd number which is one higher than the preceding odd number to the preceding number (3, 5, 7, 9).

$$1 + 3 = 4$$
$$4 + 5 = 9$$
$$9 + 7 = 16$$
$$16 + 9 = 25$$

The next number in the sequence is ($25 + 11 = 36$) = 36.

13.

The first two shapes are diagonal lines with the first one having a horizontal line at the top facing right and the second one having a horizontal line at the bottom facing left. The third shape is a diagonal line with a line at the top facing to the left. The fourth shape should be a diagonal line with a line at the bottom facing to the right, or

17. All teachers are serious
All mathematicians are serious
All mathematicians are teachers

Using Euler circles

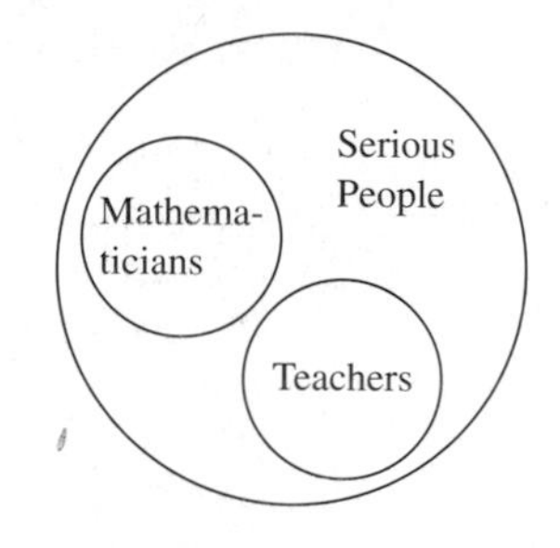

the conclusion does not follow.

21. Tom, Dick, Mary and Joan all work for the same company. One is a secretary, one a computer operator, one a receptionist, and one a mail clerk. Who is the computer operator?

1. Tom and Joan eat dinner with the computer operator.
   Therefore, the computer operator is *not* Tom or Joan.
2. Dick and Mary car pool with the secretary.
   Therefore, the *secretary* is either Tom or Joan.
3. Mary works on the same floor as the computer operator and mail clerk.
   Therefore, the computer operator is *not* Mary.

If the computer operator is *not* Tom or Joan or Mary, the computer operator *is* Dick.

GLOSSARY

This glossary provides an alphabetized listing of all the entries found in the "Key Terms" section of each Chapter Summary. For reference or further study, the corresponding section number is given at the end of each entry.

A

absolute value Absolute value is the magnitude of a number or the distance of a number from zero on a number line. **[9.1]**

acute angle An acute angle is an angle that measures between 0° and 90°. **[8.2]**

addition property of equality The addition property of equality states that the same number can be added to both sides of an equation. **[9.6]**

additive inverse The additive inverse is the opposite of a number. **[9.2]**

amount The amount in a percent problem is the part being compared with the whole. **[6.3, 6.4]**

area Area is the space taken up by a figure. **[8.3, 8.4, 8.5, 8.6]**

B

bar graph A bar graph is a graph that uses bars to show quantity or frequency. **[10.2]**

base The base in a percent problem is the entire quantity or the total. **[6.3, 6.4]**

borrowing The method of borrowing is used in subtraction if a digit is greater than the one directly above. **[1.3]**

C

cancellation When multiplying or dividing fractions, the process of dividing a numerator and denominator by a common factor is called cancellation. **[2.5]**

carrying The process of carrying is used in an addition problem when the sum of the digits in a column is greater than 9. **[1.2]** Also, the method used when the sum of the fractions of mixed numbers is greater than 1 is called carrying. Carry from the fraction to the whole number. **[3.4]**

Celsius The Celsius scale is the scale used to measure temperature in the metric system. Water boils at 100°C and freezes at 0°C on the Celsius scale. **[7.5]**

circle A circle is a figure with all points the same distance from a fixed center point. **[8.6]**

circle graph A circle graph is a circle broken up into various parts or sectors, based on percents of 360°. **[10.1]**

circumference Circumference is the distance around a circle. **[8.6]**

commission Commission is a percent of the dollar value of total sales paid to a salesperson. **[6.7]**

common factor A common factor is a number that can be divided into both the numerator and denominator to reduce the fraction. **[2.4]**

commutative property of addition The commutative property of addition states that the order of numbers in an addition problem can be changed without changing the sum. **[1.2]**

commutative propery of multiplication The commutative property of multiplication states that the product in a multiplication problem remains the same when the order of the factors is changed. **[1.4]**

comparison line graph A comparison line graph is one graph that shows how several different items relate to each other. **[10.2]**

complementary angles Complementary angles are two angles with a sum of 90°. **[8.2]**

composite number A composite number has at least one factor other than itself and 1. **[2.3]**

compound interest Compound interest is interest paid on past interest as well as on principal. **[11.1]**

compounding Interest that is compounded once each year is compounded annually, interest that is compounded twice each year is compounded semiannually, and interest that is compounded four times each year is compounded quarterly. **[11.1]**

congruent angles Congruent angles are angles that are equivalent in degrees. **[8.2]**

cross products The method of cross products, or cross multiplication, is used to determine whether a proportion is true. **[5.3]**

D

decimal point The starting point in the decimal system, called the decimal point, is the point or period that is used to separate the whole-number part from the fractional part of a number. **[4.1]**

decimals In addition to fractions, decimals are another way to show parts of a whole. **[4.1]**

degree A degree is the size of an angle that intercepts $\frac{1}{360}$th of the circumference of a circle. **[8.2]**

denominate number A denominate number is a number with a unit of measure. For example, 6 dollars, 8 feet, 4 years, and 10 pounds are denominate numbers. **[7.2]**

denominator The number below the division bar in a fraction is called the denominator. **[2.1]**

diameter Diameter is the distance across the circle, passing through the center. **[8.6]**

difference The answer in a subtraction problem is called the difference. **[1.3]**

discount Discount is often expressed as a percent of the original price; it is then deducted from the original price, resulting in the sale price. **[6.7]**

double-bar graph A double-bar graph is a graph used to show two sets of data; this graph has two sets of bars. **[10.2]**

E

endowment policy An endowment policy guarantees the payment of a fixed amount of money at a specified time while also providing life insurance during this time period. **[11.4]**

English system The English system of measurement (American system of units) is the most common system of measurement used in the United States. Common units in this system include quarts, pounds, feet, and miles. **[7.1]**

equation An equation is a statement that says two expressions are equal. **[9.6]**

estimating The process of approximating an answer to make sure the decimal is in the correct place is called estimating. **[4.5]**

expression An expression is a combination of letters and numbers. **[9.5]**

F

factors Parts of a product are called factors. For example, since $3 \times 4 = 12$, both 3 and 4 are factors of 12. **[1.4]** Also, numbers that are multiplied to give a product are factors. **[2.3]**

Fahrenheit The Fahrenheit scale is the scale used to measure temperature in the English system. Water boils at 212°F and freezes at 32°F on the Fahrenheit scale. **[7.5]**

frequency polygon A frequency polygon is the result when dots are placed at the top of the bars in a histogram and are connected with lines. **[10.3]**

H

histogram A histogram is a bar graph in which the width of each bar represents a range of numbers (class interval) and the height represents the quantity or frequency. **[10.3]**

hypotenuse The hypotenuse is the side of a right triangle opposite the 90° angle. **[8.8]**

I

improper fraction In an improper fraction, the numerator is greater than or equal to the denominator. **[2.1]**

indicators Words that suggest the operations needed for solving a problem are called indicators. **[1.7]**

integers Integers are the "counting numbers" or positive numbers, their opposites or negative numbers, and zero. **[9.1]**

interest Interest is a charge for borrowing money. **[6.8]**

interest formula The interest formula is the formula used to calculate interest (Interest = principal · rate · time or $I = p \cdot r \cdot t$). **[6.8]**

L

least common multiple Given two or more whole numbers, the least common multiple is the smallest whole number that is a multiple of these. **[3.2]**

like fractions Fractions with the same denominator are called like fractions. **[3.1]**

like terms Like terms are terms with exactly the same variable and the same exponent. **[9.7]**

line A line is a straight collection of points that goes on forever in opposite directions. **[8.1]**

line graph A line graph is a graph that uses dots connected by lines to show a trend. **[10.2]**

line segment A line segment is a part of a line bounded by endpoints. **[8.1]**

long division The process of long division is used to divide by a number with more than one digit. **[1.6]**

lowest terms A fraction is written in lowest terms when its numerator and denominator have no common factor other than 1. **[2.4]**

M

mean The mean is the sum of all the values divided by the number of values. **[10.4]**

median The median is the middle number in a group of values. It divides a group of values in half. In an evenly numbered group of values, the median is the mean of the two middle values. **[10.4]**

metric conversion line The metric conversion line is a line showing the various metric measurement prefixes and their size relationship to each other. **[7.3]**

metric system The metric system of measurement is an international system of measurement used in manufacturing, science, medicine, sports, and other fields. The system uses liters, grams, and meters. **[7.1]**

mixed number A mixed number includes a fraction and a whole number written together as one. **[2.2]**

mode The mode is the most common value in a group of values. **[10.4]**

monthly payment table A monthly payment table is a table used to find the monthly payment on a loan. **[11.3]**

multiple The product of two whole-number factors is a multiple of those numbers. **[1.4]**

multiplication property of equality The multiplication property of equality states that the same nonzero number can be multiplied or divided on both sides of an equation. **[9.6]**

N

negative numbers Negative numbers are numbers that are less than zero. **[9.1]**

numerator The number above the division bar in a fraction is called the numerator. **[2.1]**

O

obtuse angle An obtuse angle is an angle that measures between 90° and 180°. **[8.2]**

opposite of a number The opposite of a number is a number the same distance from zero as the original number but on the opposite side of zero on a number line. **[9.1]**

P

parallel lines Parallel lines are two lines that never intersect and are equidistant from each other. **[8.2]**

parallelogram A parallelogram is a four-sided figure with both pairs of opposite sides parallel. **[8.4]**

percent Percent means per one hundred. A percent is a ratio with a denominator of 100. **[6.1]**

percent equation The percent equation is percent · base = amount, and is an alternative way to solve percent problems. **[6.6]**

percent of increase or decrease Percent of increase or decrease is the amount of increase or decrease expressed as a percent of the original amount. **[6.7]**

percent proportion The proportion $\frac{a}{b} = \frac{p}{100}$ is used to solve percent problems. **[6.3, 6.4]**

perfect square A number that is the square of a whole number is a perfect square. **[1.7]**

perimeter Perimeter is the distance around a figure. **[8.3, 8.4, 8.5, 8.6]**

perpendicular lines Perpendicular lines are two lines that intersect to form a right angle. **[8.2]**

place value Place value is the value assigned to each place to the right or left of the decimal point—for example, ones, tenths, hundredths. Whole numbers are to the left of the decimal point and decimal numbers are to the right. **[4.1]**

prime factorization A factorization of a number in which every factor is prime is a prime factorization. **[2.3]**

prime number Any whole number that is not composite, except 0 and 1, is called a prime number. **[2.3]**

principal Principal is the amount of money borrowed. **[6.8]**

product The answer in a multiplication problem is called the product. **[1.4]**

proper fraction In a proper fraction, the numerator is smaller than the denominator. **[2.1]**

proportion A proportion states that two ratios are equivalent. **[5.3]**

protractor A protractor is a device (usually in the shape of a half-circle) used to measure degrees or parts of a circle. **[10.1]**

Q

quotient The answer in a division problem is called the quotient. **[1.5]**

R

radius Radius is the distance from the center to any point on the circle. **[8.6]**

rate The type of ratio used to compare unlike quantities it is a rate. An example is 22 miles per gallon to express mileage. **[5.2]**

rate of interest Often referred to as "rate," it is the charge for interest and is given as a percent. **[6.8]**

ratio A ratio is a comparison of two quantities. For example, the ratio of 6 apples to 11 apples is written as $\frac{6}{11}$. **[5.1]**

ray A ray is a part of a line that has one endpoint and extends forever in one direction. **[8.1]**

rectangle A rectangle is a four-sided figure with all sides meeting at 90° angles. **[8.3]**

remainder The remainder is the number left over when two numbers do not divide exactly. **[1.5]**

repeating decimal A repeating decimal has a digit that repeats, such as the 6 in .1666. . . . The dots indicate that the decimal does not terminate but continues to repeat. **[4.6]**

right angle A right angle is an angle that measures 90°. **[8.2]**

rounding To find a number that is close to the original number, but easier to work with, we use rounding. **[1.7]** Also, the reduction of a number with more decimals to a number with fewer decimals is called rounding. The rounded number is less accurate than the original number. **[4.2]**

S

sales tax Sales tax is a percent of the total sales charged as a tax. **[6.7]**

semicircle A semicircle is a half-circle. **[8.6]**

signed numbers Signed numbers are positive numbers, negative numbers, and zero. **[9.1]**

similar triangles Similar triangles are two triangles with the same shape but not necessarily the same size. **[8.9]**

solution The solution is a number that can be substituted for the variable in an equation, so that the equation is true. **[9.6]**

square A square is a rectangle with all four sides of equivalent length. **[8.3]**

square root A square root is one of two equal factors of a number. **[8.8]**

square root of a number The square root of a given number is the number that can be multiplied by itself to produce the given (larger) number. **[1.8]**

straight angle A straight angle is an angle that measures 180°. **[8.2]**

sum The answer in an addition problem is called the sum. **[1.2]**

supplementary angles Supplementary angles are two angles with a sum of 180°. **[8.2]**

T

term insurance Term insurance is a policy of life insurance that provides protection for a certain period of time. **[11.4]**

trapezoid A trapezoid is a four-sided figure with only one pair of opposite sides parallel. **[8.4]**

triangle A triangle is a figure with exactly three sides. **[8.5]**

true annual interest rate The true annual interest rate is the actual rate of interest. It is used to compare the rates of various loans. **[11.2]**

U

unit price The unit price is the unit rate that gives a cost per item or quantity. **[5.2]**

unit rate A unit rate is a rate with 1 unit in the denominator. **[5.2]**

unlike fractions Fractions with different denominators are called unlike fractions. **[3.1]**

V

variables Variables are letters that represent numbers. **[9.5]**

vertical angles Vertical angles are two nonadjacent angles formed by intersecting lines. **[8.2]**

volume Volume is the space inside a three-dimensional figure. **[8.8]**

W

weighted mean The weighted mean is a mean calculated so that each value is multiplied by its frequency. **[10.4]**

whole life policy A whole life policy combines life insurance and a savings plan. It is also called straight life or ordinary life. **[11.4]**

whole numbers A number made up of digits to the left of the decimal point is called a whole number. Examples are 0, 1, 2, 3, and 4. **[1.1]**

INDEX

Page numbers in **boldface type** indicate glossary entries.

E

F

G

H

I

K

L

M

N

O

P

Q

R

S

T

U

V

W

Y

Z